Feldspars and Their Reactions

NATO ASI Series

Advanced Science Institutes Series

A Series presenting the results of activities sponsored by the NATO Science Committee, which aims at the dissemination of advanced scientific and technological knowledge, with a view to strengthening links between scientific communities.

The Series is published by an international board of publishers in conjunction with the NATO Scientific Affairs Division

A Life Sciences
B Physics
Plenum Publishing Corporation
London and New York

C Mathematical and Physical Sciences
D Behavioural and Social Sciences
E Applied Sciences
Kluwer Academic Publishers
Dordrecht, Boston and London

F Computer and Systems Sciences
G Ecological Sciences
H Cell Biology
I Global Environmental Change
Springer-Verlag
Berlin, Heidelberg, New York, London, Paris and Tokyo

NATO-PCO-DATA BASE

The electronic index to the NATO ASI Series provides full bibliographical references (with keywords and/or abstracts) to more than 30000 contributions from international scientists published in all sections of the NATO ASI Series.
Access to the NATO-PCO-DATA BASE is possible in two ways:

– via online FILE 128 (NATO-PCO-DATA BASE) hosted by ESRIN, Via Galileo Galilei, I-00044 Frascati, Italy.

– via CD-ROM "NATO-PCO-DATA BASE" with user-friendly retrieval software in English, French and German (© WTV GmbH and DATAWARE Technologies Inc. 1989).

The CD-ROM can be ordered through any member of the Board of Publishers or through NATO-PCO, Overijse, Belgium.

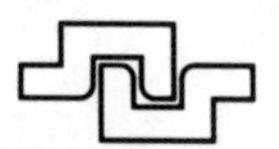

Series C: Mathematical and Physical Sciences - Vol. 421

Feldspars and Their Reactions

edited by

Ian Parsons

Department of Geology and Geophysics,
The University of Edinburgh,
Edinburgh, United Kingdom

Kluwer Academic Publishers

Dordrecht / Boston / London

Published in cooperation with NATO Scientific Affairs Division

Proceedings of the NATO Advanced Study Institute on
Feldspars and Their Reactions
Edinburgh, United Kingdom
June 29-July 10, 1993

A C.I.P. Catalogue record for this book is available from the Library of Congress

ISBN 0-7923-2722-5

Published by Kluwer Academic Publishers,
P.O. Box 17, 3300 AA Dordrecht, The Netherlands.

Kluwer Academic Publishers incorporates the publishing programmes of
D. Reidel, Martinus Nijhoff, Dr W. Junk and MTP Press.

Sold and distributed in the U.S.A. and Canada
by Kluwer Academic Publishers,
101 Philip Drive, Norwell, MA 02061, U.S.A.

In all other countries, sold and distributed
by Kluwer Academic Publishers Group,
P.O. Box 322, 3300 AH Dordrecht, The Netherlands.

Printed on acid-free paper

Printed in the Netherlands

TABLE OF CONTENTS

6. SUBSOLIDUS PHASE RELATIONS OF THE PLAGIOCLASE FELDSPAR SOLID SOLUTION **221**

M A CARPENTER, Department of Earth Sciences, University of Cambridge, Downing Street, Cambridge, CB2 3EQ, England

R J ANGEL, Department of Geological Sciences, University College London, Gower Street, London, WC1E 6BT, England

T H PEARCE, Department of Geological Sciences, Queens University, Kingston, Ontario, Canada K7L 3N6

BRUNO J GILETTI, Department of Geological Sciences, Brown University, Providence, Rhode Island 02912, U.S.A

WILLIAM L BROWN, CRPG-CNRS, BP 20, 54501 Vandoeuvre-lès-Nancy cedex, France, and IAN PARSONS, Department of Geology & Geophysics, The University of Edinburgh, Edinburgh, EH9 3JW, Scotland

PETR CERNÝ, Department of Geological Sciences, University of Manitoba, Winnipeg, MB, R3T 2N2, Canada

Preface

The title of this volume harks back to a seminal paper in mineralogy, 'Chemical reactions in crystals', written by J B Thompson more than 20 years ago (*American Mineralogist,* 1969, **54**, 341-375). The concept of intercrystalline reactions between crystals, and between crystals and melts or aqueous solutions was commonplace, but the notion that intracrystalline phenomena constituted chemical reactions, if not new, was presented in an arresting and rigorous way. Chemical reactions in crystals include a variety of styles of polymorphic transformation and exsolution phenomena which the feldspars illustrate *par excellence.* Indeed, it is true to say that the previous three NATO-sponsored volumes on Feldspars (following meetings in Oslo, 1962, Manchester, 1972, and Rennes 1983) were largely concerned with Thompson's intracrystalline reactions. The Fourth meeting, in Edinburgh, reported here, was concerned with both inter- and intracrystalline reactions, and with the unbreakable connection between them. I hope this volume is seen to stand firmly astride the boundary between, on the one hand crystallography and phase behaviour of feldspars (usually regarded as in the domain of mineralogy) and, on the other the interactions between feldspars and other crystalline phases, melts, solutions and even gases (in the domains of petrology and geochemistry). I interpret *reaction* as meaning any response of a feldspar to an *action*, be it change in P, T or X in its vicinity; and changes in X may include the minute changes in the relative concentrations of trace elements and isotopes which can be so informative in petrology.

The book sets out with reviews of crystal structure and cell parameters and leads *via* treatments of crystalline phase transitions into phase relationships involving melting and reactions at very high pressure. The non-equilibrium feature of chemical zoning provides a bridge with chapters on diffusion in feldspars, including oxygen and argon, and on the role of hydrogen in promoting feldspar reactions. From here we head into petrology, recognising the overwhelming importance of reactions between feldspar and silicate liquids in igneous petrogenesis, and the role of water activity. The special case of pegmatites is covered. Penultimately we return to aspects of structure, this time at the surface of the crystals, critical to the ultimate fate of feldspar in the weathering cycle which is the subject of the final article. Although each contribution is entirely individualistic I hope the student reader, in particular, will gain new insights, through the medium of one mineral species, into the interconnections between some of the scientific compartments to which we often adhere. Despite the specialised title, this is actually a very 'interdisciplinary' book, and this interdisciplinarity has always been one of the great strengths of the series of NATO meetings on feldspars. It is my belief that the meetings and the resulting volumes have made a really significant contribution to the development of mineralogy and indeed to petrology and geochemistry. It was the intention of the organisers that the 1993 meeting would further illuminate the connection between crystallography, thermodynamics, geochemistry and petrology and I hope the reader will find that this comes over clearly from the articles which follow.

The four volumes reporting advances in the feldspar field communicated through the medium of the NATO ASI's provide a fascinating record of the evolution of mineralogy over a thirty year period. New techniques have constantly provided new insights and the meetings have been important in their dissemination. These ten-yearly conclaves have certainly been a major factor in my own scientific development (this was my third, only Bill Brown, Paul Ribbe and Joe Smith, of this year's participants, have been to them all) and I sense that this is true for a number of the older members. I hope there will be a fifth NATO meeting, in 2003, so that old friends may meet and new younger friendships begin. Collaboration, not competition, has been the key factor in the astounding successes of science. Long may the feldspars remain a focus for such collaboration; there are many problems remaining.

Ian Parsons, Department of Geology and Geophysics, The University of Edinburgh.
5th November 1993

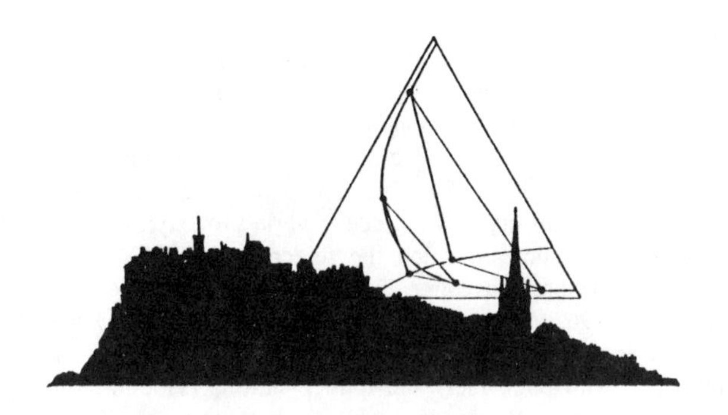

Acknowledgements

'Feldspars and Their Reactions' was largely supported by the Advanced Study Institutes programme of the North Atlantic Treaty Organisation. It was the fourth meeting on the feldspar minerals sponsored by NATO and the continuing success of these meetings confirms the constructive character of the NATO support for science, for which organisers and participants are sincerely grateful. We also thank the Mineralogical Society of Great Britain and the Royal Society of Edinburgh for contributing to the costs of scientists from outside the NATO alliance, and the Royal Society of London and International Science Foundation, New York, for support for participants from the former Soviet Union.

Sincere thanks are also due to the organising committee, W L Brown, M A Carpenter, B J Giletti and W Johannes, for support both before and during the meeting. They, and the invited lecturers, all gave excellent keynote addresses, sustaining the interest of everyone throughout the entire 10-day meeting. Participants, young and old, must be thanked for their lively involvement and good humoured enthusiasm. Peder Aspen worked extremely hard at providing practical support, and leading field excursions, and my current research colleagues, Martin Lee, Mark Hodson and Pauline Thompson were willing and efficient stewards. Aided by Adrian Finch, Kim Waldron and Ray Burgess, they were able to introduce younger members of the Institute to aspects of Scottish culture guaranteed to make the meeting a success.

Many other people made invaluable behind-the-scenes contributions. Isobel Johnstone was the Co-ordinator at Heriot Watt Conference Centre, on the outskirts of Edinburgh, where the meeting was held. Lecture facilities, accommodation and the quality of meals were widely commended by participants. Secretarial assistance of the greatest efficiency was provided by Lucian Begg, who also worked on the preparation of this volume. Ned Pegler used his computer skills to redraw some of the figures. Patricia Stewart assisted at registration, and Helena Jack helped with the production of the programme and abstracts. Heather Hooker helped with production of the final manuscript, and Peter Symms made some of the social arrangements. Last, but by no means least, I would like to thank my wife, Brenda, for her support during the build-up to the meeting, and her tolerance during what, at the end, naturally became an all-consuming event.

Ian Parsons, 1st November 1993

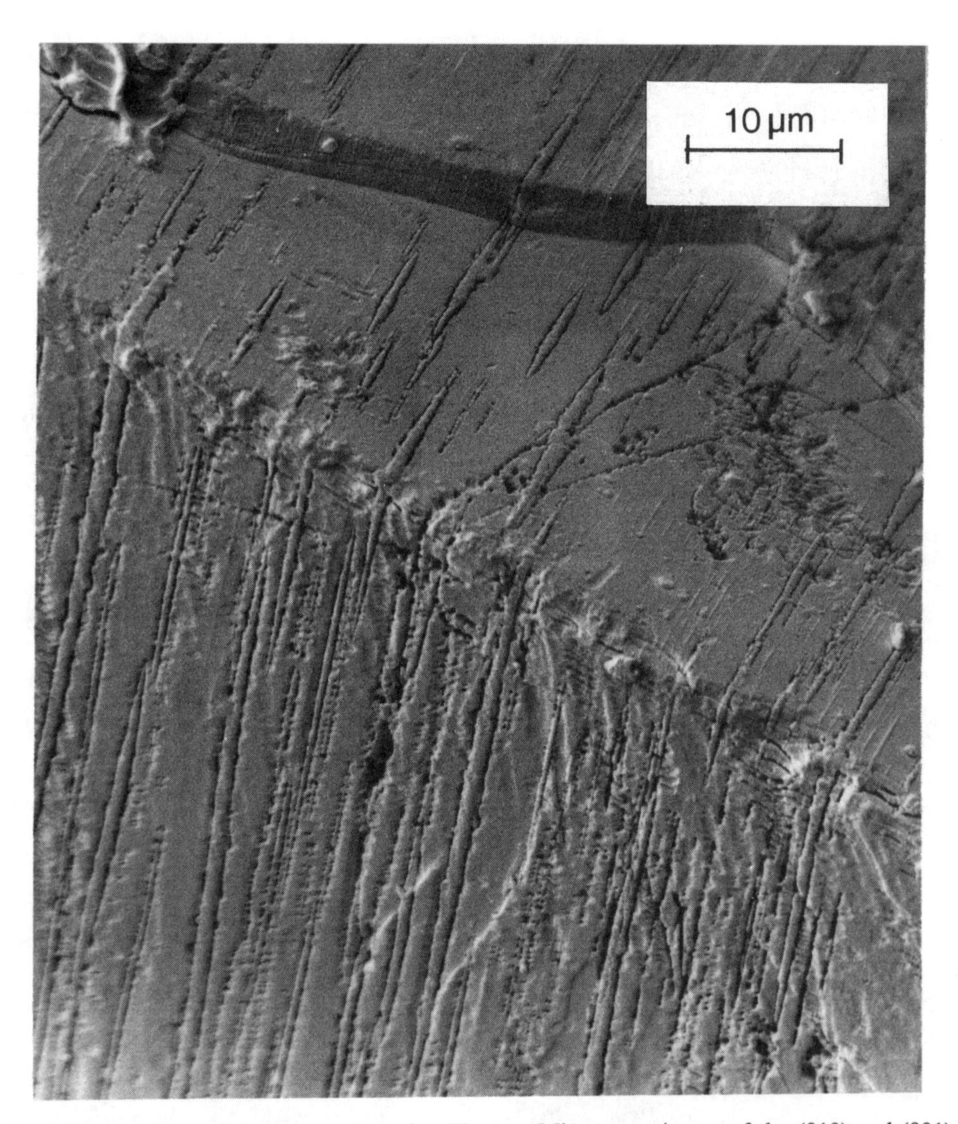

Frontispiece. Secondary electron Scanning Electron Microscope image of the (010) and (001) cleavage surfaces of a fragment of alkali feldspar from a large phenocryst in the Shap granite, northern England, lightly etched in HF vapour. The upper surface, (001), shows lenticular exsolution lamellae of albite at a variety of scales; on (010) (lower part of micrograph) the lamellae are extended sub-parallel to *c*. The edge running from left to right is the *a*-axis and the lamellae lie in the Murchison plane, near ($\bar{6}01$). Etch-pits at the interfaces of the larger lamellae are sites of dislocations; large lamellae are semicoherent while small lamellae are fully coherent. Etching has also preferentially attacked volumes of structure affected by elastic strain leading to grooves along the lamellar interfaces, and to depressions around dislocations with two-fold symmetry on (010) and mirror symmetry on (001). (Photograph by Martin Lee. See Waldron *et al.*, submitted, and Chapter 12 in the present volume).

Participants In The Nato Advanced Study Institute
'Feldspars and Their Reactions', Edinburgh, June 29 - July 10, 1993

NUMBERS REFER TO THE LISTS ON THE FOLLOWING PAGES

PARTICIPANTS IN THE NATO ADVANCED STUDY INSTITUTE
'Feldspars and Their Reactions', Edinburgh, June 29 - July 10, 1993

NUMBERS REFER TO THE GROUP PHOTOGRAPH ON THE PRECEDING PAGE

1 Aglagül, Süheyla
Graduate School of Natural & Applied Sciences, Cukurova Üniversitlsi, Fen Bilimleri Enstitüsü, Adana, Turkey

2 Angel, Ross (Invited Lecturer)
Department of Geological Sciences, University College, Gower Street, London, WC1E 6BT, England

4 Behrens, Harald
Institut für Mineralogie der Universität, Welfengarten 1, D-3000 Hannover 1, Germany

5 Blasi, Achille (Invited Lecturer)
Dipt Scienze della Terra, Universita' degli studi, via S Botticelli 23, I 20133 Milan, Italy

6 Blum, Alex (Invited Lecturer)
US Geological Survey, 345 Middlefield Road, MS 420, Menlo Park, California 94025, U.S.A.

7 Brown, Bill (Organising Committee & Invited Lecturer)
CNRS-CRPG, BP 20, 54501 Vandoeuvre-les-Nancy, cedex, France

8 Burgess, Ray
Department of Geology, The University, Oxford Road, Manchester, M13 9PL, England

9 Carpenter, Michael (Organising Committee & Invited Lecturer)
Department of Earth Sciences, The University, Downing Street, Cambridge, CB2 3EQ, England

10 Cerny, Petr (Invited Lecturer)
Department of Geological Sciences, 240 Wallace Building, University of Manitoba, Winnipeg, Manitoba, Canada, R3T 2N2

11 Chernysheva, Irene
Vernadsky Institute of Geochemistry, 19 Kosygin St, Moscow 119795, Russia

12 Colombo, Monica
Dip. di Scienze della Terra, Univ degli Studi di Milano, via S Botticelli 23, I 20133 Milan, Italy

13 Daniel, Isabelle
Institut de Geologie, Université de Rennes, 35042 Rennes cedex, France

14 De Pol Blasi, Carla
Dip. di Scienze della Terra, Univ degli Studi di Milano, via S Botticelli 23, I 20133 Milan, Italy

15 Downs, Robert
Geophysical Laboratory, Carnegie Institution of Washington, 5251 Broad Branch Road NW, Washington, DC 20015-1305, U.S.A.

16 Evangelakakis, Christos
Institut für Mineralogie, Corrensstr 24, 4400 Münster, Germany

17 Finch, Adrian
Department of Chemistry, Merton Walk, Old Aberdeen, AB9 2UE, Scotland

18 Fitz Gerald, John
Research School of Earth Sciences, Australian National University, Canberra, ACT 2061, Australia

19 Foland, Kenneth (Invited Lecturer)
Department of Geological Sciences, Ohio State University, 125 South Oval Mall, Columbus, Ohio 43210-1398, U.S.A.

20 Fountain, Kendall
1112 Turlington Hall, Department of Geology, University of Florida, Gainesville, Florida 32611, U.S.A.

21 Giletti, Bruno (Organising Committee & Invited Lecturer)
Department of Geological Sciences, Brown University, Providence, Rhode Island 02912, U.S.A.

22 Göttlicher, Jorg
Institut für Mineralogie, Corrensstr 24, 4400 Münster, Germany

23 Graham, Colin (Invited Lecturer)
Department of Geology & Geophysics, Edinburgh University, Edinburgh, EH9 3JW, Scotland

24 Grammatikopoulos, Tassos
Department of Geological Sciences, Queen's University, Kingston, Ontario, Canada, K7L 3N6

25 Grove, Martin
Université Blaise Pascal, 5 rue Kessler, 63038 Clermont- Ferrand cedex, France

26 Gualtieri, Alessandro
Istituto di Mineralogia, Modena University, via S Eufemia 19, 41100 Modena, Italy

27 Hafner, Stefan
Institute of Mineralogy, University of Marburg, Lahnberge, 3550 Marburg, Germany

28 Hanghøj, Karen
Geological Museum, Øster Voldgade 5-7, DK-1350 København K, Denmark

29 Harrison, T Mark
Department of Earth & Space Sciences, University of California, Los Angeles, CA 90024, U.S.A.

30 Hodson, Mark
Department of Geology & Geophysics, Edinburgh University, Edinburgh, EH9 3JW, Scotland

31 Hovis, Guy
Department of Geology, Lafayette College, Easton, Pennsylvania 18042-1768, U.S.A.

32 Johannes, Wilhelm (Organising Committee & Invited Lecturer)
Institut für Mineralogie, Universität Hannover, Welfengarten 1, D-3000 Hannover 1, Germany

33 Karaoglu, Nevin
Department of Geology, Dokuz Eylül University, 35100 Bornova, Izmir, Turkey

35 Kimata, Mitsuyoshi
Institute of Geoscience, The University of Tsukuba, Ibaraki 305, Japan

36 Kirkpatrick, R James
Department of Geology, University of Illinois, 245 Natural History Bldg, 1301 West Green Street, Urbana, Illinois 61801, U.S.A.

37 Klein, Jeanette
Institute of Mineralogy, Philipps University Marburg, Hans- Meerwein str, W-3550 Marburg, Germany

38 Köklü, Ünel
Istanbul Teknik Üniver, Fen-Edebiyat Fakültesi, Kimya Bölümü, 80626 Maslak, Istanbul, Turkey

39 Koroneos, Antonis
Department of Mineralogy-Petrology-Economic Geology, Aristotle University of Thessaloniki, GR 54006 Thessaloniki, Greece

40 Kotelnikov, Alexey
Institute of Experimental Mineralogy, 142432 Chernogolovka, Moscow distr, Russia

41 Kroll, Herbert
Institut für Mineralogie, Westfälische Wilhelms Univ, Corrensstrasse 24, D-4400 Münster, Germany

42 Laurenzi, Marinella
Instituto di Geochronologia e Geochimica Isotopica, via Cardinale Maffi, 36-56127 Pisa, Italy

43 Lee, Martin
Department of Geology & Geophysics, Edinburgh University, Edinburgh, EH3 9EW, Scotland

44 Lester, Alan
Department of Geological Sciences, University of Colorado, Campus Box 250, Boulder, Colorado 80309-0250, U.S.A.

45 Mannerstrand, A Maria
Department of Mineralogy and Petrology, Institute of Geology, University of Lund, Sölvegatan 13, S-223 62 Lund, Sweden

46 Martin, Robert
Department of Geological Sciences, McGill University, 3540 University Street, Montreal, Canada, H3A 2A7

47 McConnell, J Desmond
Department of Earth Sciences, University of Oxford, Parks Road, Oxford, England

48 McGuinn, Martin
Department of Geology, The University, Oxford Road, Manchester, M13 9PL, England

49 McLaren, Alex
Research School of Earth Sciences, The Australian National University, Canberra, ACT 2601, Australia

49a McLaren, Netta

50 Nater, Edward
Soil Science Department, 439 Borlaug Hall, University of Minnesota, St Paul, Minnesota 55108-6028, U.S.A.

51 Neiva, Ana
Museu e Laboratório Mineralógico e Geológico, Universidade de Coimbra, 3049 Coimbra codex, Portugal

52 Nekvasil, Hanna (Invited Lecturer)
Department of Earth & Space Sciences, State University of New York, Stony Brook, New York 11794-8200, U.S.A.

53 Ohnenstetter, Daniel
CRSCM, 1A rue de la Férollerie, 45071 Orléans cedex 2, France

54 Parsons, Ian (Director)
Department of Geology & Geophysics, Edinburgh University, Edinburgh, EH9 3JW, Scotland

54a Parsons, Brenda

55 Pavese, Alessandro
Dip. di Scienze della Terra, Univ degli Studi di Milano, via S Botticelli 23, I 20133 Milan, Italy

56 Pearce, Thomas (Invited Lecturer)
Department of Geological Sciences, Queens University, Kingston, Ontario, Canada, K7L 3N6

57 Pentinghaus, Horst
Institut für Nukleare Entsorgungs-technik, Kernforschungszentrum, Postfach 3640, W-7500 Karlsruhe 1 Germany

58 Phillips, Brian
L-219, Lawrence Liv ermore Nat Lab, PO Box 808, Libermore, CA 94551, U.S.A.

59 Pryer, Lynn
Department of Geology, University of Toronto, 22 Russell Street, Toronto, Ontario, Canada, M5S 3B1

60 Redfern, Simon
Department of Geology, The University, Oxford Road, Manchester, M13 9PL, England

61 Ribbe, Paul (Invited Lecturer)
Department of Geological Sciences, Virginia Poly Institute, 4044 Derring Hall, Blacksburg, Virginia 24061-0420, U.S.A.

62 Salje, Ekhard (Invited Lecturer)
Department of Earth Sciences, The University, Downing Street, Cambridge, CB2 3EQ, England

63 Sanchez Munoz, Luis
Museo Nacional de Ciencias Naturales, J Gutierrez Abascal 2, 28006 Madrid, Spain

64 Smith, Joseph (Invited Lecturer)
Department of Geophysical Sciences, The University of Chicago, 5734 South Ellis Avenue, Chicago, Illinois 60637, U.S.A.

65 Soldatos, Triantafyllos
Department of Mineralogy-Petrology-Economic Geology, Aristotle University of Thessaloniki, GR 54006 Thessaloniki, Greece

66 Steele, Ian
Department of Geophysical Sciences, The University of Chicago, 5734 S Ellis Avenue, Chicago, Illinois 60637, U.S.A.

67 Swierkocki, Joseph
Laboratory of Process Kinetics, Institute of Physical Chemistry, The Polish Academy of Sciences, Kasprzaka 44/52, 01-224 Warsaw, Poland

68 Teertstra, David
Department of Geological Sciences, University of Manitoba, Winnipeg, Manitoba, Canada, R3T 2N2

69 Thompson, Pauline
Department of Geology & Geophysics, Edinburgh University, Edinburgh, EH9 3JW, Scotland

70 Viswanathan, Krisnamoorthy
Abt Min und Kristal., Inst für Geowissenschaften, Postfach 3329, D-3300 Braunschweig, Germany

71 Waldron, Kim
Department of Geology, Colgate University, 13 Oak Drive, Hamilton, NY 13346-1398, U.S.A.

72 Walker, David
Corrie, Woodlands, Markinch, Fife, KY7 6HE, Scotland

73 Weger, Matthias
Mineralog-Petrogr. Inst., Ludwig Maxim Univ München, Theresienstrasse 41, W-8000 München 2, Germany,

74 Witt-Eickschen, Gudrun
Mineralogisch-Petrographisches Institut, Universität zu Köln, Zülpicherstrasse 49 b, W-5000 Köln 1, Germany

75 Wondratschek, Hans
Institut für Kristallographie, der Universität Karlsruhe, Kaiserstrasse 12, D-7500 Karlsruhe, Germany

75a Wondratschek, Hella

76 Wood, David
Department of Geology, The University, Oxford Road, Manchester, M13 9PL, England

77 Xu, Huifang
Department of Earth & Planetary Sci., The Johns Hopkins Univ, Baltimore, Maryland 21218, U.S.A.

78 McConville, Paul
Scottish Universities Research and Reactor Centre, East Kilbride, Glasgow, G75 0QU, Scotland (*Visitor*).•

THE CRYSTAL STRUCTURES OF THE ALUMINUM-SILICATE FELDSPARS

PAUL H. RIBBE
Department of Geological Sciences
Virginia Polytechnic Institute & State University
Blacksburg, Virginia, U.S.A. 24061-0420

ABSTRACT. The vast array of crystal structure refinements, lattice parameter measurements, optical and spectroscopic data, and electron microscopic observations accumulated on feldspar since W. H. Taylor unravelled the sanidine structure in 1933 has led to detailed understanding of many but by no means all of the diffusive and displacive transformations occurring in both natural and synthetic systems as functions of pressure, temperature, and composition (P,T,X). This paper reviews the basic structural variants of feldspars and summarizes the dominant structurally related observations, particularly for the compositions $(Na,K)_x(Ca,Ba)_{1-x}Al_{2-x}Si_{2+x}O_8$. The immediate effects of heating (thermal expansion) and the longer-term effects of annealing (Al,Si order-disorder) may produce first- or second-order phase transitions; these are considered together with structural changes that occur under compression and with compositional variation (by adding Rb, Li, Ag, H, Sr, Pb, Eu to the mix). Structural variables are examined as functions of the unit cell volume (or V/2 in the case of alkaline-earth feldspars with doubled **c**-axes), permitting trends in cell parameters, M-O distances, and T-O-T intertetrahedral angles to be compared among suites of feldspars with ranges of P, T, and X. An attempt to model the details of the crystal structure of low albite at 5.0 GPa, using only its lattice parameters and the principle that P, T, and X are structurally analogous variables, illustrates both the utility and the limitations of that principle.

1. Introduction

1.1. HISTORICAL PERSPECTIVE

Building on Schiebold's (1927-1930) speculations *Über den Feinbau der Feldspäte* and Machatschki's (1928) conclusion that feldspars were tetrahedral framework structures, W. H. Taylor (1933) determined the actual arrangement of Al,SiO_4 tetrahedra around the large K cation in the archetypal feldspar, monoclinic C2/m sanidine. By 1962 when he presented the introductory talk at the first NATO ASI on feldspars in Oslo, Taylor with his students and collaborators had produced only eleven crystal structure refinements but essentially had covered the range of known end-member compositions (K, Na, Ca, Ba), space group symmetries (C2/m, $C\bar{1}$, I2/c, $I\bar{1}$, $P\bar{1}$), and Al,Si order-disorder arrangements (low to high structural states in K- and Na-feldspars; partial disorder in oligoclase, andesine, and bytownite).

I. Parson (ed.), Feldspars and Their Reactions, 1–49.

In 1972 at the second NATO ASI, Megaw (1974a,b) presented her classic papers on the architecture of the feldspars in which she explained variations in structure amongst the various chemical species in terms of tilting and rotating framework tetrahedra to accommodate the large M cation, temperature changes, and various Al,Si arrangements. The first neutron diffraction results (high and low sanidine) were reported at that time, lending credence to existing interpretations of mean T-O bond lengths in terms of the relative Al/(Al+Si) content of the tetrahedral T sites. Transmission electron microscopy (TEM) was applied to imaging exsolution, twinning, and strain phenomena in alkali and plagioclase feldspars, and antiphase domains in plagioclases. By the third NATO ASI (1983) nearly 200 structure refinements had been undertaken, including some at high temperatures (low and high albite, monalbite, anorthite, sanidine) and numerous synthetic analogues containing Ge for Si, and Ga, Fe, B for Al in T sites, and Sr, Rb, NH_4 in M sites. Neutron diffraction settled the fact that Al and Si are completely ordered in low albite. High-resolution TEM provided greater insight into coherent and spinodal exsolution, strain, diffusion, incommensurate structures in intermediate *'e'*-plagioclases, antiphase domains, and phase transitions in anorthite and the K-rich feldspars.

1.2. RECENT PROGRESS AND OVERVIEW

In this chapter we will lay a base for understanding all feldspar structures and review recent progress in crystallographic studies. In the past decade these have come from neutron diffraction refinements (low albite at 13 K, a volcanic An_{48}, An_{100} at 514 K), Distance-Least-Squares (DLS) refinements (Li- and Ag-rich exchange products from K- via H-feldspar), and many conventional x-ray diffraction refinements, including several high sanidines, H-feldspar, intermediate albite, Ga- and Ge-substituted synthetics, plus a broad range of Ca-rich plagioclases – some at high pressures (see chapter by Angel, this volume). Thanks to the application of Landau theory and the accumulation of considerable thermodynamic, crystallographic, and IR and Raman spectroscopic data, the various $P\overline{1} \Rightarrow I\overline{1} \Rightarrow C\overline{1} \Rightarrow C2/m$ transitions in alkali feldspars and many plagioclases have been investigated intensively from the perspectives of spontaneous strain and Al,Si and Na,K ordering, using thermodynamic order parameters as functions of temperature and pressure (Salje, Carpenter, Angel, Navrotsky, and numerous collaborators; see other chapters). McGuinn and Redfern (1993) found a first-order ferroelastic transition $I2/c \Rightarrow I\overline{1}$ in synthetic Sr-feldspar at 3.2±0.4 GPa, and Angel reports a discontinuity in the cell parameters of low microcline at high pressure (pers. comm., 1993).

Because so many crystal structures at varying P, T, and X (composition) have been completed and the results correlated with other, more easily measured physical properties, structures often are predictable in many aspects (Al,Si distribution, bond lengths and angles, even atomic coordinates) using selected lattice parameters and/or optical properties ± composition (e.g., Su et al., 1984 et seq., on alkali feldspars). Currently the major structural unknowns include albite, microcline, and high sanidine at high pressures (the last apparently does *not* invert to $C2/m \Rightarrow C\overline{1}$ as predicted). In some respects P, T, X *are* structurally analogous variables in feldspars, but compressibility and thermal expansion have distinctively different effects on certain bond lengths and angles: T-O bonds corrected for aniostropic thermal vibration *do* increase in length with T (Downs et al., 1992), but they appear to be essentially incompressible. M-O bonds are compressible, but there must be limits; T-O-T angles (and to some lesser degree O-T-O angles, especially where T = Al) distort predictably and smoothly with P, T, and X, except at the displacive $P\bar{1} \Rightarrow I\bar{1}$ transition in anorthite at ~2.61 GPa (Angel, 1988). Predictions of high-pressure structures based on known lattice parameters and DLS refinements weighted for the force constants are informative, but intensity data collected *at* high pressures are required to complete our understanding of the full range of feldspar structures.

2. Chemistry of Feldspars

With few exceptions, natural feldspars are MT_4O_8 aluminosilicates whose structures are composed of corner-sharing AlO_4 and SiO_4 tetrahedra linked in an infinite three-dimensional array. Compounds with feldspar topology have been synthesized substituting B, Ga, and Fe^{3+} in the $[T^{3+}]_{1-2}$ sites and Ge_{3-2} and AlSiP in the $[T^{4+}]_{3-2}$ sites (Pentinghaus, 1980, et seq.; Burns and Fleet, 1990; Kroll et al., 1991; references therein), but this chapter will not be concerned with them.

Most natural feldspars occur in the K-Na-Ca ternary (cf. chapter by Nekvasil); they have the general formula $K_xNa_yCa_{1-(x+y)}Al_{2-(x+y)}Si_{2+(x+y)}O_8$, or in end-member terms, $Or_xAb_yAn_{1-(x+y)}$. Members of the K-Ba ('hyalophane') series, K-Na-Ba ternaries, reedmergnerite, ($NaBSi_3O_8$, an albite isomorph), and C2/m buddingtonite ($NH_4AlSi_3O_8$) are rarely observed. Buddingtonite has long been assumed to contain structural water (0.5 H_2O), but recent syntheses show it to be unessential (Voncken et al., 1988). Of course, minor and trace element substitutions at M as well as T sites are common.

Terrestrial feldspars with a few percent excess Al_2O_3 or SiO_2 are known (see review by Smith, 1983, p. 284). Other defect structures with the formula $M^{2+}_{1-z}\square_zAl_{2-2z}Si_{2+2z}O_8$, where $\square$ represents vacancies in Sr-, Ca-, and Eu-feldspars, have been synthesized and examined crystallographically by Grundy and Ito (1974), Bruno and Fachinelli (1974), and Kimata (1988), respectively. Bruno and Pentinghaus (1974) gave a broad overview of cation substitutions in natural and synthetic feldspars (cf. Smith and Brown, 1988, Ch. 7 for additional insight).

This chapter has been restricted to the consideration of stoichiometric $M^{1+,2+}(Al,Si)_4O_8$ feldspars, in particular those with the M cations shown in Figure 1. Encircled elements represent the primary, naturally occurring feldspar end members; solid or dashed lines join the binary series which either exist in nature or have been synthesized mostly by hydrothermal or alkali-exchange methods. The incomplete line toward Ca from K indicates the limit of K exchange for Na in natural plagioclases. Missing from the diagram is H^+, which has special properties and whose effects on the feldspar structure are unique because it can bond to only one anion at a time (see Paulus and Müller, 1988, and the chapter by Graham). Hydrogen is critical to the synthesis of Li- and Ag-rich feldspars (Deubener et al., 1991; see § 6 below), because the substitutional sequence to reach either of these compositions involves exchange of K by Na from Al,Si disordered sanidine (in molten NaCl), the exchange of Na by H (in hot, concentrated H_2SO_4), and finally the exchange of H by Li or Ag (in molten Li- or Ag-nitrate). A Cu^+ analogue could not be produced by this method. LaNa-feldspars have been synthesized, but not a Mg-feldspar, undoubtedly

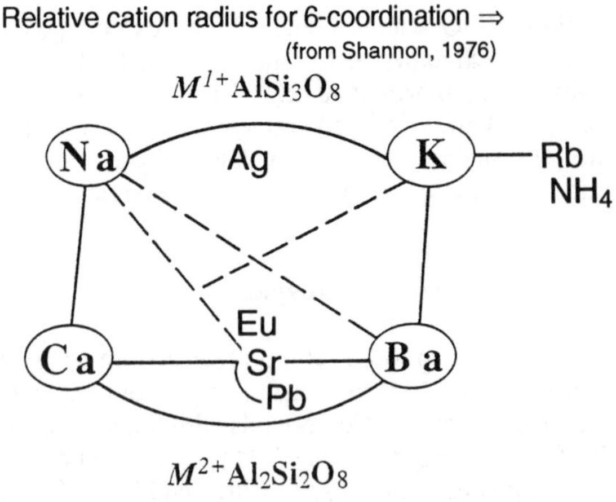

Relative cation radius for 6-coordination ⇒
(from Shannon, 1976)

$M^{1+}AlSi_3O_8$

Li Na Ag K Rb NH_4

Eu Sr Pb

Ca Ba

$M^{2+}Al_2Si_2O_8$

Figure 1. Display of the M-cations relative to the effective ionic radii (horizontal scale only) for the various aluminosilicate feldspar structures considered in this review. NH^{4+} is represented as 'larger' than Rb^+ because the unit cell volume of $NH_4AlSi_3O_8$ is slightly larger than $RbAlSi_3O_8$ (Voncken et al., 1993).

because Mg prefers tetrahedral coordination at atmospheric pressure—Murakami et al. (1992) have synthesized $Ca[Mg_{0.25}Al_{1.5}Si_{2.25}]^{iv}O_8$ with space group $C\bar{1}$.

In Figure 1 the 6-coordinated effective ionic radii range from 0.76 Å for Li^+ to twice that (1.52 Å) for Rb^+, but in fact the coordination numbers of M cations vary from as low as 5 in the former to as high as 11 in the latter, depending on how one defines coordination. Because radii are essentially bond-length predictors, their *relative* values are all that are of interest in this review. And since temperature (T) has the effect of increasing cation 'size' via thermally imparted atomic vibrations, we will find that T and X are for the most part structurally analogous variables in feldspars.

3. Topology of the Feldspar Framework

There are numerous ways to link AlO_4 and SiO_4 tetrahedra into three-dimensional framework structures of composition $M^{1+,2+}(Al,Si)_4O_8$, including monoclinic $P2_1/a$ paracelsian ($BaAl_2Si_2O_8$; Chiari et al., 1985) and slawsonite ($SrAl_2Si_2O_8$; Griffen et al., 1977), two monoclinic anorthite polymorphs (pseudo-orthorhombic $CaAl_2Si_2O_8$, which is a twinned $P2_1$ phase; pseudo-hexagonal $CaAl_2Si_2O_8$, which is a twinned C2 phase; see Iyatomi and Aoki, 1992, for a review), and a metastable orthorhombic Immm structure (Takéuchi et al., 1973)[see Abe et al (1991) and Daniel et al. (1993) for other polymorphs]. Pentinghaus (1980) reviewed the polymorphism other of compounds with feldspar-like formulas, many of which are not framework structures, and Smith (1974) summarized the topologies of minerals which are structurally related to feldspars. But the feldspar structure itself has a unique topology.

Taylor (1933) determined the structure of what H.D. Megaw called the aristotype feldspar, C2/m sanidine. His illustrations have reappeared in the literature for 60 years, with only minor embellishments (see Fig. 2). The fundamental structural unit is the so-called 'crankshaft' chain of 4-membered tetrahedral rings. Most structural accommodation of changes in P, T, and X (i.e., M and T site occupancy) can be understood in terms of thebending of T-O-T angles between TO_4 tetrahedra and, to a lesser degree, of O-T-O

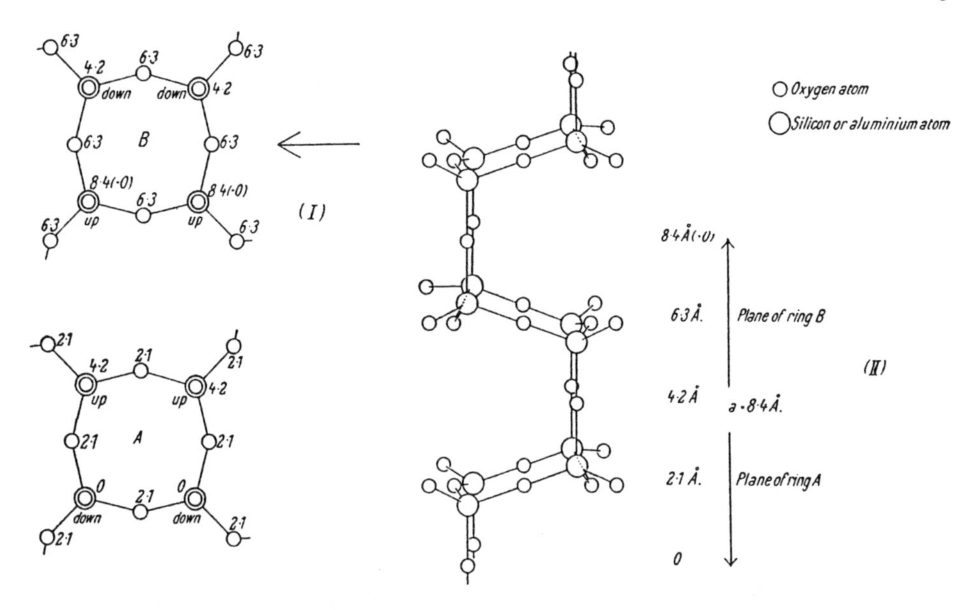

Figure 2. From Taylor (1933). "Idealized diagram of continuous chains of tetrahedron rings parallel to the a-axis. In (I) [left] the rings are viewed along the a-axis, the numbers indicating the heights of oxygen atoms above the base plane. Ring B is to lie above ring A in such a way that the oxygens at height 4.2 Å above the base plane are common to both rings. The four tetrahedra thus linked together (2 from A, 2 from B) form a new ring of four tetrahedra. In (II) the linked rings are viewed in the direction indicated by the arrow in the upper diagram."

intratetrahedral angles. Two of these chains are represented in perspective in Figure 3 in which T-atoms are at the corners of quadrilaterals (4-membered rings); most oxygen atoms are omitted. The crankshafts are mapped into one another by mirrors || (010), or ⊥**b** at $^1/_4$**b** and $^3/_4$**b**, in monoclinic C2/m feldspars. In all triclinic feldspars they will be replicated approximately by pseudo-mirrors || (010), ~⊥**b**. Megaw (1973) spoke of these structural units as double crankshaft 'slabs;' they are repeated throughout space by **c**-translations and cross-linked to one another by the OA_1 oxygens (**o** in Fig. 3; see Fig. 4a). Thus (001) is a plane of structural weakness, responsible for the good (001) cleavage in all feldspars. The OA_2 oxygens, located halfway along the dotted lines in Figure 3 (cf. Fig. 4a), are the main link between adjacent crankshaft chains. The $M^{1+,2+}$ cation sites are located between the (001) slabs and lie exactly on the (010) mirror in monoclinic feldspars or very near the pseudo-mirrors in triclinic feldspars (Fig. 4a). Because bonds to these large, 6- to 9-coordinated cations are very weak by comparison with those to the 4-coordinated Al and Si atoms, they are responsible for the excellent (010) cleavage in feldspars.

Within the crankshaft chains a variety of Al,Si distributions is possible, depending on the Al/Si ratio of the feldspar and its 'structural state,' i.e., the degree of order of Al and Si amongst the T-sites. Figure 4b is a stylized view of the shaded portion of Figure 3 and shows only the tetrahedral nodes of Figure 4a.

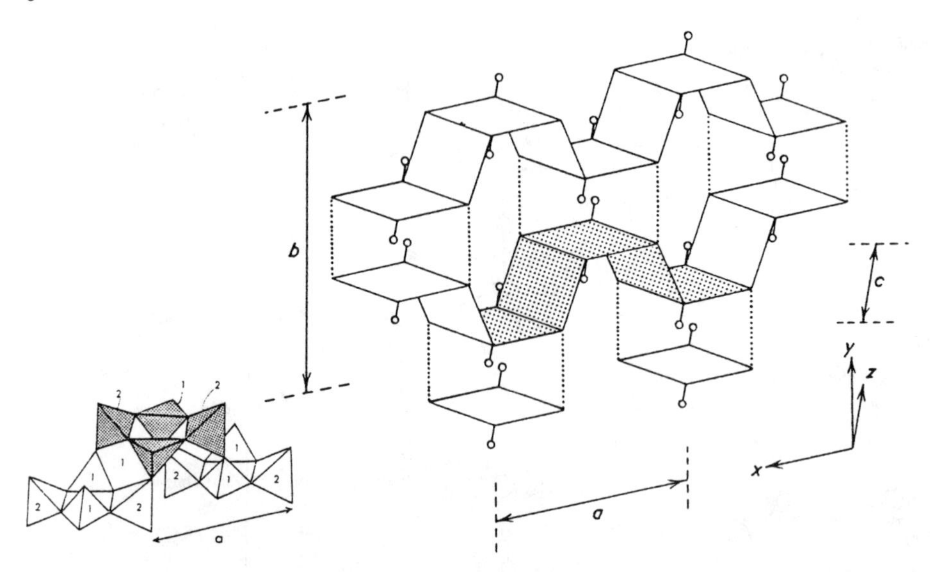

Figure 3. Lower left: Perspective view showing the linkage of (Al,Si)O_4 tetrahedral units ...T1-T2-T1-T2... in the stippled portion of the crankshaft chain (from Megaw, 1974a). Mirror-related 'double crankshaft' tetrahedral chains forming (001) 'slabs' in the feldspar structure. Details discussed in the text; modified from Megaw (1973, Fig. 11-25).

4. Patterns of Al,Si Order-Disorder: Space Groups

Figure 4b was adapted by Megaw (1974a) to display various Al,Si arrangements among feldspars. These are reproduced with projections of symmetry operators in Figure 5 for the sake of classification of the range of naturally occurring minerals, their isomorphs, and their high-P and high-T polymorphs. They also provide reference points for understanding later chapters that refer to space groups.

Space groups are easily determined from x-ray diffraction patterns. And because of the large difference in scattering cross-sections for Al and Si, direct refinement of Al/(Al+Si) contents of individual T sites in a feldspar crystal can be obtained from neutron diffraction data, although precision is reduced when the fractional occupancies of the sites approach equivalence. But most structures have been refined from x-ray diffraction data, and the x-ray scattering factors for Al and Si are too nearly alike to permit accurate site-occupancy refinements (Blasi and De Pol Blasi, this volume, can do it for ordered low microcline). Thus we are constrained to use T-O distances, calculated from atomic positional parameters in the conventional way, to estimate Al/(Al+Si) for each tetrahedral site. Precision of T-O distances is not composition-dependent. In low albite (LA), which has been refined by neutron diffraction both at 297 K and 13 K (Smith et al., 1986), the mean T-O distance for the AlO_4 tetrahedron is 1.743 Å and for the three SiO_4 tetrahedra 1.613 Å, a difference of 0.130 Å. Low microcline (LM), assumed to be fully ordered by analogy with its alkali-exchanged LA equivalent, has ⟨Al-O⟩ = 1.738 Å and ⟨Si-O⟩ = 1.613 Å, a difference of 0.125 Å. A grand mean distance, ⟨⟨T-O⟩⟩, can be calculated which would increase linearly

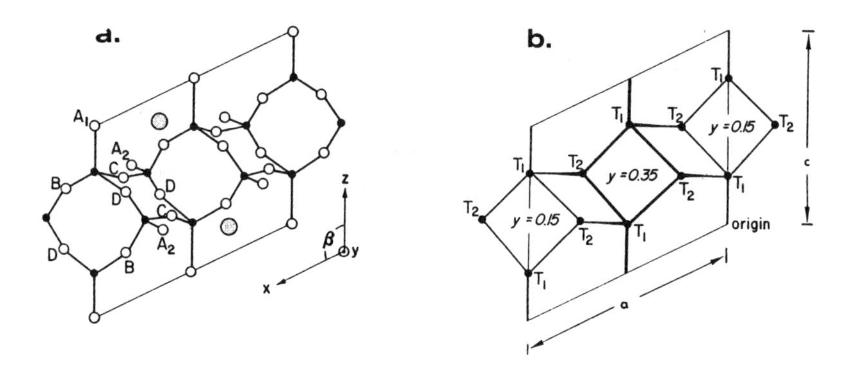

Figure 4. (a) Projection of a portion of the archetypal feldspar structure illustrated in Figures 2(II) and 3 (stippled), showing the linkage of four-fold tetrahedral rings (T atoms are dots; oxygens (O) are small circles labelled by convention, B, C, D; A_1s join T1 tetrahedra, A_2s join T2 tetrahedra [note that the OA_1 oxygens link the slabs of crankshaft chains together in the *z*-direction; cf. Fig. 3], and the M cations (large, shaded circles). (b) Stylized version of 4a showing symmetrically nonequivalent tetrahedral sites (T1 and T2) and the fractional distance along **b** for the nearly planar 4-membered rings which are ~|| (010); steeply inclined 4-membered rings joining these are the A,B rings in Figure 2(II).

with increasing total Al in the feldspar formula; for example, in $Or_{80}Ab_{16}An_4$, the 4 mol % anorthite would bring the total Al in the formula to 1.04 and add ~0.0016 Å to ⟨⟨T-O⟩⟩.

Using this information and the results of scores of modern crystal structure analyses, Kroll and Ribbe (1983) established a method for calculating the average Al content (t_i) of an individual tetrahedral site (T_i) from mean bond lengths that may be used for all but the more An-rich plagioclases at room temperature:

$$t_i = 0.25(1 + n_{An}) + (\langle T_i\text{-}O\rangle - \langle\langle T\text{-}O\rangle\rangle)/\text{constant}\,, \qquad (1)$$

where n_{An} is the mole fraction anorthite content, the "constant" term is 0.130 Å for Na-rich and 0.125 Å for K-rich feldspars; t_i is Al/(Al+Si) for site T_i, and $(1-t_i)$ is the probability of finding Si in that site.

See § 4.2.3 for an interpretation of ⟨T-O⟩ bond lengths in An-rich plagioclases.

4.1. ALKALI FELDSPARS

Alkali feldspars, $(Na,K)AlSi_3O_8$, are extremely well characterized structurally; data were compiled by Kroll and Ribbe (1983) and updated by Ribbe (1984). New refinements since then include those by Scambos et al. (1987) and Ferguson et al. (1991) - high sanidines, HS), Griffen and Johnson (1984 - intermediate microcline), Blasi et al. (1987 - LM; see also Blasi and De Pol Blasi, this volume), Phillips (pers. comm. - LM at 163 and 297 K), Phillips et al. (1989 - intermediate albite), Armbruster et al. (1990 - LA), and an important neutron diffraction refinement of LA at 13 K by Smith et al. (1986). Refinements at elevated temperatures not reported by Kroll and Ribbe include sanidine (Ohashi and Finger,

1974, 1975), three anorthoclases (Harlow, 1982), LA (Winter et al., 1977), high albite (HA; Prewitt et al., 1976), and HA, analbite (AA), and monalbite (MA) (Winter et al., 1979).

4.1.1. *Monoclinic (C2/m) K-rich Alkali Feldspars.* As we have seen, there are just two symmetrically nonequivalent T sites in C2/m feldspars (Figs. 4 and 5a), but there are four

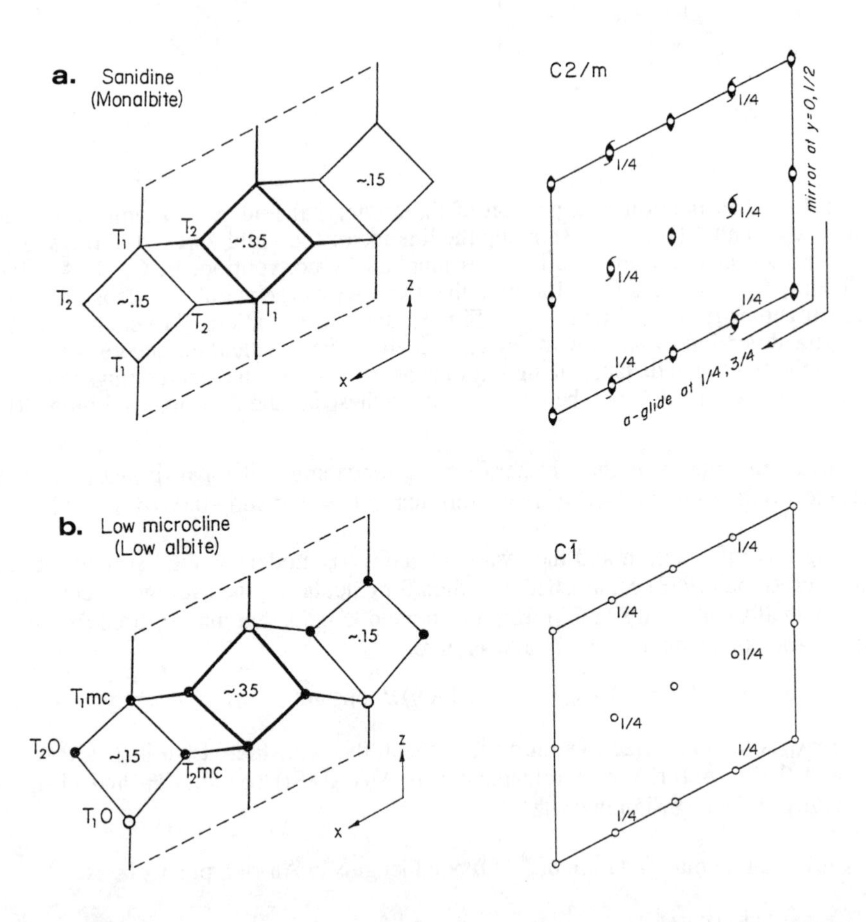

Figure 5. Projections on (010) of the symmetry elements and topology of alkali feldspars. (a) C2/m feldspars include any with **c** ≅ 7 Å whose two T1 sites (and two T2 sites) are symmetrically equivalent, related by two-folds ∥ **b**. (b) $C\bar{1}$ feldspars with **c** ≅ 7 Å have asymmetric 4-membered rings ~∥ (010), caused in the case shown by Al ordering into the T1 site. T1 must now be labelled T1O to differentiate it from its pseudo-2-fold-related T1mc site. T1mc gets its complex label from the fact that there is a pseudo-mirror-related T1m site (not shown) immediately below T1O at ~-0.15**b** which in turn is mapped by a center of symmetry (°, symbol *c*) at 1,0,$^1/_2$ into the position shown. In LA and LM, T1O contains Al; the other three T sites contain Si. Note that the central (010) ring at ~0.35**b** is related to adjacent rings by centers (◦) at $^1/_4$**b**, as well as by pseudo-*a*-glides ∥ (010).

T sites per ...T1-T2-T1-T2... ring and (1 Al + 3 Si) to fill them: thus, $2t_1 + 2t_2 = 1.0$. If complete disorder could be reached at finite temperatures, t_1 and t_2 would equal 0.25. It is not possible to have a fully ordered Al,Si distribution in the 4-membered ring: the closest to this would be to concentrate all the Al in the two T1 sites so that $t_1 = 0.5$ and $t_2 = 0.0$.

Mean T1-O and T2-O bond lengths from room temperature refinements of monoclinic K-rich feldspars (Or_{84-98}) are plotted in Figure 6a as a function of the ⟨T1-O⟩ values simply to display the known structural data. Although attempts have been made to attach a temperature scale to such a diagram, they have all failed except in the most general sense: more disordered structures are more likely to have equilibrated at higher temperatures than ordered ones, but P_{H_2O} has a profound effect (see also Martin, 1974). Using Equation (1), these structures are found to represent a range of Al,Si distributions from eseentially complete disorder [$t_1 = 0.25$] for HS (Ferguson et al., 1991) to a highly ordered low sanidine (LS) [$t_1 = 0.45$, $t_2 = 0.05$]. The latter was refined by neutron diffraction (Prince et al., 1973) and reported to be an 'ordered orthoclase' with all the Al in the two T1 sites (i.e., $t_1 = 0.50$), but the accuracy of the neutron site refinement is questioned because low sanidines are known with lattice parameters indicating even greater degrees of order (also see comments by Fitz Gerald et al., 1986, p. 1404, on neutron refinements).

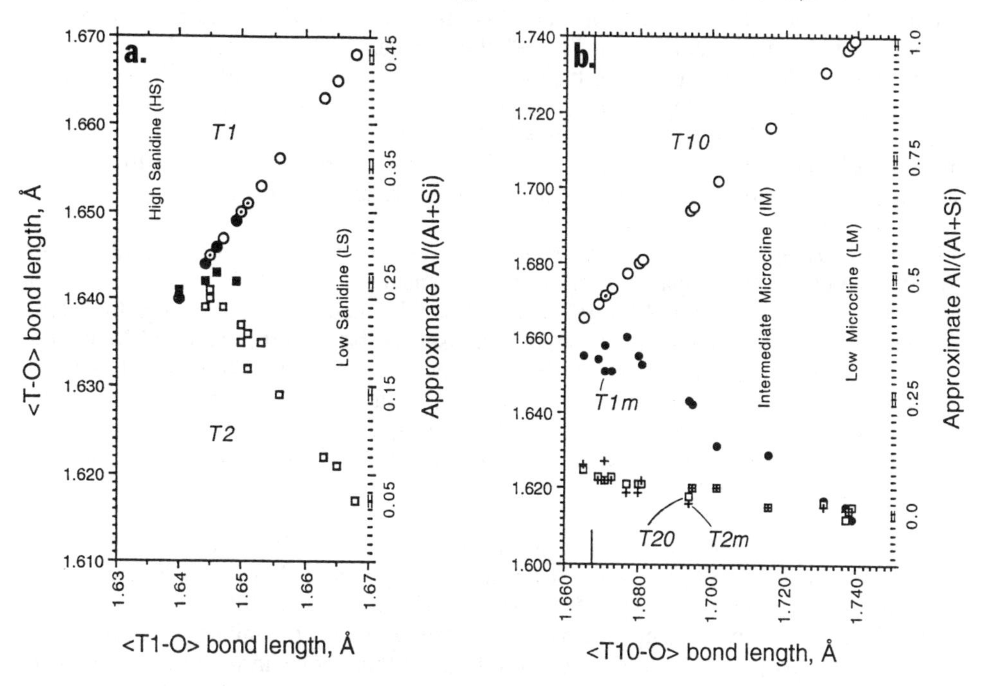

Figure 6. Plot of the mean T-O bond lengths from known crystal structures of K-rich alkali feldspars; see text for discussion. (a) Monoclinic high and low sanidines, including those feldspars commonly classified as 'orthoclases.' Dots in circles indicate two data. Note that the darkened symbols are from four refinements by Ferguson et al. (1991), all showing considerably larger than expected ⟨T2-O⟩ and ⟨⟨T-O⟩⟩ values. (b) Triclinic intermediate and low microclines.

4.1.2. T*riclinic ($C\bar{1}$) K-rich Alkali Feldspars.* The transition from monoclinic LS to triclinic intermediate microcline (IM) in nature is complex and involves the concentration of Al into one or other of the energetically equivalent T1 sites of sanidine: the T2s are already equally Si-rich. This two-step ordering process can be represented graphically from suites of natural specimens (Kroll and Ribbe, 1983, p. 94-99) and is described in detail by Ribbe (1983, p. 10-11, 21-30); it results in considerable lattice strain among adjacent, minute triclinic domains in macroscopically monoclinic 'orthoclases' (McLaren and Fitz Gerald, 1987). This ultimately produces the familiar cross-hatched twinning observed in all microclines that have passed through the $C2/m \Rightarrow C\bar{1}$ phase transition. In the process, the 4-membered ring is geometrically distorted, the two T1 (and two T2) sites lose their symmetrical equivalence, and new site designations are required. Convention has it that the site in the resulting triclinic domain that contains the most Al is T10; the other T1 site is called T1m (see Fig. 5b and its caption for explanation; the reason Al prefers T1 and then T10 sites is discussed in Ribbe, 1983). The T2 sites are likewise differentiated into T20 and T2m. Clearly, $t_10 > t_1m > t_20 \cong t_2m$.

In Figure 6b ⟨T-O⟩ distances for all four tetrahedra are plotted as a function of ⟨T10-O⟩ to display the array of known microcline structures. Examination of the data shows that $t_10 > t_1m > t_20 \cong t_2m$. It would appear that Al is 'moving' from T1m sites and, to a much lesser degree, from T20 and T2m to T10. Plotting the data in this manner calls attention to a variety of Al,Si arrangements in the more disordered intermediate microclines. It is certain that these do not represent equilibrium distributions, although the fully ordered arrangement ($t_10 = 1.00$, $t_1m = t_20 = t_2m = 0.00$; Fig. 5b) in low microcline does. Vertical lines at ⟨T10-O⟩ = 1.668 Å in Figure 6b indicate the degree to which data from C2/m structures in Figure 6a would overlap if plotted on the same scale in 6b.

4.1.3. *Room-Temperature Phase Transition in Low Microcline?* Openshaw et al. (1979) found hysteresis in heat capacity, cell parameters, and a far infrared absorption band for a martensitic-like transition at 300±10 K in a LM powder made by ion-exchange from LA. Wyncke et al. (1981) found only a similar, but lesser IR band shift in natural LM at 240-250 K. However, crystal structure refinements of natural LM at 294 K and 163 K by Phillips et al. (1988) found no hysteresis in cell parameters and opposite structural effects from those predicted by Brown et al. (1984). Thus, the question of a near-RT phase transition in LM remains unsettled; but LM does show anomalous behavior at RT and high P—see chapter by Angel).

4.1.4. *Triclinic ($C\bar{1}$) Anorthoclases, K-Analbites, and the Transition to C2/m Na-Sanidines.* Analbite (AA) is a term used to refer to any metrically triclinic Na-rich alkali feldspar which has $t_10 = t_1m \geq t_20 = t_2m$. Al,Si distributions that fulfill this condition have been called 'topochemically monoclinic,' i.e., these chemical equivalences are consistent with distributions permitted in monoclinic feldspars; thus they can invert to monoclinic symmetry upon heating. In nature, triclinic specimens with K-contents extending up to Or_{30} or higher and Ca-contents of up to $An_{\sim 20}$, are found in certain volcanic terranes. In these 'anorthoclases,' Al,Si distributions are highly disordered (Table 4 in Ribbe, 1984). Although there may be small differences among individual mean T1-O and T2-O bond lengths, some of these apparently are true K-analbites, because they readily invert from $C\bar{1}$ to C2/m to become K-monalbites (alternatively called Na-sanidines) at elevated temperatures (Harlow, 1982). Those which have 'topochemically triclinic' Al,Si distributions would be called high albites (HA) or K-high albites, and they would not invert to C2/m unless, with time at high temperatures, Al and Si diffused to a topochemically monoclinic arrangement or complete disorder. Kroll et al. (1980) have determined experimentally the equilibium diffusive (HA ⇒ AA) and displacive (AA ⇒ HS) transitions for such highly disordered phases in the composition range $Or_{40}Ab_{60}$-Ab_{100}-$Ab_{80}An_{20}$. See Kroll and Knitter (1991) for a study of Al,Si exchange kinetics.

4.1.5. T*riclinic ($C\bar{1}$) Low, Intermediate, and High Albite.* Low albite (LA) is the Al,Si-ordered isomorph of low microcline (LM); $t_10 = 1.0$, $t_1m = t_20 = t_2m = 0.0$ (see Fig. 5b). Indeed, a continuous series from $Ab_{100}Or_0 \Leftrightarrow Ab_0Or_{100}$ can be produced reversibly by alkali exchange in anhydrous molten chlorides without affecting the Al,Si distribution. When either ordered phase is heated, Al 'moves' equally from T10 into the other three T sites in what has been called a 'one-step disorder' path (see Blasi et al., 1984, Fig. 1, for LM ⇔ HS). For the $C\bar{1}$ LA ⇒ HA transition, Goldsmith and Jenkins (1985) attained reversible, equilibrium Al,Si distributions as a function of temperature at ~1.8 GPa (see Smith and Brown, 1988, Fig. 7-15). Certain of their specimens were examined optically by Su et al. (1986) and structurally by Phillips et al. (1989) and Phillips and Ribbe (unpublished). The latter found $t_10 > t_1m = t_20 = t_2m$ for all structures, (see Fig. 7), confirming the Al/(Al+Si) contents of the T sites which Su et al. had deduced from established relationships between lattice geometrical parameters and known crystal structures.

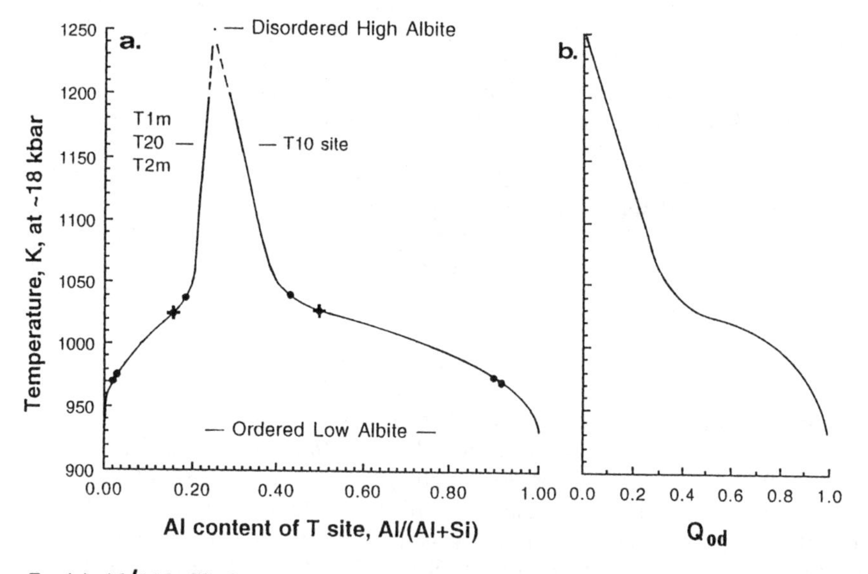

Figure 7. (a) Al/(Al+Si) for the T sites of albites intermediate in structural state between LA and HA. The curves represent Al,Si distributions derived from $\Delta 2\theta \equiv 2\theta_{131} - 2\theta_{1\bar{3}1}$ by Su et al. (1986) using data from Goldsmith and Jenkins (1985, as reported by Smith and Brown, 1988, their Fig. 7-15, uncorrected for pressure). The results from crystal structure refinements marked by symbols: + Phillips et al. (1989); • Phillips and Ribbe (unpublished). (b) The order parameter, $Q_{od} \equiv (t_10 - t_1m)/(t_10 + t_1m)$, for the same data set.

A multitude of room-temperature (RT) structure refinements of natural low albites indicated all to be essentially ordered, and this was confirmed by a neutron site refinement at 13 K (Smith et al., 1986) which yielded complete order to within ±0.004 Al at each site. Refinements at RT of high albites, formed by long-term heating of LA at T > 1050°C (see § 4.1.6 below) confirm the pattern of Al,Si distributions displayed in Figure 7.

4.1.6. *Analbite and the Transition to C2/m Monalbite.* Monalbite (MA) can exist in pure $NaAlSi_3O_8$, in which $t_10 = t_1m$ and $t_20 = t_2m$, at $T_c \cong 980°C$ or higher. T_c is the temperature at which the displacive transformation takes place: it decreases with increasing K-content (see Kroll et al., 1980), and—in the plagioclases—it increases with increasing Ca-content to the point where T_c and the (dry) melting point of the oligoclase are equal (at $An_{\sim15}$, T ≅ 1150°C; Kroll and Bambauer, 1981).

Prewitt et al. (1976) completed x-ray diffraction refinements of a synthetic HA at temperatures up to 1105°C; their crystal did not become monoclinic. Winter et al. (1979) examined a LA pre-heated at 1080°C for 60 d; it inverted to HA but did not transform to MA. Another LA preheated at 1110°C for 133 d did invert to AA and became monoclinic at ~980°C. In albite, the diffusive transition from Al,Si-ordered ⇔ Al,Si-disordered, whether LA ⇔ HA or LA ⇔ AA, can be described by an order parameter defined by Salje (1985) as

$$Q_{od} = (t_10 - t_1m)/(t_10 + t_1m)\ . \qquad (2)$$

Thus Q_{od} = 1.0 for LA and 0.0 for AA and MA. [See Fig. 7b for intermediate values associated with intermediate albites, and note that this parameter is workable for sodic plagioclases because $t_1m = t_20 = t_2m$ in all structures observed to date.] Using Landau theory, Salje introduced a displacive parameter, Q, to describe lattice distortion that occurs in the MA ⇔ AA inversion, involving the C2/m ⇔ $C\bar{1}$ phase transition, and in the $C\bar{1}$ albites from AA ⇔ HA ⇔ LA. Q is temperature-dependent and is related to elastic strain, in particular to tetrahedral tilting within and between 'crankshaft' chains of the sort reported by Prewitt et al. (1976) and Winter et al. (1979). Q_{od} and Q are 'bilinearly related' and both are necessary to describe the thermodynamic equilibrium behavior of any triclinic Na-feldspar (Salje et al., 1985, p. 96). The thermodynamically stable states of albite are shown in Figure 8a and the temperature-dependence of Q and Q_{od} in Figure 8b.

4.2. CELSIAN AND HYALOPHANE

Ribbe (1983, p. 14) discussed T-site nomenclature and symmetry relations among the 4-membered rings in I2/c celsian with **c** ≅ 14.4 Å (see Fig. 9a). This Ba-feldspar is rare in nature, and the two structure refinements that have been completed contain substantial Or-component ($Cn_{\sim82}Or_{\sim18}$, Newnham and Megaw, 1960; $Cn_{95}Or_5$, Griffen and Ribbe, 1976). Both are somewhat disordered with $t_10 \cong t_2z \cong$ 0.17-0.18 and $t_1z \cong t_20 \cong$ 0.76-0.79; see Figure 10. Heat treatment would undoubtedly increase the disorder with concomitant inversion to C2/m symmetry as the pairs of sites noted above become truly equal in Al-content. Gay and Roy (1968) studied the mineralogy of natural K-Ba feldspars, among which are found relatively disordered, K-rich 'hyalophanes' that often contain a substantial Ab-component. Four structures have been refined; data are displayed in Figure 10. They are structurally similar to HS (C2/m phases with **c** ≅ 7.2 Å; Fig. 5a), although total Al is never eactly 1.00, but (1.00 + mole fraction Cn).

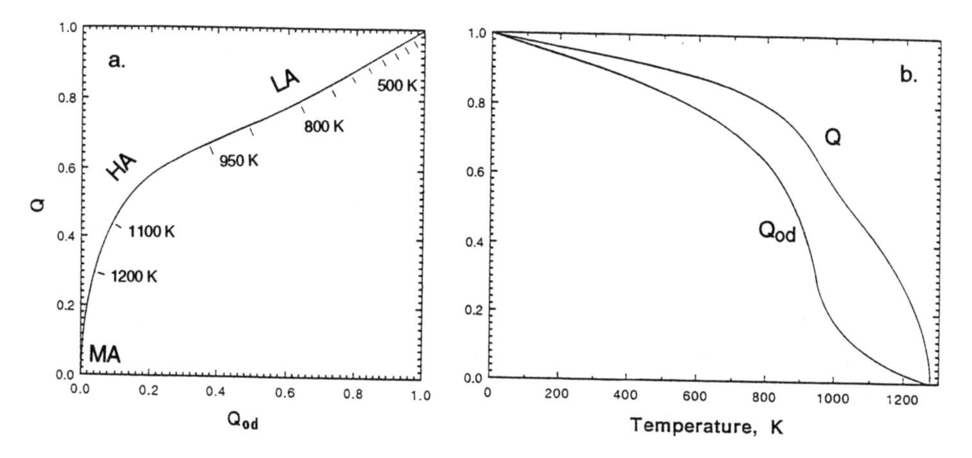

Figure 8. From Salje et al. (1985; Figs. 9, 10; caption paraphrased from pp. 104-105). (a) Thermodynamically stable states of albite. At T > transition temperature (1291 K) the only stable phase is MA [$Q = Q_{od} = 0$]. Note that this 'transition temperature' in thermal equilibrium is not identical with the transition temperature, T_c = 1251 K between AA and MA. With decreasing T, the symmetry is reduced ($Q \neq 0$; $Q_{od} \neq 0$). For T > ~950 K, the displacive parameter Q dominates over Q_{od}; this is the thermal regime of HA. For T < ~950 K, Q_{od} changes more strongly with T; this is the regime of LA. The same is seen in (b) where Q and Q_{od} are plotted as functions of T. Figure 8a has *corrected* T values.

4.3. ANORTHITE AND INTERMEDIATE PLAGIOCLASES

At temperatures very near the melting points there is essentially complete solid solution between Al,Si-disordered albite and anorthite, although space groups range from C2/m to $C\bar{1}$ to $I\bar{1}$ across the series. At lower temperatures, Al and Si partially order and a bewildering variety of exsolution textures, incommensuarte structures, and antiphase domains develop. The phase relations are intensely complex and have only reluctantly yielded information on which to base a 'phase diagram' (see chapter by Carpenter, this volume). After discussing the anorthite structures, we will attempt to summarize the known structural data for these so-called intermediate plagioclases, and later (§ 6), in a consideration of the interrelated effects of P,T,X in these and isomorphic aluminosilicate feldspars, we will investigate some of the structural variations that are observed.

4.3.1. *The 'Aluminum Avoidance' Principle and Al,Si order/disorder.* There are only two possible completely ordered Al,Si distributions in plagioclases; they occur in $C\bar{1}$ low albite (Fig. 5b) and may occur in both $P\bar{1}$ and $I\bar{1}$ anorthite, and in I2/c celsian. Figures 9a and 9b show ideal structures of Cn and P-An in schematic (010) projections, together with their respective symmetry elements. Note that the **c** cell dimensions are doubled to ~14 Å (from ~7 Å in C-centered feldspars; cf. Fig. 5) to accommodate the regular ...Al-Si-Al-Si... alternation throughout the framework. Angel et al. (1990, p. 161) suggest that the most highly ordered, P-An_{100} from Val Pasmeda (see Kirkpatrick et al., 1987) may be as much as 8% disordered, thus the alternation is really ...–Al-rich–Si-rich–... . In their neutron refinement of $I\bar{1}$ anorthite at 514 K Ghose et al. (1993) "concluded that Al/Si ordering...is virtually complete"—without specifying what they mean by "virtually."

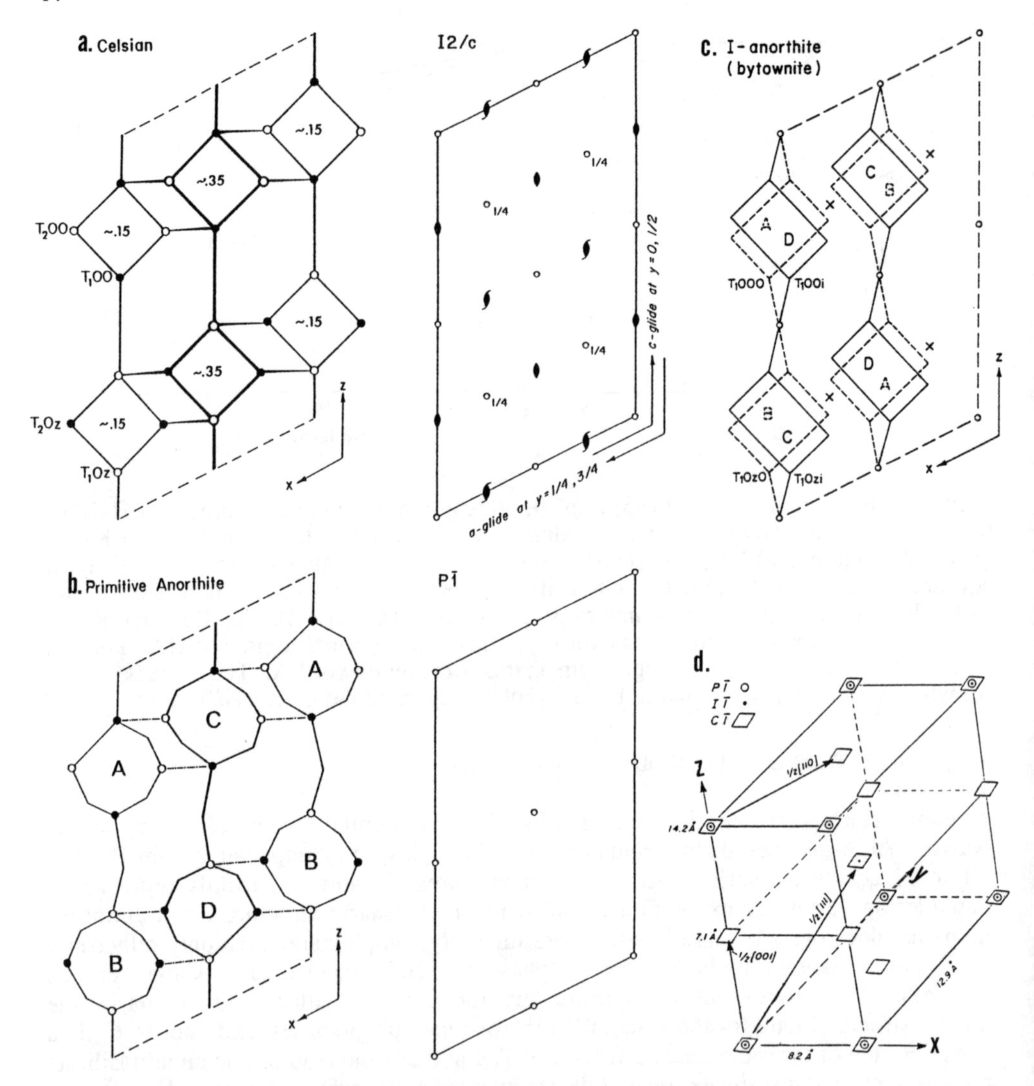

Figure 9. (a) Ideally ordered I2/c celsian and its resultant doubled **c** cell edge. Note that the ..T200,T100.. ring is pseudo-**c**-related to the ..T20z,T10z.. ring; thus z in the symbol. The monoclinic framework is held open by a large Ba cation (Sr-, Eu-, and Pb-feldspar are also I2/c). If Al,Si were disordered, the space group would be C2/m with **c** ≅ 7 Å. (b) Triclinic PĪ anorthite, whose ordered framework has collapsed around small Ca cations. Distorted rings are nonequivalt. Ring pairs A,D and B,C are geometrically similar, but their Al-Si-Al-Si sequences are different; the reverse is true of pairs A,C and B,D. Thus disordering Al,Si does not raise the symmetry unless ring geometries are modified or the diffraction geometry is 'averaged' by antiphase domain superposition, as illustrated in (c). Modified from Megaw (1974a). (c) One model for IĪ anorthite: antiphase superposition of P-An subcells of the type described by Ghose et al. (1993). (d) z = $^1/_2$[001], i = $^1/_2$[111], z+i = $^1/_2$[110] vectors interrelating lattice points of the P, I, and C unit cells of feldspars.

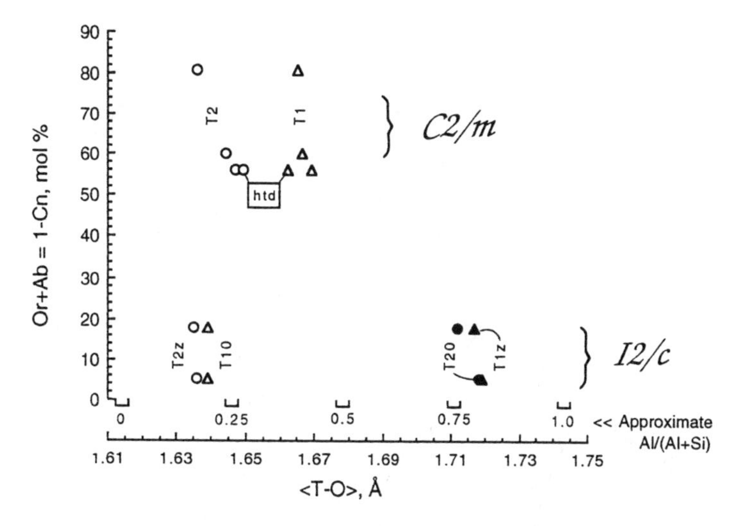

Figure 10. Display of ⟨T-O⟩ bonds as a function of Or+Ab content for the T1 and T2 tetrahedra of C2/m 'hyalophanes' and the Si-rich T2z,T10 and Al-rich T20,T1z tetrahedra of I2/c celsians. More Or-rich specimens are considered sanidines; cf. Figure 6a. Data for celsians: see text. Data for hyalophanes: Cn_{19} and Cn_{40} from Viswanathan and Brandt (1980) and Viswanathan and Kielhorn (1983a; see also 1983b); data for Cn_{43} – natural and heated natural – from De Pieri et al. (1977).

For years anorthite was simply *assumed* to be perfectly ordered, in part because of what had become an unhealthy attachment to the 'aluminum avoidance' principle of Loewenstein (1954). But Carpenter (1991a,b) demonstrated that immediately after crystallization from a glass, synthetic An_{100} is incommensurately modualted (exhibiting 'e' refelctions; see § 4.3.5 below). And Phillips et al. (1992), using ^{29}Si MAS NMR, studied An_{100} crystals annealed from a glass at 1400°C for times of 1 min to 179 h and found that the number, $N_{Al\text{-}Al}$, of Al-O-Al linkages per formula unit ranged from 0.40 to 0.19, "corresponding to large degrees of short range order, σ = 0.80–0.91. The short-range order parameter, σ, varies linearly with the square of the macroscopic order parameter [$Q_{od} = 10.1\sqrt{\varepsilon_s}$] determined by X-ray diffraction measurements on the same samples: $Q_{od}^2 = 2.66\,\sigma - 1.72$, $r = 0.99$. ...Values of $N_{Al\text{-}Al}$ also correlate well with the inverse spacing of type b and type e antiphase boundaries, indicating that much of the reduction in the number of Al-O-Al likages with short annealing time is related to the reduction in total surface area of the antiphase domains" (Phillips et al., 1992, p. 484).

4.3.2. *Primitive Anorthite.* The ordering pattern in P-An is the same as Cn; thus $\mathbf{c} \cong 14.1$ Å. But Ca^{2+} is as much smaller than Ba^{2+} as Na^+ than K^+ (see Fig. 1), and the collapse of the tetrahedral framework is dramatic. The number of unique, symmetrically nonequivalent 4-membered rings doubles from two in Cn to four in An, but the number of unique T sites quadruples from four in Cn to 16 in An, because all rotational and glide symmetry is lost, as are half of the centers of symmetry (compare Figs. 9a and 9b). This gives rise to an avalanche of site nomenclature, only half of which is shown in the detailed structure drawing of P-An in Figure 11.

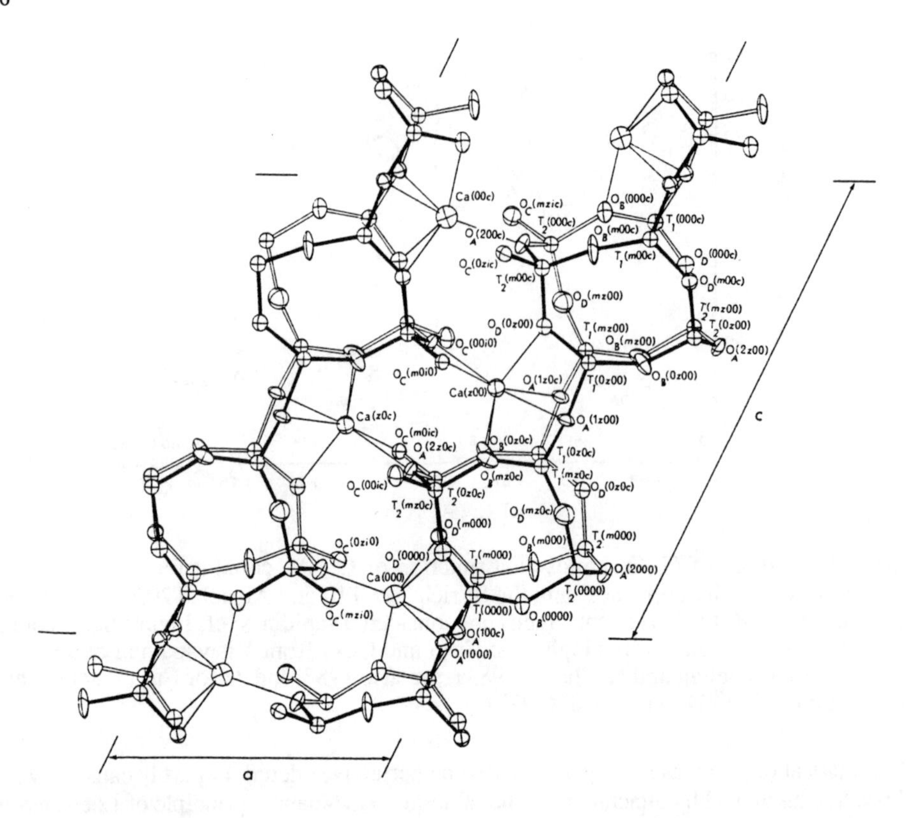

Figure 11. A portion of the structure of P-An bounded by $y = \pm 0.3$. From Wainwright and Starkey (1971).

NOTE: The space groups of crystal structures are deduced from systematic presences of classes of Bragg diffraction maxima $hk\ell$ observed in either x-ray, neutron, or electron diffraction patterns. The classification of Bown and Gay (1958) is repeated here for convenience in the subsequent discussions.

*Classes of Bragg diffraction maxima present for feldspar space groups**

Type	$P\bar{1}$, $\mathbf{c} \cong 14$ Å	$I\bar{1}$, $\mathbf{c} \cong 14$ Å	$I2/c$, $\mathbf{c} \cong 14$ Å	$C\bar{1}$ and $C2/m$, $\mathbf{c} \cong 7$ Å
a:	$h+k$ even, ℓ even	*a*	*a*	*a*; indexed on the 7 Å cell, ℓ = even *or* odd
b:	$h+k$ odd, ℓ odd	*b*	*b*	
c:	$h+k$ even, ℓ odd		For $h0\ell$, ℓ even	
d:	$h+k$ odd, ℓ even			

---

* To accommodate comparison of the various members of the feldspar group, several non-conventional space groups have been introduced, each of which could have been described by a smaller, primitive, triclinic unit cell with no easily recognizable relationship to the current choice.

4.3.3. *Pseudo-symmetry and the Average Structure Concept.* In all of the symmetry-related transitions that may occur with changes in P, T, or X among various plagioclases ($P\bar{1} \Leftrightarrow I\bar{1} \Leftrightarrow C\bar{1} \Leftrightarrow C2/m$), the pseudo-symmetrically related oxygen sites become successively irresolvable, reducing in number (from 32 to 16 to 8 to 5), likewise the T sites (from 16 to 8 to 4 to 2) and the M sites (from 4 to 2 to 1 in a general position to 1 on a mirror plane). The pseudo-symmetrical relations and derivative nomenclature for the T sites are shown in Figure 12; similar constructs are feasible for O and M atoms as well.

The concept of 'average structure' is useful for the plagioclases because of the high degree of pseudo-symmetry amongst the various structural elements. For example, one may refine the crystal structure of P-An in space group $I\bar{1}$ by using only the *a*- and *b*-type reflections (*h*+*k*+*l* even), deliberately omitting the *c* and *d* reflections whose intensities are due to slight positional shifts of atoms in the real structure from the exact body-centering operation, $^1/_2[111]$ (see Fig. 9d). The result will be two unique 4-membered rings (Fig. 9c) and two M sites in place of the four in P-An (see Fig. 13b). Every M, T, and O atom in the $I\bar{1}$ average structure would be represented by a noticeably anisotropic electron density distribution, an apparent 'thermal ellipsoid,' whose largest rms 'vibration' amplitude is related in orientation and magnitude to a 'split vector' between the atom-pair which is being averaged by artificially introducing a body-centering vector to the structure. The 'split vector' can be calculated between atoms with coordinates (x,y,z) and ($\sim^1/_2$+x,$\sim^1/_2$+y, $\sim^1/_2$+z) by subtracting *exactly* $^1/_2$ from each of the coordinates of the second atom and determining the small distance between the pair, usually in the range 0.002 to 0.5 Å (see Fig. 13a,b). Angel et al. (1990, p. 156) illustrated this with an example, and Ribbe (1983, 1984) discussed the concept in greater detail.

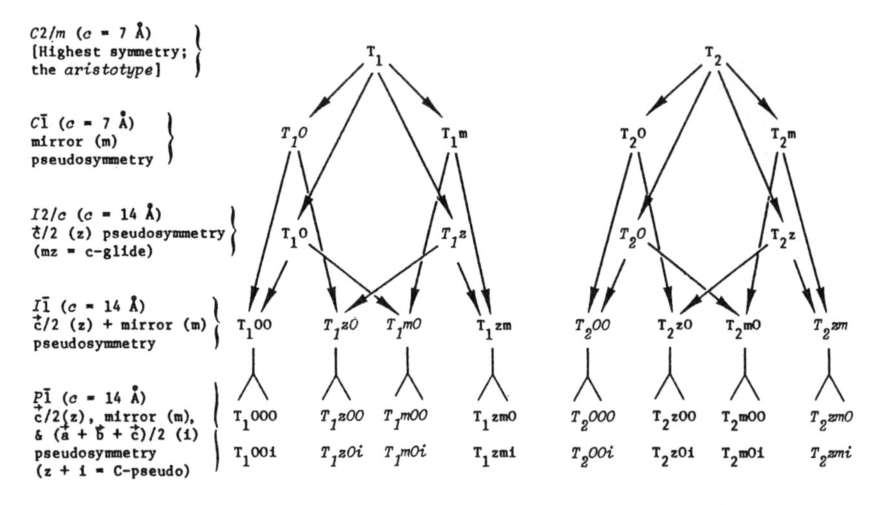

Figure 12. A 'flow chart' of T-site nomeclature for the feldspars with pseudo-symmetry indicated relative to C2/m. See Figure 11 for a site-labelled drawing of P-An and Figure 9d for the equivalent lattice points for the P and I- and C-centered lattices. Al-containing sites in ordered members are italicized. The $I2/c \Rightarrow I\bar{1}$ transformation does not occur in natural feldspars, but does occur in $SrAl_2Si_2O_8$ at 32±4 kbar and with increasing Ca-substitution in $SrAl_2Si_2O_8$ (at $An_{\sim 9}Srf_{\sim 91}$; McGuinn and Redfern, 1993b).

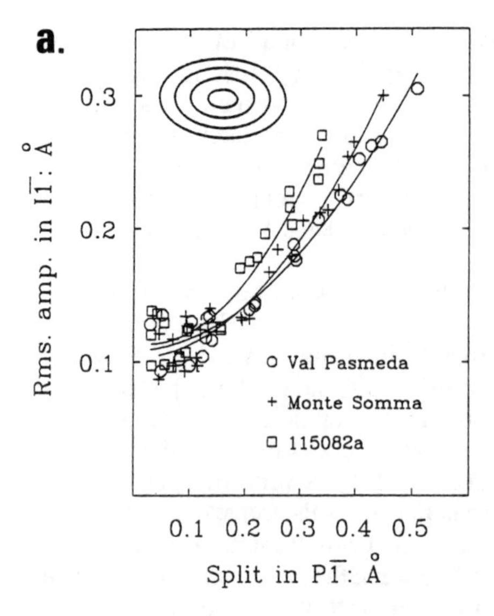

Figure 13. (a) "Comparison of the 'split vectors' for atom-pairs in P$\bar{1}$ refinements with the maximum rms amplitude of the apparent thermal ellipsoid of the corresponding atoms in I$\bar{1}$ refinements." From Angel et al. (1990, Fig. 1). Inset discussed in the text. (b) Top: Electron density maps of the four distinct Ca atoms of P$\bar{1}$ anorthite (the *y* and *z* coordinates are indicated; from Wenk and Kroll, 1984). Bottom: Another anorthite structure refined in I$\bar{1}$ at 0 GPa, which arbitrarily superposes them in pairs of highly anisotropic, 'split' atoms. The same crystal refined at 2.5 GPa is shown for comparison with that refined at 3.1 GPa, showing single, unsplit atoms, according to Angel (1988, p. 1118, his Fig. 2). Maps a,b,c = Ca(m000) site and d,e,f = Ca(mz00) site. Compare Figure 21 below; see discussion in § 5.2.4.

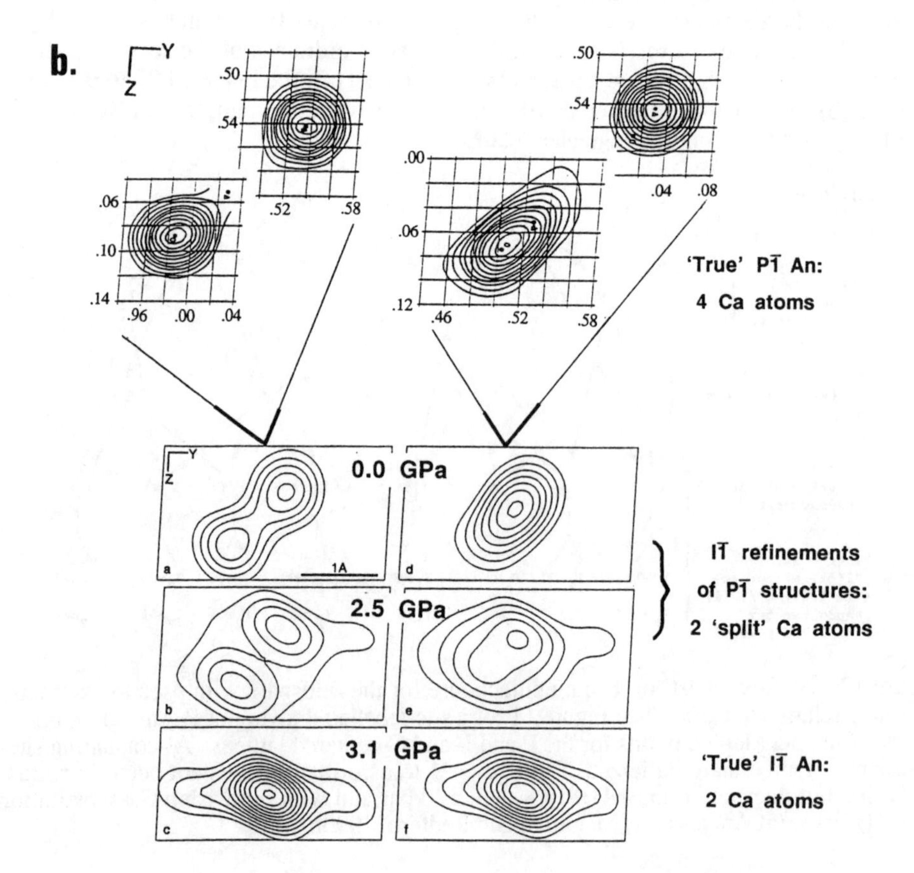

It is also possible by similar operations to refine a true $I\bar{1}$ structure (c ≅ 14 Å) in $C\bar{1}$ (**c** ≅ 7 Å), or for that matter a $P\bar{1}$ structure (**c** ≅ 14 Å) in $C\bar{1}$, using only the *a*-type reflections. In the latter case four atoms would be averaged onto each single site in the $C\bar{1}$ average structure. What is the purpose of such an exercise? The first is to make structural comparisons (of T-O-T angles or ⟨T-O⟩ distances) across a phase boundary, e.g., in the case where a $P\bar{1} \Rightarrow I\bar{1}$ transition occurs in anorthite with pressure, or temperature, or increasing Al,Si disorder. By using only $(h+k+\ell)$ even reflections, the structure appears body-centered, but each atom is found to be an anisotropic 'average atom' whose anisotropy may be attributed to the two distinct, ~$^1/_2$[111]-related atoms that would have been expected in the original P-An structure. This is particularly true of the M atoms in I-An which are not readily modeled by 'anisotropic' composite atoms, but usually require distinctly positioned half-atoms at each of the two nonequivalent sites to account for the observed electron density.

Certain physical properties, e.g., refractive indices, 2*V*, thermal expansion, compression, and of course lattice parameters, are functions of the average crystal structure, so in comparing C-, I-, and P-feldspars, it is often useful to do so in 'average structure' terms. This is done of necessity with cell dimensions determined from x-ray powder diffraction patterns, for commonly, even in P-An, no *b*, *c*, or *d* reflections are intense enough to be observed, so the 14 Å P-An appears to be 7 Å C-An. Furthermore, the feldspar may 'do' the averaging for us. This is illustrated in Figure 9c, which is a schematic representation of the structural averaging that takes place by the antiphase superposition of domains of P-An-like structure when, for example, the temperature of a P-An_{100} crystal is raised above ~240°C (see Ghose et al., 1993). Also in $C\bar{1}$ high albite, which has only *a* reflections, the single Na is so 'anisotropic' (even at RT) as to require quarter-atoms to approximate the electron density—clearly representative of positional disorder in an 'average structure' (see § 5.2 below).

4.3.4. *Body-centered Anorthite.* P-anorthite inverts to apparent $I\bar{1}$ symmetry at T_c ≅ 240°C (Brown et al., 1963). Redfern and Salje (1992) reviewed 30 years of investigation of this phenomenon. From high-T infrared studies they confirmed that T_c = 242±5°C and concluded that "on a macroscopic, thermodynamic scale the phase transition can be described as a classical displacive transition." The *framework* retains the Al,Si order of the unheated $P\bar{1}$ structure; it is "strictly $I\bar{1}$," but it is framework distortion that drives the phase transition. Ghose et al. (1993) refined an An_{100} structure *at* 514 K (241°C) by neutron diffraction, interpreting the unbalanced Ca occupancy of the two spilt-M sites in terms of "a dynamical model in which the $I\bar{1}$ phase is a statistical dynamical average of small, mobile $P\bar{1}$ domains" (as in Fig. 9c; cf. van Tendeloo et al., 1989). Each of the two Ca atoms of the $I\bar{1}$ unit cell apparently flip unequally between alternate positions not unlike those found in the lower temperature $P\bar{1}$ structure (see §5.2.4 below), violating strict $I\bar{1}$ symmetry and probably accounting for the persistence of extremely diffuse *c* and *d* reflections above T_c, as documented by Adlhart et al. (1980a,b).

Anorthite also inverts from P to an I phase with increasing pressure. Angel (1988; cf. Angel et al., 1989) fixed the first-order transition at 2.61±0.06 GPa for highly ordered An_{100} (Q_{od} = 0.92; Eqn. 3 below). Crystal structure analyses comparing $P\bar{1}$ structures refined in $I\bar{1}$ with the true $I\bar{1}$ structure above the transition "suggest that the phase transition may be described as a tilting of essentially rigid tetrahedra [by significant changes in T-O-T bond angles] so as to bring the tetrahedra related by pseudo-I symmetry [see Fig. 12] into true equivalence." A refinement at 3.1 GPa indicated single (not split!) sites for the two Ca atoms, as evidenced by electron density maps (Fig. 2 of Angel, 1988). Angel also found that increasingly disordered An (induced by heating at 1350°C < T < 1533°C) inverts at increasing pressures, e.g., $Q_{od} = 0.78$ at 4.8±0.3 GPa — a second-order transition (Angel, 1992, and this volume).

The effect of adding NaSi in place of CaAl in 'low-temperature' anorthites and bytownites is to increase the incipient disorder from $Q_{od} = 0.92$ at An_{100} to $Q_{od} = \sim 0.6$ at $An_{\sim 70}$ (Angel et al., 1990) with concomitant change from P to I symmetry. For these **c** ≅ 14 Å structures, Q_{od} is defined in terms of mean T-O bond lengths:

$$Q_{od} = K^{-1}(\langle\langle \text{Al-O} \rangle\rangle - \langle\langle \text{Si-O} \rangle\rangle) \qquad \text{(Eqn. 5 of Angel et al.)} \qquad (3)$$

with ⟨⟨Al-O⟩⟩ being the grand mean distance for the 4 Al-rich and ⟨⟨Si-O⟩⟩ for the 4 Si-rich tetrahedra in the $I\bar{1}$ structure. $K = 0.135$ Å, a value which comes from a re-examination of the relation between the grand mean distances for the entire plagioclase series:

$$\langle\langle \text{T-O} \rangle\rangle = {}^{K}/_{4}\,(n_{An}) + 1.6436$$ (from Eqn. 2 of Angel et al., 1990; see Fig. 16 below).

4.3.5. *Intermediate Plagioclases.* In addition to spinodal decomposition and exsolution in several regions of the plagioclase subsolidus, lattice strain accompanying Al,Si order in discrete phases of well annealed specimens results in incommensurate antiphase superstructures known as 'e' plagioclase. 'e' is the label given to pairs of non-Bragg diffraction maxima with coordinates in the anorthite reciprocal unit cell of $(h+\delta h, k+\delta k, \ell-\delta\ell)$ and $(h-\delta h, k-\delta k, \ell+\delta\ell)$, where $(h+k)$ and ℓ are both odd integers; 'e_2' is the label given to diffuse 'e's associated with sodic plagioclases ($An_{\sim 30\text{-}50}$) which have *a* reflections, and 'e_1' is the label given to usually sharp 'e's associated with calcic plagioclases ($An_{50\text{-}\sim 78}$) which have both *a* and diffuse *c* reflections. 'e_1's are accompanied by related, but weaker 'f' diffractions with coordinates $(h \pm 2\delta h, k \pm 2\delta k, \ell \pm 2\delta\ell)$, where $(h+k)$ and ℓ are both even integers. See Figure 14 for schematic drawings from Smith (1983). An electron diffraction pattern of the 'e' superstructure (Grove in Ribbe, 1983, p. 35) and TEM imaging thereof (Grove, 1976; Wenk and Nakajima, 1980) are helpful in visualizing this phenomenon. Several attempts to solve the structure of the 'e_1' superstructure have met with varying degrees of frustration, notably because they were model-constrained (see Horst, 1984, and a review by Smith and Brown, 1988, Ch. 5). Carpenter (1986; cf. this volume) has delineated the stability fields of the 'e_2' plagioclases relative to their more disordered $C\bar{1}$ polymorphs and the 'e_1' plagioclases relative to their $I\bar{1}$ polymorphs.

In summary of the crystal structures of $C\bar{1}$ intermediate plagioclases, ⟨T-O⟩ bond lengths are plotted vs. mole % An in Figure 15 to give an overview of the average structures refined in that space group. Smith and Ribbe (1969) noted that the most ordered of these structures (solid lines) do *not* represent a mechanical mixture of ordered Ab and An; Angel (1990) agreed.

Figure 16 is a plot of the ⟨⟨T-O⟩⟩ distances for 19 of the $C\bar{1}$ feldspars used in Figure 15, combined with calcic plagioclases refined in $I\bar{1}$ and $P\bar{1}$ before and after various heat treatments by Angel et al. (1990). In addition, ⟨⟨Al-O⟩⟩ and ⟨⟨Si-O⟩⟩ bonds for the four Al-

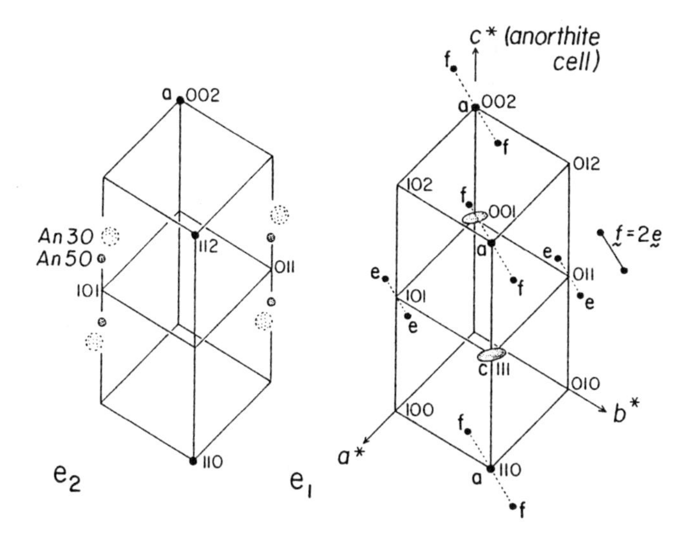

Figure 14. Schematics of the diffraction maxima observed in intermediate plagioclases. From Smith and Brown (1984, Fig. 9). Diffuse 'e_2' reflections are observed in $An_{\sim 30}$ to $An_{\sim 50}$. Sharper 'e_1's and their accompanying 'f' reflections are illustrated here for An_{70}, but they appear over the range An_{50-78}. See the chapter by Carpenter (this volume).

and four Si-rich tetrahedra of these structures are shown for eight natural specimens and those same specimens annealed at 1300 to 1370°C. The latter show decreases in ⟨⟨Al-O⟩⟩ and increases in ⟨⟨Si-O⟩⟩ that are less dramatic with increasing Al/(Al+Si). Angel et al. make the point that individual *mean* T-O distances cannot be used without adjustment to determine Al-content of T sites. Ribbe et al. (1974) had noted factors that influence *individual* T-O bonds, including not only the site's Al content, but also the Al content of the site to which the T-O oxygen is linked, a Coulombic term summed over the total M atoms coordinated to that oxygen, the T-O-T angles, and the tetrahedral distortion. Using a similar approach, Angel et al. 'corrected' individual T-Os in their structures for bonding effects to calculate adjusted ⟨T-O⟩s which they plotted vs. mol % An in their Figure 5.

5. The M Cations in Feldspar Structures

As mentioned earlier, the M^{1+} and M^{2+} cations play a significant role in feldspar structures (see Fig. 1). Larger M atoms essentially fill the cavity in the tetrahedral framework, holding it open and permitting monoclinic C2/m and I2/c symmetry to dominate if the Al,Si distribution among the T sites is topochemically monoclinic, as in Figure 17. Note that the elongation of the K anisotropic thermal ellipsoid is ⊥ to the (010) mirror (rms amplitude 0.18 Å; Scambos et al., 1987). However, framework distortion will occur if, in the course of long-term cooling, Al orders into one of the two T1 sites in the 4-membered ring or if a large cation is replaced by a smaller substituent (e.g., Na for K). Orientation of the vibration ellipsoid shifts to accommodate the distortion.

5.1. POTASSIUM IN SANIDINES AND MICROCLINES

The fragments of structure pictured in Figures 17a and 17b show effects of ordering of Al from the equivalent T1 sites (t_1 = 0.45) of LS to the T10 site (t_10 = 1.0; bonds marked |||)

Figure 15. ⟨T-O⟩ vs. mol % An for 20 $C\bar{1}$ plagioclases in the range $An_{0\text{-}66}$, including 'low' (solid line) and supposed 'high' structural states (dashed line) and a few with intermediate Al,Si disorder. [For albite, see Fig. 7.] All refinements at RT. Most data from Ribbe (1984, Tables 1 and 4); An_{38} is from Wenk and Kroll (1984); intermediate albite from Phillips et al. (1989). The structures of a ternary feldspar, $An_8Or_7Ab_{85}$, and two An_{52} labradorites which displayed distinct interference colors (an indication of exsolution) were omitted. (a) ⟨T-O⟩ bonds for T10 tetrahedra. (b) ⟨T-O⟩ bonds for the T1m tetrahedra and ⟨⟨T2-O⟩⟩ for T20 and T2m.

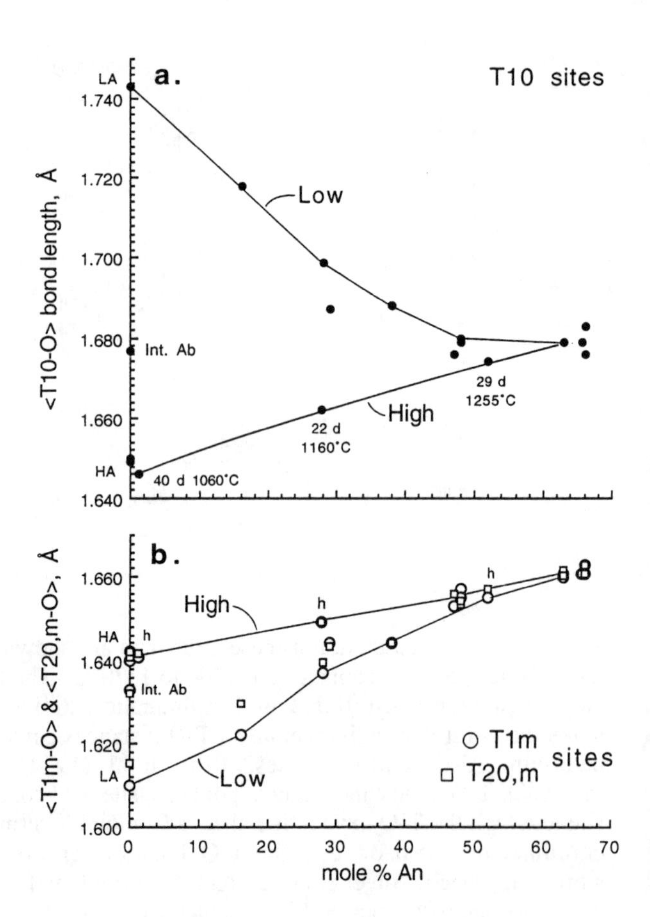

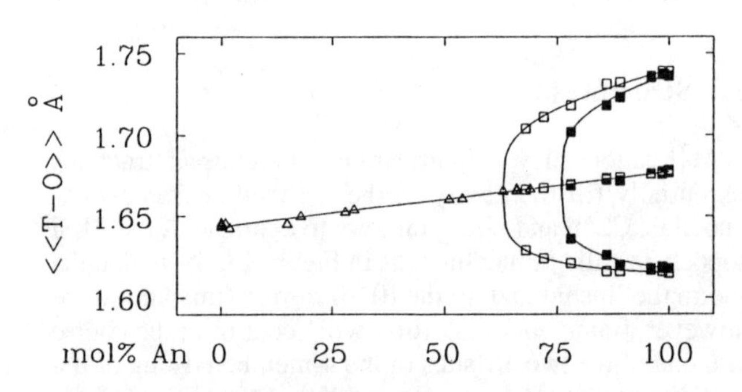

Figure 16. Grand mean T-O distances in plagioclases from Angel et al. (1990, Fig. 3). ❑ = data from their 20 $I\bar{1}$ and $P\bar{1}$ structure refinements of eight feldspars in the composition range $An_{68\text{-}100}$, Δ = data from 19 of the $C\bar{1}$ refinements plotted in Figure 15. The straight line has the equation ⟨⟨T-O⟩⟩ = $0.0338(6)n_{An} + 1.6436(4)$ (cf. §4.2.3 above) and shows no discontinuity at $An_{\sim 80}$, *contra* Kroll and Ribbe (1983, Fig. 1). Also shown are ⟨⟨Al-O⟩⟩ and ⟨⟨Si-O⟩⟩ for $I\bar{1}$ crystals: ❑ natural; ■ annealed at 1300 < T < 1370°C.

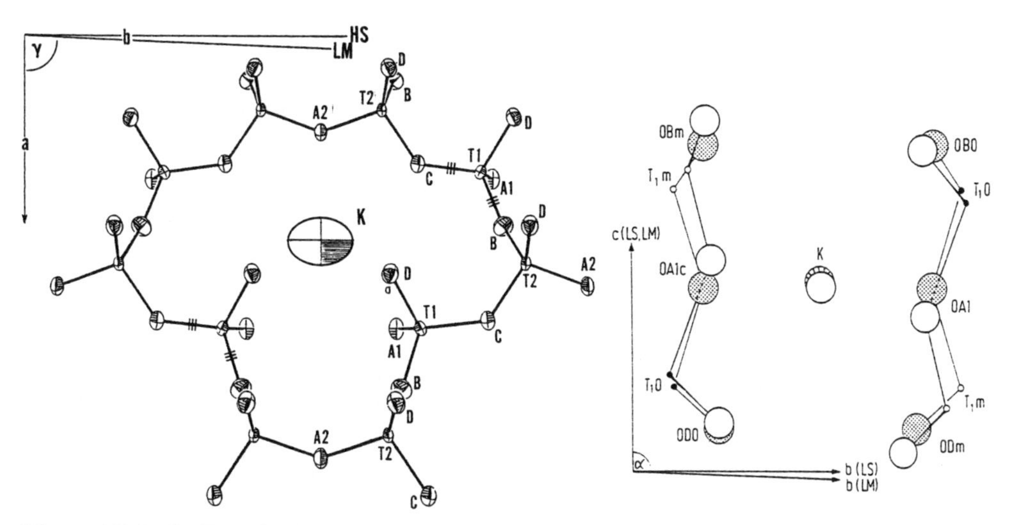

Figure 17. Left: 'Dog-face' projection of HS, labeled for C2/m symmetry (vertical mirror ⊥ **b**). ⫴ marks T1-O bonds in the 'ring' that lengthen as Al concentrates into the involved T1 sites, which are renamed T1O as the structure becomes triclinic; the other T1s become T1ms (cf. Fig. 6). Right: Structure fragment of C2/m LS (shaded) and new atom positions in $C\bar{1}$ LM (open circles) after Al enters T1O at the expense of T1m. From Kroll et al. (1980).

of LM (cf. Figs. 5a and 5b). Of course ⟨T-O⟩ bond lengths increase from ~1.67 Å to 1.74 Å (Fig. 6). The ⫴ bonds lengthen, others shorten, and the 'dog-face' projection (Fig. 17a) distorts, causing the γ angle to change from 90° in HS to 87.7° in LM. Similarly (Fig. 17b), the α angle is changed from 90° in HS and LS to 90.7° in LM. These changes and those in the **b** and **c** cell edges and/or their reciprocal lattice analogues α^*, γ^*, b^*, c^*) can be measured precisely from x-ray diffraction patterns and used to determine the actual Al,Si distribution amongst the T sites of the entire K,Na-feldspar series (Kroll and Ribbe, 1987; unit cell volume gives K/(K+Na)). Similar approaches may be applied to ternary and plagioclase feldspars (Kroll, 1983).

Although there is significant anisotropy of M in most feldspars, one may calculate an isotropic equivalent temperature factor which approximates the atom as a sphere (Stout and Jensen, 1968): $B = 8\pi^2\overline{u}^2$, where $\overline{u}^2$ is the mean-square amplitude of atomic vibration. Data for K-feldspars are plotted in Figure 18 as a function of an order parameter, Q_{od} (Eqn. 2), for sanidines and intermediate and low microclines. For IM ⇒ LM, B decreases with increasing Al,Si order. For the high and low sanidines, plotted at $Q_{od} = 0.0$, B values have a wide range (from 2 to 3 Å^2); the four highest values are all from Ferguson et al. (1991), one at 2.56 Å^2 is from a neutron refinement of HS by Brown et al. (1971), and the remainder are from C2/m K-feldspars listed in Ribbe (1984, Table 2).

The isotropic equivalent temperature factor B of monoclinic K-feldspar varies smoothly with T down to RT (Fig. 19), and, using the average RT value (2.46 Å^2, from data in Fig. 18), extrapolates to ~0.9 Å^2 at 0 K. However, Salje (pers. comm., 1993) reports that spectroscopic data indicate only slight changes in B at T < 297 K (dashed line). Positional disorder is of course expected to some degree in HS with its disordered Al,Si distribution.

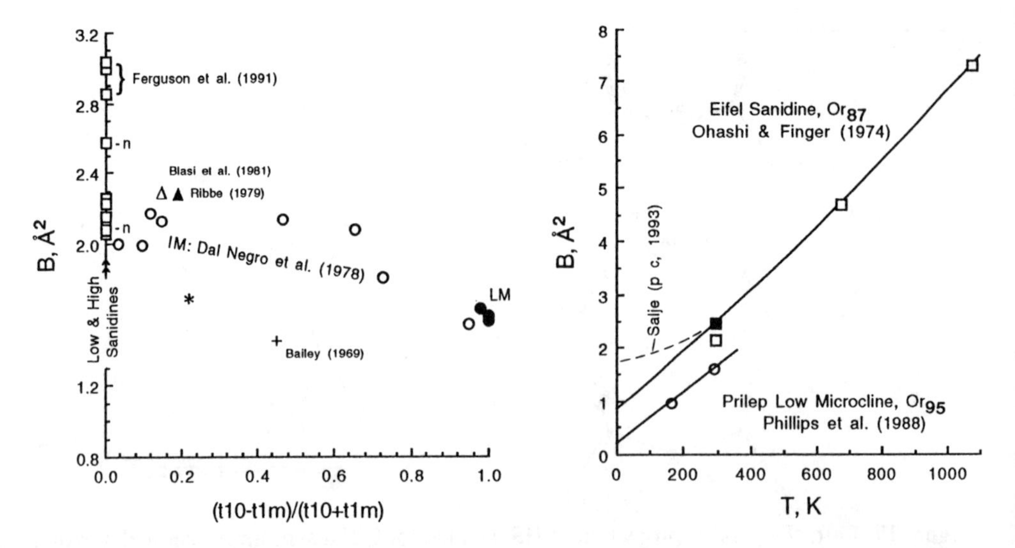

Figure 18. (left) Isotropic equivalent thermal vibration parameters B plotted as a function of $(t_10\text{-}t_1m)/(t_10+t_1m) \equiv Q_{od}$ for HS-LS at $Q_{od} = 0$, intermediate microclines (IM) with $0 < Q_{od} < 1.0$, and LM at $Q_{od} \sim 1.0$. * = aberrant value from Dal Negro et al. (1979); + is from Bailey's (1969) refinement of IM based on 1950's film data.

Figure 19. (right) B vs. T from refinements of HS (Or_{87}; ❑) and LM (Or_{95}; ○). See text.

Values of B from refinements of LM at T = 294 K and 163 K extrapolate to 0.18 $Å^2$ at 0 K. Apparently there is no significant positional disorder in LM, and it is quite likely that, within limits of error, 0.2 $Å^2$ ($\bar{u}$ = 0.05 Å) represents zero-point motion of K^+ in these ordered microclines (see discussion by Smith et al., 1986, p. 730).

5.2. SODIUM IN ALBITES

5.2.1. *Sodium in Low Albite.* One of the persistent questions concerning albite has been the nature of the Na atom. At the time of the first two-dimensional refinements at RT of albite, Ferguson et al. (1958) reported that Na showed positional disorder quantified by 'half-atom' separations of ~0.1 in LA and ~0.6 Å in HA. Using later 2-D refinements at 93 K (Williams and Megaw, 1964) and at 293, 573 and 873 K, Quareni and Taylor (1971) concluded that the Na in LA was a single atom subject to thermal vibration. This work was confirmed by Smith et al. (1986, incorporating high T studies by Winter et al., 1977), who found that the Na atom in LA was slightly anisotropic at 13 K, and that the equivalent isotropic B value extrapolated to 0 K was 0.5 $Å^2$ ($\bar{u}$ = 0.08 Å); see Figure 20. Using other near-zero-Kelvin neutron studies, Smith et al. (1986, p. 730) argued that "zero-point motion is the preferred cause rather than multiple centers-of-motion [in LA]."

5.2.2. *Sodium in High Albite and Monalbite.* Sodium in HA has always been assumed to have multiple positions in the distorted M cavity. Early 3-D studies (Ribbe et al., 1969) indicated a 0.9 Å separation of the furtherest Na 'quarter-atoms' introduced to approximate the elongate electron density distribution at RT. This was confirmed by Prewitt et al.

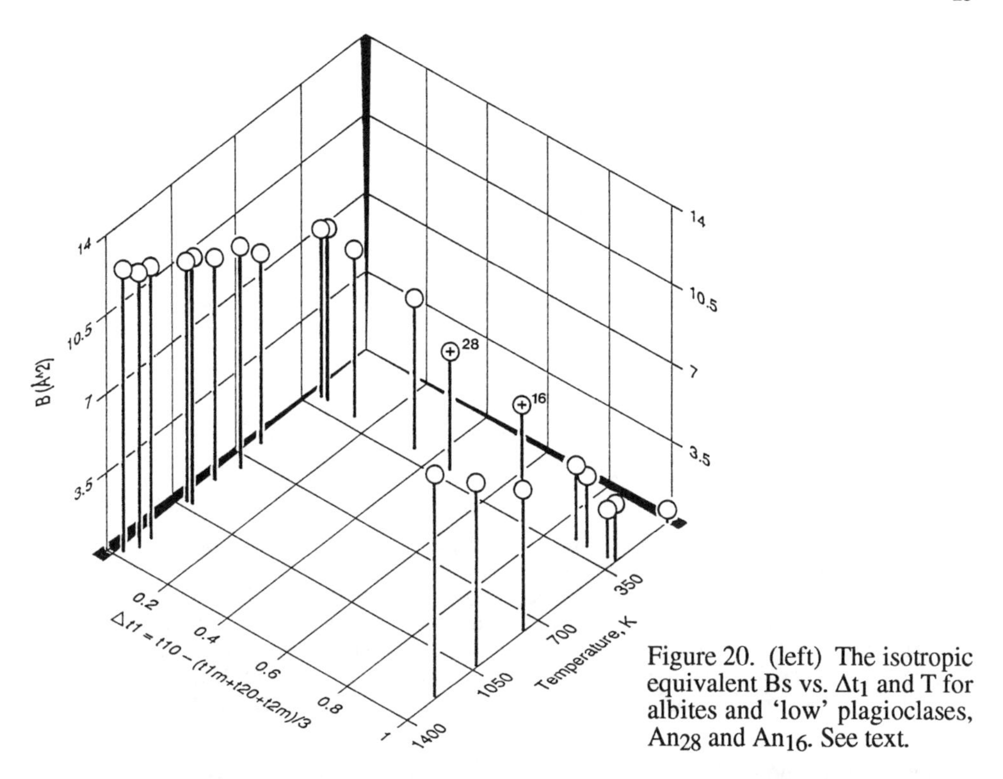

Figure 20. (left) The isotropic equivalent Bs vs. Δt_1 and T for albites and 'low' plagioclases, An_{28} and An_{16}. See text.

(1976) who used a quarter atom model for HA refinements up to 1378 K. See relative positions in Figure 21. Winter et al. (1979) repeated the study at high T and argued that the Na cavity undoubtedly had more than four low energy potential wells for Na. In other words, there were considerably more than *four* subcells required to model the Al,Si disorder of HA—a multitude of Na sites were possible, so they constrained their model to anisotropic thermal ellipsoids at all temperatures up to 1313 K. Figure 22 shows a superposition of the quarter atoms of Prewitt et al. on the thermal ellipsoids of Winter et al. at RT and ~1400 K. The isotropic equivalent Bs for the HA series are plotted to the left in Figure 20; they extrapolate to 5.85 $Å^2$ at 0 K, which is equivalent to an rms displacement of two 'half atoms' of $\bar{u} = [5.85/8\pi^2]^{1/2} = 0.27$ Å at that temperature. Notice in Figure 22b that HA at high T has a nearly monoclinic topology, much less skewed than the RT anion distribution (Fig. 22a). Had the Al,Si distribution of this albite been topochemically monoclinic, it would have already inverted to MA (at ~980°C, 1253 K). The rms amplitudes of the anisotropic displacement ellipsoid for Na in MA at 980°C are about twice those for K in HS at RT (Scambos et al., 1987, cf. Fig. 17a); both structures have C2/m symmetry.

5.2.3. *Sodium in Intermediate Albite (IA) and Na/Ca in Sodic Plagioclase.* At room temperature the anisotropy of the M cation increases regularly with increasing Al,Si disorder as measured by $\Delta t_1 \cong (t_10+t_1m)*Q_{od}$ (Fig. 20). This is expected in a transition from a simple, ordered LA structure to the average structure of disordered HA (which we have seen must be a mix of four or many more possible Al,Si arrangements in the tetrahedral framework which in turn create a range of potential energy minima for Na). The

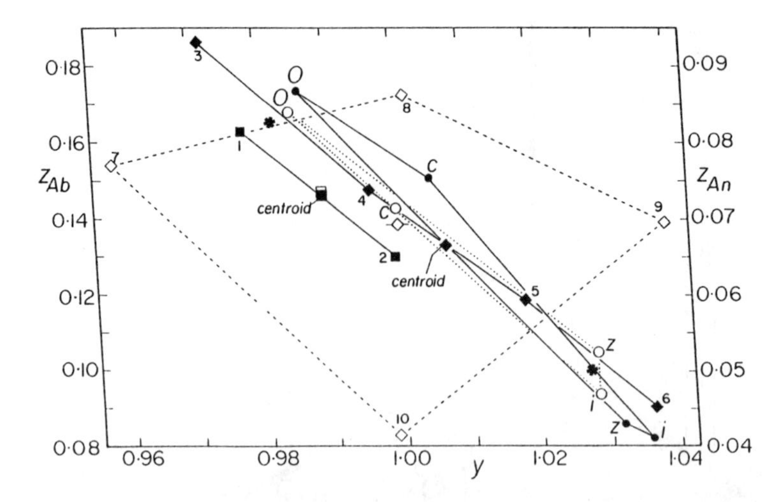

Figure 21. A *yz* projection on a generalized feldspar cell showing the relative positions of atoms, half atoms, and quarter atoms used to approximate the electron density at the M sites of LA (1,2 = half atoms; centroid indicated), of HA (3,4,5,6 and 7,8,9,10 = quarter atoms of RT and 1363 K approximations of Prewitt et al., 1976, respectively; monalbite and HA centroid from the 1243 K refinement of Winter et al., 1977), of An_{48} (*; see text), of $I\bar{1}$ An (O), and $P\bar{1}$ An (●): see caption of Figure 16 in Smith (1983) for more details and compare with similar plots of M cations in Smith and Brown (1988, Figs. 5-22 and 5-23).

fact that isotropic equivalent B values for two $C\bar{1}$ oligoclases (An_{16}, An_{28}) fit perfectly into this triaxial plot is evidence that the structural distortion expressed by incipient disorder introduced by increasing (Ca+Al) relative to (Na+Si) in low plagioclase is the same as that introduced by annealing LA at high T to produce intermediate albites (compare the changes in T-O bonds in increasingly calcic plagioclase (Fig. 15) to those in IA with increased disorder (Fig. 7a)).

It is not possible to carry the model of a 'single' anisotropic Na/Ca atom very far beyond low oligoclase compositions. In fact, refinements by x-ray and neutron methods of a volcanic andesine, An_{48}, with Q_{od} = 0.23, indicated that distinct, widely separated anisotropic M_1 and M_2 atoms (Fig. 22d and * in Fig. 21) were required to approximate the electron density distribution at the M site. A neutron site refinement by Fitz Gerald et al. (1986) estimated M_1 to be $Ca_{0.15}Na_{0.29}$ and M_2 to be $Ca_{0.23}Na_{0.33}$ to within ±0.7 atoms. This tells us that the real structure is composed of subcells with distinct geometry; T and O atoms are related by a high degree of $C\bar{1}$ pseudosymmetry, but the M cations have distinct, multiple sites within highly irregular cavities. The perseverence of diffuse 'e' reflections and small but significant Al,Si order in this volcanic andesine indicate that it has a remnant 'e' superstructure (see § 4.2.4).

5.2.4. *Ca/Na in Calcic Plagioclase.* With increase of Ca over Na, the M_1 and M_2 sites of $I\bar{1}$ plagioclases show increasing anisotropy to the point where they must be represented by M_1',M_1" and M_2',M_2" split atoms, which also may be anisotropic and unequally occupied by Na,Ca. In projection onto the yz plane of an albite cell (Fig. 21), these four half atoms of an I-An (o) are located almost exactly where the real Ca(000), Ca(z00), Ca(0i0), and Ca(zi0) atoms of P-An fall (•). Compare Figure 13 (above). It is interesting to consider the results of a neutron diffraction refinement of a P-An_{100} that had inverted to $I\bar{1}$ at 514 K. Ghose et al. (1993) have an elaborate model for the phase transition, but the M atom behavior is of particular interest: the Ca atoms are disordered on two split sites, the one split by 0.84 Å, the other by 0.49 Å; there is no reasonable rationalization for the refined occupancy of $^1/_3$ and $^2/_3$, respectively. See Ghose et al. for discussion and their Figure 4 for details of the oxygen environments of the four calcium sites. Clearly this phenomenon is critical to understanding the An-rich plagioclase structures at both high P and T; see Angel's chapter in this volume for further information.

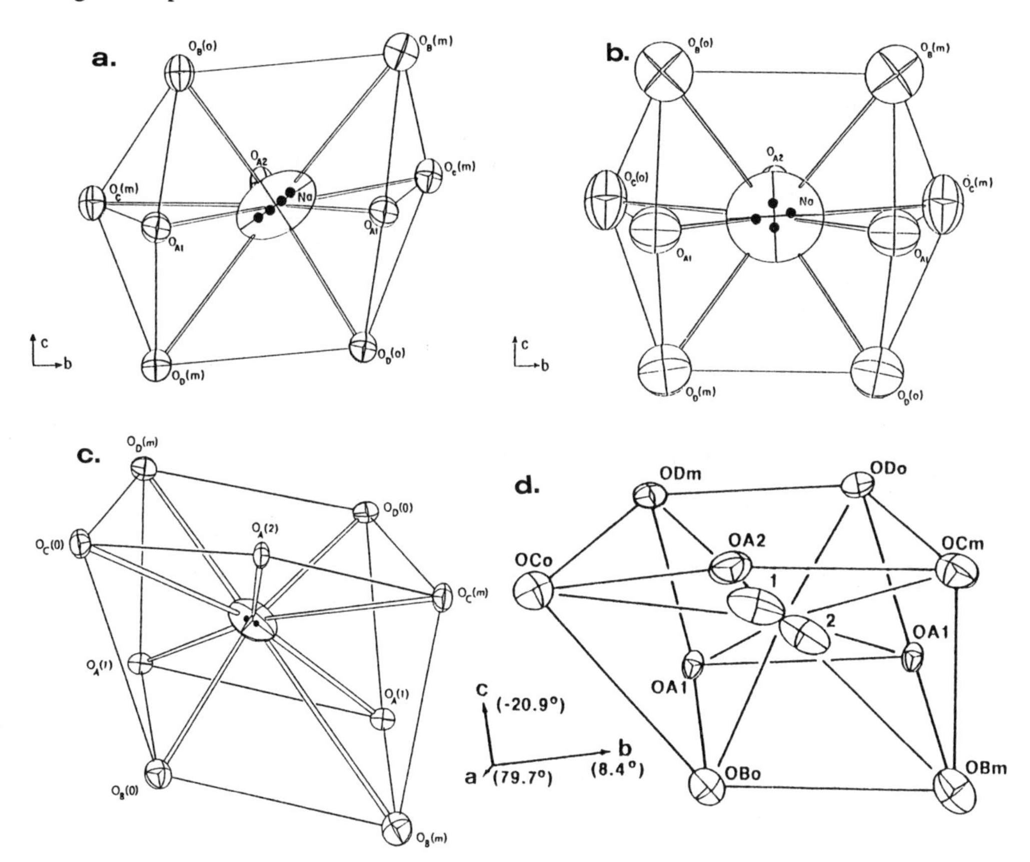

Figure 22. (a) Sodium atom configuration and oxygen neighbors in HA at RT. (b) Same, in HA at 1313 K. Both drawings from Winter et al. (1977, Fig. 9) with the approximate Na quarter-atom positions at RT and 1378 K from Prewitt et al. (1976) superimposed. These positions are also plotted on Figure 21. (c) A somewhat different perspective by Wainwright and Starkey (pers. comm.) of Na in LA at RT with the half-atom positions of Ribbe et al. (1969) superimposed. (d) Neutron diffraction refinements of half atoms $M_1 = Ca_{0.15}Na_{0.29}$ and $M_2 = Ca_{0.23}Na_{0.33}$ from a ▫ refinement of An_{48} by Fitz Gerald et al. (1986; plotted as * in Fig. 21).

6. Average Structures of Feldspars Represented by Cell Parameters: the Temperature-Pressure-Composition Analogies

To this point in our review of the feldspar structures we have confined ourselves primarily to natural phases and their behavior at higher (and a few lower) temperatures and, in the case of anorthite, at higher pressures. In 1972 Megaw (1974b, p. 112) said, "If lattice parameters can be predicted from tilt angles [of TO_4 tetrahedra], changes in lattice parameters can be expressed in terms of tilt, [and] tilt angles which vary from one feldspar to another are likely to be changed easily by physical causes other than composition, notably by temperature." She might well have added "pressure," for in the past 15-16 years numerous investigators have considered P,T,X as structurally analogous variables (see Hazen and Finger, 1982).

In the remainder of this chapter, compilations of lattice parameter data will be scrutinized. They represent stoichiometric aluminum-silicate feldspars contained in the *solid solution series* lined out between major end members in Figure 1. Cell parameters of *low-to-high-temperature series,* especially of ordered or disordered end-member feldspars, are included, as are series of *high pressure* data from LA, HS, and anorthite. The structures of anorthite are the purview of R.J. Angel (see his chapter), but they give us the only information we have on bond angles and lengths at high P. In all of this there will be no overt attempt to duplicate the close scrutinity of the thermal expansion of alkali feldspars by Kroll (1984) or their "expansion behavior...as a function of composition, temperature and pressure" based on a study of "coupled variations of cell parameters" by Brown et al. (1984; cf. Smith and Brown, 1988, Ch. 7). Some overlap is inevitable, of course, because the basic principle of framework distortion around the M-cation pervades all our observations. An excellent illustration of that principle is seen in Figure 23a,b,c, where K, Na, Li are shown with both bonded and nonbonded oxygens in *a-c* projections. The progressive distortion of the framework is cleverly represented in Figure 23d.

The broad patterns of behavior of lattice parameters with T and X in known structures will be investigated for clues to steric details in compounds whose structures have not been solved.

6.1. CHANGES IN VOLUME WITH P, T, X

6.1.1. *Compressibility and Phase Changes with Pressure.* A plot of V/V_o vs. pressure in Figure 24 shows the compressibility of LA to be similar to that of HS: the bulk moduli, $K_T = -V(\partial P/\partial V)$, are 0.70 and 0.67 Mbar, respectively. [Angel et al. (1988) present several explanations of their failure to find the C2/m-$C\bar{1}$ inversion with pressure in sanidine reported by Hazen (1976); further work is in progress.] Anorthite, with a small, divalent M cation, is stiffer. It shows a phase transition at 2.6±0.1 GPa; the lower-P phase has bulk modulus 0.94 and the high-P phase 1.06 Mbar (lower values if K' = 4.0 is used in the Murnaghan equation of state—compare calculations by Angel et al., 1988, and Hackwell and Angel, 1992). There is a distinct volume reduction (~0.2%) in anorthite accompanied by a $P\bar{1}$-$I\bar{1}$ symmetry change. Although the *c* and *d* reflections are not entirely absent from the high-P phase, Angel (1988) considers it to be $I\bar{1}$ because the unequally occupied M_1',M_1" and M_2',M_2" sites of the low-P average structure become essentially single M_1

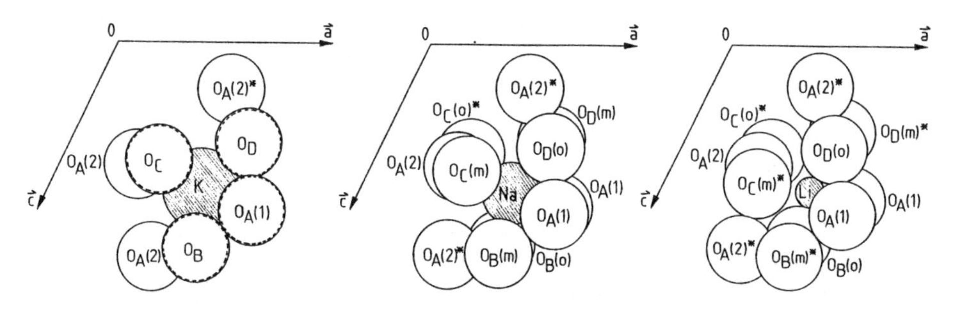

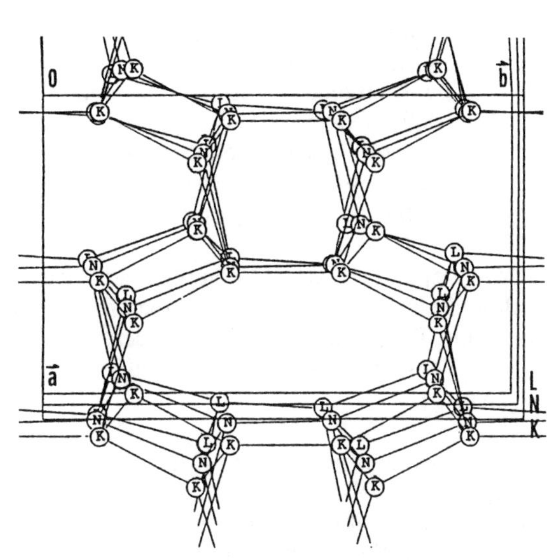

Figure 23. Oxygen environments of cations in disordered (a) K-, (b) Na-, and (c) Li-feldspars; nonbonding Os marked with *. (d) Tetrahedral nodes of the respective frameworks, labelled K, N, L, from DLS refinements of cation-exchanged powders (after Deubener et al., 1991; see § 2 above).

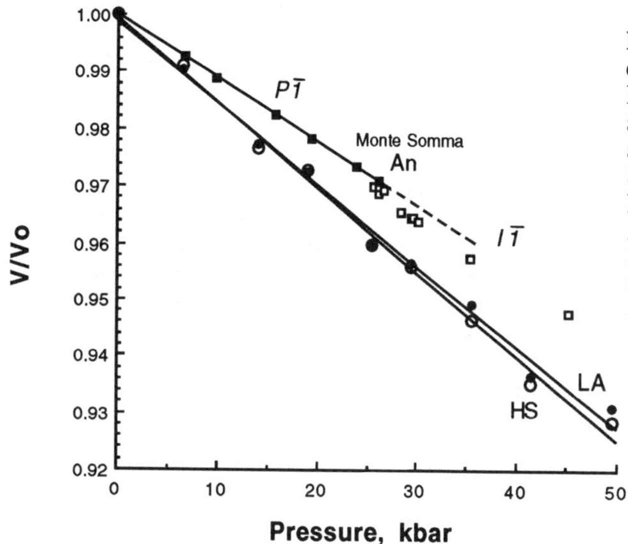

Figure 24. Change in unit cell volume with pressure for HS (high sanidine), LA (low albite) (data from Angel et al., 1988), and the Monte Somma anorthite (▪ = $P\bar{1}$, ▫ = $I\bar{1}$ structure; the data are from Angel, pers. comm., 1993). See McGuinn and Redfern (1993a) for data on synthetic Sr-feldspar.

and M_2 sites at 3.1 GPa (although, in my opinion, their significant anisotropy based on electron density distributions (Angel's Fig. 2) may indicate residual Ca site-disorder and thus account for the persistence of the $I\bar{1}$-violating reflections, $h+k+\ell$ odd). The effect of a 15% decrease in Al,Si ordering in anorthite produces a small (~5%) reduction in bulk modulus, and replacing Al in LA by B in reedmergnerite significantly stiffens the framework (see Hackwell and Angel, 1992).

6.1.2. *Thermal Expansion.* The thermal expansion curves for a number of end-member feldspars are plotted in Figure 25. Already somewhat expanded by a disordered Na cation in a highly irregular cavity in the tetrahedral framework (§ 5.2.2), the expansion of HA is slightly less than LA. There is no immediate explanation for the slight difference in curves for the data of Prewitt et al. (1976) and Winter et al. (1979). Rb-sanidine (RbS), with the most expanded framework of all the M^{1+}-feldspars, expands the least with increasing T, followed by disordered K-feldspar (HS) and ordered K-feldspar (LM). The latter exhibit the same relative expansion curves as HA and LA: the more ordered (with lower equivalent temperature factor, B, and generally smaller cation cavity) expands more readily. The Or_{38} specimen has a less expanded framework than HS due to the significant Na-component, and Or_{19} (Henderson, 1979) goes through the $C\bar{1}$-C2/m inversion at 560±10°C.

Brown et al. (1984) have compiled additional data and discuss thermal expansion of alkali feldspars in comparison with that observed by Na $\Rightarrow$ K $\Rightarrow$ Rb substitution in terms of changes in T-O-T angles, M-O distances, and individual cell parameters. See also Kroll (1984). These are comprehensive studies and there have been few new data since they were written.

At temperatures up to ~1250 K the volume thermal expansion of synthetic $BaAl_2Si_2O_8$ and $CaAl_2Si_2O_8$ (bottom, Fig. 25) are strikingly similar—a somewhat unexpected result based on the significant difference in mean Ba-O and Ca-O distances and weaker bond strengths to the 9-coordinated Ba vs. the four 6-7-coordinated calciums. High-T and high-P structure refinements of celsian would be most informative. Bruno and Gazzoni (1968) determined the cell parameters of eight Ca-Sr feldspars at 25°C and 850°C, remarking on the $I\bar{1}$-I2/c transition as functions of T and X; see forthcoming work by McGuinn and Redfern (1993).

6.1.3. *Variation of Composition in 'Solid-Solution' Series.* The effects on V/V_o of changes in M-site occupancy in binary systems are plotted in Figure 26 as a function of the mole % of the smaller of the cations. The upper graph presents the M^{1+}–M^{1+} substitutions, the middle graph the M^{2+}–M^{2+} series, and the lower graph the M^{1+}–M^{2+} series, which involve an $AlSi_3$-Al_2Si_2 substitution as well. Most were prepared by hydrothermal synthesis or fusion, except for LA-LM, AA-HS, and K-Ca, which represent cation-exchange series using natural starting materials treated in anhydrous molten chlorides. Nonideality is the rule, with the Sr-Ba solid solution being perhaps the only exception. Calcium substitution in Ba-feldspar heated to 1450°C is limited to ~25% (Bruno and Gazzoni, 1970). The squares above the Ca-Sr curve are from early fusion syntheses by Bruno and Gazzoni (1968).

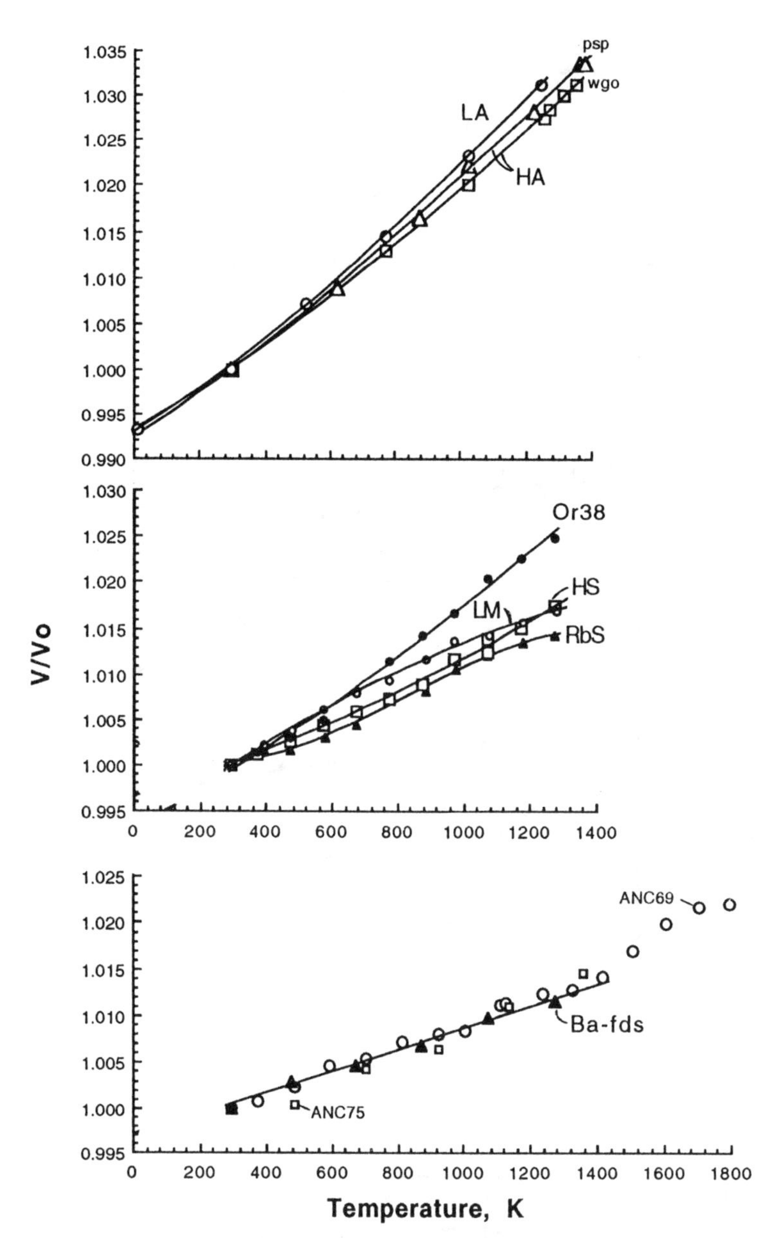

Figure 25. Volume expansion curves for low albite (○ LA; Winter et al, 1977), high albite (HA; Δ *psp* = Prewitt et al., 1976; ❑ *wgo* = Winter et al., 1979), Or_{38} and HS (● synthetic $Or_{38}Ab_{62}$ with $2t_1 = 0.61$, and ❑ synthetic Or_{100} with $2t_1 = 0.55$; Henderson, 1979; site occupancies from Kroll and Ribbe, 1987, Eqn. 5), low mirocline and Rb-sanidine (○ LM; ▲ RbS; both from Henderson, 1978), synthetic $BaAl_2Si_2O_8$ (▲ Ba-fds with fitted line; data extracted from Fig. 13 in Oehlschlegel et al., 1974), synthetic anorthite (❑ ANC75 and ○ ANC69 with $Q_{od} = 0.88$ and 0.74, respectively; data from Carpenter, 1992, Table 4).

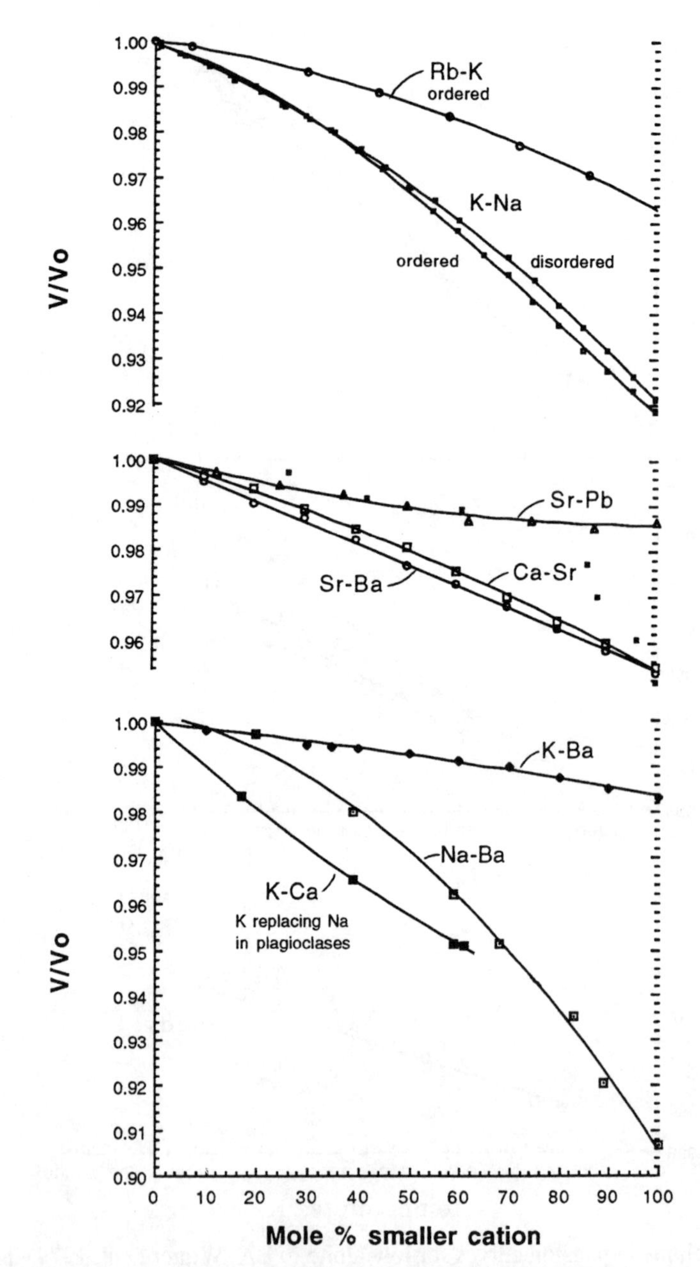

Figure 26. V/V_0 vs. mole % of the smaller cation in the Al,Si feldspar solid solution series Rb-K (McMillan et al., 1980), K-Na (Kroll et al., 1986), Sr-Pb and Ca-Sr (▪) (Bruno and Gazzoni, 1968; 1970), Ca-Sr and Sr-Ba (Bambauer and Nager, 1981), K-Ba and Na-Ba (Viswanathan, 1984; Viswanathan and Harneit, 1989; see Viswanathan and Kielhorn, 1983 for ternary K-Na-Ba feldspars), and K-Ca (K in molten chloride replacing Na in natural plagioclases; Viswanathan, 1972).

An obvious conclusion is easily confirmed: the change in effective M-cation radius (Δr) across a binary series is directly proportional to the change in volume. The only exception is Sr-Pb, whose volume change is larger than expected from Shannon (1976) radii. [The cell edges of this series depart significantly from expected trends as well; see § 6.2.2 below.] Furthermore, a line representing $dV/V_0/d\Delta r$ has a higher slope for the series in which Al:Si changes from 1:3 to 2:2 than for the line representing the two series with unchanged Al:Si ratios throughout.

6.2. CHANGES IN CELL DIMENSIONS WITH P,T,X

Four compilations of data are presented in Figures 27-29, with minimal comment on each. To keep all the data on the same scale, cell dimensions and (later) structural parameters will be plotted vs. unit cell volume, V, or $^1/_2$ V in the case of **c** ≅ 14 Å feldspars. If the path of a parameter vs. V is mimicked for P and T and X, one may safely, though cautiously assume regular variation of internal structural details, M-O distances, T-O-T angles, etc., as discussed in § 6.3.

6.2.1. *Series of Na- and K- Mixed with Ca-, Sr-, and Ba-feldspars.* Crystal chemical systematics of the plagioclases have been well established for at least a decade, and it is now possible to determine Al,Si distribution among the T sites from lattice parameters or peak positions in powder diffraction patterns, if composition is known. The same is true for most ternary (Na,K,Ca) feldspars—see review by Kroll (1983). As mentioned earlier (§ 4.2.1) the K-Ba feldspars have also been extensively investigated. At low temperatures both Na-Ca and K-Ba feldspars demonstrate exsolution and various superstructures, but at higher temperatures they form solid-solution series whose lattice parameters are plotted in Figure 27, together with synthetic Na-Sr series, several (K,Na,Ba)-ternaries, and a number of K-for-Na-substituted plagioclases formed by alkali exchange methods (Viswanathan, 1971, 1972). No high-T or -P data are included in these plots. See discussion of the various binaries in references listed in the caption of Figure 27.

6.2.2. *The M^{2+}-feldspars.* Lattice parameters of the Ca-Sr, Sr-Ba, Ca-Sr-Ba, and Sr-Pb solid-solution series at RT, anorthite (An) at high P and high T, and synthetic Ba-feldspar at high T are plotted in Figure 28. There is a notable departure from expected trends for **a**, **b**, **c** only in the Sr-Pb series (cf. Fig. 26). [This may be due to lone-pair electrons associated with lead, although I have made no attempt to follow up on this idea.] Absent from this compilation are the RT and high-P data on the Ca-Sr-feldspar series recently reported by McGuinn and Redfern (1993b; see Fig. 12 caption).

As noted in § 6.1.1 and Figure 24, there is a discontinuity of cell volume of anorthite at ~1300 Å^3, corresponding to 2.61±0.06 GPa. The accompanying decrease in **b** is less than those in **a** and **c**, but the most pronounced are discontinuities in cell angles (Fig. 28). These are due to the first-order $I\bar{1}$-$P\bar{1}$ transition (§ 4.2.3) involving changes in the framework from T-O-T angle bending and concomitant adjustments to the bonding environment of Ca atoms, which reduce in number from one 6- and three 7-coordinated calciums (not unlike those illustrated in Fig. 4 of Ghose et al., 1993; cf. Fig. 11) to two 7-coordinated calciums. Details are found in the chapter by Angel.

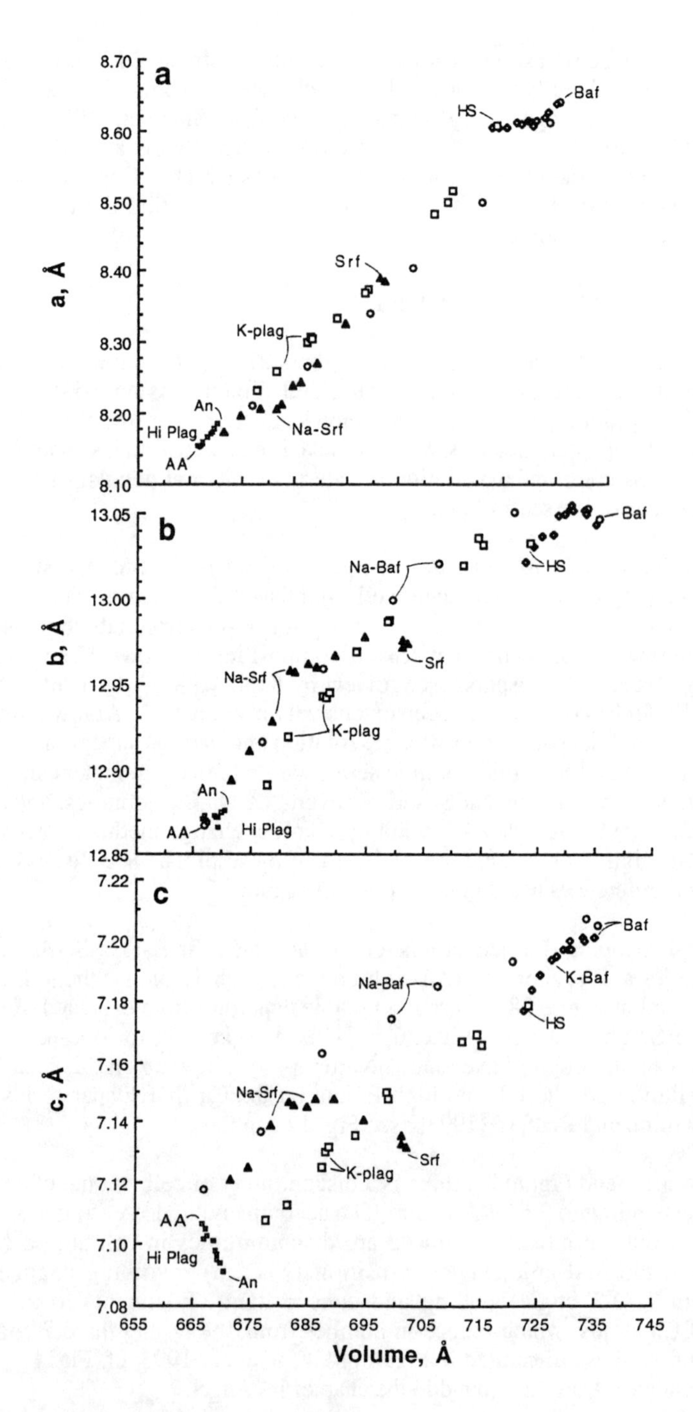

Figure 27. Full caption is on the next page.

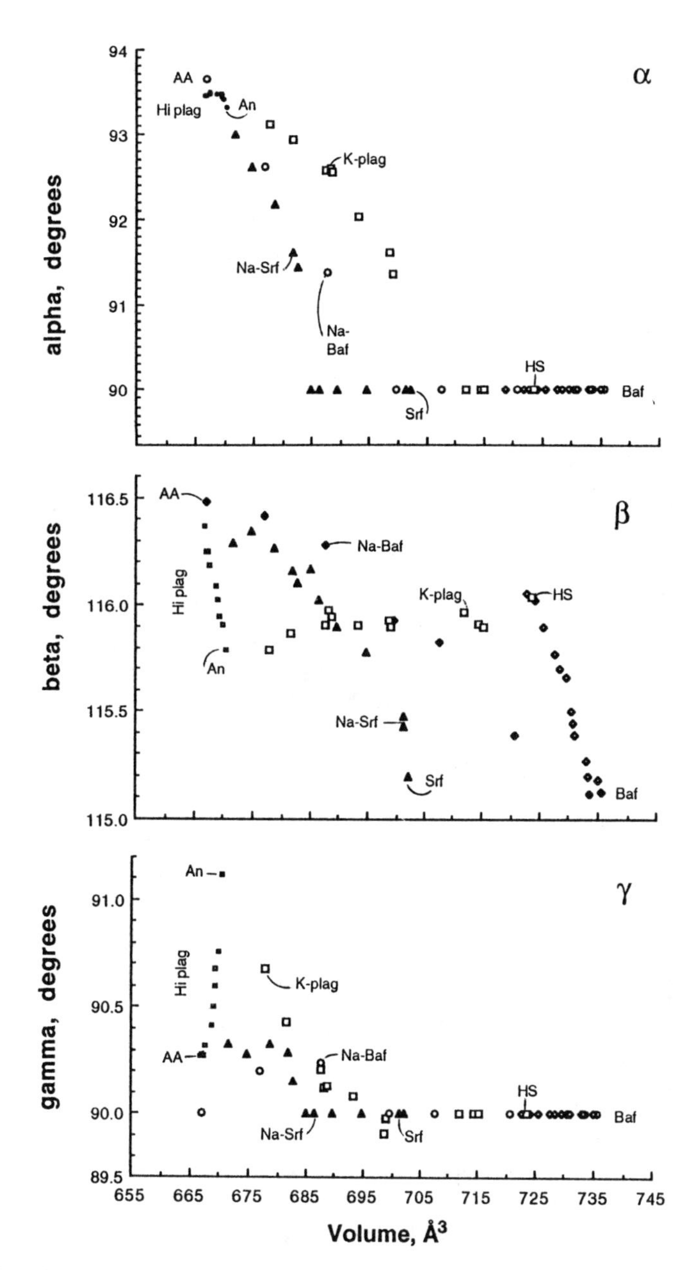

Figure 27. Compilation of lattice parameters vs. cell volume for the solid solution binaries:

□ Na-Ca (AA-An = high plagioclases) — Kroll (1971)

❑ K-Ca (HS-An = "K-plagioclases") — Viswananthan (1971,1972)

○ Na-Ba (AA-Baf = analbite-celsian series) — Viswanathan and Harneit (1989)

◊ K-Ba (HS-Baf = sanidine-celsian series: "hyalophanes") — Viswanathan (1984)

Δ Na-Srf (analbite-Sr-feldspar series) — Bambauer et al. (1984) and references therein

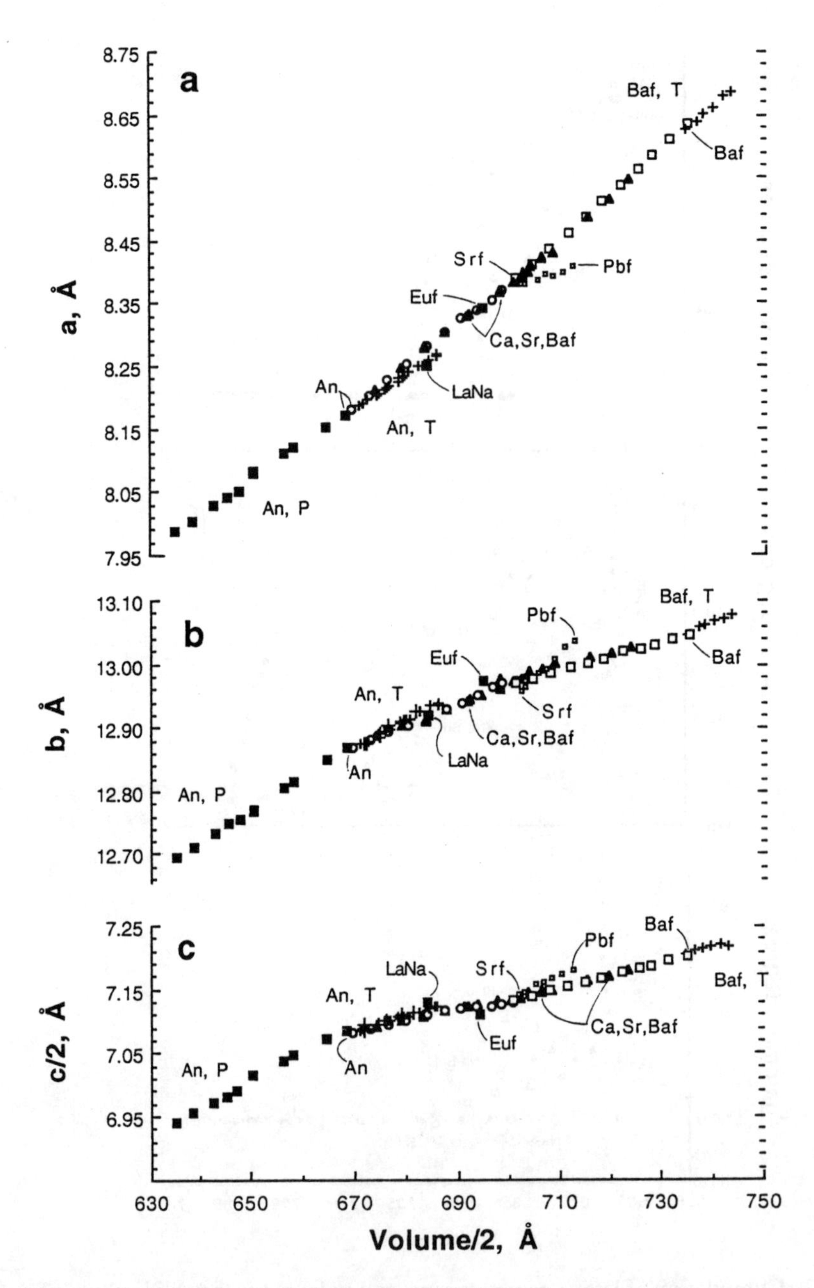

Figure 28. Cell parameters **a**, **b**, **c**/2 vs. V/2 for the M^{2+}-feldspars plus LaNa-feldspar = $LaNaAl_4Si_4O_{16}$ (from Bettermann and Liebau, 1976). Data for An@P (▪) from Angel et al. (1988); An@T from Carpenter (1992, Table 4: + = ANC75 and + = ANC69); An-Srf (○) and Srf-Baf (❑) series and the Ca,Sr,Baf ternaries (▲) from Bambauer and Nager (1981); and Euf = $EuAl_2Si_2O_8$ from Iwasaki and Kimizuka (1978). Figure continued next page.

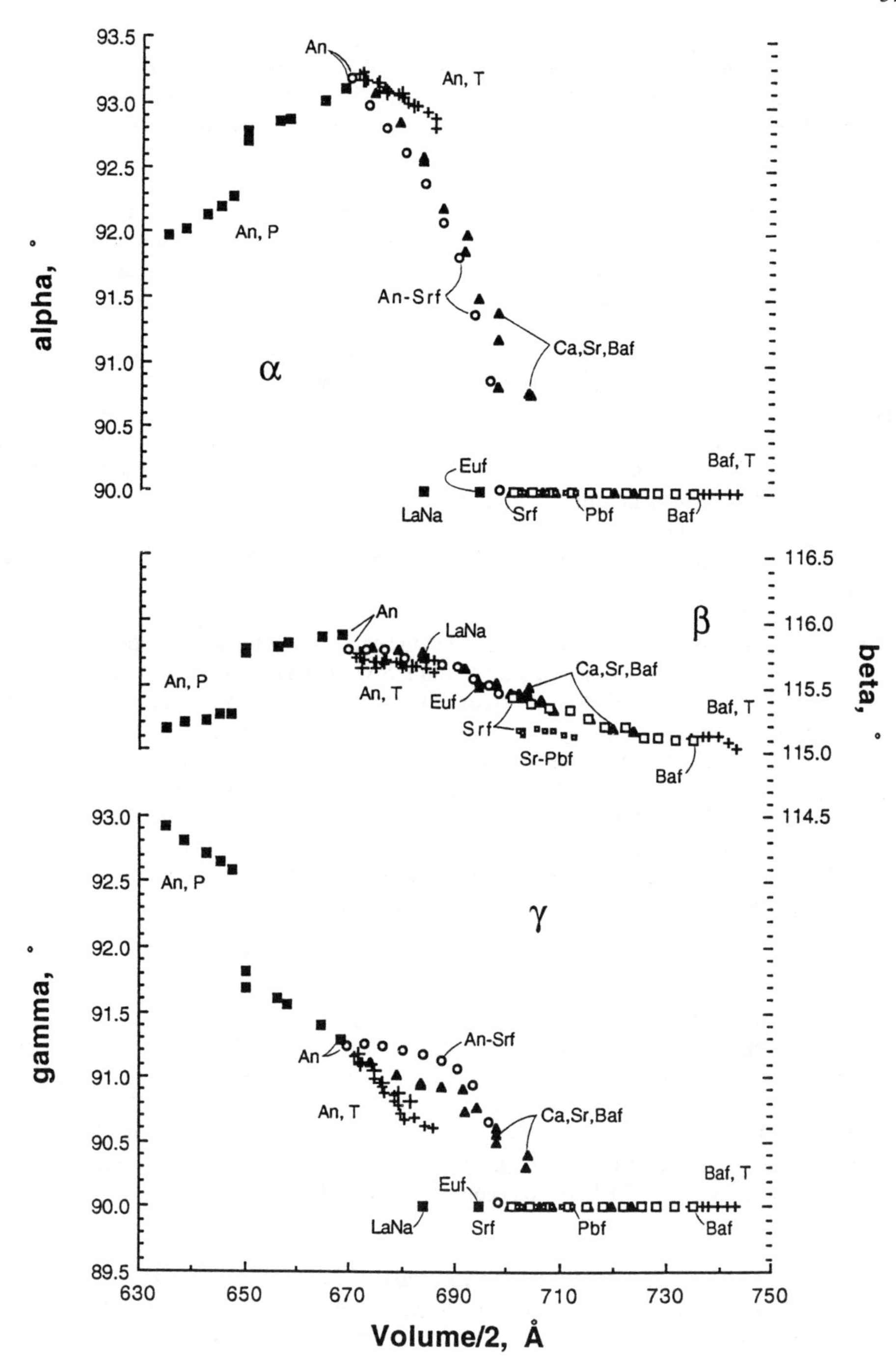

Figure 28, continued. Unit cell angles α, β, γ vs. V/2 for the M^{2+}-feldspars.

6.2.3. *Alkali Feldspars.* Most of the variations in cell parameters with T and X (plotted vs. V in Fig. 29) have been exhaustively analyzed by Brown et al. (1984; cf. Smith and Brown, 1988). Although there were some misleading high-P data on Or_{82} (from Hazen, 1976) available to Brown et al., the LA and HS 'at-pressure' data shown here are relatively recent (Angel et al., 1988).

The low-albite-at-pressure (LA@P) plots are essentially linear and continous, following nicely from the LA@T and the LM-LA compositional trends with volume, although dc/dV is noticeably greater than either [cf. An@P and HS@P], and the α angle is essentially unchanged under compression.

Except for β, changes in the lattice parameters of HS are all somewhat unexpected. First of all, there appears to be no $C2/m \Rightarrow C\bar{1}$ transition parallel to that occurring at $Or_{\sim 34}$ in the HS-AA compositonal series (at $V \cong 692$ Å^3). Why not? Clearly there is no collapse of the framework around the large K cation of the sort that accompanies Al,Si ordering (cf. Fig. 17). Furthermore there is a major departure of the HS@P line for the **c** dimension from the HS-HA trend: this may be attributed to effects of the extreme positional disorder of the small Na in HA which carries into the HA-HS solid solution series, a phenomenon which certainly is not duplicated with K in HS under pressure (cf. § 5.1 and 5.2 and accompanying figures).

Of particular interest in Figure 29 are the data representing three disordered Li-feldspars (❑) that were formed by cation exchange from HS via HA and H-feldspar (§ 2). A portion of Figure 29 has been expanded in Figure 30 to indicate that one of the Li-feldspars (+), whose structure was modeled by a DLS refinement (Deubener et al., 1991, specimen II; V = 624.6 Å^3), falls on or very near the LA@P line (equivalent pressure: ~4.0 GPa) for **a**, **b**, **c**, and β, though not α and γ (see Fig. 29 for angles). It *also* falls on or very near linear extensions of the Na-rich, triclinic segment of the AA-HS solid solution series, $Ab_{100}Or_0$-$Ab_{66}Or_{34}$, shown as a line in Figure 30. This is not an unexpected result, because this material is highly disordered with $t_1 = 0.32$. Could the DLS structure refinement of this specimen provide clues to the structure of LA at high pressures?

6.3. CHANGES IN T-O-T ANGLES AND M-O DISTANCES WITH P,T,X

6.3.1. *Disordered Alkali Feldspars.* The change in T-O-T angles with unit cell volume for the triclinic alkali feldspars, including monalbite (MA) at 1253 K, HA over the temperature range 1313 to 13 K and Li-feldspar at RT, are plotted in Figure 31. Data are arbitrarily fitted with second-order polynomials. Because the Li atom is 5-coordinated, the five shortest M-O distances for the disordered Rb-K-Na-Li-feldspars and intermediate members are plotted in Figure 32. M-OA_1 is linear across the entire range and the T-OA_1-T angle is nearly linear, as well. OA_1 is the oxygen that bridges between adjacent (001) crankshaft 'slabs' (Figs. 3 and 4).

6.3.2. *Ordered Alkali Feldspars and the Structure of Low Albite at 5 GPa.* The proximity of the cell parameters of Li-feldspar to those of LA at ~4 GPa (Figs. 29 and 30) has encouraged modelling of the 50-GPa LA structure. Cell parameters and V (= 616 Å^3)

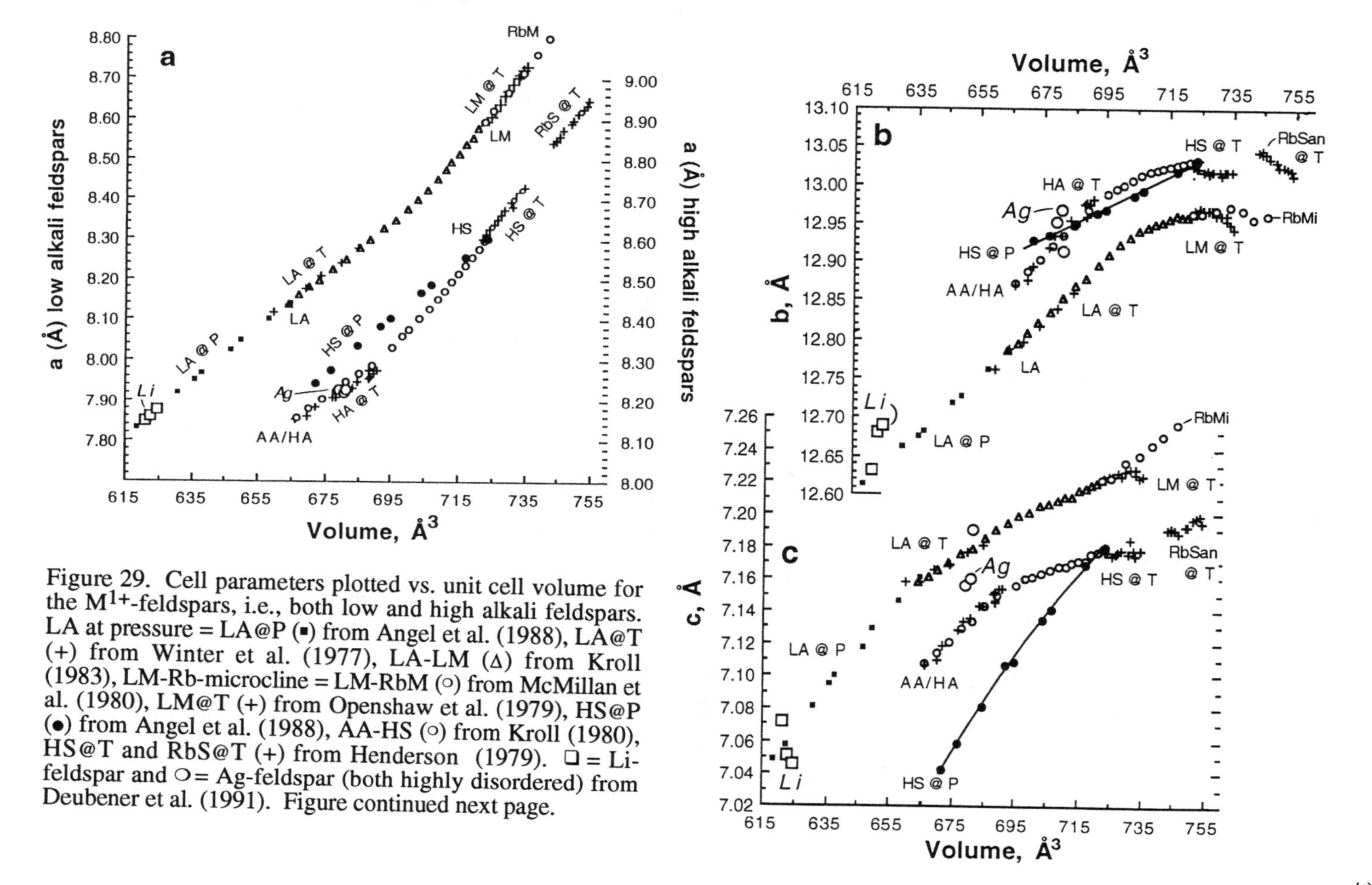

Figure 29. Cell parameters plotted vs. unit cell volume for the M^{1+}-feldspars, i.e., both low and high alkali feldspars. LA at pressure = LA@P (▪) from Angel et al. (1988), LA@T (+) from Winter et al. (1977), LA-LM (Δ) from Kroll (1983), LM-Rb-microcline = LM-RbM (○) from McMillan et al. (1980), LM@T (+) from Openshaw et al. (1979), HS@P (●) from Angel et al. (1988), AA-HS (○) from Kroll (1980), HS@T and RbS@T (+) from Henderson (1979). ❑ = Li-feldspar and ○ = Ag-feldspar (both highly disordered) from Deubener et al. (1991). Figure continued next page.

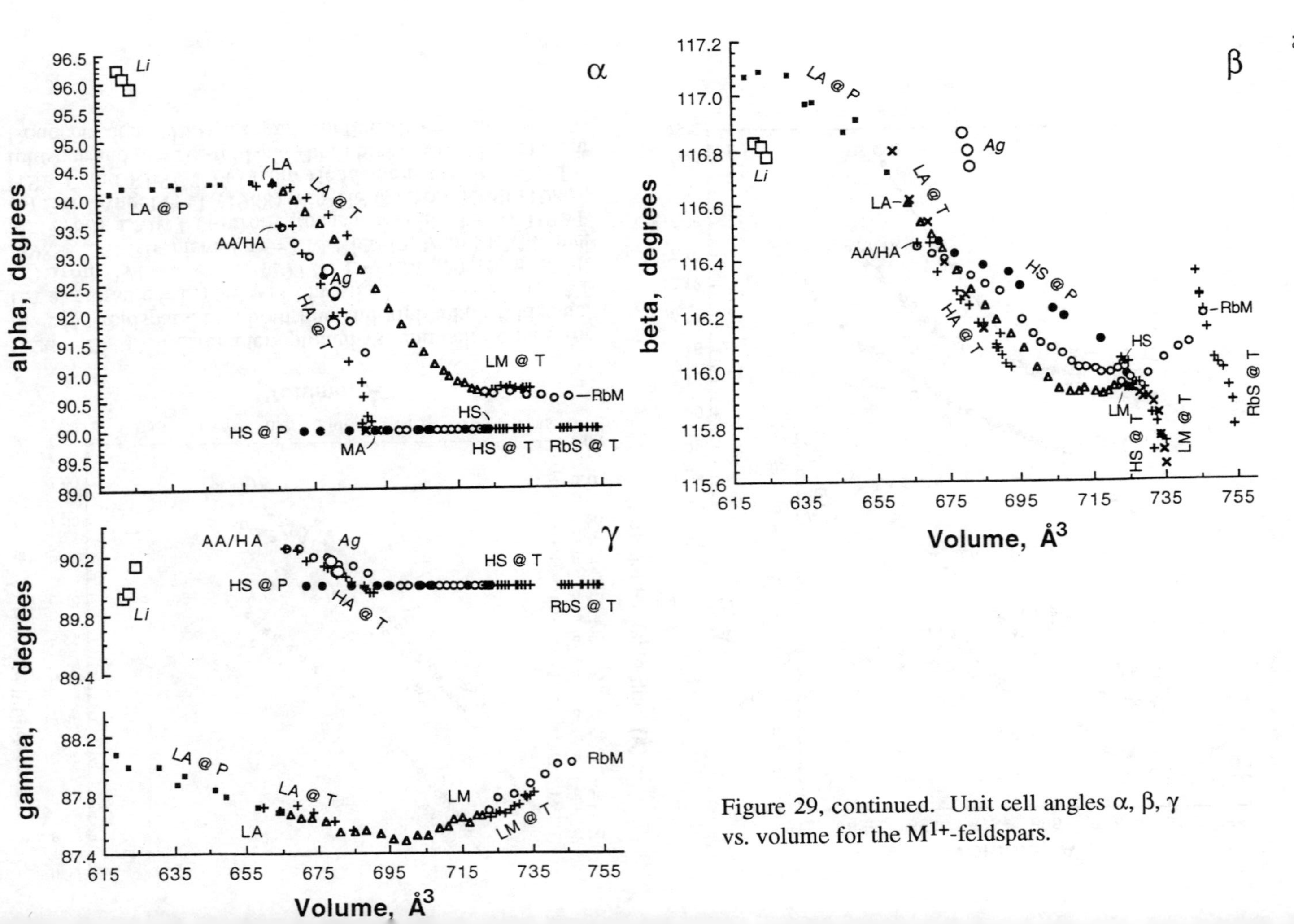

Figure 29, continued. Unit cell angles α, β, γ vs. volume for the M^{1+}-feldspars.

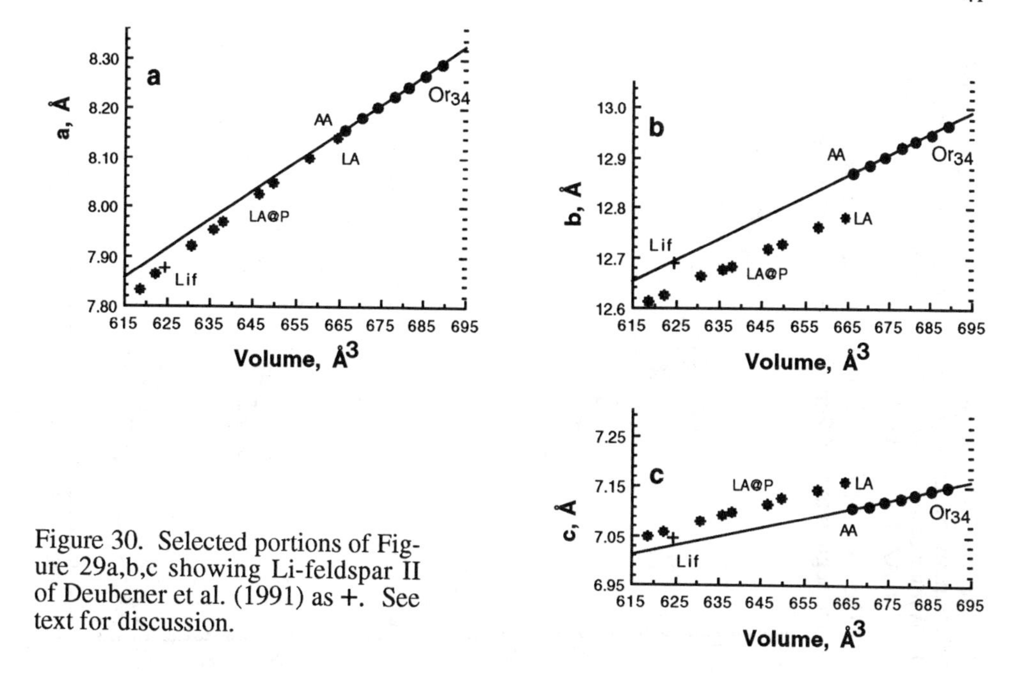

Figure 30. Selected portions of Figure 29a,b,c showing Li-feldspar II of Deubener et al. (1991) as +. See text for discussion.

were determined by linear extrapolation of the data of Angel et al. (1988). Several models were attempted.

Model 1: The x***a**, y***b**, z***c** parameters for *all* the atoms of *all* the LA@T structures were plotted and found to have remarkable linearity, so that extrapolation to 616 $Å^3$ was possible. The estimated values were divided by the 'known' **a,b,c** of the 5-GPa LA to generate a set of coordinates from which interatomic distances and angles were calculated.

Model 2: A version of DLS modified by M.B. Boisen, Jr., and G.V. Gibbs (pers. comm.) to weight the refinement for force constants involved in inter- and intra-tetrahedral bond angles and bond lengths was applied by R.T. Downs using the 'known' cell parameters and T-O, O-T-O, T-O-T 'target values' obtained from extrapolation against volume of LM and LA@T data to the 5-GPa volume of LA.

Model 3: The same program was used with a more distant set of target values, namely the structural details of the 13 K refinement of LA by Smith et al. (1986).

These models produced somewhat different structural parameters for the hypothetical 5-GPa LA. For brevity, only the T-O-T angles are plotted vs. volume in Figure 33. Notice that for most, Li-feldspar values are closest to the Model 3 approximations. These values are preferred because Model 3 was less constrained than the others. Furthermore, a plot of what I have called "T-O-T distortion," defined by the equation from Baur (1974) in Figure 34, favors Model 3 by virture of the intersecting extrapolation vs. V of the departure of these angles from the means for each structure. It is fairly sure that the parameters require further refinement, and we (R.T. Downs and this author) expect to have more certain estimates when additional force constants (for Al and Na) become available.

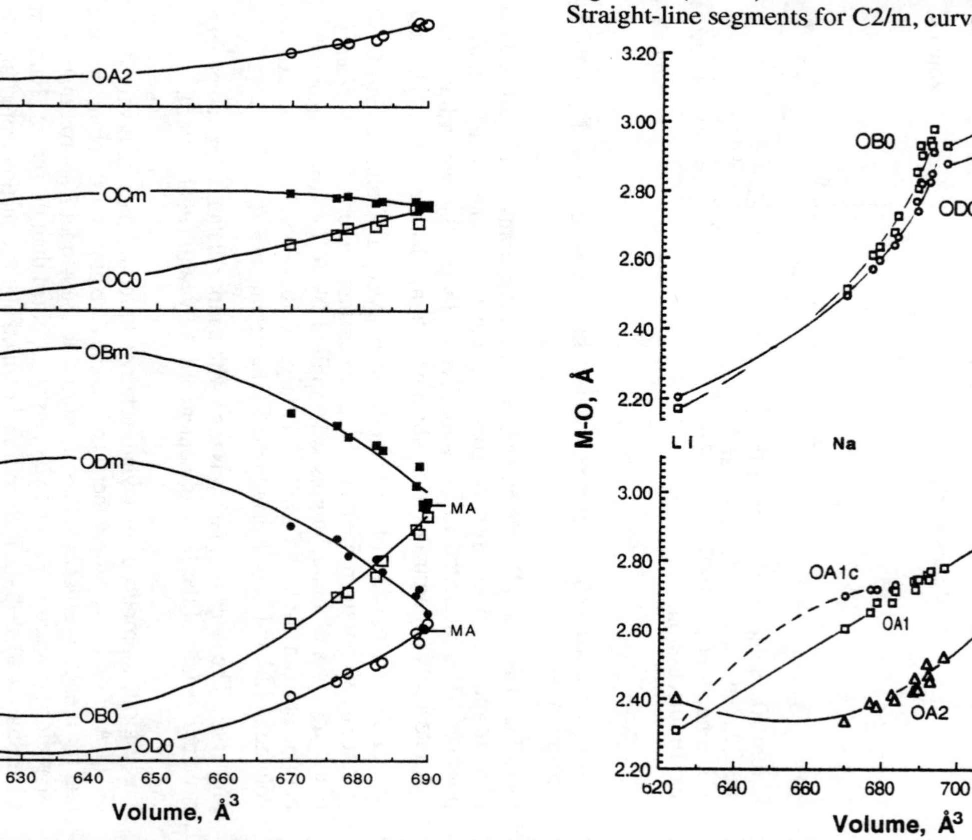

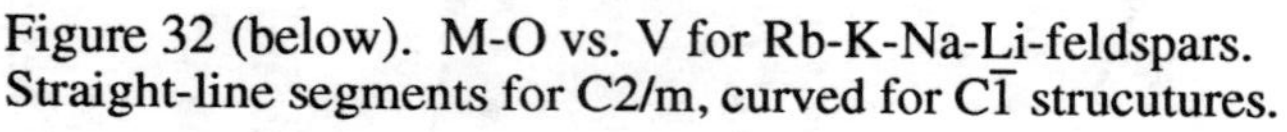

Figure 31 (left). T-O-T angles vs. V for Li-feldspar (II; new DLS refinement by Downs, pers. comm.), and HA@T and MA@T (Winter et al., 1977). "Or" = $Or_{30}Ab_{70}$ in the triclinic portion of the AA-HS solid solution series.

Figure 32 (below). M-O vs. V for Rb-K-Na-Li-feldspars. Straight-line segments for C2/m, curved for $C\overline{1}$ strucutures.

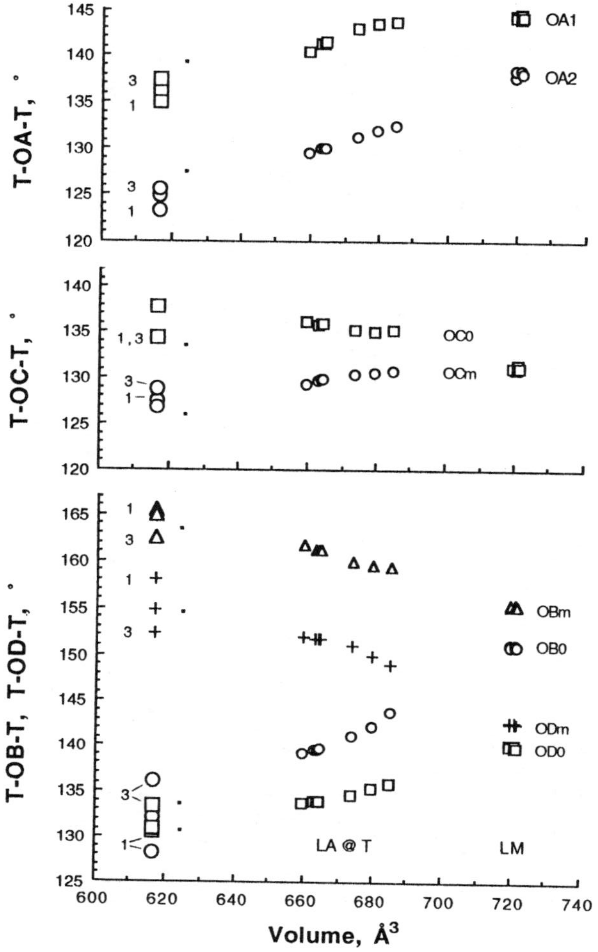

Figure 33 (left). T-O-T angles vs. volume for LM, LA@T, Li-feldspar (dots), and the results of Models 1, 2, and 3 for LA at 50 kbar.

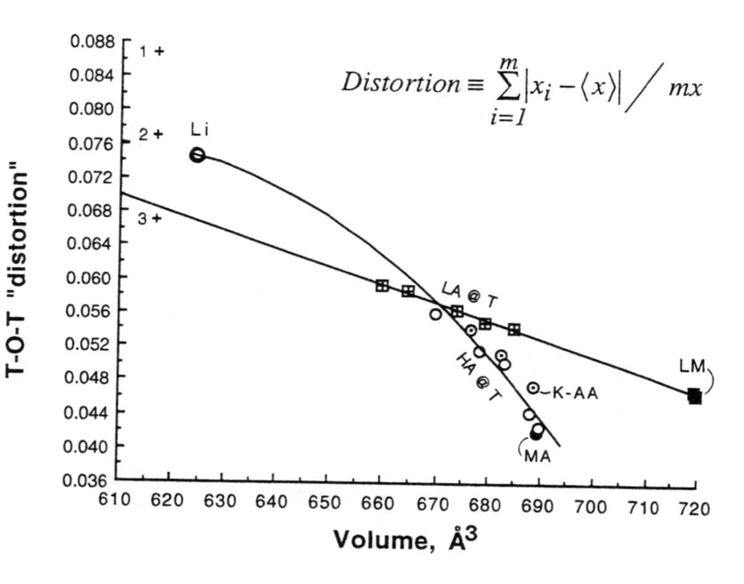

Figure 34. The "T-O-T distortion" parameter of Baur (1974; defined on the figure) vs. volume. Two trends are fitted with curves: (1) LM-LA@T, which extrapolates through the Model 3 results (+ sign). (2) MA@T–HA@T–Li-feldspar, including some RT anorthoclase data from Harlow (1982).

7. Future Work

As this book goes to press there are several studies underway that will be relevant to the structural relations among various feldspars as functions of pressure and temperature. Among them are (1) a transformation in low microcline as a function of pressure (lattice parameters only; R.J. Angel and coworker), (2) attempts by Angel and collaborators to find the monoclinic-triclinic phase transformation at high pressure in sanidine that was reported by Hazen (1976) but could not be duplicated in 1988 by Angel et al., (3) the structure of low albite at high pressures (R.T. Downs and coworkers), and (4) the studies of Sr- and Ca-Sr feldspars at pressure by McGuinn and Redfern (1993a,b–in press).

I think that the most helpful new information will proceed from high pressure structure determinations of low albite and high abite (or analbite) and low microcline and high sanidine. The structures of celsian at both high temperature and high pressure would be useful, and a general revival of feldspar structure determination *at* high temperatures might provide consistent data sets which are now spotty at best.

References

Abe T, Tsukamoto K, Sunagawa I (1991) Nucleation, growth and stability of $CaAl_2Si_2O_8$ polymorphs. Phys Chem Minerals 17:473-484

Adelhardt W, Frey F, Jagodzinski H (1980a) X-ray and neutron investigations of the $P\bar{1}$–$I\bar{1}$ transition in pure anorthite. Acta Cryst A36:450-460

Adelhardt W, Frey F, Jagodzinski H (1980b) X-ray and neutron investigations of the $P\bar{1}$–$I\bar{1}$ transition in anorthite with low albite content. Acta Cryst A36:461-470

Angel RJ (1988) High-pressure structure of anorthite. Am Mineral 73:1114-1119

Angel RJ (1992) Order-disorder and the high-pressure $P\bar{1}$–$I\bar{1}$ transition in anorthite. Am Mineral 77:923-929

Angel RJ, Carpenter MA, Finger LW (1990) Structural variation associated with compositional variation and order-disorder behavior in anorthite-rich feldspars. Am Mineral 75:150-162

Angel RJ, Hazen RM, McCormick TC, Prewitt CT, Smyth JR (1988) Comparitive compressibility of end-member feldspars. Phys Chem Minerals 15:313-318

Armbruster T, Kunz M (1990) Difference displacement parameters in alkali feldspars: effects of (Si,Al) order-disorder. Am Mineral 75:141-149

Bailey SW (1969) Refinement of an intermediate microcline structure. Am Mineral 54:1540-1545

Bambauer H-U, Nager HE (1981) Gitterkonstanten und displazive Transformation synthetischer Erdalkalifeldspäte I. System $Ca[Al_2Si_2O_8]$-$Sr[Al_2Si_2O_8]$-$Ba[Al_2Si_2O_8]$. N Jb Mineral Abh 141:225-239

Bambauer H-U, Schöps M, Pentinghaus H (1984) Feldspar phase relations in the system $NaAlSi_3O_8$-$SrAl_2Si_2O_8$. Bull Minéral 107:541-552

Baur W (1974)The geometry of polyhedral distortion. Predictive relationships for the phosphate group. Acta Cryst B30:1195-1215

Bettermann P, Liebau F (1976) Lanthanum-calcium-sodium aluminosilicates, $La_xCa_{8-2x}Na_x[Al_{16}Si_{16}O_{64}]$. Naturwiss 63:480

Blasi A, Brajkovic C, dePol Blasi C (1984) Dry-heating conversion of low microcline to high sanidine via a one-step disordering process. Bull Minéral 107:423-436

Blasi A, de Pol Blasi C, Zanazzi PF (1981) Structural study of a complex microperthite from anatexites at Mt. Caval, Argentera Massif, Maritime Alps. N Jb Mineral Abh 142:71-90

Blasi A, de Pol Blasi C, Zanazzi PF (1987) A re-examination of the Pellotsalo microcline: Mineralogical implications and genetic considerations. Can Mineral 25:527-537

Bown MG, Gay P (1958) The reciprocal lattice geometry of the plagioclase feldspar structures. Z Krist 111:1-14

Brown, GE Jr, Hamilton WC, Prewitt CT (1974) Neutron diffraction study of Al/Si ordering in a sanidine: comparison with X-ray diffraction data. In The Feldpars, MacKenzie WS, Zussman J, eds Manchester Univ Press, Manchester, UK: 68-80

Brown WL, Hoffman W, Laves F (1963) Über kontinuierliche und reversible Transformationen des Anorthits ($CaAl_2Si_2O_8$) zwischen 25 und 350°C. Naturwiss 50:221

Brown WL, Openshaw RE, McMillan PF, Henderson CMB (1984) A review of the expansion behavior of alkali feldspars: coupled variations in cell parameters and possible phase transitions. Am Mineral 69:1058-1071

Bruno E, Fachinelli A (1974) Experimental studies on anorthite crystallization along the join $CaAl_2Si_2O_8$-SiO_2. Bull Soc franc Mineral Cristallogr 97:422-432

Bruno E, Pentinghaus H (1974) Substitution of cations in natural and synthetic feldspars. In The Feldpars, MacKenzie WS, Zussman J, eds Manchester Univ Press, Manchester, UK: 574-609

Burns PC, Fleet ME (1990) Unit cell dimensions and tetrathedral site ordering in synthetic gallium albite ($NaGaSi_3O_8$) Phys Chem Minerals 17:108-116

Carpenter MA (1986) Experimental delineation of the "e"⇔$I\bar{1}$ and "e"⇔$C\bar{1}$ transformations in intermediate plagioclase feldspars. Phys Chem Minerals 13:119-139

Carpenter MA (1991a) Mechanisms and kinetics of Al/Si ordering in anorthite, I. incommensurate structure and domain coarsening. Am Mineral 76:1110-1119

Carpenter MA (1991b) Mechanisms and kinetics of Al/Si ordering in anorthite, II. Energetics and a Ginzburg-Landau rate law. Am Mineral 76:1120-1133

Chiari G, Gazzoni G, Craig JR, Gibbs GV, Louisnathan SJ (1985) Two independent refinements of the crystal structure of paracelsian, $BaAl_2Si_2O_8$. Am Mineral 70:969-974

Dal Negro A, De Pieri R, Quareni S (1978) The crystal structures of nine K feldspars from the Adamello Massif (Northern Italy). Acta Cryst B34:2699-2707

Daniel I, Gillet P, McMillan PJ (1993) A Raman spectroscopic study of $CaAl_2Si_2O_8$ at high pressure and high temperature: polymorphism, melting and amorphization. NATO ASI Feldspars and Their Reactions, Edinburgh: abstr 10

De Pieri R, Quareni S, Hall S (1977) Refinement of the structures of low and high hyalophanes. Acta Cryst B33:3073-3076

Deubener J, Sternitzke M, Müller G (1991) Feldspars $MAlSi_3O_8$ (M = H,Li,Ag) synthesized by low-temperature ion exchange. Am Mineral 76:1620-1627

Downs RT, Gibbs GV, Bartelmehs KL, Boisen MB Jr (1992) Variations of bondlengths and volumes of silicate tetrahedra with temperature. Am Mineral 77:751-757

Ferguson RB, Ball NA, Cerny P (1991) Structure refinement of an adularian end-member high sanidine from the Buck Claim pegmatite, Bernic Lake, Manitoba. Can Mineral 29:543-552

Ferguson RB, Traill RJ, Taylor WH (1958) The crystal structures of low-temperature and high-temperature albites. Acta Cryst 11:331-348

Fitz Gerald JD, Parise JB, Mackinnon IDR (1986) Average structure of an An_{48} plagioclase from the Hogarth Ranges. Am Mineral 71:1399-1408

Gasperin M (1971) Structure cristalline de $RbAlSi_3O_8$. Acta Cryst B27:854-855

Gay P, Roy NN (1968) The mineralogy of the potassium-barium feldspar series. III. Subsolidus relations. Mineral Mag 36:914-932

Ghose S, McMullan RK, Weber H-P (1993) Neutron diffraction studies of the $P\bar{1} \Rightarrow I\bar{1}$ transition in anorthite, $CaAl_2Si_2O_8$, and the crystal structure of the body-centered phase at 514 K. Z Krist 204:215-237

Goldsmith JR, Jenkins DM (1985) The high-low albite relations revealed by reversal of degree of order at high pressure. Am Mineral 70:911-923

Griffen DT, Johnson BT (1984) Strain in triclinic feldspars: a crystal structure study. Am Mineral 69:1072-1077

Griffen DT, Ribbe PH (1976) Refinement of the crystal structure of celsian. Am Mineral 61:414-418

Griffen DT, Ribbe PH, Gibbs GV (1977) The structure of slawsonite, a strontium analog of paracelsian. Am Mineral 62:31-35

Grove TL (1977) Structural characterization of labradorite-bytownite plagioclase from volcanic, plutonic and metamorphic environments. Contrib Mineral Petrol 64:273-302

Grundy HD, Ito J (1974) The refinement of a crystal sructure of a synthetic non-stoichiomentric Sr-feldspar. Am Mineral 59:1319-26

Hackwell TP, Angel RJ (1992) The comparative compressibility of reedmergnerite, danburite and their aluminum analogs. Eur J Mineral 4:1221-1227

Harlow GE (1982) The anorthoclase structures: the effects of temperature and composition. Am Mineral 67:975-996

Hazen RM (1976) Sanidine: predicted and observed monoclinic-to-triclinic reversible transformation at high pressure. Science 194:105-107

Hazen RM, Finger LW (1982) Comparative Crystal Chemistry. John Wiley & Sons, New York: 231 p

Henderson CMB (1978) Thermal expansion of alkali feldspars. II. Rb-sanidine and maximum microcline. Progress Expt Petrol 4th Progress Rpt Nat'l Environ Res Council Ser D 4:51-54

Henderson CMB (1979) An elevated temperature X-ray study of synthetic disordered Na-K alkali feldspars. Contrib Mineral Petrol 70:71-79

Horst W (1984) Modulated structures in intermediate plagioclases—history, ideas and perspectives. Fortschr. Mineral. 62:107-154

Iwasaki B, Kimizuka N (1978) Synthesis, X-ray crystallography and infrared absorption spectroscopy of $EuAl_2Si_2O_8$. Geochem J 12:1-6

Iyatomi N, Aoki Y (1992) Crystallization behavior of $P2_1$ and C2 phases of anorthite composition. N Jb Mineral Monatsh 1992:177-192

Kimata M (1988) The crystal structure of non-stoichiometric Eu-anorthite: an explanation for the Eu-positive anomaly. Mineral Mag 52:257-265

Kirkpatrick RJ, Carpenter MA, Yang Y-H, Montez B (1987) ^{29}Si Magic-angle NMR spectroscopy of low-temperature ordered plagioclase feldspars. Nature 325:236-238

Kroll H (1971) Feldspäte in System $K[AlSi_3O_8]$-$Na[AlSi_3O_8]$-$Ca[Al_2Si_2O_8]$: Al,Si-Verteilung und Gitterparameter, Phasen-Transformationen und Chemismus. Inaug-Diss Westf. Wilhelms-Univ, Münster: 119 p

Kroll H (1983) Lattice parameters and determinative methods for plagioclase and ternary feldspars. Rev Mineral 2, 2nd ed: 101-120

Kroll H (1984) Thermal expansion of alkali feldspars. In Feldspars and Feldspathoids, Brown WL, ed. NATO ASI C137:163-206

Kroll H, Bambauer H-U (1981) Diffusive and displacive transformation of (K,Na,Ca)-feldspars. N Jb Mineral Monatsh 1981:413-416

Kroll H, Bambauer H-U, Schirmer U (1980) The high albite-monalbite and analbite-monalbite transitions. Am Mineral 65:1192-1211

Kroll H, Flogel J, Breit U, Lons J, Pentinghaus H (1991) Order and anti-order in Ge-substituted alkali feldspars. Eur J Mineral 3:739-749

Kroll H, Knitter, R (1991) Al,Si exchange kinetics in sanidine and anorthoclase and modeling of rock cooling paths. Am Mineral 76:928-941

Kroll H, Ribbe PH (1983) Lattice parameters, composition and Al,Si order in alkali feldspars. Rev Mineral 2, 2nd ed: 57-99

Kroll H, Ribbe PH (1987) Determining Al,Si distribution and strain in alkali feldspars using lattice parameters and diffraction-peak positions: a review. Am Mineral 72:491-506

Kroll H, Schmiemann I, von Cölln G (1986) Feldspar solid solutions. AmMineral 71:1-16

Martin RJ (1974) Controls of ordering and subsolidus phase relations in the alkali feldspars. In The Feldpars, MacKenzie WS, Zussman J, eds Manchester Univ Press, Manchester, UK: 313-336

Machatschki F (1928) The structure and consitution of feldspars. Centr Mineral Abt A:97-104

McGuinn MD, Redfern SAT (1993a) Ferroelastic phase transition in $SrAl_2Si_2O_8$ feldspar at elevated pressure. Mineral Mag, in press.

McGuinn MD, Redfern SAT (1993b) Ferroelastic phase transition along the join $CaAl_2Si_2O_8$-$SrAl_2Si_2O_8$. Am Mineral, submitted

McLaren AC, Fitz Gerald JD (1987) CBED and ALCHEMI investigation of local symmetry and Al,Si ordering in K-feldspars. Phys Chem Minerals 14:281-292

McMillan PF, Brown WL, Openshaw RE (1980) The unit cell parameters of an ordered K-Rb alkali feldspar series. Am Mineral 65:458-464

Megaw HD (1973) Crystal Structures: A Working Approach. W.B. Saunders Co, London: 563 p

Megaw HD (1974a) The architecture of the felspars. In The Feldpars, MacKenzie WS, Zussman J, eds Manchester Univ Press, Manchester, UK: 1-24

Megaw HD (1974b) Tilts and tetrahedra in feldspars. In The Feldpars, MacKenzie WS, Zussman J, eds Manchester Univ Press, Manchester, UK: 87-113

Murakami H, Kimata M, Shimoda S, Ito E, Sasaki S (1992) Solubility of $CaMgSi_3O_8$ and $\square Si_4O_8$ endmembers in anorthite. J Mineral Pet Econ Geol 87:491-509

Newnham RE, Megaw HD (1960) The crystal structure of celsian (barium felspar). Acta Cryst 13:303-313

Oelschlegel G, Kockel A, Biedl A (1974) Anisotrope Wärmedehnung und Mischkristallbildung einiger Verbindungen des ternären Systems BaO-Al_2O_3-SiO_2, II. Glastechnische Berichte 47:31-41

Ohashi Y, Finger LW (1974) Refinement of the crystal structure of sanidine at 25° and 400°C. Carnegie Inst Washington Yrbk 73:539-574

Ohashi Y, Finger LW (1975) An effect of temperature on the feldspar structure: Crystal structure of sanidine at 800°C. Carnegie Inst Washington Yrbk 74:569-572

Openshaw RB, Hemingway BS, Robie RA, Krupka KM (1979) A room-temperature phase transition in maximum microcline. Heat capacity measurements. Phys Chem Minerals 5:83-93

Openshaw RB, Henderson CMB, Brown WL (1979) A room-temperature phase transition in maximum microcline. Unit cell parameters and thermal expansion. Phys Chem Minerals 5:95-104

Paulus H, Müller G (1988) The crystal structure of a hydrogen feldspar. N Jahrb Mineral Monatsh 11:481-490

Pentinghaus H (1980) Polymorphie in den Feldspatbildenden systemen $A^+T^{3+}T^{4+}$ Habilitationsschrift im Fachbereich Chemie, Westf Wilhelms-Univ, Münster

Phillips BL, Kirkpatrick RJ, Carpenter MA (1992) Investigation of short-range Al,Si order in synthetic anorthite by ^{29}Si MAS NMR spectroscopy. Am Mineral 77:484-494

Phillips MW, Ribbe PH, Pinkerton AA (1989) Structure of intermediate albite, $NaAlSi_3O_8$. Acta Cryst C45:542-545

Phillips MW, Zhang H, Kroll H (1988) The structure of a natural maximum microcline at 294 K and 163 K: The "room-temperature" phase transition revisited. Geol Soc Amer Abstr Program 20:225

Prewitt CT, Sueno S, Papike JJ (1976) The crystal structures of high albite and monalbite at high temperatures. Am Mineral 61:1213-1225

Prince E, Donnay G, Martin RF (1973) Neutron diffraction refinement of an ordered orthoclase structure. Am Mineral 58:500-507

Quareni S, Taylor WH (1971) Anisotropy of the sodium atom in low albite. Acta Cryst B27:281-285

Redfern SAT, Salje E (1992) Microscopic dynamic and macroscopic thermodynamic character of the phase transition in anorthite. Phys Chem Minerals 18:526-533

Ribbe PH (1974) The structure of a strained intermediate microcline in cryptoperthitic association with twinned plagioclase. Am Mineral 64:402-408

Ribbe PH (1983) Aluminum-silicon order in feldspars; domain textures and diffraction patterns. Rev Mineral 2, 2nd ed: 21-56

Ribbe PH (1984) Average structures of alkali and plagioclase feldspars: systematics and applications. In Feldspars and Feldspathoids, Brown WL, ed. NATO ASI C137:1-54

Ribbe PH, Megaw HD, Taylor WH, Ferguson RB, Traill RJ (1969) The albite structures. Acta Cryst B25:1503-1518

Ribbe PH, Phillips MW, Gibbs GV (1974) Tetrahedral bond length variations in feldspars. In The Feldpars, MacKenzie WS, Zussman J, eds Manchester Univ Press, Manchester, UK: 25-48

Salje E (1985) Thermodynamics of sodium feldspar I: Order parameter treatment and strain induced coupling effects. Phys Chem Minerals 12:93-98

Salje E, Kuscholke B, Wruck B, Kroll H (1985) Thermodynamics of sodium feldspar II: Experimental results and numerical calculations. Phys Chem Minerals 12:99-107

Scambos TA, Smyth JR, McCormick TC (1987) Crystal structure refinement of high-sanidine from the upper mantle. Am Mineral 72:973-978

Schiebold E (1927) Über den Feinbau der Feldspäte. Fortschr Mineral Krist Petr 12:78-82

Schiebold E (1928) Über den Feinbau der Feldspäte. Z Krist 66:488-493

Schiebold E (1929a) Über den Feinbau der Feldspäte. Fortschr Mineral Krist Petr14:62-68

Schiebold E (1929b) Über den Feinbau der Feldspäte. Trans Faraday Soc 25:316-520

Schiebold E (1929c) Über den Feinbau der Feldspäte. Centr Mineral Abt A:378-386

Schiebold E (1930) Über den Feinbau der Feldspäte. Z Krist 73:90-95

Shannon RD (1976) Revised effective ionic radii and systematic studies of interatomic distances in halides and chalcogenides. Acta Cryst A32:751

Smith JV (1974) Feldspar Minerals 1. Crystal Structures and Physical Properties. Springer-Verlag, Berlin: 627 p

Smith JV (1983) Some chemical properties of feldspars. Rev Mineral 2, 2nd ed: 281-296

Smith JV, Artioli G, Kvick A (1986) Low albite, $NaAlSi_3O_8$: neuton diffraction study of crystal stucture at 13 K. Am Mineral 71:727-733

Smith JV, Brown WL (1983) Feldspar Minerals 1. Crystal Structures, Physical, Chemical, and Microstructural Properties, 2nd ed. Springer-Verlag, Berlin: 828 p

Smith JV, Ribbe PH (1969) Atomic movements in plagioclase feldspars. Contrib Mineral Petrol 21:157-202

Stout GH, Jensen LH (1968) X-ray Structure Determination: A Practical Guide. Macmillan, London: 467 p

Su S-C, Bloss FD, Ribbe PH, Stewart DB (1984) Optic axis angle, a precise measure of Al,Si ordering in T1 tetrahedral sites of K-rich alkali feldspars. Am Mineral 69:440-448

Su S-C, Ribbe PH, Bloss FD (1986) Alkali Feldspars: structural state determined from composition and optic axis angle 2*V*. Am Mineral 71:1285-1296

Su S-C, Ribbe PH, Bloss FD, Goldsmith J (1986) Optical properties of single crystals in the order-disorder series low albite-high albite Am Mineral 71:1384-1392

Su S-C, Ribbe PH, Bloss FD, Warner JK (1986) Optical properties of a high albite (analbite)-high sanidine solid solution series. Am Mineral 71:1393-1398

Takéuchi Y, Haga N, Ito J (1973) The crystal structure of monoclinic $CaAl_2Si_2O_8$: a case of monoclinic structure closely simulating orthorhombic symmetry. Z Krist 137:380-398

Taylor WH (1933) The structure of sanidine and other feldspars. Z Krist 85:425-442

Taylor WH (1962) The structures of the principal felspars. Norsk Geol Tidsskr 42/2:1-24

van Tendeloo G, Ghose S, Amelinckx S (1989) A dynamical model for the $P\bar{1}-I\bar{1}$ phase transition in anorthite, $CaAl_2Si_2O_8$. I. Evidence from electron microscopy. Phys Chem Minerals 16:311-319

Viswanathan K (1971) A new X-ray method to determine the anorthite content and structural state of plagioclases. Contrib Mineral Petrol 30:332-335

Viswanathan K (1972) Kationaustausch an Plagioklasen. Contrib Mineral Petrol 37:277-290

Viswanathan K (1984) The behavior of the lattice edges of plagioclases and (Ba,K)-feldspars. Bull Minéral 107:447-454

Viswanathan K, Brandt K (1980) The crystal structure of a ternary (Ba,K,Na)-feldspar and its significance. Am Mineral 65:472-476

Viswanathan K, Kielhorn H-M (1983a) Variations in the chemical compositions and lattice dimensions of (Ba,K,Na)-feldspars from Otjosondu, Namibia, and their significance. Am Mineral 68:112-121

Viswanathan K, Kielhorn H-M (1983b) Al,Si distribution in a ternary (Ba,K,Na)-feldspar as determined by crystal structure refinement. Am Mineral 68:122-124

Viswanathan K, Harneit O (1989) Solid solution and unmixing in the feldspar system, albite ($NaAlSi_3O_8$) - celsian ($BaAl_2Si_2O_8$). Eur J Mineral 1:239-248

Voncken JHL, Konings RJM, Jansen JBH, Woensdregt CF (1988) Hydrothermally grown buddingtonite, an anhydrous ammonium feldspar ($NH_4AlSi_3O_8$). Phys Chem Minerals 15:323:328

Voncken JHL, Konings RJM, Van der Eerden AMJ, Jansen JBH, Schuiling DR, Woensdregt CF (1992) Crystal morphology and X-ray powder diffraction of the Rb-analogue of high sanidine, $RbAlSi_3O_8$. N Jb Mineral Monatsh 1993:10-16

Wainwright JE, Starkey J (1971) A refinement of the structure of anorthite. Z Krist 133:75-84

Wenk H-R, Kroll H (1984) Analysis of $P\bar{1}-I\bar{1}-C\bar{1}$ plagioclase structures. Bull Minéral 107:467-488

Wenk H-R, Nakajima Y (1980) Structure, formation, and decomposition of APB's in calcic plagioclase. Phys Chem Minerals 6:169-186

Williams PP, Megaw HD (1964) The crystal stuctures of high and low albites at -180°C. Acta Cryst 17:882-890

Winter JK, Ghose S, Okamura FP (1977) A high temperature study of the thermal expansion and the anistropy of the sodium atom in low albite. Am Mineral 62:921-931

Winter JK, Okamura FP, Ghose S (1979) A high-temperature structural study of high albite, monalbite, and the anabite-monalbite phase transition. Am Mineral 64:409-423

Wyncke B, McMillan PF, Brown WL, Openshaw R, Brehat F (1981) A room-temperature pase transition in maximum microcline. Absorption in the far infrared (10-200 cm^{-1}) in the temperature range 110-300 K. Phys Chem Minerals 7:31-34

ASPECTS OF ALKALI FELDSPAR CHARACTERIZATION: PROSPECTS AND RELEVANCE TO PROBLEMS OUTSTANDING

ACHILLE BLASI AND CARLA DE POL BLASI
Department of Earth Sciences, University of Milano,
Via Botticelli 23, 20133 Milano, Italy.

ABSTRACT. Attention will be focussed on the following topics and relevant problems concerning the characterization of alkali feldspars. *Light optics*: textural observations and determination of specific optical properties. *Composition and lattice strain*: (a) determination of composition of unstrained specimens, (b) evaluation of lattice strain caused by coherency between exsolved phases and determination of composition of strained specimens, (c) composition from K,Na site refinement of X-ray data. *Determination of tetrahedral Al contents*: (a) from lattice parameters, (b) from *T*-O distances, (c) from Si,Al site refinement of X-ray data. *Single-crystal X-ray diffraction patterns and relationships with TEM techniques*: (a) the diagonal association, (b) the antidiagonal association, (c) $P2_1/a$ structural arrangement. Unanswered questions and suggestions for future work will be underlined.

1. Introduction

Selected aspects of the characterization of alkali feldspars will be taken into consideration with regard to light optics, composition and lattice strain, tetrahedral Al contents, twins and related structures, as well as domain textures. These important subjects of the mineralogy and crystallography of alkali feldspars are most useful for interpretation of genetic conditions, rock-forming processes, and geological history. Existing work on these subjects will be reviewed, and results of new findings provided with regard to determinative methods and relevant analytical techniques.

2. Light Optics

General reviews of the optics of feldspars are given by Marfunin (1966), Bambauer (1969), Smith (1974a), Stewart and Ribbe (1983), and Smith and Brown (1988), among others. In the alkali feldspars, light optics can be employed usefully to perform textural observations and determine specific optical properties. Both represent a basic approach for unravelling mineralogical problems and the geological history of alkali feldspars to be subsequently investigated by means of more sophisticated techniques.

The familiar tartan texture of K-rich feldspar and anorthoclase indicates triclinic symmetry derived from a monoclinic precursor (cf. Laves, 1950). Patches of cross-hatching in largely untwinned K-rich feldspars may be due to transformations induced by the monoclinic-triclinic inversion. The development of the microcline structure also can give rise to undulatory extinction in K-rich feldspars in response to the formation of a variety of sub-optical textures, which

I. Parson (ed.), Feldspars and Their Reactions, 51–101.

include: (a) the tweed texture observed by TEM images in monoclinic adularia (McConnell, 1965, 1971), orthoclase (e.g., Nissen, 1967; McLaren, 1974, 1984, 1991, p. 233; Fitz Gerald and McLaren, 1982; McLaren and Fitz Gerald, 1987; Bambauer et al., 1989; Nord, 1992), and also in low microcline (Zeitler and Fitz Gerald, 1986); (b) the dimpled terrane texture illustrated by high-resolution TEM images in disordered microcline (Eggleton and Buseck, 1980); and (c) the diagonal association documented by X-ray photography in the K-rich phase of microperthite (Blasi and De Pol Blasi, 1980; Blasi et al., 1982; see also section 5.1).

Optically untwinned K-rich feldspar may be monoclinic or triclinic; symmetry can be determined by measuring the extinction angle on (001), i.e. X' to the trace of (010) on (001). Untwinned microcline may have crystallized as a triclinic or a monoclinic phase.

The textures of optically visible perthitic intergrowths are important indicators of subsolidus conditions. The regular, fine-scale lamellar textures of Na- and K-rich phases may be preserved only in favourable circumstances related to a low content of water. More commonly, coarsening can give rise to a variety of optically visible textures in some cases indistinguishable from those of non-exsolution origin (cf. Parsons, 1978; Yund, 1983; Parsons and Brown, 1984; Martin, 1988; Smith and Brown, 1988, p. 596). Optical inspection may provide a helpful guide to interpret complex alkali feldspar textures. Selected examples concern: (a) the alkali feldspar textures related to the transformation of hypersolvus mineralogy to subsolvus mineralogy in granites (e.g., Martin and Bonin, 1976), (b) the development of patch perthite from optically visible braid perthite in the Klokken syenite, South Greenland (cf. Parsons, 1978; 1980; see also section 5.1), (c) the development of bleb perthite and lamellar perthite in the Storm King granite, New York, U.S.A. (Yund and Ackermand, 1979), (d) the evolution of alkali feldspar textures during progressive metamorphism of the Scituate hypersolvus granite, Rhode Island, U.S.A. (Day and VM Brown, 1980), (e) the variety of exsolution textures in alkali feldspar from the granulite complex of Finnish Lapland (Yund et al., 1980).

A key feature of many specimens of alkali feldspar is represented by their turbidity. Much attention has been paid to this phenomenon. In some cases, it may be due to a cryptoperthitic intergrowth with units just below the limits of optical resolution (e.g., Smith, 1974b, p. 447). More commonly, the development of turbidity may be related to feldspar-fluid interactions at the subsolidus stage (e.g., Parsons, 1978, 1980; Parsons and Brown, 1984; Martin, 1988; Smith and Brown, 1988, p. 596), and especially to the presence of micropores (e.g., Folk, 1955; Martin and Lalonde, 1979; Parsons et al., 1988; Worden et al., 1990; Walker, 1990). The latter play an important role in argon-loss during geological events (Parsons et al., 1988).

Careful textural observations can reveal the coexistence of alkali feldspars of different generations in the same hand-specimen. In such cases, the choice of a feldspar crystal to be investigated by more sophisticated techniques should be done properly.

In K-rich feldspar, the optic axial angle $2V_X$ is an indicator of the total Al in T1 sites, and the extinction angle on (001) depends on $t1o - t1m$: complications arise from chemically heterogeneous intergrowths and sub-optical twins (Blasi, 1972b). Valuable estimates of structural state can be obtained from U-stage measurement of optic axial angle $2V_X$ and the orientation of the indicatrix relative to the crystallographic axes in thin sections approximately perpendicular to [100] (cf. Blasi, 1972a; Smith and Brown, 1988, p. 191). Further developments, and calibration of optical and X-ray properties, have been given by De Pieri (1979). Su et al. (1984) presented relationships expressing $\Sigma t1$, i.e. the total Al in T1 sites, as a function of $\sin^2 V_X$ in monoclinic and triclinic K-rich feldspars, respectively. Su et al. (1986a) proposed a diagram relating composition (Or), structural state ($\Sigma t1$), and optic axial angle ($2V_X$) for the entire compositional range of alkali feldspar, and gave the relevant formula to calculate $\Sigma t1$ as a function of N_{Or} and

$\sin^2 V_X$. Su et al. (1986b) found that $2V_X$ depends on both t1o + t1m and t1o − t1m in Na-rich feldspar, and gave the relevant formulae to calculate these quantities as a function of $\sin^2 V_X$. The dependence of $2V_X$ on both t1o + t1m and t1o − t1m may be due to the fact that Na-rich feldspar follows only one path of ordering. In addition, Su et al. (1986b) found that the extinction angle on (001), as well as the extinction angle on (010), i.e. *X*' to the trace of (001) on (010), depend on both t1o + t1m and t1o − t1m in Na-rich feldspar. For limitations of the optical methods, refer to Smith (1974a), Stewart and Ribbe (1983), and Smith and Brown (1988).

3. Composition and lattice strain

3.1. COMPOSITION OF UNSTRAINED FELDSPARS

Analytical techniques requiring comminution, separation and purification of the specimen cannot be used to determine the composition of the individual phases of a perthitic intergrowth because the coexisting units are usually too small in size to be separated by physical methods. In such cases, the composition is currently determined by using X-ray determinative curves. This indirect procedure is done by means of polynomial regression equations that usually express the composition in terms of mole fraction of $KAlSi_3O_8$ (N_{Or}) as a function of the cell volume *V* or cell edge *a*. If the feldspar specimen is affected by lattice strain due to coherency between exsolved phases, some complications arise, which will be considered in section 3.2.3. The determinative curves were improved in the last decade by Kroll et al. (1986) on the basis of their low microcline-low albite and high sanidine-analbite alkali-exchange series, and by Hovis (1986) on the basis of high sanidine, Eifel sanidine, orthoclase, adularia, and microcline ion-exchange series. The relevant equations for *V* and *a* are given in Table 1.

Microprobe techniques can be used to determine the composition of individual feldspar phases when their size is larger than that of the beam employed. A convenient size for the spot diameter is about 10-20 μm if investigating alkali feldspar material with a wavelength-dispersion electron microprobe, and even smaller when using an energy-dispersion system. Stewart and Wright (1974) and Blasi et al. (1984a, 1984b, 1987a), observed that microprobe-derived compositions of a K-rich phase that seems homogeneous under the optical microscope, as well as in cathodoluminescence, are commonly 2-3 mol% Or less potassic than those determined by cell volume *V*. This may be due to the fact that the volume excited by the electron beam contains small amounts of an exsolved Na-rich phase. Details of the electron microprobe investigation of feldspar minerals are described by Smith (1974b) and Smith and Brown (1988).

The chemical composition of feldspar minerals in term of trace and minor elements is rather complex. Among the 90 naturally occurring elements mentioned by Smith and Brown (1988) as possible constituents of feldspar minerals, more than 50 are actively being investigated. With regard to the alkali feldspars, in recent years, attention has been focussed on a number of elements that include Li, Be, B, F, Mg, P, Ti, Mn, Fe, Rb, Sr, Cs, Ba, and Pb, to obtain petrological information. Among them, particularly interesting elements are Rb (and Cs), Ba (and Sr) and, in special cases, Pb, Mg, Fe, P, and Ti. The analytical methods include a variety of techniques summarized by Smith (1974b, 1983) and Smith and Brown (1988). A combination of wavelength-dispersion electron microprobe and ion-microprobe techniques is highly desirable for the investigation of minor and trace elements in alkali feldspars.

The bulk composition of a perthitic intergrowth can be determined directly by chemical analysis of a representative specimen, or from the cell volume *V* after complete homogenization of the

TABLE 1. Equations to calculate N_{Or} values from cell volume V (Å^3) and cell edge a (Å)

Kroll et al. (1986)

for disordered alkali feldspars:

$$N_{Or} = -535.189 + 2.37332V - (3.52656 \times 10^{-3})V^2 + (1.75615 \times 10^{-6})V^3 \quad (1)$$

for intermediate alkali feldspars:

$$N_{Or} = -872.330 + 3.82791V - (5.61695 \times 10^{-3})V^2 + (2.75683 \times 10^{-6})V^3 \quad (2)$$

for ordered alkali feldspars:

$$N_{Or} = -1209.141 + 5.28104V - (7.70522 \times 10^{-3})V^2 + (3.75650 \times 10^{-6})V^3 \quad (3)$$

Hovis (1986)

for disordered alkali feldspars:

$$N_{Or} = -238.102 + 1.09469V - (1.69255 \times 10^{-3})V^2 + (8.79388 \times 10^{-7})V^3 \quad (4)$$

$$N_{Or} = -1362.802 + 486.4938a - 58.11977a^2 + 2.324427a^3 \quad (5)$$

for intermediate alkali feldspars:

$$N_{Or} = -129.248 + 0.610959V - (9.76800 \times 10^{-4})V^2 + (5.26832 \times 10^{-7})V^3 \quad (6)$$

$$N_{Or} = -366.3261 + 129.2335a - 15.42053a^2 + 0.6232109a^3 \quad (7)$$

for ordered alkali feldspars:

$$N_{Or} = -712.924 + 3.16601V - (4.70247 \times 10^{-3})V^2 + (2.33648 \times 10^{-6})V^3 \quad (8)$$

$$N_{Or} = -657.3958 + 239.9022a - 29.40843a^2 + 1.211078a^3 \quad (9)$$

(1) based on analbite-high sanidine solid solution series (figures -535.189 and 1.75615 from Table 9 in Kroll et al., 1986, are given as -535.186 and 1.75614, respectively, in the abstract of the same paper).
(2) based on a curve that lies halfway between the curves for analbite-high sanidine and low albite-low microcline solid solution series.
(3) based on low albite-low microcline solid solution series (figure -1209.141 from Table 9 in Kroll et al., 1986, is given as -1209.142 in the abstract of the same paper).
(4) and (5) based on two ion-exchange series for disordered topochemically monoclinic alkali feldspars (high sanidine and low sanidine series).
(6) and (7) based on two ion-exchange series for relatively ordered topochemically monoclinic alkali feldspars (orthoclase and adularia series).
(8) and (9) based on two ion-exchange series for ordered topochemically triclinic alkali feldspars (microcline series of Hovis, 1986, and Orville, 1967).

feldspar material has been achieved by heat treatment. The electron microprobe also can be used by employing broad-beam or scanning techniques (cf. Yund et al., 1980; Rollinson, 1982; Brown et al., 1983).

3.2. STRAINED FELDSPARS

3.2.1. *Background.* During exsolution in an alkali feldspar, the aluminosilicate framework expands in the regions entered by K and contracts in those enriched in Na. In spite of this, the feldspar framework may retain coherency, i.e. structural continuity, across the interfaces between compositionally different regions. The different sizes of the K and Na atoms produce coherency stresses, which render the coexisting phases elastically strained. Lattice matching across the interfaces occurs together with deformation along the directions perpendicular to them.

The coherency boundary between K- and Na-rich domains is a plane of variable orientation, and close to $(\bar{6}01)$ in the vast majority of chemically heterogeneous alkali feldspar intergrowths.

To a first approximation, the $(\bar{6}01)$ plane is parallel to *b* and *c* and normal to *a*. In going from pure $KAlSi_3O_8$ to pure $NaAlSi_3O_8$, a contraction by ~1.3% and ~0.9% occurs in *b* and *c*, respectively, and by ~5.4% in *a*. Thus, lattice matching across the $(\bar{6}01)$ interface requires that cell edges *b* and *c* undergo a small contraction in the K-rich phase and a small expansion in the Na-rich phase. This explains why a boundary near the plane that contains *b* and *c* is preferred to retain lattice coherency between the components of a perthitic intergrowth. Smith (1961, 1974a, p. 279) assumed that coherency strain occurs at a nearly constant volume in each perthitic component. This implies that cell edge *a* is forced to expand in the K-rich phase and to contract in the Na-rich phase. The change in *a* is clearly stronger compared to that in *b* and *c*, and reaches 1.6% in the most highly strained alkali feldspar found so far: the K-rich phase of cryptoperthite F97, studied by Brown and Willaime (1974) (cf. Table 2). The lattice angles α and γ seem to be little affected by coherency stresses because they are very close to those for unstrained feldspars (see also section 4.1).

The degree of coherency strain depends on the size of the exsolution lamellae. The individual components of a fine-scale perthite are indeed more strained than those of a coarser intergrowth. Perthitic lamellae thicker than a few micrometers are virtually free of strain. The amount of strain also depends on the relative proportions of the coexisting phases, the more abundant component showing less strain. In other words, as the bulk K content of a perthite increases, the amount of strain decreases in the K-rich phase and increases in the Na-rich phase. The relationship of strain to bulk composition has been documented by Stewart and Wright (1974; also Stewart, 1975), and an explanation for this behavior is given in the model of Brown and Willaime (1974).

The effects of coherency stresses can conveniently be described in terms of homogeneous and inhomogeneous strain (cf. Yund, 1974; Willaime, 1981; Yund and Tullis, 1983b; Smith and Brown, 1988, Fig. 5.2). In the first case, the elastic strain is distributed uniformly within the volume of each phase, and perfect coherency occurs across interfaces devoid of dislocations. In the second case, the elastic strain is concentrated along the interfaces, whereas partial relaxation of stresses occurs away from them (cf. Brown and Willaime, 1974; Willaime and Brown, 1974). Perfect coherency can be retained without development of dislocations at the interfaces only if the latter are of limited extent. Lamellar interfaces of larger extent require development of dislocations along them, with consequent partial loss of coherency.

In diffraction patterns, pairs of spots from lattice planes perpendicular to the coherency boundary between K- and Na-rich phases are strictly overlapping in case of homogeneous strain and slightly separated in case of inhomogeneous strain. This is because the spacings and the orientations of the relevant diffracting planes in the two coexisting phases are identical in the former case, and slightly different in the latter. The other pairs of diffraction spots show various degrees of splitting: the maximum separation occurs between pairs of spots from lattice planes parallel to the interface. The behavior of spots can best be seen in *a*b** diffraction patterns. As a result of coherency across the $(\bar{6}01)$ interface, the pairs of spots 0*k*0 for coexisting strained phases appear to be virtually overlapping (homogeneous strain) or very close to each other (inhomogeneous strain), whereas the pairs of spots *h*00 appear to be abnormally separated.

3.2.2. *Determination of lattice strain.* Evidence for coherency between K- and Na-rich phases can be obtained from various experimental techniques. In addition, an estimate of the degree of lattice strain also can be derived from several determinative methods using cell parameters.

X-ray powder diffraction patterns can usefully be employed especially in reconnaissance studies, in order to obtain information on coherency strain (cf. Parsons, 1978, 1980; Yund and Tullis, 1983a). The peaks 060 and $\bar{2}01$ are of particular interest because they depend entirely on *b* and strongly on a, respectively. Thus, in K- and Na-rich phases that are coherent across an interface close to ($\bar{6}01$), the two peaks 060 virtually overlap and the two peaks $\bar{2}01$ are abnormally separated. Further evidence for coherency also is given by the overlap of the two peaks $\bar{2}04$ due to the fact that they are largely dependent on c. The latter, however, suffer somewhat from interference from other diffraction peaks (cf. Wright, 1968; Smith, 1974a, p. 285).

More accurate information comes from the examination of spots in X-ray or electron diffraction patterns. As mentioned in the previous section, the effects of coherency stresses across an interface near ($\bar{6}01$) can be recognized from the relative positions that the pair of spots due to the two coexisting phases occupy along b^* and a^* axes. The pairs of spots $h00$ along a^* can appear to be joined by diffuse streaks, which may be due to compositional perturbations or coherency effects because a^*, as a, depends strongly on both extent of K,Na substitution and degree of strain.

Diffuse streaks between the pairs of spots $h00$ may occur in coherent intergrowths preserving early stages of exsolution in response to the relatively rapid cooling of the host rock. In such circumstances, the presence of streaking may be due to compositional perturbations. These may include residual compositional gradients within each phase, compositionally diffuse interfaces, and other kinds of compositional inhomogeneities (cf. Yund, 1974; Yund and Tullis, 1983a; Smith and Brown, 1988, p. 558). The occurrence of diffuse streaking can be explained by the fact that compositional perturbations cause part of the lattice planes parallel to the interface to have, in each of the two phases, spacings variously different from those giving rise to the pairs of main spots. Compositional disturbances in the early stages of exsolution leave the pairs of spots $0k0$ virtually overlapping along b^* (cf. Yund, 1974).

At a more advanced stage of exsolution, the compositional perturbations are expected to decrease, and the occurrence of diffuse streaking between pairs of spots $h00$ along a^* may be due to inhomogeneous strain (cf. Yund, 1974; Yund and Tullis, 1983a; Smith and Brown, 1988, p.78 and p. 154; also Brown and Willaime, 1974; Willaime and Brown, 1974). This is because in the proximity of the interface, the lattice planes parallel to it become more widely and closely spaced in the K- and Na-rich phases, respectively. Inhomogeneous strain can, in principle, be recognized because the pairs of spots $0k0$ along b^* are somewhat separated because the spacings and orientations of lattice planes normal to the interface are not exactly the same in the two coexisting phases.

Conversely, sharp pairs of spots $h00$ without any streaks joining them along a^* suggest that the individual phases are compositionally homogeneous and that strain occurs homogeneously distributed throughout the crystal (cf. Yund, 1974; Yund and Tullis, 1983a; Smith and Brown, 1988, p. 78; also Stewart and Wright, 1974, p. 374). In the case of homogeneous strain, in fact the spacings of the lattice planes parallel to the interface do not vary within an individual phase. In addition, the spacings and the orientations of lattice planes normal to the interface are the same in the two coexisting phases. Thus, homogeneous strain involves the absence of streaking between the pairs of spots $h00$ along a^* and a virtual overlap of the pairs of spots along b^*.

A combination of effects due to compositional perturbations and coherency strain may render the aspect of the diffraction patterns and their interpretation more complex. Further complications arise in the case of inhomogeneous strain because the examination of diffraction spots does not allow one to determine whether it involves perfectly coherent interfaces without dislocations or partially coherent interfaces with dislocations.

These important aspects of the strain phenomenon can be clarified by means of TEM images. Inhomogeneous strain is, in fact, revealed in TEM images by diffraction contrast, due to the non-uniform spacings and orientations of lattice planes perpendicular to the interface (cf. Robin,

1974; Yund and Tullis, 1983a; Brown and Parsons, 1984b, 1988). In addition, TEM images may show the presence of dislocations distributed along the interfaces and often typically marked by small holes due to enhanced damage by the electron beam (e.g., Brown and Parsons, 1984b; Parsons and Brown, 1984). The degree of coherency also can be evaluated in high-resolution TEM images by the aspect of lattice fringes. Brown (1983), Brown and Parsons (1984a), and Parsons and Brown (1984) have obtained high-resolution TEM images of cryptoperthites entirely coherent and dislocation-free, in which the two phases are characterized by different b and c cell dimensions. In these specimens, inhomogeneous strain does not require the development of dislocations because of the limited extent of lamellae parallel to the interface.

Currently, the degree of lattice strain due to coherency stresses is estimated from cell edges a, b, and c, or from their reciprocal counterparts, employing the determinative procedures proposed by (1) Stewart and Wright (1974): strain index Δa (Å) = a(obs) – a(est), where a(est) is obtained from a contoured bc plot; (2) Smith (1974a, Fig. 7-27): diagram of a^* *vs* $Or(b^*c^*)$, where $Or(b^*c^*)$ is derived from the contours of a b^*c^* plot; and (3) Kroll and Ribbe (1987, Fig. 8): diagram of a *vs* bc.

The Δa values for strained feldspars are assumed to be beyond the range 0.05 Å. As expected, Δa values are positive in K-rich feldspar and negative in Na-rich feldspar. The method of Smith (1974a) is conceived only for K-rich feldspar. The data points for specimens affected by coherency strain fall outside a band defined for unstrained feldspars toward lower $Or(b^*c^*)$ values. The method of Kroll and Ribbe (1987) involves a band for unstrained specimens, outside of which the data points for K- and Na-rich feldspars fall, as expected, in opposite directions.

In principle, the determination of the degree of lattice strain should be based on comparison of strained and unstrained lattice parameters for the same feldspar specimen. Thus, all three determinative methods mentioned above are vitiated by the fact that they use only strained lattice parameters (see also Yund and Tullis, 1983b, p. 156; Smith and Brown, 1988, p. 154). Brown and Willaime (1974) estimated the degree of lattice strain by comparing strained and unstrained lattice parameters, the latter being derived from composition obtained *via* the cell volume. Blasi and De Pol Blasi (1980) and Blasi et al. (1984b, 1987a) employed the difference $N_{Or}(a)$ – $N_{Or}(V)$ as a measure of the degree of lattice strain. Obviously, these two latter procedures are vitiated by the assumption of constant volume. Ideally, this weakness could be overcome by employing compositions corrected following the procedure proposed by Yund and Tullis (1983a) (see next section).

3.2.3. *Determination of composition.* Cell edge a depends principally on K,Na substitution, but it is also strongly dependent on the degree of coherency between components of perthitic intergrowths. Thus, its use to determine the composition of the coexisting phases in coherent perthites is destined to produce erroneous results. The small effect of coherency stresses on cell volume V makes it a widely employed parameter to estimate phase compositions in strained alkali feldspars (Smith, 1961, 1974a, p. 280; Wright and Stewart, 1968; Brown and Willaime, 1974; Stewart and Wright, 1974; Stewart, 1975; Parsons and Brown, 1984; Smith and Brown, 1988, p. 155). However, Robin (1974) found that N_{Or} values obtained from V are underestimated for the K-rich phase and overestimated for the Na-rich phase. This is in accordance with the findings of Tullis (1975, p. 89; also Yund and Tullis, 1983b, p. 171), who independently found that $d(\bar{6}01)$ changes by only half of the extent required to maintain constant volume.

In consideration of these results, alternative procedures to determine compositions of coherent perthitic components have been proposed by Robin (1974) and Tullis (1975). The method developed by Robin (1974) allows correct Or contents (wt%) to be obtained from the function

TABLE 2. Comparison of Or values calculated from cell edge a^* and cell volume V for strained alkali feldspars

Sample	Bulk composition (mol%)			Phase	Space group	a^* (Å^{-1})†	V (Å^3)	Or (mol%) from a^*		Or (mol%) from V		
	Or	Ab	An					1	2	3	4	5
Willaime et al. (1973)												
				K-rich(1)¶	$C\bar{1}$	0.129088	717.7§	106	87	85	86	87
L31	35	62	3	Na-rich(2)¶	$C\bar{1}$	0.137983	665.3	-13	-2	-1	1	2
				Na-rich(4)¶	$C\bar{1}$	0.137799	667.4	-10	0	1	4	6
Brown & Willaime (1974)												
Spencer P	42	58	0	K-rich	$C2/m$	0.130197	712.0	85	76	72	72	73
				Na-rich	$C\bar{1}$	0.137094	673.6	0	9	9	12	15
F99	19	68	13	K-rich	$C2/m$	0.128775	719.4	117	88	90	90	91
				Na-rich	$C\bar{1}$	0.136890	668.5	3	7	3	5	8
F91	72	26	2	K-rich	$C2/m$	0.130007	717.5	88	84	85	85	86
				Na-rich	$C\bar{1}$	0.137809	663.1	-10	8	-4	-3	-1
F97	48	49	3	K-rich	$C2/m$	0.128680	716.8	127	96	83	84	84
				Na-rich	$C\bar{1}$	0.137786	666.9	-10	3	1	3	5
L29	28	70	2	K-rich	$C\bar{1}$	0.129687	716.2	93	79	82	82	83
				Na-rich	$C\bar{1}$	0.137915	664.7	-12	-3	-2	0	1
Stewart & Wright (1974)∇												
Spencer H	71.2			K-rich	$C\bar{1}$	0.129790	718.0	91	86	86	87	87
Spencer I	64.0			K-rich	$C2/m$	0.129208	719.5	103	92	90	91	91
Spencer K	63.4			K-rich	$C2/m$	0.129407	718.8	99	89	88	89	90
Spencer M	51.0			K-rich	$C2/m$	0.129083	717.8	106	91	86	86	87
Spencer N	50.6			K-rich	$C2/m$	0.129504	718.0	97	86	86	87	87
Spencer P	43.4			K-rich	$C2/m$	0.130057	714.0	87	77	76	77	78
Spencer R	40.2			K-rich	$C2/m$	0.129164	716.3	104	87	82	82	83
G 105	59.5			K-rich	$C2/m$	0.129714	719.2	93	85	89	90	91
G 106	53.3			K-rich	$C2/m$	0.129772	719.3	92	85	89	90	91
G 107	48.2			K-rich	$C2/m$	0.129663	716.7	94	83	83	83	84
G 108	47.5			K-rich	$C2/m$	0.129680	715.9	93	82	81	81	82
G 317	43.5			K-rich	$C2/m$	0.129740	716.2	92	81	82	82	83
G 31	42.1			K-rich	$C\bar{1}$	0.130458	710.8	81	73	69	70	71
P23.4	86.3			K-rich	$C\bar{1}$	0.129338	722.2	100	96	97	98	99
M75B	30			K-rich	$C2/m$	0.130141	713.2	86	74	75	75	76
Keefer & Brown (1978)												
RC	51	48	1	K-rich	$C2/m$	0.130085	717.2	86	78	84	85	85
				Na-rich	$C\bar{1}$	0.137312	676.9	-3	8	14	17	20
Ribbe (1979)												
K-235	34◊	58	8	K-rich	$C\bar{1}$	0.128996	720.6	109	88	93	94	94
				Na-rich	$C\bar{1}$	0.138676	657.9	-27	-12	-11	-11	-11

$d(\bar{2}04) + 0.30d(\bar{2}01)$, which is not affected by elastic strain due to coherency. The procedure devised by Tullis (1975) enables apparent compositions determined from a^* to be converted into correct compositions by means of appropriate nomograms prepared for bulk compositions Or_{40}, Or_{60}, and Or_{80} mol%. Between the two methods, that of Tullis (1975) is in principle more accurate (Robin, 1974, p. 1307; Tullis, 1975, p. 84). The approach of Tullis (1975) has been

TABLE 2. Continued

Sample	Bulk composition (mol%) Or	Ab	An	Phase	Space group	a^* (Å^{-1})†	V (Å^3)	Or (mol%) from a^* 1	2	Or (mol%) from V 3	4	5
Yund et al. (1980)												
99II	35.0	56.9	8.0	K-rich	$C\bar{1}$	0.129554	719.3†	96	81	89	90	91
McCormick (1989)												
BR-2 346/133	63.0	36.1	0.9	K-rich	$C2/m$	0.130761	712.1	77	74	72	73	73
				Na-rich	$C2/m$	0.137680	672.7	-8	7	8	11	14
BR-4 086/034	63.7	35.2	1.1	K-rich	$C2/m$	0.130901	711.5	75	72	71	71	72
				Na-rich	$C2/m$	0.137489	673.3	-5	9	9	12	15
BR-6 071/050	63.4	35.7	0.9	K-rich	$C2/m$	0.131089	709.4	72	70	66	67	68
BR-2 087/032	42.7	56.3	0.9	K-rich	$C2/m$	0.130121	713.0	86	77	74	75	75
				Na-rich	$C\bar{1}$	0.137424	672.4	-4	6	8	11	13
				Na-rich	$C\bar{1}$	0.137511	673.3	-6	5	9	12	15
BR-6 044/130	40.6	58.8	0.6	K-rich	$C2/m$	0.130164	712.2	85	76	72	73	74
				Na-rich	$C\bar{1}$	0.137385	673.5	-4	6	9	12	15
				Na-rich	$C\bar{1}$	0.137316	673.7	-3	7	9	12	15
BR-7 303/186	39.9	58.5	1.6	K-rich	$C2/m$	0.130441	710.1	81	72	68	68	69
				Na-rich	$C\bar{1}$	0.137255	674.2	-2	7	10	13	16
BR-8 328/122	39.0	60.0	1.0	K-rich	$C2/m$	0.131865	703.8	62	58	55	56	57
				Na-rich	$C2/m$	0.136535	678.3	8	14	15	18	21
BR-11 250/092	40.5	58.3	1.2	K-rich	$C2/m$	0.131127	707.5	72	65	62	63	64
				Na-rich	$C\bar{1}$	0.136972	676.4	2	10	13	16	19

1 Composition uncorrected for elastic strain, calculated from the relevant cubic least-squares fit given by Luth (1974) for the sanidine-high albite series of Orville (1967). Note that the cell parameters of Orville's (1967) series were used in the method developed by Yund and Tullis (1983a) to correct apparent compositions obtained from measurements of strained a^*.
2 Composition corrected for elastic strain using the curves in Fig. 1 of Yund & Tullis (1983a). Curves for compositions out of the range Or_{20}-Or_{60} were obtained by extrapolation.
3 Composition calculated from equation (1) in Table 1.
4 Composition calculated from equation (2) in Table 1.
5 Composition calculated from equation (3) in Table 1.
† Calculated in this work.
¶ The numbers given in parentheses refer to the different feldspar regions indicated by Willaime et al. (1973) in their Fig. 5.
§ Corrected from original data (V = 716.4 Å^3). The value in the present Table was calculated using lattice constants given by Willaime et al. (1973).
∇ Ab and An mol% values were not given.
◊ Includes 1 mol% Cn.

further refined and extended by Yund and Tullis (1983a; also Yund and Tullis, 1983b; Yund, 1984) in order to yield correct compositions directly from a^* or $d(\bar{2}01)$ values for bulk compositions Or_{20}, Or_{40}, and Or_{60} mol%.

The results produced by the method of Yund and Tullis (1983a) depend, to some extent, on the soundness of the available data on elastic constants, which are at present of poor quality. In addition, the assumptions involved of monoclinic symmetry, complete coherency, and homogeneous state of strain for the coexisting phases are not valid for all natural specimens.

The situation remains complex. The strained alkali feldspars available so far (cf. Willaime et al., 1973; Brown and Willaime, 1974; Stewart and Wright, 1974; Keefer and GE Brown, 1978; Ribbe, 1979; Yund et al., 1980; McCormick, 1989) do not always fulfil the theoretical predictions of Robin (1974) and Tullis (1975), which indicate the cell volume V to yield an underestimate of K content for the K-rich phase and an overestimate of K content for the Na-rich phase (Table 2).

3.3. COMPOSITION FROM K,Na SITE REFINEMENTS OF X-RAY INTENSITY DATA

Keefer and GE Brown (1978) stressed the importance of performing K,Na site refinements of X-ray intensity data in alkali feldspar as an alternative method to determine composition. Direct K,Na site refinement performed by these authors using a sanidine in coherent cryptoperthitic intergrowth with high albite from the Rabb Canyon pegmatite, Grant County, New Mexico, U.S.A., gave $N_{Or} = 0.65$. Keefer and GE Brown (1978) thoroughly discussed all their data and concluded that a value of N_{Or} of 0.65 is the most reliable estimate of composition of their strained sanidine.

We believe that $N_{Or} = 0.65$ is an underestimated composition because cell volume V gives $N_{Or} = 0.85$ and the specimen of Keefer and GE Brown (1978) is not so strained ($\Delta a = 0.08$ Å) as to justify such a discrepancy of values. There are other reasons in favor of this assertion: (1) the crystal structure of the Rabb Canyon sanidine is virtually identical to the structure of an unstrained Eifel low sanidine #7002 of composition $N_{Or} = 0.85$ reported by Phillips and Ribbe (1973); (2) the $\langle M\text{-O}\rangle$ distance of the Rabb Canyon sanidine indicates that the composition is highly potassic, and this is confirmed, in particular, by the *M*-O*A*2 distance, which is especially sensitive to different K/Na ratios in the *M* site (cf. Fenn and GE Brown, 1977, Fig. 5); (3) the N_{Or} values estimated by Keefer and GE Brown (1978) using the method of Robin (1974) and that of Tullis (1975) are 0.92 and 0.79, respectively (cf. also Table 2).

Yund and Tullis (1983a) emphasized that the compositional data obtained by Keefer and GE Brown (1978) seem to be inconsistent with other observations related to the position of the specimen with respect to coherent solvi. Kroll and Ribbe (1983, p. 84) and Ribbe (1984, p. 42) also observed that the composition of the Rabb Canyon sanidine determined from K,Na site refinements of X-ray intensity data is not sufficiently accurate.

Site-occupancy characterization depends on the difference in scattering factors of the atoms involved (cf. Whittaker, 1977; Hawthorne, 1983). Fig. 1 shows curves relevant to X-ray scattering factors for K and Na atoms as a function of $(\sin\theta)/\lambda$. These curves express: (a) percent differences given by $\{[f(K) - f(Na)]/f(Na)\} \times 100$, $\{[f(K) - f(Na)]/\{[f(K) + f(Na)]/2\}\} \times 100$, and $\{[f(K) - f(Na)]/f(K)\} \times 100$; and (b) "absolute" differences given by $f(K) - f(Na)$. The curves for percent differences indicate that data for high angles should be used to distinguish K and Na atoms, whereas the curve for "absolute" differences indicates that low-angle data should be more appropriate.

Before attempting to perform K,Na site refinements of "difficult" specimens, as may be the case of the Rabb Canyon sanidine, it is convenient to explore the procedure in a simpler case. A specimen of a low microcline of well-known composition has been selected for this purpose, and preliminary results will be given in this section. The specimen, denoted as 7813C, is a cleavage fragment measuring $0.14 \times 0.06 \times 0.02$ mm from a white to colorless microcline overgrowth on amazonite from Pikes Peak batholith, Colorado, U.S.A. The fragment was separated from the same 14-mm^3 feldspar block employed by Blasi et al. (1984b) to obtain the two crystals 7813A and 7813B used in structure refinement.

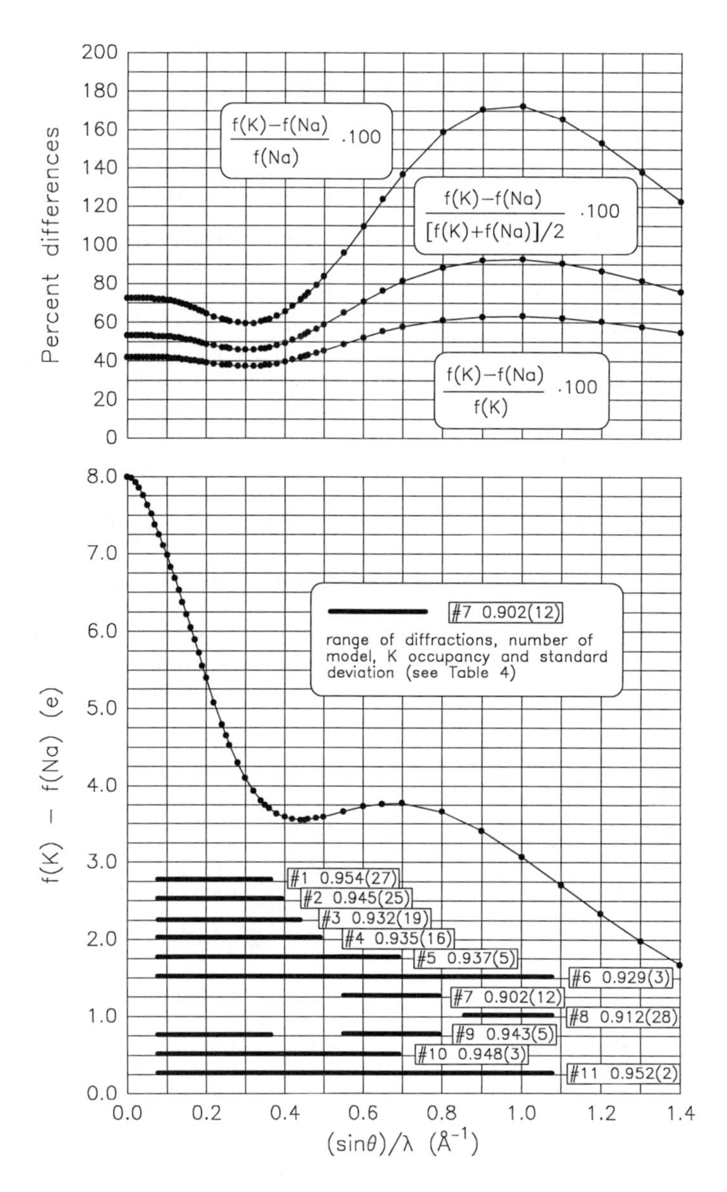

Figure 1. Percent and "absolute" differences between X-ray scattering factors for K and Na atoms as a function of (sinθ)/λ. The relevant atomic scattering factors were taken from the *International Tables for X-ray Crystallography* (1974). The heavy lines refer to the results obtained from different models of *M*-site refinements for Pikes Peak microcline 7813C.

Electron-microprobe analyses performed on several fragments from the 14-mm^3 feldspar block indicated a composition near $Or_{95}Ab_5$ (Blasi et al., 1984b; plus more recent unpublished data). Cell volume data from Blasi et al. (1984b) and this work give N_{Or} values in the range 0.96-0.97.

TABLE 3. Basic information on X-ray data collection, lattice constants and composition from V and a, standard structure refinements, and T-O distances for microcline Pikes Peak 7813C

Experimental: Single-crystal measurements at 50 kV, 50 mA and 22 °C were made with a KUMA Diffraction KM-4 computer-controlled k-axis diffractometer using graphite-monochromated Mo$K\alpha$ radiation. Lattice constants were determined by least-squares refinement of 2θ angles of 40 diffractions in the 2θ range 27-52° using Burnham's (1962) LCLSQ program as modified by Blasi (1979). Intensity data were collected using ω-2θ scan mode, scan widths $(1.5 + 0.35\tan\theta)°$ in ω, minimum and maximum scan speeds 0.005 and $0.07° s^{-1}$ in ω, respectively. A total of 14966 diffractions ($-17 < h < 17$, $-28 < k < 28$, $-14 < l < 14$) were collected in the 2θ range 3-100°. The intensities of three standard diffractions (060, 204, and $\bar{2}04$) were measured after every 50 diffractions and varied by ±1.3% from mean values. Intensities were corrected for these variations and for Lorentz and polarization effects, and were merged to give a total of 7123 unique data [R(merge) = 0.017, space group $C\bar{1}$]. Neutral-atom scattering factors corrected for the anomalous dispersion effects were taken from the *International Tables for X-ray Crystallography* (1974). Correction for absorption was not done because of the small size of the crystal (see text) and the low value of the linear absorption coefficient [$\mu(MoK\alpha) = 13.3$ cm^{-1}]. Structure refinements were performed using Sheldrick's (1976) SHELX-76 program.

Lattice constants and composition from V *and* a:

a = 8.5733(1), b = 12.9627(3), c = 7.2227(1) Å, α = 90.675(2), β = 115.934(1), γ = 87.601(1)°, V = 721.20(2) Å^3. N_{Or} from V: 0.962 (eq. 3, Table 1), 0.964 (eq. 8, Table 1); N_{Or} from a: 0.956 (eq. 9, Table 1).

Standard structure refinements [criterion for observed diffractions: $I > 3\sigma(I)$; scattering factor for M site: $0.96f(K) + 0.04f(Na)$; starting scattering factors for T sites: $0.25f(Al) + 0.75f(Si)$]:

(a) $3 < 2\theta < 60°$, $R = 0.016$ and $R_w = 0.018$, number of observed diffractions = 1788, weighting scheme $w = 1/[\sigma^2(F_o) + 0.0004F_o^2]$.

(b) $3 < 2\theta < 100°$, $R = 0.019$ and $R_w = 0.020$, number of observed diffractions = 4948, weighting scheme $w = 1/[\sigma^2(F_o) + 0.0005F_o^2]$.

T-*O distances (Å)*:

	*T*1*o* tetrahedron (a)	*T*1*o* tetrahedron (b)		*T*1*m* tetrahedron (a)	*T*1*m* tetrahedron (b)
*T*1*o*-O*A*1	1.7440(9)	1.7440(5)	*T*1*m*-O*A*1	1.5914(9)	1.5918(5)
-O*Bo*	1.7337(10)	1.7341(6)	-O*Bm*	1.6057(11)	1.6074(7)
-O*Co*	1.7364(7)	1.7373(4)	-O*Cm*	1.6234(8)	1.6242(4)
-O*Do*	1.7412(7)	1.7408(4)	-O*Dm*	1.6283(7)	1.6283(4)
Mean	1.7388(4)	1.7391(2)	Mean	1.6122(4)	1.6129(3)

	*T*2*o* tetrahedron (a)	*T*2*o* tetrahedron (b)		*T*2*m* tetrahedron (a)	*T*2*m* tetrahedron (b)
*T*2*o*-O*A*2	1.6207(7)	1.6208(4)	*T*2*m*-O*A*2	1.6416(7)	1.6412(4)
-O*Bo*	1.5838(11)	1.5841(7)	-O*Bm*	1.6293(12)	1.6284(7)
-O*Cm*	1.6275(8)	1.6273(5)	-O*Co*	1.5994(8)	1.5999(5)
-O*Dm*	1.6242(7)	1.6247(5)	-O*Do*	1.5934(7)	1.5938(5)
Mean	1.6141(4)	1.6142(3)	Mean	1.6159(4)	1.6158(3)

The value of N_{Or} = 0.95 obtained from Guinier-Hägg powder data (cf. Blasi et al., 1984b, Table III) is from feldspar material taken from a portion of the overgrowth 7813 different from that of the 14-mm^3 block. The N_{Or} values from microprobe data are slightly smaller than those from cell volume V, probably because of the reason given in section 3.1.

Basic information on X-ray data collection, lattice constants and composition from V and a, standard structure refinements, and T-O distances are given in Table 3. The values of T-O distances are very similar to the corresponding values in specimens 7813A and 7813B studied by Blasi et al. (1984b). The <*T*2*m*-O> distance in microcline 7813C, i.e. 1.6159(4) or 1.6158(3) Å (cf. Table 3), seems to be somewhat longer than in specimens 7813A and 7813B, in which it is

TABLE 4. Models of *M*-site refinements for microcline Pikes Peak 7813C

Model #	(sinθ/λ) range (Å^{-1})	No. of diffractions	*R*(%)	R_w(%)	··· *M* site ··· starting K occupancies	refined K occupancies
1	0.07-0.37	297§	3.61 3.68	4.22 4.26	0.960 0.500	0.954(27)†¶ 0.947(27)†¶
2	0.07-0.40	376§	3.54 3.54	4.32 4.31	0.960 0.500	0.945(25)∇ 0.944(25)∇
3	0.07-0.45	543§	3.28 3.33	3.94 4.01	0.960 0.500	0.932(19)◊ 0.937(20)
4	0.07-0.50	738§	2.91 2.93	3.51 3.52	0.960 0.500	0.935(16) 0.933(16)
5	0.07-0.70	1788	1.63	1.77	0.960 0.500	0.937(5) 0.937(5)
6	0.07-1.08	4948	1.86	1.99	0.960 0.500	0.929(3) 0.929(3)
7	0.55-0.80	1624	1.30	1.33	0.960 0.500	0.902(12) 0.902(12)
8	0.85-1.08	2028	2.31	1.97	0.960 0.500	0.912(28) 0.912(28)
9	0.07-0.37 & 0.55-0.80	1887	1.62	1.80	0.960 0.500	0.943(5) 0.943(5)
10	0.07-0.70	1788	1.64	1.79	0.960 0.500	0.948(3) 0.948(3)
11	0.07-1.08	4948	1.88	2.01	0.960 0.500	0.952(2) 0.952(2)

Note: See text for information about the different procedures of site refinement employed in models #1 to #9, and #10 and #11. Estimated standard errors (1σ) are given in parentheses and refer to the last decimal place.

† Not totally convergent.

¶ O*A*1, O*A*2, O*Bo*, O*Bm*, O*Co*, O*Cm*, and O*Dm* atoms are non-positive definite.

§ All diffractions obtained after merging were used.

∇ O*A*2, O*Cm*, and O*Dm* atoms are non-positive definite.

◊ O*Cm* atom is non-positive definite.

equal to 1.614(1) Å. This is, however, an idiosyncratic feature of specimen 7813C, because the same value resulted also from other data collections obtained for the same specimen with the same or a different diffractometer. Note that σ values in mean *T*-O distances are 0.001 Å in specimens 7813A and 7813B, whereas in microcline 7813C they decrease to 0.0004 Å and 0.0003 Å, with reference to the structure refinements using intensity data up to 60° and 100° 2θ, respectively (Table 3).

After the standard structure refinements presented in Table 3 were performed, several refinements of the occupancy of the *M* position were done using groups of diffractions from different ranges of (sinθ)/λ. Refinements of unconstrained site-occupancy factors for K and Na atoms gave unsatisfactory results. More promising results were obtained by constraining the sum of the site-occupancy factors to be equal to 1.

The latter technique was employed in the refinement models #1 to #9 presented in Table 4 and Fig. 1. Starting occupancies of the *M* site in these models were derived either from the composition obtained from cell volume *V*, N_{Or} = 0.96 (Table 3), or from an unrealistic composition of N_{Or} = 0.50. These two different starting compositions led to identical or very similar results in each individual model. All results are consistent with an underestimate of the refined K occupancies. In particular, the underestimate in K content decreases in structure refinements performed with diffractions at low 2θ angles, a region in which the number of data becomes insufficient to obtain satisfactory refinements of the structure and the *R* values increase (cf. Table 4). Between the two types of curves in Fig. 1, the curve for "absolute" differences in scattering factors of K and Na atoms justifies the results obtained.

Another series of refinements using the group of diffractions in the 2θ range 3-60° was done by alternating steps, which consist of: (a) refining the site-occupancy factors for K and Na atoms with the constraint that their sum be equal to 1, and fixing the displacement parameters U_{ij} for *M* position; and (b) fixing site-occupancy factors for K and Na atoms and refining displacement parameters U_{ij} for *M* position. With this technique, described by Whittaker (1977), apparent convergence was observed at the beginning of the refinement process. However, *R*(%) and R_w(%) values, if expressed with four decimal digits, reveal that convergence is slow, and that the process ends up giving the same results obtained by simultaneous refinement of site-occupancy factors and displacement parameters.

A further alternative procedure of refinement of the *M* site was explored with models #10 and #11 (Table 4 and Fig. 1) which use diffractions in the 2θ ranges 3-60° and 3-100°, respectively, in the following way: (a) starting atomic parameters were taken from a standard refinement of F_o values in the range $0.70 \leq (\sin\theta)/\lambda \leq 1.08$ Å^{-1}, and all U_{ij} displacement parameters for the *M* position were kept constant; (b) site-occupancy factors for the *M* site were constrained to be equal to 1; and (c) starting K and Na occupancies were allowed to be equal to either 0.96 or 0.50. The refined K occupancies (Table 4 and Fig.1) are very close to those obtained from low-angle data and, unlike the latter, are characterized by low values of σ, as a result of the low *R* and R_w values obtained in the relevant refinements of the structure.

4. Determination of Tetrahedral Al Contents in Alkali Feldspars

4.1. *T*-SITE OCCUPANCIES FROM LATTICE PARAMETERS

Every determinative method for estimating *T*-site occupancies from metric properties needs accurate values of lattice constants for the alkali feldspar end-members. Since 1974, complete sets of lattice parameters for the alkali feldspar end-members have been proposed by Smith (1974a; see also Smith and Brown, 1988), Stewart and Wright (1974; see also Stewart, 1975), and Kroll and Ribbe (1987; see also Kroll and Ribbe, 1983). In their set of data, Kroll and Ribbe (1983) assumed t1o and t1m for analbite, and t1 for high sanidine to be equal to 0.28. Later, Kroll and Ribbe (1987) reduced this value to 0.275.

The values of direct and reciprocal lattice constants are given in Table 5 for the three sets of data proposed by Smith (1974a), Stewart and Wright (1974), and Kroll and Ribbe (1987). The quadrilaterals joining the reference values for *b* and *c* and for *b** and *c** are represented in Fig. 2, whereas the quadrilaterals joining the reference values for α and γ and for α* and γ* are illustrated in Fig. 3. The corresponding values of α and γ or α* and γ* are very close to each other in all three sets of data (Table 5, Fig. 3), whereas the corresponding values of *b* and *c* or *b** and *c**

TABLE 5. Reference values of lattice constants for alkali feldspar end-members

End-member	Reference	a (Å) a^* (Å^{-1})	b (Å) b^* (Å^{-1})	c (Å) c^* (Å^{-1})	α (°) α^* (°)	β (°) β^* (°)	γ (°) γ^* (°)	V (Å^3) V^* (Å^{-3})
LM	Smith (1974a)†	8.5903	12.9659	7.2224	90.65	115.96	87.65	722.64¶
		0.12958	0.077190¶	0.15400	90.42	64.04¶	92.30	0.0013838§
	Stewart & Wright (1974)∇	8.597	12.964	7.222	90.637	115.933	87.683	723.24
		0.12944	0.07720	0.15397	90.419	64.071	92.267	0.001383
	Kroll & Ribbe (1987)◊	8.592	12.963	7.222	90.62	115.95	87.67	722.65
		0.129541§	0.077209§	0.153997	90.44	64.05§	92.29	0.0013838§
HS	Smith (1974a)†	8.6043	13.0289	7.1763	90	116.03	90	722.89
		0.12934	0.076752	0.15508	90	63.97	90	0.0013833§
	Stewart & Wright (1974)∇	8.610	13.033	7.174	90	116.017	90	723.45
		0.12924	0.07673	0.15511	90	63.983	90	0.001382
	Kroll & Ribbe (1987)◊	8.605	13.031	7.177	90	116.00	90	723.32
		0.129297§	0.076740§	0.155023	90	64.00§	90	0.0013825§
LA	Smith (1974a)†	8.138	12.785	7.158	94.26	116.60	87.68	664.06
		0.13743	0.078436	0.15655	86.40	63.49	90.46	0.0015059§
	Stewart & Wright (1974)∇	8.134	12.781	7.160	94.317	116.617	87.667	663.57
		0.13752	0.07847	0.15654	86.340	63.475	90.449	0.001507
	Kroll & Ribbe (1987)◊	8.135	12.785	7.158	94.27	116.60	87.68	663.81
		0.137481§	0.078437§	0.156552	86.39	63.49§	90.46	0.0015065§
HA	Smith (1974a)†	8.160	12.873	7.110	93.52	116.43	90.27	667.11
		0.13694	0.077879	0.15746	85.93	63.50	87.95	0.0014990§
	Stewart & Wright (1974)∇	8.160	12.871	7.110	93.517	116.400	90.250	667.19
		0.13690	0.07789	0.15742	85.949	63.529	87.972	0.001499
	Kroll & Ribbe (1987)◊	8.156	12.871	7.108	93.52	116.44	90.26	666.44
		0.137019§	0.077890§	0.157516	85.94	63.49§	87.96	0.0015005§

Note: LM low microcline, HS high sanidine, LA low albite, HA high albite.

† See also Smith & Brown (1988, Table 7.1). Their data reproduce those given by Smith (1974a). In Smith & Brown (1988) c value for low microcline was misprinted as 0.72232 nm, a, b, and c values for low microcline and high sanidine were rounded to three decimals, and V value for low microcline and high sanidine was rounded to one decimal.

¶ Corrected from original data (V = 722.62 Å^3, b^* = 0.77200 nm^{-1}, β^* = 64.05°). The values in the present Table were calculated from direct lattice constants given by Smith (1974a).

§ Calculated in this work.

∇ See also Stewart (1975).

◊ See also Kroll and Ribbe (1983, Table 4). In their set of data, b and V values for low microcline were 12.962 Å and 722.60 Å^3; a, β and V values for high sanidine were 8.606 Å, 116.03°, and 723.22 Å^3.

are slightly different (Table 5, Fig. 2). The data available in the last decade indicate that major uncertainties occur in the case of lattice constants for high sanidine end-member.

A highly disordered sanidine, obtained by Blasi et al. (1984a) by dry heating a low microcline from Bedford County pegmatite district, Virginia, U.S.A. to 1050 °C for 150 days, gave b = 13.0319(8) Å and c = 7.1742(5) Å. These data, obtained from least-squares refinements of 45 uniquely indexed diffractions collected using X-ray powder diffractometer methods, led Blasi et al. (1984a) to suggest that the reference values chosen by Smith (1974a) for high sanidine end-member should be amended slightly (cf. Table 5). Another specimen of high sanidine produced by dry annealing of a Bedford low microcline for 200 days at 1050 °C was employed in structure refinement by Blasi et al. (1987b). Lattice refinements based on 40 independent diffractions ob-

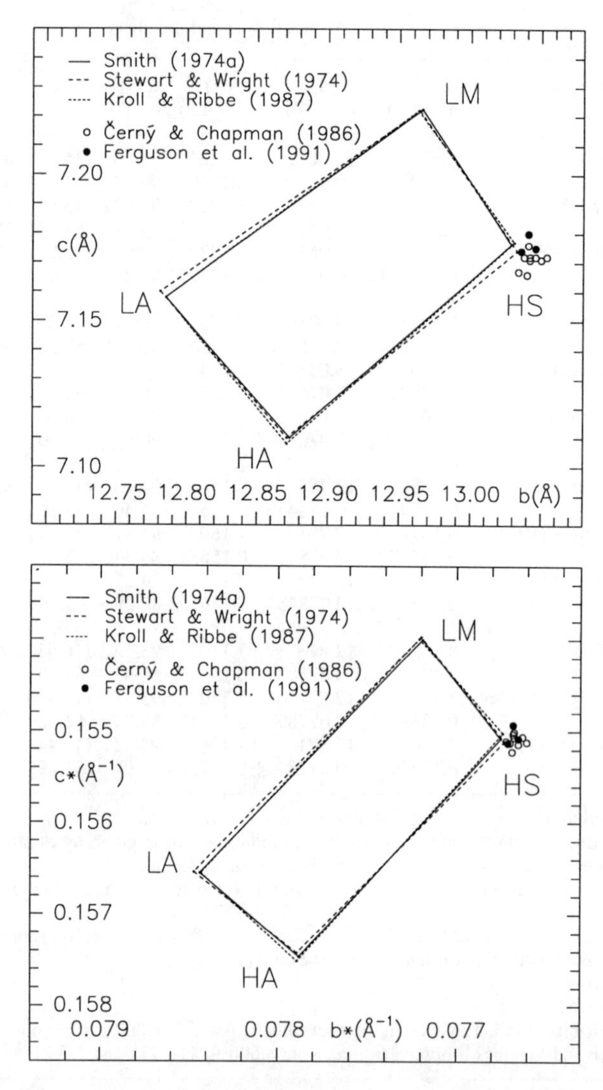

Figure 2. Relation between b and c, and b^* and c^*, for alkali feldspars. The quadrilaterals join the reference values for low microcline (LM), high sanidine (HS), high albite (HA), and low albite (LA) proposed by the authors indicated in the inset. Open and closed circles represent data obtained for adularia specimens by means of X-ray powder patterns and single-crystal X-ray diffractometry, respectively.

tained by single-crystal X-ray diffractometry gave b = 13.0334(3) Å and c = 7.1747(2) Å, two values that confirm the earlier data obtained with powder methods and that agree very well with the data proposed by Stewart and Wright (1974) for their high sanidine end-member (cf. Table 5). The composition of the two specimens of high sanidine mentioned is close to $Or_{91}Ab_9$ mol%.

Specimens of adularia from a variety of granitic pegmatites and hydrothermal vein deposits were studied by Černý and Chapman (1984, 1986) by X-ray powder diffractometer methods and

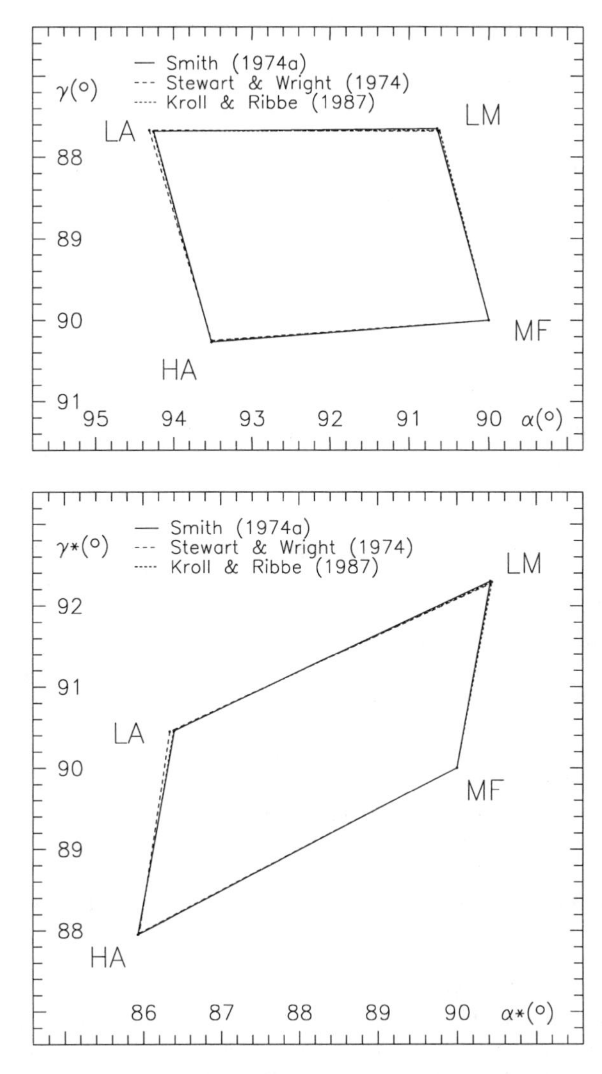

Figure 3. Relation between α and γ, and α* and γ*, for alkali feldspars. The quadrilaterals join the reference values for low microcline (LM), monoclinic feldspars (MF), high albite (HA), and low albite (LA) proposed by the authors indicated in the inset.

infrared absorption techniques, indicating unusually high Si,Al disorder. The chemical compositions of these specimens closely approach that of the pure end-member. The relevant data, including those for another specimen of high sanidine [the Eagle Harbor sanidine studied by Stewart and Wright (1974) and reported by Černý and Chapman (1986) in their Fig. 3], are presented in Fig. 2. Recently, Ferguson et al. (1991) refined the structure of four specimens of adularia studied by Černý and Chapman (1984) from a granitic pegmatite at Buck Claim, Bernic Lake, Manitoba, Canada. In three of these specimens, the cell parameters determined with a 4-circle single-crystal X-ray diffractometer indicate strong Si,Al disorder, which is, however, less pronounced than that determined by Černý and Chapman (1984). The relevant data are plotted in Fig. 2. It is surprising that the most disordered specimen in terms of *T*-O distances appears to be the least disordered in terms of lattice constants. The *R* values of all three structure refinements are rather high (4-6%). Ferguson et al. (1991) attributed these results to the poor quality of the crystals, which show marked wavy extinction.

Černý and Chapman (1986) reported that H. Kroll examined some of their specimens of adularia with a Guinier camera and found many diffraction lines doubled, indicating the presence of two monoclinic phases. Subsequent examinations by Černý and Chapman (1986) in those parts of the hand specimen from which the material studied by H. Kroll was separated, revealed structural heterogeneity. Single-crystal photographs and TEM techniques could be used to clarify the nature of the structural heterogeneity in the feldspar material investigated by Černý and Chapman (1984, 1986) and Ferguson et al. (1991). These feldspars are certainly highly disordered, and further determinations of their lattice parameters are desirable in order to properly revise the reference values for the high sanidine end-member.

At present, the lattice constants reported in Table 5 for the alkali feldspar end-members can be considered good reference values for estimating Al contents in alkali feldspars. There is the possibility of adopting the data that give rise to the largest quadrilaterals in Figs. 2 and 3. In this case, the data proposed by Stewart and Wright (1974) for high sanidine and low albite, and those proposed by Kroll and Ribbe (1987) for high albite (analbite) could be employed. The three sets of data for low microcline are very close to each other (Table 5, Figs. 2 and 3), and one of them, or their average, could be used.

To a first approximation, complete Si,Al order could be assumed in low microcline and low albite end-members in Table 5, and complete disorder could be assumed for high albite (analbite) and high sanidine end-members in Table 5. The latter assumption, which was not shared by Kroll and Ribbe (1987), seems to be supported by the following data. The specimen of high sanidine studied by Blasi et al. (1987b) has $\langle T1\text{-O}\rangle = 1.6428(8)$ Å and $\langle T2\text{-O}\rangle = 1.6398(9)$ Å, and the most disordered adularia of Ferguson et al. (1991) has $\langle T1\text{-O}\rangle = 1.640(1)$ Å and $\langle T2\text{-O}\rangle =$ 1.642(1) Å. Both sets of data are, therefore, very close to the expected values for a fully disordered high sanidine. Note that cell edges *b* and *c* for both specimens, i.e. 13.0334(3) Å and 7.1747(2) Å for the former and 13.036(4) Å and 7.174(2) Å for the latter, are very close to the reference values proposed by Stewart and Wright (1974) for their high sanidine end-member (Table 5).

Over the last two decades, the most widely employed procedure for estimating *T*-site occupancies in alkali feldspars from metric properties has been that based on a model, the Δ method for short, that uses the lattice parameters *b* and *c*, b^* and c^*, α and γ, and α^* and γ^* in the structural indicators $\Delta(bc)$ (Stewart and Ribbe, 1969; Stewart and Wright, 1974; Stewart, 1975; Blasi, 1980, 1982), $\Delta(b^*c^*)$ (Smith, 1968, 1970, 1974a; Blasi, 1980, 1982), $\Delta(\alpha\gamma)$ (Blasi, 1978, 1980, 1982), and $\Delta(\alpha^*\gamma^*)$ (Smith, 1968, 1970, 1974a; Stewart and Ribbe, 1969; Stewart and Wright, 1974; Stewart, 1975; Blasi, 1978, 1980, 1982). In $C\bar{1}$ alkali feldspars, $\Delta(bc)$ or $\Delta(b^*c^*)$, and $\Delta(\alpha\gamma)$ or $\Delta(\alpha^*\gamma^*)$, give an estimate of the sum and difference, respectively, of the Al occupancies in the *T*1*o* and *T*1*m* sites. In *C*2/*m* alkali feldspars, $\Delta(bc)$ or $\Delta(b^*c^*)$ give an estimate of 2t1. The values of the relevant structural indicators can be derived graphically from Figs. 2 and 3, or by using the equations proposed by Blasi (1977). The soundness of the Δ method using the structural indicators $\Delta(bc)$ or $\Delta(b^*c^*)$, normalized between 0.5 and 1.0, and $\Delta(\alpha\gamma)$ or $\Delta(\alpha^*\gamma^*)$, normalized between 0.0 and 1.0, is reconfirmed here to estimate *T*-site occupancies in alkali feldspars.

Further alternative determinative procedures using metric parameters have been devised by Blasi and Blasi De Pol (1977), Kroll (1973, 1980), Kroll and Ribbe (1983, 1987), and Hovis (1986, 1988). The c_K-Z method proposed by Hovis (1986) has been criticized by Kroll and Ribbe (1987), and the relevant reply can be found in Hovis (1988). Simple determinative methods using selected diffraction-peak positions have been developed by Kroll and Ribbe (1987) and Hovis (1989). We consider below the procedures devised by Kroll and Ribbe (1983, 1987) because these are related to the above mentioned Δ method.

Modifications to the Δ method have been proposed by Kroll and Ribbe (1983; see also Ribbe, 1984) by giving separate plots of *b vs c* for topochemically monoclinic and topochemically triclinic alkali feldspars, and assuming $t1o + t1m$ in the analbite (high albite) and $2t1$ in high sanidine to be equal to 0.56. Subsequently, Kroll and Ribbe (1987) abandoned the two *bc* plots and proposed the use of two separate plots of *b vs c** for topochemically monoclinic and topochemically triclinic alkali feldspars, reducing the value of $t1o + t1m$ in the analbite (high albite) and the value of $2t1$ in high sanidine to 0.55. It should be noted that the method based on the diagrams of *b vs c** for *alkali* feldspars is largely dependent on structural data obtained from monoclinic and triclinic *K-rich* feldspars.

The use of direct and reciprocal cell edges *b* and *c** was done by Kroll and Ribbe (1987) with the purpose of (a) linearizing alkali exchange paths that are curved on the plot of *b vs c*, and (b) reducing the departure of the results given by their equations from the actual values that can be derived graphically.

However, all equations proposed by Kroll and Ribbe (1987) show deviations from the values that can be obtained graphically from the relevant plots. According to the authors (their Table 6), these deviations in the diagrams of *b vs c** reach +0.007 and –0.006 in topochemically monoclinic and triclinic K-rich feldspars, respectively. In the α^* *vs* γ^* diagram, these deviations reduce to 0.002. The equations developed by Blasi (1977), which correspond to a rigorous treatment, were not used by Kroll and Ribbe (1987) because they considered all the deviations mentioned above to be negligible.

Reference values and feldspar names for all quadrilateral corners in the diagram of *b vs c** for topochemically monoclinic feldspars were not given by Kroll and Ribbe (1987), and the same applies to the two corners for disordered feldspars in the diagram of *b vs c** for topochemically triclinic feldspars. It might be unclear to some readers whether these corners, free of reference values and names, correspond to ideal feldspars, to feldspars occurring in nature, or to those obtainable only in the laboratory.

The fact that in principle it is not possible to plot in a single diagram both monoclinic and triclinic alkali feldspars may represent a drawback in several cases. This means that specimens with different topochemistry but similar chemistry, e.g. monoclinic and triclinic K-rich feldspars from the same geological suite, cannot be compared. The same applies to specimens with different chemistry and topochemistry, as is the case with two alkali feldspars coexisting in perthitic or cryptoperthitic intergrowths.

With regard to strained alkali feldspars, the determination of Al contents from metric properties is, at present, quite uncertain. In fact, the number of strained alkali feldspars used in structure refinement that can be employed as reference specimens in order to check the methods based on metric properties is completely inadequate. In practice, only two specimens are available, (a) the intermediate microcline K-235 from the Eastern border group of the Kûngnât syenites, SW Greenland, with $\Delta a = 0.30$ Å, studied by Ribbe (1979), and (b) a high albite from the Rabb Canyon pegmatite, with $\Delta a = -0.23$ Å, investigated by Keefer and GE Brown (1978) (see section 3.3 and Table 2). With reference to these two specimens, the results obtained from the structural indicators based on cell edges b, c, b^*, and c^* are confusing. In contrast, the results obtained from the structural indicators based on cell angles α, γ, α^*, and γ^* agree well with the data derived from $\langle T\text{-O}\rangle$ distances, confirming that these cell angles are little affected by coherency stresses (cf. section 3.2.1).

Cell dimensions used in the determinative methods for estimating T-site occupancies should be very accurate. The simulated diffractometer patterns of alkali feldspars given by Borg and Smith (1969) and the indexing guides prepared by Wright and Stewart (1968), Ribbe (1983b), Blasi (1984), Kroll et al. (1986) and Hovis (1989) provide a helpful tool to a correct use of diffractions in refinements of unit-cell parameters.

4.2. *T*-SITE OCCUPANCIES FROM *T*-O DISTANCES

Smith (1974a, pp. 56-59 and p. 70) reviewed the basic principles concerning the determination of Al contents of T sites from $\langle T\text{-O}\rangle$ distances in $AlSi_3$ feldspars. He cautioned that whatever determinative calibration curve is used, serious problems might arise; in desperation (sic), he decided to convert all $\langle T\text{-O}\rangle$ distances to Al occupancy with a single relationship, which consists of a straight line from 1.612 Å for 0.0 Al to 1.676 Å for 0.5 Al, and a second straight line therefrom to 1.745 Å at 1.0 Al. The equations of the two straight lines are

$$t_i = 0.5(\langle T_i\text{-O}\rangle - 1.612)/(1.676 - 1.612) \quad (1)$$

and

$$t_i = 0.5 + 0.5(\langle T_i\text{-O}\rangle - 1.676)/(1.745 - 1.676). \quad (2)$$

Obviously, equations (1) and (2) are to be used with $\langle T_i\text{-O}\rangle$ distances shorter and longer than 1.676 Å, respectively. The sum of Al contents derived from Smith's (1974a) method can be different from 1.0.

Ribbe (1975, p. R-22) proposed a method based on simple proportionation of bond length differences, e.g. $\langle T1o\text{-O}\rangle - \langle T1m\text{-O}\rangle$, to the difference in $\langle T\text{-O}\rangle$ lengths of pure AlO_4 and pure SiO_4 tetrahedra. The value chosen for this difference was 0.130 Å. Ribbe's (1975) method leads to the following equations in $AlSi_3$ feldspars:

$$t1o - t1m = (\langle T1o\text{-O}\rangle - \langle T1m\text{-O}\rangle)/\text{const}, \quad (3)$$

$$t1o - t2o = (\langle T1o\text{-O}\rangle - \langle T2o\text{-O}\rangle)/\text{const}, \quad (4)$$

$$t1o - t2m = (\langle T1o\text{-O}\rangle - \langle T2m\text{-O}\rangle)/\text{const}, \quad (5)$$

where const = 0.130 Å. Equations (3), (4), and (5) can be solved to obtain t1o, t1m, t2o, and t2m by letting t1o + t1m + t2o + t2m = 1. The latter equation implies that the sum of Al contents determined by Ribbe's (1975) method will be equal to 1.0.

Kroll and Ribbe (1983, p. 67; see also Ribbe 1984, p. 28) claimed that Smith's (1974a) method had limited success and proposed a "new model" considering the differences between average individual and grand mean tetrahedral distances, i.e. $\langle T_i\text{-O}\rangle - \langle\langle T\text{-O}\rangle\rangle$, rather than size differences among individual tetrahedra, as in Ribbe's (1975) model. The relevant equation proposed by Kroll and Ribbe (1983) for $AlSi_3$ feldspars is

$$t_i = 0.25 + (\langle T_i\text{-O}\rangle - \langle\langle T\text{-O}\rangle\rangle)/\text{const}, \quad (6)$$

where const, the difference between Al-O and Si-O distances, was estimated to be equal to 0.125 Å in K-rich feldspars and to 0.13 Å in Na-rich feldspars. It will be shown in the following that the method of Kroll and Ribbe (1983) is not new, rather it is a compact reformulation of Ribbe's

(1975) model. If solved as a function of t1o, t1m, t2o, and t2m, Ribbe's (1975) equations (3), (4), and (5) become

$$\mathrm{t1o} = (1 + A + B + C)/4 = 1/4 + (A + B + C)/4, \quad (7)$$

$$\mathrm{t1m} = (1 - 3A + B + C)/4 = 1/4 + (A + B + C)/4 - A = \mathrm{t1o} - A, \quad (8)$$

$$\mathrm{t2o} = (1 + A - 3B + C)/4 = 1/4 + (A + B + C)/4 - B = \mathrm{t1o} - B, \quad (9)$$

$$\mathrm{t2m} = (1 + A + B - 3C)/4 = 1/4 + (A + B + C)/4 - C = \mathrm{t1o} - C, \quad (10)$$

where

$$A = (<T1o\text{-O}> - <T1m\text{-O}>)/\mathrm{const},$$

$$B = (<T1o\text{-O}> - <T2o\text{-O}>)/\mathrm{const},$$

$$C = (<T1o\text{-O}> - <T2m\text{-O}>)/\mathrm{const}.$$

Since

$$\begin{aligned}(A + B + C)/4 &= 3<T1o\text{-O}>/4\mathrm{const} - (<T1m\text{-O}> + <T2o\text{-O}> + <T2m\text{-O}>)/4\mathrm{const} \\ &= <T1o\text{-O}>/\mathrm{const} - (<T1o\text{-O}> + <T1m\text{-O}> + <T2o\text{-O}> + <T2m\text{-O}>)/4\mathrm{const} \\ &= (<T1o\text{-O}> - <<T\text{-O}>>)/\mathrm{const},\end{aligned}$$

equations (7), (8), (9), and (10) can be rewritten as

$$\mathrm{t1o} = 1/4 + (<T1o\text{-O}> - <<T\text{-O}>>)/\mathrm{const}, \quad (11)$$

$$\begin{aligned}\mathrm{t1m} &= 1/4 + (<T1o\text{-O}> - <<T\text{-O}>>)/\mathrm{const} - (<T1o\text{-O}> - <T1m\text{-O}>)/\mathrm{const} \\ &= 1/4 + (<T1m\text{-O}> - <<T\text{-O}>>)/\mathrm{const}, \quad (12)\end{aligned}$$

$$\begin{aligned}\mathrm{t2o} &= 1/4 + (<T1o\text{-O}> - <<T\text{-O}>>)/\mathrm{const} - (<T1o\text{-O}> - <T2o\text{-O}>)/\mathrm{const} \\ &= 1/4 + (<T2o\text{-O}> - <<T\text{-O}>>)/\mathrm{const}, \quad (13)\end{aligned}$$

$$\begin{aligned}\mathrm{t2m} &= 1/4 + (<T1o\text{-O}> - <<T\text{-O}>>)/\mathrm{const} - (<T1o\text{-O}> - <T2m\text{-O}>)/\mathrm{const} \\ &= 1/4 + (<T2m\text{-O}> - <<T\text{-O}>>)/\mathrm{const}, \quad (14)\end{aligned}$$

and then condensed into the form

$$t_i = 1/4 + (<T_i\text{-O}> - <<T\text{-O}>>)/\mathrm{const},$$

which corresponds to equation (6) of Kroll and Ribbe (1983). As in the original formulation proposed by Ribbe (1975), the sum of Al contents obtainable from equation (6) turns out to be equal to 1.0 by definition. In fact, by summing equations (11), (12), (13), and (14) we obtain

$$\Sigma t_i = \mathrm{t1o} + \mathrm{t1m} + \mathrm{t2o} + \mathrm{t2m} = 1 + (4<<T\text{-O}>> - 4<<T\text{-O}>>)/\mathrm{const} = 1.$$

If equations involving $<T\text{-O}>$ and $<<T\text{-O}>>$ distances are used, such as in the case of equation (6), it is mandatory to employ individual *T*-O distances as starting data, rather than mean *T*-O distances and grand mean *T*-O values derived from mean *T*-O data. This procedure was suggested by Blasi et al. (1987a, Table 7 at p. 534) to avoid truncation and rounding errors that could amount to 4 units in the third decimal place of Al contents usually published with three decimal digits. Such errors could even amount to 8 units in the third decimal place if rounding is made in $<<T\text{-O}>>$ distances derived from $<T\text{-O}>$ distances.

A careful analysis of data reported by Kroll and Ribbe (1983) in their Tables 2 and 3 (the same Tables were also reported by Ribbe, 1984) shows a number of rounding and misprint errors in $<T\text{-O}>$ distances and Al contents. The occurrence of a misprint error in their Prilep microcline was the key that focussed our attention and led to a careful interpretation of the significance of equation (6).

Consider just two sites of Prilep microcline: $T1o$ and $T1m$. Kroll and Ribbe (1983, Table 3) gave t1o = 0.995 and t1m = 0.000 instead of 0.994 and –0.006, respectively. The latter represent the correct values obtainable from equation (6) when using the relevant data of Prilep microcline $<T1o\text{-O}> = 1.738$ Å, $<T1m\text{-O}> = 1.613$ Å, and $<<T\text{-O}>> = 1.645$ Å. The same values of 1.738 Å and 1.613 Å were assumed by Kroll and Ribbe (1983, p. 67; also Ribbe 1984, p. 23) for pure AlO_4 and pure SiO_4 tetrahedra in low microcline. Therefore, in the absence of other assumptions, the Al contents of $T1o$ and $T1m$ sites of Prilep microcline would be expected to be equal to 1.000 and 0.000, respectively.

The relevant underestimate of Al contents of $T1o$ and $T1m$ sites in the Prilep microcline can be explained by the following interpretation of equation (6). With reference to $AlSi_3$ feldspars, Fig. 4a allows us to derive the equation of the straight line passing through the points of coordinates ($<\text{Si-O}>$, 0.0) and ($<\text{Al-O}>$, 1.0):

$$(<T_i\text{-O}> - <\text{Si-O}>)/(<\text{Al-O}> - <\text{Si-O}>) = (t_i - 0.0)/(1.0 - 0.0),$$

from which

$$t_i = - <\text{Si-O}>/\text{const} + <T_i\text{-O}>/\text{const}, \qquad (15)$$

where const = $<\text{Al-O}> - <\text{Si-O}>$. The equation of the straight line parallel to that represented by equation (15) and passing through one of the points of coordinates ($<<T\text{-O}>>$, $t_{<<T\text{-O}>>}$) is given by the relationship:

$$t_i = (t_{<<T\text{-O}>>} - <<T\text{-O}>>/\text{const}) + <T_i\text{-O}>/\text{const}. \qquad (16)$$

It is: slope eq. (15) = slope eq. (16) = 1/const, and, with reference to the t axis, intercept eq. (15) = $- <\text{Si-O}>/\text{const}$ and intercept eq. (16) = $(t_{<<T\text{-O}>>} - <<T\text{-O}>>/\text{const})$.

If we assume that in the points of coordinates ($<<T\text{-O}>>$, $t_{<<T\text{-O}>>}$) the value of $t_{<<T\text{-O}>>}$ remains fixed at 0.25, whereas the value of $<<T\text{-O}>>$ can vary, equation (16) becomes formally identical to equation (6) of Kroll and Ribbe (1983). In this respect, equation (6) can be regarded as the equation of the infinite straight lines passing through the infinite points of coordinates ($<<T\text{-O}>>$, 0.25) and parallel to a "reference straight line" represented by equation (15), which joins the points of coordinates ($<\text{Si-O}>$, 0.0) and ($<\text{Al-O}>$, 1.0).

Turn now to the analysis of equation (6) in the case of K-rich feldspars. If we let, in equation (6), const = 0.125 Å and either

$$t_i = 0.0 \text{ and } <T_i\text{-O}> = 1.613 \text{ Å} \quad \text{or} \quad t_i = 1.0 \text{ and } <T_i\text{-O}> = 1.738 \text{ Å},$$

then $<<T\text{-O}>>$ will be equal to 1.64425 Å. The latter can be assumed as the value of the "ideal" grand mean T-O distance in low microcline end-member.

In this specific case, equation (6) represents the alignment condition of the three points of coordinates (1.613, 0.0), (1.64425, 0.25), and (1.738, 1.0). Therefore, the same equation (6) also can be expressed by the relationship:

$$t_i = (<T_i\text{-O}> - 1.613)/\text{const}, \qquad (17)$$

which corresponds to equation (15) and represents the simplest determinative curve for esti-

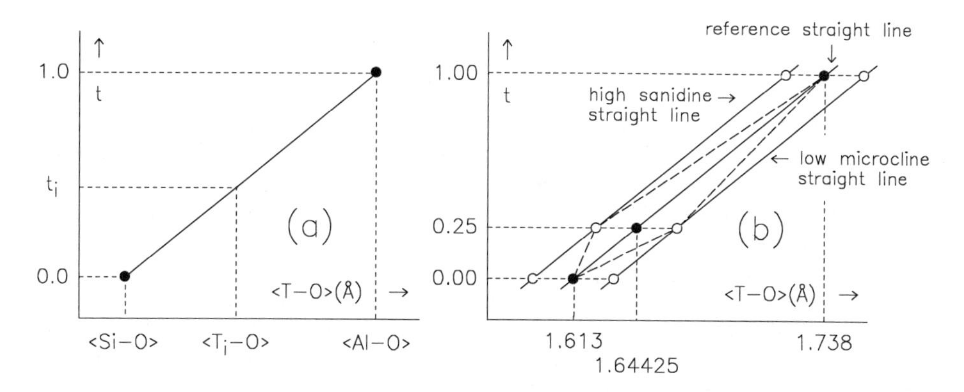

Figure 4. Schematic diagrams to illustrate the relation between mean *T*-O distances, <*T*-O>, and Al contents of *T* sites, t, in (a) alkali feldspars and (b) K-rich feldspars. See text for explanation.

mating Al contents in K-rich feldspars. Unlike equation (6), the sum of Al contents calculable from equation (17) is not constrained to be equal to 1.0.

The actual grand mean *T*-O distance decreases almost systematically passing from low microcline (~1.645 Å) to high sanidine (~1.641 Å). This means that with equation (6), Al contents will be determined for each feldspar with a different determinative straight line characterized by a particular <<*T*-O>> value.

With respect to the "reference straight line" as defined by equation (17), Al contents of specimens of low microcline will be underestimated. This is because the values of their <<*T*-O>> distances are higher than the "ideal" value of 1.64425 Å, and, therefore, their straight lines will be on the right of the "reference straight line". This is the case of Prilep specimen. If one uses <<*T*-O>> = 1.64425 Å in equation (6), the Al contents of $T1o$ and $T1m$ sites of Prilep microcline will be equal to 1.000 and 0.000, respectively, as expected. This procedure is equivalent to that based on the use of equation (17). On the other hand, if one uses <<*T*-O>> = 1.645 Å, i.e. the actual grand mean *T*-O distance of the Prilep specimen, the determinative curve will be a straight line parallel to that represented by equation (17) and passing through the point of coordinates (1.645, 0.25). This implies the above-mentioned underestimate of Al contents, namely $t1o$ = 0.994 and $t1m = -0.006$.

In constrast, the straight lines for specimens of high sanidine will be placed to the left of the "reference straight line" represented by equation (17). Thus, the relevant Al contents in specimens of high sanidine will be overestimated with respect to the same "reference straight line".

The behavior described above for specimens of low microcline and high sanidine is clearly shown in Fig. 4b, in which the deviations of the values of <<*T*-O>> distances for low microcline and high sanidine from 1.64425 Å are grossly exaggerated. Examples of the behavior of selected specimens of K-rich feldspar are given in Table 6. All K-rich feldspars having a <<*T*-O>> distance longer or shorter than 1.64425 Å show a behavior similar to that described for low microcline and high sanidine, respectively.

An alternative procedure to that of employing a straight line parallel to the "reference straight line" (as defined above) could be the use of a straight line from point (1.613, 0.0) to point (<<*T*-O>>, 0.25) and a second straight line therefrom to point (1.738, 1.0) (cf. Fig. 4b). The relevant equations are:

$$t_i = 0.25(<T_i\text{-O}> - 1.613)/(<<T\text{-O}>> - 1.613) \qquad (18)$$

and

$$t_i = 0.25 + 0.75(<T_i\text{-O}> - <<T\text{-O}>>)/(1.738 - <<T\text{-O}>>). \qquad (19)$$

Equations (18) and (19) are to be used with $<T_i\text{-O}>$ distances shorter and longer than $<<T\text{-O}>>$, respectively. The sum of Al contents obtained from equations (18) and (19) is not constrained to be equal to 1.0. Al contents estimated from equations (18) and (19) for selected specimens of K-rich feldspar can be compared with the corresponding values obtained from equation (6) and equation (17) in Table 6 .

The above procedure based on equations (18) and (19) resembles the one proposed by Smith (1974a), also based on two straight lines. However, the two methods are somewhat different. The intersection between the two straight lines represented by equations (18) and (19) is given by a point of variable coordinates ($<<T\text{-O}>>$, 0.25), whereas the intersection between the two straight lines represented by equations (1) and (2) is given by a fixed point (1.676, 0.5).

Unlike the methods of Ribbe (1975) and of Kroll and Ribbe (1983), the procedure based on equations (18) and (19) does not give rise, in principle, to underestimates in Al contents in fully ordered K-rich feldspar with respect to the "reference straight line" represented by equation (17). Obviously, overestimates can be produced in specimens of low microcline if their $<T\text{-O}>$ values are higher than 1.738 Å or smaller than 1.613 Å and their $<<T\text{-O}>>$ values are higher than 1.64425 Å. This happens in the case of the *T1o* site of Pikes Peak 7813C low microcline (cf. Table 6).

On the other hand, equations (18) and (19) give rise to overestimates in Al contents of specimens of high sanidine with respect to the "reference straight line" mentioned earlier. These overestimates are virtually smaller than those produced by equation (6). It should be clear from Fig. 4b and the data in Table 6 that the overestimates produced by equations (18) and (19) in specimens of high sanidine are due to the fact that in these feldspars the values of $<T\text{-O}>$ distances are close to the actual values of their $<<T\text{-O}>>$ distances.

At present, it is therefore difficult to establish whether the intersection point ($<<T\text{-O}>>$, 0.25) between the straight lines represented by equations (18) and (19) can be considered a valid choice. On the other hand, the reference values adopted by Smith (1974a) for his straight lines would require appropriate revision.

Table 6 shows that the differences among the corresponding values obtained from equation (6), equation (17), and equations (18) and (19) are small and reduce to zero where the $<<T\text{-O}>>$ distance of a given specimen becomes equal to the "ideal" grand mean *T*-O distance of low microcline end-member (cf. intermediate microcline RC20C). However, the conceptual differences among the three models are significant. Unlike equation (6), equation (17) or equations (18) and (19) have the advantage that the sum of Al contents is not constrained to 1. Deviations from 1 give an indication of the inadequacy of the model in estimating Al contents. Equation (6) gives information on such inadequacy only in the extreme cases in which the values of Al contents become negative or higher than 1.

J.V. Smith abandoned equations (1) and (2) in the book by Smith and Brown (1988), and underlined (p. 42) that no attempt was made to convert *T*-O distances into Al contents. He used a graphical calibration to evaluate approximate Al contents in K-rich feldspars in a diagram of $<T_i\text{-O}>$ distances *vs* $<T1o\text{-O}>$ distance (his Fig. 3.6, p. 43). The same procedure was used by Ribbe (1983a, Fig. 1 at p. 23; 1984, Fig. 19 at p. 39). This graphical calibration corresponds to a relationship similar to that expressed by equation (17).

A new approach was proposed by Alberti and Gottardi (1988) to estimate tetrahedral Al contents in framework silicates. The method involves various structural parameters other than

TABLE 6. Comparison of Al contents of *T* sites calculated from different equations, given in section 4.2, using <*T*-O> distances of selected specimens of K-rich feldspar

#	Specimen Reference	<*T*1*o*-O> <*T*2*o*-O>	<*T*1*m*-O> <*T*2*m*-O>	<<*T*-O>>	Equation(s)	t1o	t1m	t2o	t2m	tot
		.... ångströms	....							
1	Pikes Peak 7813C	1.7391	1.6129	1.64551	(6)	0.998	-0.011	0.000	0.013	1.000
	low microcline	1.6142	1.6158		(17)	1.008	0.000	0.010	0.023	1.040
	this work (cf. Table 3)				(18) & (19)	1.009	0.000	0.009	0.022	1.039
2	Pellotsalo	1.738	1.613	1.64456	(6)	0.996	-0.005	0.006	0.004	1.000
	low microcline	1.614	1.614		(17)	0.998	-0.002	0.008	0.006	1.010
	Blasi et al. (1987a)				(18) & (19)	0.998	-0.002	0.008	0.006	1.010
3	Prilep	1.738	1.613	1.645	(6)	0.994	-0.006	0.002	0.010	1.000
	low microcline	1.614	1.615		(17)	1.000	0.000	0.008	0.016	1.024
	Strob (1983)				(18) & (19)	1.000	0.000	0.008	0.016	1.023
4	CA1E	1.731	1.618	1.64544	(6)	0.937	0.031	0.017	0.017	1.000
	low microcline	1.616	1.616		(17)	0.946	0.040	0.026	0.026	1.038
	Dal Negro et al. (1978)				(18) & (19)	0.945	0.039	0.025	0.025	1.034
5	RC20C	1.717	1.630	1.64431	(6)	0.832	0.132	0.018	0.020	1.000
	intermediate microcline	1.615	1.616		(17)	0.832	0.132	0.018	0.020	1.002
	Dal Negro et al. (1978)				(18) & (19)	0.832	0.132	0.018	0.020	1.002
6	Spencer U	1.694	1.643	1.64281	(6)	0.662	0.248	0.056	0.036	1.000
	intermediate microcline	1.619	1.616		(17)	0.650	0.236	0.044	0.024	0.954
	Bailey (1969)				(18) & (19)	0.655	0.247	0.046	0.025	0.974
7	Mt. Caval 273	1.677	1.660	1.64381	(6)	0.516	0.380	0.054	0.052	1.000
	high microcline	1.619	1.619		(17)	0.512	0.376	0.050	0.048	0.986
	Blasi et al. (1981)				(18) & (19)	0.514	0.379	0.051	0.049	0.993
8	7007	1.665		1.64325	(6)	0.426		0.074		1.000
	adularia	1.621			(17)	0.418		0.066		0.968
	Phillips & Ribbe (1973)				(18) & (19)	0.424		0.068		0.985
9	Spencer C	1.656		1.64188	(6)	0.359		0.141		1.000
	orthoclase	1.628			(17)	0.340		0.122		0.924
	Colville & Ribbe (1968)				(18) & (19)	0.356		0.132		0.977
10	Roberts Victor	1.644		1.64163	(6)	0.269		0.231		1.000
	high sanidine	1.639			(17)	0.248		0.210		0.916
	Scambos et al. (1987)				(18) & (19)	0.268		0.229		0.995
11	Bedford†	1.643		1.64125	(6)	0.262		0.238		1.000
	high sanidine	1.640			(17)	0.238		0.214		0.904
	Blasi et al. (1987b)				(18) & (19)	0.262		0.237		0.997
12	Buck Claim I	1.640		1.64113	(6)	0.241		0.259		1.000
	high sanidine	1.642			(17)	0.216		0.234		0.900
	Ferguson et al. (1991)				(18) & (19)	0.240		0.259		0.997

Note: The values of <*T*-O> and <<*T*-O>> distances were calculated from individual *T*-O distances. The values of the Al contents and their sum were calculated without performing any rounding during the various steps of calculations. The values indicated by tot may be different from the total of Al contents calculable from the values of t1o, t1m, t2o and t2m given in the present Table, because the latter involve rounding. The values of <<*T*-O>> distances are given with 5 decimals in order to compare them with the value of the ideal grand mean *T*-O distance in low microcline (1.64425 Å).

† Low microcline converted to high sanidine after dry heating for 200 days at 1050 °C.

T-O distances. There are still some problems with this approach applied to feldspars, as relatively high disorder is inferred in specimens of low microcline and low albite that are considered to be fully ordered or nearly so. However, the method proposed by Alberti and Gottardi (1988) is capable of further improvement and should be taken in due consideration by feldspar investigators.

4.3. *T*-SITE OCCUPANCIES FROM Si,Al SITE REFINEMENTS OF X-RAY INTENSITY DATA

Direct Si,Al site refinements using neutron intensity data have been successfully performed in sanidine (GE Brown et al., 1974), orthoclase (Prince et al., 1973), and low albite (Harlow and GE Brown, 1980; Smith et al., 1986), because the difference in the scattering lengths of Si and Al allows the two atoms to be easily distinguished.

X-ray intensity data, which have the advantage of being "home-collected" and obtainable from very small crystals in comparison with neutron diffraction data, give poor results in separating Si and Al by site-refinement techniques because of the similarity in the scattering factors of the two atoms.

Fischer and Zehme (1967) were able to refine T-site occupancies of a Pellotsalo microcline using X-ray data measured by BE Brown and Bailey (1964). Their results, however, might have some bias: in fact, the X-ray intensities of a Pellotsalo microcline remeasured by Blasi et al. (1987a) showed that the data collected by BE Brown and Bailey (1964) were anomalous, being consistent with the structure of a hyper-ordered microcline.

Phillips and Ribbe (1973) and GE Brown et al. (1974) were unsuccessful in their attempts to refine T-site occupancies in specimens of Eifel sanidine. Dal Negro et al. (1978) also made unsuccessful attempts to refine T-site occupancies of a suite of nine specimens of K-rich feldspar from the Adamello Massif. Wenk and Kroll (1984) found that Al and Si atoms were easily separated in a Cazadero low albite, even though the uncertainty in T-site occupancies was rather high (±0.1). Armbruster et al. (1990) found that in a Roc Tourné low albite, a data set with diffractions in the range $0.70 \leq (\sin\theta)/\lambda \leq 0.90$ Å^{-1} yielded populations of 1.00(2) for all tetrahedrally coordinated sites. However, they did not give any detail on the site-refinement procedure employed. Limitations of the method have been discussed by Whittaker (1977), Hawthorne (1983) and Kroll and Ribbe (1983, p. 58).

Fig. 5 gives a representation of curves relevant to X-ray scattering factors for Si and Al atoms as a function of $(\sin\theta)/\lambda$. These curves express: (a) percent differences given by $\{[f(\text{Si}) - f(\text{Al})]/f(\text{Al})\} \times 100$, $\{[f(\text{Si}) - f(\text{Al})]/\{[f(\text{Si}) + f(\text{Al})]/2\}\} \times 100$, and $\{[f(\text{Si}) - f(\text{Al})]/f(\text{Si})\} \times 100$; and (b) "absolute" differences given by $f(\text{Si}) - f(\text{Al})$. The curves for percent differences indicate that high angle-data should be used to distinguish Si and Al atoms, whereas the curve for "absolute" differences indicates that intermediate-angle data should be more appropriate.

Direct Si,Al site refinements were performed for the Pikes Peak microcline 7813C using the intensity data mentioned in Table 3. The groups of diffractions employed were selected in the following ranges of $(\sin\theta)/\lambda$: (1) 0.07-0.20 plus 0.45-0.80 Å^{-1}, (2) 0.45-0.80 Å^{-1}, (3) 0.70-0.90 Å^{-1}, (4) 0.90-1.08 Å^{-1}, and (5) 0.70-1.08 Å^{-1} (cf. Fig. 5). Three different types of T-site occupancy refinements were performed in the following way:

(a) starting atomic parameters were taken from a standard refinement of F_o values in the relevant range of $(\sin\theta)/\lambda$; site-occupancy factors of T sites were replaced by those for ideal high sanidine or ideal low microcline, and were used either unconstrained or constrained to be equal to 1.0;

(b) same technique as in (a), except for starting atomic parameters, which were taken from a

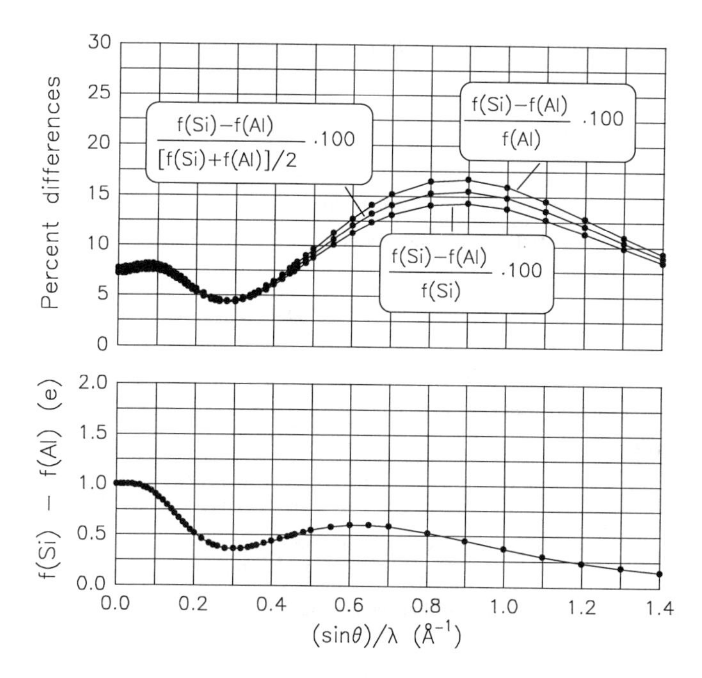

Figure 5. Percent and "absolute" differences between X-ray scattering factors for Si and Al atoms as a function of (sinθ)/λ. The relevant atomic scattering factors were taken from the *International Tables for X-ray Crystallography* (1974).

standard refinement of F_o values in the range $0.70 \leq (\sin\theta)/\lambda \leq 1.08$ Å^{-1}, and for all U_{ij} displacement parameters for *T* positions, which were kept fixed;

(c) same technique as in (b) but with alternating steps consisting of *T*-site refinement with fixed U_{ij} values, and U_{ij} refinement with fixed *T*-site occupancies.

Refinements using starting *T*-site occupancy factors of ideal low microcline, either unconstrained or constrained, were in general straightforward, whereas refinements using starting *T*-site occupancy factors of ideal high sanidine were definitely more difficult. The latter, however, present the best evidence to demonstrate that Si and Al atoms can actually be distinguished.

Table 7 summarizes the results obtained from selected models of *T*-site refinements. Brief information on the various types of site refinements performed will be given in the following.

With regard to site refinements of type (a), site-occupancy factors seem to give better results if they are unconstrained. Starting site occupancies of ideal high sanidine required an extremely high number of least-squares cycles of site refinement to see Al contents increase in the *T*1*o* site and decrease in the other three sites. Si and Al atoms were more easily distinguished with the use of low- plus intermediate-angle data, i.e. those in the (sinθ)/λ ranges 0.07-0.20 plus 0.45-0.80 Å^{-1}. The curve for "absolute" differences in scattering factors between Si and Al atoms in Fig. 5 seems to justify this behavior. The variations in Al and Si contents of *T* sites obtained from 100 cycles of least-squares site refinement in the (sinθ)/λ ranges mentioned are given in Table 8 and shown in Fig. 6.

A somewhat different behavior resulted from site refinements of type (b). Proper results were obtained in several cases within a few cycles of site refinement. Refinements based on uncon-

TABLE 7. Selected models of T-site refinements for microcline Pikes Peak 7813C

No. of diffractions	No. of least-squares cycles	$R(\%)$	$R_w(\%)$		*T*-site occupancies			
					$T1o$	$T1m$	$T2o$	$T2m$
Refinement of type (a)								
Starting T-site occupancy factors of ideal high sanidine (unconstrained)†								
$(\sin\theta)/\lambda$ range: 0.07-0.20 plus 0.45-0.80 Å^{-1}								
2068	100	1.36	1.42	Si	0.098(25)	0.984(25)	1.022(25)	0.987(25)
				Al	0.918(26)	0.045(26)	-0.004(26)	0.036(26)
Refinements of type (b)								
Starting T-site occupancy factors of ideal high sanidine (unconstrained)¶								
$(\sin\theta)/\lambda$ range: 0.45-0.80 Å^{-1}								
2024	4	1.37	1.42	Si	0.036(24)	0.966(24)	1.024(24)	1.029(24)
				Al	0.982(27)	0.067(27)	0.002(27)	-0.005(27)
do.	5	1.50	1.58	Si	0.005(24)	0.978(24)	1.036(25)	1.037(24)
				Al	1.017(27)	0.055(27)	-0.012(27)	-0.014(27)
Starting T-site occupancy factors of ideal high sanidine (constrained)								
$(\sin\theta)/\lambda$ range: 0.70-0.90 Å^{-1}								
1611	4-10§	1.63	1.57	Si	0.061(42)	1.054(0)	1.057(0)	1.057(0)
$(\sin\theta)/\lambda$ range: 0.90-1.08 Å^{-1}								
1583	6	2.45	2.06	Si	-0.097(0)	1.007(0)	1.020(0)	0.999(114)
$(\sin\theta)/\lambda$ range: 0.70-1.08 Å^{-1}								
3190	4	2.00	1.81	Si	0.007(22)	1.004(0)	1.007(0)	1.006(0)
Starting T-site occupancy factors of ideal low microcline (unconstrained)								
$(\sin\theta)/\lambda$ range: 0.70-1.08 Å^{-1}								
3190	2	2.00	1.82	Si	0.001(18)	1.002(18)	1.001(18)	1.000(18)
				Al	0.999(21)	-0.001(21)	0.001(21)	0.001(21)
Starting T-site occupancy factors of ideal low microcline (constrained)								
$(\sin\theta)/\lambda$ range: 0.70-1.08 Å^{-1}								
3190	4	2.00	1.81	Si	-0.0011(3)	1.0011(4)	1.0015(3)	1.0014(3)
Refinement of type (c)								
Starting T-site occupancy factors of ideal high sanidine (unconstrained)∇								
$(\sin\theta)/\lambda$ range: 0.45-0.80 Å^{-1}								
2024	10	1.30	1.35	Si	0.003(23)	0.964(23)	1.002(23)	1.005(23)
				Al	1.020(26)	0.069(26)	0.024(26)	0.020(26)

Note: Estimated standard errors (1σ) are given in parentheses and refer to the last decimal place.
† See Table 8 for details.
¶ See Table 9 for details.
§ No significant variations were observed in this range of least-squares cycles.
∇ See Table 10 and footnote # for details: the data reported here correspond to those given under #6 in Table 10.

TABLE 8. T-site refinements of type (a) using 2068 observed diffractions in the $(\sin\theta)/\lambda$ ranges 0.07-0.20 & 0.45-0.80 Å^{-1} for microcline Pikes Peak 7813C

#	R(%)	R_w(%)	T-site occupancies											
			$T1o$			$T1m$			$T2o$			$T2m$		
			Si σ	Al σ	tot	Si σ	Al σ	tot	Si σ	Al σ	tot	Si σ	Al σ	tot
10	1.3546	1.4175	0.52445	0.46600	0.99045	0.83781	0.19968	1.03749	0.90489	0.11985	1.02474	0.83955	0.19148	1.03103
			0.02483	0.02619		0.02483	0.02618		0.02483	0.02619		0.02488	0.02618	
20	1.3516	1.4136	0.37995	0.61931	0.99926	0.87885	0.15637	1.03522	0.98326	0.03671	1.01997	0.88465	0.14419	1.02884
			0.02477	0.02612		0.02476	0.02611		0.02475	0.02612		0.02481	0.02611	
30	1.3668	1.4285	0.28325	0.72184	1.00509	0.91148	0.12191	1.03339	1.02213	-0.00449	1.01764	0.91784	0.10932	1.02716
			0.02499	0.02634		0.02497	0.02633		0.02496	0.02634		0.02502	0.02633	
40	1.3654	1.4268	0.21788	0.79106	1.00894	0.93510	0.09689	1.03199	1.02191	-0.00428	1.01763	0.94114	0.08476	1.02590
			0.02496	0.02630		0.02493	0.02629		0.02492	0.02630		0.02498	0.02629	
50	1.3645	1.4258	0.17385	0.83767	1.01152	0.95157	0.07944	1.03101	1.02176	-0.00413	1.01763	0.95712	0.06792	1.02504
			0.02494	0.02628		0.02491	0.02627		0.02490	0.02628		0.02496	0.02627	
60	1.3639	1.4251	0.14441	0.86884	1.01325	0.96309	0.06724	1.03033	1.02167	-0.00404	1.01763	0.96811	0.05633	1.02444
			0.02493	0.02626		0.02490	0.02626		0.02489	0.02626		0.02494	0.02625	
70	1.3636	1.4247	0.12484	0.88955	1.01439	0.97115	0.05870	1.02985	1.02161	-0.00398	1.01763	0.97572	0.04831	1.02403
			0.02493	0.02626		0.02489	0.02625		0.02488	0.02626		0.02494	0.02625	
80	1.3634	1.4245	0.11187	0.90328	1.01515	0.97683	0.05268	1.02951	1.02158	-0.00395	1.01763	0.98100	0.04275	1.02375
			0.02492	0.02625		0.02488	0.02624		0.02488	0.02625		0.02493	0.02624	
90	1.3633	1.4244	0.10331	0.91234	1.01565	0.98082	0.04846	1.02928	1.02156	-0.00393	1.01763	0.98466	0.03889	1.02355
			0.02492	0.02625		0.02488	0.02624		0.02488	0.02625		0.02493	0.02624	
100	1.3633	1.4243	0.09767	0.91831	1.01598	0.98363	0.04548	1.02911	1.02155	-0.00392	1.01763	0.98723	0.03618	1.02341
			0.02492	0.02625		0.02488	0.02624		0.02487	0.02625		0.02493	0.02624	

Note: The high number of digits for all quantities is given to follow better their behavior during the course of site refinements.
refers to the number of least-squares cycle.

TABLE 9. T-site refinements of type (b) using 2024 observed diffractions in the range $0.45 \leq (\sin\theta)/\lambda \leq 0.80$ Å^{-1} for microcline Pikes Peak 7813C

#	R(%)	R_w(%)	*T*-site occupancies											
			T1o			*T1m*			*T2o*			*T2m*		
			Si	Al	tot	Si	Al	tot	Si	Al	tot	Si	Al	tot
			σ	σ		σ	σ		σ	σ		σ	σ	
1	1.9394	2.6898	0.45254	0.53701	0.98955	0.84866	0.18205	1.03071	0.87942	0.14694	1.02636	0.88429	0.14079	1.02508
			0.42484	0.47232		0.42011	0.47027		0.42174	0.46987		0.41961	0.46821	
2	1.4479	1.5102	0.24076	0.75234	0.99310	0.90649	0.13600	1.04249	0.94378	0.09188	1.03566	0.95392	0.08021	1.03413
			0.03943	0.04382		0.03985	0.04407		0.04008	0.04425		0.03888	0.04315	
3	1.3562	1.4341	0.10004	0.91092	1.01096	0.94626	0.09088	1.03714	1.00130	0.02792	1.02922	1.00841	0.01909	1.02750
			0.02702	0.03013		0.02687	0.03007		0.02699	0.03006		0.02686	0.02996	
4	1.3656	1.4218	0.03615	0.98215	1.01830	0.96644	0.06738	1.03382	1.02395	0.00161	1.02556	1.02874	-0.00454	1.02420
			0.02428	0.02706		0.02419	0.02702		0.02432	0.02703		0.02412	0.02686	
5	1.5039	1.5753	0.00493	1.01742	1.02235	0.97792	0.05451	1.03243	1.03590	-0.01179	1.02411	1.03703	-0.01384	1.02319
			0.02445	0.02727		0.02437	0.02724		0.02450	0.02724		0.02433	0.02710	
6	1.5455	1.6169	-0.01047	1.03467	1.02420	0.98308	0.04861	1.03169	1.03589	-0.01192	1.02397	1.03624	-0.01308	1.02316
			0.02700	0.03011		0.02692	0.03008		0.02706	0.03008		0.02687	0.02992	
7	1.6276	1.7208	-0.01791	1.04294	1.02503	0.98543	0.04581	1.03124	1.03510	-0.01117	1.02393	1.03546	-0.01235	1.02311
			0.02752	0.03069		0.02745	0.03066		0.02760	0.03066		0.02738	0.03048	
8	1.6817	1.7917	-0.02133	1.04671	1.02538	0.98652	0.04449	1.03101	1.03440	-0.01050	1.02390	1.03487	-0.01177	1.02310
			0.02904	0.03237		0.02897	0.03234		0.02913	0.03235		0.02888	0.03213	
9	1.7102	1.8279	-0.02290	1.04845	1.02555	0.98705	0.04385	1.03090	1.03412	-0.01021	1.02391	1.03463	-0.01154	1.02309
			0.03001	0.03345		0.02994	0.03342		0.03011	0.03343		0.02983	0.03320	
10	1.7250	1.8463	-0.02365	1.04929	1.02564	0.98734	0.04352	1.03086	1.03405	-0.01015	1.02390	1.03457	-0.01148	1.02309
			0.03054	0.03403		0.03047	0.03401		0.03064	0.03402		0.03036	0.03378	

Note: The high number of digits for all quantities is given to follow better their behavior during the course of site refinements.
refers to the number of least-squares cycle.

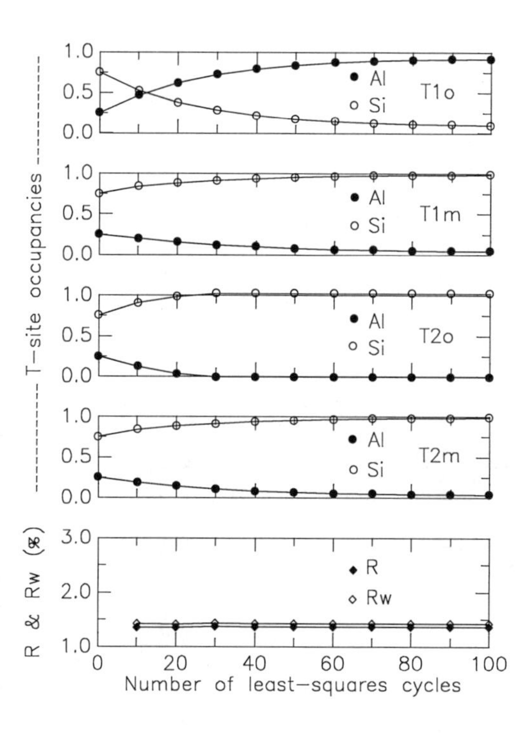

Figure 6. Variation in Al and Si contents of T sites during 100 cycles of T-site refinements of type (a) (see text), performed by using 2068 observed diffractions in the $(\sin\theta)/\lambda$ ranges 0.07-0.20 plus 0.45-0.80 Å^{-1} for microcline Pikes Peak 7813C.

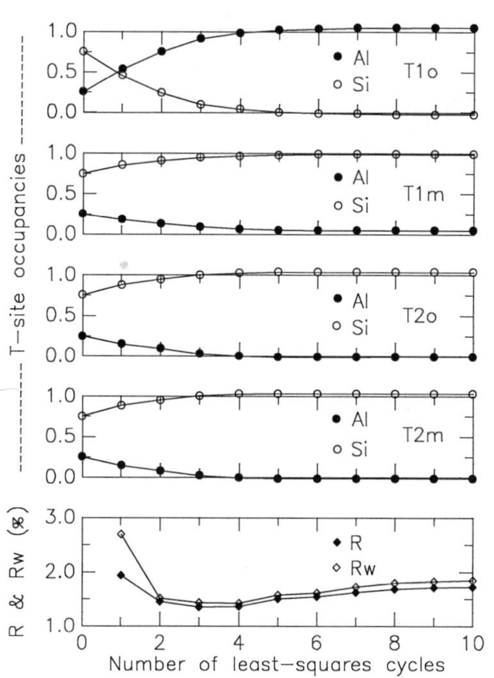

Figure 7. Variation in Al and Si contents of T sites during 10 cycles of T-site refinements of type (b) (see text), performed by using 2024 observed diffractions in the $(\sin\theta)/\lambda$ range 0.45-0.80 Å^{-1} for microcline Pikes Peak 7813C.

TABLE 10. *T*-site refinements of type (c) using 2024 observed diffractions in the range $0.45 \le (\sin\theta)/\lambda \le 0.80$ Å^{-1} for microcline Pikes Peak 7813C

#	R(%)	R_w(%)	T-site occupancies: T1o Si σ	T1o Al σ	T1o tot	T1m Si σ	T1m Al σ	T1m tot	T2o Si σ	T2o Al σ	T2o tot	T2m Si σ	T2m Al σ	T2m tot
3	1.3042	1.3562	0.11085	0.89929	1.01014	0.93683	0.09992	1.03675	0.97862	0.05063	1.02925	0.98681	0.04103	1.02784
			0.02404	0.02680		0.02392	0.02675		0.02403	0.02674		0.02389	0.02662	
4	1.3003	1.3494	0.04873	0.96895	1.01768	0.95338	0.08092	1.03430	0.99425	0.03267	1.02692	1.00075	0.02496	1.02571
			0.02356	0.02627		0.02346	0.02623		0.02357	0.02623		0.02343	0.02611	
5	1.2976	1.3458	0.01816	1.00328	1.02144	0.96013	0.07273	1.03286	1.00060	0.02487	1.02547	1.00469	0.01978	1.02447
			0.02357	0.02629		0.02348	0.02626		0.02360	0.02626		0.02346	0.02614	
6	1.2976	1.3457	0.00333	1.01994	1.02327	0.96354	0.06884	1.03238	1.00177	0.02355	1.02532	1.00477	0.01969	1.02446
			0.02345	0.02616		0.02337	0.02613		0.02348	0.02612		0.02334	0.02601	
7	1.3243	1.3774	-0.00622	1.03028	1.02406	0.96287	0.06915	1.03202	1.00214	0.02274	1.02488	1.00426	0.01988	1.02414
			0.02354	0.02626		0.02346	0.02623		0.02357	0.02623		0.02343	0.02611	
8	1.3750	1.4433	-0.01101	1.03566	1.02465	0.96253	0.06952	1.03205	1.00056	0.02456	1.02512	1.00244	0.02197	1.02441
			0.02391	0.02668		0.02383	0.02665		0.02395	0.02665		0.02381	0.02653	
9	1.2979	1.3455	0.05189	0.97438	1.02627	0.94899	0.08254	1.03153	0.98795	0.03673	1.02468	0.98899	0.03506	1.02405
			0.02368	0.02641		0.02360	0.02639		0.02371	0.02639		0.02357	0.02627	
10	1.2977	1.3455	0.07837	0.94463	1.02300	0.94283	0.08958	1.03241	0.98177	0.04386	1.02563	0.98304	0.04194	1.02498
			0.02348	0.02618		0.02339	0.02616		0.02350	0.02615		0.02336	0.02604	

	Starting U_{ij}† for T sites (Cycle # 1): 11	22	33	12	13	23	Final U_{ij}† for T sites (Cycle # 10): 11	22	33	12	13	23
T1o	802	683	587	-176	322	-45	837	709	637	-203	349	-55
T1m	767	677	565	91	312	15	766	657	556	70	314	19
T2o	673	549	709	-95	273	-34	674	519	710	-117	286	-38
T2m	659	544	725	7	264	5	653	521	719	-13	265	7

Note: The high number of digits for all quantities is given to follow better their behavior during the course of site refinements.

\# refers to the number of least-squares cycle in which U_{ij} values for *T* sites were kept constant. In particular, the first two cycles were performed with constant U_{ij} for *T* sites, the second two cycles with constant *T*-site occupancies, the third again with constant U_{ij} for *T* sites, and so on. The first two cycles, denoted as #1 and #2, are given in Table 9. The third two cycles, denoted as #3 and #4, are given in this Table along with the fifth two, seventh two and ninth two cycles denoted as #5 and #6, #7 and #8, #9 and #10, respectively.

† U_{ij} ($\times 10^5$) are expressed in Å^2 and appear in the anisotropic displacement factor as: $\exp\{-2\pi^2[h^2a^{*2}U_{11} + k^2b^{*2}U_{22} + l^2c^{*2}U_{33} + 2hka^*b^*U_{12} + 2hla^*c^*U_{13} + 2klb^*c^*U_{23}]\}$.

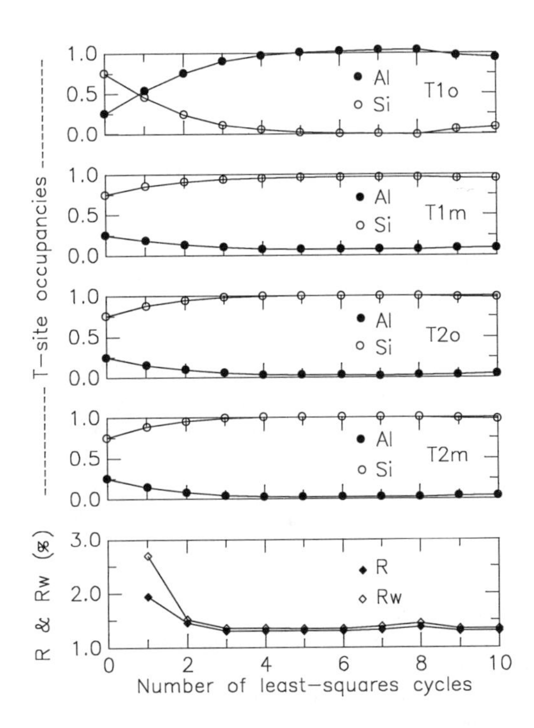

Figure 8. Variation in Al and Si contents of T sites during 18 cycles of T-site refinements of type (c) (see text), performed by using 2024 observed diffractions in the $(\sin\theta)/\lambda$ range 0.45-0.80 $Å^{-1}$ for microcline Pikes Peak 7813C. The number of least-squares cycles indicated on the abscissa refers to the cycles in which U_{ij} values for T sites were kept constant (see footnote # in Table 10).

strained site-occupancy factors of ideal high sanidine gave better results with low- plus intermediate-angle data, and especially intermediate-angle data. Table 9 presents data pertinent to the latter case [$(\sin\theta)/\lambda$ range = 0.45-0.80 $Å^{-1}$], and Fig. 7 gives a representation of the behavior of Al and Si atoms with an increasing number of least-squares cycles of site refinement. Refinements based on constrained site-occupancy factors of ideal high sanidine gave the best results with high-angle data, especially in the $(\sin\theta)/\lambda$ region 0.70-1.08 $Å^{-1}$ (see Table 7). The behavior of site refinements of type (b) is not yet understood, in terms of the relationships with percent and "absolute" differences in scattering factors between Si and Al atoms in Fig. 5. In fact, the unconstrained technique is consistent with "absolute" differences, whereas the constrained technique agrees with percent differences.

Site refinement of type (c) has been explored using F_o values in the $(\sin\theta)/\lambda$ range 0.45-0.80 $Å^{-1}$. The data presented in Table 10 and the relevant curves illustrated in Fig. 8 show that Al and Si contents reached extreme values and then inverted their trends.

5. Single-Crystal X-ray Diffraction Patterns and Relationships with TEM Techniques

In the past, a specialized application of the X-ray oscillation method developed by Smith and MacKenzie (1955) and the X-ray precession technique have made the best contribution to our knowledge of fine-scale intergrowths in chemically homogeneous and heterogeneous alkali feldspars. TEM techniques have made it possible to confirm and extend the findings previously obtained by means of these methods. However, single-crystal X-ray photography cannot be considered superseded. The oscillation method represents the most rapid technique to select feldspar crystals suitable for X-ray structure refinement or to identify very fine intergrowths, such as twins and exsolution textures, which are not visible with an optical microscope and require further investigation by TEM for a complete mineralogical characterization. Unlike the precession technique, the oscillation method allows precious information to be obtained even though the crystal is not perfectly oriented. Thus, considering that 15 min or so are enough to obtain good photographs, it is possible to examine many crystals in a very short time and to decide if it is convenient to continue the investigation of a given feldspar specimen or to abandon it. Heating experiments requiring step-by-step checks of the specimen to see whether an exsolved phase has been partially or completely resorbed, can be advantageously complemented by the employment of this technique.

The oscillation method allowed us to identify, in specimens of K-rich feldspar from peraluminous granites from the Mojave Desert, California, U.S.A., and Calabria, southern Italy, very small amounts of an exsolved Na-rich phase showing Pericline twin-type superstructures. This is an uncommon phenomenon in K-rich feldspar from such geological environments, and was very difficult to see by TEM. This is because the presence of such small amounts of Na-rich phase is statistically very rare in the small volumes that can be explored by TEM.

Another example showing the usefulness of the X-ray oscillation technique is that of the diffuse extra diffractions *hkl* with $(h + k)$ odd, which occur in K-rich feldspar having small $P2_1/a$ domains (see section 5.3). These weak extra diffractions are more clearly visible in oscillation photographs than in precession patterns. This may be due to the fact that the spots of type $21\bar{4}$ or $2\bar{1}\bar{4}$, which appear in oscillation photographs, are particularly strong.

Ideally, the information obtained by the oscillation method should be complemented by the data gathered by the precession technique, which has the great advantage of showing an undistorted picture of the reciprocal lattice. The precession technique has recently been relaunched for the characterization of phases and twins in alkali feldspars by McGregor and Ferguson (1989). Unfortunately, their paper contains a high number of conceptual and pagination errors; the latter were corrected in an Erratum (Can. Mineral., 1990, vol. 28, p. 184).

5.1. THE DIAGONAL ASSOCIATION

The diagonal association consists of two units, D1 and D2, that are related by a rotation of 180° about an irrational direction intermediate between the *b* and *b** axes. The composition plane is an irrational plane normal to the twin axis. With reference to the positions occupied in diffraction patterns by the spots from the A1, A2, P1, and P2 components of the ideal M-type association, the D1 diffractions from the diagonal association lie between the A1 and P1 spots, and the D2 diffractions between the A2 and P2 spots (Fig. 9). Thus, the twin axis of the diagonal association coincides with the twin axes *b** and *b* of the Albite and Pericline components, respectively, of the M-type association, and the composition plane of the diagonal association coincides with the (010) composition plane for the two units related by the Albite law.

Only in the ideal diagonal association does D1 lie midway between A1 and P1, and D2 midway between A2 and P2 (Fig. 9). The positions for D1 and D2 can rotate from the orientations for Pericline twinning to those for Albite twinning. Very commonly, the positions of D1 and D2 deviate only slightly from those for Albite twinning. Thus, the diagonal association in many occurrences can be regarded as a distorted Albite twinning.

The diagonal association was discovered by Smith and MacKenzie (1959; also 1954, 1955, and MacKenzie and Smith, 1955) by using the X-ray oscillation method, and so called because of the diagonal directions relating pairs of spots in diffraction patterns. Marfunin and Rykova (1961, Russian original dated 1960; also Marfunin, 1966, pp. 97-100, Russian original dated 1962)

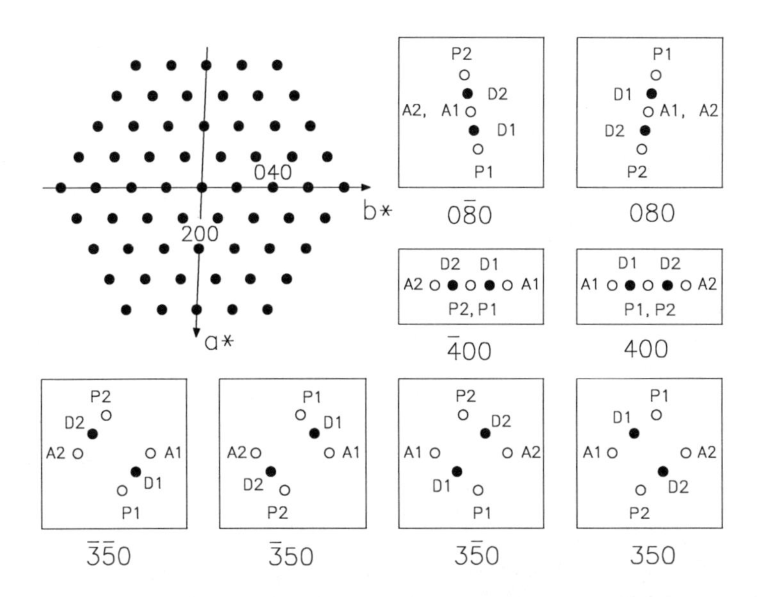

Figure 9. Schematic representation of the relationships between the diagonal association and the ideal M-type association in microcline, as shown by clusters of spots in *hk*0 precession photographs. The relevant clusters of spots are referred to a triclinic reference lattice a^*b^*. The diffractions due to the D1 and D2 units are related by the diagonal association. The diffractions due to the A1 and A2 units are related by Albite twinning, and those due to the P1 and P2 units are related by Pericline twinning, as in the ideal M-type association. The D1 diffractions lie midway between A1 and P1 spots, and the D2 diffractions midway between A2 and P2 spots. The directions joining P1 and P2 spots are slightly tilted because the axes b^*(P1) and b^*(P2) are up and down the plane of the paper, respectively.

independently described the diagonal association with the name "irrational twinning", in reference to the irrational relationships between the two components involved. From then on, numerous occurrences of the diagonal association have been reported (e.g., MacKenzie and Smith, 1962; Smith, 1962; Parsons, 1965; Wright, 1967; Brown et al., 1972; Lorimer and Champness, 1973; Willaime et al., 1973, 1976; Brown and Willaime, 1974; Champness and Lorimer, 1976; Willaime, 1977, 1981; Blasi and De Pol Blasi, 1980; Carter and Champness, 1980; Blasi et al., 1982; Brown, 1983, 1987; KL Smith and McLaren, 1983; Brown and Parsons, 1984a; Parsons and Brown, 1984; Krause et al., 1985; Kroll et al., 1991; see also Smith, 1974a, 1974b; Yund, 1983; Smith and Brown, 1988).

The diagonal association has been found only in triclinic K-rich feldspar. It occurs in the K-rich phase of mesoperthitic to perthitic intergrowths, as well as in nonperthitic, compositionally pure microcline.

However, the occurrence of the diagonal association most frequently reported is in the K-rich phase of mesoperthitic intergrowths. TEM images of such intergrowths show spectacular microtextures in which diagonally associated microcline, forming parallel-sided zig-zag lamellae, coexists with an Albite-twinned Na-rich phase in a variable form, from pinch-and-swell lenses to well-defined lozenges (e.g., Brown et al., 1972; Lorimer and Champness, 1973; Willaime et al., 1973, 1976; Brown and Willaime, 1974; Champness and Lorimer, 1976; Willaime, 1977, 1981; Carter and Champness, 1980; Brown, 1983, 1987; Brown and Parsons, 1984a; Parsons and Brown, 1984).

Diffraction patterns of such intergrowths show that pairs of spots from the diagonally associated microcline are joined by streaks. The streaks correspond to gradual changes in the lattice from the orientation of one diagonal component to the other (cf. Carter and Champness, 1980), as is shown in high-resolution TEM images by the bending of the 110 and $1\bar{1}0$ fringes on passing through the composition plane (Brown, 1983; Brown and Parsons, 1984a). According to Carter and Champness (1980), the composition plane between the two diagonal units is not coherent because the direction of the splitting of the relevant diffraction spots is not invariant. By contrast, Brown (1983) and Brown and Parsons (1984a) showed with high-resolution TEM images that the 110 and $1\bar{1}0$ fringes, though bent, are continuous through the composition plane, indicating complete coherency between the diagonally associated domains.

Each of the two spots from the microcline in such mesoperthitic intergrowths is joined by a streak to the corresponding spot for the Na-rich phase. The directions of these two streaks are invariant in the reciprocal space, and each of them is perpendicular to one of the two interfaces between the two phases. This configuration corresponds to complete coherency between the two phases (cf. Carter and Champness, 1980; also Robin, 1974) and is further documented by high-resolution TEM images that show continuity of fringes on passing from one phase to the other (Brown, 1983; Brown and Parsons, 1984a). Thus the adoption of the diagonal association can be regarded as a mode to retain coherency between exsolved phases in mesoperthitic intergrowths (cf. Carter and Champness, 1980; Brown, 1983; Brown and Parsons, 1984a).

The most comprehensive and detailed reconstruction of the development and evolution of the diagonal association in the K-rich phase of fine-scale mesoperthitic intergrowths has been reported by Brown and Parsons (1984a; see also Parsons and Brown, 1984; Brown, 1987; Smith and Brown, 1988, Fig. 19.14). This reconstruction is based on a suite of geologically related specimens from a layered syenite sequence from Klokken, South Greenland. It extends and refines the model developed by Brown and Willaime (1974) and Willaime et al. (1976) and supersedes that of Brown et al. (1972). Significant aspects of the interpretation of Carter and Champness (1980; see also a critique by Smith and Brown, 1988, p. 585, justified in Brown, 1983) were given earlier. The reconstructions of Lorimer and Champness (1973; also Champness and Lorimer, 1976) and Fleet (1984, 1985; also 1982) deal with general aspects of the development of diagonally associated mesoperthite, and do not provide a definite explanation for the formation of the diagonal association in the K-rich phase.

On the basis of the reconstruction proposed by Brown and Parsons (1984a), it is now quite certain that the diagonal association develops from original long-period Albite twins of high microcline coexisting in coherent straight lamellar mesoperthitic intergrowths with short-period Albite twins of Na-rich phase. As ordering proceeds, the obliquity of the triclinic K-rich phase increases, and the original long-period Albite twinning transforms into the diagonal association. This involves a rotation of the average exsolution boundary from near $(\bar{6}01)$ toward $(\bar{6}\bar{6}1)$, as the obliquity of the K-rich component increases to that of low microcline. As a result, the exsolution boundaries become progressively wavy to zig-zag shaped, and the microtexture assumes the characteristic aspect of diagonally associated mesoperthite. Further evolution and coarsening of such microtextures ultimately lead to the formation of optically visible braid perthites.

The rotation of the exsolution boundary can be explained by the minimization of the elastic strain energy required to maintain coherency between the coexisting phases (Brown and Willaime, 1974). The interface orientations observed are indeed in good agreement with those predicted by the coherent elastic theory (Willaime and Brown, 1974; see also Fleet, 1982, 1984, 1985, and Willaime and Brown, 1985).

A few percent An tends to slow down the alkali interdiffusion between the two coexisting phases (e.g., Mardon and Yund, 1981; Yund, 1983, 1984; Smith and Brown, 1988, p. 587), as well as the ordering and the increase of obliquity in the K-rich component (e.g., Brown, 1987). The rotation of the exsolution boundary, which seems to be promoted by these processes, is therefore blocked, and the strain energy is reduced by the development of periodic dislocations (Brown and Parsons, 1984b, 1988; Parsons and Brown, 1984; Brown, 1987; Smith and Brown, 1988, p. 591). In addition, where the bulk Or content is higher than 67 mol%, the amount of the Na-rich phase decreases to such extent as to form isolated lenses rather than continuous lamellae in the dominant K-rich phase. In such conditions, the rotation of the exsolution interface is not observed, and the K-rich phase develops a tweed texture on cooling rather than a diagonal association (Brown and Parsons, 1984a, 1984b, 1988; Parsons and Brown, 1984; Brown, 1987; Smith and Brown, 1988, pp. 587-590). Thus large amounts of Ca-poor Na-rich phase are considered to exert the necessary control to promote the development of the diagonal association in the K-rich component of mesoperthitic intergrowths (Brown and Parsons, 1984b; Parsons and Brown, 1984).

It would therefore seem surprising to find the development of the diagonal association in compositionally pure microcline, which is not in direct contact with an exsolved Na-rich phase. KL Smith and McLaren (1983) discovered such a diagonal association in tiled microcline crystals from a nepheline syenite from Ilimaussaq, West Greenland. The specimens are formed by dominant, compositionally pure, low microcline in a tile pattern coexisting with small areas of almost pure Na-rich feldspar with a less sharp tile microtexture. The areas of different compositions are clearly observable by optical microscopy. The complex tiled microtexture could be the result of metasomatic processes. TEM investigations performed by KL Smith and McLaren (1983) on the Ilimaussaq specimens showed that the microcline consists of a mosaic of domains, most of which are slightly misoriented, without any crystallographic relationship, whereas others are related by the Albite law, the Pericline law, and the diagonal association. The pairs of spots in the electron diffraction pattern of the diagonally associated microcline are very sharp and apparently free of streaks joining them, and show perfectly regular geometric relationships consistent with a 180° rotation about an axis midway between b and b^*. No other spots are visible in the diffraction pattern except those from diagonally associated microcline. The origin of this example of the diagonal association was discussed by KL Smith and McLaren (1983), with reference to the formation of the M-type association in microcline and the subsequent degeneration of the Pericline-twin units. The development of the diagonal association in some chess-board microtextures could represent intermediate stages during this process of degradation (McLaren, 1978; Fitz Gerald and McLaren, 1982). However, the diagonal association observed in the tiled microcline from Ilimaussaq is quite distinct, and could be due to metasomatism of microcline that crystallized with triclinic symmetry (McLaren, 1984).

Further occurrences in which the development of diagonally associated microcline cannot be attributed to the guiding action of the exsolved Na-rich phase were described by Wright (1967), Blasi and De Pol Blasi (1980), Blasi et al. (1982), Krause et al. (1985), and Kroll et al. (1991).

The examples of the diagonal association investigated by Wright (1967) were developed in the K-rich phase of microperthitic intergrowths from pegmatites in the contact aureole of the Eldora stock, Colorado, U.S.A. Optically, these microperthites show the tartan pattern, but only two optical orientations were detected, which are related by the Albite law. X-ray oscillation photographs showed that the K-rich phase is in a diagonal association close to an Albite-twin relationship, and that the Na-rich phase is Albite twinned. In several specimens, a monoclinic K-rich phase also was found to be present. Smith (1974b, p. 335) interpreted the absence of the M-type association in the diffraction patterns of Wright's (1967) specimens to be due to a probable recrystallization of Pericline-twinned units into the Albite orientation without obliteration of the tartan texture. The reason for such recrystallization is not clear. However, Smith (1962; also 1974b, p. 336) observed that the diagonal association might occur under special conditions whereby reduction of strain between the components is important. In fact, the maximum angular misfit between the two components of the diagonal association is less than that between either the components of Albite twinning or those of Pericline twinning (cf. Smith, 1962; 1974b, p. 336; also Smith and MacKenzie, 1959, Fig. 3).

The examples of the diagonal association studied by Blasi and De Pol Blasi (1980) and Blasi et al. (1982) occur in the K-rich phase of microperthites from migmatites in the Argentera Massif, Maritime Alps. These feldspar specimens are optically untwinned but show a strong shadowy extinction. X-ray oscillation and precession photographs show that the K-rich phase is in the diagonal association, whereas the subordinate Na-rich phase may commonly be in the M-type association. An increasing number of such occurrences has been found recently. Fig. 10 is a precession photograph of one of the specimens investigated, which shows pairs of spots D1 and D2 slightly unbalanced in their intensity. In other examples, pairs of spots of the diagonal association tend to be elongated in the directions joining the diffractions A1 and P1, as well as A2 and P2, of the ideal M-type association.

A similar situation was described by Marfunin and Rykova (1961) and Marfunin (1966, pp. 97-100, see also Tables 4 and 6, pp. 84, 88, and 92 for pertinent data and description of specimens) in their examples of diagonally associated microcline. Two specimens, Kapustino and Kirovograd, are from trachytoid granites, two others, Voinovka and Adabash, from charnockites, and one, Karelia, from a pegmatite. The first four specimens show optical properties characteristic of submicroscopically twinned microcline. Their bulk compositions (Kapustino Or_{63}, Kirovograd Or_{65}, Voinovka Or_{64}, Adabash Or_{72} mol%) straddle the value of Or_{67} mol% suggested by Brown (1987) as the limit below which the Na-rich phase is expected to control the development of the diagonal association in mesoperthites. In contrast, the Karelia specimen is a distinctly cross-hatched microcline with bulk composition Or_{78} mol%. In the above specimens, the Na-rich phase shows the Albite superstructure (Kapustino, Kirovograd, Voinovka), both the Albite and Pericline superstructure (Adabash), and simple Albite twinning (Karelia). According to Marfunin (1966, p. 99), the development of the diagonal association (his "irrational twinning") may be due to an increase in the volume of the feldspar material with intermediate orientations between those of components A1 and P1, as well as A2 and P2 of the M-type association. From his findings, Marfunin (1966, p. 99) deduced that the observed occurrence of only two optical orientations in distinctly cross-hatched microcline could represent either Albite twins or diagonally associated units.

The specimen investigated by Kroll et al. (1991; cf. also Krause et al., 1985) is from contact-metamorphosed quartzites from the Ballachulish Igneous Complex, Scotland. A TEM image of the specimen shows "irregular microcline" exhibiting tweed areas, ill-defined twin boundaries and subgrain boundaries. An electron diffraction pattern shows that grain and subgrain of microcline are related by the diagonal association. The relevant D1 and D2 spots are sharp and clearly separated, and no other diffractions are visible.

In conclusion, the types of occurrences described suggest that the diagonal association should be considered as a case of phenomenological convergence in alkali feldspars. The development of the diagonal association in the K-rich phase of mesoperthitic intergrowths seems clearly to be controlled by the large amount of exsolved Na-rich phase. The control on the diagonal associa-

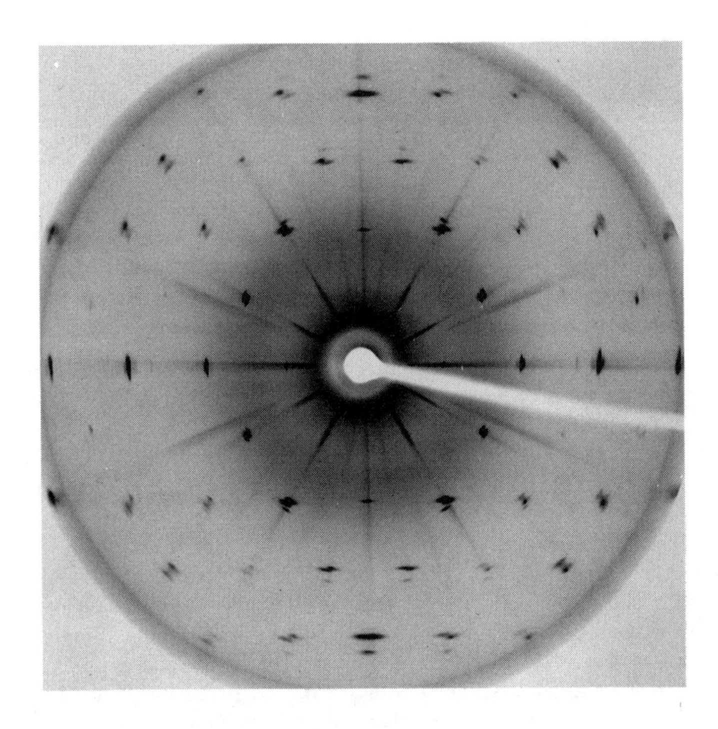

Figure 10. X-ray precession photograph (Ni-filtered Cu-radiation) of a specimen of perthitic K-rich feldspar in the diagonal association from migmatite of the Argentera Massif, Maritime Alps (specimen #3870). The vertical axes, a^*(D1) and a^*(D2), are in the plane of the film, whereas the horizontal axes, b^*(D1) and b^*(D2), are tilted up and down, respectively (cf. Fig. 9). As shown especially by the pairs of spots 240, 260, and 350, the two D1 and D2 units are slightly unbalanced. The coexisting Na-rich phase is in the M-type association, even though in this photograph it shows apparently single spots.

tion cannot be the same in nonperthitic, compositionally pure microcline and in the K-rich component of perthitic intergrowths. In many of these examples, the occurrence of the diagonal association could be related to the development of the cross-hatched texture of microcline.

5.2. THE ANTIDIAGONAL ASSOCIATION

The antidiagonal association was observed by X-ray oscillation and precession photography in the triclinic K-rich phase of K-rich perthites from migmatites in the Argentera Massif, Maritime Alps. It is so called because the directions relating pairs of spots of type *hk*0 in the a^*b^* plane are perpendicular to those observed in the diagonal association. The precession photograph in Fig. 11 shows the typical pattern of the antidiagonal association in a perthitic K-rich feldspar from the Argentera Massif. The pairs of related spots, e.g. the pair of 350 diffractions, are unbalanced in intensity and have the same Bragg angle.

The antidiagonal association is interpreted as a strongly unbalanced M-type association in which the components A1 and P1 (or A2 and P2) are virtually absent, whereas the components A2 and P2 (or A1 and P1) give rise to a distinct 2-spot pattern. With the orientation given to the pattern in Fig. 11, the two components forming the antidiagonal association are A1 and P1 (cf.

Fig. 9). A 180° rotation of the pattern in Fig. 11 about the vertical axis produces a new orientation in which the two spots related by the antidiagonal association become A2 and P2.

When superimposing the photograph in Fig. 11 on a precession pattern of a K-rich feldspar showing the M-type association, there is a perfect coincidence of the two spots of the antidiagonal association with the spots A1 and P1 of the M-type association. This confirms the soundness of the interpretation given above for the antidiagonal association. As in the case of the diagonal association occurring in K-rich perthite, the antidiagonal association seems to be related to the development of the cross-hatched texture of microcline.

5.3. $P2_1/a$ STRUCTURAL ARRANGEMENT

Diffuse X-ray diffractions elongated in the direction of the *b* axis and centered on positions *hkl* with $(h + k)$ odd have been documented by Laves (1950, Fig. A in Plate 2) in a monoclinic adularia from St. Gotthard by means of oscillation photography. The same specimen showed the main diffractions with $(h + k)$ even to be crossed by diffuse streaks in the direction of spots for the M-type association of microcline, the streaks for Albite twinning being clearly more intense than those for Pericline twinning.

Other occurrences of the extra diffractions *hkl* with $(h + k)$ odd have been described in crystals of the Spencer B adularia from St. Gotthard (Smith and MacKenzie, 1959; Colville and Ribbe, 1968; also Smith and Brown, 1988, Fig. 6.4b) and in the K-rich phase of the Spencer U microperthite from the mica-bearing pegmatite of Kodarma, Bihar, India (Smith and MacKenzie, 1959; Smith, 1974a, Fig. 6-13). In these specimens, the main spots with $(h + k)$ even obey monoclinic symmetry and are associated with streaks involving incipient M-type association, with the Albite units much more developed than the Pericline units. For the Spencer U specimen, this is an unfamiliar aspect because this feldspar is well-known as a distinct example of intermediate microcline from the structure determination by Bailey and Taylor (1955; also Bailey, 1969). The structural state of the K-rich phase of Spencer U microperthite has been found to be rather variable (Smith and MacKenzie, 1959; Smith, 1974a, pp. 67 and 197).

The diffuse diffractions *hkl* with $(h + k)$ odd violate *C*-face centering and are indicative of a primitive $c \sim 7$ Å structure. They were interpreted by Laves and Goldsmith (1961, Figs. 4 and 6) as the result of the formation of $P2_1/a$ domains consisting of regularly repeated out-of-step boundaries of twinned microcline. Laves and Goldsmith (1961) suggested that in an extreme case the domains may be essentially of unit-cell size, but Megaw (1961; also 1962) argued against the concept of domains on such a fine scale. Nevertheless, according to Smith (1974b, p. 335 and caption to Fig. 18-18), the diffuseness of the extra diffractions *hkl* with $(h + k)$ odd indicates that the $P2_1/a$ domains should be very small. Colville and Ribbe (1968) were able to observe such diffractions in the Spencer B adularia only by means of a *b*-axis oscillation photograph. However, they ignored them and refined the crystal structure of their specimen in the space group $C2/m$.

An alternative interpretation of the extra diffractions *hkl* with $(h + k)$ odd has been presented by Ribbe (1983a) on the basis of the model proposed by McConnell (1965, 1971) to explain the modulated patterns observed in TEM images of monoclinic adularia. The model devised by McConnell (1965, 1971, Fig. 2) involves the formation of two orthogonal systems of waves with transverse distortions in the lattice of monoclinic adularia. These systems of waves account for the occurrence in this material of weak streaks that cross the main diffractions with $(h + k)$ even in the direction of spots for the M-type association of microcline. The model of distortion waves of McConnell (1971, Fig. 2) implies the emergence of 2_1 screw axes and *a* glide planes parallel

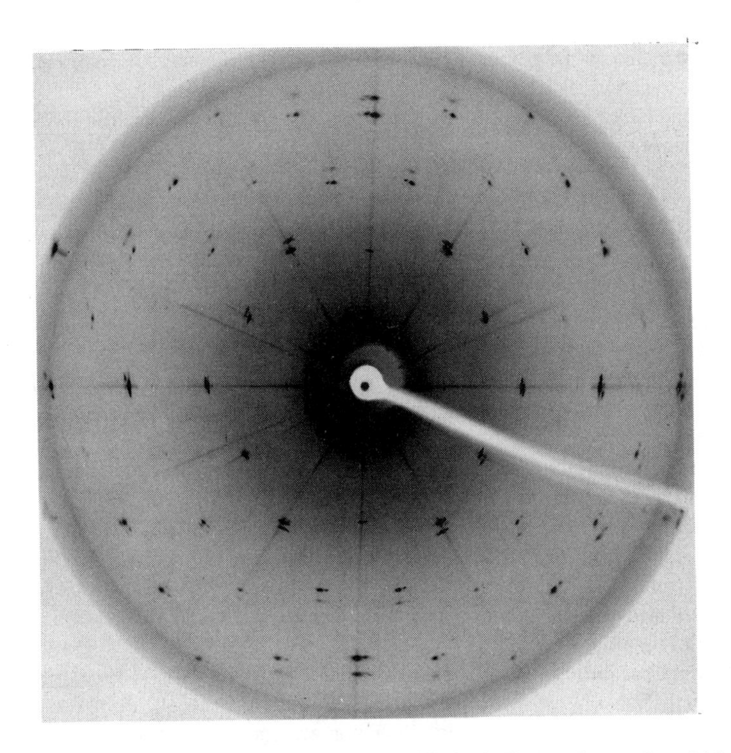

Figure 11. X-ray precession photograph (Ni-filtered Cu-radiation) of a specimen of perthitic K-rich feldspar in the antidiagonal association from migmatite of the Argentera Massif, Maritime Alps (specimen #3866). The vertical axes are a^*(A1) and a^*(P1), and the horizontal axes are b^*(P1) and b^*(A1), the b^*(P1) axis being tilted above the plane of the film (cf. Fig. 9). As shown, especially by the 350 diffractions, the directions relating pairs of spots of type *hk*0 are perpendicular to those observed in the diagonal association (cf. Fig. 9). The Na-rich phase gives rise to weaker spots which simulate, along with the spots of the K-rich phase, a distorted M-type association. This is shown especially by the cluster of diffractions 260.

to the original axes and planes of symmetry, respectively, in the $C2/m$ structure. According to Ribbe (1983a), this configuration may give rise to weak diffuse streaks centered on positions *hkl* with $(h + k)$ odd, which are consistent with a feldspar with space group $P2_1/a$.

The interpretation given by Ribbe (1983a, Fig. 4) seems to be consistent with the available experimental observations. The diffractions violating *C*-face centering indeed coexist with those related to incipient M-type association in the same feldspar specimen. In addition, the extra diffractions are elongated in the direction of the *b* axis as if they were due to the formation of a rudimentary Albite twin-type superstructure. This is consistent with the pre-eminent development of streaks in the direction of spots for the Albite components of the incipient M-type association commonly observed in specimens of $P2_1/a$ feldspar.

Laves (1950) documented the occurrence of the extra diffractions disobeying *C*-face centering only in monoclinic adularia. He reported that these extra maxima have not been observed in untwinned microcline nor in plagioclase; therefore they seem to be restricted to feldspar material with the state of "monoclinic" orthoclase. However, not all examples of adularias show such a phenomenon (e.g., Zilczer, 1981, p. 20). Smith (1974b, p. 455) reported that such maxima were not found in specimens out of granites or plutonic rocks. The Spencer U specimen, whose monoclinic K-rich phase shows such extra diffractions, is from a pegmatite.

Recently, we have found that such extra diffractions systematically occur in the K-rich phase of perthites from peraluminous granites from the Mojave Desert, California, U.S.A., and Calabria, southern Italy. The extra maxima are very weak but clearly distinct. They occur in either the monoclinic or the untwinned triclinic K-rich phase, both of them usually showing weak streaks indicative of incipient M-type association, with the Albite components more developed than the Pericline components. The distinctive habit of adularia has not been observed in these specimens.

Fig. 12 shows a portion of a *b*-axis oscillation photograph that exhibits elongated diffuse $21\bar{4}$ and $\bar{2}1\bar{4}$ diffractions from an untwinned triclinic K-rich feldspar from the Mojave Desert. The lattice parameters of this specimen [a = 8.5666(8) Å, b = 12.9620(11) Å, c = 7.2105(9) Å, α = 90.286(9)°, β = 116.034(14)°, γ = 89.069(8)°], determined with a KUMA Diffraction KM-4 computer-controlled k-axis diffractometer using graphite-monochromated Mo$K\alpha$ radiation, are consistent with the structural state of an intermediate microcline. Preliminary data obtained with the same instrumentation indicate that the extra diffractions are detectable, but centering is rendered difficult by their extreme diffuseness.

The occurrence of the extra diffractions in the triclinic K-rich phase of perthites from the peraluminous granite from California seems to be consistent with either the interpretation of Laves and Goldsmith (1961) or that of Ribbe (1983a). In fact, small domains of $P2_1/a$ material with a structural arrangement as described by either interpretation may occur in a dominant untwinned triclinic matrix.

Many specimens exhibiting the weak streaks indicative of incipient M-type association do not show the extra diffraction *hkl* with $(h + k)$ odd. With reference to Ribbe's (1983a, Fig. 4) interpretation, this could be explained by either non-emergence or loss of 2_1 axes and a glides due to imperfect formation of the ideal modulated structure or to deviations therefrom because of later transformations. In the interpretation of Laves and Goldsmith (1961), the out-of-step boundaries of twinned microcline (see their Fig. 6) may act as sites from which recrystallization of coarser microcline domains can take place, with consequent disappearance of the extra diffractions. According to Laves and Goldsmith (1961), this does not imply that the formation of microcline must take place from a parent $P2_1/a$ structural arrangement.

With reference to the Spencer B specimen, Smith and Brown (1988, p. 44) observed that out-of-step domains with local symmetry $P2_1/a$ undoubtedly are the result of metastable processes involved in the low-temperature history of adularias; the stronger the extra diffractions *hkl* for which $(h + k)$ odd, the more a K-rich feldspar should deviate from the stable state (cf. Smith 1974a, p. 422). The preservation of the $P2_1/a$ structural arrangement in monoclinic and triclinic specimens of K-rich feldspar from the Mojave Desert and Calabria may be due to special conditions related to the peraluminous environment.

6. Unanswered questions and suggestions for future work

The relationship of the undulatory extinction to sub-optical textures in K-rich feldspar is not clear. The monoclinic adularia and orthoclase, which usually display undulatory extinction, typically show the tweed texture in TEM images. However, the tweed texture and the diagonal association have been found by TEM investigations and by X-ray photography, respectively, in specimens of low microcline showing undulatory extinction (cf. section 2). Such unusual occurrences indicate that the crystallographic and genetic significance of undulatory extinction

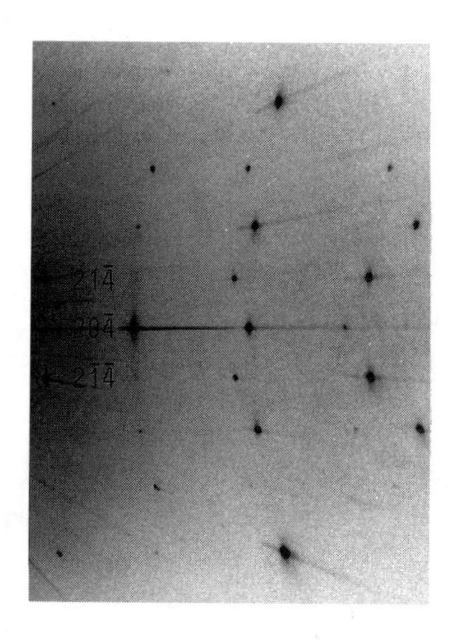

Figure 12. Portion of *b*-axis X-ray oscillation photograph (Ni-filtered Cu-radiation) of a K-rich feldspar from peraluminous granite from the Mojave Desert, California, U.S.A. (specimen #SW8). The photograph shows $21\bar{4}$ and $2\bar{1}\bar{4}$ diffuse extra diffractions, which are elongated in the direction of the *b*-axis and violate *C*-face centering. Weak streaks indicative of incipient Albite twinning cross the main spots with $(h + k)$ even.

deserves much more investigation in K-rich feldspar from underformed and deformed rocks, especially by TEM and single-crystal X-ray diffraction techniques.

The determination of composition in strained alkali feldspars poses severe problems. At present, it is not clear whether the compositions determined by the method of Yund and Tullis (1983a) are more reliable than those determined by cell volume *V* (cf. section 3.2.3 and Table 2). Yund and Tullis (1983b, p. 159) considered that further improvements of their method with regard to the values of the cell parameters and elastic constants would provide refinements of the results. We believe that this is an essential condition for abandoning the simple method based on cell volume *V*.

In principle, K,Na site refinements from X-ray data could be of particular value to estimate the composition of strained alkali feldspars. However, serious problems have been encountered in refining K,Na site in a "simple" alkali feldspar structure, as is the case of the Pikes Peak microcline 7813C (cf. section 3.3). These difficulties justify the poor results obtained by Keefer and GE Brown (1978) for the Rabb Canyon sanidine, which is undoubtedly a more "difficult" structure. In spite of this, the procedures used in the refinement models #10 and #11 for microcline 7813C (cf. section 3.3) are encouraging. They are capable of improvement and should be checked in estimating the composition of strained alkali feldspars.

The determination of structural state from cell parameters by means of the Δ method is a standard routine in alkali feldspars with normal cell dimensions, but problems seem to arise with strained specimens (cf. section 4.1). Until new structure refinements of strained feldspars are performed, to allow a better evaluation of data, the situation will remain unclear.

While waiting for further refinements of relevant procedures (cf. section 4.2), we propose that the determination of *T*-site occupancies by means of *T*-O distances be done in K-rich feldspar by means of equations (18) and (19), or by equation (17), and in Na-rich feldspars by equations based on the same principles.

The small differences in scattering factors of Si and Al atoms remain the most crucial problem in refinements of *T*-site occupancies using X-ray data. However, the refinement technique leading to the results presented in Fig. 7 and Table 9 for microcline 7813C (cf. section 4.3) seems to be promising, and should be applied to feldspar structures with an intermediate structural state. In this respect, the experimental conditions of data collection, including scan width and scan speed, could be better investigated, in spite of the fact that the standard procedure of structure refinements of microcline 7813C gave very low R and R_w values (cf. Table 3).

Twin-related structures such as the diagonal association and the antidiagonal association (see sections 5.1 and 5.2) are a matter for future consideration. Whereas the development of the diagonal association in the K-rich phase of mesoperthites is clearly controlled by the exsolved Na-rich phase, the nature of the diagonal association and the antidiagonal association occurring in the low microcline of K-rich perthites requires further investigation. In particular, whether the diagonal association as well as the antidiagonal association in these occurrences could be related to an earlier or to a later stage in the formation of the M-type association should be clarified.

The presence of out-of-step domains with local symmetry $P2_1/a$ in either the monoclinic or the untwinned triclinic K-rich feldspars occurring in peraluminous granites from the Mojave Desert, California, U.S.A., and Calabria, southern Italy, is a challenging problem, considering that these specimens do not show the habit of adularia (cf. section 5.3). Further work is needed on these specimens to clarify subtle crystallographic aspects of $P2_1/a$ domains and their genetic significance.

ACKNOWLEDGEMENTS

We thank G. Artioli, W.L. Brown, R.F. Martin, P.H. Ribbe, J.V. Smith and P.F. Zanazzi for fruitful discussions. We also thank I. Parsons for his careful editorial work and helpful suggestions. The manuscript benefitted from readings by R.F. Martin, P.F. Zanazzi and referees P.H. Ribbe and J.V. Smith. We acknowledge financial support by MURST, through grants denoted as 40% and 60%, CNR, through the Centro di Studi per la Geodinamica Alpina e Quaternaria, CARIPLO, through the Università degli Studi di Milano, and NATO, through the Advanced Study Institute on Feldspars and their Reactions in Edinburgh.

REFERENCES

Alberti A, Gottardi G (1988) The determination of the Al-content in the tetrahedra of framework silicates. Z Kristallogr 184: 49-61

Armbruster TH, Bürgi HB, Kunz M, Gnos E, Brönnimann ST, Lienert CH (1990) Variation of displacement parameters in structure refinements of low albite. Am Mineral 75: 135-140

Bailey SW (1969) Refinement of an intermediate microcline structure. Am Mineral 54: 1540-1545

Bailey SW, Taylor WH (1955) The structure of a triclinic potassium felspar. Acta Crystallogr 8: 621-632

Bambauer HU (1969) Feldspat-Familie $R_x{}^{I}R_{1-x}{}^{II}[Al_{2-x}Si_{2+x}O_8]$, $0 \leq x \leq 1$. In: Tröger WE Optische Bestimmung der gesteinsbildenden Minerale (Teil 2, Textband, 2. Auflage). E Schweizerbart'sche Verlagsbuchhandlung, Stuttgart, pp 645-762

Bambauer HU, Krause C, Kroll H (1989) TEM-investigation of the sanidine/microcline transition across metamorphic zones: the K-feldspar varieties. Eur J Mineral 1: 47-58

Blasi A (1972a) Identificazione delle varianti strutturali del K-feldspato in sezioni sottili di rocce. Nota I and Nota II. Rend Acc Naz Lincei, serie 8, 52: 773-782 and 934-945

Blasi A (1972b) "Iso-microclino" ed altre varianti strutturali del K-feldspato coesistenti in uno stesso cristallo nei graniti del Massiccio dell'Argentera (Alpi Marittime). Rend Soc It Mineral Petrol 28: 375-411

Blasi A (1977) Calculation of *T*-site occupancies in alkali feldspar from refined lattice constants. Mineral Mag 41: 525-526

Blasi A (1978) Structural explanation for $\Delta(\alpha^*\gamma^*)$ from α^* vs γ^* in alkali feldspar. Tschermaks Mineral Petrogr Mitt 25: 47-56

Blasi A (1979) Mineralogical applications of the lattice constant variance-covariance matrices. Tschermaks Mineral Petrogr Mitt 26: 139-148

Blasi A (1980) Different behavior of $\Delta(bc)$ and $\Delta(b^*c^*)$ in alkali feldspar. Neues Jahrb Mineral Abh 138: 109-121

Blasi A (1982) Appraisal of the Ferguson method and the linear model using $\Delta(bc)$, $\Delta(b^*c^*)$, $\Delta(\alpha\gamma)$, $\Delta(\alpha^*\gamma^*)$ to estimate tetrahedral Al-contents in alkali feldspar. Mineral Mag 46: 465-468

Blasi A (1984) The variation of 2θ angles in powder diffraction patterns of one- and two-step K-rich feldspars. Bull Minéral 107: 437-445

Blasi A, Blasi De Pol C (1977) Role and convenience of lattice elements for deriving Si,Al distribution in alkali feldspar. Rend Soc It Mineral Petrol 33: 497-509

Blasi A, Brajkovic A, De Pol Blasi C (1982) Highly ordered monoclinic K-feldspars from Haut Boréon anatexites, Argentera Massif, Maritime Alps. Tschermaks Mineral Petrogr Mitt 29: 241-253

Blasi A, Brajkovic A, De Pol Blasi C (1984a) Dry-heating conversion of low microcline to high sanidine via a one-step disordering process. Bull Minéral 107: 423-435

Blasi A, Brajkovic A, De Pol Blasi C, Foord EE, Martin RF, Zanazzi PF (1984b) Structure refinement and genetic aspects of a microcline overgrowth on amazonite from Pikes Peak batholith, Colorado, USA. Bull Minéral 107: 411-422

Blasi A, De Pol Blasi C (1980) Highly ordered triclinic K-feldspars from Mt. Pélago anatexites (Argentera Massif, Maritime Alps). Bull Minéral 103: 209-216

Blasi A, De Pol Blasi C, Zanazzi PF (1981) Structural study of a complex microperthite from anatexites at Mt. Caval, Argentera Massif, Maritime Alps. Neues Jahrb Mineral Abh 142: 71-90

Blasi A, De Pol Blasi C, Zanazzi PF (1987a) A re-examination of the Pellotsalo microcline: mineralogical implications and genetic considerations. Can Mineral 25: 527-537

Blasi A, De Pol Blasi C, Zanazzi PF (1987b) Structure refinement of a high sanidine produced by dry annealing a Bedford low microcline for 200 days at 1050 °C. Rend Soc It Mineral Petrol 42: 325 (abstr)

Borg IY, Smith DK (1969) Calculated X-ray powder patterns for silicate minerals. Geol Soc Am Memoir 122: 1-896

Brown BE, Bailey SW (1964) The structure of maximum microcline. Acta Crystallogr 17: 1391-1400

Brown GE, Hamilton WC, Prewitt CT, Sueno S (1974) Neutron diffraction study of Al/Si ordering in sanidine: a comparison with X-ray diffraction data. In: MacKenzie WS, Zussman J (eds) The feldspars. Manchester University Press, Manchester, pp 68-80

Brown WL (1983) Étude par microscopie électronique en haute résolution de la texture cohérente des cryptoperthites en association diagonale et origine de certaines macles du microcline de type-M. C R Acad Sci Paris 296 Série II: 143-148

Brown WL (1987) Interprétation des microstructures d'exsolution des feldspaths ternaires. Utilisation de la microscopie électronique par transmission combinée avec des données de terrain. In: Willaime C (ed) Initiation à la microscopie électronique par transmission. Minéralogie, Sciences des matériaux. Soc Franç Minéral Cristallogr, Paris, pp 197-212

Brown WL, Becker SM, Parsons I (1983) Cryptoperthites and cooling rate in a layered syenite pluton: a chemical and TEM study. Contrib Mineral Petrol 82: 13-25

Brown WL, Parsons I (1984a) Exsolution and coarsening mechanisms and kinetics in an ordered cryptoperthite series. Contrib Mineral Petrol 86: 3-18

Brown WL, Parsons I (1984b) The nature of potassium feldspar, exsolution microtextures and development of dislocations as a function of composition in perthitic alkali feldspars. Contrib Mineral Petrol 86: 335-341

Brown WL, Parsons I (1988) Zoned ternary feldspars in the Klokken intrusion: exsolution microtextures and mechanisms. Contrib Mineral Petrol 98: 444-454

Brown WL, Willaime C (1974) An explanation of exsolution orientations and residual strain in cryptoperthites. In: MacKenzie WS, Zussman J (eds) The Feldspars. Manchester University Press, Manchester, pp 440-459

Brown WL, Willaime C, Guillemin C (1972) Exsolution selon l'association diagonale dans une cryptoperthite: étude par microscopie électronique et diffraction des rayons X. Bull Soc Franç Minéral Cristallogr 95: 429-436

Burnham CW (1962) Lattice constant refinement. Carnegie Inst. Wash Year Book 61: 132-135

Carter S, Champness PE (1980) The diagonal association in alkali feldspars. In: Brederoo P, Boom G (eds) Electron microscopy 1980. Vol 1. Physics. Seventh European Congress on Electron Microscopy Foundation, Leiden, pp 454-455

Černý P, Chapman R (1984) Paragenesis, chemistry and structural state of adularia from granitic pegmatites. Bull Minéral 107: 369-384

Černý P, Chapman R (1986) Adularia from hydrothermal vein deposits: extremes in structural state. Can Mineral 24: 717-728

Champness PE, Lorimer GW (1976) Exsolution in silicates. In: Wenk H-R (ed) Electron microscopy in mineralogy. Springer-Verlag, Berlin, pp 174-204

Colville AA, Ribbe PH (1968) The crystal structure of an adularia and a refinement of the structure of orthoclase. Am Mineral 53: 25-37

Dal Negro A, De Pieri R, Quareni S, Taylor WH (1978) The crystal structures of nine K feldspars from the Adamello Massif (Northern Italy). Acta Crystallogr B34: 2699-2707

Day HW, Brown VM (1980) Evolution of perthite composition and microstructure during progressive metamorphism of hypersolvus granite, Rhode Island, USA. Contrib Mineral Petrol 72: 353-365

De Pieri R (1979) Cell dimensions, optic axial angle and structural state in triclinic K-feldspars of the Adamello Massif (Northern Italy). Mem Sci Geol 32: 1-17

Eggleton RA, Buseck PR (1980) The orthoclase-microcline inversion: a high-resolution transmission electron microscope study and strain analysis. Contrib Mineral Petrol 74: 123-133

Fenn PM, Brown GE (1977) Crystal structure of a synthetic, compositionally intermediate, hypersolvus alkali feldspar: evidence for Na,K site ordering. Z Kristallogr 145: 124-145

Ferguson RB, Ball NA, Černý P (1991) Structure refinement of an adularian end-member high sanidine from the Buck Claim pegmatite, Bernic Lake, Manitoba. Can Mineral 29: 543-552

Fischer K, Zehme H (1967) Röntgenographische Untersuchung der Si-Al-Verteilung in einem Mikroklin durch Verfeinerung des atomaren Streuvermögens. Schweiz Mineral Petrogr Mitt 47: 163-167

Fitz Gerald JD, McLaren AC (1982) The microstructures of microcline from some granitic rocks and pegmatites. Contrib Mineral Petrol 80: 219-229

Fleet ME (1982) Orientation of phase and domain boundaries in crystalline solids. Am Mineral 67: 926-936
Fleet ME (1984) Orientation of feldspar intergrowths: application of lattice misfit theory to cryptoperthites and e-plagioclase. Bull Minéral 107: 509-519
Fleet ME (1985) Orientation of phase and domain boundaries in crystalline solids: reply. Am Mineral 70: 130-133
Folk RL (1955) Note on the significance of "turbid" feldspars. Am Mineral 40: 356-357
Harlow GE, Brown GE Jr (1980) Low albite: an X-ray and neutron diffraction study. Am Mineral 65: 986-995
Hawthorne FC (1983) Quantitative characterization of site-occupancies in minerals. Am Mineral 68: 287-306
Hovis GL (1986) Behavior of alkali feldspars: Crystallographic properties and characterization of composition and Al-Si distribution. Am Mineral 71: 869-890
Hovis GL (1988) Enthalpies and volumes related to K-Na mixing and Al-Si order/disorder in alkali feldspars. J Petrol 29: 731-763
Hovis GL (1989) Effect of Al-Si distribution on the powder-diffraction maxima of alkali feldspars and an easy method to determine *T*1 and *T*2 site occupancies. Can Mineral 27: 107-118
International tables for X-ray crystallography (1974) Vol IV. The Kynoch Press, Birmingham
Keefer KD, Brown GE (1978) Crystal structures and compositions of sanidine and high albite in cryptoperthitic intergrowth. Am Mineral 63: 1264-1273
Krause C, Bambauer HU, Kroll H, McLaren AC (1985) TEM-investigation of metamorphic K-feldspars: the sanidine-microcline transition. 3rd Meeting of the European Union of Geosciences, Strasbourg 1985. Terra Cognita 5: 222 (abstr)
Kroll H (1973) Estimation of the Al,Si distribution of feldspars from the lattice translations Tr[110] and Tr[$1\bar{1}0$]. I. Alkali feldspars. Contrib Mineral Petrol 39: 141-156
Kroll H (1980) Estimation of the Al,Si distribution of alkali feldspars from lattice translations tr[110] and tr[$1\bar{1}0$]. Revised diagrams. Neues Jahrb Mineral Mh 1980: 31-36
Kroll H, Krause C, Voll G (1991) Disordering, re-ordering and unmixing in alkali feldspars from contact-metamorphosed quartzites. In: Voll G, Töpel J, Pattison DRM, Seifert F (eds) Equilibrium and kinetics in contact metamorphism. Springer-Verlag, Berlin, pp 267-296
Kroll H, Ribbe PH (1983) Lattice parameters, composition and Al,Si order in alkali feldspars. In: Ribbe PH (ed) Feldspar mineralogy (2nd edition). Mineral Soc Am, Rev Mineral 2: 57-99
Kroll H, Ribbe PH (1987) Determining (Al,Si) distribution and strain in alkali feldspars using lattice parameters and diffraction-peak positions: A review. Am Mineral 72: 491-506
Kroll H, Schmiemann I, von Cölln G (1986) Feldspar solid solutions. Am Mineral 71: 1-16
Laves F (1950) The lattice and twinning of microcline and other potash feldspars. J Geol 58: 548-571
Laves F, Goldsmith JR (1961) Polymorphism, order, disorder, diffusion and confusion in the feldspars. Cursillos y Conferencias del Instituto "Lucas Mallada" 8: 71-80
Lorimer GW, Champness PE (1973) The origin of the phase distribution in two perthitic alkali feldspars. Philos Mag 28: 1391-1403
Luth WC (1974) Analysis of experimental data on alkali feldspars: unit cell parameters and solvi. In: MacKenzie WS, Zussman J (eds) The Feldspars. Manchester University Press, Manchester, pp 249-296
MacKenzie WS, Smith JV (1955) The alkali feldspars: I. Orthoclase-microperthites. Am Mineral 40: 707-732
MacKenzie WS, Smith JV (1962) Single crystal X-ray studies of crypto- and micro-perthites.

Norsk Geol Tidsskr 42, No 2: 72-103

Mardon D, Yund RA (1981) The effect of anorthite on exsolution rates and the coherent solvus for sanidine-high albite. EOS, Transactions, Am Geophys Union 62: 411 (abstr)

Marfunin AS (1966) The feldspars: phase relations, optical properties, and geological distribution. (Translation from the Russian edition, 1962). Israel Prog Sci Translations, Jerusalem

Marfunin AS, Rykova SV (1961) On the irrational twinning of potash feldspars. (Translation from the Russian edition, 1960). Doklady Acad Sci USSR 134: 1007-1009

Martin RF (1988) The K-feldspar mineralogy of granites and rhyolites: a generalized case of pseudomorphism of the magmatic phase. Rend Soc It Mineral Petrol 43: 343-354

Martin RF, Bonin B (1976) Water and magma genesis: the association hypersolvus granite-subsolvus granite. Can Mineral 14: 228-237

Martin RF, Lalonde A (1979) Turbidity in K-feldspars: causes and implications. Geol Soc Am, Abstracts with programs 11: 472-473

McConnell JDC (1965) Electron optical study of effects associated with partial inversion in a silicate phase. Philos Mag 11: 1289-1301

McConnell JDC (1971) Electron-optical study of phase transformations. Mineral Mag 38: 1-20

McCormick TC (1989) Partial homogenization of cryptoperthites in an ignimbrite cooling unit. Contrib Mineral Petrol 101: 104-111

McGregor CR, Ferguson RB (1989) Characterization of phases and twins in alkali feldspars by the X-ray precession technique. Can Mineral 27: 457-482. Erratum: Can Mineral, 1990, 28:184

McLaren AC (1974) Transmission electron microscopy of the feldspars. In: MacKenzie WS, Zussman J (eds) The Feldspars. Manchester University Press, Manchester, pp 378-423

McLaren AC (1978) Defects and microstructures in feldspars. In: Roberts MW, Thomas JM (eds) Chemical physics of solids and their surfaces, Vol 7. A Specialist Periodical Report. The Chemical Society, London, pp 1-30

McLaren AC (1984) Transmission electron microscope investigations of the microstructures of microclines. In: Brown WL (ed) Feldspars and feldspathoids: structures, properties and occurrences. D. Reidel Publishing Co, Dordrecht, pp 373-409

McLaren AC (1991) Transmission electron microscopy of minerals and rocks. Cambridge University Press, Cambridge

McLaren AC, Fitz Gerald JD (1987) CBED and ALCHEMI investigation of local symmetry and Al,Si ordering in K-feldspars. Phys Chem Minerals 14: 281-292

Megaw HD (1961) Discussion following the paper by Laves, Goldsmith (1961, p 80)

Megaw HD (1962) Order and disorder in felspars. Norsk Geol Tidsskr 42, No 2: 104-137

Nissen H-U (1967) Direct electron-microscopic proof of domain texture in orthoclase ($KAlSi_3O_8$). Contrib Mineral Petrol 16: 354-360

Nord GL Jr (1992) Imaging transformation-induced microstructures. In: Buseck PR (ed) Minerals and reactions at the atomic scale: transmission electron microscopy. Mineral Soc Am, Rev Mineral 27: 455-508

Orville PM (1967) Unit-cell parameters of the microcline-low albite and the sanidine-high albite solid solution series. Am Mineral 52: 55-86

Parsons I (1965) The feldspathic syenites of the Loch Ailsh Intrusion, Assynt, Scotland. J Petrol 6: 365-394

Parsons I (1978) Feldspars and fluids in cooling plutons. Mineral Mag 42: 1-17

Parsons I (1980) Alkali-feldspar and Fe-Ti oxide exsolution textures as indicators of the distribution and subsolidus effects of magmatic 'water' in the Klokken layered syenite intrusion, South Greenland. Trans R Soc Edinburgh Earth Sci 71: 1-12

Parsons I, Brown WL (1984) Felspars and the thermal history of igneous rocks. In: Brown WL (ed) Feldspars and feldspathoids: structures, properties and occurrences. D. Reidel Publishing Co, Dordrecht, pp 317-371

Parsons I, Rex DC, Guise P, Halliday AN (1988) Argon-loss by alkali feldspars. Geochim Cosmochim Acta 52: 1097-1112

Phillips MW, Ribbe PH (1973) The structures of monoclinic potassium-rich feldspars. Am Mineral 58: 263-270

Prince E, Donnay G, Martin RF (1973) Neutron diffraction refinement of an ordered orthoclase structure. Am Mineral 58: 500-507

Ribbe PH (1975) The chemistry, structure, and nomenclature of feldspars. In Ribbe PH (ed) Feldspar mineralogy. Mineral Soc Am, Short Course Notes 2: R1-R52

Ribbe PH (1979) The structure of a strained intermediate microcline in cryptoperthitic association with twinned plagioclase. Am Mineral 64: 402-408

Ribbe PH (1983a) Aluminum-silicon order in feldspars; domain textures and diffraction patterns. In: Ribbe PH (ed) Feldspar mineralogy (2nd edition). Mineral Soc Am, Rev Mineral 2: 21-55

Ribbe PH (1983b) Guides to indexing feldspar powder patterns. In: Ribbe PH (ed) Feldspar mineralogy (2nd edition). Mineral Soc Am, Rev Mineral 2: 325-341

Ribbe PH (1984) Average structures of alkali and plagioclase feldspars: systematics and applications. In: Brown WL (ed) Feldspars and feldspathoids: structures, properties and occurrences. D. Reidel Publishing Co, Dordrecht, 1-54

Robin P-Y (1974) Stress and strain in cryptoperthite lamellae and the coherent solvus of alkali feldspars. Am Mineral 59: 1299-1318

Rollinson HR (1982) Evidence from feldspar compositions of high temperatures in granite sheets in the Scourian complex, N.W. Scotland. Mineral Mag 46: 73-76

Scambos TA, Smyth JR, McCormick TC (1987) Crystal-structure refinement of high sanidine from the upper mantle. Am Mineral 72: 973-978

Sheldrick GM (1976) SHELX 76. Program for crystal structure determination. University of Cambridge, England.

Smith JV (1961) Explanation of strain and orientation effects in perthites. Am Mineral 46: 1489-1493

Smith JV (1962) Genetic aspects of twinning in feldspars. Norsk Geol Tidsskr 42, No 2: 244-263

Smith JV (1968) Cell dimensions b^*, c^*, α^*, γ^* of alkali feldspars permit qualitative estimates of Si,Al ordering: albite ordering process. Geol Soc Am, Abstracts for 1968, Special Paper 121: 283

Smith JV (1970) Physical properties of order-disorder structures with especial reference to feldspar minerals. Lithos 3: 145-160

Smith JV (1974a) Feldspar minerals, Vol 1. Crystal structure and physical properties. Springer-Verlag, Berlin

Smith JV (1974b) Feldspar minerals, Vol 2. Chemical and textural properties. Springer-Verlag, Berlin

Smith JV (1983) Some chemical properties of feldspars. In: Ribbe PH (ed) Feldspar mineralogy (2nd edition). Mineral Soc Am, Rev Mineral 2: 281-296

Smith JV, Artioli G, Kvick Å (1986) Low albite, $NaAlSi_3O_8$: Neutron diffraction study of crystal structure at 13 K. Am Mineral 71: 727-733

Smith JV, Brown WL (1988) Feldspar minerals, Vol 1. Crystal structures, physical, chemical, and microtextural properties. Springer-Verlag, Berlin

Smith JV, MacKenzie WS (1954). Further complexities in the lamellar structure of alkali

feldspars. Acta Crystallogr 7: 380
Smith JV, MacKenzie WS (1955). The alkali feldspars: II. A simple *X*-ray technique for the study of alkali feldspars. Am Mineral 40: 733-747
Smith JV, MacKenzie WS (1959). The alkali feldspars V. The nature of orthoclase and microcline perthites and observations concerning the polymorphism of potassium feldspar. Am Mineral 44: 1169-1186
Smith KL, McLaren AC (1983). TEM investigation of a microcline from a nepheline syenite. Phys Chem Minerals 10: 69-76
Stewart DB (1975) Lattice parameters, composition, and Al/Si order in alkali feldspars. In: Ribbe PH (ed) Feldspar mineralogy. Mineral Soc Am, Short Course Notes 2: St1-St22
Stewart DB, Ribbe PH (1969) Structural explanation for variations in cell parameters of alkali feldspar with Al/Si ordering. Am J Sci, Schairer Vol 267-A: 444-462
Stewart DB, Ribbe PH (1983) Optical properties of feldspars. In: Ribbe PH (ed) Feldspar mineralogy (2nd edition). Mineral Soc Am, Rev Mineral 2: 121-139
Stewart DB, Wright TL (1974) Al/Si order and symmetry of natural alkali feldspars, and the relationship of strained cell parameters to bulk composition. Bull Soc Franç Minéral Cristallogr 97: 356-377
Strob W (1983) Strukturverfeinerung eines Tief-Mikroklins, Zusammenhange zwischen <T-O> Abstanden und Al,Si-Ordnungsgrad und metrische Variation in einer Tief-Albit/Tief-Mikroklin-Mischkristallreihe. Diplomarbeit, Inst Mineralogie, Westf. Wilhelms-Univ, Münster (not seen; extracted from Kroll and Ribbe 1983, Ribbe 1984)
Su SC, Bloss FD, Ribbe PH, Stewart DB (1984) Optical axial angle, a precise measure of Al,Si ordering in T1 tetrahedral sites of K-rich alkali feldspars. Am Mineral 69: 440-448
Su SC, Ribbe PH, Bloss FD (1986a) Alkali feldspars: structural state determined from composition and optic axial angle 2V. Am Mineral 71: 1285-1296
Su SC, Ribbe PH, Bloss FD, Goldsmith JR (1986b) Optical properties of single crystals in the order-disorder series low albite-high albite. Am Mineral 71: 1384-1392
Tullis J (1975) Elastic strain effects in coherent perthitic feldspars. Contrib Mineral Petrol 49: 83-91
Walker FDL (1990) Ion microprobe study of intragrain micropermeability in alkali feldspars. Contrib Mineral Petrol 106: 124-128
Wenk H-R, Kroll H (1984). Analysis of $P\bar{1}$, $I\bar{1}$ and $C\bar{1}$ plagioclase structures. Bull Minéral 107: 467-487
Whittaker EJW (1977) Determination of atomic occupancies. In Fraser DG (ed) Thermodynamics in geology. D. Reidel Publishing Co, Dordrecht, pp 99-113
Willaime C (1977). Apports de la microscopie électronique a l'étude des exsolutions dans les minéraux. J Microsc Spectrosc Electron 2: 405-424
Willaime C (1981) Les exsolutions dans les minéraux. In: Gabis V, Lagache M (eds) Les transformations de phases dans les solides minéraux. Vol 2. Soc Franç Minéral Cristallogr, Paris, pp 499-532
Willaime C, Brown WL (1974) A coherent elastic model for the determination of the orientation of exsolution boundaries: application to the feldspars. Acta Crystallogr A30: 316-331
Willaime C, Brown WL (1985). Orientation of phase and domain boundaries in crystalline solids: discussion. Am Mineral 70: 124-129
Willaime C, Brown WL, Gandais M (1973) An electron-microscopic and X-ray study of complex exsolution textures in a cryptoperthitic alkali feldspar. J Mater Sci 8: 461-466
Willaime C, Brown WL, Gandais M (1976) Physical aspects of exsolution in natural alkali

feldspars. In: Wenk H-R (ed) Electron microscopy in mineralogy. Springer-Verlag, Berlin, pp 248-257

Worden RH, Walker FDL, Parsons I, Brown WL (1990) Development of microporosity, diffusion channels and deuteric coarsening in perthitic alkali feldspars. Contrib Mineral Petrol 104: 507-515

Wright TL (1967). The microcline-orthoclase transformation in the contact aureole of the Eldora stock, Colorado. Am Mineral 52: 117-136

Wright TL (1968) X-ray and optical study of alkali feldspar: II. An X-ray method for determining the composition and structural state from measurement of 2θ values for three reflections. Am Mineral 53: 88-104

Wright TL, Stewart DB (1968) X-ray and optical study of alkali feldspar: I. Determination of composition and structural state from refined unit-cell parameters and 2*V*. Am Mineral 53: 38-87

Yund RA (1974) Coherent exsolution in the alkali feldspars. In: Hofmann AW, Giletti BJ, Yoder HS Jr, Yund RA (eds) Geochemical transport and kinetics. Carnegie Inst Wash, Publication 634: 173-183

Yund RA (1983) Microstructure, kinetics and mechanisms of alkali feldspar exsolution In: Ribbe PH (ed) Feldspar mineralogy (2nd edition). Mineral Soc Am, Rev Mineral 2: 177-202

Yund RA (1984) Alkali feldspar exsolution: kinetics and dependence on alkali interdiffusion. In: Brown WL (ed) Feldspars and feldspathoids: structures, properties and occurrences. D. Reidel Publishing Co, Dordrecht, pp 281-315

Yund RA, Ackermand D (1979) Development of perthite microstructures in the Storm King granite, N.Y. Contrib Mineral Petrol 70: 273-280

Yund RA, Ackermand D, Seifert F (1980) Microstructures in the alkali feldspars from the granulite complex of Finnish Lapland. Neues Jahrb Mineral Mh 1980: 109-117

Yund RA, Tullis J (1983a) Strained cell parameters for coherent lamellae in alkali feldspars and iron-free pyroxenes. Neues Jahrb Mineral Mh 1983: 22-34

Yund RA, Tullis J (1983b) Subsolidus phase relations in the alkali feldspars with emphasis on coherent phases. In Ribbe PH (ed) Feldspar mineralogy (2nd edition). Mineral Soc Am, Rev Mineral 2: 141-176

Zeitler PK, Fitz Gerald JD (1986) Saddle-shaped $^{40}Ar/^{39}Ar$ age spectra from young, microstructurally complex potassium feldspars. Geochim Cosmochim Acta 50: 1185-1199

Zilczer JA (1981) Adularia and orthoclase: X-ray diffraction and crystal structure analyses of diffuse streaking and aluminum/silicon ordering in natural potassium feldspars. Ph D dissertation, The George Washington Univ, Washington DC

PHASE TRANSITIONS AND VIBRATIONAL SPECTROSCOPY IN FELDSPARS

E.K.H. SALJE
Department of Earth Sciences
University of Cambridge
Downing Street
Cambridge CB2 3EQ
England

ABSTRACT. Structural changes of the feldspar framework result in significant changes of the phonon frequencies. Optically active lattice vibrations show systematic changes of their absorption or scattering profiles due to displacive structural shear (e.g. during displacive phase transitions), Al,Si ordering and exsolution phenomena. Systematic studies of alkali feldspars and anorthite lead to an empirical calibration of these structural and microstructural parameters using hard-mode spectroscopy. An important advantage of this technique is its short characteristic length scale. In contrast to diffraction experiments, structural changes on a scale of some 10 Å are detected without interference from the coherent lattice. This allows the analysis of heterogeneous and modulated structures. A typical example is orthoclase with strain modulation on a 100 Å scale. This material has no equivalent modulation of its Al,Si degree of order. Kinetically Al,Si (dis-)ordered Na-feldspar and anorthite are used as further examples for the discussion of the experimental technique and the theoretical analysis of spectrocopic data.

1. Introduction

Although stability relationships and kinetic processes in feldspars are complex, much can be understood starting from a surprisingly simple idea. Feldspars, like many other minerals, have to reconcile two apparently contradictory aspects of their crystal structure. Firstly, we may ask what structure will have the minimum internal energy. The answer is: a very low symmetric ($P\bar{1}$) structure without defects, twins or other microsctructures, ordered Al,Si sites and exsolved K, Na and Ca atoms when miscibility gaps occur. Our second question is: what is the structure of the topologically most symmetrical state, i.e. the "ideal" feldspar structure. The answer is now a high symmetric structure (C2/m) with randomly distributed Al,Si and K, Na, Ca, respectively. At any short time interval (i.e. a 'snapshot' of the structure) this state is often riddled with embryos of low-symmetry material, textured and tremendously heterogeneous. Over sufficiently long time intervals all this averages out to the typical monoclinic paraphase structure of all feldspars.

These two situations represent, indeed, two structural states of physical significance. The first is the ground state, i.e. the structure which we expect at absolute zero temperature. The second, high symmetry state represents the parastructure which is stabilized by entropic effects and appears at high temperatures. Let us now focus on this high symmetry structure (at, hypothetically, infinitely high temperatures). We also idealize the situation by ignoring melting, facetting and other effects not related to the infinite solid state. Our "ideal" structure has topologically monoclinic symmetry, i.e. the

I. Parson (ed.), Feldspars and Their Reactions, 103–160.

structural crankshafts are all parallel, and all interchangeable atoms are randomly distributed. Let us now cool this structure down in a thought-experiment. Cooling from infinitely high temperatures will, at first, result in Al,Si ordering between the energetically most different sites, namely T_1 and T_2. This ordering does not change the monoclinic symmetry and, thus, some degree of order, described by the order parameter Q_t, must be present at any finite temperature. To put emphasis on the absence of a critical temperature above which total disorder is achieved, the term "non-convergent" ordering is normally used to characterize this process.

Cooling further, several structural changes may occur, each leading to a structural phase transition. Remember that our symmetry is still monoclinic C2/m. Possible instabilities leading to phase transitions are:

1. Tilt of the structural crankshafts in a displacive phase transition. The order parameter of this transition is usually called Q.

2. Ordering of Al and Si on tetrahedral sites. Various ordering schemes are possible, e.g. maintaining the structural unit cell C2/m - C$\overline{1}$ as in Na-feldspar. The order parameter is Q_{od}. Other ordering schemes, e.g. in anorthite, change the size of the unit cell (I2/c or P2/c). The order parameter Q^I_{od} or Q^P_{od}

3. Further displacive transitions can occur after prior Al,Si ordering such as the P$\overline{1}$ - I$\overline{1}$ transition in anorthite. The order parameter is Q^{I1}, whereby the index is often omitted if there is no danger of confusion with Q of C2/m $\rightarrow$ C$\overline{1}$.

4. Exsolution of Na - K or Na - Ca occurs for binary feldspars with the order parameters η (Na,K) and η (Na,Ca), respectively.

No further instabilities are known and it would appear very simple to analyse each of these structural phase transitions individually. It will, indeed, be argued later that all these transitions fall into the most trivial class of phase transitions, namely those following the mean field behaviour with some kind of a Landau-Ginzburg type free energy. The complexity of the actual behaviour of feldspars does not come from any "anomalies" of these transitions but stems mainly from two effects:

1. The phase transitions are not independent of each other. Their mutual influence is described by the coupling between the relevant order parameters.

2. Feldspars are often not in thermodynamic equilibrium. This leads to specific structural states which are typical for the "kinetics" of structural phase transitions.

These two effects can result in heterogeneous states of feldspars which make their analysis by traditional diffraction methods rather difficult (or even impossible). New analytical techniques were necessary for the characterisation of these states. One of the most successful approaches is vibrational Hard Mode Spectroscopy, as commonly used in the fields of research on ferroelectrics or semiconductors. It is the aim of this paper to explain this technique and show how it can be applied to feldspars.

The paper is organized as follows. In Part 2 some of the essentials of the order parameter treatment of the structural phase transitions in feldspars are reviewed. Several fundamental questions, concerning the correlation between atomic potentials and thermodynamic Gibbs free energies are treated in Appendix 1. The thermodynamic irrelevance of fluctuations in Γ - point transitions and similar phenomena are discussed in Appendix 2.

In Part 3 kinetic rate behaviour and related microstructures are discussed. In Part 4 the general concept of Hard Mode Spectroscopy is introduced. Part 5 contains several applications of the technique to specific problems related to phase transitions in feldspars.

There is a large amount of literature published on the subject of this paper. The reader is referred to Smith and Brown (1988) and references given therein for a review on the structural states of feldspars. The thermodynamics of phase transitions were treated more generally by Salje (1993a-b), and the specific way that phonon spectroscopy is applied was previously reviewed by Bismayer (1990) and Salje (1992).

2. Phase transitions and order parameters

2.1 THE ORDER PARAMETER

In order to illustrate the meaning of the order parameter - the parameter which we wish to measure - we return to our thought-experiment. The monoclinic "ideal" feldspar has a crystal structure with an average distribution of atoms which can be described by a density function $\rho_O(r)$. This is essentially the same density function used for ordinary structure analysis. Cooling this crystal leads to structural changes with $\rho(T,r) = \rho_O(r) + \Delta\rho(T,r)$. We see immediately that all of the information about structural distortions, Al,Si ordering, and even exsolution phenomena is contained in $\Delta\rho(T,r)$. Let us now use as an example one of these processes, say the rotations of the crankshafts in Al,Si disordered anorthoclase. The function $\Delta\rho(T,r)$ can easily then be constructed. As we consider only one mechanism, we simply compare the structure ρ^* with the largest effect (e.g. the maximum rotation at absolute zero temperature) with the density function of the monoclinic structure $\rho^*(r) = \rho_O(r) + \Delta\rho(r)_{max}$. Any intermediate state is then given by $\Delta\rho(r,T) = Q(T) \cdot \Delta\rho(r)_{max}$ where the amplitude $Q(T)$ describes "how much happens" and $\Delta\rho_{max}$ describes the pattern of "what happens". More strictly speaking, we have separated the thermodynamic function Q, which is independent of all crystallographic properties, and the crystallographic features which are summarized in $\Delta\rho(r)_{max}$. Note that $\Delta\rho_{max}$ is *independent* of T, P, N. It is obvious that Q is a function of all external thermodynamic quantities, e.g. temperature T, pressure P, chemical composition N. We choose T as an example for all of them. The difference-density $\Delta\rho(r)$ can represent any transformation pattern, e.g. Al,Si ordering, a displacive transition or an exsolution process. The subsequent treatment is in all cases identical.

So far we have reduced the treatment of the thermodynamics of the phase transition to the point that all relevant thermodynamic information is contained in Q, whereas $\Delta\rho(r)$ can be dismissed as crystallographic ballast (as far as the thermodynamic treatment is concerned). The parameter Q is called the "order parameter". Before we describe Q in more detail, let us return to a simple thermodynamic argument. The idealised monoclinic feldspar structure is determined by a Gibbs free energy G_O (as it must in order to be a thermodynamically stable phase). G_O depends still on T, P and N but not on Q because this structure is already fully determined by the fact that all Al,Si positions are randomly occupied; K, Na, Ca are randomly distributed and the symmetry is C2/m. If any of these features changes during cooling, it is trivial to see that $G(T, P, N)$ will necessarily change. The more fundamental question is: is $G(T, P, N)$ still fully defined? The answer is clearly: No! In fact, we have to specify which Gibbs free energy we mean. Firstly, there is the one for the undercooled monoclinic structure. Then there is the equilibrium structure with either triclinic symmetry or, at least, non-symmetric Al,Si occupancies on T_1 and T_2 sites. Let us consider again, as an example, the triclinic $C\bar{1}$ structure in anorthoclase with tilted crankshafts. It is clearly important to state how much these crankshafts are tilted, i.e. what the value of the order parameter is. This means that G depends not only on T, P and N but also on Q : $G = G(T, P, N, Q)$. If this function is known, the equilibrium order parameter Q is easily obtained because it must correspond to a minimum of G:

$$\partial G/\partial Q = 0 \tag{II.1}$$

There is always one trivial solution which is $G(T, P, N) = 0$ for one phase. This represents the ideal monoclinic structure. All other quantities are now measured as excess quantities (just as $\Delta\rho(r,T)$) with respect to this ideal structure.

The remaining problem is now to derive expressions for $\Delta G(T, P, N, Q) = \Delta G(T, Q)$ for the relevant phase transitions. Rather little is known about the exsolution features from a "first-principles" viewpoint but for structural transitions the forms of $\Delta G(T, Q)$ are known or can be anticipated from some fundamental physical pictures (Appendices 1 and 2). Both the theoretical approaches and the empirical experimental determinations of $G(T, Q)$ show that this function can be written (for temperatures above room temperature) in a Landau-type potential (Appendix 1)

$$\Delta G(T,Q) = \frac{1}{2}A(T-T_c)Q^2 + \frac{1}{4}BQ^4 + \frac{1}{6}CQ^6 - HQ \tag{II.2}$$

It appears that this general approach holds for all structural phase transitions in feldspars. Often it can be simplified with $H = 0$. This parameter H is the conjugated field which is non-zero only for non-convergent ordering phenomena defect fields etc. (Lebedev et al., 1983; Levanyuk et al., 1979, Salje et al., 1991a, Carpenter et al., 1993, Carpenter and Salje 1993a,b). The parameters A, B and C can be determined experimentally for each of the transitions (Salje 1993). Before we can discuss specific solutions for various feldspars, it is imperative to consider the interaction between different phase transitions.

2.2. COUPLING BETWEEN ORDER PARAMETERS

It was noted in the introduction that feldspar structures show a variety of structural phase transitions in their phase diagram. It is now important to understand that these phase transitions are not independent events but strongly influence each other. The most extreme case occurs in Na-feldspar. This material is monoclinic at $T > 1251$K. On cooling, the symmetry is reduced to triclinic by two interacting processes: the displacive tilt of the structural crankshafts and ordering of Al and Si atoms between the relevant tetrahedral sites of the feldspar structure. The order parameter for the displacive process was called Q, that for the Al,Si ordering was called Q_{od} (Salje 1985). Both transitions would result in the same symmetry change but the transition temperature of Q is ca. 1251K, i.e. much higher than the equivalent temperature for Q_{od}, namely ca. 983K (Salje et al., 1985; Kroll et al., 1980; Brown and Parsons 1989; Carpenter 1988). Each of these transition processes should occur independently if the other transition did not occur. If, for example, the displacive transition failed to occur, the Al,Si ordering would still take place, albeit at a different temperature. As the displacive transition in Na-feldspar does occur, however, and reduces the symmetry to triclinic, there is no further possibility for the Al,Si ordering to create an additional phase transition and the Al,Si ordering is now limited to modifications of the displacive transition mechanism. These modifications are, nevertheless, essential for the thermodynamic stability of the structure and it is impossible to understand the physical behaviour of Na-feldspar if either of these mechanisms is ignored.

Another instructive example is the transition $I\bar{1}$ - $P\bar{1}$ in anorthite (Salje 1987; Redfern and Salje, 1987). This transition occurs at 514.5K for fully Al,Si ordered anorthite and its bare Landau potential is again the same as introduced above with a small 4th-order term. In the case of $B = 0$, the transition is called "tricritical" with certain consequences for its thermodynamic behaviour. The most interesting aspect of this transition is that the transition mechanism is heavily influenced by other phase transitions (e.g. Al,Si ordering), chemical variations (increase of albite component) and pressure. Let us focus first on the effect of Al,Si ordering. A transition sequence C2/m - $C\bar{1}$ - $I\bar{1}$ can be postulated for end member anorthite to occur at temperatures above the melting point. The order parameter describing Al,Si ordering is again called Q_{od}. This order parameter might be expected not to interact with $Q(I\bar{1} - P\bar{1})$ for two reasons : 1. the two critical temperatures are some 2000K apart, i.e. the two order parameters act on widely different energy scales. 2. The symmetry of the two order parameters is different so that a simple (bi)linear interaction as in Na-feldspar is not possible. More indirect coupling mechanisms are possible, however. These are (Salje 1993a):

(i) both transitions produce a lattice distortion (spontaneous strain e_{ik} which is proportional to Q^2 and Q_{od}^2 in this particular case). The superposition of both strains leads to excess elastic energy which is proportional to $Q^2 Q_{od}^2$.

(ii) Direct coupling via structural modifications. This mechanism leads the same effective coupling energy as in (i).

(iii) Inhomogeneous structural variations break the translational symmetry of the crystal, and coupling between Q and the spatial gradient of Q_{od}, and vice versa, might occur.

How strong the interaction can be is demonstrated by the fact that well Al-Si ordered anorthite from the Val Pasmeda locality shows a tricritical phase transition whereas slightly disordered material (by heat treatment or from another locality) shows a second order transition. Such behaviour is typical for certain types of indirect coupling mechanisms, as discussed by Salje (1993a).

Having introduced some ideas about coupling mechanisms, we can now formalize this approach. The concept is simply based on the fact that Gibbs free energies are additive quantities. Each phase transition is then described by our appropriate Landau potential L as introduced before:

$$G_o(Q_1,Q_2) = L(Q_1) + L(Q_2) \qquad \text{(II.3)}$$

The only energy missing so far is the interaction energy. The form of this interaction energy is determined by symmetry and only three cases are relevant:

(i) Bilinear coupling, with $G_{coupling} = \lambda_1 Q_1 Q_2$, occurs when Q_1 and Q_2 have the same symmetry properties.

(ii) Biquadraric coupling, with $G_{coupling} = \lambda_2 Q_1^2 Q_2^2$, can occur under all symmetry conditions and is a very common type of coupling.

(iii) Linear-quadratic coupling, with $G_{coupling} = \lambda_3 Q_1 Q_2^2$, appears to be a rather rare case and occurs particularly for nonconvergent ordering in feldspar.

Which coupling is symmetry-allowed follows simply from the argument that $G_{coupling}$ is a scalar quantity so that the relevant combination of Q_1 and Q_2 has to transform as the identity representation of the common paraphase. It is not necessary to go into details of group theory because all the important combinations of symmetry relationships have been discussed previously in great detail (Tolédano and Tolédano 1976, 1977; Stokes and Hatch 1988). In the case of feldspars, bilinear coupling occurs between Q and Q_{od} in alkali feldspars (as described for the case of Na-feldspar). All other interaction phenomena are well described by bi-quadratic coupling. Linear-quadratic coupling may occur if one coupling partner is Q_t and coupling between an order parameter and the spatial gradient of a second order parameter is important for plagioclase structures. The consequences of all these coupling energies were analysed in great detail and the reader is referred to Salje (1993a) and references given therein for further details.

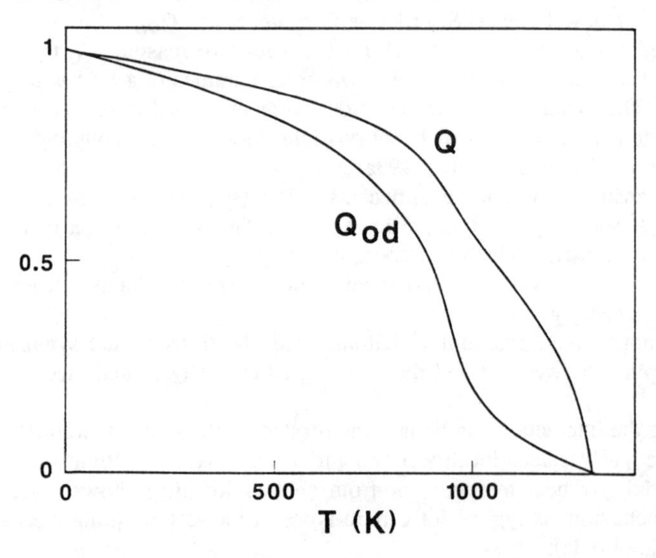

Figure 1. Temperature evolution of the displacive (Q) and order/disorder ($Q_{\rm od}$) order parameter in Na-feldspar. Monoclinic Na-feldspar is thermodynamically stable for $Q = Q_{\rm od} = 0$. The phase transition occurs when both order parameters become non-zero during cooling. A crossover between a "high" albite and a "low" albite occurs when $Q_{\rm od}$ increases sharply when the sample is cooled further.

2.3. SOME TYPICAL LANDAU POTENTIALS

So far we have established the tools for the quantitative description of phase transitions of feldspars. Our main assumption is that the crystal is uniformly in one and the same thermodynamic state, i.e. we have not yet introduced domain boundaries, phase mixtures and defects. This assumption of uniform crystals with little or no disturbance from heterogeneities is, in fact, rather appropriate for many feldspars. In particular, very pure alkali-feldspars can be found in nature which display the Landau-type phase transitions in an (almost) ideal way. It is not surprising, therefore, that the historically first quantitative treatment of a phase transition in feldspars was in 1985 on Na-feldspar (Salje et al., 1985). The Landau potential (i.e. the Landau type Gibbs free energy) of Na-feldspar was determined to be:

$$G=\frac{1}{2}5.479(T-1251)Q^2+\frac{1}{4}6854Q^4$$
$$+\frac{1}{2}41.62(T-824)Q_{od}^2-\frac{1}{4}9301Q_{od}^4 \tag{II.4}$$
$$+\frac{1}{6}43600Q_{od}^6+\lambda QQ_{od}$$

with $\lambda = -2.171 - 3.043T - 0.0016\,T^2 + 2.1.10^{-6}\,T^3$ (II.5)

where all the energies are in units of J/mol and all temperatures are in K. The temperature dependences of the two order parameters Q and Q_{od} are shown in Fig. 1, and the evolution of the thermodynamic functions is shown in Fig. 2.

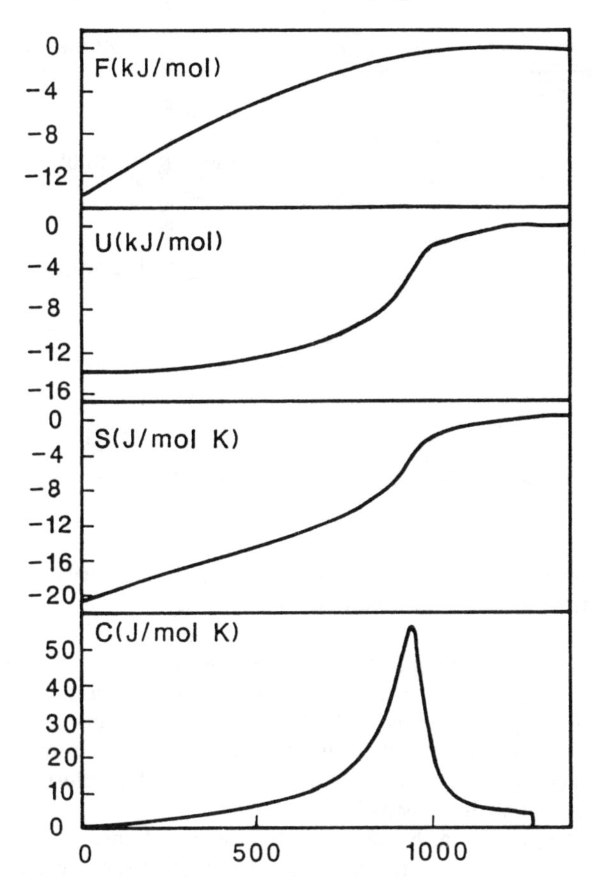

Figure 2. The temperature dependence of the thermodynamic excess quantities F (excess Gibbs free energy), the internal energy U, the entropy S and the specific heat C for Na-feldspar under equilibrium conditions as derived from Landau theory (from Salje et al., 1985).

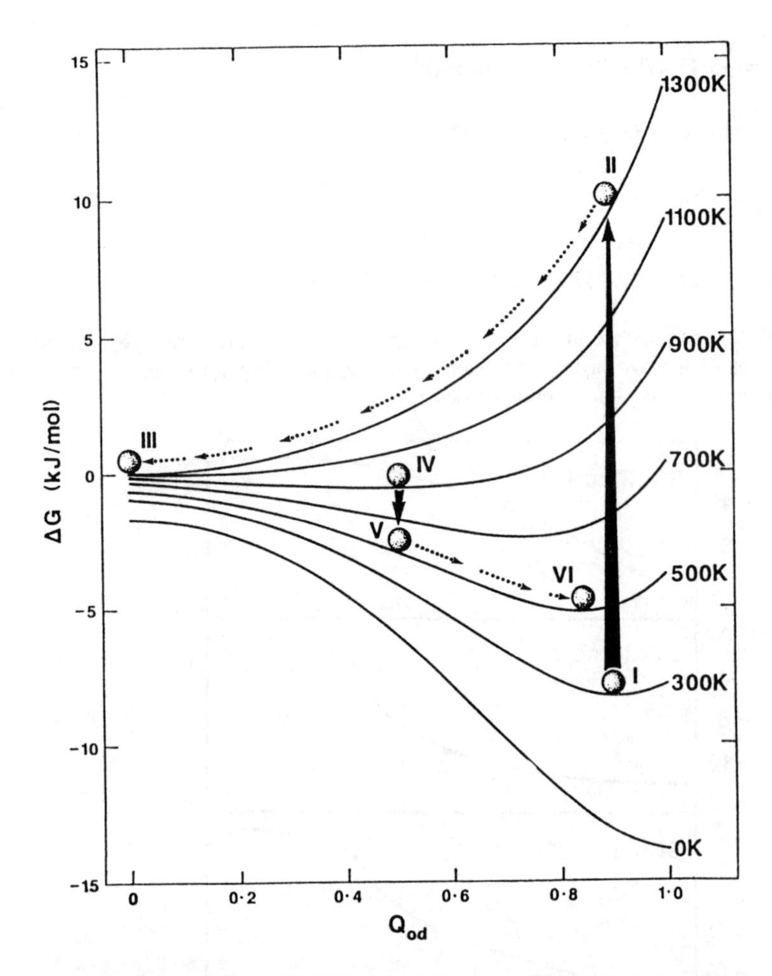

Figure 3. The Landau potentials of Na-feldspar as a function of Q_{od} with Q relaxed for each value of Q_{od}. These potentials govern the kinetics of Al,Si ordering in Na-feldspar.

The main part in G is given by Q_{od}. For further discussions in part 3, it is convenient to reduce $G(Q_{od}, Q)$ to $G(Q_{od})$ in such a way that Q is always in thermodynamic equilibrium for each value of Q_{od}. The way to do that is to calculate numerically for each value of Q_{od} (and each temperature T) the equilibrium value Q by minimization of $G(Q_{od}, Q)$ with respect to Q. Such a projection is shown in Fig. 3 which illustrates well the typical shape of Landau potentials in feldspars.

Let us finally note that, in case of Na-feldspar, the coupling strength λ is an (almost) linear function of temperature. This indicates that it is not a constant static distortion of the structure which transmits the interaction between Q and Q_{od} but a process which is itself temperature dependent. A typical candidate for such T-dependent coupling is phonon-lattice interactions or other dynamical effects.

The second example for a Landau potential is that of K-feldspar. The main differences between K-feldspar and Na-feldspar are

(i) In K-feldspar the symmetry breaking transition relates to Q_{od} whereas the influence of Q is small. In Na-feldspar, the symmetry is broken by Q with Q_{od} coupling with Q.

(ii) The nonconvergent ordering is strong in sanidine (i.e. monoclinic K-feldspar) because its stability field expands to rather low temperatures (Salje and Kroll 1991). In Na-feldspar, this effect also exists in monalbite, although the stability field is small and the gradual effect of T_1, T_2 ordering is harder to observe. The relevant order parameter is Q_t. The Landau potential now contains the symmetry breaking order parameter Q_{od} (equivalent to Q_{od} in Na-feldspar) and Q_t, whereas Q is ignored. The resulting Landau potential is (Carpenter and Salje 1993b)

$$\begin{aligned} G &= 1.650Q_t + 11.14(T-586)Q_t^2 + 2.45Q_t^6 \\ &+ 20.55(T-239)Q_{od}^2 + 1636Q_{od}^6 - 20.13TQ_tQ_{od}^2 \end{aligned} \qquad \text{(II.6)}$$

The coupling is again dependent on temperature. The equilibrium order parameters show a first order transition at 753K with breaks of Q_t and Q_{od} of 0.4 → 0.95 and 0 → 0.94, respectively.

Instead of coupling between different order parameters, one commonly observes a coupling with the chemical composition in binary or ternary feldspars. This effect was first mentioned for the $P\bar{1}$ - $I\bar{1}$ transition in anorthite although the experimental data are not yet well enough constrained to derive the quantitative expression for the Landau potential. Instead, we may consider the $C\bar{1}$ - $I\bar{1}$ ordering transition in anorthite. The first determination was (Carpenter 1992)

$$\begin{aligned} G &= \frac{1}{2}37.6(T-2283+2525X_{Ab})Q^2 \\ &+ \frac{1}{4}(198371-412954X_{Ab})Q^4 \\ &+ \frac{1}{6}(198371-412954X_{Ab})Q^6 \end{aligned} \qquad \text{(II.7)}$$

The phase transition is first order for endmember anorthite *(X_{Ab} = 0)*. The jump of the order parameter decreases with increasing albite-component X_{Ab}, becomes tricritical at ca. An_{78} and second order for more Ab-rich plagioclases.

A new determination of equilibrium data together with kinetic data by Salje et al. (1993) for a first order transition leads to a very similar potential for An_{100} with

$$G = \frac{1}{2}40.7(T-2325)Q^2 - \frac{1}{4}107600Q^4 + \frac{1}{6}202200Q^6 \qquad \text{(II.8)}$$

Further potentials for second order and tricritical transitions were also discussed by Salje et al (1993).

These three examples are typical for the energies and types of potentials involved in phase transitions of feldspar. They represent a good starting point for further research into the transition behaviour of alkali feldspars and plagioclases which still awaits a quantitative treatment.

3. Kinetic and other non-equilibrium behaviour

The treatment of phase transitions in feldspars in Part 2 of this paper considered only structural states in thermodynamic equilibrium. Recent research on feldspars has extended this scheme to non-equilibrium states which are encountered during kinetic experiments. We will now turn to the description of the relevant kinetic rate laws.

Before such a description becomes meaningful, we have to dispose of one red herring which has led to some confusion in the mineralogical literature. Thermodynamic quantities such as order parameters and fluctuations are defined as time averages and, with some caution, as space averages of local state parameters. Imagine, for example, Al,Si ordering in a feldspar. The order parameter, Q_{od}, measures the average degree of order, where the term "average" describes a clearly defined procedure of standard thermodynamics (e.g. Landau and Lifshitz 1980). Local states on a short time scale, on the other hand, do not represent a thermodynamic quantity. This clear distinction is sometimes blurred when phenomena at temperatures close to T_C in a second order phase transition are discussed. In this case, the correlation length between state parameters increases, whereas the time scale for (dis-) ordering decreases. This is true for displacive and order/disorder phase transitions alike. It is nonsense, however, to assume that these correlated regions inevitably form longlived clusters causing any displacive transition to become automatically of the "order/disorder type" at temperatures near the transition point (or, more precisely speaking, that there is a measurably wide Ginzburg interval around T_C). It is, of course, even worse to postulate this for first order transitions. As long as fluctuations are Gaussian, no such effects occur in the defect free crystal (Larkin and Pikin 1969; Levanyuk and Sobyanin 1970; Cowley 1976; Folk et al., 1979). When non-Gaussian fluctuations appear, i.e. in the Ginzburg interval, we can no longer use Landau type theories altogether and have to resort to Renormalization Theory. It can be proved, however, that ferroelastic and co-elastic phase transitions with Γ - point instabilities have no Ginzburg interval at all, so that the whole discussion becomes irrelevent for such materials (Appendix II). Note however, that defect structures lead to order parameter behaviour which is very similar to predictions of standard Renormalization Theory although they follow, in fact, Landau-type theories when the influence of defects are properly included. Spurious fluctuations and cluster formation of this kind have nothing to do with the intrinsic behaviour of phase transitions and are, thus, excluded from the following discussion. In kinetic experiments, we have to give some margin to the averaging process. Nevertheless, we always consider here a situation in which spatial variations are smooth, with characteristic length scales of a few unit cells. The relevant time scale is always long compared to phonon times and any attempt-time for Al,Si (dis-) ordering.

Most non-equilibrium states in feldspars are the result of changes of the thermodynamic conditions of the mineral on a timescale which is faster than that of global Al,Si ordering or those of exsolution processes. A typical example is Na-feldspar in which the order parameter Q reacts on a phonon timescale to external changes. Al,Si ordering, and hence Q_{od}, relate to thermally activated processes and react only on a much longer timescale. Kinetic behaviour is determined by Q_{od}, whereas Q follows all changes almost immediately. In the language of kinetics, Q is "slaved" by Q_{od}. The question arises as to how to quantify such behaviour. If the crystal is in a non-equilibrium situation, it will tend to lower its energy by undergoing structural changes. These changes define one or several "kinetic pathways" and it is the task of kinetic theories to identify these pathways (Dattagupta et al., 1991; Salje 1988; Salje 1993a,b; Marais et al., 1991; Carpenter and Salje 1989). In the case of feldspar structures (and other highly correlated systems) one finds that, after averaging over equivalent configuration coordinates, only one such pathway exists. Each step on this pathway is then correlated with changes of a multitude of structural parameters, and we must first identify the "pattern function" which is equivalent to $\Delta\rho(r)$ in the case of the equilibrium phase transition. Glauber (1963), Kawasaki (1966), Salje (1988) and others have argued that, by a judicious selection of parameters, kinetic expressions can be formulated in terms of a "kinetic order parameter". This kinetic order parameter is, in general, the amplitude function of the pattern in the same way as we defined the equilibrium order parameter. The experimental studies of Na-feldspar, K-feldspar and anorthite have shown that the structural meaning of the kinetic and the equilibrium order parameter are identical within experimental resolution so that we will no longer insist on their, more philosophical, difference.

The second main question concerns the nature of the kinetic driving force. In analytical theories, we can always define such a "force" as the gain of some (kinetic) Gibbs free energy, G_{kin}, per step of the order parameter. It was argued that, in cases with long range interactions and a multitude of intermediate structural states, the most general kinetic rate law can be written as (Salje 1993a,b and references given there),

$$\frac{dQ}{dt} = \frac{\tau^{-1}}{RT}\left(1 - \frac{\xi_c^2}{\xi^2}\frac{\sinh(\xi\nabla)}{\xi\nabla}\right)\frac{\delta G_{kin}}{\delta Q} \tag{III.1}$$

This rate law was first introduced by Salje (1988, 1993b) and subsequently discussed in a series of papers. Here it is only important to note that τ is a general time constant ($\tau = \tau_0 \exp(\Delta G^*/RT)$), ∇ is the gradient operator and ξ is a correlation length related to the kinetic process. The factor ξ_c^2/ξ^2 indicates the importance of order parameter conservation : $\xi_c = 0$ is the fully non-conserved case, $\xi_c = \xi$ is the fully conserved (Cahn-Hilliard) case. The physical picture describing ξ and ξ_c is still very much a matter of discussion (see Salje 1993b for a review) although some aspects can be visualized rather easily. The parameter ξ sets the length scale of the gradient operator similar to the length scale of the Ginzburg energy which determines the thickness of domain boundaries etc.. This length is usually long compared with the unit cell. The length ξ_c indicates the distance over which correlated exchanges Al,Si occur. If such exchanges are limited to nearest neighbours, one expects $\xi_c << \xi$. The diffusion length of Al in plagiclase at 1673K during the kinetic time interval τ is some 50 Å so that ξ_c is, indeed, thought to be smaller than ξ albeit not zero. The analysis of Al,Si ordering in feldspar so far has been based on the assumption of $\xi_c << \xi$. Furthermore, the quantitative expression for G_{kin} was found to be identical with the excess Gibbs free energy of the phase transition, i.e. G_{kin} = G (G_{kin} is also identical with the Lyapunov potential in the field of pattern formation, e.g. Cross and Hohenberg 1994 for a review).

Within the scope of this paper, the most striking consequence of the analysis of the general rate law is that uniform kinetic states should occur only in exceptional circumstances. The reasons for this prediction are twofold. Firstly, all thermal noise ($\propto kT$) will be amplified in relation to the slope of $G(Q)$. The order parameter distribution depends explicitly on the distribution function of the noise and the effective order parameter susceptibilities. All solutions have in common that the uniform crystal will break down into domains with domain boundaries as high energy features in the structure. These domain boundaries are albite and pericline twins in alkali-feldspars, with additional APBs in plagioclases. Early stages of kinetic runs are characterised by a tweed pattern with two main directions parallel to the walls in albite and pericline twins. Note that these tweed patterns occur also in systems such as Na-feldspar during kinetic runs where such patterns do not occur in quasi equilibrium. After long annealing times (or, say, in natural materials) the material becomes uniform again. Tweed patterns were first observed in Na-feldspar by Wruck et al. (1991). Rather similar patterns are well known for orthoclase in K-rich feldspars (Bambauer et al., 1984; Eggleton and Busek 1980; FitzGerald and McLaren 1982; Krause et al., 1986; McConnell 1965, 1971; McLaren 1974; 1984; McLaren and FitzGerald 1987; Nissen 1967; Ribbe 1983).

The second reason for kinetic heterogeneities is due to the possible admixture of conservation into the general rate behaviour. It was recently found that any ordering which contains more than 1/12 kinetic steps being conserved while 11/12 steps are non-conserved (as measured by the ratio ξ_c/ξ) will always form a periodic pattern even when the transformation process is fast (i.e. on the timescale of acoustic wave propagation). Details of this mechanism are discussed by Salje (1993b) and Van Saarloos and Hohenberg (1990). Both mechanisms have been studied in great detail by computer simulation.

In summary, the global kinetic process of Al,Si ordering in feldspars can be described by a general rate equation in which the driving force is related to the excess Gibbs free energy of the (dis-) ordering process. Crystals will tend to form heterogeneous states in a kinetic process, with the characteristic sequence: uniform initial state → tweed → stripes → uniform final state. In Na-feldspar, the tweed pattern survives the early kinetic stages, whereas the stripe phase seems to be less prominent. In the case of partial conservation, pattern formation can occur via oscillating wave propagation.

4. Hard Mode Phonon Spectroscopy

4.1. INTRODUCTORY REMARKS ON OPTICAL SPECTROSCOPY

The analysis of equilibrium phase transitions and kinetic behaviour of feldspars clearly defines a formidable task for the experimentalist who wants to measure the order parameter. Let us first define the requirements we encounter for the ideal experiment. These are:

1. The order parameters should be measured as intrinsic properties and not as defect related phenomena. If possible, doping samples with impurities etc. (e.g. for EPR measurements) should be avoided.
2. The technique should be selective if several order parameters occur simultaneously. Theoretical models are clearly helpful to disentangle such multi-order parameter behaviour (e.g. as measured from the spontaneous strain), but a more direct approach could be helpful.

3. The order parameter should be measured using a technique with a short characteristic length scale. This requirement avoids the phase problem of X-ray and neutron diffraction in samples which undergo kinetic pattern formation.

One particular technique which, to a large extent, fulfils all three requirements is hard mode phonon spectroscopy. Its basic idea is the following. Phonon signals, such as observed in infrared absorption, transmission, reflection or emission spectroscopy, or in Raman spectroscopy, respond to any structural change of a feldspar. This change can be generated by chemical variations, e.g. cation exchange, cation ordering, displacive phase transitions etc.. The nature of the phonon response can be, mostly in the cases of low frequency or heavily damped phonons, rather complex so that a simple, phenomenological interpretation is not always possible. The reason for this complexity is not directly related to the rather complex crystal structure but stems from the physical fact that only the most simple, harmonic or quasi-harmonic lattice vibrations can be calculated using off-the-shelf lattice dynamical routines. Non-harmonic vibrations, on the other hand, usually cannot be calculated this way, and one convenient approach to understanding the whole phonon spectrum starts from the following thought experiment. Imagine that we could exert arbitrary forces f inside a crystal. The response of the crystal to such a force will always follow the application of the force. Any local state X (e.g. a distortion, site occupancy etc.) will then be determined by

$$X(t) - X_o = \Delta X(t) = \int_o^\infty \varepsilon(\tau) f(t-\tau) d\tau \quad \text{(IV.1)}$$

where X_o is the undisturbed state, $X(t)$ is the state at time t, τ measures the time before t, and ε measures the strength of the response ΔX to the perturbation $f(t-\tau)$. In the frequency domain, one finds easily after Fourier transformation that this relationship simplifies to

$$X_\omega = \varepsilon(\omega) \,.\, f_\omega \quad \text{(IV.2)}$$

where

$$\varepsilon(\omega) = \int_o^\infty \varepsilon(t) e^{i\omega t} dt \quad \text{(IV.3)}$$

The important parameter is $\varepsilon(\omega)$ which is called the susceptibility and plays in phonon spectroscopy a similar role to the order parameter susceptibility (χ) for phase transitions. ε is a complex quantity

$$\varepsilon(\omega) = \varepsilon'(\omega) + i\varepsilon''(\omega) \quad \text{(IV.4)}$$

where ε' and ε'' can be calculated from each other using the Kramers-Kronig relationship.

Let us now relate this dissipative response of a system to spontaneous fluctuations of the crystal. These fluctuations, $(X^2)_\omega$ are thermally created and will operate with the same frequency distribution as the dissipation. In frequency representation this correlation is particularly simple and became known in 1951 as the "dissipation-fluctuation theorem", the starting point for most of the more rigorous lattice dynamical calculations (for detailed discussion see Landau and Lifshitz 1980)

$$(X^2)_\omega = \hbar\varepsilon \cosh\frac{\hbar\omega}{2kT} \quad \text{(IV.5)}$$

The mean square of the fluctuation quantity is then

$$< X^2 >= \frac{\hbar}{\pi} \int_0^{\infty} \varepsilon''(\omega) \cosh \frac{\hbar\omega}{2kT} d\omega \tag{IV.6}$$

For high temperatures, the cosh-function becomes linear with ($kT >> \hbar\omega$)

$$(X^2)_\omega = \frac{2kT}{\omega} \varepsilon(\omega) \tag{IV.7}$$

The functional form of ε'' can be derived for the typical situation in a feldspar with quasi-harmonic lattice vibrations and diffusive transport movements as follows. The phonon susceptibility of an oscillator is

$$\varepsilon_{phonon} = (\omega_0^2 - \omega^2 + 2i\omega\Gamma)^{-1} \tag{IV.8}$$

where Γ is the damping constant and ω_O is the eigenfrequency of the phonon. The contribution from diffusive modes is usually described by a Debye relaxation so that

$$\varepsilon_{total} = (\omega_0^2 - \omega^2 + 2i\omega\Gamma - \frac{\delta^2}{1 - i\omega\tau})^{-1} \tag{IV.9}$$

where τ is the Debye time constant and δ is the coupling constant. Although Γ and δ can, in principle, depend on the frequency ω, no such observation in feldspars is known to the author. It appears to be a safe assumption therefore, then Γ and δ may be considered as numerical constants.

For small values of Γ, the two damping terms can be separated so that ε'' contains two features, namely a phonon peak with

$$\varepsilon'' \propto \frac{2\omega\Gamma}{(\omega_0^2 - \omega^2)^2 + 4\omega^2\Gamma^2} \tag{IV.10}$$

and a relaxational peak at $\omega_O = 0$ with

$$\varepsilon'' \propto \frac{\delta^2 \omega\tau}{1 + \omega^2\tau^2} \tag{IV.11}$$

Additional dynamical excitations at frequencies well below those of typical IR or Raman active phonons in feldspars appear, for example, in anorthite at temperatures close to T_c of the P $\bar{1}$ - I $\bar{1}$ transition. Such excitations lead to modifications of $\varepsilon''(\omega)$, mainly as an increase of the damping constant Γ. The formal treatment is similar to that of a Debye relaxation with

$$\varepsilon''_{phonon}(\omega) = \frac{2\omega\Gamma}{(\omega_0^2 - \omega^2)^2 + 4\omega^2\Gamma^2} \tag{IV.12}$$

and $$\varepsilon''_{total}(\omega) = \int \varepsilon''_{phonon}(\omega')\varepsilon''_{fluctuation}(\omega' - \omega) d\omega' \tag{IV.13}$$

where the convolution in frequency space is a simple multiplication in time space (i.e. the reverse process as used to derive $\varepsilon(\omega)$ in equation IV.7). Fluctuations are often described by a Gaussian "noise" function

$$\varepsilon''_{fluctuation} \propto \exp\left[-\left(\frac{\omega-\omega_f}{\sigma}\right)^2\right] \tag{IV.14}$$

so that ε''_{total} shows for Γ, $\sigma << \omega_f < \omega_o$ a broadening directly proportional to ω_f. In this way, even "slow" fluctuations ($\tau < 10^{-10}$secs) can be detected as changes of ε'' inside the fluctuation regime.

Direct coupling with solitonic excitations, flip motions etc. leads to similar effects via quasi-harmonic mode coupling. Line broadening also occurs from heterogeneous breaking of the translational symmetry and the correlated optical activity of phonons with wavevectors (in the uniform state) outside the immediate vicinity of the Γ-point of the Brillouin zone (i.e. the "density of states" effect). These effects can be described in a similar way as in equation (IV.14) and the interested reader is referred to Bruce et al.(1980) and Salje et al.(1983) for a detailed treatment of such effects.

4.2. SOFT MODES *VERSUS* HARD MODES

Let us now combine the results of parts 3. and 4.1. The phase transition is described by a Landau potential $L(Q,T)$. The dissipation is then calculated as the response of the structure to a hypothetical perturbation symbolized by an external field H_{ex}.

The Gibbs free energy (Appendix I for $T >> \theta_s$) is

$$G(Q, T, H_{ex}) = \tfrac{1}{2}A(T-T_c)\,Q^2 + \tfrac{1}{4}BQ^4 + \tfrac{1}{6}CQ^6 - QH_{ex} \tag{IV.15}$$

The field H_{ex} modifies Q by δQ with $\delta Q = \chi\ H_{ex}$ where χ is the *(field)* susceptibility. If $Q = Q_o$ for $H_{ex} = 0$ and $Q = Q_o + \delta Q$ for $H_{ex} \neq 0$, we can develop G in a power series of δQ and find, for $\delta\Gamma/\delta Q = 0$, in lowest order

$$\chi = \frac{1}{A|T-T_c|} \qquad \text{for } T > T_c \tag{IV.16}$$

$$\chi = \frac{1}{2A|T-T_c|} \qquad \text{for } T < T_c \tag{IV.17}$$

For small variations of Q, we find, therefore,

$$L(\delta Q, T) = \tfrac{1}{2}\chi^{-1}(\delta Q)^2 \tag{IV.18}$$

which is identical to the energy of a phonon with the amplitude $u = \delta Q$ and the square of the frequency ω_s^2 equal to χ^{-1}. Phonons with $u = \delta Q$ and $\omega_o = \omega_s$ are called "soft modes" and appear as elementary excitations in displacive phase transitions ($\Delta = 0$ in Appendix I).

There are two phase transitions in feldspars which, in principle, might have qualified for a soft-mode type phase transition. Firstly, the C2/m - C $\bar{1}$ displacive transition allows for an optically active soft mode with B_{1g} - symmetry. Raman spectroscopic investigations by Salje (1986) did not find such a soft mode and it seems to be commonly accepted that the relevant transition mechanism is the softening of an acoustic phonon related to the elastic constant $\{C_{44}C_{66} - C_{46}^2\}$. The second example is the P$\overline{1}$ - I$\overline{1}$ transition in anorthite. The soft mode, if it were to exist, would condense at the Z-point on the surface of the first Brillouin zone. No experimental evidence for this soft mode is known to the author.

None of the other phase transitions can relate to a proper soft mode behaviour because δQ describes in these transitions occupational probabilities ($\Delta >> 0$ in Appendix I) and not a phonon amplitude. The only possible manifestation of the phase transition is then the relative change of $\varepsilon''(\omega)$ for modes other than the soft modes. These modes are called "hard modes", a name probably first used by the Prague group around Petzelt and Dvorak (1976). The surprising sensitivity and the rather straightforward analysis of experimental data for hard modes is now introduced.

The starting point is analogous to the treatment of order parameter coupling in part 2.2, with the difference that we now couple an order parameter with a phonon degree of freedom. Let us consider the order parameter Q and the phonon amplitude Q_i. The Gibbs free energy is in this case the sum of the Landau potential $L(Q)$, the energy of the undisturbed phonon (i.e. the phonon energy or 'self energy' for $Q = 0$), $\frac{1}{2}\sum \chi_{ij}^{-1} Q_i Q_j$ and, finally, the coupling energy. Note that we assume the phonon self energy to be harmonic. This assumption is justified by the experimental observation that all optically active phonons in feldspars are well described by a form of ε'' as in equation (IV.10).

The coupling energy depends on the symmetry properties of the phonon coordinates Q_i and the order parameter Q. As a rule, only those products $Q_i Q_j Q^m$ ($m \geq 1$) which are symmetry allowed would also appear in a Landau potential (so that the same rules can be applied as far as the construction of coupling energies between different order parameters). The total energy is then for a simple example:

$$G = \frac{1}{2}A(T - T_c)Q^2 + \frac{1}{4}BQ^4 + \frac{1}{6}CQ^6 + \frac{1}{2}\sum \chi_{ij}^{-1} Q_i Q_j + \sum \delta_{ij}^m Q_i Q_j Q^m \qquad \text{(IV.19)}$$

Minimizing G with respect to the phonon coordinates ($\partial G/\partial Q_j = 0$) leads to the new self energy ($i = j$)

$$\chi_{ii}^{-1} Q_i + 2\sum \delta_{ij}^m Q_i Q^m = 0 \qquad \text{(IV.20)}$$

Thus the renormalized phonon susceptibility (i.e. the value changed for $Q \neq 0$) becomes

$$\chi_{ii}^{*-1} = \chi_{ii}^{-1} + \sum_m \delta_{ii}^m Q^m \qquad \text{(IV.21)}$$

The change in the phonon frequency is proportional to the change of χ^{-1}, i.e.

$$\Delta(\omega^2) \propto \chi_{ii}^{*-1} - \chi_{ii}^{-1} = \sum_m \delta_{ii}^m Q^m \qquad \text{(IV.22)}$$

Symmetry arguments show that m is, in lowest order, either 1, 2 or 2, 4 so that the Q-dependence of $\Delta(\omega^2)$ is either

$$\Delta(\omega^2) = \delta_1 Q + \delta_2 Q^2 \qquad \text{(IV.23)}$$

or

$$\Delta(\omega^2) = \delta_3 Q^2 + \delta_4 Q^4 \qquad \text{(IV.24)}$$

No higher order dependences have yet been documented. The dependence in (IV.23) applies to nonconvergent ordering in feldspars only, and the equation (IV.24) to all symmetry breaking phase transitions in feldspars. The change of the amplitude I of ε'' (and, hence, all correlated quantities) can be calculated using a similar line of argument. It is found that in feldspars the amplitude renormalises for symmetry breaking processes as

$$\Delta I \propto a\,Q^2 + b\,Q^4 \qquad \text{(IV.25)}$$

where one expects $a = 0$ for modes which are not IR or Raman active in the paraphase.

Finally, the damping constant

$$\Gamma = \Gamma_o + \Delta\Gamma(Q) \quad (\Gamma_o = A_\Gamma T + B_\Gamma T^2) \qquad \text{(IV.26)}$$

for a single mode (i.e. a profile described by eq. (VI.10)) shows a Q-dependence of $\Delta\Gamma$ similar to $\Delta(\omega^2)$, with the additional observation that the maximum $\Delta\Gamma$ for Al,Si ordering is of the same order of magnitude as the maximum shift $\Delta\omega$ of the mode frequency. The increase of the intrinsic value of $\Delta\Gamma$ during a displacive phase transition in feldspars is very small (≤ 1 cm^{-1}) although it is often increased by the influence of defects.

In summary, we find that all relevant parameter, $\Delta(\omega^2)$, ΔI and $\Delta\Gamma$ depend for symmetry breaking phase transitions in feldspars on the order parameter Q via the simple relationship:

$$\Delta(\omega^2),\ \Delta I,\ \Delta\Gamma \propto AQ^2 + BQ^4 \qquad \text{(IV.27)}$$

whereas nonconvergent ordering also allows a term proportional to Q. Note that for small changes of ω one finds $\Delta(\omega^2) = (\omega^2 - \omega_o^2) \approx 2\omega_o(\omega - \omega_o) \propto \Delta\omega$. The coefficients A and B have to be determined experimentally. It appears from previous results that all phase transitions in feldspars display the expected rapid convergency of the expansion in equation (IV.22), with $A >> B$ so that all higher order terms may be ignored. Although this preliminary result would indicate an extremely simple way to analyse spectra (i.e. all parameters are proportional to Q^2 for symmetry breaking transitions), it may be too early to jump to this conclusion yet.

4.3 THE CHARACTERISTIC LENGTH SCALE

Hard modes "see" order parameters on a rather local scale. This makes the technique of Hard Mode Spectroscopy comparable to NMR and EPR and contrasts it to more long range probes such as coherent scattering techniques and soft mode spectroscopy.

The characteristic length can be derived from the dissipation-fluctuation theorem in equation (IV.7). The dispersion of a hard mode in the long-wavelength limit is approximately:

$$\omega^2 = \omega_o^2 + \alpha^2 q^2 \qquad \text{(IV.28)}$$

where q is the wavevector in the appropriate direction and α is a measure of the strength of the dispersion (Einstein oscillators would have $\alpha = 0$). The phonon amplitude is for high temperatures

$$\left\langle Q_i^2 \right\rangle \propto \frac{kT}{\omega_o^2 + \alpha^2 q^2} \tag{IV.29}$$

The equivalent correlation function in real space is described by Q_i at the site i and Q_j at the site j by

$$\left\langle Q_i Q_j \right\rangle \propto \frac{1}{|r_i - r_j|} \exp\left(\frac{-\omega_o (r_i - r_j)}{\alpha} \right) \tag{IV.30}$$

This correlation decays exponentially (besides the trivial prefactor $1/|r_i - r_j|$) with a characteristic length ξ where

$$\xi = \frac{\alpha}{\omega_o} \tag{IV.31}$$

This length decreases rapidly with increasing ω_o even if $\alpha \neq 0$. Expressing α by the phonon band width $\Delta\omega = \omega_o$ and a lattice constant $a = 10$Å one finds

$$\xi \approx \frac{a}{\pi} \sqrt{\frac{2\Delta_\omega}{\omega_o}} \approx 5\text{Å} \tag{IV.32}$$

Hard modes at sufficiently high frequencies are therefore not very different from independent Einstein oscillators. Local structural changes are detected by these modes on a level of one or two tetrahedra with little influence from other structural effects further apart. This means that the phonons provide us with a "true picture" of structural variations with little disturbance by the measuring process itself. Its interpretation can be based on a simple superposition of signals ignoring spatial averaging by the phonons.

4.4 HOW TO DO IT

The essence of hard mode spectrocopy is to measure small changes of phonon signals and to correlate these changes with the thermodynamic order parameter. This correlation becomes the more simple the smaller the change of the phonon signal is. The best phonon is always the one which (i) is isolated in a spectrum so that it does not overlap with other modes, (ii) has a narrow line profile, and (iii) shows at first glance no or little effect when the phase transition takes place. Such a mode is then measured in high resolution and the small changes of its line profile (peak freqency, intensity, line width) is measured with the highest possible accuracy. In virtually all cases, one finds that phase transitions create line shifts of between 0.5 cm^{-1} and 6 cm^{-1}, intensity changes of some % and changes of line widths in order-disorder transitions of the same magnitude as the frequency shift (and much less for displacive transitions). These small changes have been the bane of spectroscopists in the past because they were difficult to quantify. The pioneering work of Laves and Hafner (1956) was one of the most daring attempts to use hard mode spectroscopy before more advanced experimental techniques developed around 1970.

Small as the effects may appear, the experimental resolution in Raman and Infrared spectroscopy is largely sufficient for their investigation. Modern Raman spectrometers allow the determination of peak positions and line widths with a resolution of better than

0.1 cm^{-1}. The measurement of absolute intensities still varies considerably between different instruments. Good experimental arrangements on large enough single crystals allow a reproducibility of intensity measurements to within 15% on small amounts of material. Infrared powder spectrometry is still one of the best ways to measure phonon intensities. Independently prepared samples of feldspars show spectra which agree on an absolute scale within 15%. Relative changes on the same sample (e.g. due to changes of temperature) can be measured within an accuracy of 8%. Frequency changes (peak positions and line widths) are reproducibly determined within a resolution of better than 0.1 cm^{-1} for isolated peaks. Emission and reflection measurements have similar virtues although the control of the sample environment is often the major problem.

The challenge to the experimentalist is to measure as many phonon signals as possible as a function of temperature, pressure etc.. The individual line profiles can then be directly correlated with the dielectric function. In the case of Raman spectrocopy, the profile of a one-phonon Stokes signal is determined by the time correlation function of the polarisation coordinates P:

$$I(\omega) = \frac{1}{2\pi N} \int_{-\infty}^{+\infty} \langle P^*(O)P(t)\rangle e^{i\omega t} dt \propto (n(\omega)+1)P^2 \varepsilon''(\omega) \qquad \text{(IV.33)}$$

where $n(\omega) = 1/(\exp(\hbar\omega/kT)-1)$ and $\varepsilon''(\omega)$ is as given in equation (IV.12). If P depends directly on the order parameter Q, one finds the predicted dependence of ΔI on Q^2. All changes of ω_O and Γ relate directly to changes of $\varepsilon''(\omega)$ as a function of the order parameter. Further effects are discussed by Bruce et al. (1980) and Salje et al. (1983).

Infrared spectroscopy using powder samples is particularly useful for the investigation of feldspars because it requires only a small amount of well-characterised starting material (say 20 mg). Its disadvantages are, firstly, that powder samples in KBr or NaCl cannot easily be heated in vacuum at temperatures beyond 700°C. Secondly, the powder spectrum is only approximately related to $\varepsilon''(\omega)$. In the case of feldspars, however, this second obstacle is much less serious than in other, more polar, materials. A very good approximation for the frequency shift from the peak in $\varepsilon''(\omega)$ for a sample with grain sizes of ca. 0.5 μm for feldspar embedded in a matrix of KBr is

$$\frac{\omega^2_{\text{experiment}}}{\omega_o^2} = \frac{\varepsilon'_o + 2\varepsilon'_m}{\varepsilon'_\infty + 2\varepsilon'_m} \qquad \text{(IV.34)}$$

where $\varepsilon'_m \approx 2.3$ is the dielectric constant of KBr near 600 cm^{-1}, ε'_∞ is the dielectric constant of feldspar at frequencies above ω_O and ε'_o the same parameter at frequencies below ω_O. For peaks with small intensities (i.e. small LO-TO splitting) one finds $\varepsilon'_o / \varepsilon'_\infty \approx 1.1$. A reasonable estimate for ε'_o in the mid-IR region is ca. 4, so that the unwanted shift of $\omega_{experiment}$ from ω_O is only some 2%. The additional relative shift due to the phase transition is virtually uneffected by the embedding because no macroscopic change of ε'_o appears to occur in feldspars. Under these conditions we find that the absorption spectrum of sufficiently fine powder samples is well approximated by $\varepsilon''(\omega)$ (without further temperature dependent prefactors such as in the Raman spectrum).

Reflection spectra on single crystals are analysed using standard procedures but no systematic work on phase transitions in feldspars seems to have been undertaken, possibly due to the lack of appropriate sample material. Although there are new developments in

IR emission and stroboscopic techniques, it is too early to report on their application to feldspars in this paper.

Let us finally mention a useful trick for the numerical analysis of spectra. An important constraint on the possible intensity variations follows from the so-called "sum rule" which states that the integrated intensities of bands of phonons which do not interact with other phonons remain constant (under conditions met by feldspars). This feature is particularly impressive to observe when the temperature evolution of IR powder spectra is investigated. As such spectra are essentially proportional to ε'' without additional temperature dependences, one sees that phonon bands with gaps to other phonon bands rearrange the intensities inside each band but keep the same integral intensities of a band at all temperatures. This feature can also be used to correct for spurious temperature effects. One possible method is to integrate the intensities of the stretching modes between 830 cm^1 and 1450 cm^{-1}, and normalize the intensity scale for each absorption spectrum with respect to these integrated intensities.

5. Applications and examples

We now discuss specific applications of hard mode spectroscopy for specific phase transitions and for the determination of degrees of Al,Si order in feldspars. The examples are chosen to represent typical applications rather than to review the full breadth of work done in this field.

5.1 TEMPERATURE DEPENDENCE OF HARD MODES IN ALBITE

Al,Si ordered albite shows, when heated rapidly, no structural phase transition. This material is ideal, therefore, for the investigation of the temperature dependence of phonon spectra which occur as a background to feldspars with no superimposed signals due to phase transitions. A first study was carried out by Salje (1986) using a highly stabilized Raman spectrometer which allowed heating experiments without mechanical changes to the scattering geometry. Note that this is an essential pre-requisite for hard mode spectroscopy (more in Raman spectroscopy than in IR spectroscopy), because each sample movement (e.g. due to thermal expansion of the heater) leads to changes of the line profiles which may obscure the effects one wants to observe. Raman spectra for two scattering geometrics are shown in Fig. 4. Several strong peaks are easily identified, namely near 293 cm^{-1}, 482 cm^{-1}, 510 cm^{-1} and, outside the depicted spectral range, at 1104 cm^{-1}. No soft modes or even low frequency modes below 40 cm^{-1} were seen.

There is little temperature dependence of the mode frequencies and also, more importantly, of the integral scattering intensities (Fig. 5). Na-feldspar without phase transitions does not show any phonon anomalies which could be mistaken for the finger print of a phase transition.

The damping constants Γ are measured as full width at half maximum (FWHM) for each peak, with $A_\Gamma >> B_\Gamma$ in eq. (IV.26). Their temperature dependence in Fig. 6 closely follows the predicted dependence for Γ_o. Note that, as a rule of thumb, Al,Si ordered Ab,Or and An feldspars show for most phonon lines a Γ_o at absolute zero temperature of less than 1 cm^{-1}, and a characteristic value of 5-7 cm^{-1} at room temperature. We can compare this result for albite with that of anorthoclase with low degree of Al,Si order and ca 31% replacement of Na by K (Or_{31}). The line widths shown in Fig. 7 are now much

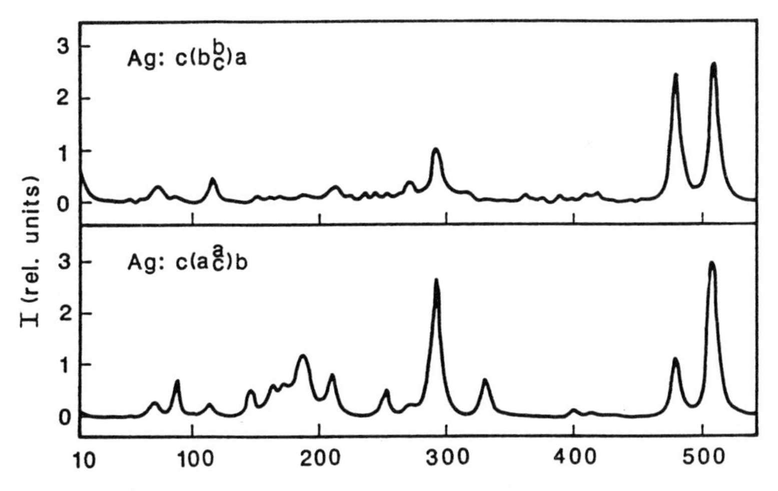

Figure 4. Raman spectra on Na-feldspar (natural sample from the Amelia locality) for two scattering geometries.

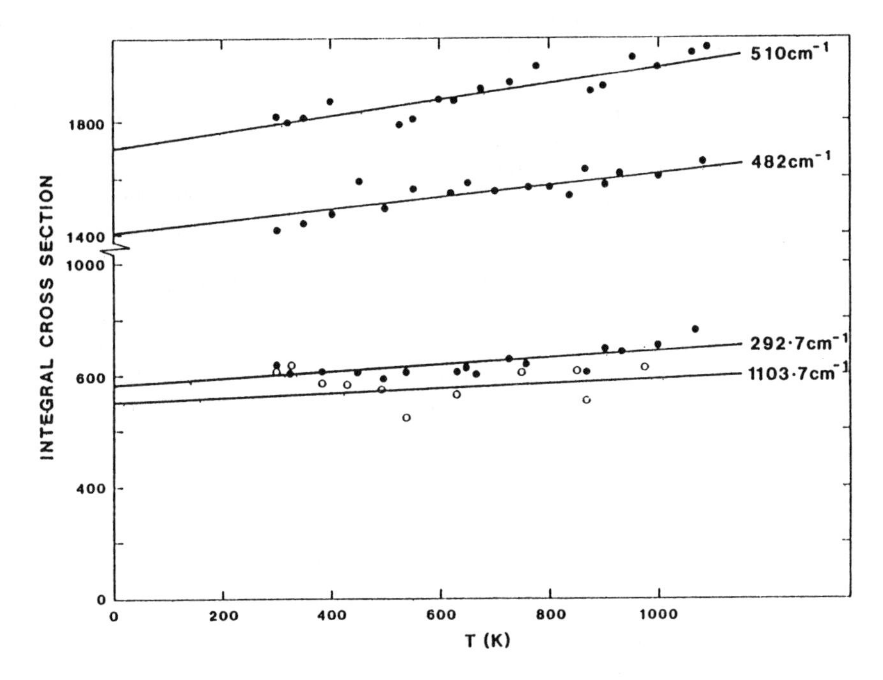

Figure 5. Temperature dependence of the integral Raman cross section for 4 signals in the spectrum. None of these lines shows a strong temperature dependence which could be taken as indicative for a phase transition.

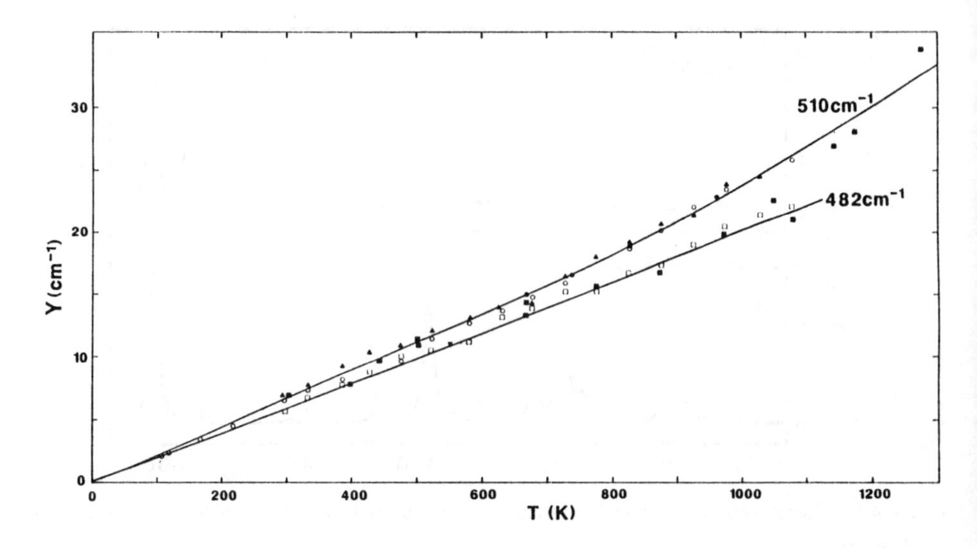

Figure 6. Increase of the thermal line width of Raman signals for Na-feldspar ($Q_{od} = 1$) following $\Gamma = AT + BT^2$ (details in Salje, 1986).

broader, with a minimum value of 10-16 cm^{-1} at 0K. These values are comparable with the spread of phonon frequencies due to Al,Si disordering enhanced by additional effects from (quasi) random K-Na distributions. The total line width is deconvoluted (eq. IV.13) into a temperature independent part (from static disorder) and thermal effects. The thermal effect is of the same order of magnitude as in albite (i.e. some 2 cm^{-1} per 100K).

5.2. EFFECT OF THE C2/m-C$\overline{1}$ TRANSITION IN ANORTHOCLASE

The best example of a displacive phase transition in feldspars is the C2/m-C$\overline{1}$ transition in Na-rich alkali feldspars with no Al,Si order ($Q_{od} = 0$). Raman active phonon lines show little or no coupling between the structural order parameter Q and the mode frequencies. The scattering intensities, on the other hand, are enhanced in the triclinic phase. Fig. 8 shows a temperature independent cross section (i.e. the integrated intensity divided by the Bose-Einstein factor $(n(\omega) + 1)$ at $T > T_c$. At T_c, this value starts to increase linearly with decreasing temperature. Using eq. IV.27 (with $B = 0$) we find

$$\Delta I \propto Q^2 \propto |T - T_c|$$

or

$$Q = \left(\frac{T_c - T}{T_c} \right)^{1/2}$$

indicating a second order phase transition with a Landau potential of the form

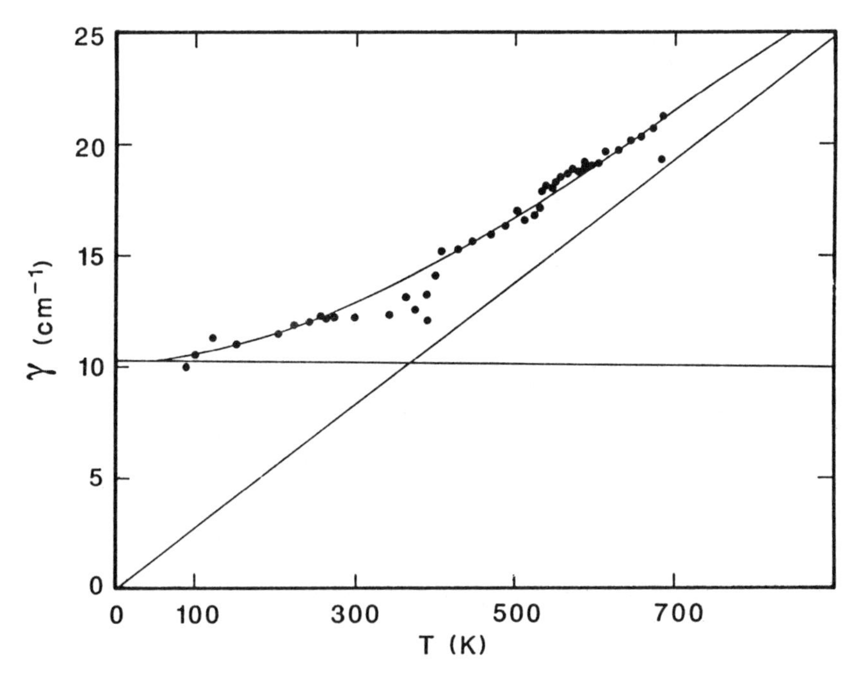

Figure 7. The temperature dependence of the line width of Raman signals of hypersolvus alkali feldspars shows two components : a temperature independent part (ca 10 cm^{-1}) and a thermal broadening similar to the results in Na-feldspar in Fig. 6.

$$L(Q) = \frac{1}{2}A(T-T_c)Q^2 + \frac{1}{4}BQ^4$$

where T_c = 523K and $AT_c = B$ (because Q = 1 at T = 0K). No anomalous line broadening was observed in good quality single crystals, indicating that no heterogeneous transitions or large scale fluctuations occur. Deviations from such an ideal, displacive transition is seen in extremely fine-grained material, as sometimes used for ion-exchange synthesis of binary feldspars. This suggestion of a more defect-influenced transition in fine grained material could indicate that some care should be taken when ion exchanged samples are used for the investigation of transition mechanisms in feldspars.

5.3. A DISPLACIVE PHASE TRANSITION WITH SOME FLUCTUATIONS: $P\bar{1}$ - $I\bar{1}$ IN ANORTHITE

This phase transition was first observed by Brown et al (1963). Its fundamental nature and the driving force behind the transition has been investigated in a number of structural and thermodynamic studies. Their results can be summarized in terms of three models:

5.3.1. *Space Average Model.* Czank et al. (1972) first linked their electron microscopic observations of *c*-domain structure with the persistence of *c*-intensity using a static disorder model for the $I\bar{1}$ structure. In this model, they suggested that the observed

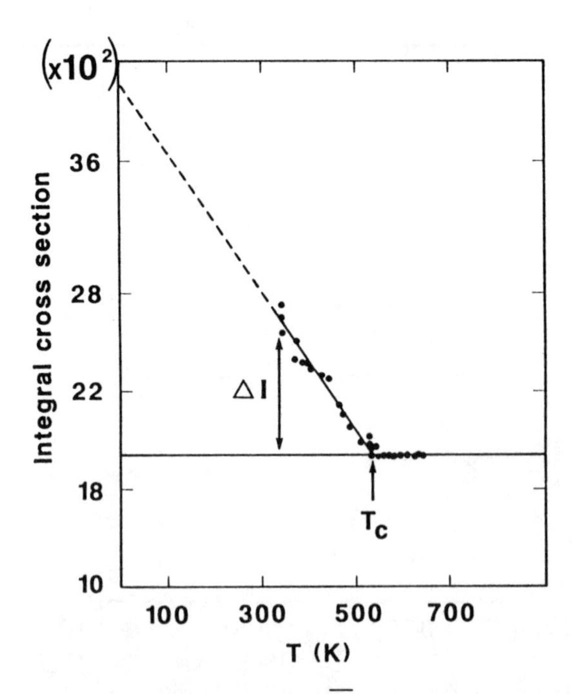

Figure 8. The phase transition C2/m - $C\bar{1}$ in hypersolvus alkali feldspar appears as break in the intensity *versus* temperature curves. The excess intensity at low temperatures is proportional to Q^2.

average structure was due to the space averaging of near-body-centred domains of anorthite. Although this could account for persistent elastic scattering at the *c*-reflection position above T_c, it is in complete contradiction with the NMR results of Staehli and Brinkmann (1974) who observed an increase in Al site-symmetry above T_c on a dynamic time scale.

5.3.2. *The Ca-Jumping Model.* Megaw et al.'s (1962) first description of the anorthite structure drew attention to the rigid nature of the aluminosilicate framework compared with the weak bond strength at the calcium atoms. The calcium jumping model draws on the assumed detachment of these two structural features and presumes that the coupling between the calcium atoms and the framework is weak. In the positional disorder model of the $I\bar{1}$ phase associated with Megaw et al.'s (1962) description, Ca atoms jump between two positions in a deep double well potential. The time average of these two $P\bar{1}$ positions gives rise to the split-atom refinements of the Ca position. Staehli and Brinkmann (1974) rationalised their NMR results within this model by suggesting that the jump frequency was greater than the precession frequency of the local field around the Al atom (of the order of 6 x 10^3 Hz) and that their results were themselves a time average of the high-temperature dynamic structure. Under such a model any *c*-intensity at high-temperature must be due to inelastic scattering. High-temperature neutron scattering experiments (Frey et al. 1977; Adlhart et al. 1980a, b) were inconclusive, although Adlhart et al. (1980a) identified low-energy inelastic scattering above T_c, from which they suggested any dynamic process must occur at a frequency less than 10^{12}Hz.

5.3.3. *Soft-Mode Model.* Adlhart et al (1980a) favoured a soft-mode model for the phase transition, in which there is strong corelation between the behaviour of the framework and the calcium atoms. They identified critical phonons at the Brilloin zone boundary and suggested that the transition was driven by softening of phonons specifically associated with the calcium atoms. They suggested that dynamic in-phase and antiphase domains of $P\bar{1}$ symmetry may exist in the high-temperature structure, separated by boundaries of zero-energy. This model accounts for some of the neutron scattering results but does not explain the observation of split calcium positions in the high-temperature phase. Salje (1987) argued that this "soft mode" might not actually condense for specific structural reasons although he supported the general idea of this approach.

Ghose et al. (1988) added some thoughts on the high-temperature behaviour, elaborating on the manner in which the framework first moves towards $I\bar{1}$ symmetry as T_c is approached from below. In what seems to be a re-working of the soft-mode model they suggested that, close to T_c, "breathing motion type lattice vibrations of the framework cause the structure of the Ca sites to interchange dynamically" between body-centre-related primitive sites. At the same time they suggested that an order-disorder mechanism involving the Ca atom configurations finally drives the transition. The supposed order-disorder nature of this process is then linked to the speculative assertion that the transition is actually first-order in all anorthites at T_c and that the high-temperature phase is a disordered average structure.

In an approach to describe the transition in terms of a Landau type Gibbs free energy, Salje (1987) developed an expression which explained the co-elastic strain, the excess entropy and the influence of the Al,Si ordering associated with Q_{od}. Furthermore, the incorporation of Ab-component was considered. Coupling between Al/Si order (Q_{od}) and the order parameter of the $I\bar{1}$-$P\bar{1}$ transition (Q_o) accounts for the observed change in transition character (as a function of temperature) from tricritical for pure ordered An_{100} to second order for slightly disordered $An_{98}Ab_2$ (Wruck 1986; Redfern and Salje 1987; Redfern et al. 1988). Subsequently, the same model has been used to explain the high-pressure behaviour of anorthite, in which the $I\bar{1}$ phase appears to be stable above 2.5 GPa (Angel 1988, 1992; Angel et al. 1989, 1992), and the transition becomes first-order. Interestingly, Ca appears to sit on the true body-centred position in the high-pressure $I\bar{1}$ anorthite structure (Angel 1988), and there is no evidence of any deviation from body-centred symmetry for either the lattice or the alkali site. For further discussions the reader is referred to the article by R. Angel in this book.

A distinction between the various models is possible from the in-situ study of hard modes in anorthite. Such a study was carried out by Redfern and Salje (1992) at temperatures up to 715 K on a sample of anorthite from the Monte Somma locality (where it occurs in rapidly cooled limestone ejecta). This sample has been characterised by Carpenter et al. (1985), who determined a composition of around $An_{98}Ab_2$. The $I\bar{1}$-$P\bar{1}$ transition in Monte Somma anorthite had previously been studied using single-crystal X-ray and neutron diffraction (Adlhart et al. 1980b), by DSC (Wruck 1986) and by high-temperature powder diffraction (Redfern et al. 1988). The results of these experimental investigations suggest that the $I\bar{1}$-$P\bar{1}$ transition in Monte Somma anorthite is second-order as a function of temperature and behaves according to a Landau-type model. Redfern et al. (1988) also reported significant low-temperature saturation of the order parameter (Q_o) below 200 K, which agrees with the results of the generalised Landau theory developed by Salje et al. (1991).

Redfern and Salje (1992) used powder hard mode infrared spectroscopy in a high-temperature study. Monte Somma anorthite was crushed and milled in an agate mortar in a Spex micro ball mill for 35 minutes. The even-sized fine-grained anorthite powder was kept in a drying oven at 120° C to ensure it was totally anhydrous prior to the preparation of KBr pellets. The standard KBr technique was employed by diluting anorthite powder in spec-pure KBr in the ratio 1:300 and die-pressing under vacuum to produce an optically transparent 13mm pellet.

Spectra were recorded under vacuum at temperatures between room temperature and 715 K using the Bruker 113v FTIR instrument. The sample was positioned within a cylindrical platinum-wound furnace in the sample compartment and the sample temperature was measured using a Pt-Rh thermocouple held in contact with the sample pellet. The sample temperature was controlled using a Eurotherm temperature control system to a stability of better than $\pm$ 1 K. All spectra were collected using a liquid-nitrogen cooled MCT detector and KBr beamsplitter at a resolution of 2 cm^{-1} over the range 450-1500 cm^{-1}. Transformed spectra were analysed by least-squares fitting of fixed-shape mixed Gaussian-Lorentzian peak profiles to the recorded absorption with an R-factor typically less than 0.05. Errors in measurements of peak positions and widths are of the order of $\pm$ 0.2 cm^{-1} by this technique, and 10% for integrated intensities.

The temperature evolution of the absorption spectra are shown in Figs. 9 and 10. Most hard modes show breaks of their frequency *versus* temperature curves at the transition point. In all cases a linear excess frequency $\Delta\omega$ was found with

$$\Delta\omega \propto q^2 \propto (T_c - T), \qquad (T < T_c = 515 \pm 5\ K)$$

Here q represents the "local" order parameter equivalent to the thermodynamic parameter Q. The order parameter shows the same behaviour as in our previous example of anorthoclase so that the phase transition in Monte Soma anorthite is, indeed, second order as found by calorimetric and diffraction experiments.
We now focus on the question of displacive or order/disorder character of the transition. This question can be solved by the analysis of the line profiles of the phonon signals as discussed before (see also Matsushita, 1976). The mode showing the greatest change in $\partial\omega/\partial T$ is that around 582 cm^{-1}. The linewidth of this, the strongest mode in the region of interest, is shown in Fig. 11. The theoretically expected temperature dependence of the damping factor, Γ, follows from multi-phonon interactions and may be expressed in the form of eq. (IV.26).

The residual line width at 0 K reflects principally the broadening due to density of states effects. The expected increase in Γ as a function of temperature due to multiphonon damping effects, $\Gamma_o(T)$, is shown by the solid line for the absorption band at around 582 cm^{-1}. The low temperature extrapolation of this behaviour gives a line width (full width half maximum height) of 11.6 cm^{-1}, which is slightly higher than might be expected for a perfect crystal, and probably reflects the inherent static Al/Si disorder and Na-inhomogeneity due to the 2 mol% Ab content of Monte Somma anorthite. The multi-phonon damping from Fig. 11 is $\Gamma_o(T)$ = 11.6 + 2.55 x 10^{-2} T (ignoring the quadratic term). In addition to this broadening, there is an obvious transition-related line broadening

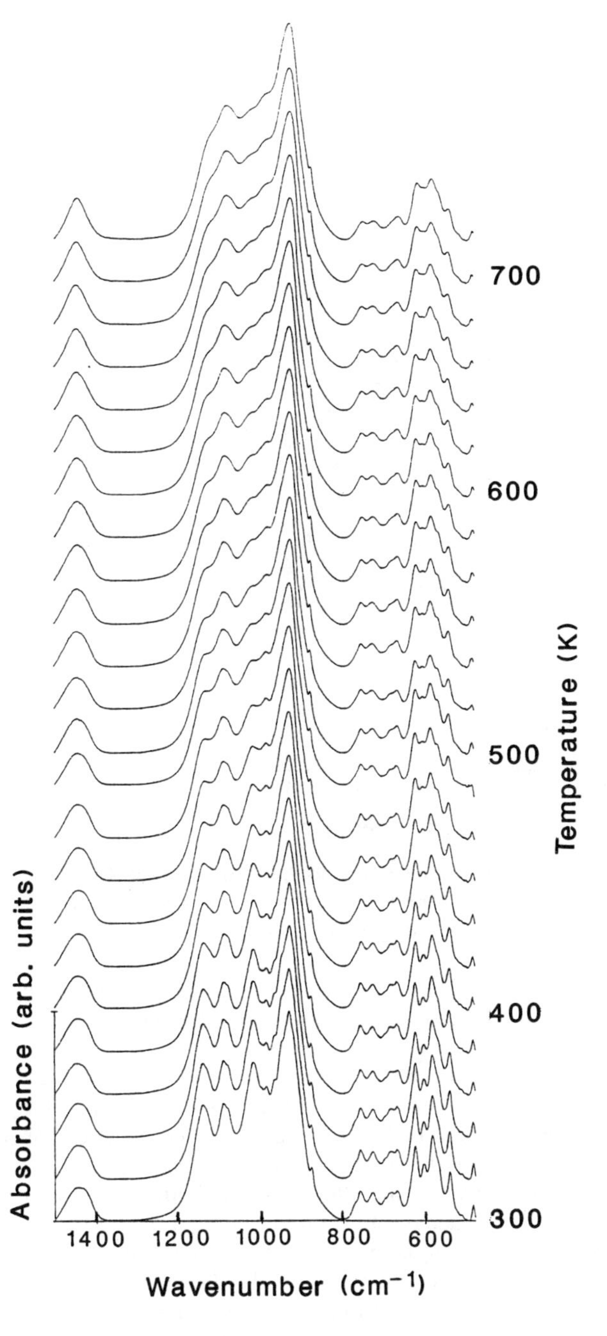

Figure 9.a Temperature evolution of infrared absorption spectra in anorthite.

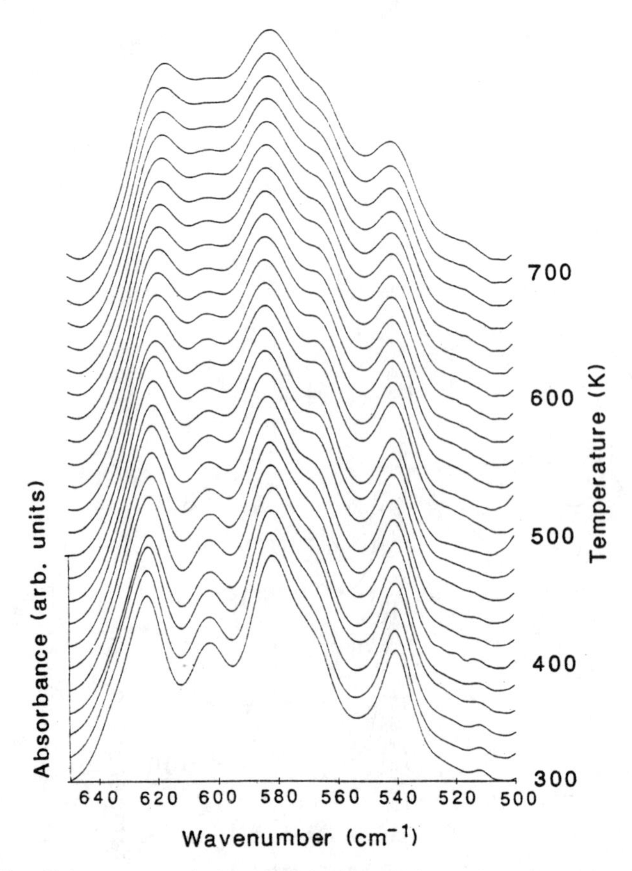

Figure 9.b Temperature evolution of infrared absorption spectra in anorthite. Note the well-separated peaks near 540cm^{-1}, 620cm^{-1} which are well suited for hard mode spectroscopy. The integral intensities of the spectra are virtually temperature independent as predicted theoretically.

shown by the dashed line. The time-scale of the dynamic fluctuation is given by the excess broadening above that expected from pure multi-phonon anharmonic effects, Γ-Γ_0. This excess is of the order of 2 cm^{-1}, corresponding to dynamic fluctuations on a time scale of 6x10^{10} Hz. The temperature interval over which these fluctuations are discernible by infrared absorption is some 70 K or so either side of T_c and there is no evidence for their persistence at high-temperature. Indeed, none of the spectra appear to provide evidence of order-disorder phenomena deep in the I $\overline{1}$ phase.

It is tempting to correlate the line broadening in anorthite with the observed antiphase boundaries (Van Tendeloo et al. 1989). Similarly, we could envisage a model in which the excess line broadening is linked to dynamic behaviour at the Ca-site. In either case it is most important to note, however, that the thermodynamic transition mechanism (as seen by the temperature evolution of the mean phonon frequencies) remains essentially uninfluenced by these fluctuations. Our present results also show that the mean order parameter follows a classical second-order phase transition, with no indication of a

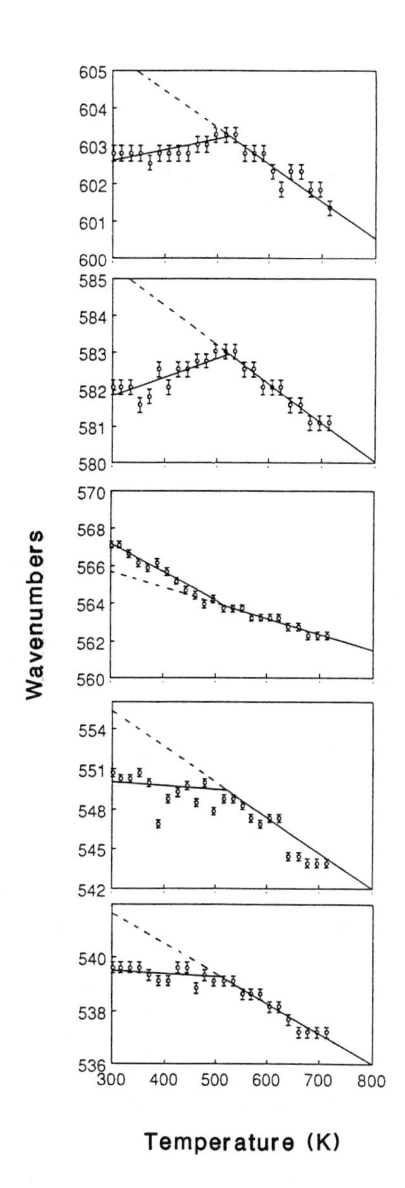

Figure 10. Shift of phonon frequencies due to the phase transition P $\overline{1}$ - I $\overline{1}$ in anorthite. T_c is the temperature of the break on the slope dω/dT. The extrapolated frequencies of the high symmetry phase are indicated by the dotted lines. The difference between the dotted line and the data points is $\Delta\omega$ which is proportional to Q^2 (after Redfern and Salje, 1992).

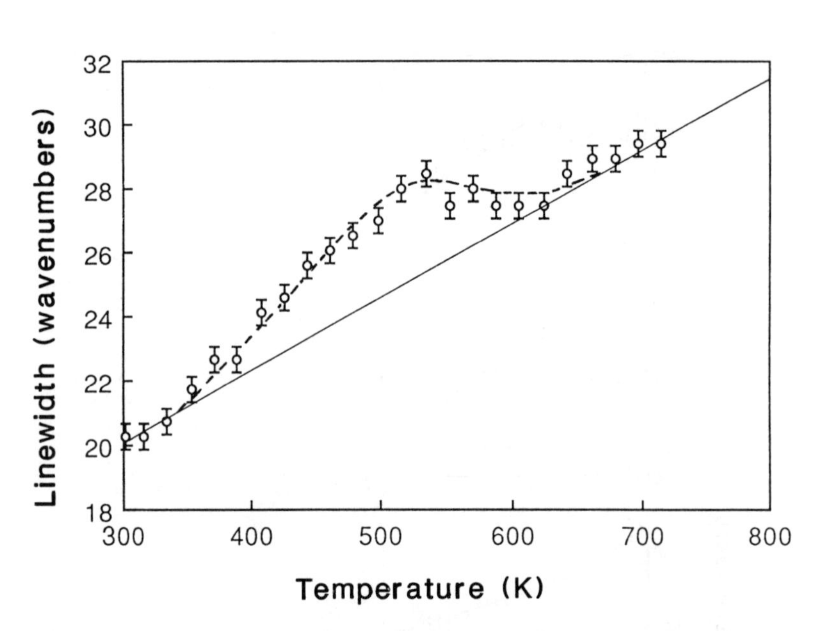

Figure 11. Excess line broadening of the 582-1 phonon due to "fluctuations" near the P 1 - I 1 transition in anorthite (after Redfern and Salje, 1992).

stepwise behaviour. The vanishing excess linewidth above T_c allows us to reject confidently an order-disorder model for the high-temperature phase since no large volume proportion of $P\bar{1}$ phase exists at $T >> T_c$, although minor components (for example those related to lattice imperfections or chemically pinned defects) cannot be ruled out.

In summary, the results from hard mode spectroscopy are in agreement with the idea that the phase transition is driven by a lattice distortion similar to the model proposed by Adlhart et al. (1980a). The lattice distortion shows additional fluctuations which are beyond a soft-mode model in its most simple form. Redfern and Salje attribute these fluctuations to weak coupling to the Ca-flip motions, which we expect to stabilize the soft mode. Thermodynamically (with respect to the behaviour of the space and time averaged order parameter) this coupling is not relevant, however. The fluctuations decay rapidly as T becomes greater than T_c, and do not support an order-disorder model for the $I\bar{1}$ phase. On a macroscopic, thermodynamic scale the phase transition can well be described as a classical displacive transition with some fluctuations which do not, however, destroy the thermodynamic second order character of the transition. This does not mean that such fluctuations do not change the stability field of the $P\bar{1}$ phase, e.g. its T_c or P_c. It is very likely that the application of pressure will limit the Ca position to one single site rather than a split site. In this case the $P\bar{1}$ phase would become more stable because the entropic stabilisation of the $I\bar{1}$ phase via jumps is suppressed. The expectation is then that, within the framework of the theory by Salje (1987), the extrapolated T_c for the high pressure part of the P_cT_c diagram should be as high as some 1300 K. This extrapolated T_c is then reduced for $P = 0$ to 515 K (see also the paper by Angel in this book).

5.4. DETERMINATION OF THE DEGREE OF Al,Si ORDER IN Na-FELDSPAR

In the last two examples we have discussed the effect of displacive phase transitions on the Raman and IR spectra of feldspars. The example in this section is chosen to illustrate that hard mode spectroscopy is an equally powerful tool for the investigation of ordering processes. The crucial advantage of hard mode spectroscopy is its short characteristic length scale. This means that the peak positions indicate the most probable local configuration, while the peak profiles indicate the distribution function. It is irrelevant for the application of this technique whether or not the sample displays microstructures (e.g. tweed structures), because the characteristic length scale of hard modes is, in general, smaller than the length scales of such patterns.

The dependence of hard modes on the degree of Al,Si ordering in Na-feldspar was measured on samples from the Amelia locality. The lattice parameters and the results of the chemical analysis (microprobe) are given in Carpenter et al. (1985). The albite samples were annealed isothermally at 1353 K for various times. Each run was performed with fresh starting material from the same batch. All samples were subsequently analysed using X-ray powder diffraction, and transmission electron microscopy.

The annealed Na-feldspar samples were then crushed and milled using a spex-mill (micro mill 35 min). Infrared spectra of all samples were obtained by using the standard KBr-pellet technique. Each pellet had a weight of 200 mg and contained 0.67 mg of Na-feldspar. All pellets were prepared under vacuum and were optically transparent. Extreme care was taken to obtain exactly reproducible conditions for sample preparation. Pellets from the same experimental runs showed band shifts of corresponding peaks not larger than ± 0.2 cm^{-1} and variations in their absolute intensities not larger than 8 percent.

A sequence of 12 samples of Na-feldspar with different annealing times at an annealing temperature of 1353 K is shown in Fig. 12. The samples here held at 77 K while the spectra were collected. The spectrum of the starting material shows rather sharp absorption peaks. As the annealing time is increased, the peaks become broader and the number of resolved peaks decreases. This is particularly apparent in the ranges 700-800 cm^{-1} and 900-1300 cm^{-1}. Only in the range of 500-700 cm^{-1} do the bands remain very distinct, and the analysis of the fine structure, as employed in HMIS, was focussed on these modes.

With increasing annealing times, the band at 615 cm^{-1} becomes less prominent and, after an annealing time of 290 h, this mode appears only as a small 'shoulder' of the band at 590 cm^{-1}. An additional band (545 cm^{-1}) appears on the high energy side of the 535 cm^{-1} band after 40 h annealing, the absorption signal of the 535 cm^{-1} band decreasing simultaneously. A 5-band fit was used for the analysis of this spectral region. The variation of the intensities of the two bands at 535 and at 545 cm^{-1} is shown in Figure 13.

Absolute frequencies of the phonon bands can be obtained from the spectra with an accuracy of ± 0.2 cm^{-1}. The observed frequency shifts due to the annealing process are below 5 cm^{-1}. The shift of the band around 535 cm^{-1} was previously used by Hafner and Laves (1957) as a measure of the degree of Al,Si order, because these authors believed that an actual line shift of 14 cm^{-1} would occur. This assignment is erroneous, however, since two bands with a gap of 8 cm^{-1} overlap in this band, and each of them shifts by only ca. 3cm^{-1}. The large apparent line shift is mimicked by the relative change of the intensities of these two bands which could not be resolved by Hafner and Laves (1957).

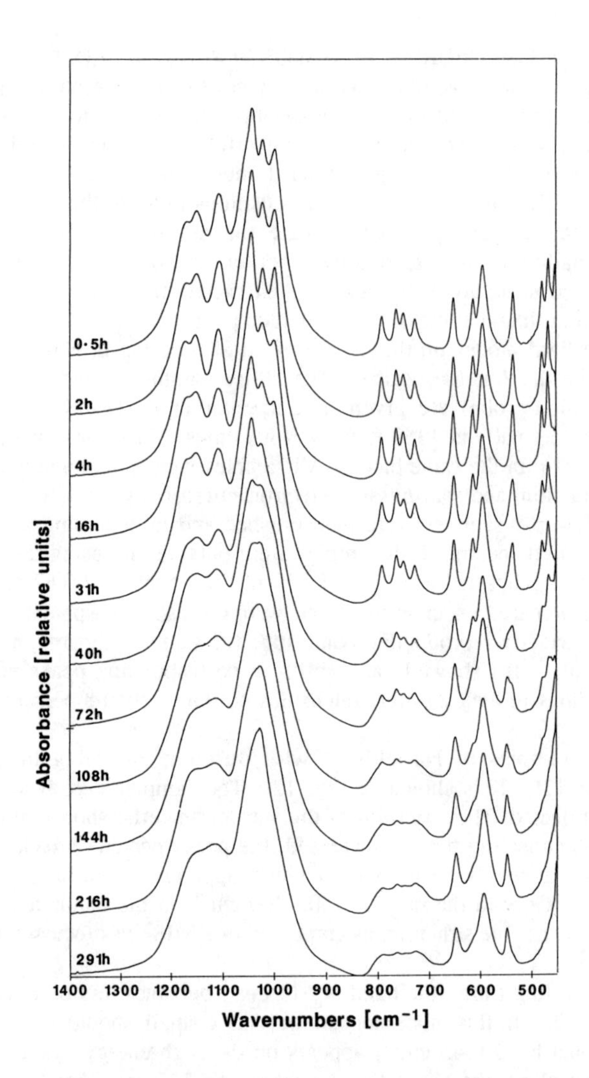

Figure 12.a Infrared absorption spectra of kinetically disordered Na-feldspar. The annealing time at 1353K is indicated. The spectra were measured at 77K.

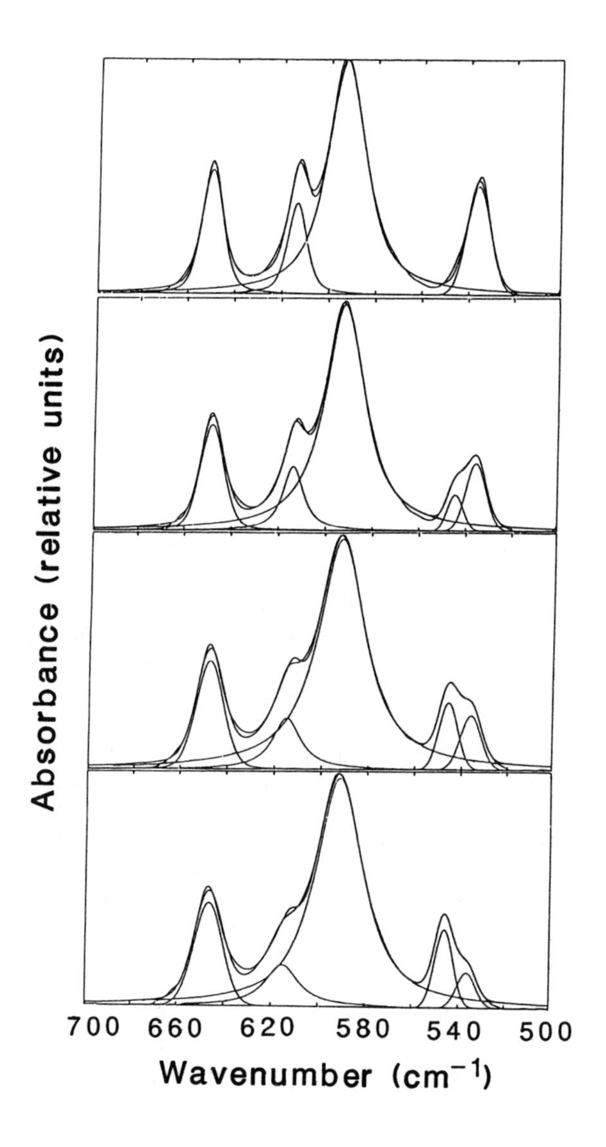

Figure 12.b Analysis of peak profiles of modes between 500 cm^{-1} and 700cm^{-1} for the determination of ω, Γ and I.

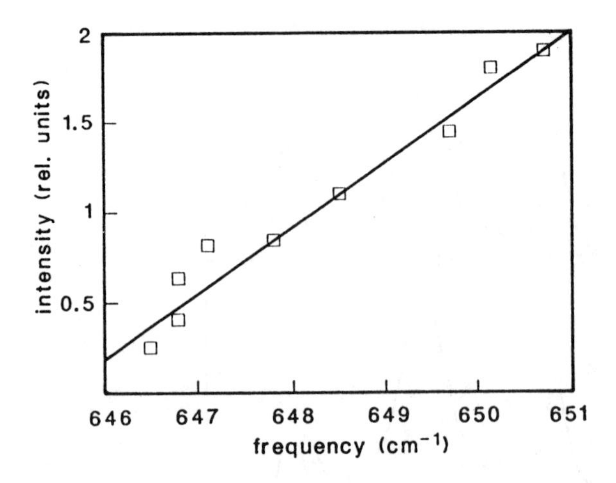

Figure 13. Correlation between the integrated intensity and the frequency for the 650 cm^{-1} mode.

The frequencies of the strongest peaks can be determined with the highest accuracy, and only these need to be used for the subsequent analysis. All other lines show within their data scatter exactly the *same* behaviour as the phonon bands between 500 cm^{-1} and 700 cm^{-1}. For the sake of clarity, we will discuss in the following the behaviour of the 650 cm^{-1} line *as an example.*

We now analyse the results in terms of the proportionalities in equation IV.27 with $B = 0$. As a first step we test the internal consistency by plotting the integral peak intensity versus the frequency shift and the linewidth Γ (Fig. 14). We find that all these parameters are, indeed, proportional to each other, although the data scatter is typically largest for ΔI, less for $\Delta\Gamma$ and smallest for $\Delta\omega$. Note that Γ for the disordered sample (14.2 cm^{-1}) is 5.2 cm^{-1} larger than that of the most ordered sample. This increment $\Delta\Gamma = 5.2$ cm^{-1} is in the same order of magnitude as the frequency shift between these samples (which is 4.2 cm^{-1}). This behaviour is to be contrasted with that of displacive transitions where $\Delta\Gamma$ is much smaller than $\Delta\omega$.

In a second step, we test the relationship between $\Delta\omega$ and the structural order parameter as determined by powder X-ray diffraction. It is known that the shift of the 2θ value of the $1\,\bar{3}\,1$ reflection is directly proportional to Q_{od}, and we expect $\sqrt{\Delta\omega} \propto Q_{od} \propto \Delta(2\theta_{1\bar{3}1})$. This relationship is, indeed, found in Fig. 15, which justifies the approximation $B = 0$. Applications of this study for the investigation of kinetic rate laws of Al,Si disordering in Na-feldspars are described by Wruck et al (1991).

5.5 STRUCTURAL MODULATIONS IN NATURAL POTASSIUM FELDSPAR

The structural behaviour of K-feldspar is similar to that of Na-feldspar, with the difference that T_c of Q is now at (extrapolated) negative absolute temperatures, whereas T_c of Q_{od} is of the same order of magnitude as in Na feldspar (Kroll et al. 1980; Carpenter 1988; Hovis 1974; Bambauer and Bernotat 1982). The large stability field of the monoclinic phase (i.e. sanidine) allows a careful study of the non-convergent ordering, Q_t (Salje and Kroll 1991). The thermodynamic behaviour of K-feldspar involving Q_t and

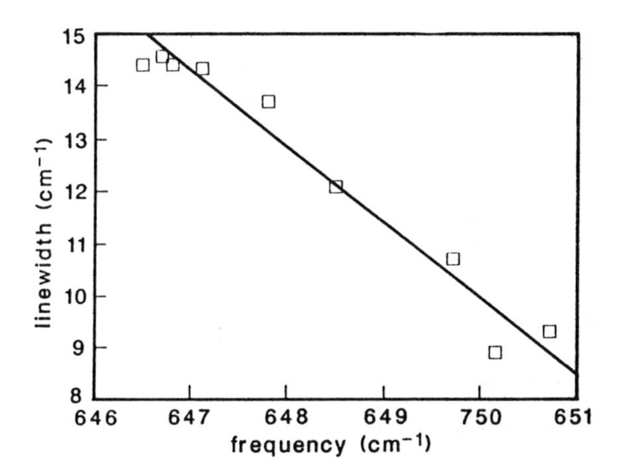

Figure 14. Correlation between Γ and ω showing that the maximum value of Γ is in the same order of magnitude as the total frequency shift. ($\Delta\Gamma_{max} = \Delta\omega_{max}$). This observation is typical for Al,Si ordering in feldspars.

Q_{od} was analysed by Carpenter and Salje (1993b). The effect of Q_t and Q_{od} on hard modes was investigated by Harris et al (1989).

The microstructural features of K-feldspar are of particular interest within the context of this paper, because K-feldspar sometimes appears as a natural material with a tweed texture. Indeed, much of recent research on such microstructures was stimulated by the early observation of cross-hatched patterns in adularia and orthoclase by McConnell (1965, 1971) and McLaren (1984), and their analysis by ALCHEMI by McLaren and FitzGerald (1987). For a discussion of the generic aspect of tweed in K-feldspar, the reader is referred to Putnis and Salje (1993) and Salje (1993b).

Before we discuss the application of hard mode spectroscopy to modulated and tweed structures (Harris and Salje, 1992), let us first sharpen up the theoretical question of the nature of tweed pattern (Putnis and Salje 1993).

The early mineralogical analysis was based on the idea that competition between different Al,Si ordering schemes with well defined symmetries was at the origin of tweed formation. In the ideas promoted by McConnell (1971, 1985, 1988) two orthogonal waves of ordering intersect and form a sinusoidal chessboard pattern with 50% of the volume of the intersections taken by the thermodynamically unstable phase. This model was never fully analysed, although the one dimensional model of crest riding periodons by Barsch and Krumhansl (1988) addressed a similar question. In this case only the strain variable forms a modulated structure (possibly on the background of a "rippled" background of Al,Si ordering, Houchmandzadeh et al., 1992a). This model was found to be unstable unless the mutual attraction and annihilation of two stable regions in the pattern is prevented by matching areas which are not sinusoidal, however (Horovitz et al., 1991).

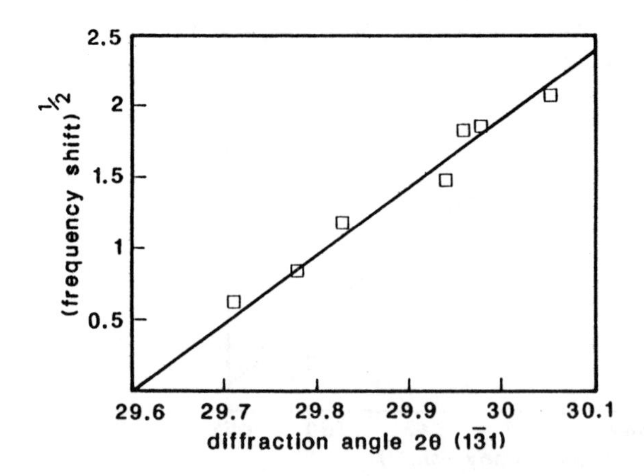

Figure 15. Linear correlation between ω and the 2θ angle of the $1\bar{3}1$ diffraction signal. Both quantities are proportional to Q_{od}.

Work on pattern formation and molecular dynamics studies of tweed draw a different picture altogether (Khatchaturyan 1983, Marais et al. 1991a,b, Bratkovsky et al. 1993a,b,c; Lajzerowicz 1981, Salje 1993a, Gordon 1983, Houchmandzadeh et al. 1991,1992a,b; Magyari and Thomas 1986, Salje and Parlinski 1991, Parlinski et al. 1993a,b, Semenovskaya and Khatchaturyan 1991). In all of these studies the tweed was found to consist of patches of material with approximately the average degree of order of the sample separated by walls which have all the typical qualities of segments of twin walls. These walls intersect, and typical features of tweed are the high density of such junctions and a (large) junction energy.

It is not possible to distinguish between the relevance of various models by transmission electron microscopy because they all predict the same star-shaped diffraction pattern. The crucial question is: is the distribution of Q_{od} bimodal as suggested by McConnell and in the various theories proposed by Barsch and Krumhansl, or is the distribution single-peaked with some (possibly small) broadening due to "walls"? In the language of incommensurate phase transitions, the question can be formulated as: How great is the soliton density? If it is unity, the pattern is roughly sinusoidal, if it is less than unity, the pattern contains well-defined walls (Jacobs, 1985).

Harris et al (1989) addressed this question using hard mode spectroscopy for the analysis of the local degree of order. Infrared spectra of a number of natural K-feldspars (table 1) were measured, and a systematic shift of several mode frequencies and line-widths found (Figs. 16,17). The shifts were due to Al,Si ordering because the typical relationship $(\Delta\omega)_{max} \approx \Delta\Gamma_{max}$ was found for the two characteristic phonon lines at 540 cm^{-1} and 645 cm^{-1} (Fig. 16). Their line profiles are single-peaked with a Gaussian distribution. No indications of bi-modal or Van-Hove singularities were found in K-feldspar although they exist in incommensurate phases of other materials (Harris and Salje 1992). This observation rules out any model based on a sinusoidal chess-board pattern of Q_{od}.

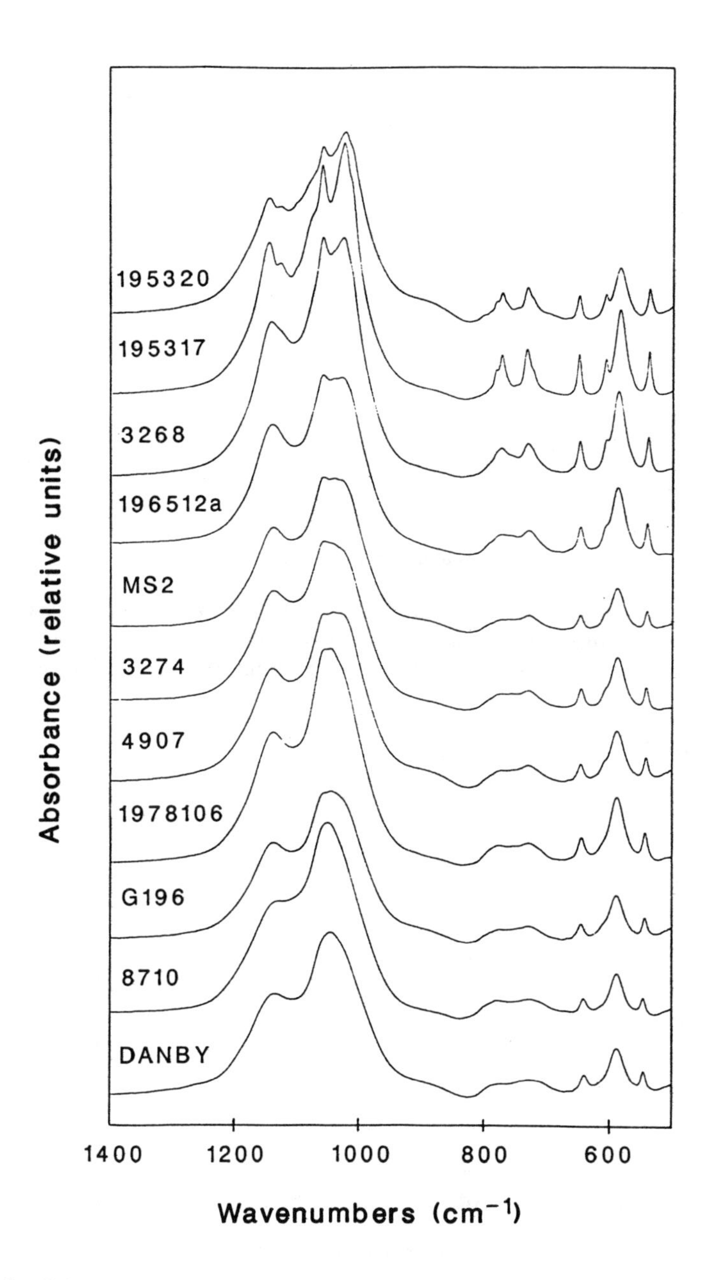

Figure 16. Infrared spectra of natural K-feldspars measured at 77K. The numbers refer to Tab. 1, the most ordered sample is at the top, the most disordered sample is at the bottom of the figure.

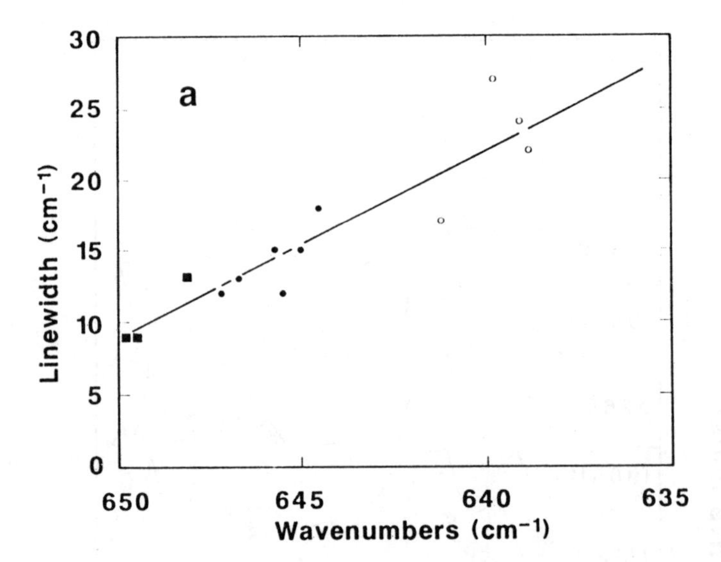

Figure 17.a Correlation between line width and frequency shift for two phonon signals (microcline: filled squares, orthoclase : filled circles, sanidine : open circles). The data scatter in Γ is due to chemical inhomogeneities, and the experimental resolution is equivalent to the size of the symbols (after Harris et al. 1989).

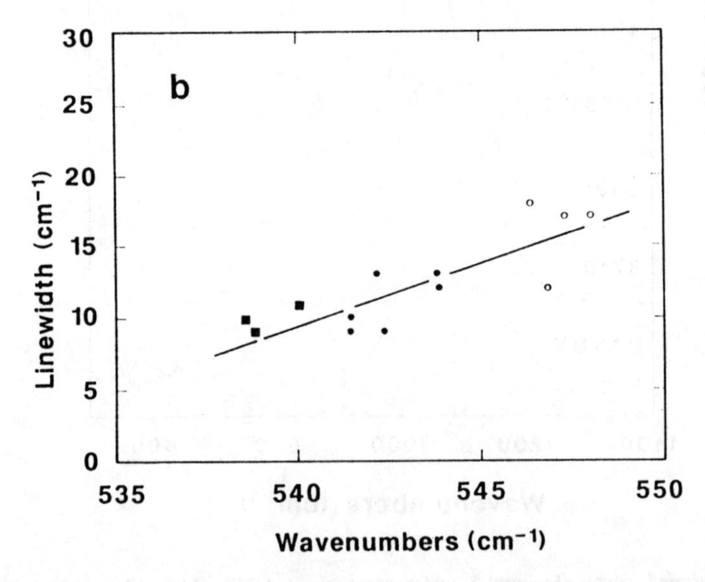

Figure 17.b Correlation between line width and frequency shift for two phonon signals (microcline: filled squares, orthoclase : filled circles, sanidine : open circles). The data scatter in Γ is due to chemical inhomogeneities, and the experimental resolution is equivalent to the size of the symbols (after Harris et al. 1989).

TABLE 1

Sample numbers (Harris et al. 1989), Al content on T_1 sites in monoclinic K-feldspars, and Al content of T_{10} + T_{1m} sites in triclinic K-feldspars as used for the spectroscopic analysis.

Sample no.	$2t_1$	$t_{t10} + t_{1m}$	Classification	TEM observation
8710z	0.535		high sanidine	
DANBYz	0.542		high sanidine	
DANBY	0.596		high sanidine	homogeneous
8710	0.608		high sanidine	homogeneous
G196	0.749		orthoclase	modulated
1978106	0.752		orthoclase	modulated
4907	0.828		orthoclase	modulated
3274	0.831		orthoclase	modulated
MS2	0.835		orthoclase	modulated
196512a	0.848		orthoclase	modulated
3268		0.957	intermediate microcline	twinned
195317		0.997	low microcline	twinned
195320		1.009	low microcline	twinned

Let us now concentrate on the determination of the mean value of Q_t in modulated K-feldspar. Q_t follows Q_{od} at low temperatures and is non-zero in sanidine. Its equilibrium behaviour was determined by Carpenter and Salje (1993b). The advantage of working with Q_t rather than Q_{od} is that Q_t can be determined directly from lattice parameter measurements (Kroll and Ribbe 1983). Typical values of Q_t in sanidine are below 0.38. Modulated K-feldspar appears to develop with Q_t between ca. 0.38 and values around 0.9. Microcline has Q_t values typically between 0.9 and 1. The question arises as to whether these values of Q_t are also valid on a local scale. Alternatively, the lattice parameters may simply measure some average structure, and the local patches of K-feldspar could, in fact, be much better ordered. Using hard mode spectroscopy we find that the frequency shifts of modulated feldspar are intermediate between those of microdine and sanidine. When these shifts are plotted as a function of $\tau = Q_t^2$, we find an approximately linear relationship for microcline and modulated K-feldspar and a kink between these structures and sanidine (Fig. 18). If the same data are plotted as a function of Q_t, the kink becomes less pronounced. (Note that Q_t is non-symmetry breaking and $\Delta\omega \propto Q_t$ is possible in lowest order). The answer to our question is identical for both ways of analysing the data: the degree of Al,Si order as measured macroscopically by Q_t using X-ray diffraction is essentially the same as seen on a local scale. This means that it is not the degree of Al,Si ordering (as seen by Q_t) which is modulated, but the structural strain. This interesting phenomenon has not been elucidated by further theoretical studies,

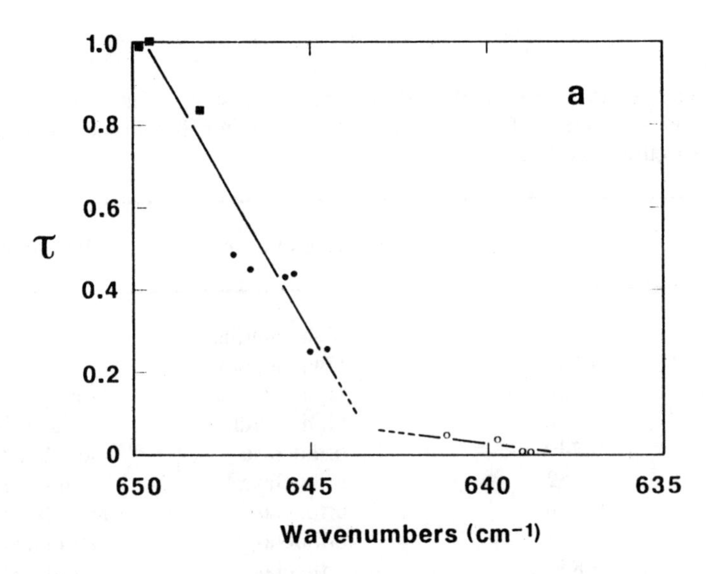

Figure 18.a Correlation between the phonon frequencies and the degree of short range order $\tau = \langle Q_t^2 \rangle$ (microcline : filled squares, orthoclase : filled circles, sanidine : open circles).

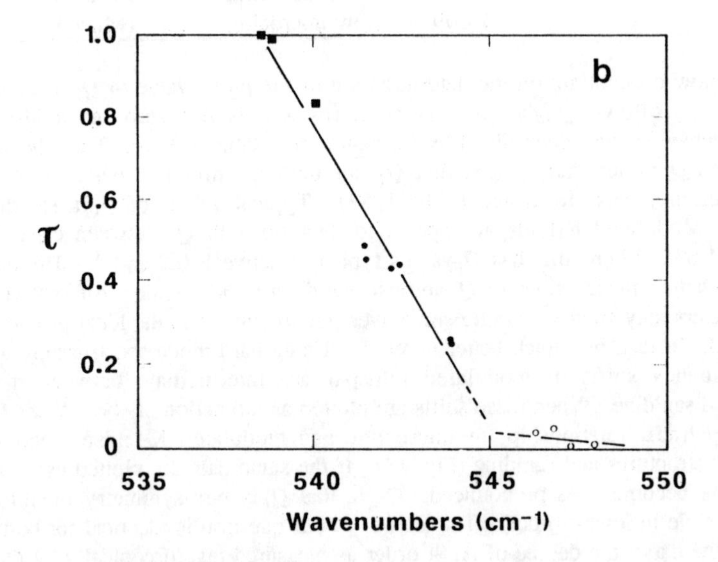

Figure 18.b Correlation between the phonon frequencies and the degree of short range order $\tau = \langle Q_t^2 \rangle$ (microcline : filled squares, orthoclase : filled circles, sanidine : open circles).

although the thermodynamical viability of such a state was asserted by Carpenter and Salje (1993b). Further work using computer simulation techniques to investigate the structural features of the tweed pattern is under way.

5.6 ON THE EFFECT OF Na-K MIXING AND THE STUDY OF EXSOLUTION: AN OUTLOOK

Although vibrational spectroscopy is commonly used for the identification of chemical phases, etc., rather little systematic work seems to have been done on the effect of exsolution processes in feldspars (Farmer 1974). The formal analysis of exsolution is identical with that of structural phase transitions in feldspar. The order parameter for Ab-Or exsolution is called $\eta_{Ab\text{-}Or}$, where we use a Greek letter to signify that the exsolution process is described by a conserved order parameter, in contrast with Q, as described so far in this paper, which is a non-conserved order parameter (Binder 1987, Cahn and Hilliard 1958, Langer et al. 1975, Binder and Stauffer 1976, Dattagupta et al. 1991). A conserved order parameter is contrainted so that the sum over all local states is always the same (e.g. equal to zero). In this case any change of the order parameter somewhere in a crystal has to be compensated by another change in the opposite direction somewhere else in the structure. Typical examples are exsolution processes where each Na leaving a site has to be replaced by K because of the constraint of having a constant chemical composition. No such constraints exist for Q or Q_{od} when the uniform sample has the same value of the order parameter everywhere. The terms "conserved" and "non-conserved" are used here for thermodynamic equilibrium, but kinetic processes do not necessarily follow the same conservation behaviour (see Salje 1993b and references given there). The Gibbs free energy of exsolution is identical to that discussed in Appendix 1. As a consequence, the treatment of hard mode behaviour is the same as discussed throughout this paper.

The applicability of hard mode spectroscopy for the study of local Ab-Or exsolution is illustrated in Fig. 19. Sequences of spectra for two series of crystals with a fully ordered Al,Si distribution and a fully Al,Si disordered distribution are shown.
Firstly, we see that the ordered series shows phonon signals with small Γ, and every spectral feature is well resolved. The disordered series shows broad peaks due to the more random Al,Si distribution as discussed before. Peak frequencies shift in each series as a function of the chemical composition. This effect is clearly seen when the low frequency part of the spectrum is enlarged (Fig. 20). The narrow line near 330 cm^{-1} is a typical example of a useful mode for hard mode spectroscopy because its frequency shifts as a function of chemical composition in a similar manner to what we saw before for the changes due to Q or Q_{od}. This particular mode relates to a bond-bending SiO_4 vibration which couples with the overall chemical effect in the sample. In a simple physical picture, this coupling can be depicted as the long-ranging lattice strain due to the larger size of K than Na. Replacing Na by K will somewhat squeeze the SiO_4 tetrahedra. The simple shift of the mode frequency indicates that this squeeze penetrates the entire structure and is not confined to the immediate vicinity of a K-site. This is to be expected, due to the long-ranging strain interaction (Appendix 2), for all internal tetrahedra modes.

Vibrations of Na or K with respect to the surrounding framework cannot be treated in this way. The modes with frequencies below 318 cm^{-1} "see" the moving Na or K atom and react on the basis of their mass difference and any difference of the effective bond to the lattice, inducing the local lattice relaxations. A quantitative analysis requires for these

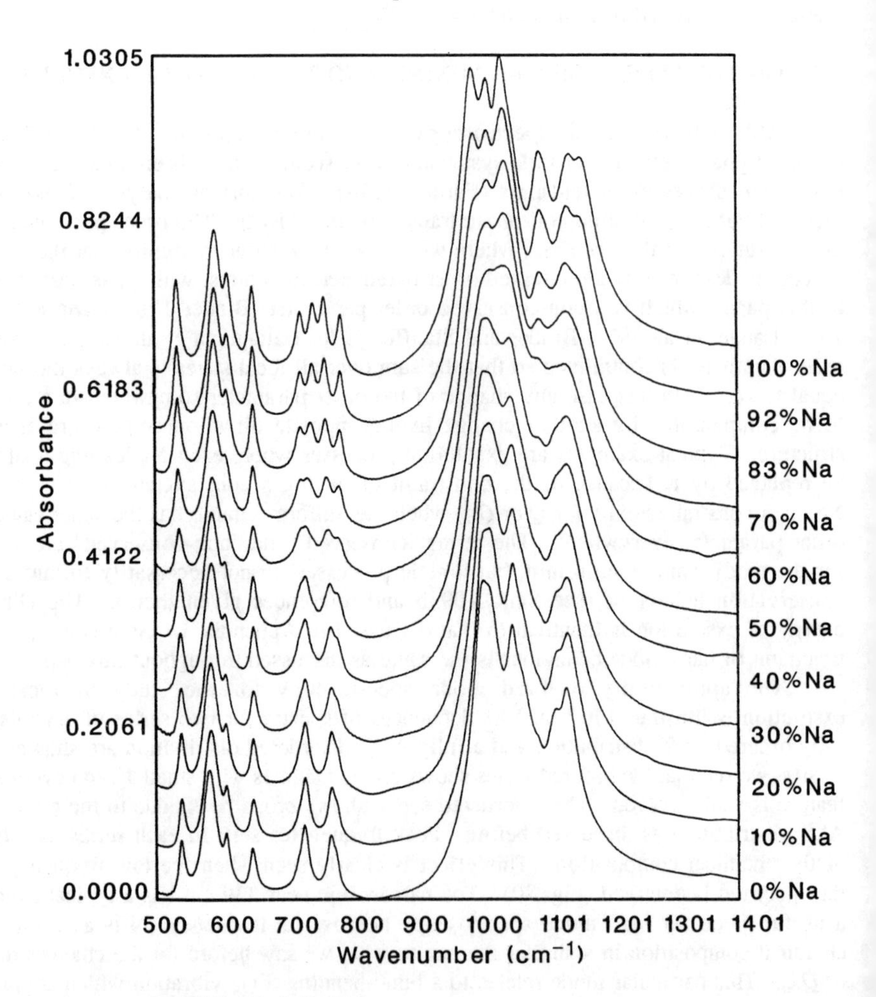

Figure 19.a IR spectra of alkali feldspars for high degrees of Al,Si ordering. The amount of Ab component is indicated for each spectrum. Note the rather continuous changes in the spectral region between $250cm^{-1}$ and $680cm^{-1}$ which is ideal for the application of hard mode spectroscopy.

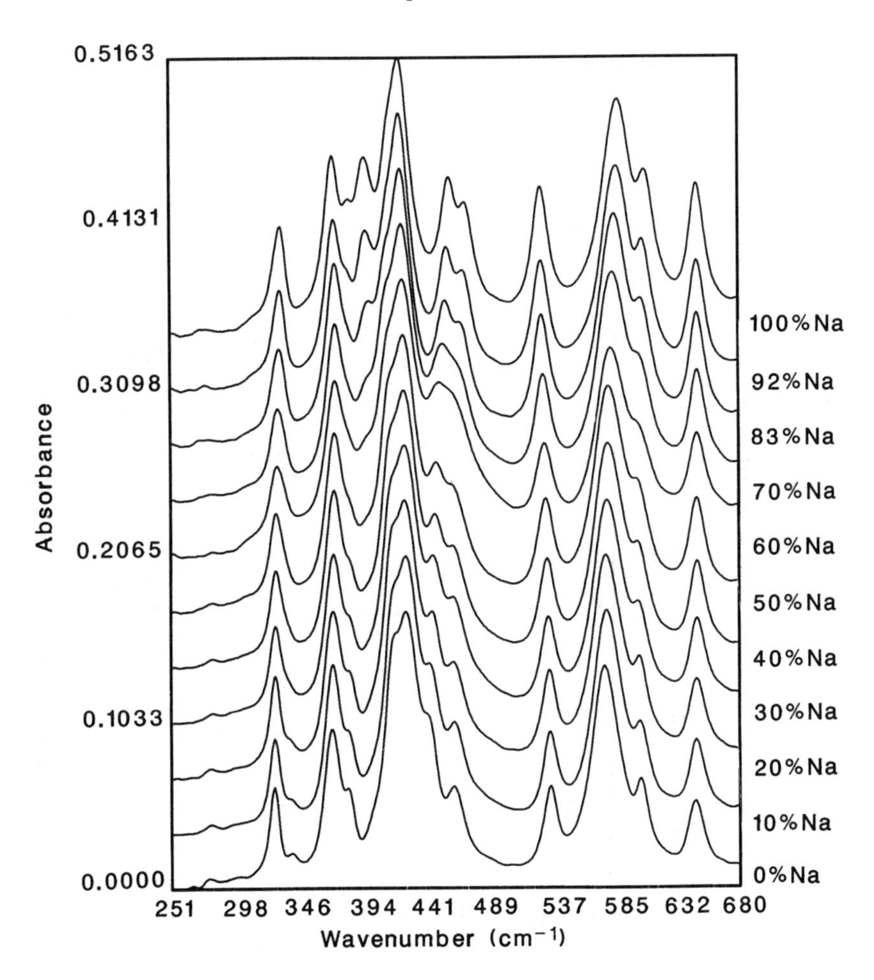

Figure 19.b IR spectra of alkali feldspars for high degrees of Al,Si ordering. The amount of Ab component is indicated for each spectrum. Note the rather continuous changes in the spectral region between 250cm^{-1} and 680cm^{-1} which is ideal for the application of hard mode spectroscopy.

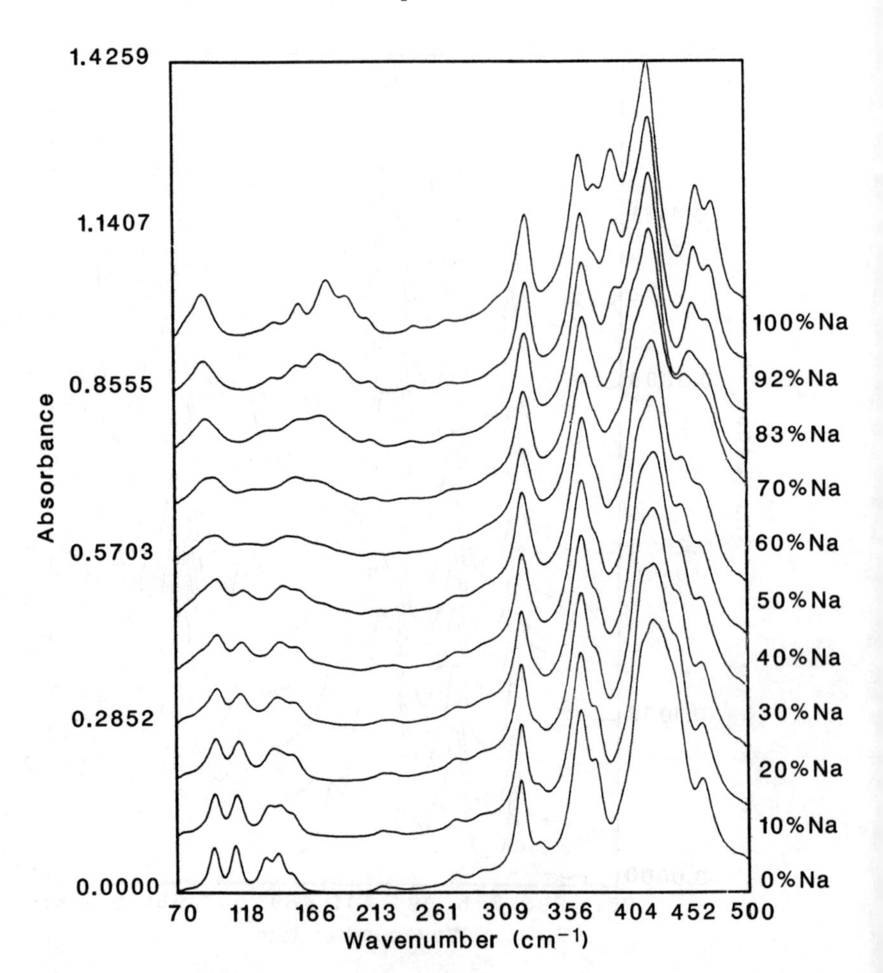

Figure 19.c IR spectra of alkali feldspars for high degrees of Al,Si ordering. The amount of Ab component is indicated for each spectrum. Note the rather continuous changes in the spectral region between 250cm^{-1} and 680cm^{-1} which is ideal for the application of hard mode spectroscopy.

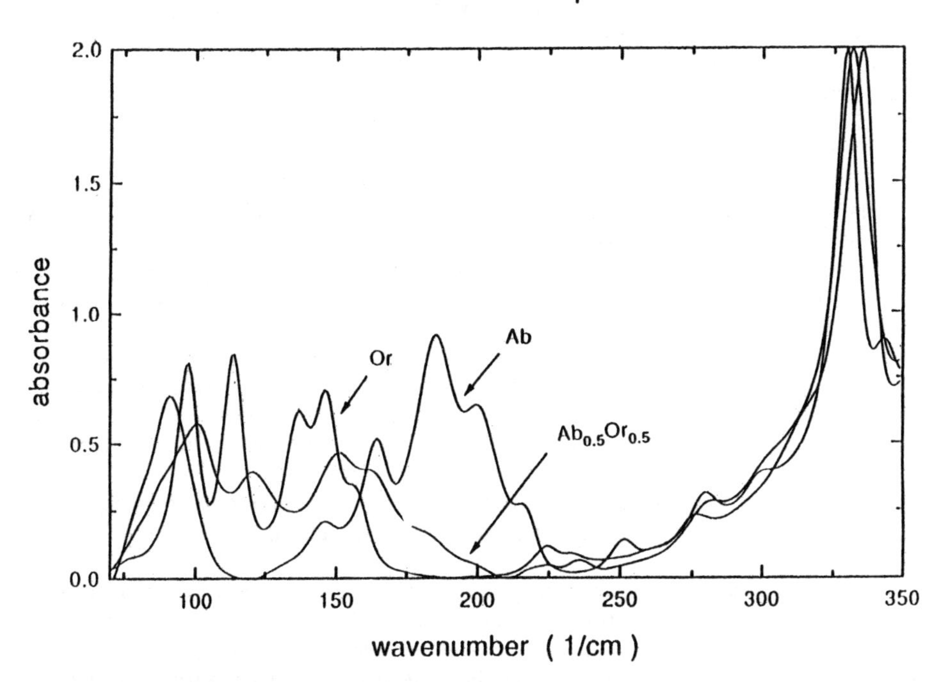

Figure 20. Enlarged part of the IR spectra of Ab, Or and $Ab_{0.5}$ $Or_{0.5}$ (high degree of Al,Si order). The mode near near 325cm^{-1} shifts continuously and can easily be used for the calibration of the chemical composition. Correlation effects at low wave numbers lead to large variations for vibrations directly involving the large cations.

modes the full lattice dynamical treatment (and not a simple perturbation approach as valid for hard modes discussed so far). Some conclusions can be reached without going through the full theory, however. Probably the most interesting aspect is related to the question of whether Na and K are randomly distributed in hypersolvus anorthoclase. The answer can easily be found using the following argument. If K and Na were to cluster, the spectra would consist of a superposition of those of pure Na- and K feldspar ($Q_{od} = 0$). This is clearly not the case (Fig. 20) because none of the dominant peaks of the endmember material between 70 cm^{-1} and 220 cm^{-1} is also a peak position in the Or_{50}. If, on the other hand, the Na,K distribution were random, the spectrum should contain a broad distribution of peaks roughly between those of the two endmembers. This is also not the case (Fig. 20). In fact new phonon lines (e.g. near 101 cm^{-1}, 120 cm^{-1} etc.) appear with linewidth slightly larger than those of the endmembers. This clearly indicates that some short range K,Na order exists in hypersolvus anorthoclase, although not in the sense of simple segregation of Na and K rich areas. Similar conclusions were already drawn from the profile analysis of Raman signals by Salje (1986). Futher work on the kinetics of exsolution processes in alkali feldspars is under way (Wruck and Ming 1993).

CONCLUSION. Hard mode spectroscopy is a potent tool for the characterisation of feldspar structures. Most of the necessary work to calibrate changes in the phon spectra with structural changes has been accomplished so that phonon spectroscopy can be used, on the most basic level, as an empirical indicator for the degree of Al,Si ordering, lattice distortion etc.. The key for the analysis is that frequency shifts, changes in intensities and linewidth are all proportional to the square of the symmetry breaking order parameter.

Beyond the most basic aplication of Hard Mode Spectroscopy, a more refined aspect related to its short characteristic length. Each signal is measured essentially on an atomistic length scale, the total signal is simply the sum of all local signals. This is obviously not true for diffraction techniques where the crux of the interpretation relates to phase factors etc. These difficulties do not exist for the methods explained in this paper which make them particularly useful for the investigation of incommensurate structures, tweed structures, short range order etc..

Once these principles are clearly established, further work can now proceed. I have three particular directions in mind. Firstly, all work done on temperature effects can be extended to include also pressure as external variable. The theoretical analysis is very similar to that presented here when the parameter P is introduced via the appropriate coupling energy (e.g. $\lambda Q^2 P$).

Secondly, kinetic studies of tweed formation and exsolution in alkali feldspar can be tackled on the correct (local) length scale so that the relevant structural states can be identified.

Thirdly, structural states with mesoscopic repetition units (such as c-plagioclases) or very fine lamellar structures (perthites) can be analysed to determine their thermodynamic behaviour. This can extent to the analysis of non-ideality of mixing where the phonon response of hypersolvus alkali feldspars show already short range correlations which do not coincide with those of the exsolved structures.

All these projects are now possible to be carried out and will produce a wealth of new insight into the structural behaviour of feldspars - far beyond our present knowledge.

ACKNOWLEDGEMENT. Work supported by NERC. The author is grateful for fruitful discussions with colleagues in Cambridge, in particular with B. Wruck, M.A. Carpenter and A. Putnis. Recent experimental work was carried out by M. Zhang.

Appendix 1

DEVIATION OF THE GIBBS FREE ENERGY FOR A PHASE TRANSITION WITH CONTINUOUS LOCAL POTENTIALS

The experimental investigations of phase transitions in minerals, and even more so in all ferroelastic and ferroelectric materials, over the last two decades have produced a wealth of empirical data which can be rationalized in terms of Landau Ginzburg theories. Theorists have argued that this is not an accident, but found that Landau Ginzburg type Gibbs free energies are, indeed, very often the underlying excess energies which lead to a phase transition (at least in a very good approximation). In this appendix we give a line of arguments which shows under which conditions the application of some generalized Landau theory is justified (Salje et al. 1991a,b).

The starting point is a local potential "seen" by the atoms. As all framework structures locally relax when atomic exchanges occur, or if the transition is displacive, we write the potential as a continuous function of the local state parameter Q_l:

$$V(Q_l)=\frac{1}{2}M\Omega_o^2Q_l^2+\frac{1}{4}uQ_l^4 \tag{A1.1}$$

The quadratic term is written in the tradition of phonon potentials (Ω_o would be a hypothetical phonon frequency) although the approach is general and does not depend on the actual physical process which determines the parameters. If the sides of the local potentials are steeper, one can replace the Q^4-term by a Q^6-term or the Slater term Q^P where $p \to \infty$. The subsequent treatment of the potentials is similar in all cases.

We now add the energy of the intensive interaction $\frac{1}{2}\Sigma\, \upsilon_{ll'} Q_l Q_{l'}$ i.e. the interaction between the site l and the site l'. This energy is identical to the leading term in Ising type models. Adding the momentum related energy $\Sigma\frac{1}{2M}P_l^2$ we finally construct the Hamiltonian

$$H=\sum_l\left[\frac{1}{2M}P_l^2+\frac{1}{2}M\Omega_o^2Q_l^2+\frac{1}{4}uQ_l^2\right]-\frac{1}{2}\sum_{l\neq l'}\upsilon_{ll'}Q_lQ_{l'} \tag{A1.2}$$

This most general Hamiltonian cannot (so far) be translated into a Gibbs free energy, but the essentials of the solution for dominating local potentials and longranging interactions were derived by Salje et al. (1991a).

Let us now calculate the transition behaviour of a system described by this Hamiltonian. First we Fourier-transform the interaction term as

$$\upsilon=\sum_{l'}\upsilon_{ll'}\exp(iqR_{ll'}). \tag{A1.3}$$

where q is the wavevector of the strutural instability and $R_{ll'}$ is the distance vector between the sites l and l'. The Gibbs free energy is

$$G=\mathrm{Tr}(\rho H+k_BT\rho\ln\rho),$$

where ρ is the density matrix and Tr means the trace of the matrix. Mean-field theory (MFT) is appropriate for the further treatment because the relevant interactions have long correlation lengths and critical cluster formation is rare in minerals investigated so far. The Boltzmann expression for such a density function is, in good approximation

$$\rho = (1/Z^{tr})\exp(-H^{tr}/k_BT), \tag{A1.4}$$

where the index *tr* indicates that this is a trial function. The simplest trial function is Gaussian, when we choose

$$H^{tr} = \sum\left(\frac{1}{2M}P_l^2 + \frac{1}{2}M\Omega_o^2(Q_l - \overline{Q}_l)^2\right) \tag{A1.5}$$

with two variational parameters per site to be determined by minimizing G: an effective single-site frequency Ω_l and the order parameter $\overline{Q}_l = \langle Q_l \rangle$ of the transition.

The quadratic function of H^{tr} in Q_l is clearly a good approximation if the true local potentials have only one minimum. The transition temperature for a purely parabolic potential $M\Omega_o^2 = \upsilon$ is absolute zero tempeature. Even for $M\Omega_o^2 \leq \upsilon$, the parabolic approximation is good. All these cases are called "displacive" with $T_C \rightarrow 0$ K defined as the "displacive limit". For order-disorder phase transitions the approach is still valid for situations where the local potential has a weak double minimum ($M\Omega_o^2 << \upsilon$) and, interestingly enough, also for the extreme limit of deep double-well potentials at low temperature (Salje et al., 1991b). The reason for the latter argument is that the Gaussian trial function is valid for each side of the double potential provided that fluctuations between the two sides of the potentials are sufficiently rare (i.e. their volume contribution disappears in the thermodynamic limit).

The thermal fluctuations are

$$\sigma = \left\langle (Q_l - \overline{Q}_l)^2 \right\rangle \tag{A1.6}$$

and can be expressed via the fluctuation-dissipation theorem as

$$\sigma = (\hbar / 2M\Omega)\coth(\hbar\Omega / 2k_BT) \tag{A1.7}$$

where Ω is the renormalized frequency which has to be determined self-consistently. Evaluating the various terms in the thermodynamic limit,

$$\left\langle P_l^2 \right\rangle = (M\Omega)^2\sigma \qquad \left\langle Q_l^2 \right\rangle = Q^2 + \sigma \qquad \left\langle Q_l^4 \right\rangle = Q4 + 6Q^2 + 3\sigma^2 \tag{A1.8}$$

one finds for the Gibbs free energy

$$G = -k_BT\ln Z^{tr} + \left\langle H - H^{tr} \right\rangle \tag{A1.9}$$

the expression

$$G = N\left\{\frac{1}{2}(M\Omega_0^2 - \upsilon + 3u\sigma)Q^2 + \frac{1}{4}uQ^4 + \frac{1}{2}(M\Omega_0^2\sigma + \frac{3}{4}u\sigma^2 \right.$$
$$\left. -\frac{1}{2}\hbar\Omega\coth(\hbar\Omega / 2k_BT) + k_BT\ln[2\sinh(\hbar\Omega / 2k_BT)]\right\} \tag{A1.10}$$

where N is the number of particles. Minimisation with respect to Q and Ω yields the mean-field equations ("self-consistency equations") (Salje 1993a).

$$(M\Omega_0^2 - \upsilon + 3u\sigma + uQ^2)Q = 0 \qquad M\Omega^2 = M\Omega_0^2 + 3u(\sigma + Q^2) \tag{A1.11}$$

These equations can be made more transparent if we define two dimensionless parameters. The first, $\Delta = (\upsilon / M\Omega_o^2) - 1$, describes the distance from the displacive limit $\Delta = 0$ when the on-site and intersite potentials are equal with opposite sign. The second, $\eta = \hbar / 2MQ_o\sigma_c$, is a measure of the quantum influence ($\hbar = 0$ yields "classical" thermodynamic results). The order parameter then becomes

$$Q = \sqrt{3(\sigma_c - \sigma)} \tag{A1.12}$$

the fluctuation amplitude is

$$\sigma = \sigma_c(\eta\Omega_o / \Omega)\coth(\eta\Omega / \Omega_o x) \tag{A1.13}$$

and the effective local mode frequency is

$$\Omega^2 = \Omega_o^2[1 + \Delta(3 - 2\sigma / \sigma_c)] \tag{A1.14}$$

The temperature is measured by

$$x = k_B T / M\Omega_o^2\sigma_c \tag{A1.15}$$

It is important to note at this point that it is impossible to derive a simple analytical expression for the Gibbs free energy. In fact, the general case has to be solved numerically, which is computationally easy [eqs. A1.12-A1.15]. Although a general form for the Gibbs free energy cannot be derived at this point, it is possible, nevertheless, to find such expressions for "displacive" transitions with $\Delta << 1$. Note that the term 'displacive' is chosen for the reason that all soft mode related transitions show $\Delta << 1$. Al,Si ordering in feldspars and rotational ordering of CO_3 groups in calcite are not 'displacive' in the geometrical sense of the term, but they still show $\Delta < 1$ so that the following arguments remain valid in these cases. In this case, we can write

$$\sigma = (k_B\theta_s / M\Omega_o^2)\coth(\theta_s / T) \tag{A1.16}$$

$$k_B\theta_s = -\frac{1}{2}\hbar\Omega_o, \qquad A = 3nk_B / M\Omega_o^2, \qquad B = u, \tag{A1.17}$$

$$T_C = \theta_S \operatorname{arctanh}(M\Omega_o^2 Q_o^2 / 3k_B\theta_s). \tag{A1.18}$$

We recover the generalised Landau expression

$$G = \frac{1}{2}A\theta_s[\coth(\theta_s / T) - \coth(\theta_s / T_c)]Q^2 + \frac{1}{2}BQ4 + \ldots \tag{A1.19}$$

We see at this point that the phase transition is, in fact, characterised by *two* temperatures T_C and θ_S. T_C is a measure of the collective forces which have to be compensated by thermal forces in order to achieve the high symmetry form at $T > T_C$. θ_s is a measure of the anharmonicity of the system which, when frozen out, does not allow changes of Q. We find that at $T < T_S = \frac{1}{2}\theta_S$, the order parameter becomes (almost) independent of temperature. If θ_s is determined experimentally, one can analyse which anharmonicities drive the phase transition. In the case of a soft mode, in which no coupling to other thermodynamic degrees of freedom exists, and thus, where only one frequency scale occurs, one expects $T_s \approx (\hbar / 2k_B)\Omega_{ls}$ where Ω_{ls} is the soft mode frequency at T_C. This case was found in an approximation for calcium chloride, $CaCl_2$ (Unruh et al. 1992). If the soft mode couples strongly with other modes one finds $T_s > (\hbar / 2k_B)\Omega_{ls}$ and, in the limit of coupling with virtually all phonon degrees of freedom, one expects $T_s \approx \frac{1}{4}\theta_E$ where θ_E is the thermodynamic Einstein temperature. This latter case was found in quartz (Salje et al. 1991b).

At high temperatures, $T > \theta_S$, this expression becomes the classical Landau-type Gibbs free energy, i.e. independent of θ_S.

$$G = \frac{1}{2}A(T - T_c)Q^2 + \frac{1}{4}BQ^4 + ... \tag{A1.20}$$

So far, we have only considered the case of a second-order phase transition with a local potential of the form $\frac{1}{2}A'Q^2 + \frac{1}{4}BQ^4$. It can be shown that the same arguments apply for potentials which include a 6th -order term. A particular case is a "tricritical" potential of the type $\frac{1}{2}A'Q^2 + \frac{1}{6}CQ^6$. The thermodynamic treatment is the same as before when we replace Q^2 by Q^4 in the relevant equations (Salje et al. 1991a).

Appendix 2

ON THE INTERCORRELATION OF AL,SI ORDERING AND THE GENERATION OF SPONTANEOUS STRAIN

In a first attempt to describe Al,Si ordering in a feldspar, one might be tempted to use a simple mapping procedure, which is sometimes successful in other phase transitions involving atomic ordering. In this mapping each site is given a spin variable *S (R)*, which can have the values + 1 or -1 depending on its occupancy with Al or Si. The solution for this problem is then simply identical to that of an appropriate Ising-type model. We know, however, that tetrahedra containing Al are different from those containing Si and that the lattice geometry changes considerably if the Al,Si order is changed (Kroll and Ribbe, 1983). Let us now describe what happens when we take these simple elastic forces into account. Let *R* be a set of lattice coordinates and *F(R)* the relevant spin-lattice forces. In the harmonic approximation, which is the most simple possible model, we can describe the spin-lattice interaction by the Hamiltonian (Marais et al. 1991, Bratkovsky et al. 1993a,b,c):

$$H = H_o + \frac{1}{2}\sum_{RR'} A^{ij}(R-R')u^i(R)u^j(R) - \frac{1}{2}\sum_{RR'} F^i(R-R')u^i(R)s(R') \qquad \text{(A2.1)}$$

Where A^{ij} is a dynamical matrix, *u(R)* describes the displacement of atoms from ideal lattice sites, defined for all $F = 0$ (1/2 is included in the definition of F for convenience). There is a complete analogy between the problem described by this Hamiltonian for spin-lattice coupling and those for elastic effects in solid solutions (in which the set of *F* is called Kanzaki forces) and both problems can be handled in the same manner (Khachaturyan, 1983). In both Hamiltonians, the lattice plays the role of the background and acts as a "reservoir" of elastic forces, mediating interactions between (discrete) variables *s(R)*. We eliminate the noncritical variables *u(R)* in the effective Hamiltonian (Wilson and Kogut, 1974):

$$\exp(-H_{eff}(\{s\})/T) = \int \mathrm{D}u \exp(-H(\{s,u\})/T) \qquad \text{(A2.2)}$$

where D*u* denotes the functional integration over ion coordinates. The exclusion of ionic motion can be also obtained by a simple shift of variables (Marais et al., 1991) with the result

$$H_{eff} = H_o - \sum_{RR'} J(R-R')s(R)s(R') \qquad \text{(A2.3)}$$

$$J(R-R') = \sum_{R''R'''} F^i(R-R'')\left[A^{-1}(R''-R''')\right]^{ij} F^j(R'''-R') \qquad \text{(A2.4)}$$

The effective Hamiltonian (A2.3) can easily be interpreted in the continuous (elastic) limit of the effective spin-spin coupling. In the ordered state, when all spins have the same sign, a lattice deforms homogeneously. In this case ions will form a low-symmetry lattice with ionic coordinates displaced from the high-symmetry positions by vectors

$$u^i = \varepsilon_{ij} R^j \qquad \text{(A2.5)}$$

where ε_{ij} is a usual homogenous strain tensor transforming under group G. Substituting (A2.5) into eq. (A2.1) we find the classical internal energy

$$E = \overline{E} = V(\frac{1}{2}C_{iklm}\varepsilon_{ik}\varepsilon_{lm} - \lambda_{ik}\varepsilon_{ik}Q) \tag{A2.6}$$

where $V = N\upsilon_c$ is the total volume of the sample, υ_c is the volume of the unit cell, C_{iklm} is the tensor of elastic moduli, Q is the average spin of the system and we introduced a set of microscopic parameters

$$\lambda_{ij} = \frac{1}{4\upsilon_c}\sum_R F^i(R)R^j + F^j(R)R^i \tag{A2.7}$$

Minimising the energy E for the case of free boundary conditions we get the following expression for the homogeneous strain and the elastic energy

$$\varepsilon_{ik} = Qu^o_{ik}, \qquad u^o_{ik} = S_{iklm}\lambda_{lm} \qquad \overline{E}/V = -\frac{1}{2}C_{iklm}u^o_{ik}u^o_{lm}Q^2 \tag{A2.8}$$

where S_{iklm} is the usual compliance tensor. The elastic energy in (A2.8) is a harmonic (quadratic) function of the order parameter Q. Such harmonic systems show no phase transitions in the thermodynamic limit if fluctuations are ignored. Simple calculations (Bratkovsky et al., 1993a,b,c) show that fluctuations lead to an effective Hamiltonian

$$H_{eff} = H_o + \frac{1}{2N}\sum_k J(k)s(k)s(-k) \tag{A2.9}$$

$$J(k) = -\frac{1}{4}F^i(k)G^{ij}(k)F^{j*}(k) \tag{A2.10}$$

where $G^{ij}(k)$ is the lattice Green function.

We see that in the general case $J(k)$ is a nonanalytic function of k with a non-analyticity at $k \to 0$ because $F(k)$ and $G(k)$ are explicit functions of k. These functions depend on the orientation of k as well as on its absolute value. This means that, for $k \to 0$, $J(k)$ assumes different values depending on the direction along which the point $k = 0$ is approached.

This non-analytical behaviour is crucial for the understanding of the transition behaviour and it is also the physical reason why fluctuations in feldspars often appear as mesoscopic tweed. The phase transition occurs when kT has the same value as the maximum of $J(k)$ (which is for C2/m - C$\overline{1}$ at the Γ point ($k = 0$)). There are two directions in k-space which link with this maximum value. Texture and fluctuations occur only in these directions. Any other direction has a different limiting value of $J(k)$ when $k \to 0$. Fluctuations in these directions cost additional energy and are effectively suppressed in real systems (Cowley 1976, Folk et al., 1979). The reduction of fluctuations to one-dimensional subsystems makes all critical fluctuations impossible and the system has to follow mean field behaviour. Furthermore, as $J(k)$ is approximately parabolic along the relevant directions, we find that the Hamiltonian has the same general form as used in Landau Ginzburg type theories.

References

Adlhart W, Frey F, Jagodzinski H (1980a) X-ray and neutron investigation of the P$\bar{1}$ -I$\bar{1}$ transition in pure anorthite. Acta Crystallogr A 36 : 450-460

Adlhart W, Frey F, Jagodzinski H (1980b) X-ray and neutron investigation P$\bar{1}$-I$\bar{1}$ of the transition in anorthite with low albite content.. Acta Crystallogr A 36 : 461-470

Angel RJ (1988) High pressure structure of anorthite. Amer Mineralogist 73 : 1114-1119

Angel RJ (1992) Order-disorder and the high-pressure P$\bar{1}$-I$\bar{1}$ transition in anorthite. Amer Mineralogist 77 : 923-929

Angel RJ, Redfern SAT, Ross NL (1989) Spontaneous strain below the I$\bar{1}$-P$\bar{1}$ transition in anorthite at pressure. Phys Chem Minerals 16 : 539-544

Angel RJ, Ross NL, Wood IG, Woods PA (1992) Single-crystal X-ray diffraction at high pressures with diamond-anvil cells. Phase Transitions 39 : 13-32

Bambauer HK, Bernotat WH, Kroll H, Voll G (1984) Structural states of K-feldspars from regional and contact metamorphic regions: Central Swiss Alps and Scottish Highlands. Bull Mineralogie 107 : 385-386

Bambauer HU, Bernotat WH (1982) The microcline/sanidine transformation isograd in metamorphic regions I. Composition and structural state of alkali feldspars from granitoid rocks in two N-S transverses across the Aar Massif and Gotthard "Massif", Swiss Alps. Schweizerische Mineralogische und Petrologische Mitteilungen 62 : 185-230

Barsch GR, Krumhansl JA (1988) Nonlinear and nonlocal continuum model of transformation precursors in martensites. Metallurgical Trans AIME A19 (2) : 761-775

Binder J, Stauffer D (1976) Statistical theory of nucleation, condensation and coagulation. Adv Phys 25 : 343-396

Binder K (1987) Theory of first order transitions. Rep Prog Physics 50 : 783-859

Bismayer U (1990) Hard mode Raman spectroscopy and its application to ferroelastic and ferroelectric phase transitions. Phase Transitions 27 : 211-267

Bratkovsky AM, Marais SC, Heine V, Salje EKH (1993a) Theory of fluctuations and texture embryos in structural phase transitions. J Physics: Condensed Matter, in preparation

Bratkovsky AM, Salje EKH, Heine V, (1993b) Theory and simulation of the origin. form and coarsening of tweed texture. Europhysics Letters, in preparation

Bratkovsky AM, Salje EKH, Marais SC, Heine V, (1993c) Theory and computer simulation of tweed texture. Phase Transitions, in press

Brown WL, Hoffman W, Laves F (1963) Über kontinuierliche und reversible Transformation des Anorthites ($CaAl_2Si_2O_8$) zwischen 25 und 350° C. Naturwissenschaften 50 : 221

Brown WL, Parsons I (1989) Alkali feldspars: ordering rates, phase transformations and behaviour diagrams for igneous rocks. Mineral Mag 53 : 25-42

Bruce AD, Taylor W, Murray AF (1980) Precursor order and Raman scattering near displacive phase transitions. J Phys C 13 : 483-504

Cahn JW, Hilliard JE (1958) Free energy of a non-uniform system I: interfacial free energy. J Chem Phys 28 : 258-267

Carpenter MA (1988) Thermochemistry of aluminium/silicon ordering in feldspar minerals. In: EKH Salje (ed) Physical properties and thermodynamic behaviour of minerals NATO ASI series C 225 : 265-323. Reidel, Dordrecht

Carpenter MA (1992) Equilibrium thermodynamics of Al/Si ordering in anorthite. Phys Chem Minerals 19 : 1-24
Carpenter MA, McConnell JCD, Navrotsky A (1985) Enthalpies of ordering in the plagioclase feldspar solid solution. Geochim Cosmochim Acta 49 : 947-966
Carpenter MA, Powell R, Salje EKH (1993) Thermodynamics of non-convergent cation ordering in minerals, I: an alternative approach. Am Mineralogist, submitted
Carpenter MA, Salje EKH (1989) Time-dependent Landau theory for order/disorder processes in minerals. Mineral.Mag 53 : 483-504
Carpenter MA, Salje EKH (1993a) Thermodynamics of non-convergent cation ordering in minerals, II: spinels and the orthopyroxene solid solution. Am Mineralogist, submitted
Carpenter MA, Salje EKH (1993b) Thermodynamics of non-convergent cation ordering in minerals, III: order parameter coupling in K-feldspar. Am Mineralogist, submitted
Cowley RA (1976) Acoustic phonon instabilities and structural phase transitions. Phys Rev B 13 : 4877-4885
Cross MC, Hohenberg PC (1993) Pattern formation outside equilibrium. Rev. Modern Physics, to be published
Czank M, Landuyt J van, Schulz H, Laves F, Amelinckx S (1972) Temperature dependence of domains in anorthite. Naturwissenschaften 59 : 646
Dattagupta S, Heine V, Marais S, Salje E (1991) Rate equations for atomic ordering in mean field theory II: general considerations. J Phys Condensed matter 3 : 2975-2984
Eggleton RA, Buseck PR (1980) The orthoclase-microline inversion: a high rsolution transmission electron microscope study and strain analysis. Contrib Mineral Petrol 74 : 123-133
Farmer VC (1974) The infrared spectra of minerals. Mineralogical society monograph 4, Adlard & Son Ltd, Dorking, Surrey
FitzGerald JD, McLaren AC (1982) The microstructure of microcline from some granitic rocks and pegmatites. Contrib Mineral Petrol 80 : 219-229
Folk R, Iro H, Schwabl F (1979) Critical dynamics of elastic phase transitions. Phys Rev B 20 : 1229-1244
Frey F, Jagodzinski H, Prandl W, Yelon WB (1977) Dynamical character of the primitive to body-centred phase transition in anorthite. Phys Chem Minerals 1: 227-231
Ghose S, Van Tendeloo G, Amelinckx S (1988) Dynamics of a second-order phase transition: $P\bar{1}$ to $I\bar{1}$ phase transition in anorthite, $CaAl_2Si_2O_8$. Science 242 : 1539-1541
Glauber RJ (1963) Time dependent statistics of an Ising model. J Math Phys 4 : 294
Gordon A (1983) Tricritical phase transitions in ferroelectrics. Physica B 122 : 321-332
Hafner S, Laves F (1957) II. Variation der Lage und Intensität einiger Absorptionen von Feldspaten zur Struktur von Orthoklas und Adular. Z Kristallogr 109 : 204-225
Harris MJ, Salje EKH (1992) The incommensurate phase of sodium carbonate: an infrared absorption study. J Phys Condensed Matter 4 : 4399-4408
Harris MJ, Salje EKH, Güttler BK, Carpenter MA (1989) Structural states of natural potassium feldspar: an infrared spectroscopic study. Phys Chem Minerals 16 : 649-658
Horovitz B, Barsch GR, Krumhansl JA (1991) Twin bands in martensites: statics and dynamics. Phys Rev B 43 : 1021-1033
Houchmanzadeh B, Lajzerowicz J, Salje EKH (1991) Order parameter coupling and chirality of domain walls. J Phys Condens Matter 3 : 5163-5169

Houchmanzadeh B, Lajzerowicz J, Salje EKH (1992a) Interfaces and ripple states in ferroelastic crystals - a simple model. Phase Transitions 39 : 77-87

Houchmanzadeh B, Lajzerowicz J, Salje EKH (1992b) Relaxations near surfaces and interfaces for first, second and third neighbour interactions: theory and applications to polytypism. J Phys Condens Matter 4 : 9779-9794

Hovis GL (1974) A solution caloritmetric and X-ray investigation of Al-Si distribution in monoclinic potassium feldspars. In: Mackenzie WS, Zussmann J (eds) The Feldspars , Manchester University Press, Manchester, UK, pp 114-144

Jacobs AE (1985) Solitons of the square-rectangular martensitic-transformation. Phys Rev B 31 : 5984-5989

Kawasaki K (1966) Diffusion constants near critical point for time dependent Ising models 3 : Self-diffusion constant, Phys. Rev. 145 : 224-

Khatchaturyan AG (1983) The theory of structural transformation in solids. Wiley, New York

Krause C, Kroll H, Breit U, Schmiemann I, Bambauer H-U (1986) Formation of low-microcline twinning in tweed-orthoclase and its Al,Si-order estimated by ALCHEMI and X-ray powder methods. Zeitschrift für Kristallographie 174 : 123-124

Kroll H, Bambauer H-U, Schirmer U (1980) The high albite-monalbite and analbite - monalbite transitions. Am Mineralog 65 : 1192-1211

Kroll H, Ribbe PH (1983) Lattice parameters, composition and Al,Si order in alkali feldspars. Minerl Soc of America Reviews in Mineralogy 2 : 57-99

Lajzerowicz J (1981) Domain wall near a 1st order phase transition - role of elastic forces. Ferroelectrics 35 : 219-222

Landau LD, Lifshitz EM (1980) Statistical Physics Oxford: Pergamon Press, UK

Langer JS, Bar-on M, Miller HD (1975) New computational method in the theory of spinodal decomposition. Phys Rev A11 : 1417-1429

Larkin AL, Pikin SA (1969) ZH Eksp Teor Fiz 56 : 540-543

Laves F, Hafner S (1956) Ordnung/Unordnung und Ultrarotabsorption I. (Al,Si)-Verteilung in Feldspäten. Z Kristallogr 108 : 52-63

Lebedev NI, Levanyuk AP, Morozov AI, Signow AS (1983) Defects near phase transition points: approximation of quasi-isolated defects. Fizika Tverdogo tela 25: 2975-2978

Levanyuk AP, Osipov VV, Sigov AS, Sobyanin AA (1979) Evolution of defect structure near the phase transition points and the anomalies in the properties of solids due to the defect s (1976). Zh Eksp Teo 76: 345-368

Levanyuk AP, Sobyanin AA (1970) Second order phase transitions without divergences in second derivatives of thermodynamic potential. JETP Lett 11 : 371-

Magyari E, Thomas H (1986) Stability of solitons. Helv Acta Phys 59 : 845-858

Marais S, Heine V, Nex C, Salje EKH (1991a) Phenomena due to strain coupling in phase transitions. Phys Rev Lett 66 : 2480-2483

Marais S, Padlewski S, Salje EKH (1991b) On the origin of kinetic rate equations: Salje-Glauber-Kawasaki. J Phys Condens Matter 3 : 6571-6577

Matsushita M (1976) Anomalous temperature dependence of the frequency and damping constants of phonons near T_l in ammonium halides. J Chem Phys 65 : 23-28

McConnell JDC (1965) Electron optical study of effects associated with partial inversion in a silicate phase. Philosophical Magazine 11 : 1289-1301

McConnell JDC (1971) Electron-optical study of phase transformations. Mineral Mag 38 : 1-20

McConnell JDC (1985) Symmetry aspects of order-disorder and the application of Landau theory. In: Kieffer SW, Navrotsky A (eds) Macroscopic to microscopic (Reviews in Mineralogy 14). American Mineralogical Society, pp 165-186

McConnell JDC (1988) Symmetry aspects of petransformation behaviour in alloys. Metallurgical Trans AIME A19 (2) : 159-167

McLaren AC (1974) Transmission electron microscopy of the feldspars. In: MacKenzie WS, and Zussman J (eds) The Feldspars, Manchester University Press, pp 378-423

McLaren AC (1984) Transmission electron microscope investigations of the microstructures of microlines. In: Brown WL (ed) Feldspars and Feldspathoids, NATO ASI series C 137, Reidel, Dordrecht, pp 373-409

McLaren AC, Fitz Gerald JD (1987) CBED and ALCHEMI investigation of local symmetry and Al,Si ordering in K-feldspars. Physics and Chemistry of Minerals 14 : 281-292

Megaw HD, Kempster CJE, Radoslovich EW (1962) The structure of anorthite, $CaAl_2Si_2O_8$ II: Description and discussion. Acta Crystallogr 15 : 1017-1035

Nissen H-U (1967) Direct electron-microscopic proof of domain texture in orthoclase ($KAlSi_3O_8$). Contributions to Mineralogy and Petrology 16 : 354-360

Parlinski K, Heine V, Salje EKH (1993a) Origin of tweed texture in the simulation of a cuprate superconductor. J Phys Condensed matter 5 : 497-518

Parlinski K, Salje EKH, Heine V, (1993b) Annealing of tweed tmicrostructure in high T_c superconductors studies by computer simulation. Acta metall mater 41 : 839-847

Petzelt J, Dvorak V (1976) Changes of infrared and Raman-spectra induced by structural phase transitions I: general considerations. J Phys C 9 : 1571-1586

Putnis A, Salje EKH (1993) Tweed microstructures: experimental observations and theoretical approaches. Phase Transitions, in press

Redfern SAT, Graeme-Barber A, Salje EKH (1988) Thermodynamics of plagioclase III: Spontaneous strain at the $P\bar{1}$ -$I\bar{1}$ phase transition in Ca-rich plagioclase. Phys Chem Minerals 16 : 157-163

Redfern SAT, Salje EKH (1987) Thermodynamics of plagioclase II: Temperature evolution of the spontaneous strain at the $P\bar{1}$ -$I\bar{1}$ phase transition in anorthite. Phys Chem Minerals 14 : 189-195

Redfern SAT, Salje EKH (1992) Microscopic dynamic and macroscopic thermodynamic character of the $P\bar{1}$ -$I\bar{1}$ phase transition in anorthite. Phys Chem Minerals 18 : 526-533

Ribbe PH (1983) Aluminum-silicon order in feldspars: domain textures and diffraction patterns. In: Mineralogical Society of America Reviews in Mineralogy, 2, second edition, pp 21-55

Salje EKH, Devarajan V, Bismayer U, Guimaraes DMC (1983) Phase transitions in $Pb_3(P_{1-x}As_xO_4)_2$: influence of the central peak and flip mode on the Raman scattering of hard modes. J Phys C 16 : 5233

Salje EKH (1985) Thermodynamics of sodium feldspar I: order parameter treatment and strain induced coupling effects. Phys Chem Minerals 12 : 93-98

Salje EKH, Kuscholke B, Wruck B, Kroll H (1985) Thermodynamics of sodium feldspar II: experimental results and numerical calculations. Phys Chem Minerals 12 : 99-107

Salje EKH (1986) Raman spectroscopic investigation of the order parameter behaviour of hypersolvus alkali feldspar: displacive phase transition and evidence for Na-K site ordering. Phys Chem Minerals 13 : 340

Salje EKH (1987) Thermodynamics of plagioclase I: theory of the $P\overline{1}$-$I\overline{1}$ phase transition in anorthite and Ca-rich plagioclases. Phys Chem Minerals 14 : 181-188
Salje EKH (1988) Kinetic rate laws as derived from order parameter theory I: theoretical concepts. Phys Chem Minerals 15 : 336-348
Salje EKH, Kroll H (1991) Kinetic rate laws derived from order parameter theory III: Al,Si ordering in sanidine. Phys Chem Min 17 : 563-568
Salje EKH, Parlinski K (1991) Microstructures in high T_c Superconductors. Supercond Science and Technology 4 : 93-97
Salje E, Wruck B, Thomas H (1991a) Order parameter saturation and a low-temperature extension of Landau theory. Z Phys B Condensed Matter 82 : 399-404
Salje EKH, Bismayer U, Wruck B, Hensler J (1991b) Influence of lattice imperfections on the transition temperatures of structural phase transitions: the plateau effect. Phase Transitions 35 : 61-74
Salje EKH (1992a) Application of Landau theory for the analysis of phase transitions in minerals, Physics reports 215 : 49-99
Salje EKH (1992b) Hard Mode Spectroscopy: experimental studies of structural phase transitions. Phase Transitions 37 : 83-110
Salje EKH, Wruck B, Graeme-Barber A, Carpenter MA (1993) Experimental test of rate equations: time evolution of Al,Si ordering in anorthite $CaAl_2Si_2O_8$. J Physics Condensed Matter 5 : 2961-2968
Salje EKH (1993a) Phase Transitions in Ferroelastic and Co-elastic Crystals. Cambridge University Press, England
Salje EKH (1993b) On the kinetics of partially conserved order parameters: a possible mechanism for pattern formation. J Phys Condensed Matter, 5 : 4775-4784
Semenovskaya S, Khatchaturyan AG (1992) Structural transformations in nonstoichiometric $YBa_2Cu_3O_{6+d}$. Phys Rev B 46 : 6511-6534
Smith JV, Brown WL (1988) Feldspar minerals I: crystal structures, physical, chemical and microstructural properties. Springer Verlag, Berlin
Staehli JL, Brinkmann D (1974) A nuclear magnetic resonance study of the phase transition in anorthite, $CaAl_2Si_2O_8$. Z Kristallogr 140 : 360-373
Stokes HT, Hatch DM (1988) Isotropy subgroups of the 230 crystallographic space groups, Singapore: World Scientific
Toledano P, Toledano JC (1976) Order parameter symmetries for improper ferroelectric non-ferroelastic transitions. Phys Rev B14 : 3097-3109
Toledano P, Toledano JC (1977) Order parameter symmetries for phase transitions of non-magnetic secondary and higher order ferroics. Phys Rev B16 : 386-407 and references given there
Unruh HG, Muhlenberg D, Hahn C (1992) Ferroelastic phase transition in $CaCl_2$ studied by Raman-spectroscopy. Z Physik B 86 : 133-138
Van Tendeloo G, Ghose S, Amelinckx S (1989) A dynamic model for the $P\overline{1}$-$I\overline{1}$ phase transition in anorthite, $CaAl_2Si_2O_8$ I. Evidence from electron microscopy. Phys Chem Minerals 16 : 311-319
Van Saarloos W, Hohenberg PC (1990) Pulses and fronts in the complex Ginzburg-Landau equation near a subcritical bifurcation. Phys Rev Letters 64 : 749-753
Wilson KG, Kogat J (1974) Physics Reports 12: 75

Wruck B (1986) Einfluss des Na-Gehaltes und der Al,Si Fehlordnung auf das thermodynamische Verhalten der Phasenumwandlung $P\bar{1}$ -$I\bar{1}$. In: Anorthit Dissertation, Universität Hannover

Wruck B, Salje EKH, Graeme-Barber A (1991) Kinetic rate laws derived from order parameter theory IV:kinetics of Al,Si disordering in Na-feldspars. Phys Chem Minerals 17 : 700-710

Wruck B, Ming X (1993) Infrared spectroscopic investigation of exsolution in alkali feldspars, in preparation

PARTIAL MELTING REACTIONS OF PLAGIOCLASES AND PLAGIOCLASE-BEARING SYSTEMS

W. JOHANNES, J. KOEPKE and H. BEHRENS
Institut für Mineralogie
Universität Hannover
Welfengarten 1
D-30167 Hannover
Germany

ABSTRACT. The kinetics of partial melting were studied in the systems:

Ab-An	(P = 1 atm)
Ab-An-H_2O	(P = 5 kbar)
Qz-Ab-An-H_2O	(P = 2 kbar)
Qz-Ab-An-Al_2O_3-H_2O	(P = 2 - 5 kbar)
Qz-Or-Ab-An-H_2O	(P = 2 kbar, X_{H2O} = 0.2 - 1.0)
Qz-Or-Ab-An-Bio-H_2O	(P = 2 kbar, X_{H2O} = 0.2 - 1.0)
Qz-Or-Ab-An-MgO-H_2O	(P = 2 kbar)

Starting materials were single crystals of natural plagioclases which in some of the experiments were embedded in powder of various materials.

At 1 atm pressure melting nucleated at the surface as well as at defects in the interior of the plagioclase. In contrast, at elevated water pressures the melting reaction exclusively began at the surface of the crystal and at cracks connected to the surface.

New plagioclase was always formed by crystallographically orientated solution/reprecipitation mechanisms. The propagation of melting was controlled by the interface between old plagioclase and melt as well as by transport in the melt. Diffusion rates within the plagioclase were too small to give a significant contribution to the formation of new plagioclases.

Under dry conditions the melting reaction was very sluggish. Even after a run duration of 1000 h, equilibrium conditions could not be attained at 1420 °C. In contrast plagioclase was completely recrystallized within 25 h at 1000 °C and at P_{H2O} = 5 kbar.

Addition of Al_2O_3 to quartz/plagioclase assemblages as well as the addition of biotite or MgO to $Qz_{60}Or_{40}$ glass/plagioclase assemblages resulted in a considerable enhancement of the melting process. New Al_2O_3-bearing phases were formed on the surface (corundum, biotite, phlogopite) and partially in the interior of the crystal (corundum). This suggests that the transport distance of Al_2O_3 in the melt is very important for the progress of melting. Furthermore, decreasing Qz-activity, and especially decreasing H_2O-activity results in lowering of the reaction rates. Models for partial melting of plagioclase at various conditions are proposed.

1. Introduction

Fractional crystallization and partial melting are important processes for the formation of magmatic rocks. Leucogranites, end products of granitic magmatism, are rich in sodium and potassium and poor in calcium as compared to their ultimate source, the Earth's mantle. This is due to partial melting and fractionation of components between melt and crystals.

I. Parson (ed.), Feldspars and Their Reactions, 161–194.

The most abundant minerals in the Earth's crust are feldspars. Therefore, a special interest exists in melting reactions in feldspar-bearing system. The determination of phase equilibria in such systems can greatly help in understanding the development of rocks. Despite a number of investigations performed on this subject, our understanding of the interaction of feldspars with melts is poor. This is due mainly to the slow kinetics of the melting reactions, especially if plagioclase is involved. Often metastable products are formed and chemical equilibrium cannot be achieved in experiments.

The main purpose of the present study was to elucidate the melting kinetics in plagioclase-bearing systems. Partial melting of plagoclase crystals was investigated under dry (1 atm) and wet (P_{H2O} = 5 kbar) conditions. In Qz-bearing systems the influence of additional components (Or, Al_2O_3, biotite, MgO) on the melting processes was studied. Melting textures and electron microprobe analyses of the run products were taken as the basis for the development of models for partial melting of plagioclases.

2. Previous work on plagioclase-melt interaction

Dry melting of plagioclases at P = 1 atm was studied by Bowen (1913) and by Tsuchiyama and Takahashi (1983). Bowen used synthetic starting materials whereas Tsuchiyama and Takahashi used natural single crystals. Solidus and liquidus temperatures determined in both studies are in fairly good agreement. Tsuchiyama and Takahashi (1983) suggested from their observations, that the progress of partial melting of plagioclase under these conditions is controlled by chemical diffusion in the crystal. In further studies Tsuchiyama investigated the influence of diopside on partial melting of plagioclase (Tsuchiyama, 1985a) and the dissolution kinetics of plagioclase in melts of the system diopside-albite-anorthite (Tsuchiyama, 1985b).

Solidus and liquidus temperatures of plagioclases are considerably decreased by high water pressures. Yoder et al. (1957) determined solidus and liquidus temperatures of the system Ab-An-H_2O for P_{H2O} = 5 kbar. The liquidus curve could be reproduced well by Johannes (1978), but the solidus was found to be at up to 60 °C degrees lower temperatures. Johannes observed that melting kinetics in this system are too slow at temperatures ≤ 800 °C to achieve equilibrium compositions.

The presence of Qz results in an additional decrease of melting temperatures. Liquidus temperatures and the cotectic line in the tonalite system Qz-Ab-An-H_2O were determined by Yoder (1968) for P_{H2O} = 5 kbar. The solidus surface for the same conditions was established by Johannes (1978). Cotectic melting at P_{H2O} = 2 kbar was studied by Johannes (1989) using single crystals of natural plagioclase. The cotectic line obtained and the solidus temperatures agree well with data of Tuttle and Bowen (1958) and of Stewart (1967) for the systems Qz-Ab-H_2O and Qz-An-H_2O, respectively.

In the granite system Qz-Or-Ab-An-H_2O several attempts were made to obtain chemical equilibrium, but equilibrium compositions could not be proved in these attempts (for summary see Johannes 1980, 1985). The cotectic temperatures are too low for an experimental investigation. Johannes and Holtz (1992) suggested that equilibrium compositions in plagioclase-bearing systems can be observed only at temperatures ≥ 830 °C.

TABLE 1: Compositions of starting plagioclases (normalized to 100%)

	An_{58}	An_{60}	An_{66}	An_{68}
analyses	29	9	79	11
	content of oxides (wt%)			
SiO_2	53.79	53.80	51.36	50.88
Al_2O_3	29.22	29.03	30.82	31.14
Fe_2O_3	0.46	0.42	0.47	0.48
CaO	12.07	12.28	13.43	13.82
Na_2O	4.68	4.35	3.80	3.58
K_2O	0.34	0.35	0.12	0.12
	content of Feldspar components (mol%)			
Ab	40.6	38.3	33.6	31.7
Or	1.9	2.0	0.7	0.7
An	57.5	59.7	65.7	67.6

All iron is given as Fe_2O_3.

3. Starting materials and experimental procedure.

The starting plagioclases were single crystals of gemstone quality from Lake County, Oregon. A detailed description of plagioclases from this locality is given by Stewart et al. (1966). The crystals used in the experiments were clear with slightly yellow color due to dissolved Fe^{3+}. No fluid inclusions or other phases were present as indicated by optical microscope and scanning electron microscope. The plagioclases were very homogeneous in composition. The deviation in contents of the elements was always within the uncertainty of electron microprobe analyses (1 % for main elements). Typically, the crystals contained about 0.5 wt% Fe_2O_3. The Or content was between 0.7 and 2.0 mol%. Chemical analyses are given in table 1. The Al/Si distribution was almost disordered as indicated by the Δ-131 value obtained from X-ray powder measurements.

The plagioclases exhibited only a few fractures parallel to (001) and only a few twin boundaries parallel to (010). These macroscopic defects were used for orientation of the crystals. Approximately rectangular pieces with sizes between 1 x 1 x 2 and 2 x 2 x 3 mm were cut from crack-free parts of the crystals. Unfortunately, the crystals were too small to give material for all the experiments. But the different plagioclases used were very similar in composition and structural state and, therefore, the results should be comparable.

Dry melting of plagioclase was performed at 1 atm in an horizontal tube furnace. Samples were inserted in platinum capsules welded tight at one side and shut with help of pliers at the other. This arrangement was believed to prevent significant evaporization of alkalies. The capsules were mounted on a corundum sample holder. Temperature was measured about 2 mm from the capsule using a PtRh6/PtRh30 thermocouple. Temperature was calibrated at the sample position using the melting point of Li_2SiO_3 (1204 °C) and is believed to be accurate within ± 2 °C at run conditions. For quenching, the sample holder was pulled out of the furnace and the capsule was thrown onto a metal plate, cooling the sample to room temperature within a few seconds.

Studies on wet melting of plagioclases were performed at 5 kbar and at temperatures from 950 to 1100 °C in an internally heated pressure vessel. Argon was used as the pressure medium. Samples were sealed in platinum tubes with 10 to 100 wt% (relative to the mass of the solid) of double-distilled water.

The influence of various materials (e.g. quartz, Al_2O_3, mullite, sillimanite, biotite, MgO) on the melting kinetics was investigated at temperatures between 820 and 880 °C and at 2 or 5 kbar confining pressure. Samples were surrounded by powder of these materials and sealed in gold tubes with 10 to 15 wt% water added. The runs were performed in cold-seal pressure vessels of Tuttle type using water as the pressure medium.

The temperatures were measured in the internally heated pressure vessel by a Ni/NiCr thermocouple located directly beside the samples. The actual temperatures at the sample positions were determined by melting of gold and silver wires. In the cold-seal pressure vessels temperatures were measured by external Ni/NiCr thermocouples. These thermocouples were calibrated against internal thermocouples at run conditions. In both types of apparatus, the actual temperatures are believed to be accurate within ± 10 °C. Pressures were determined using a strain-gauge manometer (accuracy ± 20 bar).

Runs in the internally heated pressure vessel were terminated by switching off the heating power. Pressure was held constant by automatic pumping during the quench. The sample cooled down from 1000 to 600 °C within 2 minutes and to 200 °C within 5 minutes. Quenching in the cold-seal pressure vessels was performed by using a flow of compressed air. The temperature dropped from 850 to 200 °C in less than 4 minutes. Constant pressure during the quench was adjusted in the cold seal pressure vessels by manual pumping.

Polished thin sections of the charges were prepared for analysis by optical microscope and electron microprobe. Two different orientations of the plagioclases, one perpendicular to the a-axis and one parallel to (001), were prepared for analyzing experimental products in the system Ab-An and Ab-An-H_2O.

Chemical analyses of crystals and quenched melts (glasses) were carried out using a CAMECA CAMEBAX microprobe. Operation parameters were 15 kV accelerating voltage and 18 nA beam current. A counting time of 10 seconds and a beam diameter of about 1 μm was used for solids. The quantitative analyses of glasses show a very sensitive dependence on the diameter of the applied beam and on the counting time. Using a focused beam, the signal of Na decreases quickly while the count rates of Si and Al increase. This is particularly important for hydrous glasses. In order to get stable conditions for quantitative analyses, the beam was defocused to approximately 10 to 20 μm and the counting time was reduced to 2 seconds for Na and K and to 5 seconds for Ca, Si, Al and Fe. The raw data were corrected by the CAMECA "PAP" program according to Pouchou and Pichoir (1984). Qualitative line-scans were performed with an accelerating voltage of 10 kV and a beam current of 10 nA leading to a local resolution of approximately 1 μm. Backscattered electron (BSE) images were taken to characterize the phase assemblages of the run products.

4. Experimental results and discussion

4.1. DRY MELTING OF PLAGIOCLASE AN_{68} AT 1420 °C

Beginning of melting. Partial melting of the plagioclase was observed after a run time of only 1 minute. Melting was observed not only at the surface but also in the interior of the crystal (Fig. 1A,B). The melt exhibited either drop-like or plate-like shape with a preferred orientation parallel to the main cleavage, (001). In both sections, perpendicular to the a-axis and parallel to (001), an arrangement of drops of melt along lines was visible(Fig. 1 A-D). It is assumed that this phenomenon is due to nucleation of melt at dislocations in the crystal. In Fig. 1D it can be seen that twin boundaries also act as nucleation sites for melts. Plates of melt are formed at the albite twin boundaries (parallel to (010)).

Our findings differ from the observations of Tsuchiyama and Takahashi (1983). These authors observed, at the same temperature, the preferential beginning of melting at the surface and penetration of melt into the crystal with the formation of dendritic and platy patterns. Nucleation at the surface was not dominant in our experiments and the patterns observed by Tsuchiyama and Takahashi (1983) could not be detected. The differences are assumed to be due to different starting materials. The same crystal from Chihuahua, Mexico used by Tsuchiyama and Takahashi (1983) in their melting experiment was also used by us in a study on cation diffusion in ternary feldspars (Behrens et al., 1990). For this crystal, the Na-diffusivity was 2 to 4 times slower and the activation energy for Na-diffusion was 6 % higher when compared to a plagioclase of same composition from Lake County. Furthermore, the Chihuahua plagioclase was considerably more resistant during mechanical treatment (sawing, polishing) than the Lake County plagioclase. The findings in the melting experiments suggests, that the differences are at least partially due to the number of dislocations in the crystals. The number of dislocations is probably higher in the Lake County plagioclase.

The formation of new plagioclase with a higher An content than the starting material is already observed after 6 min. (Fig. 1C,D). The new plagioclase surrounds the melt areas or forms thin plates with a preferred orientation parallel to (010). At the twin boundaries plates of new plagioclases were found often between melt areas.

Melting textures. The textures of run products obtained after run times ≥ 1 h were similar to those reported by Tsuchiyama and Takahashi (1983). The BSE images showed no distinct differences in melting textures in the two sections prepared perpendicular to the a-axis and parallel to (001). Therefore, in Fig. 2A-D only one section is presented (perpendicular to the a-axis). Analyses of plagioclase and melt compositions are given in table 2.

Small drops of melt with sizes below 1 μm and only very thin rims of new plagioclase were present in all run products. This suggests that either melt drops can be nucleated in the late stage of the reaction or melt drops can be hindered in growth by the surrounding crystal or both. The volume change by melting produces a local pressure in the melt drop and in the surrounding plagioclase. The growth of a melt nucleus will be stopped when the pressure exceeds a critical value. Melting can proceed further, only if paths are available for the melt to reach the surface of the crystal.

Even after a run duration of 1000 hours small volumes of the old plagioclase with an An content of 68 mol% were detectable. BSE images illustrate that the contacts between the old and the new plagioclase are still sharp (Fig. 2D). Line-scans for elements demonstrate step-like transitions in composition when crossing the boundary between old and new plagioclase (Fig. 3).

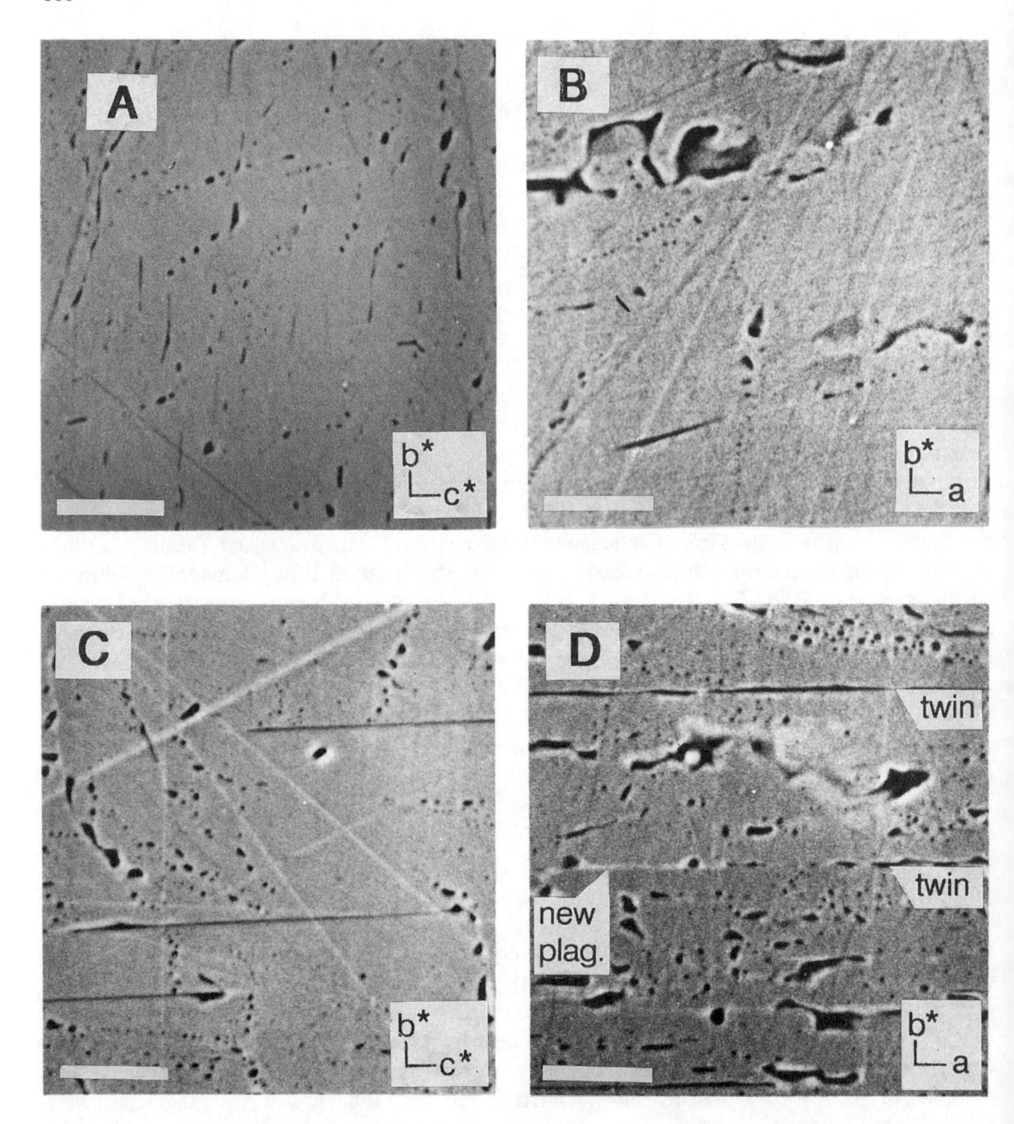
A
b*
c*
B
b*
a
C
b*
c*
D
twin
twin
new
plag.
b*
a

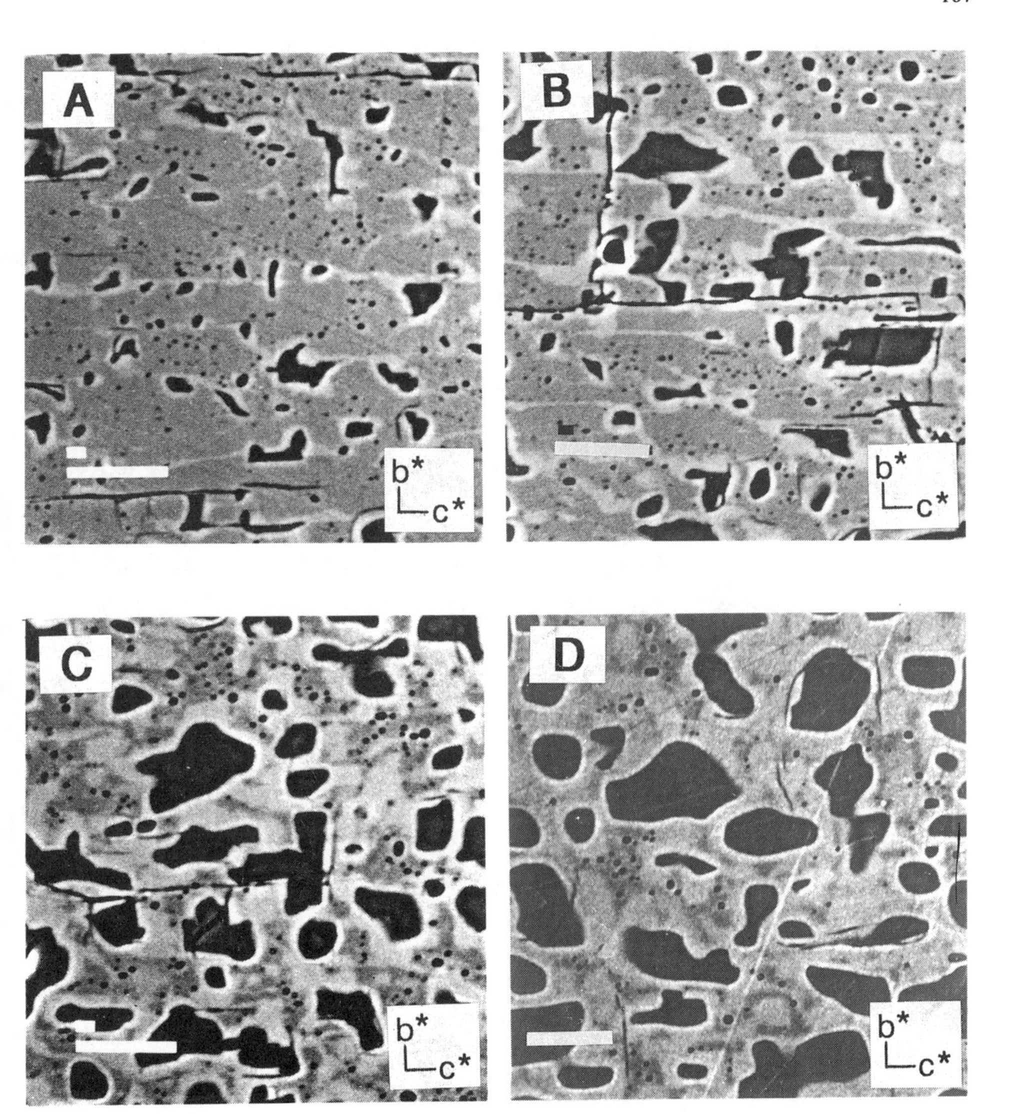

Fig. 2. BSE images of melting textures of a plagioclase An_{68} heated for longer times at 1420 °C and at 1 atm. All sections are orientated perpendicular to the a-axis and are taken in the center of the samples. Dark areas: quenched melt, grey areas: old plagioclase, light grey areas: new plagioclase. Bar: 10 μm.

(**A**) 1 hour, (**B**) 10 hours, (**C**) 100 hours, (**D**) 1000 hours

TABLE 2: Results of dry melting of plagioclase. P = 1 atm, T = 1420 °C

run duration		An mol% (1σ)	Ab mol% (1σ)	Or mol% (1σ)	Fe_2O_3 wt% (1σ)
	pl_{st}	67.6 (0.5)	31.7 (0.5)	0.7 (0.1)	0.47 (0.04)
1 h	pl_{re}	67.8 (0.5)	31.8 (0.5)	0.4 (0.1)	0.50 (0.03)
	pl_n	77.0 (0.9)	22.7 (0.9)	< 0.3	0.15 (0.04)
	m_r	33.7 (0.9)	61.5 (0.6)	4.8 (0.3)	1.88 (0.05)
10 h	pl_{re}	68.9 (0.7)	30.7 (0.7)	0.4 (0.1)	0.44 (0.03)
	pl_n	78.5 (0.9)	21.5 (0.9)	< 0.3	0.21 (0.02)
	m_c	37.0 (0.8)	60.0 (0.8)	3.0 (0.1)	1.50 (0.03)
	m_r	43.0 (1.1)	54.6 (1.1)	2.4 (0.1)	0.92 (0.06)
100 h	pl_{re}	68.0 (0.5)	31.5 (0.6)	0.5 (0.1)	0.30 (0.02)
	pl_m	79.1 (1.2)	20.9 (1.3)	< 0.3	0.24 (0.02)
	m_c	39.8 (0.9)	57.7 (0.8)	2.5 (0.1)	1.25 (0.05)
	m_r	43.0 (0.8)	54.9 (0.8)	2.1 (0.2)	0.60 (0.07)
1000 h	pl_{re}	69.0 (0.6)	30.6 (0.7)	0.4 (0.1)	0.17 (0.04)
	pl_m	81.2 (1.1)	18.8 (1.1)	< 0.3	0.19 (0.04)
	m_c	41.7 (0.6)	56.5 (0.6)	1.8 (0.2)	0.80 (0.05)
	m_r	43.2 (0.8)	55.1 (0.7)	1.7 (0.2)	0.59 (0.03)

All iron is given as Fe_2O_3. pl_{st}: starting plagioclase, pl_{re}: residual plagioclase, pl_n: new plagioclase, m_c: melt in the center of the crystal, m_r: melt in the rim around the crystal.

After a run duration of one hour a segregation of melt at the surface of the plagioclase was observed. This is attributed to the volume increase by melting. The rim of melt at the surface became larger with increasing run time. Both the new plagioclase and the segregated melts reach a size which enables quantitative analyses after a run duration of 1 hour. After 10 hours the volumes of melt in the interior of the crystal are large enough to be analyzed.

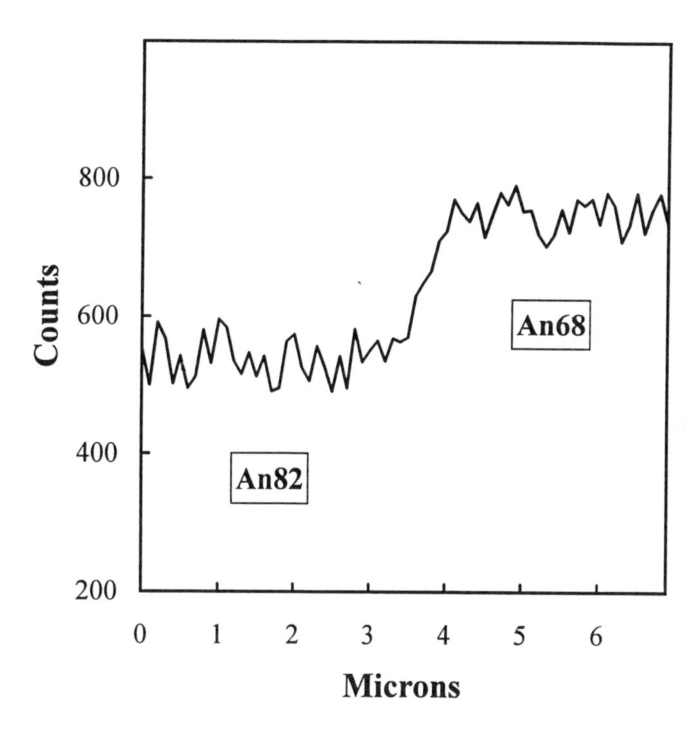

Fig. 3. Microprobe line scan for Na across the boundary between new (An_{82}) and old (An_{68}) plagioclase. Run conditions: T = 1420 °C, P = 1 atm, t = 1000 h.

Development of chemical compositions. The An content of both the new plagioclase and the melt inside the crystal, increase with run time (new plagioclase: 1 h - An_{76-79}, 10 h - An_{78-79}, 100 h - An_{78-80}, 1000 h - An_{80-83}; melt in the interior: 10 h - An_{36-38}, 100 h - An_{39-41}, 1000 h - An_{41-43}). Even after 100 hours there is a change in composition of these phases. In contrast to this the An content of the melt in the segregation rim reached a stable value after 10 hours (An_{42-45}). Considering both the development of the melt and of the crystal compositions (Fig. 4), we assume that local equilibrium between melt and new plagioclase is almost established after 1000 hours. But there still are relics of unchanged plagioclase indicating that the run time was insufficient to get an equilibrium state in the whole sample. For both the melt and the new plagioclase the maximum An contents are slightly higher than determined by Bowen (1913) for the pure system Ab-An (Fig. 5). The composition of the new plagioclase indicated a solidus curve approximately 30 °C below that given by Bowen (1913). The differences are attributed to the small amounts of Or component and of Fe present in the starting plagioclase.

The Or content of the relics of the starting plagioclase decreased very quickly (within 1 hour) from 0.7 to 0.4 mol% and remained almost constant in runs of longer duration. Initially, the new plagioclase had a low Or content (0.3 mol%), but after 100 hours potassium could be detected no more. This corresponds to the increase of melt with time and to the favoured incorporation of potassium in the melt. The Or content in both, the melt in the core, and in the segregation rim decreased with run duration due to the increase in melt volume (Fig. 6).

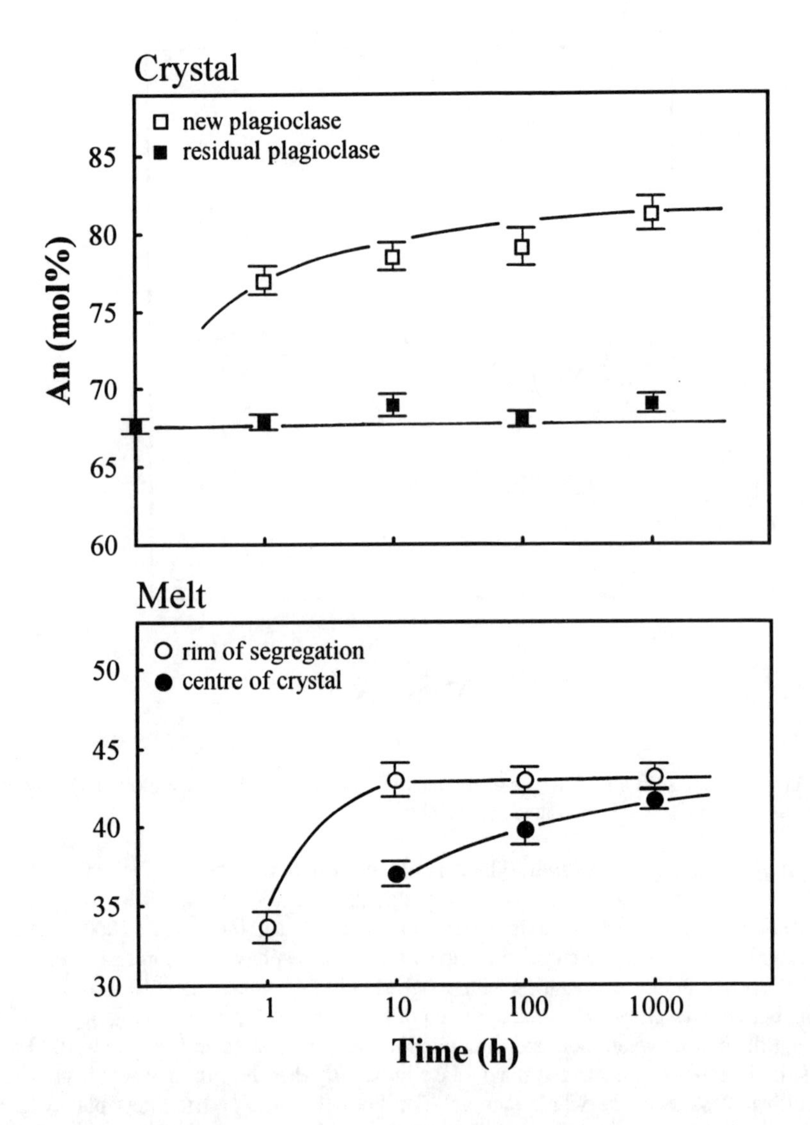

Fig. 4. Chemical compositions (An content) of plagioclases and of quenched melts obtained from melting experiments at 1420 °C and at 1 atm. Starting material: An_{68}. Error bars correspond to 1σ errors.

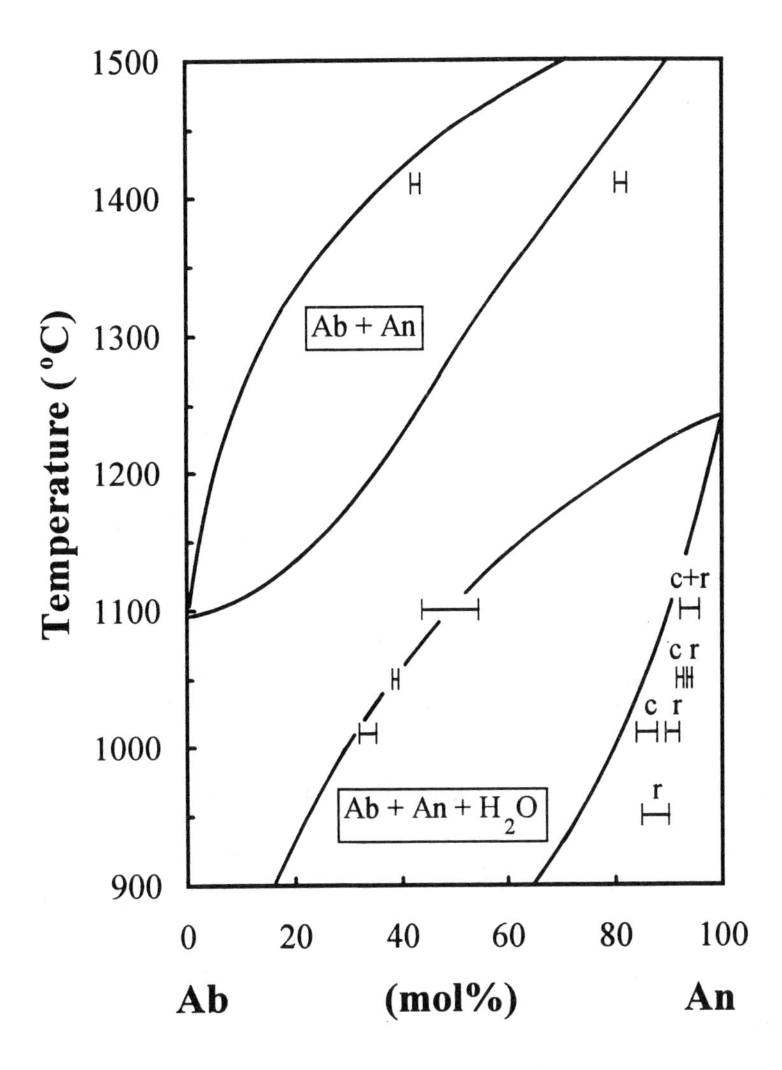

Fig. 5. An/(An+Ab) ratios of plagioclases and coexisting melts in the systems Ab-An (P = 1 atm) and Ab-An-H_2O (P = 5 kbar). The bars correspond to the average composition $\pm$ 1σ determined by microprobe analyses (bars on the right side: plagioclase; bars on the left side: melt). Solidus and liquidus lines of the system Ab-An were determined by Bowen (1913). For the system Ab-An-H_2O (P = 5 kbar) the liquidus was defined by Yoder et al. (1957) and the solidus by Johannes (1978). For P = 1 atm the results of the 1000 h run only are plotted. The data for P_{H2O} = 5 kbar were obtained from completely reacted samples except of those for 950 °C. At this low temperature small reaction layers could only be produced in the sample. For T = 1000 °C the data result from two experiments, otherwise from one. Plagioclase compositions obtained in the center (c) and in the rim (r) of the crystals are given for the system Ab-An-H_2O (P = 5 kbar).

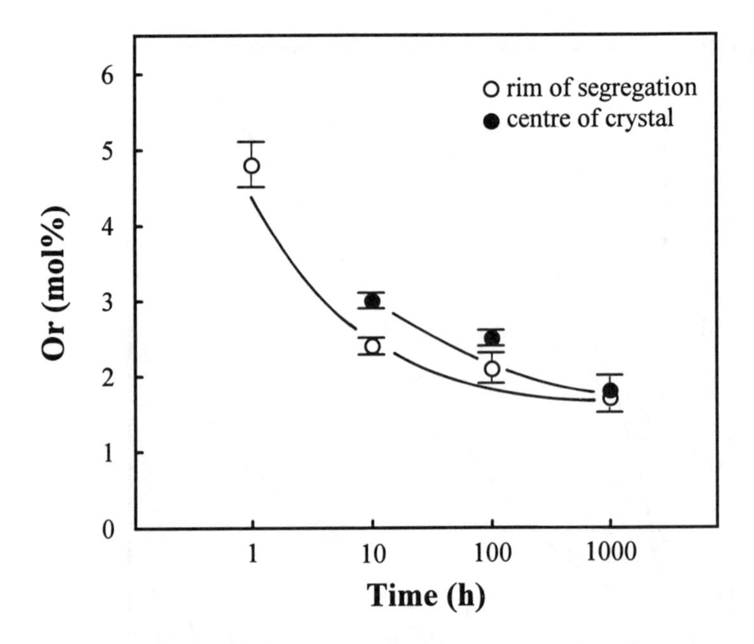

Fig. 6. Or content of quenched melt in the center of the crystal and in the rim of segregation as a function of run duration. Run conditions: An_{68}, T = 1420 °C, P = 1 atm. Error bars correspond to 1σ errors.

Iron, like the Or component, was preferentially dissolved in the melt. The iron content of the both the new and the residual plagioclase became smaller with run duration (Fig. 7). But the exchange of Fe between crystal and melt was significantly slower than that of Or component. This can be explained by considering the diffusivities of the cations involved in the exchange process in the feldspars (the diffusivities in the melt are much higher and, therefore, are not rate-controlling). The exchange of Or component only requires an interdiffusion of K and Na, whereas in case of the iron exchange network-forming cations are involved. The diffusivities of alkali cations are much higher than those of Fe, especially at high oxygen fugacity (Foland, 1974; Giletti et al., 1974; Kasper 1974; Christoffersen et al., 1983; Behrens et al. 1990).

Model for dry melting of plagioclases at 1 atm. The development of melt and crystal compositions as well as the BSE pictures and the line-scans of elements do not agree with the melting model proposed by Tsuchiyama and Takahashi (1983). According to these authors, the compositional exchange between melt and plagioclase and, therefore, the progress of partial melting is controlled by chemical diffusion (NaSi-CaAl) in the plagioclase. If this is true, then a diffusion profile should be developed between new and residual plagioclase. At high temperatures as in the experiments, there is a complete miscibility of the plagioclases and no step-like transition in the diffusion profile (due to an immiscibility gap) should occur.

According to a diffusion coefficient D of 10^{-12} cm^2/s as deduced by Tsuchiyama and Takahashi (1983) from their melting experiments, a diffusion length $(2Dt)^{1/2}$ of 27 μm would be expected for the 1000 hours run. Even if the much smaller diffusivity (10^{-15} cm^2/s) obtained by Grove et al.

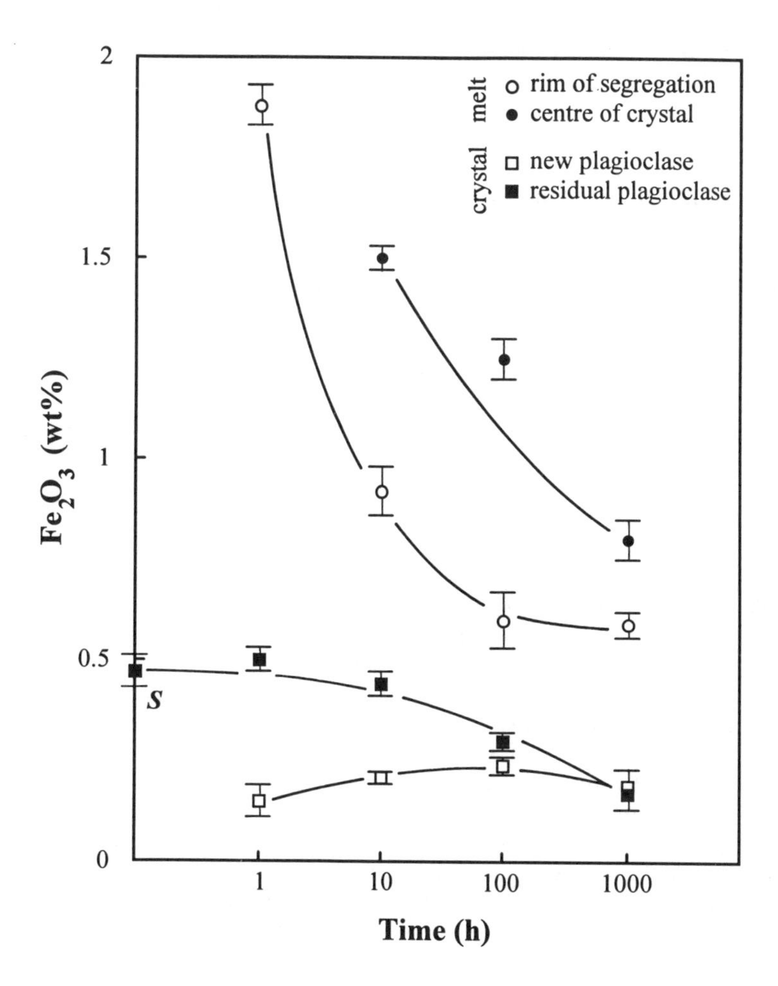

Fig. 7. Fe_2O_3 content of plagioclases and quenched melt as a function of run duration. Run conditions: An_{68}, T = 1420 °C, P = 1 atm. Error bars correspond to 1 σ errors.

(1984) from homogenization experiments is used for the estimation, a diffusion length of 1 μm should occur after this time. The total length of the diffusion profile would be in the latter case about 4 μm and this distance should be resolvable by electron beam scanning. But in no case we were able to detect a continuous change in composition between new and residual plagioclase. Taking the resolution of the electron microprobe into account the transitions are step-like in all stages of our melting experiments. This suggests that under dry conditions diffusion within the plagioclase is too slow to give a significant contribution to the formation of new plagioclase. Chemical diffusion within the plagioclase only is important to get equilibrium compositions of the new plagioclases after the melting reaction.

We conclude from our experiments that new plagioclases are formed by a solution/ reprecipitation mechanism. The old plagioclase is dissolved in the melt and new plagioclase is precipitated epitactically on the surfaces of old and new plagioclase. The solution reaction is critically dependent on the interface area between old plagioclase and melt. The interface is covered continuously by newly formed plagioclase. This prevents the further dissolution of the old plagioclase. In this situation, the only pathways for the melt to the old plagioclase are grainboundaries in the new plagioclase. Higher reaction rates are possible if cracks are formed in the plagioclase by local stresses during the melting process.

A further influence on the kinetics of partial melting of plagioclases can be the transport in the melt. Under dry conditions (1 atm) the diffusivities of network-forming cations, Al and Si, are at 1420 °C in the order of 10^{-8} - 10^{-9} cm^2/s as determined by Baker (1990) for dacite and rhyolite melts. From these data it can be concluded, that the homogenization of melts on a scale of 1 mm in the dry melting experiments requires more than 100 h.

On the basis of our findings, the melting reaction can be described by the following model:

(1) Melt is nucleated on the surface and at extended defects in the interior of the crystal (cracks, twins, dislocations).
(2) Ab-rich melt and An-rich plagioclase are formed in local equilibrium after short times; both melt and plagioclase are lower in An than the equilibrium compositions of the whole system.
(3) Formation of new plagioclase take place by a solution/reprecipitation mechanism. The progress of melting is controlled by the interface between old plagioclase and melt and/or by transport of components in the melt.
(4) Finally, equilibrium compositions are attained by diffusion in the crystal and in the melt after disappearance of the residual plagioclase.

4.2. WET MELTING OF PLAGIOCLASE AT P = 5 KBAR.

The kinetics of wet melting of plagioclase at a confining pressure of 5 kbar were investigated at 1000 °C using a plagioclase An_{66}. In contrast to dry melting at 1 atm, the reaction starts exclusively from the surface and from cracks connected to the surface (Fig. 8A). Reaction rates and melting textures were found to depend on the size of the plagioclase samples. For small crystals (size: 1 x 1 x 2 mm) the reactivity is much higher than for larger ones (size: 2 x 2 x 3 mm) (Table 3). In both, small and large crystals, melting progresses faster in direction parallel to the a-axis than perpendicular to this. But for small samples the dependence of the melting rates on the crystallographic orientation is essentially more significant (Fig. 8A,B).

Melting textures. The high reaction rates in the small samples are due to thin needles of melt growing into the crystal (Fig 8A). The needles were orientated parallel to the a-axis, which is the intersection of the main cleavage planes (001) and (010). They were too thin for a quantitative analyses by the microprobe, but qualitatively an Ab-rich composition could be deduced from the analyses. In the large crystals such needles were not observed. Instead of this, melting in the direction of the a-axis produced platy melt patterns (Fig. 8B). In all samples melt planes either parallel to (001) or to (010) were formed by the reaction in directions perpendicular to the a-axis. Obviously, the structure of the plagioclase (low concentration of chemical bonds across (010) and (001)) controls the propagation of melt. The effect of the size of the crystals is believed to be related to mechanical stress induced in the samples. This stress can produce microcracks and can facilitate cleaving of the crystal. The smaller samples probably are more strained by the

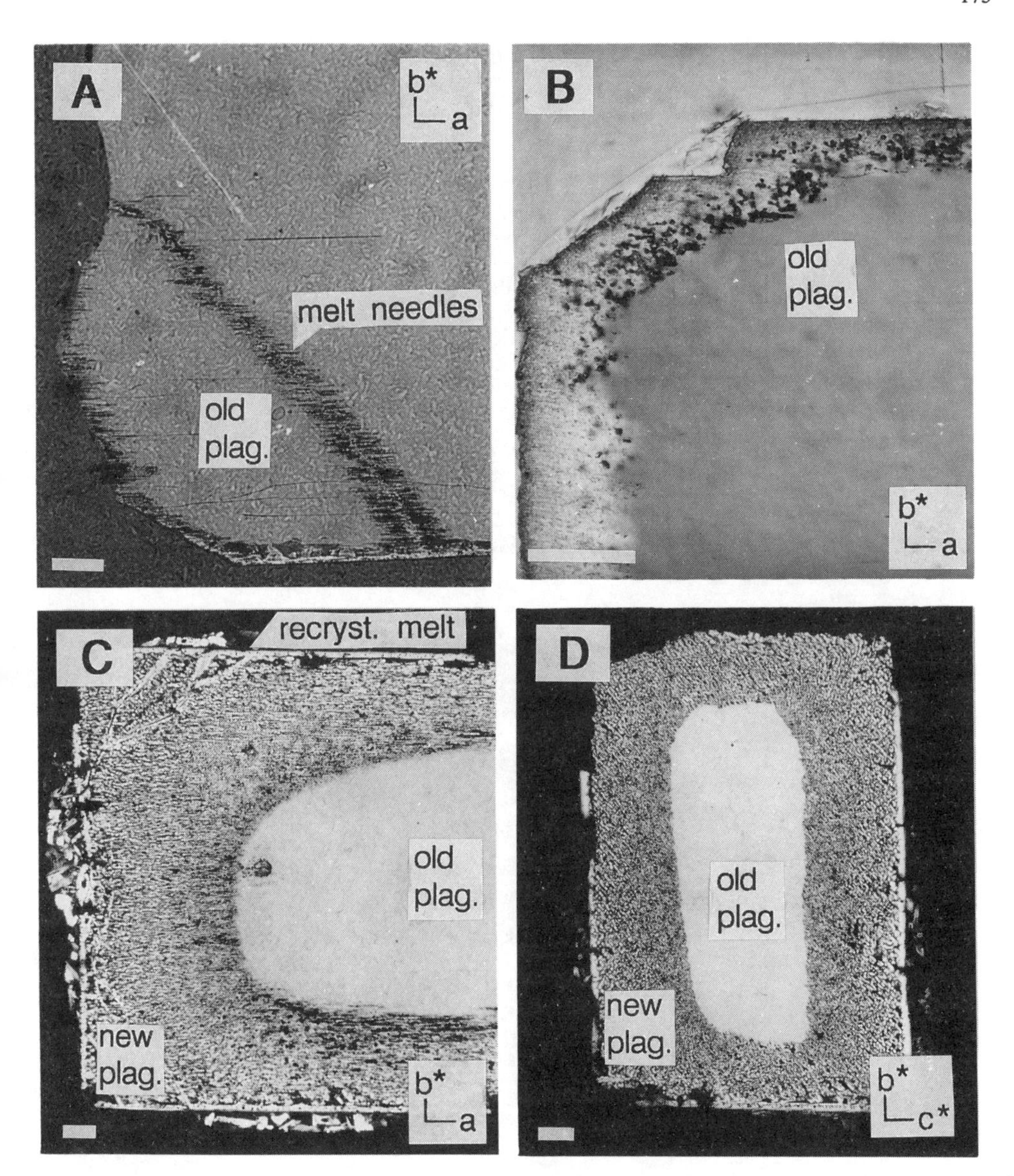

Fig. 8. Photographs taken by a microscope of run products of melting experiments performed at T = 1000 °C and at P_{H2O} = 5 kbar. Starting material: An_{66}. Bar: 100 μm.

(**A**) Run conditions: small crystal, t = 5 min. Section: parallel to (001)
(**B**) Run conditions: large crystal, t = 60 min. Section: parallel to (001)
(**C**) Run conditions: small crystal, t = 30 min. Section: parallel to (001)
(**D**) Run conditions: see (C). Section: perpendicular to the a-axis.

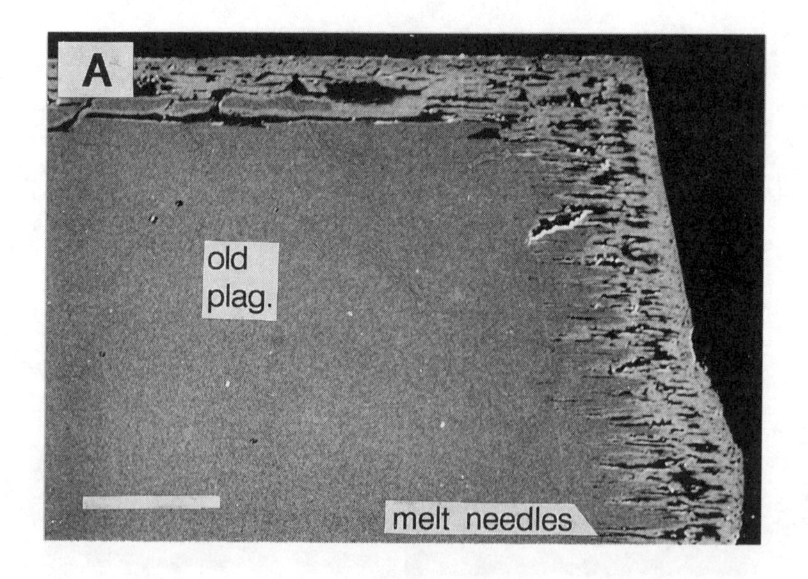

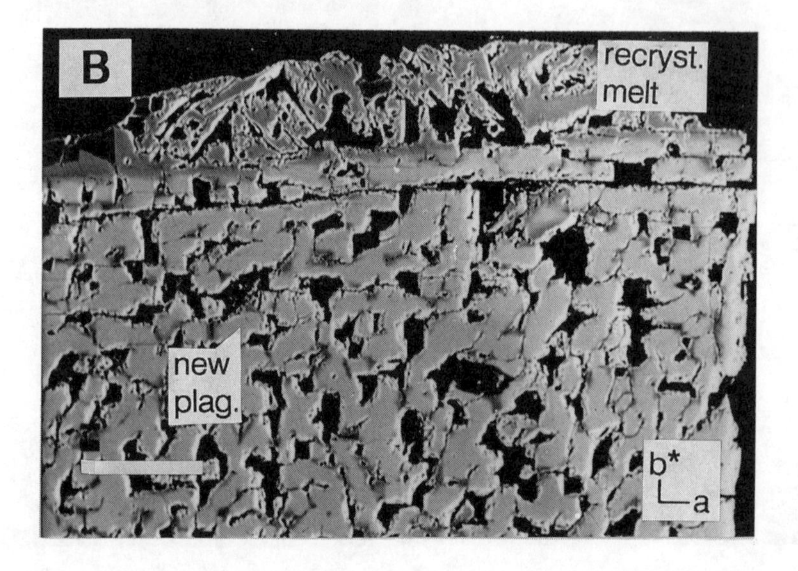

Fig. 9. BSE images of run products of melting experiments performed at T = 1050 °C and at P_{H2O} = 5 kbar. Starting materials: small crystals of An_{66}. Bar: 100 μm.

(**A**) Run duration: 20 min.
(**B**) Run duration: 45 h.

TABLE 3: Results of kinetic studies on the melting of plagioclase.
P_{H2O} = 5 kbar, T = 1000 °C

No.	time (min)	wt% water added	thickness of the layer (μm)	composition of new plagioclases (mol% An)	
				rim	core
1 sc	3	10	40	n.d.	n.d.
3 sc	300	10	30	83-89	n.d.
5 sc	1500	10	completely reacted	89-93	83-87
6 sc	5	50	80	88-94	-
8 sc	10	50	80	90-94	-
9 sc	20	50	270	91-96	87
7 sc	30	50	530	92-95	84-89
10 sc	1800	50	completely reacted	91-96	85-88
12 lc	60	50	110	92-95	-
13 lc	15	100	60	94-96	-
14 lc	45	100	80	94-96	-
15 lc	60	100	90	96-97	-

sc and lc denotes small and large crystals as the solid starting material, respectively. The thickness of the reaction layer was always determined for the a-direction of the plagioclase.

preparation (cutting with a diamond saw) and by the heating experiment (non-hydrostatic pressure by collapsing of the capsule during the build-up of pressure) than the larger ones.

The melting zone continuously changed to a zone of new plagioclases forming a mosaic texture (Fig. 8C,D, 9B). Both Fe and K contents are strongly reduced in the new plagioclases as compared to the starting plagioclase (Table 4). Relics of unchanged plagioclase were not detected behind the melting zone in any of the experiments. This suggests the formation of the new plagioclases by a solution/reprecipitation mechanism.

TABLE 4: Results of melting plagioclase. P_{H2O} = 5 kbar.

temp. °C	run time		An mol% (1σ)	Ab mol% (1σ)	Or mol% (1σ)	Fe_2O_3 wt% (1σ)
-	-	pl_{st}	65.7 (0.3)	33.6 (0.3)	0.7 (0.1)	0.48 (0.04)
950	60 h	pl_{re}	66.5 (0.4)	32.7 (0.4)	0.8 (0.1)	0.49(0.07)
		pl_r	87.5 (2.5)	12.5 (2.5)	< 0.3	0.09 (0.03)
1000	25 h	pl_c	85.9 (1.9)	14.1 (1.8)	< 0.3	0.20 (0.03)
		pl_r	90.7 (1.3)	9.3 (1.4)	< 0.3	0.11 (0.05)
		m_{gl}	31.9 (1.5)	63.1 (1.5)	5.0 (1.7)	0.40 (0.12)
		m_{cr}	34.2 (2.6)	65.1 (1.4)	0.6 (0.2)	0.40 (0.02)
1000	30 h	pl_c	85.9 (1.3)	14.0 (1.3)	< 0.3	0.24 (0.10)
		pl_r	92.5 (3.2)	7.4 (7.4)	< 0.3	0.17 (0.04)
		m_{gl}	35.4 (3.2)	62.5 (2.7)	2.1 (0.7)	0.21 (0.09)
1050	45 h	pl_c	92.5 (1.0)	7.5 (1.0)	< 0.3	0.12 (0.04)
		pl_r	93.7 (0.8)	6.3 (0.8)	< 0.3	0.07 (0.05)
		m_{gl}	32.7 (0.5)	51.6 (0.6)	15.7 (0.8)	0.08 (0.02)
		m_{cr}	35.3 (3.0)	64.0 (4.5)	0.7 (0.3)	0.04 (0.02)
1100	40 h	pl_c	94.1 (1.8)	5.9 (1.8)	< 0.3	0.08 (0.07)
		pl_r	95.1 (0.3)	4.8 (0.4)	< 0.3	0.03 (0.02)
		m_{cr}	49.0 (5.4)	50.4 (5.2)	0.6 (0.2)	0.08 (0.05)

All iron is given as Fe_2O_3. pl_{st}: starting plagioclase, pl_{re}: residual plagioclase, pl_r: new plagioclase close to the rim of the crystal, pl_c: new plagioclase in the center of the crystal, m_{gl}: melt quenched to glass, m_{cr}: melt crystallized during the quench.
50 wt% of water, relative to the mass of the plagioclase, were added in all runs except of that hold for 25 h at 1000 °C. All samples had completely reacted except of those from the run at 950 °C. The glass-plagioclase assemblage obtained from the run at 1000 °C (30 h) was heated for a few minutes in a dry argon stream at 1000 °C in order to remove water dissolved in the glass.

The newly formed plagioclases were of rectangular shape, often nearly cubic, with an edge length of about 50 μm. The surfaces of the new crystals were orientated nearly parallel to those of the starting material. Furthermore, an investigation of the samples by polarized light microscopy showed that the new plagioclase had almost the same crystallographic orientation as the old. This suggests that the new plagioclases are epitactically grown on the surfaces of the old plagioclase.

The space between the new plagioclases was empty. In no case glass (quenched melt) was detected in this zone. Glass was observed only in the melting zone and in a layer segregated at the surface. It is assumed that the empty space was filled with the fluid during the experiment. The separation of new plagioclase and melt suggests, that the surface tension of plagioclase and fluid is lower than that of plagioclase and melt.

Complete melting of a small crystal of plagioclase could be obtained at temperatures $\geq$ 1000 °C within 25 h. At 950 °C the reaction was too slow to give a significant melting zone. After 60 hours melting extended only over a few microns from the initial surface of the starting plagioclase.

All the fully reacted samples consisted of newly formed plagioclase with mosaic texture and a rim of Ab-rich quenched melt with a thickness of up to several 100 μm. With increasing run temperature the melts became more and more recrystallized during the quench of the sample. The recrystallized plagioclases could be easily distinguished from the new plagioclase formed by the melting reaction (Fig. 9B). The recrystallized plagioclases are preferentially nucleated on the surface of the original crystal and grow into the melt. The size of the recrystallized plagioclases can reach up to 100 μm.

Development of chemical compositions. The compositions of the new plagioclases formed by the melting reaction decreases in An content from the rim to the core of the sample. The maximum An contents detected near the initial surface are correlated to the fluid/solid mass ratio: 0.1 - An_{89-93}, 0.5 - An_{90-95}, 1.0 - An_{94-97} for the experiments at 1000 °C. This observation can be explained by considering the semi-quantitative phase diagram of the system Ab-An-H_2O at 1000 °C and at 5 kbar (Fig. 10). The first step of the reaction is the dissolution of feldspar components in the fluid. For simplicity the dissolution is assumed to be congruent for the plagioclase as observed for albite, but in principle this restriction is not necessary to explain the observed phenomena. At the beginning of the experiment, only a small amount of the plagioclase participate in the reaction and the fluid/solid ratio is almost infinite. The Ab component preferentially dissolves in the fluid and, therefore, the plagioclases formed at this stage have very high An contents. With progressing dissolution reaction the amount of plagioclase participating in the reaction increases. The composition of the reacting system moves on the connection of H_2O to An_{66} in direction of the plagioclase and both the new plagioclase and the fluid becomes more Ab rich. Melt is formed when the composition of the system reaches the three-phase field. Then the composition of the plagioclase is thermodynamically fixed by the three-phase equilibrium.

The chemical diffusion within the feldspars at 1000 °C is too slow to equilibrate the plagioclase with the changing chemical environment (Grove et al., 1984; Yund, 1986; Liu and Yund, 1991; Baschek and Johannes, 1992) and the initial compositions of the new plagioclases are almost preserved in the run. At temperatures above 1000 °C the differences in plagioclase composition between rim and core become smaller. In runs at 1050 and 1100 °C the new plagioclase near to the initial surface were only 1 mol% higher in An than in the center of the sample (Table 2). Possibly, at these temperatures, the diffusion rates within the feldspars are high enough to change their compositions.

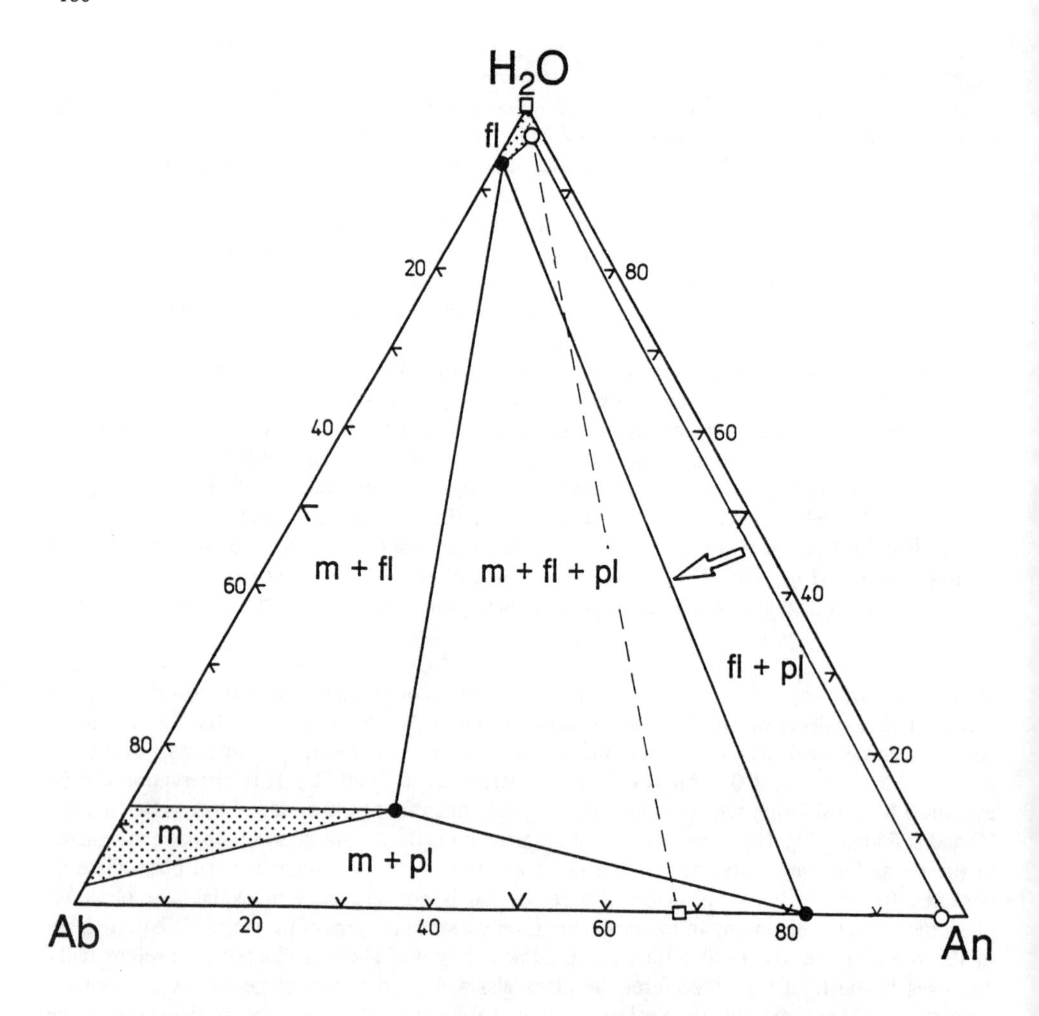

Fig. 10 Phase diagram of the system Ab-An-H_2O at T = 1000 °C and at P = 5 kbar (pl: plagioclase, m: melt, fl: fluid). Open squares: starting composition. Open circles: fluid and plagioclase compositions at the beginning of the experiment. Filled circles: three phase equilibria. The arrow denotes the development of fluid and plagioclase compositions in the first stage of the reaction.

Data used for the construction of the diagram: Solidus composition: An_{82} (Johannes, 1978). Liquidus composition without consideration of the water content: An_{30} (Yoder et al., 1957). Maximum water contents in the melts determined by Karl-Fischer titration: 11.5 wt% (Ab) and 10.9 wt% (An_{35}, recalculated from an experiment at 1200 °C). Water content of Ab melt in equilibrium with crystallin Ab determined by experiments using Ab glass and various amounts of water: 3.3 ± 1.0 wt%. Solubility of Ab in the fluid determined by the weight loss of a glass piece encapsulated with 50 wt% of water: 6.9 wt%. Microprobe analyses of globules quenched from the fluid indicates that Ab dissolved almost congruently in the fluid. Solubility of a natural plagioclase An66 in the fluid determined by the weight loss of a single crystal: 3.5 wt% (run for 30 h at 1000 °C, see table 4). Microprobe analyses of globules quenched from the fluid of this run indicates that the Ab (and Or) component is enriched in the fluid relative to the An component.

In conclusion, only the new plagioclases in the center of the original sample have a composition close to that of the three-phase equilibrium. Considering these data, the obtained solidus temperatures are approximately 30 °C below those of the pure system Ab-An-H_2O as determined by Johannes (1978) (see Fig. 6). As in the case of dry melting this is attributed to the Or content and to the iron present in the natural starting plagioclase.

A reliable determination of the melt composition was possible only for runs at 1000 °C displaying a small degree of recrystallization. In runs at temperatures above 1000 °C a large part of the melt was recrystallized during the quench (e.g. see Fig. 9). The recrystallized plagioclases show a large variation in An-content. Typically, the An-contents of the recrystallized plagioclases decrease with distance to the nucleation sites. The Or component is enriched during recrystallization in the residual melt.

Because of this recrystallization phenomenon, a quantitative determination of the melt composition is difficult. The approximate An/Ab ratio was obtained by the average of analyses of the recrystallized melt (1050 °C: $An_{36}Ab_{66}$; 1100 °C: $An_{49}Ab_{51}$), but these estimates have large uncertainties. Nethertheless, the values are in good agreement with the liquidus determined by Johannes (1978) for the pure system Ab-An-H_2O (see Fig. 5).

Both the new plagioclases and the melt were reduced in iron content as compared to the starting plagioclase. The iron content of the new plagioclases increases from rim to core. This can be explained in the same way as for the main components. In the beginning of melting iron is leached away from the plagioclase by the fluid. Thus, the iron activity is low in the beginning and the first plagioclases contain relatively small amounts of iron (0.11 wt% at 1000 °C). As melting processes the part of iron dissolved in the limited volume of fluid decrease and consequently the iron content of the new crystals becomes higher (0.20 wt% at 1000 °C).

Model of wet melting of plagioclase. On the basis of our experimental findings the following model of the melting reaction in presence of an hydrous fluid is developed:

(1) Feldspar components are dissolved in the hydrous fluid, An-rich plagioclase is reprecipitated from the fluid.
(2) Melt is nucleated at the surface of the crystal and at cracks connected to the surface.
(3) Melt volumes grow into the crystals. This process is controlled by the dissolution kinetics of the old plagioclase.
(4) New plagioclases grow epitactically on the surfaces of old and new plagioclases by reprecipitation from the melt. The new crystals are in local equilibrium during their growth, but they are not in equilibrium with respect to the whole sample.
(5) After the melting process is finished, equilibrium compositions in the plagioclases are established by solid state diffusion.

The rate-controlling step of the melting reaction is the movement of the melt/crystal interface into the crystal (step 3). This process seems to be strongly influenced by the stress state of the crystal and by extended defects within the crystal.

4.3. THE INFLUENCE OF ADDITIONAL COMPONENTS AND PHASES

Solidus and liquidus temperatures of plagioclase-bearing systems are strongly lowered by quartz. Various assemblages containing quartz as a mineral or SiO_2 as a component of minerals, glasses or gels were used in order to study melting processes at almost natural conditions. Results of melting in the tonalite system Qz-Ab-An at P_{H2O} = 2 kbar have been reported earlier (Johannes, 1989; Johannes and Holtz, 1992) and will be briefly summarized. The effect of the Qz-activity on partial melting of plagioclases was systematically investigated in the tonalite system using materials with various SiO_2/Al_2O_3 mass ratios. The dependence of partial melting of plagioclases on pressure and on water activity was studied in the granite system Qz-Or-Ab-An-H_2O. In some of these experiments biotite was added to the starting assemblages. In others, the effect of the Al_2O_3 activity in the melt on melting kinetics was tested.

The melting textures observed in SiO_2-oversaturated systems (with respect to feldspar compositions) are similar to those found for wet melting of plagioclase without additional components. The melting reaction progresses most rapidly into the plagioclase in the direction of the crystallographic a-axis. For this direction the reaction rates are up to two times higher than for directions perpendicular to the a-axis. Interfaces between old plagioclase and melt often are orientated parallel to the main cleavage planes, (001) and (010) (e.g. Fig. 11). Furthermore, melt channels or intergrowths formed between the new plagioclases are preferentially orientated parallel to these cleavage planes.

4.3.1. Melting reactions in the system Qz-Ab-An-H_2O

In this set of experiments plagioclases An_{60} were embedded in powdered natural quartz from Göschenen, Switzerland. Runs were carried out in the T-range 820 - 880 °C at 2 kbar water pressure. Reaction zones in the plagioclases were small even after a run duration of 19 d at 880 °C (maximum thickness: 100 μm). The melt and the new plagioclase had equilibrium compositions at the beginning of melting. Plagioclase components were strongly fractionated between melt and new plagioclase (Fig. 5). For example at 850 °C the mole fraction An/(An + Ab) in the melt was 0.3 and that in the plagioclase was 0.8. Both iron and Or component were enriched in the melt compared to the new plagioclase.

4.3.2. Melting reactions in the system Qz-Ab-An-Al_2O_3-H_2O

The basic intention for this set of experiments was to test how the melting kinetics are influenced by the Qz-activity. The solidus temperatures in the system Qz-Ab-An-H_2O increase with decreasing Qz-activity. For a given temperature this shift of the solidus results in smaller proportions of plagioclase dissolving in the melt. This was believed to a have marked influence on the kinetics of melting.

The Qz-activity was systematically varied in the experiments by using materials with different SiO_2/ Al_2O_3 mass ratios for surrounding the plagioclase. These materials were quartz, quartz/Al_2O_3 gel 3:1, quartz/sillimanite 1:1, quartz/Al_2O_3 gel 2:1, quartz/mullite 1:1, sillimanite, mullite or Al_2O_3 (arranged in decreasing Qz-activity). Sillimanite powder was prepared from a natural crystal from Mjödö, Sweden. The mullite composition was introduced either as a gel or as a crystalline material obtained after heating the gel for 12 h at 1200 °C in air.

Runs were performed at water pressures of 2 or 5 kbar using plagioclases An_{58} or An_{60}. At P_{H2O} = 2 kbar, two temperatures (835 and 880 °C) with different degrees of partial melting were chosen. At P_{H2O} = 5 kbar, it was not possible to operate at subliquidus conditions in the system Qz-Ab-An-H_2O. At the temperatures necessary for this, the melting kinetics are too slow to give a significant reaction within acceptable times. Experiments were, therefore, conducted at a temperature above the liquidus, but the amount of SiO_2 was always less than that necessary for complete melting of the plagioclase.

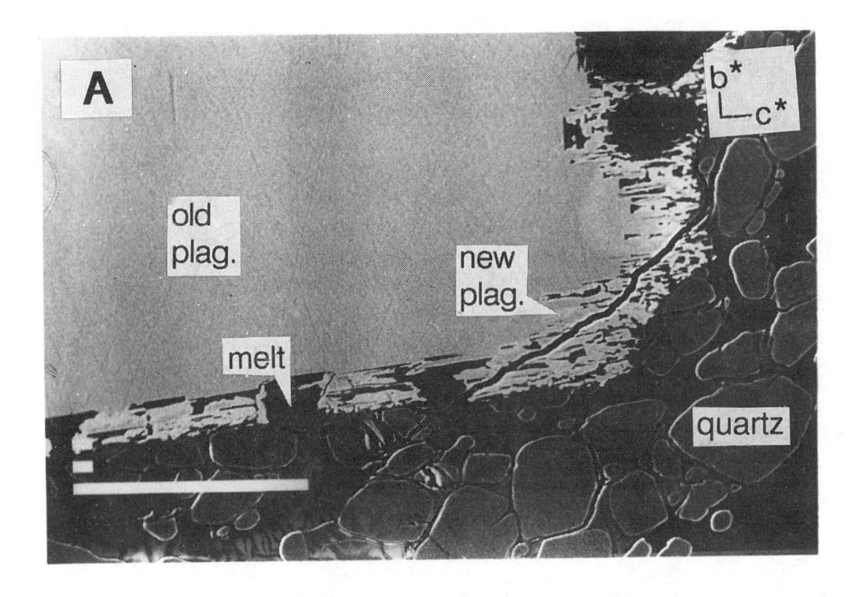

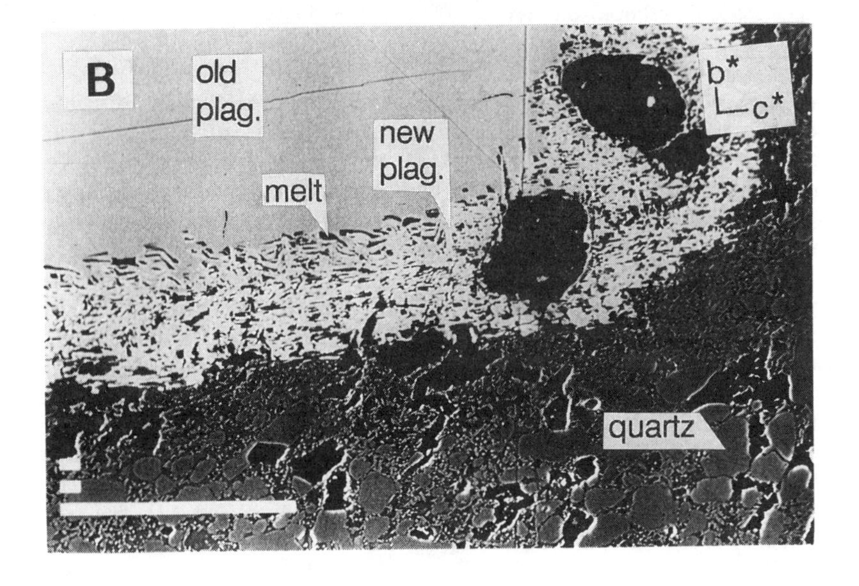

Fig. 11. BSE images of run products of melting experiments performed in the system Qz-Ab-An-Al_2O_3-H_2O. Run conditions: An_{58}, T = 880 °C, P = 2 kbar. Run duration: 6 d. Bars: 100 μm.

(**A**) Crystal embedded in quartz
(**B**) Crystal embedded in quartz and Al_2O_3 gel, mass ratio 3:1.

Fig. 12. BSE image of a run product obtained in the system Qz-Ab-An-Al_2O_3-H_2O. Run conditions: An_{60} embedded in quartz/Al_2O_3 gel mass ratio 3:1, T = 835 °C, P = 5 kbar, t = 12 h. Bar: 100 μm.

Experiments at P_{H2O} = 2 kbar and at T = 835 °C. After 7 days run duration, reaction zones in plagioclase An_{60} were always very small. If Al_2O_3 or sillimanite surrounded the plagioclase, dissolution textures were visible only at the surface of the plagioclase crystals. The Qz activity was too low in these runs to initiate melting. Neither new plagioclase nor melt could be detected. If the plagioclases were embedded in quartz the Qz activity was sufficiently high to initiate melting, but the reaction layer had a maximum thickness of only 50 μm.

Experiments at P_{H2O} = 2 kbar and at T = 880 °C. In this set of experiments the plagioclases An_{58} were embedded in quartz, quartz/Al_2O_3 gel 3:1, sillimanite or mullite. Run duration was always 6 days. No melting was observed for the plagioclase in contact with sillimanite or with mullite under these conditions. In the sample containing quartz a reaction layer is formed with a thickness comparable to that of the run at 835 °C. This indicates that the ratio of melt to solid within the plagioclase was not controlling the melting process. At the higher temperature a significant larger amount of the plagioclase was dissolved in the melt (see Johannes (1989), Fig. 7). The melting reaction was considerably enhanced if Al_2O_3 was added to the quartz (Fig. 11). The reaction rate was approximately twice that found when the plagioclase was embedded exclusively in quartz.

Experiments at P_{H2O} = 5 kbar and at T = 835 °C using plagioclase single crystals. In all assemblages containing free quartz, the Qz-activity was very high at the beginning of the run and a large zone of completely molten plagioclase was formed (Fig. 12). The Qz-activity decreased at the interface between old plagioclase and melt by propagation of melting and new, An-rich plagioclases became stable.

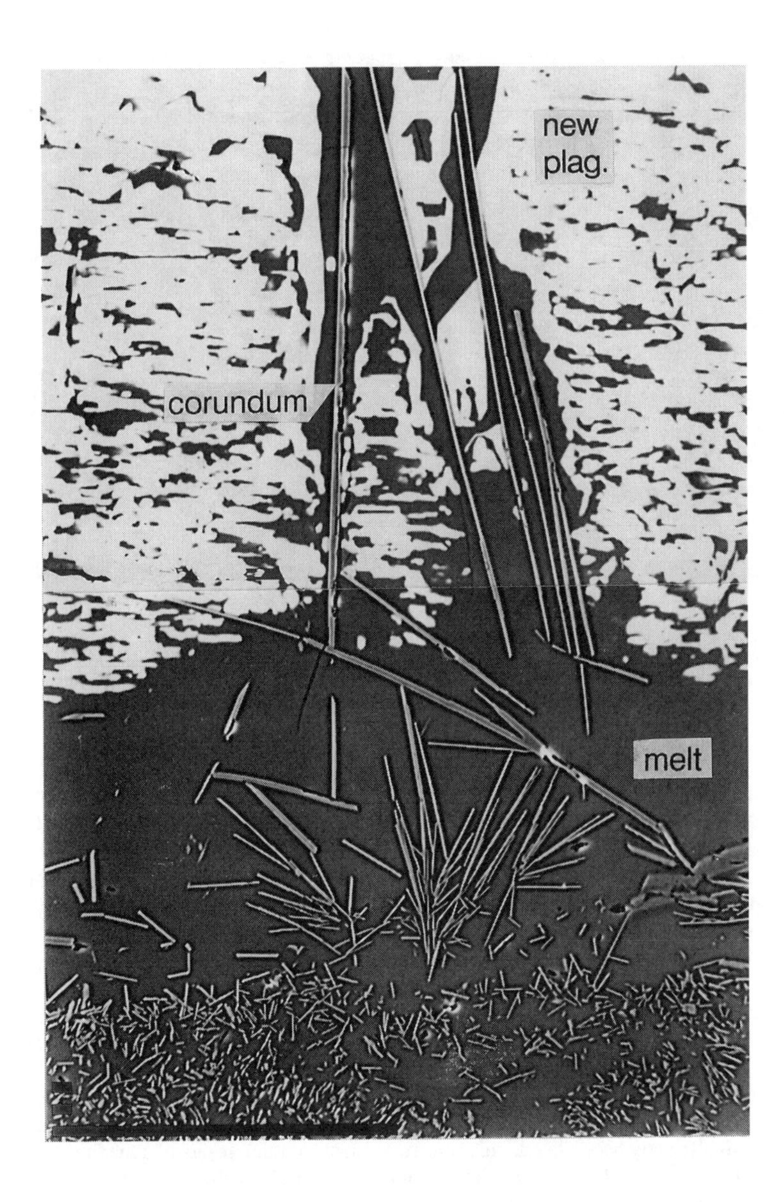

Fig. 13. BSE image of an experimental product obtained from an An_{60} crystal surrounded by quartz/Al_2O_3 gel, mass ratio 2:1. Run conditions: $T = 835$ °C, $P_{H2O} = 5$ kbar, $t = 14$ d. Bar: 100 μm.

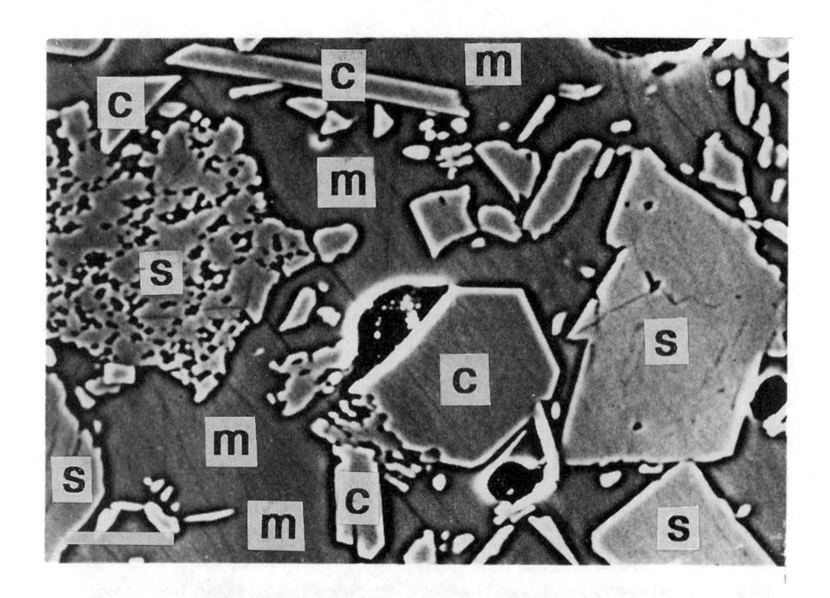

Fig. 14. BSE image of the surrounding material after a melting experiment using an An_{60} crystal and sillimanite powder. Run conditions: T = 835 °C, P_{H2O} = 5 kbar, t = 3d. Phases identified by microprobe analyses (c: corundum, s: sillimanite, m: melt). Bar: 10 μm.

Crystals in the form of thin plates were present in all runs performed at 5 kbar using powders with normative corundum. Quantitative determination of the composition of this plates by electron microprobe was difficult due to the small size of these crystals, but a very high Al_2O_3 content (> 90 wt%) could be proved. X-ray powder methods clearly identified this plates as α-corundum. Corundum was formed in the surrounding powder as well as in the interior of the plagioclase. Typically, corundum crystals within the feldspar were much larger than outside, but the amount of crystals was smaller. This was probably due to nucleation kinetics. Within the plagioclases the corundum is surrounded by extended melt channels (see Fig. 13).

Mullite was formed metastably in runs with high silica contents (quartz/Al_2O_3 gel 3:1, quartz/sillimanite 1:1). Sillimanite employed as starting material shows dissolution textures resulting from the reaction with the melt (Fig. 14).

The melting reaction was very fast in the Qz-rich systems. For example in a quartz/Al_2O_3 mixture 3:1, the crystal had completely reacted within 12 h. New plagioclase and melt was formed, indicating that the Qz-activity was significantly below one in this run. In contrast, reaction layers of a maximum thickness of 200 μm were formed within 14 d in plagioclase + mullite gel or plagioclase + mullite crystals. In these assemblages the Qz activity is much lower than in the example given before and only a small part of the plagioclase is dissolved in the melt.

Experiments at P_{H2O} = 5 kbar and at 835 °C using plagioclase powder. An important question in the experiments using plagioclase single crystals was that of the thermodynamic stability of the run products. The formation of corundum from sillimanite or mullite may be due to the kinetics of the reaction. In order to test this a set of experiments was performed using a fine-grained plagioclase (hydrothermally synthesized An_{60}, grain size < 2 μm). The transport distances become smaller and the reaction can progress faster.

TABLE 5: Results of experiments using powders as solid starting materials. Run conditions: T = 835 °C, P_{H2O} = 5 kbar, t = 6 d. 12.5 wt% of water was added (relative to the mass of all solids).

No.	mass of solids (mg)				$m(SiO_2)$/	run products
	plag	sill	mull	qz	m(plag)	
ps3	10.07	3.01	-	7.00	0.811	m + fl + plag + sill
ps2	10.5	5.08	-	5.04	0.691	m + fl + plag + sill
ps6	10.15	3.05	4.06	3.06	0.525	m + fl + plag + sill + c
ps1	10.07	10.08	-	-	0.370	m + fl + plag + sill + c
ps4	10.17	5.10	5.07	-	0.331	m + fl + plag + sill + c
ps5	10.16	5.08	-	-	0.185	m + fl + plag + sill + c

m: melt, fl: fluid, plag: plagioclase, sill: sillimanite, mull: mullite, c: corundum.
$m(SiO_2)$ denotes the total mass of SiO_2 present in the various components of the surrounding powder.

Sillimanite was preserved in all run products and corundum was formed in the assemblages with a mass ratio of SiO_2 (in the added material) to plagioclase $\leq$ 0.52 (Table 5). Mullite was observed in none of the run products even if it was present in the starting material. In assemblages containing mullite and sillimanite, the mullite was preferentially dissolved. This suggests that under these conditions mullite is not a stable phase between corundum and sillimanite in the system Al_2O_3-SiO_2. This is consistent with observations in the nature. Mullite is found in contact metamorphism of the sanidinite facies, but not in low-temperature metamorphism (Deer et al., 1982). Experimental studies on the stability of mullite are difficult due to slow reaction rates even at high temperatures (Klug et al.,1991) and no reliable data exist for temperatures below 1000 °C.

From thermodynamic considerations, the presence of both sillimanite and corundum is an indication for an invariant point in the system Qz-Ab-An-Al_2O_3-H_2O at the experimental conditions (5 components, 5 coexisting phases: fl., m., plag., sill., c.). The melt composition should be unambiguously fixed and corundum/sillimanite should as a buffer for SiO_2. But corundum and sillimanite were found in experiments with a large variation in composition of the starting materials. We assume that in the runs with low $m(SiO_2)$/m(plag) the run duration was not long enough to remove all sillimanite and only corundum should be the stable at these conditions. Further experiments are necessary to test this, but this was out of the scope of the present study.

The plagioclases in the run products were much larger (length up to 30 μm) than the starting ones. This is a strong indication that the new plagioclases are formed by a solution/reprecipitation mechanism. The An content of the new plagioclases increases with increasing $m(SiO_2)$/m(plag) of the starting material: An_{64-70} for 0.37 (ps1) , An_{82-87} for 0.69 (ps2), An_{90-96} for 0.81 (ps3).

4.3.3. Melting reactions in the system Qz-Or-Ab-An-H_2O

In this type of experiment a plagioclase An_{66} was surrounded by $Qz_{40}Or_{60}$ glass powder. Run temperatures were between 840 and 880 °C. The total pressure was always 2 kbar. The water activity in the fluid phase was varied by using H_2O/CO_2 mixtures. CO_2 was generated in the experiments by thermal decomposition of $Ag_2C_2O_4$.

Effect of water activity. No partial melting of the plagioclase was observed in this temperature range when the fluid contained less than 80 mol% of water. A narrow rim occurred on the plagioclase in which Na was substituted by K due to Na/K interdiffusion. A small reaction layer of 20 μm thickness was formed in the plagioclase after 28 d at 880 °C using a fluid with 80 mol% of water. New plagioclases within this layer had compositions $Or_7Ab_{11}An_{82}$. Partially, overgrowth of metastable Or-rich plagioclases (composition: $Or_{34}Ab_4An_{62}$) could be detected on the surface of the original crystal. Much deeper reaction zones in the plagioclases were obtained after the same run duration at identical P/T conditions if the fluid was pure water. Intergrowths of alkali feldspar, $Or_{92}Ab_3An_5$, and new plagioclase, $Or_4Ab_6An_{90}$, were formed (Fig. 15).

The influence of water activity on partial melting of plagioclase can be due to the lowering of the solidus temperatures with increasing water activity as well as to faster transport of components in melts of higher water contents. Lowering of the solidus temperature gives a higher degree of dissolution of the plagioclase. Thus, the ratio of melt to new plagioclase is increased. If the melting reaction is exclusively controlled by the dissolution and the reprecipitation kinetics of the plagioclase, this process should be enhanced by higher proportions of solvent within the crystal. This is not the case, if transport in the melt has a significant influence on the progress of partial melting. Higher degrees of melting requires higher amounts of SiO_2 which must be transported to or produced at the reaction front. As shown by Baker (1991) the diffusivity of SiO_2 in aluminosilicate melts depends strongly on the water content of the melt and, therefore, on the water activity. Melting experiments in the system Qz-Ab-An at p_{H2O} = 2 kbar have shown no effect of the melt/plagioclase ratio within the crystal on the melting kinetics (see 4.3.2.). Therefore, it is concluded that the main effect of water activity is on the transport properties of the melt.

Partial melting in biotite-bearing systems. Addition of synthetic biotite to the $Qz_{40}Or_{60}$ glass powder gave a strong enhancement of the melting reaction. The added biotite was dissolved close by the plagioclase and new Al_2O_3-rich biotites were formed at the outer surfaces of the plagioclase. The parallel arrangement of the new biotites suggests growth in the direction of the plagioclase (Fig. 16). This indicates that the growth of the biotites was controlled by diffusion of components from the melting front in the plagioclase to the biotite. Pyroxene was found in the melt far away from the plagioclase (> 500 μm), suggesting a low Al_2O_3-activity. Obviously, the Al_2O_3-activity decreases in this system from the melting front within the plagioclase to the melt outside of the plagioclase.

Partial melting in MgO-bearing systems. Mixing of MgO and quartz (mass ratio 1:4) to the $Qz_{40}Or_{60}$ glass also gave a considerable increase in reaction rate compared to runs using only the glass. Phlogopite was formed at the outer surfaces of the plagioclase, acting as a sink of Al_2O_3 (Fig. 17). In contrast, no effect of MgO on the melting kinetics was observed if the surrounding material contains no Or component. In this case, no Al_2O_3 consuming phase can be formed.

Fig. 15. BSE image of a reaction layer performed in a melting experiment using a plagioclase An_{66} embedded in $Qz_{40}Or_{60}$ glass. Run conditions: T = 880 °C, P_{H2O} = 2 kbar, t = 28 d. Bar: 100 μm. Section: parallel to (001). Black areas: melt, dark grey areas: old plagioclase, medium grey areas: Or-rich alkali feldspar, light grey areas: new plagioclase.

Fig. 16. BSE image of the run product obtained from a reaction of a plagioclase An_{66} with a powder composed by 80 wt% of $Qz_{40}Or_{60}$ glass and 20 wt% of synthetic biotite. Run conditions: T = 880 °C, P_{H2O} = 2 kbar, t = 1 d. Bar: 100 μm. Section: parallel to (001). Black areas: melt, dark grey areas: old plagioclase, medium grey areas: new plagioclase with intergrowth of Or-rich alkali feldspar. A rim of Al_2O_3 rich biotites (white colour) is formed around the plagioclase.

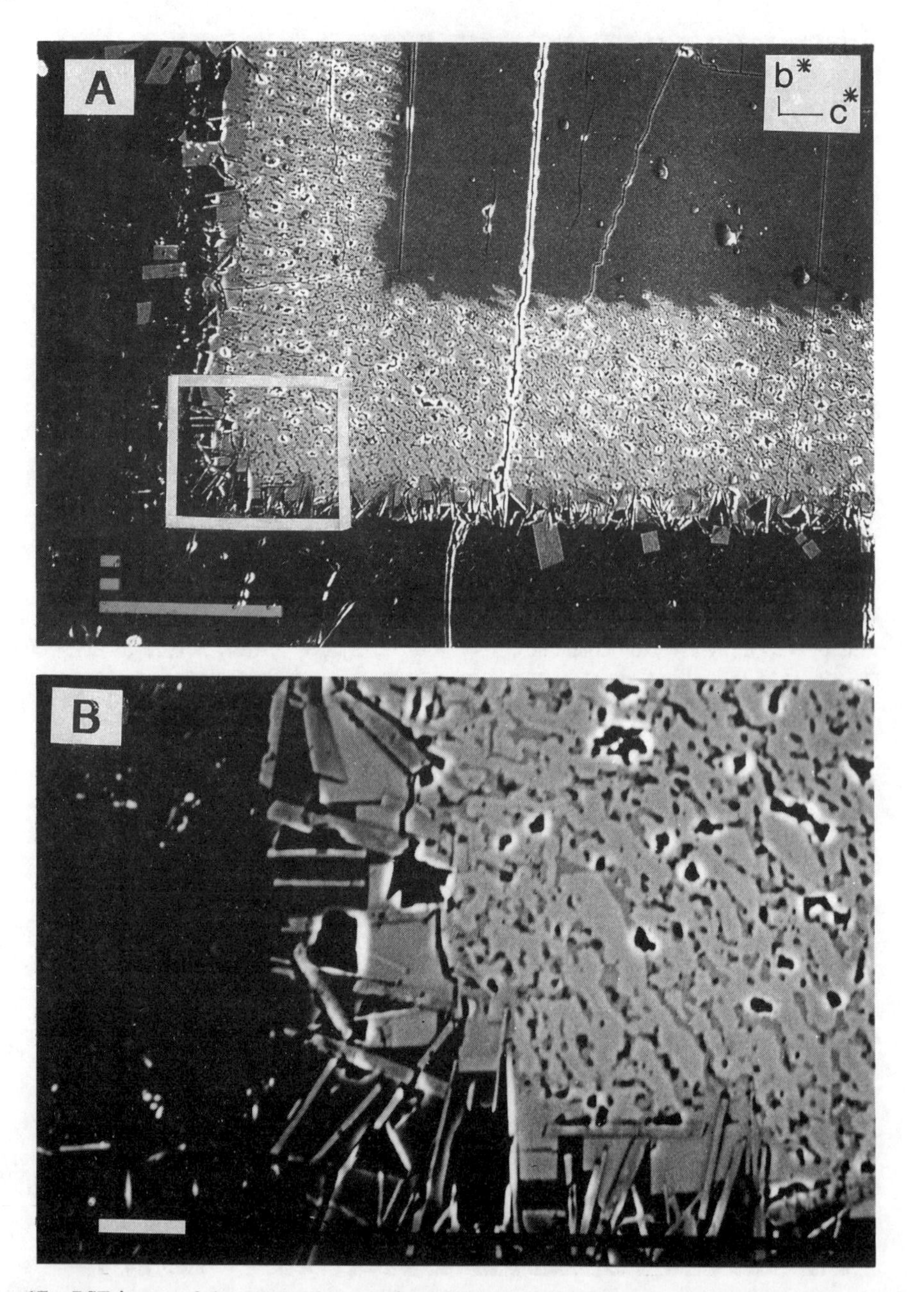

Fig. 17. BSE image of the run product obtained from a reaction of a plagioclase An_{66} with a powder composed of 90 wt% $Qz_{40}Or_{60}$ and 10 wt% of 3 MgO : 4 SiO_2. Run conditions: T = 880 °C, P_{H2O} = 2 kbar, t = 2 d. (**A**) Overview. Bar: 100 μm. (**B**) Detail. Bar: 10 μm. The reaction layer in the plagioclase consists of new plagioclase (light grey), intergrowth of Or-rich alkali feldspar (medium grey) and melt (dark). A zone of alkali feldspar (plates) and phlogopite (needles) is formed around the plagioclase.

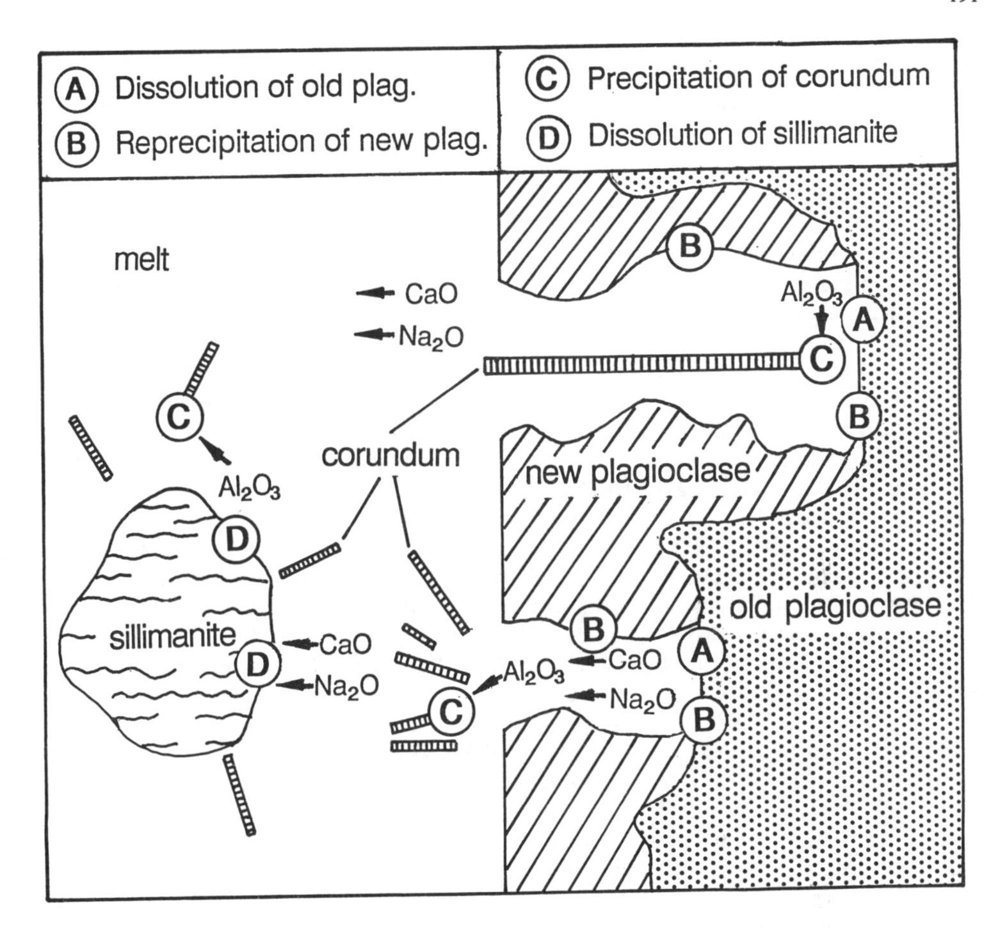

Fig. 18. Schematic illustration of partial melting of plagioclase in presence of sillimanite. Explanations see text.

4.3.4. Kinetics of partial melting in systems with additional SiO_2-containing phases beside feldspars.

Partial melting of plagioclase in systems containing quartz or other SiO_2-bearing phases is different from the first two processes (dry and wet melting of plagioclase) in that SiO_2 must be available at the interface between melt and old plagioclase. This can be achieved in two ways. Either SiO_2 diffuses to the reaction sites, or the feldspar itself acts as the source of SiO_2. In the latter case, Na_2O, CaO and Al_2O_3 must diffuse away from the interface or must be locally bonded. Which of the mechanisms operates, depends on the diffusivities of the involved components and on the presence of sinks for the components.

In the system with normative Al_2O_3, corundum is formed in the interior of the plagioclases. There is no driving force for a transport of Al_2O_3 into the crystal. Therefore the formation of corundum must be the result of a high Al_2O_3-activity produced by the melting reaction. A high Al_2O_3-activity of the melt within the plagioclase also is indicated in other types of experiments, in particular in those where Al_2O_3-rich biotite and phlogopite were formed at the outer surfaces of the plagioclase.

On the basis of these results it is concluded that the SiO_2 necessary for melting is produced by decomposition of the feldspar if a sink for Al_2O_3 is available. In this case neither SiO_2 nor Al_2O_3 must be transported in the melt for long distances. Possible sinks for Al_2O_3 are corundum and other peraluminous minerals like biotite or phlogopite.

A scheme of the melting reaction for sillimanite as the surrounding material is given in Fig. 18. Reaction of plagioclase with mullite or quartz/Al_2O_3 are very similar to the reaction with sillimanite and are not specially described. The lower part of Fig. 8 illustrates the more common situation that the sink for Al_2O_3 is on the outer surface of the plagioclase. Plagioclase is dissolved in the melt at the interface melt/old plagioclase. New plagioclase is precipitated on the surfaces of the new plagioclase as well as on those of the old plagioclase. Al_2O_3 is transported to the outer surface of the plagioclase and it is bound there in corundum crystals. The longest transport distances are for Na_2O and CaO. Both must diffuse out of the plagioclase to the sillimanite. There they react with the sillimanite to form $NaAlO_2$ and $CaAl_2O_4$ units in the melt. The amount of Al_2O_3 produced by decomposition of the sillimanite is higher than its solubility in the melt and corundum is precipitated also close to the sillimanite. The upper part of fig. 18 describes the situation in that corundum crystals grow into the plagioclase. In this situation the transport distance for Al_2O_3 is very short and the melting reaction proceeds much faster into the plagioclase than in the case where the sink is located at the outer surface of the plagioclase.

Partial melting in the systems Qz-Or-Ab-An-Bio-H_2O and Qz-Or-Ab-An-MgO-H_2O is very similar to the reaction illustrated in the lower part of Fig. 18. The difference is that components (e.g., K_2O, Fe_2O_3 in case of the biotite-containing system, MgO in case of MgO-containing system) must diffuse from the added minerals through the melt to the outer surface of the plagioclase. But these processes obviously are not rate-controlling for the melting process under the conditions of the experiments. It should be mentioned that in the MgO-containing system there there is an excess of alkali and alkaline earth cations with respect to aluminum ($\sum Al_2O_3 < \sum Na_2O + K_2O + CaO + MgO$). The reaction rates are similar to those of the biotite-containing system in which aluminum is compensated by alkali and alkaline earth cations ($\sum Al_2O_3 = \sum Na_2O + K_2O + CaO$). At the beginning of melting the reaction rates in the MgO-containing system may be higher due to the dissolution of Al_2O_3 in the melt initially undersaturated in this component. But the amount of Al_2O_3 produced by decomposition of the feldspar is much higher than its solubility in the melt and must be transported away from the reaction front.

In the experiments in which the plagioclase is surrounded exclusively by quartz, no sinks for Al_2O_3 are available and either SiO_2 must be transported from the quartz to the reaction front or Al_2O_3 must be transported out of the plagioclase. In both cases there are large transport distances for components of low mobility. The transport distances become even larger with progressing of the melting reaction and, therefore, the amounts of components transported away from the reaction front becomes smaller. Consequently, the dissolution rates of the old plagioclase become smaller. If the reprecipitation of new plagioclase on the surface of the old plagioclase becomes faster than the dissolution, the old plagioclase is covered by a layer of new plagioclase and further progress of the melting reaction is prevented. This can explain identical reaction zones in the experiments in the system Qz-Ab-An-H_2O at $P = 2$ kbar despite of largely different run times (4 d and 18 d).

5. Summary

Partial melting of plagioclase was studied under various conditions. At all experimental conditions the transformation of the old plagioclase to a new, An-rich plagioclase took place by solution/reprecipitation mechanisms. Chemical diffusion within the plagioclase was always too slow to give a significant contribution to the melting reaction. The kinetics of partial melting were controlled by the dissolution of the old plagioclase as well as by transport in the melt.

Under dry conditions at 1 atm partial melting of plagioclase was very sluggish despite of the high temperature. This can be attributed to the growth of new plagioclase on the surfaces of the old plagioclase, resulting in small dissolution rates of the old plagioclase due to small interface areas between old plagioclase and melt.

The kinetics of partial melting under wet conditions depend on the mechanical state of the crystal. Reaction rates of small crystals was observed to be much higher than those of large crystals. The higher reactivity of the small crystals can be explained by higher dissolution rates of the old plagioclase due to mechanical stress.

The kinetics of partial melting in Qz-bearing systems under wet conditions are strongly influenced by the presence of sinks for Al_2O_3. If such sinks are available at the surface or within the crystal the transport distances for Al_2O_3 becomes short and the reaction is considerably enhanced.

Acknowledgements

This work was supported by the DFG (SFB 173 and Jo63-26). Thanks are given to G. Bartels and A. Exner for performing some of the experiments. Furthermore we thank I. Parsons and T. Pearce for reviewing the manuscript and for helpful comments.

References

Baker DR (1990) Chemical interdiffusion of dacite and rhyolite: anhydrous measurements at 1 atm and 10 kbar, application of transition state theory, and diffusion in zoned magma chambers. Contrib Mineral Petrol 104: 407-423

Baker DR (1991) Interdiffusion of hydrous dacitic and rhyolitic melts and the efficacy of rhyolite contamination of dacitic enclaves. Contrib Mineral Petrol 106: 462-473

Baschek G, Johannes W (1992) Chemische Diffusion in Plagioklasen. Europ J Mineral Beih 4: 16

Behrens H, Johannes W, Schmalzried H (1990) On the mechanisms of cation diffusion processes in ternary feldspars. Phys Chem Minerals 17: 62-78

Bowen NL (1913) The melting phenomena of the plagioclase feldspars. Am J Sci 35: 577-599

Christoffersen R, Yund RA, Tullis J (1983) Inter-diffusion of K and Na in alkali feldspars: diffusion couple experiments. Am Mineral 68: 1126-1133.

Deer WA, Howie RA, Zussman J (1982) Rock-forming minerals, Volume 1A, Orthosilicates (2nd ed) Longman Group Limited, London, New York.

Foland KA (1974) Alkali diffusion in orthoclase. In: Hofmann AW,Giletti BJ, Yoder HS jr, Yund RA (eds) Geochemical transport and kinetics. Carnegie Inst, Washington 77-98

Giletti BJ, Semet MP, Kaspar RB (1974) Selfdiffusion of postassium in low-albite using ion

microprobe. Geol Soc Am Abstr Progr 68: 1126-1133
Grove TL, Ferry JM, Spear FS (1984) Phase transitions and decomposition relations in calcic plagioclase. Am Mineral 68: 41-59
Johannes W (1978) Melting of plagioclase in the system Ab-An-H_2O and Qz-Ab-An-H_2O at P_{H2O} = 5 kbars, an equilibrium problem. Contrib Mineral Petrol 66: 295-303
Johannes W (1980) Metastable melting in the granite system. Contrib Mineral Petrol 86: 264-273
Johannes W (1984) Beginning of melting in the granite system Qz-Or-Ab-An-H_2O. Contrib Mineral Petrol 86: 264-273
Johannes W (1985) The significance of experimental studies in the formation of migmatites. In Ashworth JR (ed) Migmatites. Blackie and Son, Glasgow 36-85
Johannes W (1989) Melting of plagioclase-quartz assemblages at 2 kbar water pressure. Contrib Mineral Petrol 103: 270-276
Johannes W, Holtz F (1992) Melting of plagioclase in granite and related systems: composition of coexisting phases and kinetic observations. Trans R Soc Edingburgh Earth Sci 83: 417-422
Kaspar RB (1974) Cation diffusion in low albite. Ph D thesis, Brown University, Providence, USA
Klug FJ, Prochazka S, Doremus RH (1990) Alumina silica phase diagram in the mullite region. Ceram Trans 6: 15-44
Liu M, Yund RA (1991) NaSi-CaAl interdiffusion in plagioclase. Am Mineral 77: 275-283
Pouchou JL, Pichoir F (1984) A new model for quantitative x-ray microanalysis. Part I: Application of the analysis of homogeneous samples. La Recherche Aérospatiale 3: 13-38
Stewart DB (1967) Four phase curve in the system $CaAl_2Si_2O_8$-SiO_2-H_2O between 1 and 10 kilobars. Schweiz Mineral Petrog Mitt 47: 35-39
Stewart DB, Walker GW, Wright TL, Fahey JJ (1966) Physical properties of calcic labradorite from Lake County, Oregon. Am Mineral 51: 177-197
Tsuchiyama A (1985a) Partial melting kinetics of plagioclase-diopside pairs. Contrib Mineral Petrol 91: 12-23
Tsuchiyama A (1985b) Dissolution kinetics of plagioclase in the melt of the system diopside-albite-anorthite, and origin of dusty plagioclase in andesites. Contrib Mineral Petrol 89: 1-16
Tsuchiyama A, Takahashi E (1983) Melting kinetics of plagioclase feldspar. Contrib Mineral Petrol 84: 345-354
Tuttle OF, Bowen NL (1958) Origin of granite in the light of experimental studies in the system $NaAlSi_3O_8$-$KAlSi_3O_8$-SiO_2-H_2O. Geol Soc Am Mem 74: 153p.
Yoder HS (1968) Albite-anorthite-quart-water at 5 kb. Carnegie Inst Washington Yearb 56: 477-478
Yoder HS, Stewart DB, Smith JR (1957) Feldspars. Carnegie Inst Washington Yearb 56: 206-214
Yund RA (1986) Interdiffusion of NaSi-CaAl in peristerite. Phys Chem Minerals 13: 11-16

TERNARY FELDSPAR/MELT EQUILIBRIA: A REVIEW

H. NEKVASIL
Department of Earth and Space Sciences
State University of New York
Stony Brook, NY 11794-2100
U.S.A.

ABSTRACT. The compositions of ternary feldspars coexisting in equilibrium with melts are constrained by the subsolidus characteristics in the system Ab-An-Or. The presence of a solvus in the ternary feldspar system indicates that there is limited solubility of the three components at magmatically reasonable temperatures and pressures. The shape of the solvus in T-X space indicates that this solubility increases significantly with temperature but only slightly with pressure, thereby laying the foundation for ternary feldspar geothermometry. Several thermodynamic models describing the mixing behavior of components in ternary feldspars have been developed for use in geothermometry. However, because of the paucity of data at high temperatures and in Ab-poor regions, there are significant variations in the predicted high-temperature characteristics of the solvus. Nonetheless, when combined with models for silicate melts in the feldspar system, most of the models for the solid solutions yield the same general trends in compositional evolution of feldspars during crystallization or melting (although the specific temperatures and pressures at which various changes in behavior are manifested vary with the model chosen). The major differences between the evolutionary histories of ternary feldspars undergoing equilibrium or fractional crystallization at high or low temperatures and under H_2O-unbuffered or H_2O-buffered conditions provide important criteria for the interpretation of the histories of natural igneous rocks.

1. Introduction

Natural feldspars are commonly ternary solid solutions of the three main endmembers $NaAlSi_3O_8$ (Ab), $KAlSi_3O_8$ (Or), and $CaAl_2Si_2O_8$ (An). The ternary nature of feldspars is of particular importance in high-temperature felsic to intermediate igneous rocks such as syenites and trachytes as well as in high-grade metamorphic rocks such as granulite. If the effects of composition (X), temperature (T), and pressure (P) on ternary solution in feldspars were understood, ternary feldspars would provide an invaluable means of interpreting the genetic history of many high-temperature rocks. This paper focuses specifically on two uses of ternary feldspar stability relations: geothermometry and the prediction of crystal/liquid crystallization paths. For the purposes of this paper, all feldspars exhibiting measureable ternary solubility will be termed ternary feldspars.

I. Parson (ed.), Feldspars and Their Reactions, 195–219.

2. Ternary Feldspar Geothermometry

Ternary feldspar solvus relations form the basis for ternary feldspar geothermometry. Experimental investigations of the ternary feldspar solvus have most commonly involved use of feldspars synthesized from gels and analysis of run products using powder X-ray techniques (Iiyama 1966; Seck 1971a, 1971b; Johannes 1979). These experiments, however, are not definitive in their location of the ternary solvus in P-T-X space. The experiments of Seck (1971a, 1971b), for example, suffered from two main problems. First, the experiments were synthesis experiments and were not based on the direction of compositional changes in crystalline feldspars. Second, the compositions of coexisting plagioclase were computed from the bulk composition of the gels and the composition of the alkali feldspar. However, Seck (1971a) tested the assumption of equilibrium by reacting two-feldspar pairs that had the same bulk composition as the ternary gels. From the agreement of the alkali feldspar compositions for each set of experiments, his results were at least consistent with attainment of equilibrium although no proof of this was possible (see Johannes 1979 for further discussion). More recent data (Elkins and Grove 1990) were obtained using natural and K-exchanged crystalline feldspars and analyzing the products using the electron microprobe. These workers attempted to bracket their compositional results by using both ternary and binary feldspar starting materials. Their problems, however, included the formation of zoned feldspars and thus changes in the apparent bulk composition of the "equilibrium" assemblage.

None of the available data shed much light on the Al-Si ordering in ternary feldspars. It is likely that the synthetic feldspars are disordered. As summarized by Kroll et al. (1993 in press), the result of increasing order in ternary feldspars is to widen the ternary solvus [as occurs in the binary Ab-Or solvus (Bachinsky and Muller 1971)]. The use of the metastable disordered solvus means that the solubility of An in ordered natural alkali feldspar and Or in plagioclase will be overestimated or, equivalently, the temperatures estimated from coexisting natural pairs will be low.

2.1. ACTIVITY MODELS FOR TERNARY FELDSPAR SOLID SOLUTIONS

When two feldspars coexist at equilibrium, the chemical potentials of like components in each feldspar (i.e., plagiocase, Pl or alkali feldspar, Af) must be equivalent, i.e., $\mu_{Ab}{}^{Pl} = \mu_{Ab}{}^{Af}$; $\mu_{An}{}^{Pl} = \mu_{An}{}^{Af}$; $\mu_{Or}{}^{Pl} = \mu_{Or}{}^{Af}$. Since the standard state for each of the components is its pure state at P and T, these requirements may be restated that the activities of like components in each feldspar must be equal, i.e., $a_{Ab}{}^{Pl} = a_{Ab}{}^{Af}$; $a_{An}{}^{Pl} = a_{An}{}^{Af}$; $a_{Or}{}^{Pl} = a_{Or}{}^{Af}$. This is a simplification, however, that contains hidden implications. Theoretically, the standard state of a component in a solution is the pure component in the same structure as the solution (e.g., for a disordered $C\bar{1}$ plagioclase, the standard state of anorthite component should be pure anorthite in the $C\bar{1}$ structure). As the ordered standard state is used for anorthite component, the $RT\ln a_{An}$ term includes the energy of the transformation from the $I\bar{1}$ to the $C\bar{1}$ structural states [see Carpenter and Ferry (1984) for further discussion]. The error introduced by the use of mixed standard states should, however, be small relative to the experimental uncertainty.

The activity of each component (i) in each phase (Pl or Af, written generically as Pl(Af)) is given by the simple expression

$$a_i^{Pl(Af)} = \gamma_i^{Pl(Af)}\ a_{IDi}^{Pl(Af)} \tag{1}$$

where $\gamma_i^{Pl(Af)}$ is the activity coefficient for component i in either plagioclase or alkali feldspar that takes into account the energetics of non-ideal mixing of component i with all other components. The ideal activity term $a_{IDi}^{Pl(Af)}$ term can be expressed on either a molecular mixing or a site mixing basis. In the former case, random mixing of fully charged balanced molecular units of each component is considered and $a_{IDi}^{Pl(Af)} = X_i^{Pl(Af)}$. This approach has been used by Nekvasil and Burnham (1987), Lindsley and Nekvasil (1989) and Elkins and Grove (1990). In contrast, Ghiorso (1984), Green and Usdansky (1986), and Fuhrman and Lindsley (1988) considered random site mixing and obtained the following expressions

$$a_{ID_{Ab}}^{Pl(Af)} = 4 \cdot X_{Na}^{oct} \cdot X_{Al}^{T1} \cdot X_{Si}^{T1} \cdot \left(X_{Si}^{T2}\right)^2$$

$$a_{ID_{Or}}^{Pl(Af)} = 4 \cdot X_{K}^{oct} \cdot X_{Al}^{T1} \cdot X_{Si}^{T1} \cdot \left(X_{Si}^{T2}\right)^2$$

$$a_{ID_{An}}^{Pl(Af)} = X_{Ca}^{oct} \cdot \left(X_{Al}^{T1}\right)^2 \cdot \left(X_{Si}^{T2}\right)^2$$

where $X_{Si}^{T2} = 1$ for strict Al-avoidance (for further discussion see Price 1985).

Expressions can be derived for the activity coefficients of each component in the ternary solution based on Margules formalism for both binary and ternary interactions from the Gibbs free energy of mixing :

$$\begin{aligned}
RT\ln a_{Ab}^{Pl(Af)} = {} & W_{Or,Ab}\,[2X_{Ab}X_{Or}\,(1-X_{Ab}) + X_{Or}X_{An}\,(1/2-X_{Ab})] \\
& + W_{Ab,Or}\,[X_{Or}^2\,(1-2X_{Ab}) + X_{Or}X_{An}\,(1/2-X_{Ab})] \\
& + W_{Or,An}\,[X_{Or}\,X_{An}(1/2-X_{Ab}-2X_{An})] \\
& + W_{An,Or}\,[X_{Or}\,X_{An}(1/2-X_{Ab}-2X_{Or})] \\
& + W_{Ab,An}\,[X_{An}^2\,(1-2X_{Ab}) + X_{Or}X_{An}\,(1/2-X_{Ab})] \\
& + W_{An,Ab}\,[2X_{Ab}X_{An}\,(1-X_{Ab}) + X_{Or}X_{An}\,(1/2-X_{Ab})] \\
& + W_{OrAbAn}\,[X_{Or}X_{An}\,(1-2X_{Ab})]
\end{aligned}$$

$$\begin{aligned}
RT\ln a_{Or}^{Pl(Af)} = {} & W_{Or,Ab}\,[X_{Ab}^2\,(1-2X_{Or}) + X_{Ab}X_{An}\,(1/2-X_{Or})] \\
& + W_{Ab,Or}\,[2X_{Ab}X_{Or}\,(1-X_{Or}) + X_{Ab}X_{An}\,(1/2-X_{Or})] \\
& + W_{Or,An}\,[X_{An}^2\,(1-2X_{Or}) + X_{Ab}X_{An}\,(1/2-X_{Or})]
\end{aligned}$$

$$+ W_{An,Or} [2X_{Or}X_{An} (1-X_{Or}) + X_{Ab}X_{An} (1/2-X_{Or})]$$

$$+ W_{Ab,An} [X_{Ab} X_{An}(1/2-X_{Or}-2X_{An})]$$

$$+ W_{An,Ab} [X_{Ab} X_{An}(1/2-X_{Or}-2X_{Ab})]$$

$$+ W_{OrAbAn} [X_{Ab}X_{An} (1-2X_{Or})]$$

$$RT \ln a_{An}^{Pl(Af)} = W_{Or,Ab} [X_{Ab}X_{Or} (1/2-X_{An}-X_{Ab})]$$

$$+ W_{Ab,Or} [X_{Ab} X_{Or}(1/2-X_{An}-2X_{Or})]$$

$$+ W_{Or,An} [2X_{Or}X_{An} (1-X_{An}) + X_{Ab}X_{Or} (1/2-X_{An})]$$

$$+ W_{An,Or} [X_{Or}^2 (1-2X_{An}) + X_{Ab}X_{Or} (1/2-X_{An})]$$

$$+ W_{Ab,An} [2X_{Ab}X_{An} (1-X_{An}) + X_{Ab}X_{Or} (1/2-X_{An})]$$

$$+ W_{An,Ab} [X_{Ab}^2 (1-2X_{An}) + X_{Ab}X_{Or} (1/2-X_{An})]$$

$$+ W_{OrAbAn} [X_{Ab}X_{Or} (1-2X_{An})$$

The W terms in these equations refer to the Margules parameters W_G such that $W_{G,i,j} = W_{Hi,j} - T W_{Si,j} + P W_{Vi,j}$. Values of these parameters are in part available from measured data. For the Ab-Or join, W_H and W_S from Thompson and Hovis (1979) have been used by Ghiorso (1984) and Burnham and Nekvasil (1987); those of Haselton et al. (1983) have been used in all of the other models. Several models (e.g., Nekvasil and Burnham 1987; Elkins and Grove 1990) use a symmetric value for $W_{VAb,Or}$ from Hovis (1988); others (e.g., Fuhrman and Lindsley 1986; Lindsley and Nekvasil 1989) used values for $W_{VOr,Ab}$ similar to that of Thompson and Hovis (1979) but obtained by regression of volume data from Kroll et al. (1986) and Newton et al. (1980). Such regression also yields significant negative excess volumes for the ternary and for one of the An-Or binaries that counteract the positive ones for Ab-Or. The net result is little change in the ternary feldspar solvus with pressure. This result is still maintained even when the more recent data of Hovis et al. (1991) for the Ab-Or solvus are considered (Kroll et al., in press).

Margules parameters for the Ab-An interactions used in models based on Al-avoidance entropies are those from Newton et al. (1980). In contrast, models based on molecular mixing of the feldspar components have used two different approaches to obtain these parameters. Nekvasil and Burnham (1987) used ion-exchange data from Seil and Blencoe (1979 and pers. comm.) to obtain their values. Lindsley and Nekvasil (1989) adopted the values of Nekvasil and Burnham (1987). Elkins and Grove (1990) obtained their parameters by regression of experimental data concommitent with their regression for the An-Or and ternary interaction parameters. Importantly, the models based on molecular mixing of An and Ab components still provide good agreement with the enthalpy of solution data of Newton et al. (1980). Elkins and Grove (1990) demonstrated this specifically for their model, comparing it with a model for Ab-An interaction from Newton et al. (1980) based on the Kerrick and Darken (1975) Al-avoidance entropies. Elkins and Grove (1980) suggest that this agreement is a result of the relative contributions of the $-TS^{config}$ and G^{ex} terms to the total free energy of mixing in each model. In the Kerrick and Darken (1975) approximation, the $-TS^{config}$ term adds a large negative contribution that is compensated for by a large positive G^{ex} term. In the molecular (single-site) model, the $-TS^{config}$ term is smaller and the contribution from G^{ex} is also smaller, thereby resulting in similar free energies of mixing using both modeling approaches. [Use of the molecular mixing model, however, appears to give much better results when used to predict crystal/liquid data; perhaps, as suggested by Kroll et al. (in press) this is because the Al-avoidance expressions do not take into account charge balance restrictions.] All of the models used the experimental solvus data to solve for the parameters of the An-Or and Or-An interactions as well as for the ternary interaction term (when used).

The ternary feldspar solvus can be readily calculated using the activity expressions above and equating the activities of each component in both feldspars. Figure 1 shows the ternary feldspar solvus calculated at P=5 kbar using the model of Elkins and Grove (1990). The increasing mutual solubility with increasing temperature is evident as is the maintenance of an immiscibility gap even at temperatures as high as 1500°C. Importantly, the tielines flatten with decreasing temperature. This flattening results in an increase in Or content of alkali feldspar and a slight increase in Ab content of plagioclase with lower temperature.

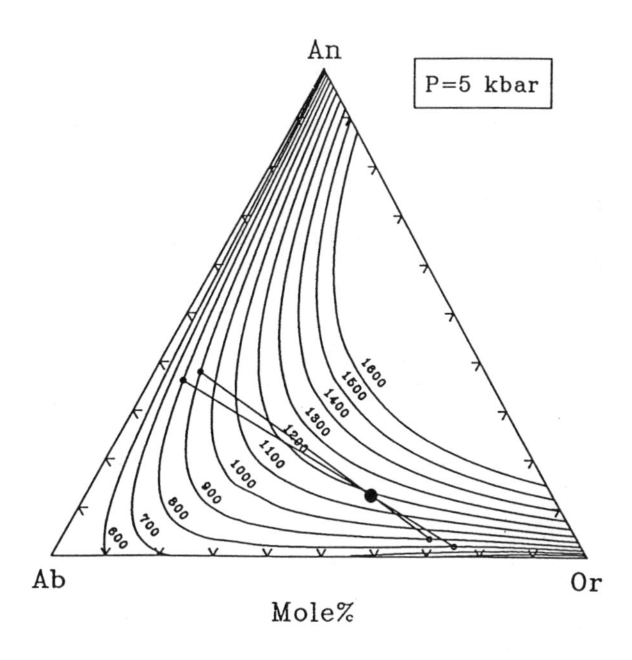

Figure 1. Calculated solvus projection at 5 kbar using the ternary feldspar model of Elkins and Grove (1990). The tielines indicate the compositions of coexisting feldspars at two temperatures for the bulk composition given by the solid circle.

Figure 2 shows calculated isothermal solvus sections at 5 kbar and (a) 700 °C and (b) 1200 °C using several solid solution models. As all of the models were calibrated using predominantly low-pressure and moderate-temperature data, they all agree reasonably well at low temperatures (Fig. 2a). The region of greatest extrapolation lies at high temperatures (and pressures). As seen in Figure 2b, under these conditions the models show significant disagreement.

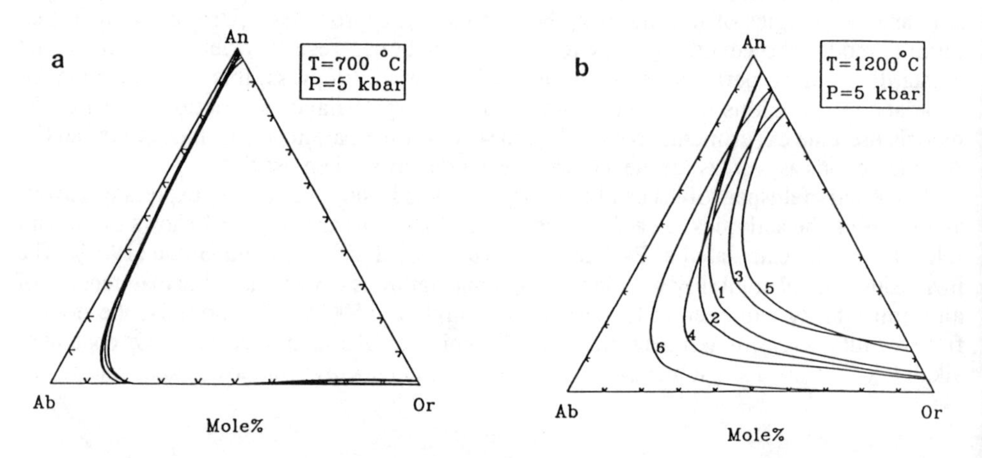

Figure 2. Calculated isothermal solvus sections at 1 kbar and (a) 600 °C and (b) 1200 °C using six available solid solution models. [Symbols: 1 (Elkins and Grove 1990); 2 (Fuhrman and Lindsley 1988); 3 (Ghiorso 1984); 4 (Green and Usdansky 1986); 5 (Lindsley and Nekvasil 1989; 6 (Nekvasil and Burnham 1987)]

TABLE 1. Calculated coexisting anorthite (An) and sanidine (Sa) compositions in the system An-Or at 11.3 kbar, 1280 °C (from Nekvasil and Carroll 1993)

Source	X_{Or}^{An}	X_{An}^{Sa}
Experimental data (Nekvasil and Carroll 1993)	< 0.11	< 0.09
Ghiorso (1984)	0.18	0.04
Nekvasil and Burnham (1987)	0.07	0.05
Fuhrman and Lindsley (1988)	0.22	0.17
Lindsley and Nekvasil (1989)	0.13	0.19
Elkins and Grove (1990)	0.06	0.08

There are few experimental data available to evaluate the accuracy of the models. Nekvasil and Carroll (1993) obtained anhydrous liquidus data in the system An-Or at 11.3 kbar and temperatures in excess of 1250 °C. They were also able to determine maximum limits on the mutual solubility of An in alkali feldspar and Or in plagioclase. Table 1 gives a comparison of calculated coexisting Ab-free feldspar pairs (in mole fractions of the components) for 1280°C and 11.3 kbar. Only the models of Nekvasil and Burnham (1987) and Elkins and Grove (1990) yield pairs that are consistent with the data. The model of Green and Usdansky (1986) was not used as it predicts decreasing solubility of An in sanidine and Or in anorthite with increasing temperature and essentially no mutual solubility at these conditions.

2.2. GEOTHERMOMETRY: CALCULATION METHODOLOGY AND RESULTS

Geothermometric calculations are similar to those for the solvus. For such calculations compositions (e.g., X_{Ab}^{Pl}) are given for a feldspar pair presumed to have been at equilibrium at some stage of their crystallization or metamorphic history. The activity equations above with the criteria that $a_{Ab}^{Pl} = a_{Ab}^{Af}$, $a_{An}^{Pl} = a_{An}^{Af}$, and $a_{Or}^{Pl} = a_{Or}^{Af}$ for two ternary feldspars coexisting in equilibrium, permit the calculation of three independent temperatures. Ideally, these temperatures will be the same (i.e., concordant). Fuhrman and Lindsley (1988) suggested that analytical uncertainties, however, may give rise to discordant temperatures. In order to circumvent this potential problem they developed a geothermometric calculation methodology that takes into account some of the uncertainty by searching in successively larger increments around each feldspar composition for the pair of compositions that minimizes the sum of differences

$$f(X_{Or},X_{Ab},X_{An} \text{ in 2 feldspars}) = ||T_{Ab}| - |T_{Or}|| + ||T_{Or}| - |T_{An}|| + ||T_{An}| - |T_{Ab}|| \quad .$$

Table 2 shows temperatures calculated using the Fuhrman and Lindsley (1988) approach and the 6 solid solution models of Figure 1 for two experimental feldspar pairs presumed to represent at least an approach to equilibrium by Johannes (1979). It is evident that even for intermediate temperatures the models result is significant temperature differences.

The Fuhrman and Lindsley (1988) approach to geothermometric calculations differs from that of Ghiorso who utilized a least squares method to find the temperature that best satisfies all the activity equations. Green and Usdansky's (1986) geothermometric methodology does not require an independent pressure estimate. They determine, instead, a pressure at which the differences between temperatures are minimized; thus the compositions are not modified during calculation.

Upon application of Fuhrman and Lindley's (1988) compositional correction scheme to take into account possible analytical uncertainties, many compositions still have 40 °C or more difference among the three temperatures at the 2% analytical uncertainty level. Based on random compositional variations of a feldspar pair, Fuhrman and Lindsley (1988) determined that high or low discordant anorthite temperatures commonly reflect overestimation or underestimation of An in alkali feldspar. Underestimation of Or in plagioclase typically yields T_{Or} very much lower than the two other temperatures. That this occurs so frequently in natural samples is an indication of the ease of K-Na exchange relative to the Al-Si exchange. Kroll et al. (in press) have suggested that the simple corrections of the feldspar pair compositions by Fuhrman and Lindsley (1988) may be insufficient and suggest a systematic path of change that may form the basis of a modified

TABLE 2. Comparison of calculated temperatures and adjusted compositions using various ternary feldspar solution models and the Fuhrman and Lindley (1988) (M) and the Kroll et al. (in press) (K) geothermometric approaches.

	Plagioclase			Alkali feldspar						
	An*	Ab	Or	Or	Ab	An	TAb	TOr	TAn	TAv**
Experimental										
Johannes (1979) 1 kbar	40.0	54.0	6.0	75.0	24.0	1.0	800	800	800	800
Calculated										
E&G (M)	40.0	55.5	4.5	74.2	24.0	1.8	768	766	768	767
E&G (K)	40.0	57.0	3.0	79.4	19.6	1.0	674	673	675	674
F&L (M)	40.0	54.6	5.4	75.0	22.5	2.5	767	768	769	768
F&L (K)	40.0	57.3	2.7	82.2	16.8	1.0	631	628	630	630
G (M)	40.0	54.1	5.9	75.0	22.5	2.5	794	794	795	794
G (K)	40.0	57.4	2.6	85.0	14.0	1.0	621	585	579	595
G&U (M)	40.0	53.0	7.0	74.0	25.0	1.0	887	886	885	886
G&U (K)	40.0	53.0	7.0	74.0	25.0	1.0	887	886	885	886
L&N (M)	41.7	54.0	4.3	74.0	24.0	2.0	726	724	724	725
L&N (K)	40.0	57.2	2.8	79.3	19.7	1.0	622	623	622	623
N&B (M)	40.0	53.6	6.4	74.6	24.4	1.0	786	786	786	786
N&B (K)	40.0	53.6	6.4	74.6	24.4	1.0	786	786	786	786

*all compositions in mole %; all temperatures in °C ** TAv (average temperature)
E&G (Elkins and Grove 1990); F&L (Fuhrman and Lindsley 1988); G (Ghiorso 1984); G&U (Green and Usdansky 1986); L&N (Lindsley and Nekvasil 1989); N&B (Nekvasil and Burnham 1987)

calculation methodology. Brown and Parsons (1985 unpublished) and Kroll et al. (in press) suggest that in slowly cooled granulite-facies rocks compositions of feldspars may be changed by intercrystalline Na-K exchange at constant An content, thus shifting the compositions of the feldspars in opposite directions parallel to the Ab-Or join and therefore off the equilibrium tieline. Such change may be followed by exsolution and intracrystalline exchange within alkali feldspar crystals while plagioclase crystals continue to exchange K for Na with alkali feldspar crystals. The intracrystalline exchange in alkali feldspar crystals will lead to continuous changes in lamellar and matrix compositions with decreasing temperature. This change should result in increasing Or and decreasing An contents of the matrix and decreasing Or and increasing An contents of the lamellae. Brown and Parsons (1985 unpublished) and in Kroll et al. (in press), however, noted that for all of the feldspar crystal pairs that they investigated, the alkali feldspar lamellae did not have lower An contents than the plagioclase crystals, but instead had essentially the same An contents. They suggested that this is a result of the simultaneous change in the bulk composition of the alkali feldspar to more Or-rich compositions (thereby resulting in steeper tie lines) because of the intercrystalline K-Na exchange, while the lamellae shift to more Or-poor compositions (and hence flatter tielines as a result of temperature drop). The net effect would be only a slight change in the An content of the lamellae.

Table 2 presents a comparison of the temperatures obtained using the Fuhrman and Lindsley (1988) and Kroll et al. (in press) geothermometric approaches. For the latter calculations, the An content of each feldspar is required to remain constant during the search and there is no other limit on the range of the search. As seen in Table 2, the temperatures obtained by these two approaches can differ greatly. These differences indicate the importance of using the Kroll et al. (in press) approach only for rocks in which there is direct evidence for subsolidus reaction.

In summary, uncertainties in two-feldspar geothermometry arise from numerous sources, many of which will remain inherent in the use of ternary feldspars for geothermometry [e.g., lack of attainment of an equilibrium (compositional, and/or ordering), formation of zoned crystals, subsolidus exsolution and exchange processes]. However, with advances in our understanding of the thermodynamic properties of feldspars mixing relations, phase equilibria, and importantly, subsolidus processes, the reliability of ternary feldspar geothermometry will be improved.

3. Ternary Feldspar/Melt Equilibria

Experimental investigations of crystal/melt equilibria of ternary feldspars are very scarce because of the kinetic problems associated with feldspar nucleation and growth. Based on experimental data on the phase relations in the bounding binaries of the system Ab-An-Or and the few experimental data within the ternary (e.g., Yoder et al. 1957), however, it is possible to conclude that liquidus relations in this system are characterized by the presence of a two-feldspar curve separating the dominant plagioclase + L field from the smaller alkali feldspar + L field. Importantly, in reflection of the low-temperature eutectic and high temperature azeotrope in the subsystem Ab-Or, this curve reaches the Ab-Or binary system only at low temperatures. In contrast, at high temperatures the curve terminates within the ternary system. As concluded by Tuttle and Bowen (1958), termination of the 2 feldspar +

L curve within the feldspar system requires that this curve becomes a reaction curve somewhere along its length (i.e., the reaction m (melt)$\rightarrow$ Pl + Af is replaced by m + Pl$\rightarrow$Af or m + Af - Pl during crystallization). Because of the experimental problems associated with the equilibration of ternary feldspars, geometric considerations and thermodynamic calculations have played an important role in the development of our understanding of the compositional evolution of ternary feldspars during crystallization and melting in this complex region.

Solid solution models for ternary feldspars can be combined with mixing models for silicate liquids in order to compute crystal/melt equilibria through the requirement that for a solid and liquid in equilibrium $\mu_i^{PL(AF)} = \mu_i^m$ for any component i. Any one of the available ternary feldspar solution models discussed above can be used in such calculations. Thermodynamic models describing feldspar liquids, however, are scarce. Nekvasil (1986) and Burnham and Nekvasil (1986) presented a model for melts within the ternary feldspar system that basically involves correction of standard state thermochemical data to permit agreement with the congruent portions of the single component melting curves and ideal mixing of components based on these revised standard states. As this model permits reliable prediction of phase relations within the bounding binaries and even of the few ternary data available (see Nekvasil and Burnham 1987 for details) it represents a good first-order approximation. Calculated phase relations and crystallization paths in the feldspar system at high temperatures are much more sensitive to the solid solution model chosen than the melt model. As is to be expected, the greatest disagreement between the phase relations calculated with different solid solution models arises in the An-rich region at high temperatures. A limited but useful basis for evaluating the effects of the various solid solution models on calculated liquidus relations is provided by the liquidus data of Nekvasil and Carroll (1993) in the anhydrous system An-Or at 11.3 kbar. Only the Elkins and Grove (1990) and Nekvasil and Burnham (1987) models reliably predict the phase relations in this system. Nonetheless, all of the solid solution models (with the exception of Green and Usdansky 1986) yield the same general trends and can be used to evaluated crystallization pathways.

3.1. LIQUIDUS RELATIONS IN THE FELDSPAR SYSTEM

Figure 3 shows the calculated liquidus surface under H_2O-saturated conditions at 5 kbar (Fig. 3a) and for very low water contents [a_w (activity of water in the melt) = 0.1)] (Fig.3b) in the vicinity of the 2 feldspar + L curve. In addition to the differences in temperatures of the liquidus surfaces and in the location of the terminations of the PAL curves, the curvature of the isotherms in the alkali feldspar + L field indicates the cotectic nature of the PAL curve at low temperatures and the peritectic nature of the curve at high temperatures. In the peritectic region, the liquidus temperatures in the bounding subsystem Ab-Or(-H_2O) are lower than those along the PAL curve and liquids are no longer constrained to follow this curve at lower temperatures during equilibrium crystallization.

The transition between the peritectic-like and eutectic-like character of the PAL curve occurs at a "neutral point" (Abbott 1978) where the reaction (given by three-phase triangles) changes from even to odd and the backtangent to the PAL curve giving the change in liquid composition intersects one of the alkali feldspar compositions. This neutral point lies at increasingly higher Ab contents with increasing water content. In the ternary feldspar-H_2O system, this gives rise to a PAL surface shown schematically in Figure 4 in an anhydrous projection. A locus of neutral points separates the cotectic-like and peritectic-like regions of this surface.

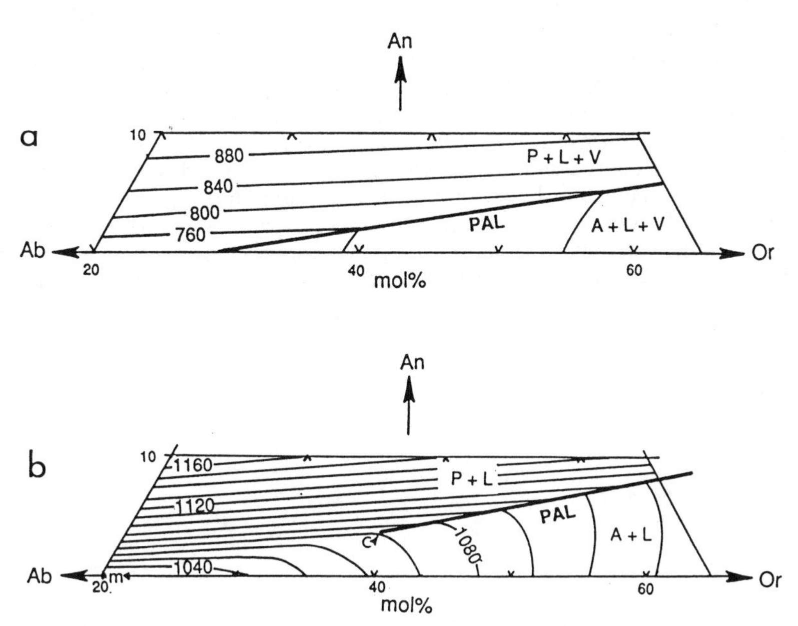

Figure 3. Calculated liquidus surface in the An-poor portion of the feldspar system at (a) 5 kbar, H_2O-saturated and (b) 2 kbar for a_w= 0.1 (from Nekvasil 1990). Solid model of Lindsley and Nekvasil (1989) and melt model of Burnham and Nekvasil (1986) were used in the calculations. [Symbols: P (plagioclase); A (alkali feldspar); L (melt); V (aqueous fluid)].

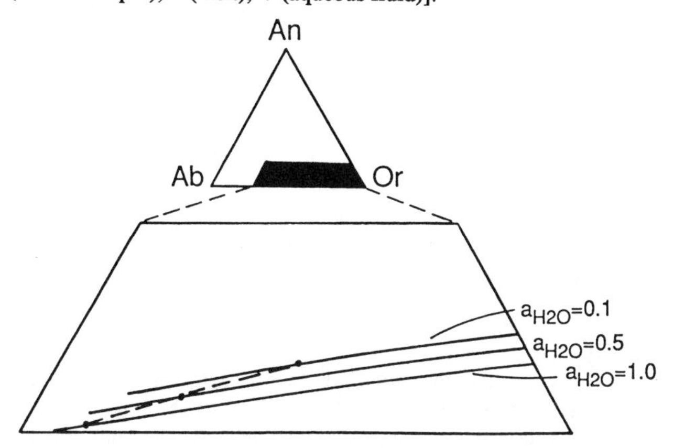

Figure 4. Schematic diagram of the PAL surface (shown by three isoactivity contours) in the feldspar-H_2O system projected into the anhydrous feldspar system. The dashed line connects the neutral points and outlines the peritectic-like (on the side closest to the Ab-An sideline) and cotectic-like regions. (From Nekvasil 1990)

Whether or not a bulk composition will contain two feldspars at the end of equilibrium crystallization is dependent upon whether or not it lies within the solvus intersection at the solidus temperature. If not, then only a single feldspar will be present (plagioclase if the bulk composition is above the PAL curve, and alkali feldspar if it is in the An-poor field below this curve). Equilibrium crystallization paths of bulk compositions with a constant water activity that will reach equilibrium with two feldspars and melt at some stage may show very different evolutionary histories depending upon whether the bulk compositions lie in regions A, B, C, D, or E in Figure 5. Crystallization of bulk compositions plotting in region A will include coprecipitation of two feldspars to the solidus temperature once two feldspars have been stabilized. Bulk compositions plotting in regions B and C will instead go through a stage in which coprecipitation of two feldspars will be replaced by a reaction relation between melt and one feldspar resulting in partial resorption of the reacting feldspar (plagioclase for field B and alkali feldspar for field C) with decreasing temperature.

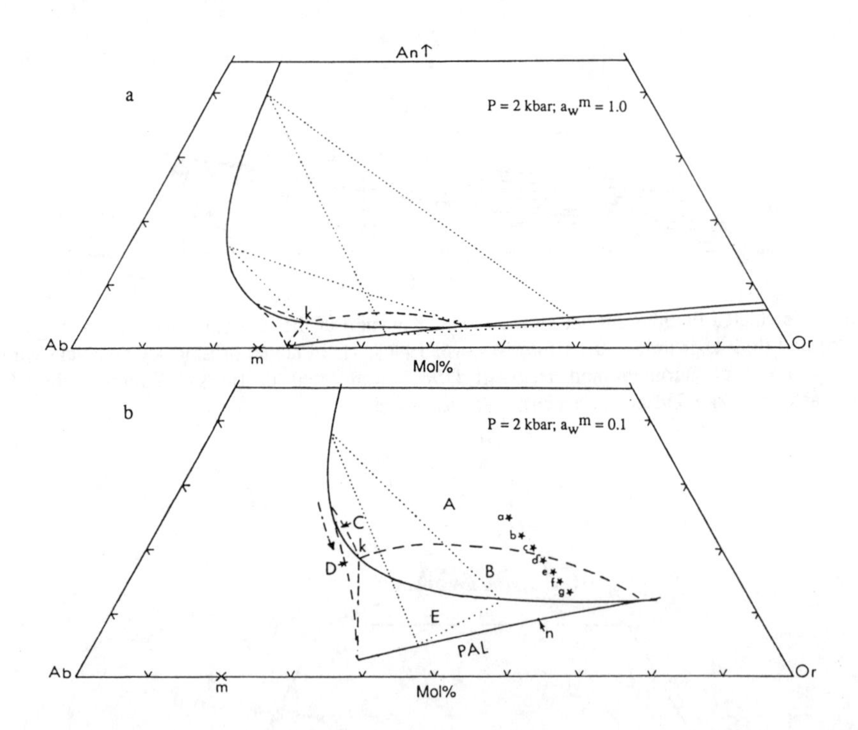

Figure 5. Calculated phase relations in the system Ab-An-Or -(H2O) at 2 kbar for (a) a_w= 1.0 and (b) a_w= 0.1 modified from Nekvasil (1990) showing the PAL curve and corresponding polythermal solvus intersection (which indicates the locus of compositions of feldspars in equilibrium with melt along the PAL curve). The calculated feldspar composition at the consolute point on the polythermal solvus is given by *k*; the location of the neutral point is indicated by *n*. The double arrowheads indicate the location of the thermal minimum (m) in the bounding Ab-Or-(H2O) subsystem. Compositions a-g are plotted for reference to Fig. 6. The fields A-E are discussed in the text.

Complete resorption of the reacting feldspar will occur for bulk compositions plotting in fields D or E (alkali feldspar in field D and plagioclase in field E). These fields can be readily outlined by cystallization path calculations conducted analytically by combining the chemical potential equalities for each component in the feldspars and melt with mass balance constraints (e.g., Nekvasil 1990). The calculations also indicate that these fields shrink markedly with increasing water content (and decreasing temperature), as can be clearly seen by comparison of Figures 5a and 5b.

Figure 6 shows calculated variations in plagioclase proportions during equilibrium crystallization along the PAL curve for bulk compositions a-g lying in fields A and B of Figure 5. At high bulk An contents, once the PAL curve is reached, two feldspars will coprecipitate to the solidus temperature (as indicated by the positive slope of curve a). As the bulk An content is decreased, bulk compositions are obtained for which coprecipitation changes to resorption. The arrows in Figure 6 indicate the locations along the PAL curve at which resorption begins. The lower the bulk An content, the earlier (with respect to evolution of the melt composition along the PAL curve) is the onset of resorptional behavior. Such calculations for several bulk compositional series indicate a locus of points describing the curved boundary of the region B in which the high proportion of alkali feldspar in the system and the more An-rich nature of this feldspar relative to the melt requires the resorption of some plagioclase during equilibrium crystallization. Compositions more An-rich than this boundary have a small enough proportion of alkali feldspar at the solidus temperature, so that the melt can provide the An requirements for both feldspars during crystallization. Importantly, as indicated in Figure 6, reaching the neutral point is a necessary but insufficient condition for the onset of resorptional behavior.

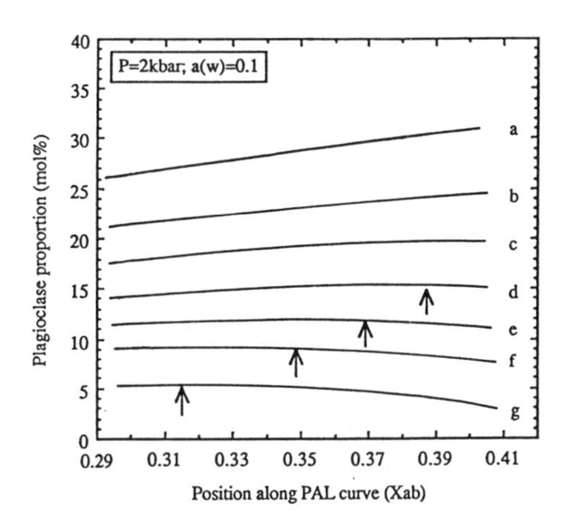

Figure 6. Calculated variations in plagioclase abundance (mol% of system) during equilibrium crystallization along the PAL curve (indicated by Xab in the melt) for the compositions a-g of Figure 5. The arrows indicate the point of onset of resorptional behavior. (From Nekvasil 1990)

The termination point *K* of the PAL curve and the topologic features of the liquidus surface around *K* have never been investigated experimentally; however, through calculation and geometric analysis, it is possible to shed light on the behavior of compositions in this region. Figure 7 shows the detailed calculated geometry (Nekvasil and Lindsley 1990) near the termination at P=2 kbar and a_w=0.1 including selected calculated three-phase triangles.

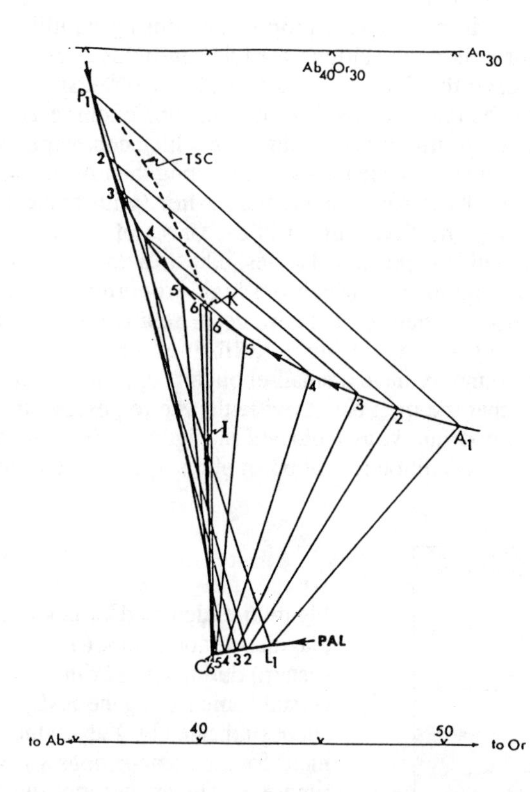

Figure 7. Detailed calculated geometry of plagioclase (P), alkali feldspar (A), and melt (L) compositions in the vicinity of the termination (C) of the PAL curve at P=2 kbar and a_w=0.1. K is the consolute point on the polythermal solvus section (curve P1-K-A1) corresponding to the temperatures along the PAL curve. (Modified from Nekvasil and Lindsley 1990)

Variations of phase proportions of bulk compositions in fields D and E (Figure 5) can be readily calculated from the three-phase triangles constructed through mass balance and feldspar/melt equilibrium relationships. Figure 8 shows the phase proportions and variations in component abundances during equilibrium crystallization along the PAL curve for the bulk composition $Ab_{53}An_{13}Or_{34}$ which lies very close to composition I on the *K-C* line at 2 kbars and a_w=0.1 but on the alkali feldspar side of this line (in field E of Fig. 5). Figure 8a shows that for this composition the amount of plagioclase remains essentially constant with dropping temperature as the proportion of liquid decreases and the abundance of alkali feldspar increases until three-phase triangle 9 (not shown) is reached. At lower temperatures plagioclase reacts with liquid to produce alkali feldspar in a reaction relation that results in a sharp decrease of plagioclase abundance and concommitant increase in alkali feldspar abundance. The smooth decrease in liquid abundance does not reflect this change in behavior. Figure 8b indicates that while the total plagioclase abundance remains essentially constant over much of the crystallization path, the total amount of An component tied up in plagioclase (i.e, the composition of plagioclase weighted by the amount of plagioclase in the system) decreases throughout the crystallization interval. Once resorption begins the abundances of all three components in plagioclase decrease. Figure 8c shows that the abundances of all components in alkali feldspar increases. Importantly, the complex behavior of components in the two feldspars is not reflected in the abundance of the melt components (Fig. 8d).

A bulk composition in field D will show similar behavior to one in field E. Coprecipitation of plagioclase and alkali feldspar will occur until alkali feldspar undergoes a reaction with melt to produce plagioclase and the resorption of alkali feldspar goes to completion (see Nekvasil and Lindsley, 1990 for a detailed example). The abundance of An component in bulk plagioclase decreases until alkali feldspar begins to undergo resorption and the total abundance of plagioclase increases dramatically. In contrast, abundances of all components in alkali feldspar increase until resorption of alkali feldspar occurs.

The termination of the PAL curve has received a significant amount of attention. Stewart and Roseboom (1962) considered the possible terminations of the PAL curve depending upon the relative positions of the consolute point of the solvus section (*K*), the termination of the PAL curve (*C*), and the nature of the three-phase triangles. Henderson (1984) incorrectly applied an unpublished graphical construction developed by Greig to evaluate the geometry of the region close to the termination using hypothetical three-phase triangles and a hypothetical consolute point. He concluded that the PAL curve must be tangent to the line *K*-*C* at the termination; therefore, the PAL curve must turn sharply to lower An contents near the termination. Nekvasil and Lindsley (1990) have shown, however, that although the instantaneous alkali-feldspar and plagioclase compositions indicate precipitation of both phases, the material added to plagioclase is negative in An content for bulk compositions on the *K*-*C* line. This imples that bulk An component is being removed from plagioclase even as Ab and Or components are being added. This change in An content is with respect to the fraction of the system that consists of each component in each phase (i.e., the An content of the plagioclase weighted by the amount of plagioclase in the system). The phenomenon of decreasing bulk An or Or content during precipitation of a feldspar (and hence negative values of the material added to plagioclase) is a result of more rapid decrease in the An content of plagioclase or Or content in alkali feldspar than increase in amount of the phase precipitating. This occurs only in the region of the solvus characterized by major changes in the plagioclase and alkali feldspar compositions (i.e. around the nose of the solvus). It is the negative An content of the material added to plagioclase that indicates that the melt will not simply move away from the solid composition as the consolute point is approached and a sharp downturn of the PAL curve near the termination is not required. (This is because the total mass of a feldspar can change composition both by addition of material and subtraction of a component.)

A bulk composition along the *K*-*C* line will crystallize both plagioclase and alkali feldspar to the solidus temperature. However, the abundance of An in plagioclase continuously decreases with dropping temperature. The abundance of Or in alkali feldspar also decreases but not until close to the solidus temperature.

The resorptional behavior summarized above will be a feature only of high temperature magmas because of shrinkage of fields B-E (Figure 4) with higher water contents as the PAL curve extends further and further towards the Ab-Or(-H_2O) subsystem. Any melt

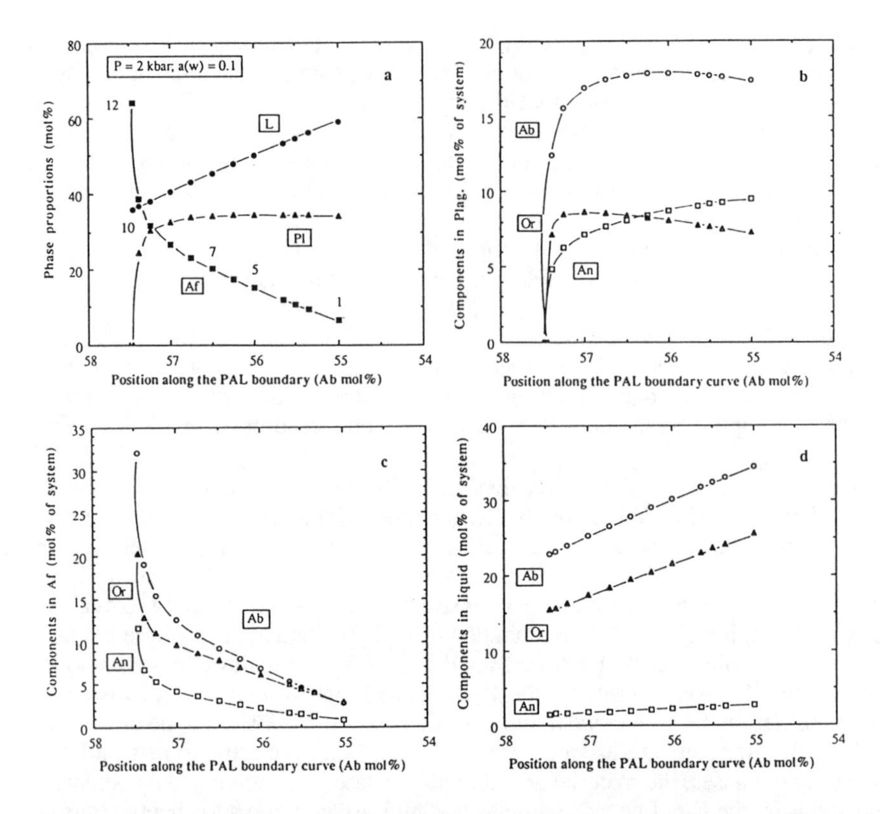

Figure 8. Calculated phase proportions and abundances during equilibrium crystallization of the bulk composition $Ab_{53}An_{13}Or_{34}$ (in field E of Fig. 5) along the PAL curve at P=2 kbar and a_W=0.1. (a) Variations in phase proportions. The numbers refer to the three-phase triangles of Figure 7. (b) Variation in component abundance in bulk plagioclase (i.e., the composition of plagioclase weighted by the amount of plagioclase present at each three phase triangle). (c) Variation in component abundance in bulk alkali feldspar. (d) Variations in component abundance in the melt. (From Nekvasil and Lindsley 1990)

components that mainly dilute the system (e.g., quartz) will have an effect similar to H_2O.

3.2. COMPOSITIONAL EVOLUTION OF FELDSPARS DURING CRYSTALLIZATION

3.2.1 Equilibrium Crystallization. During equilibrium crystallization under H_2O-buffered conditions (i.e., the activity of water is fixed) the temperature interval of crystallization is small and feldspar compositional changes follow paths subparallel to an isothermal solvus section. Figure 9 shows a typical path for low activities of water and hence for high temperatures. With dropping temperature, the plagioclase becomes enriched in both Ab and Or while the alkali feldspar is enriched in Ab and An. For this composition and activity of water, resorption of plagioclase begins as soon as alkali feldspar appears and continues until all plagioclase has redissolved into the melt.

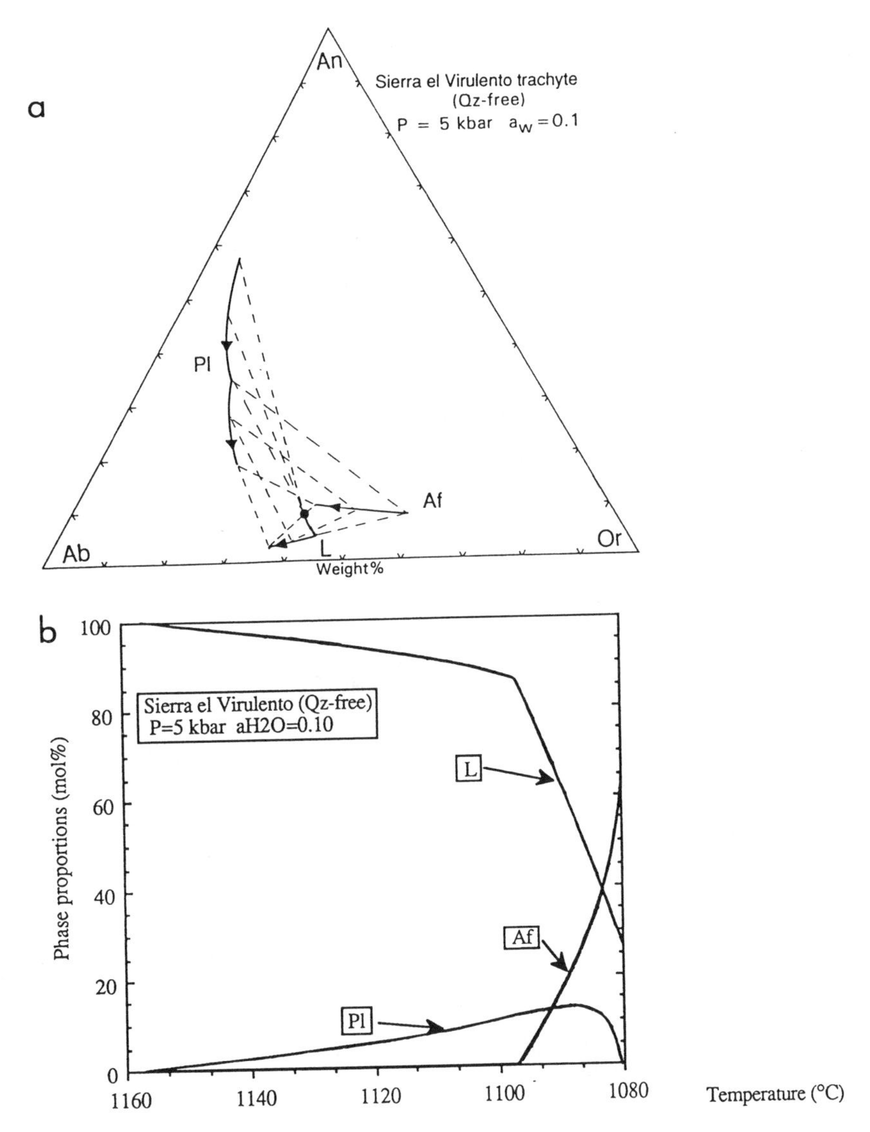

Figure 9. Portion of the calculated H_2O-buffered equilibrium crystallization path for a model Sierra el Virulento trachyte [$Or_{39.3}Ab_{52.0}An_{8.67}$ (wt%)] at P=5 kbar and a_W=0.1. (a) Melt, plagioclase and alkali feldspar compositions projected into the anhydrous system. This path will continue into the alkali feldspar + L field once all of the plagioclase has been resorbed. (b) Variations in proportions of phases vs. temperatures for the path in (a). All phase proportions sum to $(1-X_W) \times 100$.

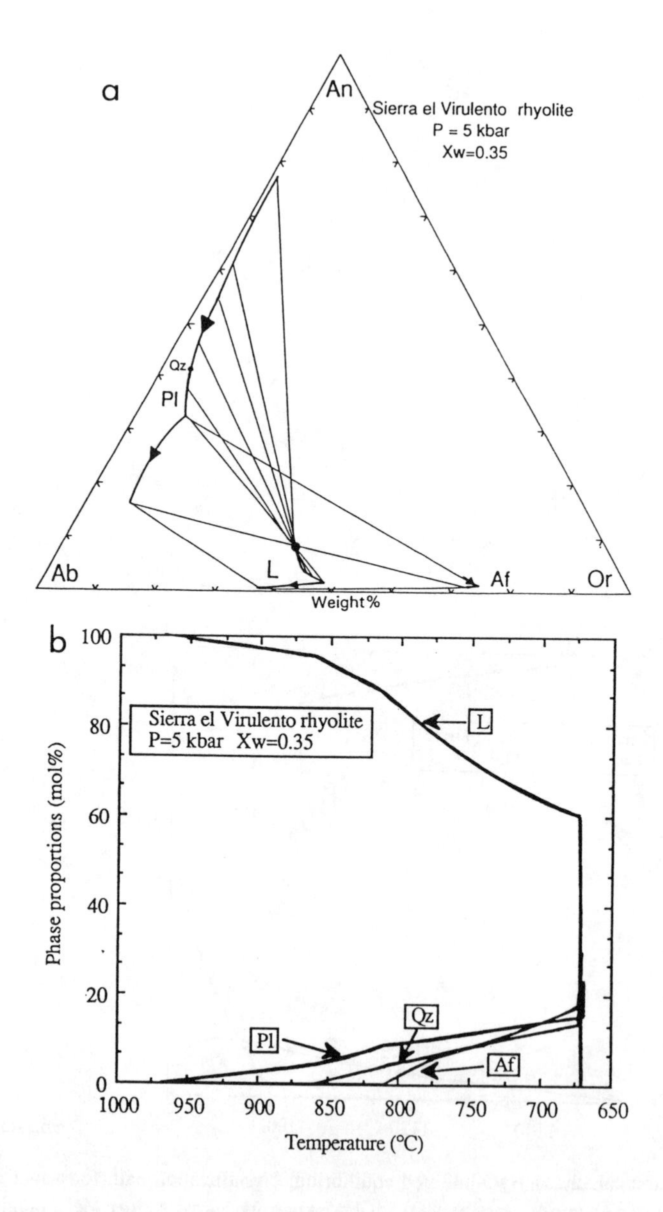

Figure 10. Calculated H_2O-unbuffered equilibrium crystallization path for a Qz-normative Sierra el Virulento trachyte ($Or_{28.5}Ab_{37.7}An_{6.27}Qz_{27.6}$ wt%) at 5 kbar and 3.5 wt% H_2O (X_w=0.35). (a) Melt, plagioclase, and alkali feldspar compositions projected into the anhydrous system. (b) Variations in proportions of phases vs. temperatures for the path in (a). From Nekvasil (1992)

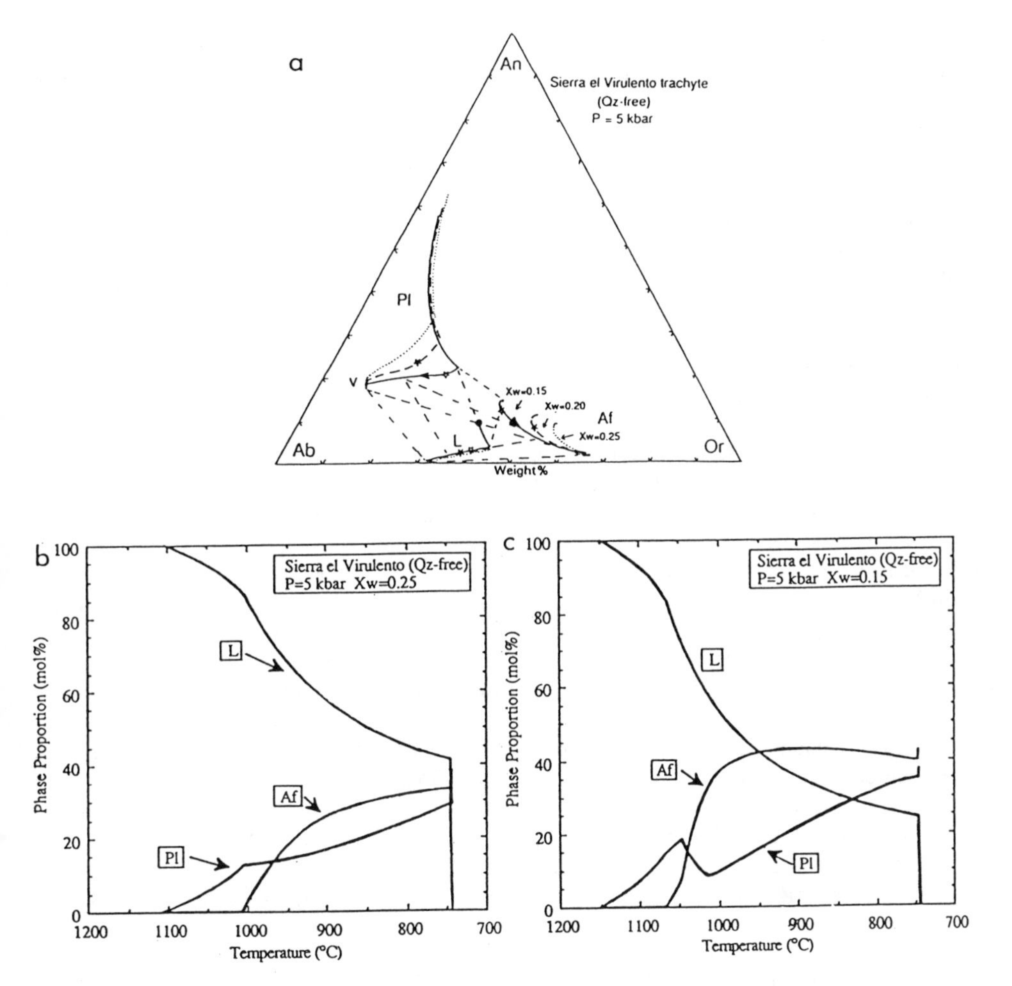

Figure 11. Calculated H_2O-unbuffered equilibrium crystallization paths for the Qz-free model Sierra el Virulento composition of Figure 10 at 5 kbar and bulk H_2O content of 2.2 wt% (X_W=0.25; dotted curve), 1.7 wt% (X_W=0.20; dashed curves) and 1.2 wt% (X_W=0.15; solid curves). (a) Melt (L), plagioclase (Pl), and alkali feldspar (Af) compositions projected into the feldspar system for each bulk H_2O content. Arrowheads point to the direction of decreasing temperature. V indicates when H_2O-saturation is attained. For all of these paths resorption begins once the solvus is intersected by the solidus. Solid stars (X_W=0.2) and open stars (X_W=0.15) indicate the compositions of all phases when partial resorption of plagioclase is replaced by coprecipitation of both feldspars. (b) Variations in proportions (mol%) of each phase along the path for Xw=0.15. (c) Variations in proportions (mol%) of each phase along the path for Xw=0.20. From Nekvasil (1992)

When the H_2O content is allowed to build up in the melt because of the crystallization of anhydrous feldspar phases, the interval of melting is much greater than for H_2O-buffered conditions. However, if the region of the solvus traversed is steep in T-X space the changes in feldspar composition will also be subparallel to an isothermal solvus section as is the solidus trend for H_2O-buffered crystallization. Figure 10a shows an equilibrium crystallization path for X_w (mole fraction of water equal to 0.35 for the bulk system), for a quartz trachyte composition projected into the feldspar system. This composition reaches H_2O-saturation when 40 mol% of the system is crystalline. From this point to the solidus temperature the H_2O-content is buffered. Although the Ab content of the plagioclase increases markedly during crystallization alkali feldspar compositions change only slightly (around ~Or_{78}). Such compositional trends characterize feldspars found in low-temperature hydrous granites. As seen in Figure 10b, there is no stage of the crystallization path in which partial resorption of either feldspar occurs.

The increase in melt water content during crystallization of compositions with low bulk H_2O-contents under H_2O-unbuffered conditions results in a solidus path that reflects a major expansion of the solvus in response to the large temperature interval of crystallization. If the expansion of the solvus is large compared with the change in bulk feldspar compositions during crystallization, compositional changes very different from those of H_2O-buffered paths might be anticipated. The tielines in Figure 1 indicate that alkali feldspar compositions, for example, may become progressively more potassic instead of more sodic with decreasing temperature.

Figure 11a shows three H_2O-unbuffered crystallization paths at P=5 kbar and X_w=0.25, X_w=0.20, and X_w=0.15. Importantly, the low H_2O contents and the increasing water content results in alkali feldspar evolving mainly by increasing Or content and plagioclase mainly by decreasing Or content. The magnitude of this effect diminishes at higher bulk water contents (cf. curves for X_w=0.25 with X_w=0.15).

During crystallization under H_2O-unbuffered conditions, many melts that enter the resorptional region at high temperatures may leave it again as the expansion of the solvus with dropping temperature causes the resorptional regions to shrink (Fig. 5a and 5b); in other words, the melt composition traverses the PAL surface (Fig. 4) from the odd region into the even region. This is certain to be the case for pressures at which the system Ab-Or-H_2O contains a eutectic under H_2O-saturated conditions. When this occurs, the period of resorptional behavior is followed by coprecipitation unlike for H_2O-buffered crystallization. Figure 11b shows the variations in phase proportions for the H_2O-unbuffered crystallization path for X_w=0.15. For this path the negative slope of the plagioclase curve indicates the region of resorption. In contrast, for X_w=0.25 (Fig. 13c), the slope of the plagioclase curve is never negative although entrance into the resorptional field does result in diminished precipitation of plagioclase until the even region is reached.

Natural crystallization paths are not likely to strictly follow equilibrium crystallization paths. Instead, natural paths may mimic stepwise fractional crystallization paths in which, for example, feldspars become zoned periodically during cooling, thereby changing the effective bulk composition in a stepwise manner. Such paths are intermediate between ideal equilibrium and ideal fractional crystallization paths.

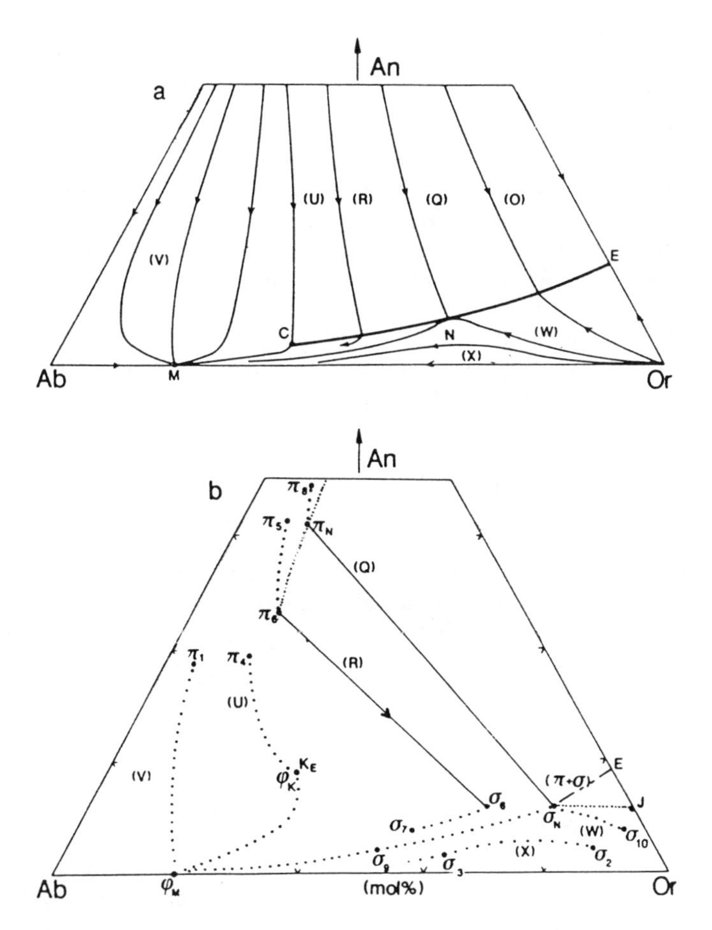

Figure 12. Schematic compositional evolution of (a) melt and (b) feldspar compositions during fractional crystallization under H_2O-buffered conditions and low water activities. N indicates the location of the neutral point on the PAL curve and M the minimum in the subsystem Ab-Or(-H_2O). Letters in parenthesis refer to specific fractionation paths in both a and b. Plagioclase (p) and alkali feldspar (s) compositions along solidus paths correlating to the lettered melt paths are given schematically in b. (From Brown 1993)

3.2.2 *Fractional Crystallization.* Ideal fractional crystallization paths involve the continual removal of crystals during crystallization. In the ternary feldspar system, as is the case for ideal equilibrium crystallization, this results in significantly different paths of liquid and solid compositional evolution depending upon the water content and whether or not the system is buffered with respect to H_2O.

Abbott (1978) and Brown (1993) considered geometrically possible paths taken by melt and crystal compositions during fractional crystallization. Figure 12a shows schematically the basic types of liquid fractionation paths for H_2O-buffered conditions and low water activities constructed by Brown (1993) based on geometric considerations and the calculated equilibrium crystallization paths of Nekvasil (1990).

Melt fractionation paths beginning in the plagioclase field will either converge on the PAL curve (paths O, Q, R, U) or on the minimum (M) in the system Ab-Or(-H_2O) (paths such as V). As the melt evolves towards the PAL curve, the plagioclase compositions will generally be changing along paths such as $\pi 8$-πN or $\pi 5$-$\pi 6$ in Figure 12b. Melts evolving on paths such as O intersect the PAL curve on the even side of the neutral point N crystallizing both plagioclase and alkali feldspar. In such cases, the melt will remain on the PAL curve until N is reached and the plagioclase composition has evolved to π_N and the alkali feldspar composition to σ_N (Fig. 12b). At this point the melt will leave the curve and move along a "special line" (Abbott 1978) towards M and will crystallize only an alkali feldspar that will evolve along a path such as σ_N-M.

Paths such as Q, R and U reach the PAL curve at N or on the odd side of N. In such cases, the melt composition will immediately leave the PAL curve and move towards the special line and eventually to M. Only at the temperature of intersection of the PAL curve by paths such as R will two feldspars be stable. Paths such as U and V will never precipitate two feldspars.

Melt fractionation paths beginning in the alkali feldspar field will behave similarly to those in the plagioclase field except for the fractionation of alkali feldspar both during the early and late stages of crystallization. If intersection of the PAL curve occurs on the even side of N, the melt will evolve to N as coprecipitation of two feldspars continues. Further evolution will be in the single alkali feldspar field once again. Along the unique path that intersects the PAL curve at N (path W) alkali feldspar will be the only feldspar stable. Paths such as X will not intersect the PAL curve and will only precipitate alkali feldspar. Note, however, the initial increase in An content of the alkali feldspar and later decrease in An content during the later stages of fractionation.

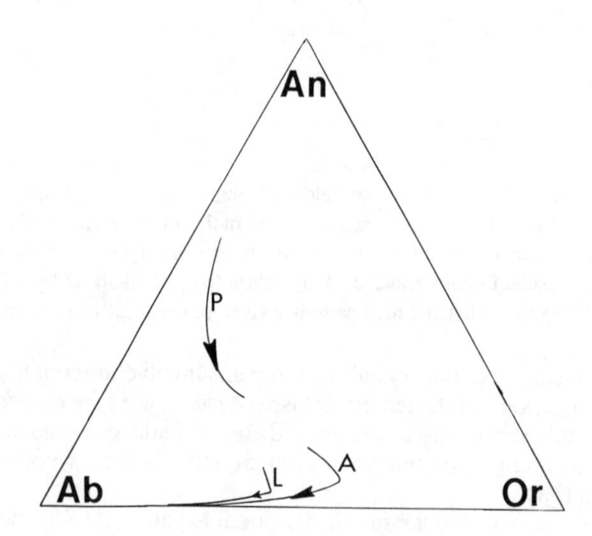

Figure 13. Calculated stepwise fractional crystallization path (removal of crystals at 1% increments) of the Qz-free El Virulento trachyte at 5 kbar, X_w=0.15. (P: plagioclase; A: alkali feldspar; L: Melt)

Fractionation paths for H_2O-unbuffered crystallization have many of the features of H_2O-buffered paths as the melt will eventually move towards the minimum or eutectic in the Ab-Or(-H_2O) subsystem. Figure 13 shows the calculated stepwise fractional crystallization path (along which equilibrium crystallization was permitted to produce 1% crystals before they were removed from the bulk composition) for the Qz-free Sierra el Virulento composition at 5 kbars and X_w=0.15. For this composition under these conditions, the melt path intersects the PAL curve but on the odd side of the neutral point. Two feldspars are thus stable only at the intersection. The melt immediately leaves the PAL curve and moves towards the H_2O-saturated eutectic composition in the binary while crystallizing only alkali feldspar. Note that there is a slight enrichment of Or in alkali feldspar before the composition begins evolving similarly to the H_2O-buffered case (Fig. 12).

4. Conclusions

Understanding of ternary feldspar/melt relations has progressed rapidly due to the development of solid-solution models of the solvus relations and melt models for quartzo-feldspathic melts. Calculations of feldspar differentiation trends indicate that magmatic feldspar zoning characteristics may prove very sensitive indicators of crystallization conditions. Many of the unusual zoning patterns and partial resorptional features discussed above are found in natural trachytes, syenites, and even high temperature rhyolites (see Nekvasil, 1992 for a summary). Through continued combined experimental and theoretical efforts to understand the controls on feldspar zoning the use of feldspar zoning as a powerful interpretational tool for natural systems may soon be realized.

5. Acknowledgements

The support of this review by the 1993 NATO Advanced Study Institute on feldspars is greatfully acknowledged. The careful reviews of Bill Brown and Ian Parsons resulted in significant improvements of the manuscript.

6. References

Abbott RN Jr (1978) Peritectic relations in the system An-Ab-Or-Qz-H_2O. Can Mineral 16: 245-256

Bachinsky SW, Müller G (1971) Experimental determination of the microcline-low albite solvus. J Petrol 12: 329-356

Brown WL (1993) Fractional crystallization and zoning in igneous feldspars: ideal water-buffered liquid fractionation lines and feldspar zoning paths. Contrib Mineral Petrol 113: 115-125

Burnham CW, Nekvasil H (1986) Equilibrium properties of granite pegmatite magmas. Am Mineral 71: 239-263

Carpenter MA, Ferry JM (1984) Constraints on the thermodynamic mixing properties of plagioclase feldspars. Contrib Mineral Petrol 87: 138-148

Elkins LT, Grove TL (1990) Ternary feldspar experiments and thermodynamic models. Am Mineral 75: 544-559

Fuhrman ML, Lindsley DH. (1988) Ternary-feldspar modeling and thermometry. Am Mineral 73: 201-215

Ghiorso MS (1984) Activity/composition relations in the ternary feldspars. Contrib Mineral Petrol 87: 282-296

Green NL, Usdansky SI (1986) Ternary-feldspar mixing relations and thermobarometry. Am Mineral 71: 1100-1108

Haselton HT Jr, Hovis GL, Hemingway BS, Robie RA (1983) Calorimetric investigation of the excess entropy of mixing in analbite-sanidine solid solutions: Lack of evidence for Na, K short-range order and implications for two feldspar thermometry. Am Mineral 68: 398-413

Henderson CMB (1984) Graphical analysis of phase equilibria in An-Ab-Or. In Progress in Experimental Petrology, NERC Publication No 25, Series D: 70-78

Hovis GL (1988) Enthalpies and volumes related to K-Na mixing and Al/Si order/disorder in alkali feldspars. J Petrol 29: 731-763

Hovis GL, Delbove F, Bose MR (1991) Gibbs energies and entropies of K-Na mixing for alkali feldspars from phase equilibrium data: Implications for feldspar solvi and short-range order. Am Mineral 76: 913-927

Iiyama JT (1966) Contribution a l'étude des équilibres subsolidus du systēme ternaire orthoclase-albite-anorthite a l'aide des réactions d'échange d'ions Na-K au contact d'une solution hydrothermale. Bull Soc Fr Mineral Crist 89: 442-454

Johannes, W (1979) Ternary feldspars: Kinetics and possible equilibria at 800 °C. Contrib Mineral and Petrol 68: 221-230

Kerrick DM, Darken LS (1975) Statistical thermodynamic models for ideal oxide and silicate solid solutions in plagioclase. Geochim Cosmochim Acta 39: 1431-1442

Kroll H, Schmiemann I, von Cölln G (1986) Feldspar solid solutions. Am Mineral 71: 1-16

Kroll H, Evangelakakis E, Voll G. (in press) Two-feldspar geothermometry: A review and revision for slowly cooled rocks.

Lindsley DH, Nekvasil H (1989) A ternary feldspar model for all reasons. EOS 70: 506

Nekvasil H (1986) A theoretical thermodynamic investigation of the system Ab-Or-An-Qz(-H_2O) with implications for melt speciation. Ph.D. Dissertation, The Pennsylvania State University.

Nekvasil H (1990) Reaction relations in the granite system: Implications for trachytic and syenitic magmas. Am Mineral 75: 760-771

Nekvasil H, Burnham CW (1987) The calculated individual effects of pressure and H_2O content on phase equilibria in the granite system. in B. Mysen (ed.) Magmatic Processes: Physicochemical Principles, Geochemical Society Special Pub 1, The Geochemical Society: 433-446

Nekvasil H, Carroll W (1993) Experimental constraints on the high-temperature termination of the anhydrous 2 feldspar + L curve in the feldspar system at 11.3 kbar. Am Mineral 78: 601-606

Nekvasil H, Lindsley DH (1990) Termination of the 2 feldspar + L curve in the system Ab-Or-An-H_2O at low H_2O contents. Am Mineral 75: 1071-1079

Newton RC, Charlu TV, Kleppa OJ (1980) Thermochemistry of high structural state plagioclases. Geochim Cosmochim Acta 44: 933-941

Price JG (1985) Ideal site-mixing in solid solutions, with applications to two-feldspar geothermometry. Am Mineral 70: 696-701

Seck HA (1971a) Koexistierende Alkalifeldspäte und Plagioklase im System $NaAlSi_3O_8$-$KAlSi_3O_8$-$CaAl_2Si_2O_8$-H_2O bei Temperaturen von 650 °C bis 900 °C. N Jahrb Mineral Abhan 115: 315-342

Seck HA (1971b) Der Einfluss des Drucks auf die Zussamensetzung koexistierende Alkalifeldspäte und Pagioklase im System $NaAlSi_3O_8$-$KAlSi_3O_8$-$CaAl_2Si_2O_8$-H_2O. Contrib Mineral Petrol 31: 67-86

Seil MK, Blencoe JG (1979) Activity-composition relations of $NaAlSi_3O_8$-$CaAl_2Si_2O_8$ feldspars at 2 kbar, 600-800°C. Geol Soc Am, Prog Abstr 11: 513

Stewart DB, Roseboom EH Jr (1962) Lower temperature terminations of the three-phase region plagioclase-alkali feldspar-liquid. Journal of Petrology 3, part 2: 280-315

Thompson JB, Hovis GL (1979) Entropy of mixing in sanidine. Am Mineral 64: 57-65

Tuttle OF, Bowen NL (1958) Origin of granite in light of experimental studies in the system $NaAlSi_3O_8$-$KAlSi_3O_8$-SiO_2-H_2O. Geol Soc Am Mem 74, pp 1-153

Yoder HS, Stewart DB, Smith, JR (1957) Ternary feldspars. Carnegie Inst Wash Yearb 55: 206-214

SUBSOLIDUS PHASE RELATIONS OF THE PLAGIOCLASE FELDSPAR SOLID SOLUTION

M. A. CARPENTER
Department of Earth Sciences
University of Cambridge
Downing Street
Cambridge CB2 3EQ
England

ABSTRACT. Experimental, microstructural and calorimetric evidence for the stability relations among plagioclase feldspars is reviewed, and a new phase diagram proposed. The incommensurate "e" structure is now believed to have a field of equilibrium stability at intermediate compositions in the solid solution. Two phase fields are interpreted as arising from breaks in the ordering behaviour as a function of composition. The Huttenlocher gap may extend up to ~800°C and is presumed to occur between the stability fields of e_1 and $I\bar{1}$ ordering. A spinodal mechanism for exsolution can be explained if the $I\bar{1} \rightleftharpoons e_1$ transition is taken to be thermodynamically continuous. The Bøggild gap may also extend up to ~800°C as a consequence either of a break between $C\bar{1}$ and e_1 structures or of a break between e_1 and e_2 structures at ~An_{50}. Again, a thermodynamically continuous $C\bar{1} \rightleftharpoons e_1$ transition would allow a spinodal exsolution mechanism. As in previous reviews, the peristerite gap is ascribed to the influence of an increase in the equilibrium degree of Al/Si order in pure albite below ~650-700°C. Incommensurate ordering is now known to occur also in anorthite under metastable conditions and this has implications for the nature of the e structure in crystals with intermediate compositions. Finally, schematic free energy curves and selected Landau free energy expansions are shown to provide a basis for future thermodynamic models of the overall mixing and ordering behaviour.

1. Introduction

Progress in understanding the plagioclase feldspar solid solution can be measured conveniently by referring to the 10 year intervals between NATO feldspar conferences. At the time of the 1962 conference, the principal structure types and their variations with composition had been characterised using single crystal X-ray diffraction. Some experimental heat treatments of natural samples, to explore the stability of the different structures as a function of temperature, had also been undertaken, and the importance of Al/Si ordering and displacive effects was understood. Gay (1962) summarised relationships between the structures in a manner similar to that shown in Figure 1. At high temperatures, the solid solution has the high albite structure for most of the composition range, but adopts the body centered anorthite structure at anorthite-rich compositions and becomes monoclinic at albite-rich compositions. At low temperatures, low albite gives way to the intermediate structure and then to the primitive anorthite structure with increasing anorthite content. Anorthite-rich crystals with diffuse reflections associated with the change from $I\bar{1}$ to $P\bar{1}$ symmetry were referred to as transitional anorthite. Exsolution in the peristerite gap had also been identified, and Ribbe (1962) included the now familiar link between exsolution and ordering

I. Parson (ed.), Feldspars and Their Reactions, 221–269.

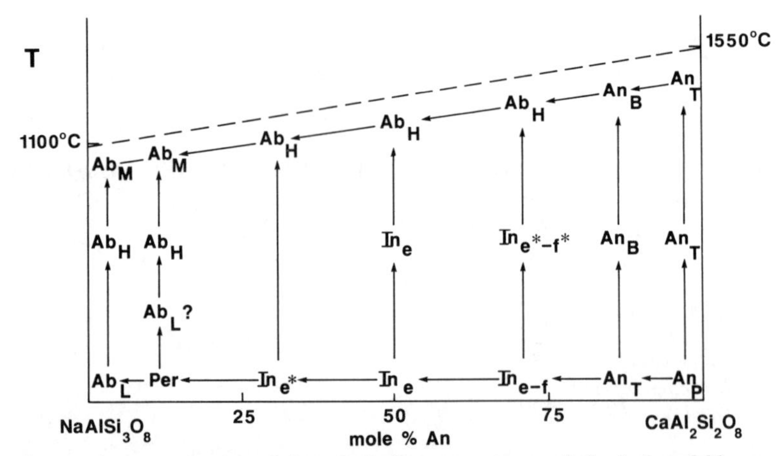

Figure 1. Schematic representation of the principal structure types of plagioclase feldspars, modified slightly after Gay (1962). The dashed line indicates approximately the onset of melting. Ab_L = low albite, Ab_H = high albite, Ab_M = monoclinic albite, In_{e*} = intermediate structure with diffuse e reflections, In_e = intermediate structure with sharp e reflections, $In_{e\text{-}f}$ = intermediate structure with sharp e, f reflections, An_T = transitional anorthite (diffuse c, d reflections), An_P = primitive anorthite, An_B = body centred anorthite, Per = peristerite exsolution. These structural relationships are essentially as we understand them now, with the exception that in 1962 it was thought that the stable state of anorthite at high temperatures was An_T. There was also only limited experimental evidence for the nature of structural states of crystals with intermediate compositions at intermediate temperatures. Gay did not discuss the Bøggild and Huttenlocher miscibility gaps.

processes in a schematic phase diagram for albite-rich compositions which is reproduced in Figure 2.

In the following decade, the advent of transmission electron microscopy as a routine technique for examining submicroscopic intergrowths in minerals led to the characterisation of the Bøggild and Huttenlocher miscibility gaps and of the accompanying changes in structure type across the two phase fields. A summary of the approximate composition limits for the miscibility gaps was given at the 1972 conference by Nord et al. (1974), and their schematic diagram is reproduced in Figure 3. McConnell (1974a) proposed a more specific relationship between Al/Si ordering transitions and the miscibility gaps, in which the onset of exsolution is directly related to the onset of ordering on the basis of the albite, $I\bar{1}$, or incommensurate "e" structures (Fig. 4). At this stage the e structure was considered to represent a non-equilibrium state favoured over more stable ordered structures as a consequence of kinetic factors (see, also, Wenk et al., 1980).

The following decade appears to have been one of consolidation. More TEM observations of microstructural relationships in natural samples were completed, further experiments were undertaken to delineate stability fields of different structures, and new observations were made of coexisting plagioclases in metamorphic rocks. Smith (1984) summarised the status of the known structure types at the 1983 conference in two separate diagrams (Fig. 5). The issues for discussion had become the nature of the phase transitions, i.e. whether they are thermodynamically first order or second order in character (see, also, Grove et al., 1983, 1986), and the stability of the intermediate structure with respect to alternative ordered states or with respect to two phase mixtures. The experimental data then available were not adequate to distinguish between several alternative topologies for the phase diagram, however.

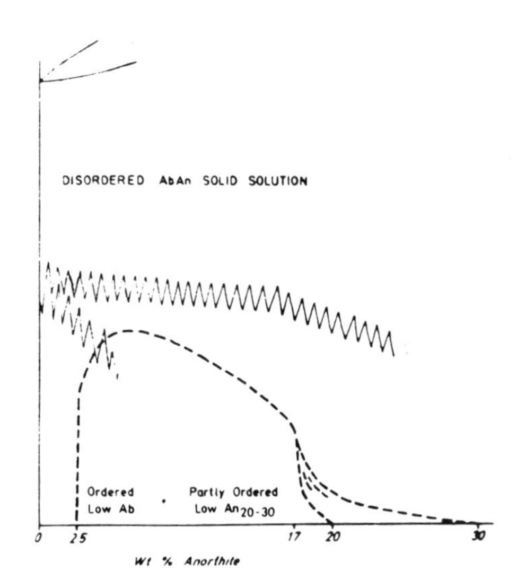

Figure 2. Phase relations for the peristerite gap, as proposed by Ribbe (1962), with the now accepted interrelationship between ordering and exsolution processes. The jagged lines represent possible positions for high ⇌ low albite ordering as a function of temperature and composition, based on the analysis of experimental data for ordering in albite by McConnell and McKie (1960). Reproduced by kind permission of P.H.Ribbe.

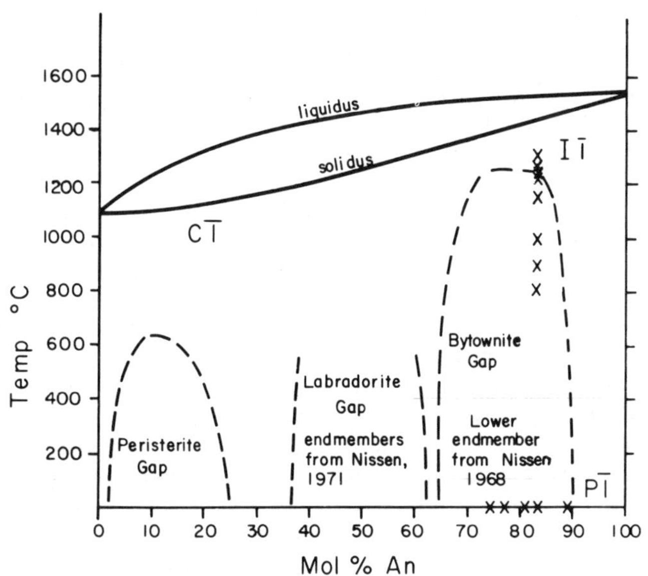

Figure 3. Simplified phase diagram for plagioclase showing only the miscibility gaps, as envisaged by Nord et al. (1974). Reproduced by kind permission of G.L.Nord, Jr.

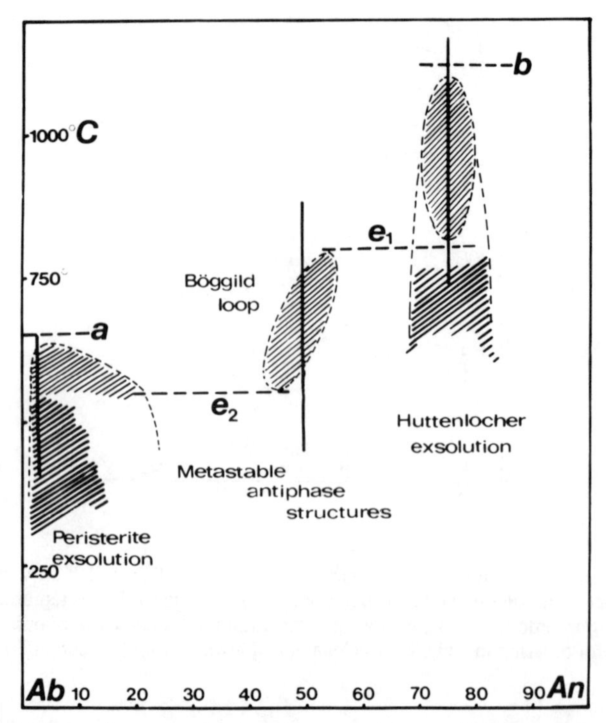

Figure 4. Temperature-composition plot showing the major features which govern the time-temperature-transformation behaviour of the plagioclase system, as proposed by McConnell (1974a). The miscibility gaps (shaded) are shown as being related to breaks in the ordering behaviour as a function of composition. Dashed lines indicate the estimated onset temperatures for ordering to give the different structure types. Reproduced by kind permission of J.D.C.McConnell.

Since then, new information has accrued. A large number of experiments have been completed to define the positions of at least some of the phase transitions. Calorimetric data have provided a measure of the enthalpy variations associated with Al/Si ordering at different compositions, and are consistent with the view that the e structure might have an equilibrium field of stability. Incommensurate ordering, previously thought to be restricted to natural crystals with intermediate compositions, has also been found in synthetic crystals of pure anorthite. At the same time, advances have been made in treating the thermodynamics of displacive and Al/Si ordering transitions in framework silicates. While it is not yet possible to calculate the complete phase diagram, some of the phase transitions in plagioclase crystals can be treated in a truly quantitative manner, and a clearer view of the relative stabilities of the different structure types is emerging.

The purpose of the present review is twofold. It is intended both to present a plausible and self-consistent version of the subsolidus phase relations, taking account of the most recent observations and experimental data, and to identify critical issues which now need to be addressed. The format of the review consists of three main sections. Firstly, an alternative phase diagram, shown in Figure 6, is introduced. The lines of evidence used to provide support for or against this version are then critically reviewed. Finally, schematic free energy-composition curves and selected Landau series expansions are used to illustrate lines of enquiry that might become more

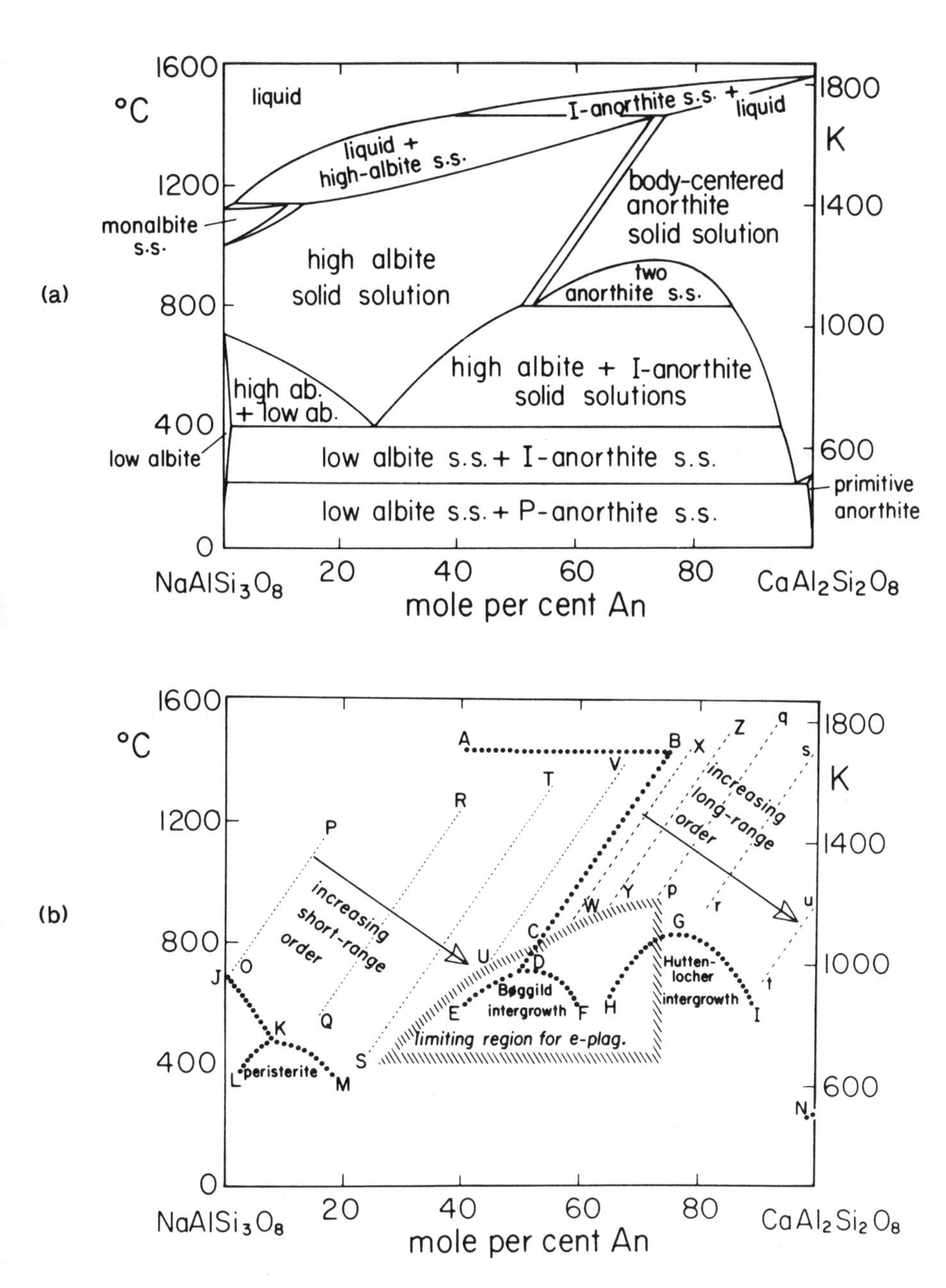

Figure 5. Phase diagrams for plagioclase at low pressure, as proposed by Smith (1984). In (a) all the phase transitions are shown as being first order in character. In (b) possible positions for the miscibility gaps are shown in relation to transitions which are thermodynamically continuous. In both cases, the e structure is presumed to be metastable with respect to other ordered phases or two phase assemblages. Reproduced by kind permission of J.V.Smith.

quantitative. An underlying philosophy is that ordering and exsolution in the plagioclase solid solution are so intimately connected that a proper understanding of the ordering behaviour should automatically lead to a proper understanding of the miscibility gaps. In this regard, pressure may not be a crucial variable and attention is concentrated on the roles of temperature and composition.

It is not the intention to provide a comprehensive bibliography for work on the plagioclase system. There are already excellent reviews of the feldspar literature and readers are referred for background information to Ribbe (1983a), Brown (1984) and Smith and Brown (1988). Similarly, for the thermodynamic treatments, reviews by Carpenter (1988, 1992a) and Salje (1990, 1994) should provide an adequate introduction into the literature.

2. A possible phase diagram

At high temperatures there appears to be complete solid solution between albite (Ab) and anorthite (An). No direct evidence of a miscibility gap, even due to the $C\bar{1} \rightleftharpoons I\bar{1}$ transition at anorthite-rich compositions, has been reported. There are structural changes in this sequence, however. Pure albite is monoclinic (C2/m) above ~978°C, but the $C2/m \rightleftharpoons C\bar{1}$ transition temperature increases steeply with increasing anorthite content. For much of the composition range the high albite structure is stable, before Al/Si ordering occurs to give the $I\bar{1}$ anorthite structure (Fig. 6).

The phase relations at low temperatures are more controversial. There is no disputing the existence of three miscibility gaps, universally referred to as the peristerite, Bøggild and Huttenlocher gaps. The qualitative behaviour of pure albite and pure anorthite is also not in dispute. A high degree of Al/Si order occurs in $C\bar{1}$ low albite and a comparably high degree of order also occurs in $I\bar{1}$ anorthite. Below ~240°C the $I\bar{1}$ structure of pure anorthite gives way to the $P\bar{1}$ structure by a largely displacive mechanism. The proposed phase diagram, Figure 6, shows the incommensurate structure to have stability fields at intermediate compositions which are labelled e_1 and e_2 (following McConnell, 1974a). These occur as a consequence of a phase transition from the $C\bar{1}$ structure, for the e_2 field, and by a phase transition from the $I\bar{1}$ structure, for the e_1 field. The transition line for the latter has been determined with more certainty than the transition line for the former. It is not known with any degree of certainty how these transition lines meet, however, or how they are related to the miscibility gaps. The low albite structure has very limited solid solution towards anorthite, whereas ordered anorthite has a much wider range towards albite. The phase diagram becomes even less certain below ~400°C, and the assemblage albite plus anorthite could become the stable assemblage at all intermediate compositions.

This phase diagram differs from previously proposed versions primarily because it includes stability fields for the e structures, separated by the Bøggild gap. The e_1 and e_2 structures were originally distinguished by their diffraction patterns: e_1 structures typically give sharp e and f reflections whereas the e reflections in diffraction patterns from e_2 crystals tend to be diffuse and the f reflections absent. With more significant thermodynamic implications, the e_1 stability field occurs to the $I\bar{1}$ side of the $C\bar{1} \rightleftharpoons I\bar{1}$ transition line and the e_2 stability field occurs on the $C\bar{1}$ side. Diffuse intensity has been reported in diffraction patterns from crystals with the high albite structure, indicating some short range ordering in the $C\bar{1}$ field, but this has not been characterised in any systematic manner.

As will become clear from the analysis which follows, the peristerite gap is believed to occur as a consequence of increasing Al/Si order in pure albite, the Bøggild gap as a consequence of either a $C\bar{1}/e_1$ break or an e_1/e_2 break, and the Huttenlocher gap as a consequence of an $I\bar{1} \rightleftharpoons e_1$ ordering transition.

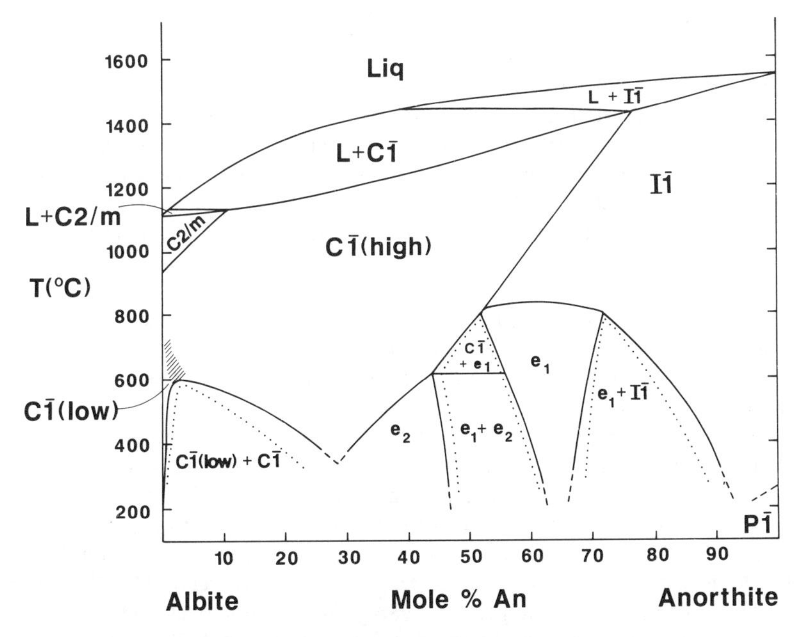

Figure 6. Proposed equilibrium phase diagram for plagioclase as low pressure. The topology is based on experimental data for some of the order/disorder transitions, including the conclusion that the e_1 structure has an equilibrium field of stability. The topology of how the various transition lines meet and their exact relationships to the miscibility gaps are matters for speculation. The form shown here does not represent a unique solution. It has the $I\bar{1} \rightleftharpoons e_1$ transition joined to the peaks of the Bøggild and Huttenlocher gaps. The $C\bar{1} \rightleftharpoons e_2$ transition is known to occur below ~750°C, but its position is otherwise poorly constrained. If it is restricted to below ~600°C, as shown here, there may be a field of coexisting $C\bar{1}$ + e_1 crystals in the Bøggild gap. The hatched zone shows the assumed composition/temperature region for high $\rightleftharpoons$ low albite ordering. Dotted lines show possible topologies for the limbs of conditional spinodals.

3. Lines of evidence

Many sources of evidence have been used to constrain the possible subsolidus phase relations of plagioclase feldspars. They all involve a measure of interpretation, but, in the end, must all be reconciled in some self-consistent manner. Each line of argument is summarised critically in this section.

3.1. ORDER/DISORDER EXPERIMENTS

The most direct evidence for the stability relations between different structure types has come from annealing experiments on natural samples. Crystals are heat treated under dry or hydrothermal conditions and changes in their states of order detected by transmission electron microscopy or X-ray diffraction. Above ~1050°C, significant variations in cation order can be induced on time scales of less than a few weeks. Under hydrothermal conditions with $P_{H_2O} \lesssim 1$ kb, the lower temperature limit for producing comparable ordering or disordering effects is ~750-800°C in run times of up to ~6 months. This lower limit can be extended by using higher pressures (Christoffersen and Yund, 1984; Goldsmith and Jenkins, 1985).

The phase transitions which have been observed in this way are: $C\bar{1} \rightleftharpoons I\bar{1}$, $I\bar{1} \rightleftharpoons e_1$ and $e_2 \rightarrow C\bar{1}$. An arrow in the last transition is used to signify that the reversal, $C\bar{1} \rightarrow e_2$, has not been achieved experimentally. Structural changes associated with the displacive monoclinic $\rightleftharpoons$ triclinic ($C2/m \rightleftharpoons C\bar{1}$) transition can be observed in situ at high temperatures using X-ray diffraction (Kroll et al., 1980; Kroll and Bambauer, 1981; Redfern et al., 1988; Redfern, 1992; Carpenter, 1992b). Only the most recent and pertinent results are summarised here. For older work see Smith (1974) and Smith and Brown (1988).

3.1.1. *C2/m* $\rightleftharpoons$ *$C\bar{1}$ transition.* Albite-rich plagioclase crystals with complete Al/Si disorder (other than the non-convergent ordering between T1 and T2 sites) will undergo a monoclinic $\rightleftharpoons$ triclinic transition at high temperatures which is largely displacive in character. The transition temperature is ~1238K in pure albite but increases with increasing anorthite content (Fig. 7a; data from Kroll et al., 1980; Kroll and Bambauer, 1981; Carpenter, 1988). For most of their range of solid solution, plagioclase crystals would melt before they become monoclinic, but the extrapolated line in Figure 7a gives $T_c \approx 3{,}000$K for the transition in pure anorthite (Carpenter, 1992b). Under equilibrium conditions the transition also involves a small component of Al/Si ordering, and coupling between order parameters for the displacive and Al/Si ordering effects causes the equilibrium transition temperature to increase. In pure albite, this increase in T_c with respect to fully disordered crystals is ~40°C (Salje et al., 1985; Carpenter, 1988), but has not been determined for other compositions. Limited experimental data for anorthite suggest that Al/Si ordering on the basis of $I\bar{1}$ symmetry does not influence the monoclinic $\rightleftharpoons$ triclinic transition in the same way (Carpenter, 1992b). The transition is thermodynamically continuous in albite and is shown on the equilibrium phase diagram as a single line (Fig. 6).

3.1.2. *$C\bar{1} \rightleftharpoons I\bar{1}$ transition.* The development of long range Al/Si ordering at the $C\bar{1} \rightleftharpoons I\bar{1}$ transition is revealed by sharp type b reflections (h + k = odd, l = odd) in single crystal diffraction patterns. Carpenter and McConnell (1984) bracketed the transition by heat treating natural crystals with a variety of compositions. Both ordering and disordering directions were followed and a linear compositional dependence for the transition temperature, $T_c(C\bar{1} \rightleftharpoons I\bar{1})$, is reasonably consistent with the experimental brackets (Fig. 7b). Carpenter (1986) has given the dependence as: $T_c(C\bar{1} \rightleftharpoons I\bar{1}) = 2283 - 2525X_{Ab}$, where T_c is in Kelvin and X_{Ab} is the mole fraction of albite component in the solid solution.

No experimental evidence for a two phase field associated with the transition has been reported, and, from the apparently continuous change in lattice parameters through the transition in crystals quenched from high annealing temperatures, the transition appears to be thermodynamically continuous (Carpenter and McConnell, 1984; Carpenter, 1988). Available thermodynamic data are also consistent with the transition being continuous at intermediate compositions in the plagioclase solid solution (Carpenter et al., 1985; Carpenter, 1988, 1992b).

3.1.3. *$I\bar{1} \rightleftharpoons e_1$ transition.* The steep variation with composition of the $C\bar{1} \rightleftharpoons I\bar{1}$ transition temperature contrasts with the approximately constant temperature of the $I\bar{1} \rightleftharpoons e_1$ transition for crystals in the range ~An_{55}-An_{70}. This transition has been reversed, i.e. following $e_1 \rightarrow I\bar{1}$ and $I\bar{1} \rightarrow e_1$ changes, at two compositions (Fig. 7c,d), constraining the equilibrium transition temperature to be between 800 and 850°C (Carpenter, 1986). Again, no experimental evidence has been found for two phase fields associated with the transition. The thermodynamic data are inadequate

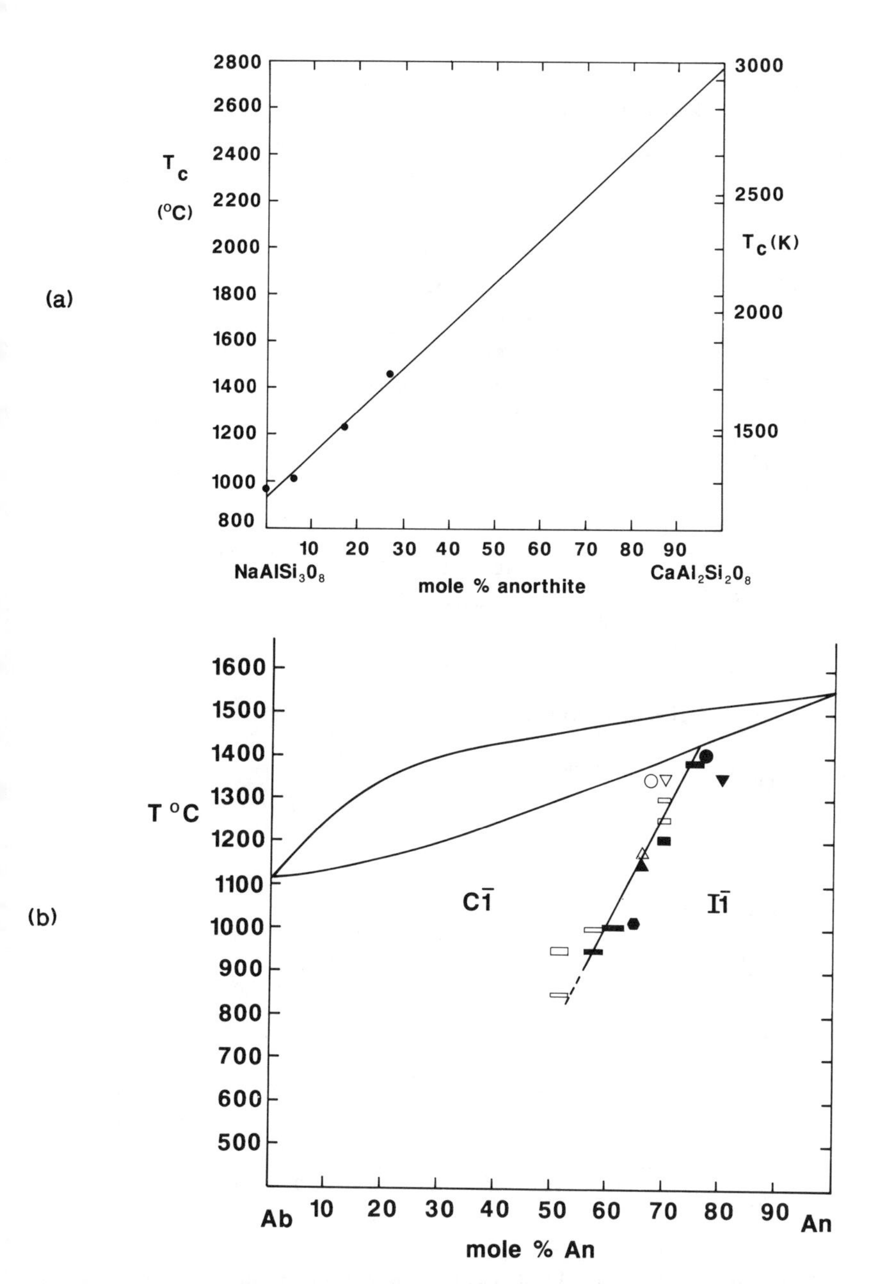

Figure 7. Experimental delineation of phase transitions in the plagioclase solid solution. (a) Data from Kroll and Bambauer (1981) and Carpenter (1988) for the C2/m ⇌ C$\bar{1}$ displacive transition in crystals with disordered Al/Si distributions (Carpenter, 1992b). (b) Bracketing of the C$\bar{1}$ ⇌ I$\bar{1}$ transition, from Carpenter and McConnell (1984).

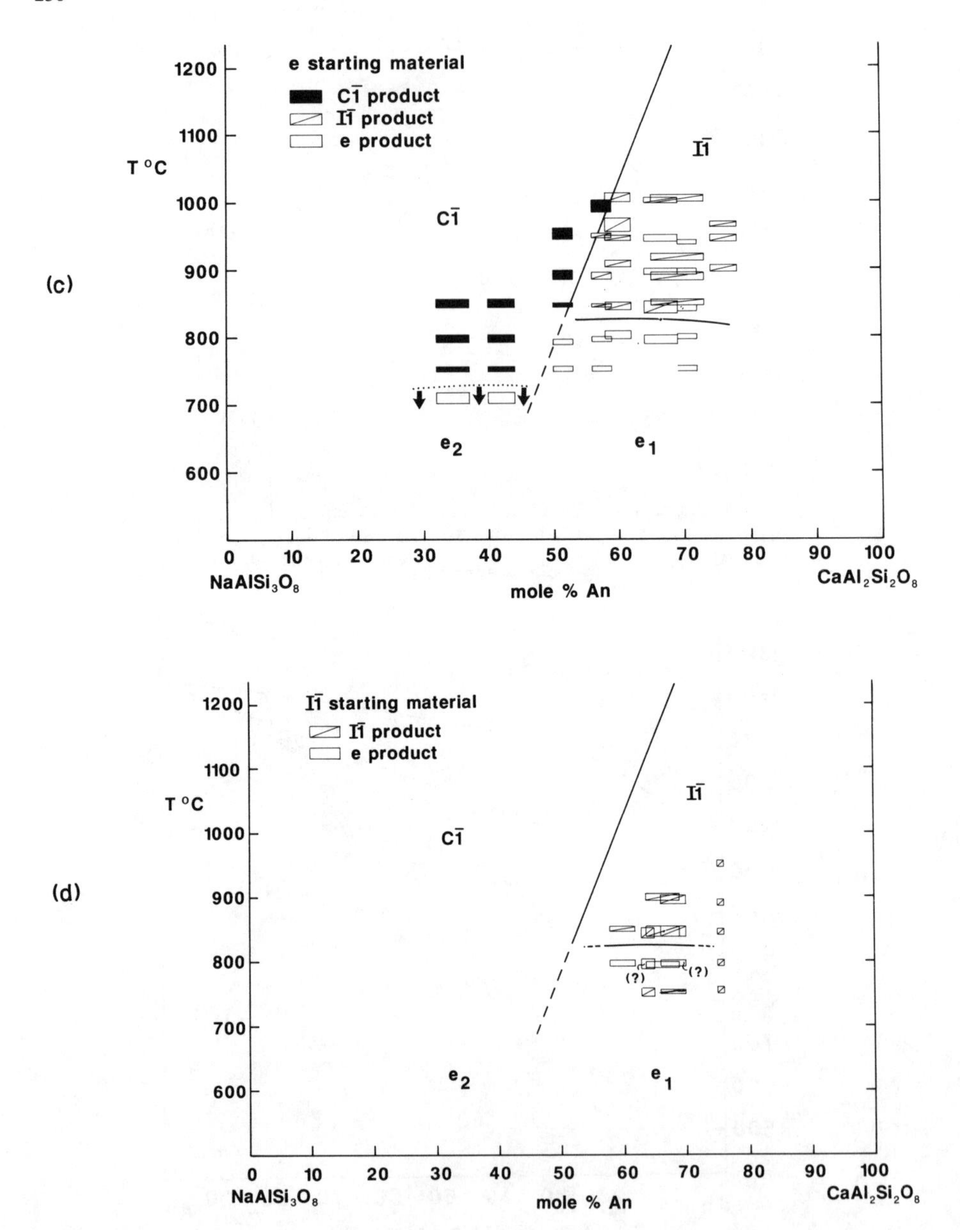

Figure 7 (cont.). (c) Results of prolonged hydrothermal annealing of starting crystals with diffraction patterns containing e reflections. The upper temperature limit for a C$\bar{1} \rightleftharpoons e_2$ transition is constrained to be below 750°C (dotted line). (d) Results of hydrothermal annealing of crystals with diffraction patterns containing b reflections. Note that crystals with the e structure were produced at 800°C.

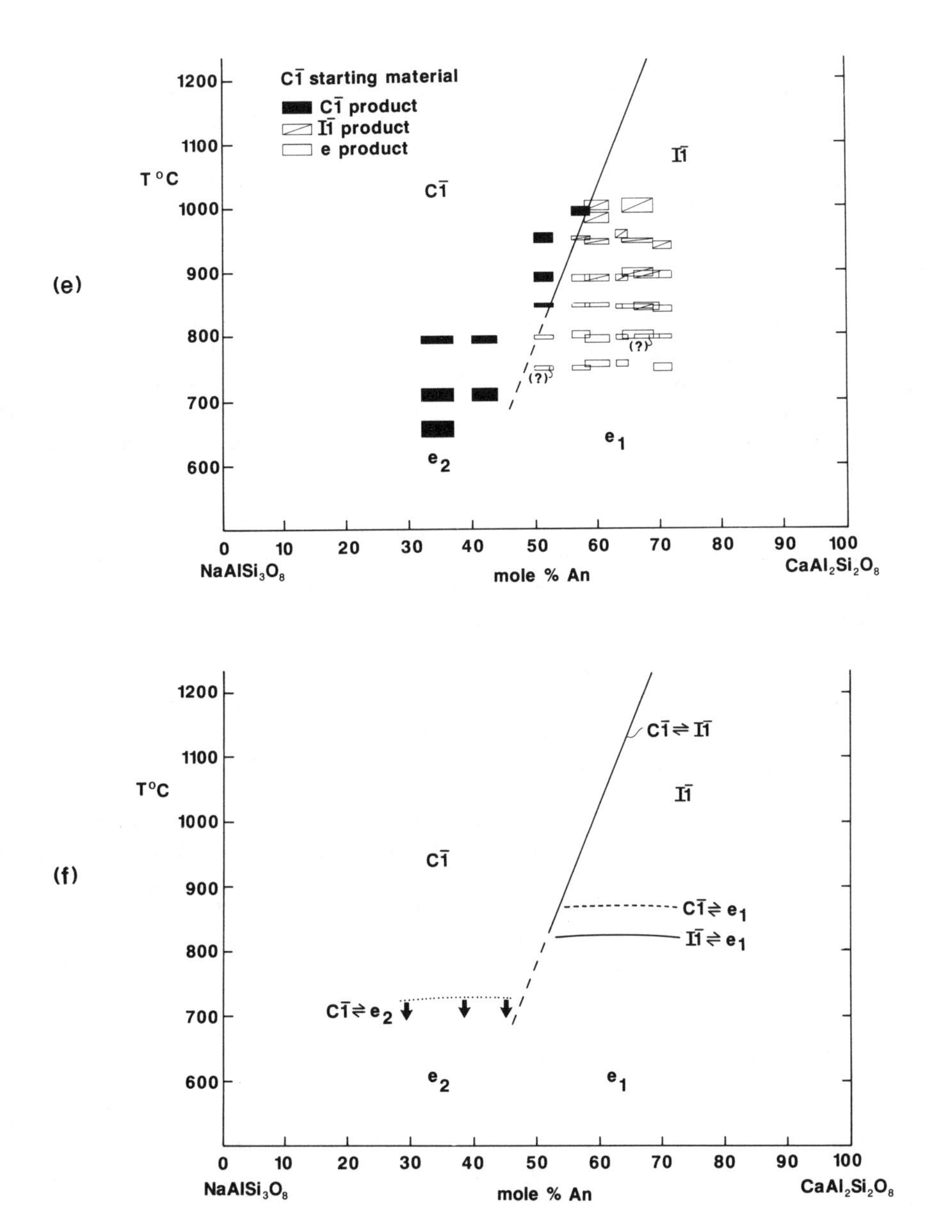

Figure 7 (cont.). (e) Results from starting crystals with the $C\bar{1}$ high albite structure. (f) Summary of results of hydrothermal annealing experiments (Carpenter, 1986). Note that the $I\bar{1} \rightleftharpoons e_1$ transition has been reversed. The dashed line represents either a metastable $C\bar{1} \rightleftharpoons e_1$ transition or a kinetic cut-off temperature below which e_1 ordering occurs more rapidly than $I\bar{1}$ ordering.

to discriminate between a first order transition and a continuous transition in this case. As will be argued later, continuous behaviour would be consistent with the spinodal microstructures observed in crystals that cooled into the Bøggild and Huttenlocher gaps. The transition is therefore shown on the phase diagram as a single line.

Starting with $C\bar{1}$ structures, e ordering can be induced at temperatures above the $I\bar{1} \rightleftharpoons e_1$ line (Fig. 7e). Thus there are circumstances in which e ordering that is metastable with respect to $I\bar{1}$ order can occur for kinetic reasons. The sequence $e_1 \rightarrow C\bar{1}$, to constrain a metastable $C\bar{1} \rightleftharpoons e_1$ transition within the $I\bar{1}$ stability field (Fig. 7f) has not been observed, however.

3.1.4. *$C\bar{1} \rightleftharpoons e_2$ transition.* The disordering reaction $e_2 \rightarrow C\bar{1}$ has been induced in natural crystals with compositions between ~An_{30} and ~An_{45} down to ~750°C (Fig. 7c). Although increases in the degree of order of disordered crystals can be induced experimentally at 700°C (e.g. Christoffersen and Yund, 1984), the development of e reflections has not yet been observed (Carpenter, 1986). The $C\bar{1} \rightleftharpoons e_2$ transition is thus constrained only to occur at some temperature below ~750°C (Fig. 7f).

3.1.5. *Ordering within the $C\bar{1}$ stability field.* In addition to the T1/T2 nonconvergent ordering of both monoclinic and triclinic crystals, two types of ordering can develop in crystals equilibrated within the $C\bar{1}$ stability field. Natural low albite crystals have a high degree of long range Al/Si order (Smith and Brown, 1988, and references therein). From experimental studies this is known to decrease as a function of increasing temperature, with a particularly marked, but continuous reduction in the interval ~650-700°C at low pressures (Goldsmith and Jenkins, 1985; Salje et al., 1985). The temperature at which this change from high albite to low albite occurs is generally presumed to decrease with increasing anorthite content because the Al/Si configuration in low albite precludes the addition of excess Al without placing Al in adjacent tetrahedra (hatched line in Figure 6).

In addition, more anorthite-rich crystals which have been annealed experimentally in the $C\bar{1}$ field often show diffuse intensity in their single crystal diffraction patterns. This diffuse intensity is located at and around the h + k = odd, l = odd positions, and could be interpreted as being due to diffuse b or diffuse e reflections (Kroll and Müller, 1980; Carpenter and McConnell, 1984; Christoffersen and Yund, 1984; Carpenter, 1986), indicating a degree of local ordering based on the $I\bar{1}$ structure or on the incommensurate e structure. The e reflections in diffraction patterns from e_2 crystals are also characteristically diffuse. The extent and nature of this short range ordering has not been determined systematically, but it presumably varies with temperature and composition, and must have some influence on the thermodynamic properties of the $C\bar{1}$ solid solution.

3.1.6. *$I\bar{1} \rightleftharpoons P\bar{1}$ transition.* The $I\bar{1} \rightleftharpoons P\bar{1}$ transition in pure anorthite is largely displacive in character, with a transition temperature of ~240°C in samples with a high degree of Al/Si order. Its thermodynamic character has recently been analysed in detail by Redfern and co-workers (Redfern and Salje, 1987, 1992; Redfern et al., 1988; Redfern , 1992; see, also, Ghose et al., 1988), but, because it occurs at too low a temperature to influence the ordering and exsolution processes within the solid solution, it is not reviewed here.

3.1.7. *Summary.* The important results extracted from the experimental data are summarised in Figure 7f. The transition lines for the $C\bar{1} \rightleftharpoons I\bar{1}$ and $I\bar{1} \rightleftharpoons e_1$ transitions are constrained to intersect between ~800 and ~ 850°C in the interval ~An_{50}-An_{54}. The data do not define the nature of the intersection, however. The anorthite-rich limit of the $I\bar{1} \rightleftharpoons e_1$ transition line is between ~An_{70} and

~An_{75} if the experimental results for $An_{75\text{-}76}$, in which the only state of order encountered was $I\bar{1}$, are considered to have approached equilibrium (Fig. 7d). The transition line must presumably turn down at this point or, as discussed below, give way to a two phase field. The dashed line for $C\bar{1} \rightleftharpoons e_1$ may be a metastable transition or it may represent an upper temperature limit for the kinetically favoured development of e ordering in place of $I\bar{1}$ ordering.

A clear distinction can be made between e_1 and e_2 ordering, aside from any structural distinctions. The former occurs to the $I\bar{1}$ side of the $C\bar{1} \rightleftharpoons I\bar{1}$ transition, while the latter occurs to the $C\bar{1}$ side. The former also develops at markedly higher temperatures than the latter and the experimental data are consistent with a substantial temperature difference between the onset of e_1 and e_2 ordering, as proposed by McConnell (1974a).

3.2. MICROSTRUCTURES

The most direct evidence for the nature and extent of miscibility gaps has come from electron-optical observations of ordering and exsolution microstructures in natural plagioclase crystals with different thermal histories. These observations have been reviewed thoroughly by Ribbe (1983b), Smith (1983, 1984) and Smith and Brown (1988). Only certain specific features are discussed here to highlight their possible implications for the form of the equilibrium phase diagram.

3.2.1. *Peristerite gap.* Summarising from Smith and Brown (1988), the maximum extent of the peristerite two phase field is generally taken to be ~An_1 to ~An_{25}. A variety of exsolution structures has been described in natural samples, some of which are consistent with having developed by a spinodal decomposition mechanism (e.g. Fig. 8, from Nord et al., 1978). The sodic component is ordered low albite while the calcic component appears to be less well ordered oligoclase. In some cases the oligoclase has short range order on the basis of the incommensurate structure, as implied by the presence of diffuse e reflections in single crystal diffraction patterns. The currently accepted hypothesis is that the driving force for exsolution is due primarily to the free energy of ordering in albite, and hence that the existence of the miscibility gap is conditional upon the Al/Si ordering taking place. The peak of the asymmetric solvus is usually placed approximately in, or just below the temperature interval over which the equilibrium degree of order in albite changes markedly, and is shown at ~600°C in Figure 6.

3.2.2. *Bøggild gap.* Exsolution textures in natural labradorite crystals with different compositions define the Bøggild miscibility gap as existing between ~An_{46} and ~An_{60} (summarised by Smith and Brown, 1988). The coherent microstructures are also consistent with a spinodal decomposition mechanism. Specimens in which the scale of exsolution is sufficient to cause an optical schiller are derived from slowly cooled anorthosites, which are believed to have had an igneous origin. The schiller specimens typically have $\gtrsim 2$ mole% orthoclase component in solid solution (Nissen et al., 1967), but Bøggild exsolution can also occur in crystals which have lower potassium contents. For example, Carpenter et al. (1985) reported exsolution on a scale of ~2,000Å in crystals with composition ~$An_{49}Ab_{51}Or_0$ from the granulite metamorphic rocks of Broken Hill, Australia.

It has long been postulated that the miscibility gaps arise as a consequence of changes in the ordering behaviour as a function of composition. In this context it is unlikely to be a coincidence that the intercept of the $C\bar{1} \rightleftharpoons I\bar{1}$ and $I\bar{1} \rightleftharpoons e_1$ transition lines at ~$An_{50\text{-}54}$ falls almost exactly in the middle of the Bøggild range. The volume proportions of albite-rich and anorthite-rich lamellae are also equal at about An_{50} (Nissen, 1971; Miura, 1978), signifying that the miscibility gap is

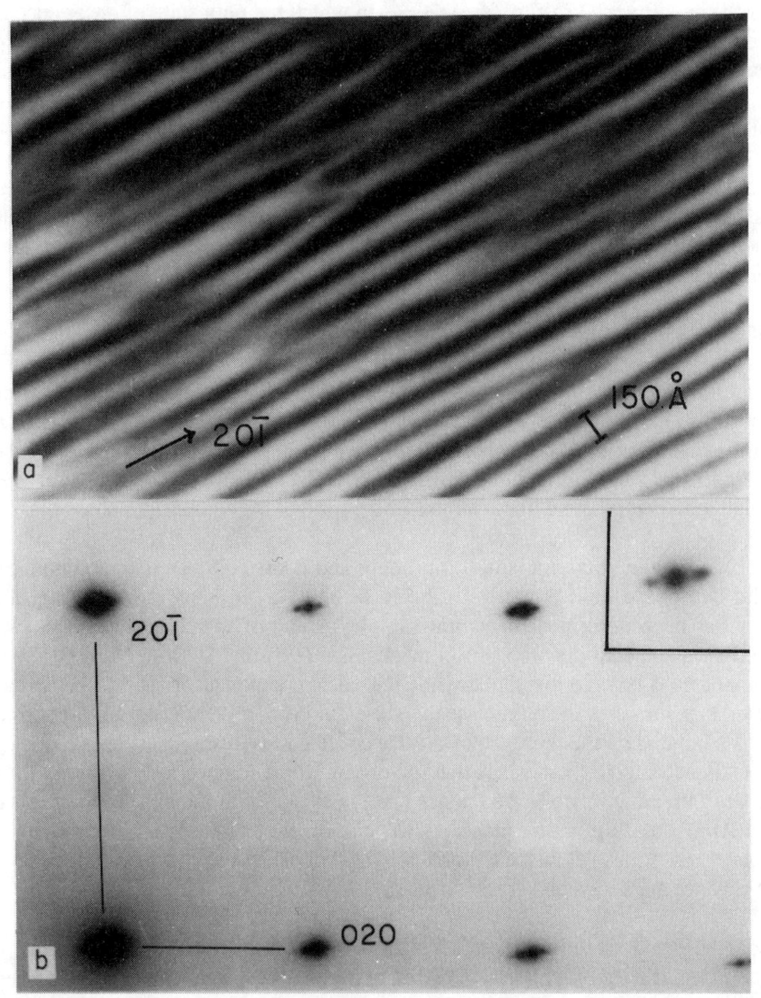

Figure 8. Electron micrograph and diffraction pattern of an exsolution texture in a crystal with an albite-rich composition that fell within the peristerite gap. (Courtesy of G.L.Nord, Jr., from Nord et al., 1978). The branching lamellar texture and the well developed satellite reflections (e.g. inset) are characteristic of a spinodal mechanism of decomposition.

approximately symmetric about this point. From the topology of these transitions (Figs. 6, 7f), it is clear that crystals with compositions to the albite side of the intersection should have a quite different ordering history during slow cooling than those to the anorthite side. The latter should first order to an $I\bar{1}$ state and then to an e_1 state, while the former should order directly from $C\bar{1}$ to either e_1 or e_2 structures, depending on the position of the $C\,\bar{1} \rightleftharpoons e_2$ transition line. Evidence for this difference might be preserved in the ordering microstructures of natural samples and, indeed, two distinct types of texture are observed. In one group, e or f type antiphase boundaries are observed to vary systematically with composition across the exsolution lamellae (Fig. 9a; McConnell, 1974b; Wenk and Nakajima, 1980; Müller, in Ribbe, 1983b). In a second group, the

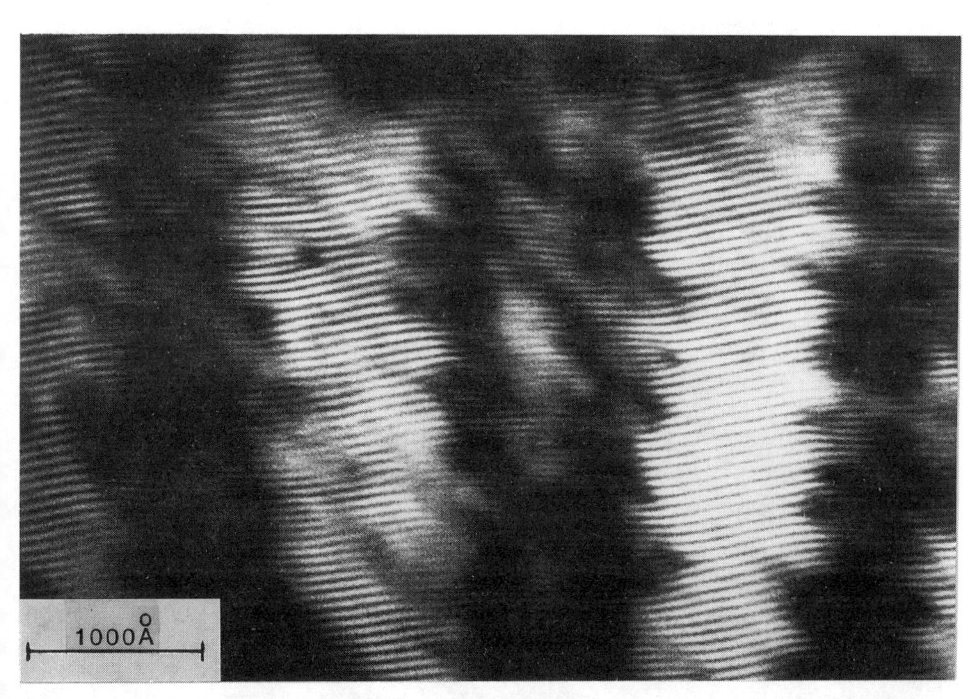

Figure 9. Some characteristic exsolution textures in natural crystals from the Bøggild and Huttenlocher gaps. (a) Dark field image of Bøggild exsolution lamellae showing the variation in spacing and orientation of type e APB's with composition. Bulk composition: An_{53}. (Courtesy of H.-R.Wenk, from Wenk and Nakajima, 1980).

type e or f antiphase boundaries retain approximately constant orientations and spacings across the lamellae (Fig. 9b; Hashimoto et al., 1976; Olsen, 1977). Compositions reported for each group are An_{55}, An_{53}, An_{52} and An_{54}, An_{51}, respectively. These compositions span the critical range, but there is overlap between the two groups. It may be that insufficient care has been taken to determine accurate compositions for the exact area of crystal examined by TEM, though other factors, such as pressure or orthoclase content, could influence the position of the intersection point in relation to a given crystal as it cools. A possible explanation of the two types of microstructures is offered in a later section on the systematics of e and f reflections.

A third possible group of Bøggild microstructures has e or f type APB's appearing in only one component of the Bøggild exsolution (McLaren, 1974; McLaren and Marshall, 1974). This is a surprising observation given that all homogeneous plagioclase crystals from slowly cooled rocks so far examined which have compositions between ~An_{30} and ~An_{65} have the incommensurate e structure. Figure 10 of McLaren (1974), from a crystal with composition An_{52}, appears to be consistent with the second group of microstructures (no change in orientation and spacing of e or f APB's) with only a small portion of one component out of contrast. Some artefact, such as the preferential beam degradation of the albite-rich component might have caused the loss of contrast. Figure 6b of McLaren and Marshall (1974), also from a crystal with composition An_{52}, is less easy to explain away and the possibility of coexisting $C\bar{1}$ and e_1 phases across the Bøggild gap may be implied. Further systematic TEM studies are needed to resolve these uncertainties.

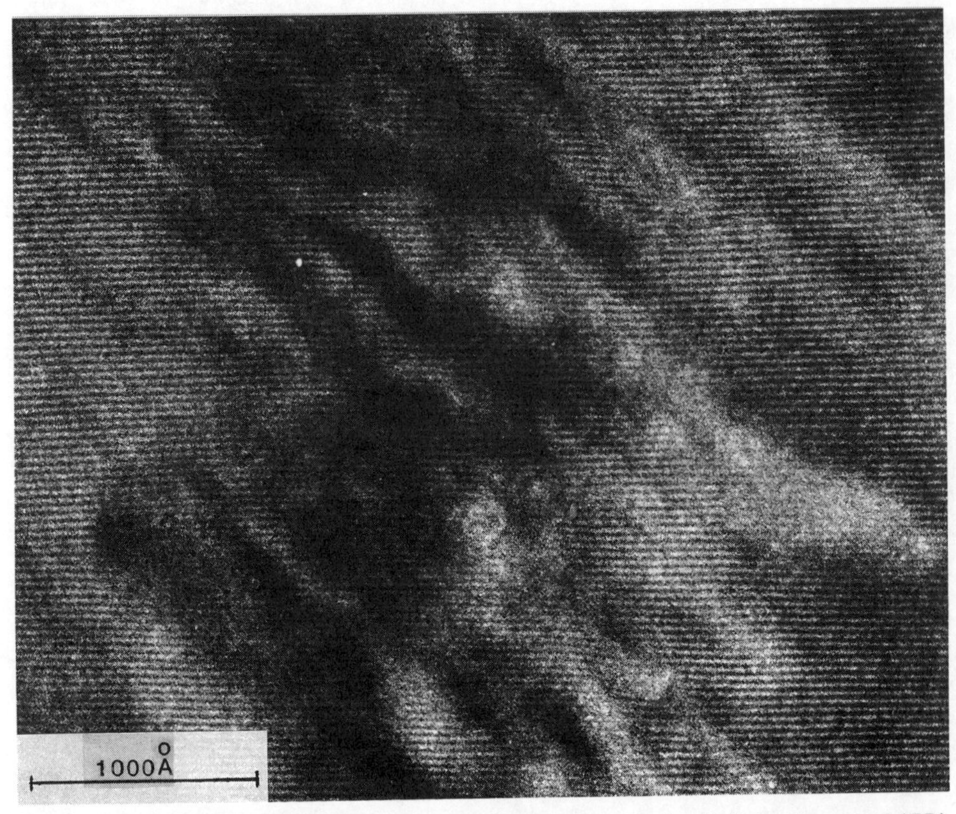

Figure 9 (cont.). (b) Dark field image of Bøggild exsolution lamellae (vertical) showing type f APB's (horizontal) with no change of orientation or spacing. Bulk composition: $An_{53}Ab_{46}Or_1$. The darker vertical band is an albite-rich lamella. (Courtesy of H.Hashimoto, from Hashimoto et al., 1976).

Extended time at relatively high temperatures appears to be required for Bøggild exsolution. For example, incipient exsolution has not been observed in crystals with composition ~An_{50-53} from Skaergaard (with ~1 mole% Or) or in crystals with composition ~An_{56-59} from Duluth (with ~2 mole% Or), (Carpenter, 1986). This presumably reflects a relatively faster cooling rate for high level igneous intrusions, in comparison with the deeper crustal origin of ancient anorthosites. Similarly, Grove et al. (1983) appear not to have found lamellar exsolution textures in metamorphic plagioclase crystals of composition An_{50} which were subjected to a peak temperature of only ~530°C. The role of K^+ also remains somewhat perplexing (see discussion in Smith and Brown, 1988). Its presence in solid solution is not an essential prerequisite for exsolution to occur, but it does appear to enhance the exsolution process in some way.

3.2.3. *Huttenlocher gap*. Again summarising from Smith and Brown (1988), the limiting compositions for exsolution in the Huttenlocher miscibility gap are ~An_{67} and ~An_{90}. As with peristerite and Bøggild exsolution, many of the microstructures are consistent with a spinodal decomposition mechanism. Coexisting phases across the two phase field have the e_1 and $I\bar{1}$ structures, with diffuse type c and d reflections due to local $P\bar{1}$ distortions. The coarsest

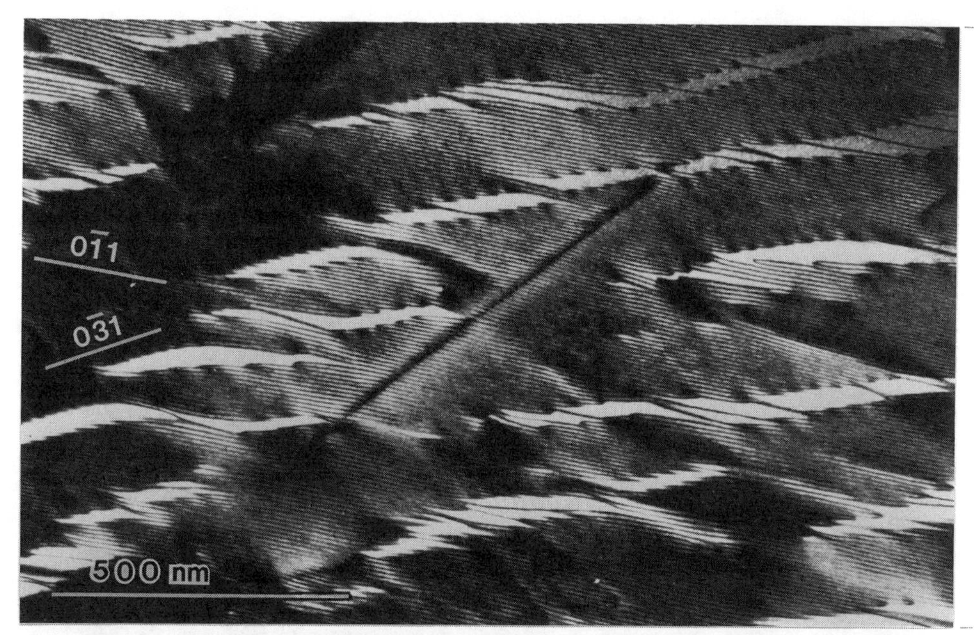

Figure 9 (cont.). (c) Dark field image of exsolution texture from the Huttenlocher gap revealing the distribution of APB's. The lozenge shaped regions with few APB's are interpreted as anorthite-rich plagioclase with $I\bar{1}$ structure which exsolved from a matrix with type e APB's. Bulk composition: An_{75}. (Courtesy of H.-R.Wenk, from Wenk and Nakajima, 1980).

exsolution, just visible in an optical microscope, occurs predominantly in the range ~An_{69}-An_{76} (Nissen, 1974; Grove, 1976, 1977a; Smith and Brown, 1988). This places constraints on the composition at which the miscibility gap reaches its highest temperature, suggesting that the solvus is slightly asymmetric.

It is unlikely to be coincidental that the anorthite rich limit of e_1 ordering determined experimentally (Fig. 7) falls within the expected range of the solvus crest. If the solvus does not extend above ~800°C, as shown in Figure 6, the microstructures of exsolved natural samples should fall into two groups. Slow cooling of crystals with compositions to the albite-rich side of the e_1 limit should order first on the basis of the $I\bar{1}$ structure and then on the basis of the e_1 structure before entering the miscibility gap. The microstructures might be expected to imply exsolution of an $I\bar{1}$ phase from an e_1 host (Fig. 9c). On the other hand, during slow cooling of crystals with compositions to the anorthite-rich side of the $I\bar{1}$ limit, the two phase field would be entered directly from the stability field of $I\bar{1}$ crystals. A characteristic microstructure of natural plagioclases in this range is shown in Figure 9d. It involves relict zig-zag type b APB's, derived from $C\bar{1} \rightarrow I\bar{1}$ ordering, with fine scale exsolution of the e_1 phase spread throughout. From a systematic survey of published electron micrographs, the composition which marks the divide between these two types of microstructures appears to be in the range ~An_{72}-An_{75} (Heuer et al., 1972; Nord et al., 1974; Grove, 1977a; Wenk and Nakajima, 1980; Grove et al., 1983).

The unique sample of Phillips et al. (1977) which was reported to show optically visible exsolution lamellae in the composition range ~An_{64}-An_{67} requires further examination (Smith, 1984).

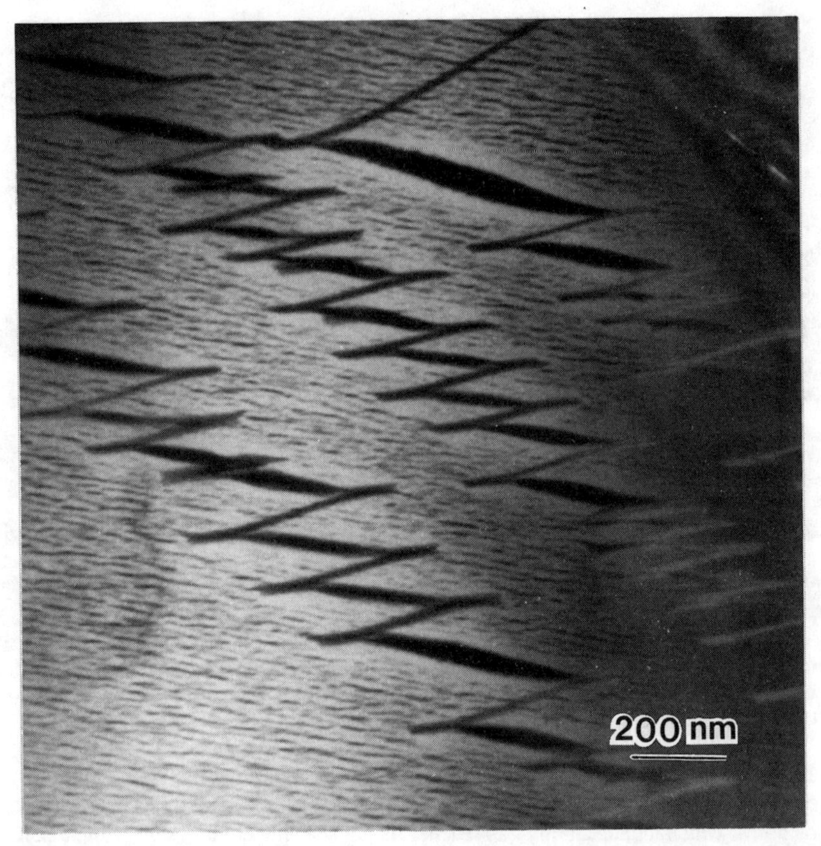

Figure 9 (cont.). (d) Zig-zag type b APB's in a crystal with a fine scale exsolution texture observed in dark field using a type b reflection. The crystal is interpreted as having first ordered on the basis of the $I\bar{1}$ structure before entering the Huttenlocher gap. Bulk composition: An_{82}. (Courtesy of G.L.Nord, Jr., from Nord et al., 1974).

3.3. COMPOSITIONS OF COEXISTING CRYSTALS

There are many reports of coexisting plagioclase crystals with different compositions occurring in metamorphic rocks (reviews: Goldsmith, 1982; Smith and Brown, 1988). In principle, such assemblages should carry information concerning the limits of miscibility gaps in the plagioclase solid solution. Frequently they are mutually inconsistent, however, implying that factors other than equilibrium stability can control their development. The issues are illustrated most clearly for crystallisation of sodic plagioclase below the peak temperature of the peristerite gap. Crystallisation behaviour associated with the Bøggild and Huttenlocher gaps appears to be less predictable.

3.3.1. *Coexisting phases across the peristerite gap.* In many greenschist and amphibolite facies rocks two feldspars, albite and oligoclase, appear to have crystallised contemporaneously or with a core/rim relationship (Maruyama et al., 1982; Grove et al., 1983; Grapes and Otsuki, 1983;

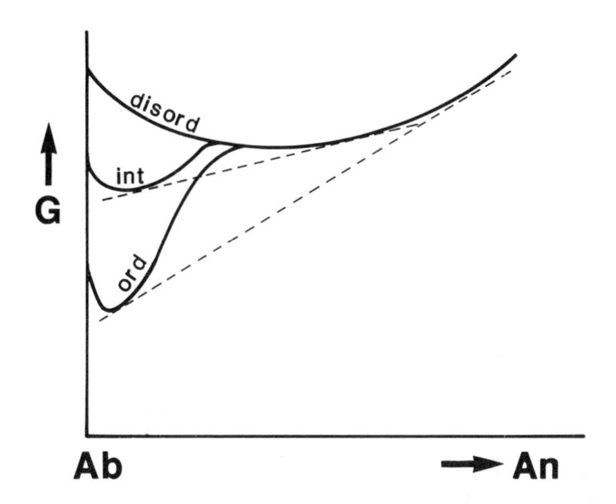

Figure 10. Schematic free energy (G) - composition curves defining the peristerite gap for different degrees of Al/Si order in albite (from Carpenter and Putnis, 1985). The equilibrium limits of the two phase field are given by the common tangent to the curves for the ordered albite solid solution (ord) and the disordered solid solution (disord). If crystals grow metastably with an intermediate degree of order (int), the effective miscibility gap would be smaller. In the absence of any Al/Si ordering occuring during crystal growth, there would be no miscibility gap.

Ashworth and Evirgen, 1985a; see Goldsmith, 1982, for older references). Inconsistencies in the apparent composition limits of the peristerite gap defined by the coexisting crystals are found (Crawford, 1966; Cooper, 1972), and crystallisation of phases with compositions well within the miscibility gap can occur (Nord et al., 1978). Equilibrium evidently does not always prevail and a possible explanation arises from considerations of the kinetics of crystal growth (Carpenter and Putnis, 1985).

The entropy of activation for nucleation and growth of new phases in a rock acts to favour the development of crystals with disordered cation distributions rather than crystals with ordered distributions, even when the latter are thermodynamically stable. If there is a miscibility gap that exists only when one of the end member phases has a high degree of cation order, then the compositions of coexisting crystals that nucleate and grow will also depend on the rate of crystal growth. In the limiting case illustrated in Figure 10, kinetically enhanced growth of disordered crystals would give a homogeneous feldspar with some intermediate composition, when equilibrium growth would give low albite + oligoclase. From the converse argument, the development of two plagioclase assemblages with a regular pattern of composition variations as a function of metamorphic temperature would imply that the crystals grew sufficiently slowly that they acquired a rather high initial degree of order. Data from Spear (1980), Grove et al. (1983) and Ashworth and Evirgen (1985a) are shown in Figure 11 to illustrate a set of self-consistent limits to the peristerite gap. Maruyama et al. (1982) have shown that different limits, also self-consistent for a given association, can be extracted from rocks metamorphosed at different pressures. An equilibrium pressure effect for the miscibility gap is possible but a kinetic pressure effect, through the activation volume for nucleation and growth, might also contribute (Carpenter and Putnis, 1985). Inevitably, questions concerning equilibrium versus metastable crystal growth are associated with these observations.

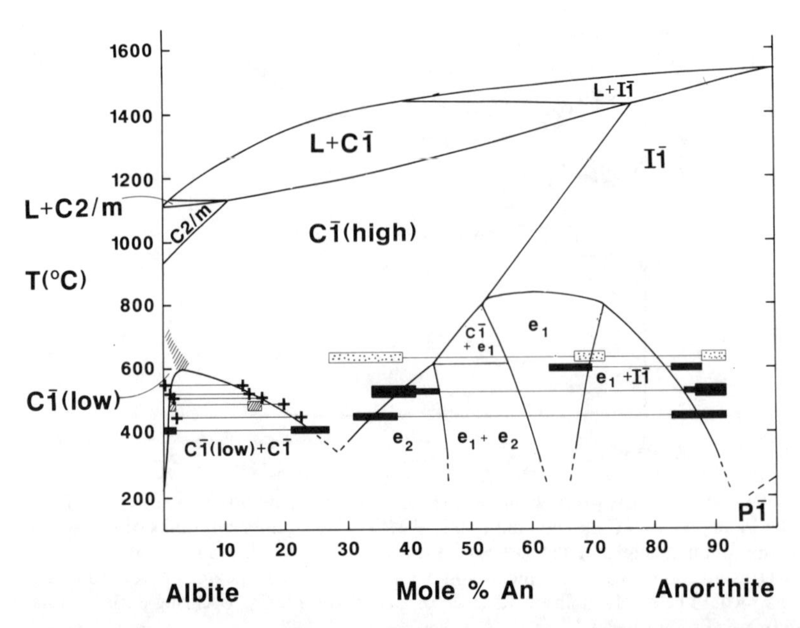

Figure 11. Relationship between selected two plagioclase assemblages and the proposed phase diagram. A qualitatively self-consistent delineation of the peristerite gap is provided by observations of coexisting albite and oligoclase by Ashworth and Evirgen (1985a), crosses, Grove et al. (1983), filled rectangles, and Spear (1980), hatched rectangles. Coexisting anorthite-rich plagioclases do not give such a well defined view of the Bøggild and Huttenlocher miscibility gaps (shaded rectangles from Wenk et al., 1975, Wenk and Wenk, 1977, and Wenk et al., 1980; filled rectangles from Grove et al., 1983). In the interpretation discussed here, nucleation and growth of crystals with e_1 order would not be kinetically favourable at low temperatures so that alternative, but metastable assemblages might develop instead. Note that if all data for coexisting plagioclases reported in the literature were shown on this diagram there would be little or no consistent correlation with the positions of the miscibility gaps known from exsolution textures.

3.3.2. *Coexisting plagioclases with intermediate compositions.* Almost any combination of coexisting plagioclase crystals with intermediate compositions can be found in metamorphic rocks (Crawford, 1972; Wenk et al., 1975; Wenk and Wenk, 1977, 1984; Garrison, 1978; E.Wenk, 1979, 1983a,b; H.-R.Wenk, 1979a; Wenk et al., 1980, 1991; Spear, 1980; Grove et al., 1983; and see review by Goldsmith, 1982). Possible causes unrelated to miscibility gaps in the solid solution include discontinuous zoning caused by the release or uptake of components in other metamorphic reactions (Goldsmith, 1982), or ion-exchange between crystals and a later fluid phase (Mora and Ramseyer, 1992).

At the very least, there should be textural or chemical evidence for equilibrium before a given two-feldspar assemblage can be taken as delimiting the Bøggild or Huttenlocher miscibility gaps. For example, coexisting andesine (An_{37-42}) and bytownite (An_{87-92}) described by Spear (1980) show a regular partitioning of elements with amphiboles. The coexisting compositions described by Wenk et al. (1975, 1980), Wenk and Wenk (1977), Spear (1980) and Grove et al. (1983) have been discussed most frequently in this context and are shown in Figure 11. Taken at face value they are consistent with the e_1 structure having a stability field above ~550-600°C, giving way to andesine + bytownite at lower temperatures. However, given the degree of long range order required in the e_1 ordering scheme, it is likely that the nucleation and growth of either homogeneous disordered crystals or of some assemblage of crystals with smaller entropies of activation will be kinetically favoured (Carpenter and Putnis, 1985). Thus the andesine + bytownite assemblage might show evidence of textural and chemical equilibrium with other phases but still be metastable with respect to e_1 crystals with compositions ~An_{60}-An_{67}. Only permissive evidence rather than proof of the form of the equilibrium subsolidus phase diagram is provided, therefore. At the heart of the problem is the likely causal relationship between Al/Si ordering and the development of miscibility gaps at low temperatures in the solid solution, and the sensitivity of the former to kinetic factors. Wenk et al. (1991) have recently shown that all plagioclase compositions are possible in a metamorphic area in which individual rocks may contain two plagioclases. According to the kinetic argument, this reflects variable nucleation and growth rates under different local conditions of supersaturation.

3.3.3. *The low temperature assemblage albite + anorthite.* The apparent breakdown of single plagioclase crystals to a mixture of albite + anorthite in natural assemblages has been described by Voll (1971) and H.-R.Wenk (1979a). If the textures are correctly interpreted as representing an equilibrium phenomenon, they provide the most direct evidence for the form of the phase diagram at low temperatures. The temperature of this lower stability limit for homogeneous crystals with e_1 or e_2 order is not known, however.

3.4. PROPERTIES OF THE INCOMMENSURATE "e" STRUCTURE

Until recently, the incommensurate plagioclase structure was only known to occur in natural crystals with compositions between ~An_{25} and ~An_{75}. A rather similar ordered structure has now been found in pure anorthite (Carpenter, 1991a), and the new observations provide a different perspective for understanding the nature, properties and behaviour of the e structure. It has also commonly been assumed that the e ordering scheme develops as a consequence of kinetic effects and may not therefore be stable even with respect to an $I\bar{1}$ ordering scheme at the same composition. This contrasts with experience of other materials in which incommensurate ordering gives rise to structural states with significant equilibrium fields of stability (McConnell, 1988, 1991, 1992, and references therein). In this section, the incommensurate structure of pure anorthite is briefly described. A qualitative description of the structure is then presented in terms of order parameters and order parameter coupling. The intention is to consider the factors which influence the stability and thermodynamic behaviour rather than to address details of the crystal structure. Comprehensive reviews of current thinking on the latter have been given by Horst (1984), Kitamura and Morimoto (1984), Jagodzinski (1984), McConnell (1988) and Smith and Brown (1988).

3.4.1. *Metastable incommensurate structure of anorthite.* The first crystals to form on heating anorthite glass at temperatures of ~1100-1400°C have a high degree of metastable Al/Si disorder and essentially $C\bar{1}$ symmetry. With further annealing, an apparently continuous increase in Al/Si order causes the development, initially, of an incommensurate structure which has diffraction characteristics which are remarkably similar to those arising from the more familiar e structure, and then of the stable $I\bar{1}$ structure (Carpenter, 1991a,b). Type e reflections in diffraction patterns from crystals annealed for the shortest times are weak and diffuse. They become stronger and sharper as the total degree of order increases, while their orientation and spacing also change systematically (Fig. 12a-e). There is complete overlap with the orientations and spacings of e reflections in natural crystals with compositions of ~An_{30} - ~An_{55} in the early stages of ordering, but as the e

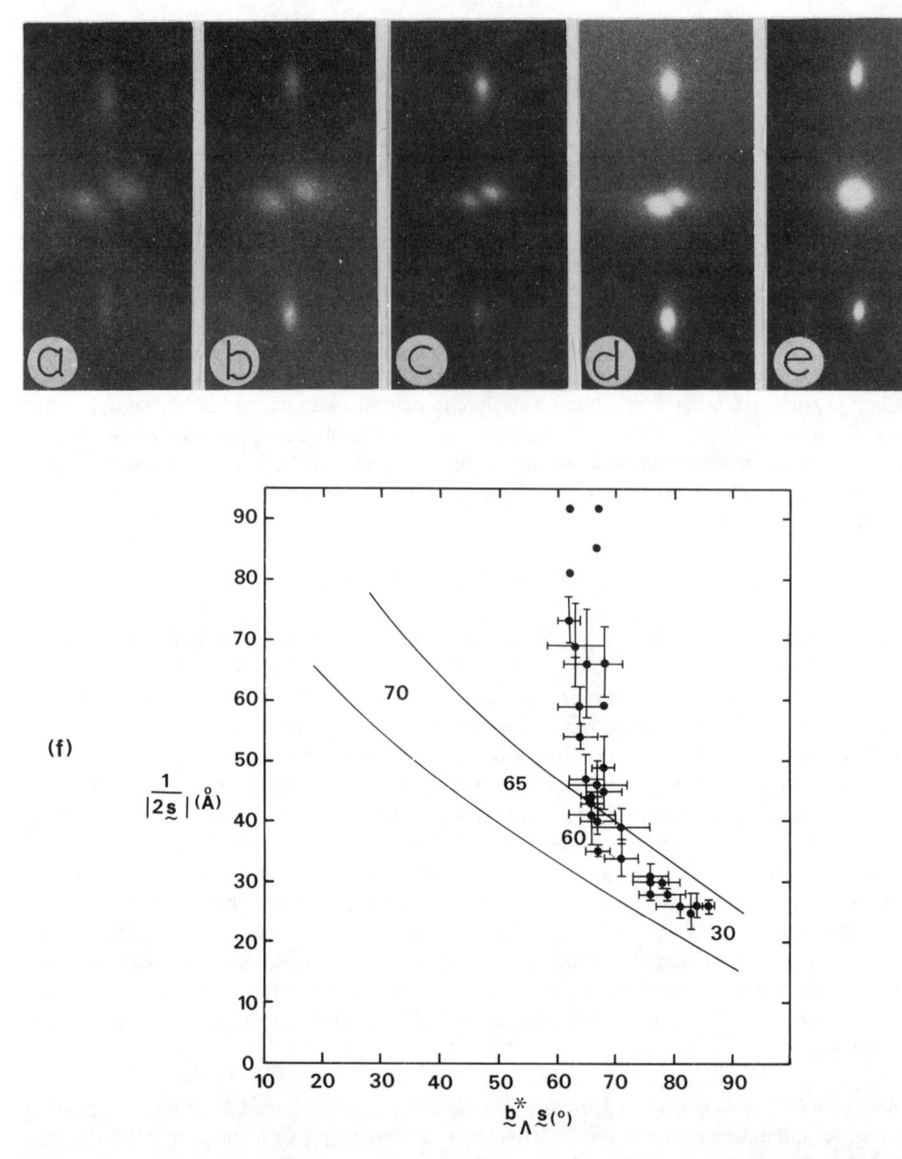

Figure 12. Characteristic features of the metastable incommensurate structure of anorthite (from Carpenter, 1991a). (a-e) Representative selection of portions of electron diffraction patterns showing e and b reflections (central) and elongate c reflections (top and bottom). With increasing Al/Si order (left to right) the diffuse e reflections sharpen, rotate and become closer, before merging in an apparently continuous manner to form relatively sharp b reflections. (f) Variation of spacing between type e APB's ($1/|2\underset{\sim}{s}|$) with orientation ($b^* \wedge \underset{\sim}{s}$), where $2\underset{\sim}{s}$ is the vector separating the individual e reflections belonging to a given pair, and b^* is a vector of the reciprocal lattice. Continuous lines represent the approximate limits of the equivalent variation for natural intermediate plagioclases with compositions indicated in mole% An.

reflections become closer together their orientations diverge from the trend of natural crystals with compositions of ~An_{55} - ~An_{75} (Fig. 12f). The size of the first type b APB's appears to be determined by the spacing of type e APB's in the last stages of the evolving incommensurate state.

In spite of the close similarities, the incommensurate anorthite structure is not identical to the e structure of natural crystals in that the type f satellites around type a reflections (h+k = even, l = even) have not been observed in the former. There is, however, a correlation between the type e and type c reflections (h+k = even, l = odd) which is comparable with trends in the natural crystals. Anorthite crystals with the weakest and most diffuse e reflections also have the weakest and most diffuse c reflections. As the e reflections sharpen up, so do the c reflections. Type d reflections (h+k = odd, l = even) can be detected at about the point where e reflections give way to b reflections. The sequence therefore seems to mirror, at least qualitatively, the sequence of ordered states with increasing An content in natural crystals.

3.4.2. *Structure of e plagioclase.* Many descriptions of the incommensurate plagioclase structure make use of slabs of material with different structural or compositional states, which are put together in a sequence that generates the incommensurate repeat. In the simplest representation of these models, the layers might repeat in the sequence: -Ab-An-Ab-An*-Ab-An-Ab-An*-, where An and An* represent anorthite-like layers with an antiphase relationship to each other and Ab represents layers with an albite-like structure (Smith and Brown, 1988, and references therein). Given that ordered structures with closely analogous diffraction characteristics can develop in pure anorthite, compositional variations between the slabs may not represent an essential condition for e plagioclase formation. Without loss of generality, therefore, the structure might be better represented as a sequence of slabs such as -$C\bar{1}$-$I\bar{1}$-$C\bar{1}$-$I\bar{1}$*-$C\bar{1}$-$I\bar{1}$-$C\bar{1}$-$I\bar{1}$*-, where $C\bar{1}$ and $I\bar{1}$ refer to the symmetry of the slabs. $I\bar{1}$ and $I\bar{1}$* again have an antiphase relationship to each other. This now concentrates attention exclusively on the cation ordering in individual slabs and the mechanisms by which ordering in one slab couples with ordering in the next.

In general, incommensurate structures owe their stability to interactions between at least two ordering processes (see reviews by Bruce and Cowley, 1981; Heine and McConnell, 1984; McConnell, 1988, 1991, 1992). Because of the low symmetry of the plagioclase system and the known behaviour of many end-member phases, the choice of possible processes, as described in terms of order parameters, appears to be restricted to three. One component must be Q_{od}^{I1} for the $C\bar{1} \rightleftharpoons I\bar{1}$ Al/Si ordering transition, but the second could be either Q for the $C2/m \rightleftharpoons C\bar{1}$ displacive transition or Q_{od}^{C1} for the $C2/m \rightleftharpoons C\bar{1}$ Al/Si ordering transition. As found in many other materials, the interaction or coupling mechanism is likely to involve a common strain; a change in one order parameter causes a lattice distortion which then influences the second order parameter, and vice versa (see Carpenter, 1988, 1992a; Salje, 1990). As discussed in more detail by Carpenter (1991a), the strain common to Q and Q_{od}^{I1} is small, but the strain accompanying both Q_{od}^{C1} and Q_{od}^{I1} is predominantly in the x_6 component (~cosγ) giving a significant common strain. Furthermore, x_6 tends to be positive ($\gamma < 90°$) for $C\bar{1}$ ordering and negative ($\gamma > 90°$) for $I\bar{1}$ ordering (Fig. 13) so that the only manner in which favourable interactions can occur between the two ordering schemes is for distinct regions of crystal with $C\bar{1}$ and $I\bar{1}$ order to coexist.

This qualitative picture of the incommensurate plagioclase structure can be formalised in terms of variations in Q_{od}^{C1} and Q_{od}^{I1} as a function of distance in the crystal. The slabs will not be rigid units, but, rather, would make up a structure with more or less continuous variations in the local ordering. Two possible configurations are illustrated in Figure 14. In the locally $I\bar{1}$ regions, Q_{od}^{I1} will have positive or negative values (corresponding to the antiphase relationship) and Q_{od}^{C1} will be zero. In the locally $C\bar{1}$ regions there might be gradients in both Q_{od}^{C1} and Q_{od}^{I1} (Fig. 14a) or there

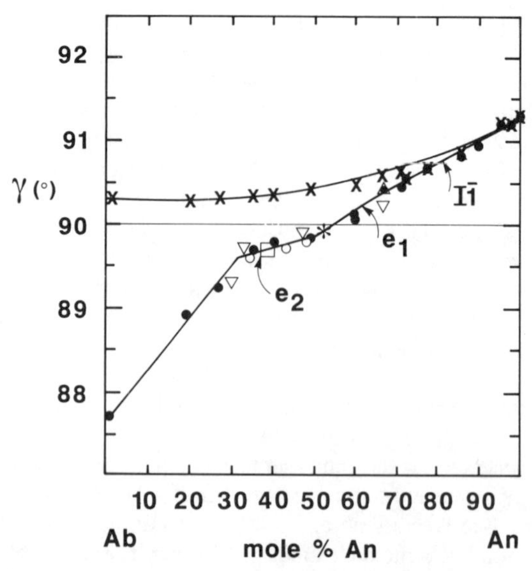

Figure 13. Variation of lattice angle γ with composition at room temperature for natural ordered plagioclases (filled circles, Carpenter et al., 1985; open circles, Christoffersen and Yund, 1984; open squares, Steurer and Jagodzinski, 1988; inverted open triangles, H.-R.Wenk, 1979b; open triangle, Wenk et al., 1980; star, Horst et al., 1981), and experimentally disordered plagioclases (crosses, Carpenter et al., 1985). Separate trends for e_1 and e_2 structures are shown with a break in slope at ~An_{52}. Note that $I\bar{1}$ ordering gives $\gamma > 90°$ and $C\bar{1}$ ordering gives $\gamma < 90°$, while the e_1 and e_2 structures have γ close to 90°.

might be segments with relatively constant $Q_{od}^{C\bar{1}}$ and zero $Q_{od}^{I\bar{1}}$ (Fig. 14b). The energy due to ordering constitutes three parts, that due to $C\bar{1}$ ordering, that due to $I\bar{1}$ ordering and that due to interactions between the two ordering schemes. If one of the first two contributions were to dominate, then a commensurate structure with either $C\bar{1}$ or $I\bar{1}$ order would be expected to develop, but if both gave comparable energy contributions, the coupling might tilt the energetic balance in favour of the incommensurate structure. $Q_{od}^{C\bar{1}}$ is shown as having only one sign in Figure 13 since regions of crystal with positive and negative values would have a twin relation, as in albite. Such twins develop characteristically as the consequence of transformation from a monoclinic parent crystal, while the incommensurate structures develop in crystals which are already triclinic.

Since each of $Q_{od}^{C\bar{1}}$ and $Q_{od}^{I\bar{1}}$ is coupled to strain, there would be a variation in $\cos\gamma$ (~ x_6) accompanying the variations in $Q_{od}^{C\bar{1}}$ and $Q_{od}^{I\bar{1}}$. This is illustrated schematically in Figure 14c for the ordering behaviour represented by Figure 14b. First of all, it is apparent that if the x_6 strains are alternatively positive and negative, the net strain would be close to zero. The γ angle of incommensurate plagioclase crystals at room temperature is indeed close to 90°, i.e. $\cos\gamma \approx 0$ (Fig. 13). Secondly, the repeat distance of the strain modulation is exactly half that of the full ordering modulation. The complete strain depends on the remaining five strain components as well, but if this involves only a shear for both $C\bar{1}$ and $I\bar{1}$ strains, then the overall strain modulation will also involve only shearing. If, however, there is an element of volume strain which differs between the $C\bar{1}$ and $I\bar{1}$ regions, the structure will automatically have a density modulation with a wavelength of half that of the order modulation. At room temperature, $I\bar{1}$ ordering in anorthite gives a volume strain of ~3‰ (Carpenter, 1991b), while the room temperature volume difference between natural

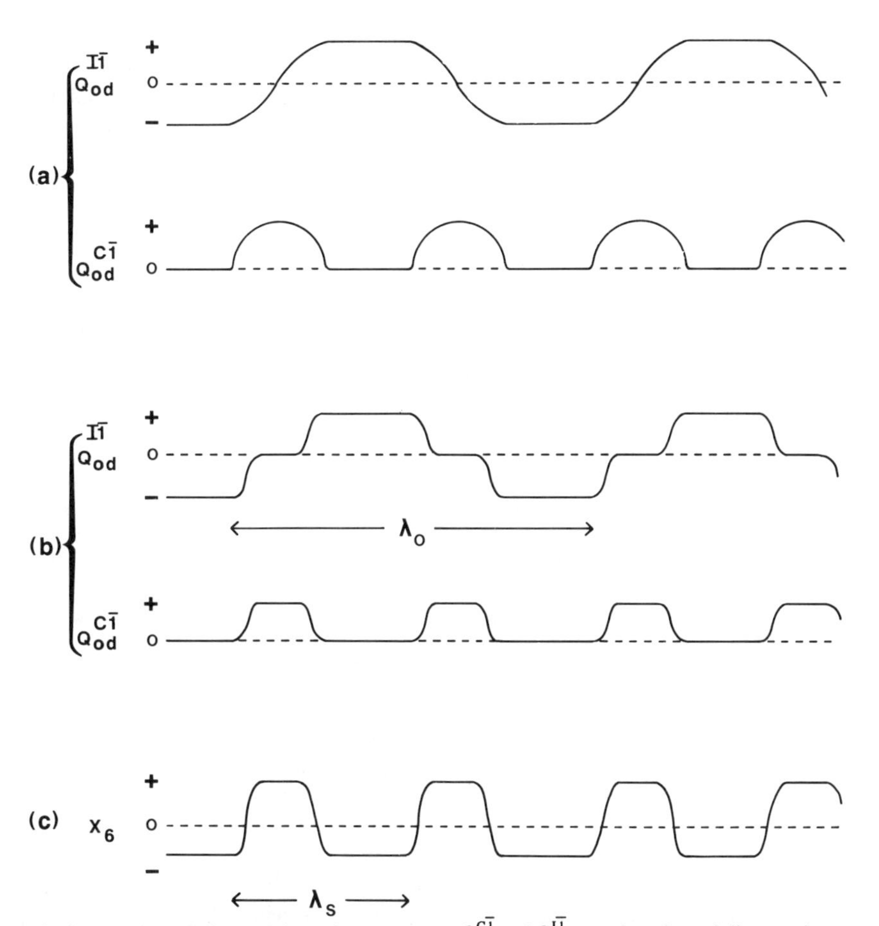

Figure 14. Schematic variations of the order parameters $Q_{od}^{C\bar{1}}$ and $Q_{od}^{I\bar{1}}$ as a function of distance in a crystal with e ordering, to illustrate possible coupling behaviour. In (a) $I\bar{1}$ APD's have $Q_{od}^{C\bar{1}} = 0$ but within the APB's there are gradients in both $Q_{od}^{C\bar{1}}$ and $Q_{od}^{I\bar{1}}$. In (b) the APB's are represented as having a core region with $Q_{od}^{I\bar{1}} = 0$ and $Q_{od}^{C\bar{1}}$ = constant. In (c) a schematic variation for the strain component x_6 accompanying the order modulations represented in (b) is shown. Since x_6 is given approximately by $\cos\gamma$ and is positive for $C\bar{1}$ ordering and negative for $I\bar{1}$ ordering, the net strain will be small, but only zero if the strain contributions from $Q_{od}^{C\bar{1}}$ and $Q_{od}^{I\bar{1}}$ exactly match. A strain modulation provides a mechanism for interactions between the two order parameters. Note that the wavelength of the strain modulation, λ_s, is exactly half that of the order modulation, λ_o.

ordered albite and fully disordered albite is ~4‰ (using data of Carpenter et al., 1985). The difference in volume strain would thus appear to be small in the pure phases, but at intermediate compositions it could, perhaps, be sufficient to cause the appearance of type f superlattice reflections in single crystal diffraction patterns. Compositional variations might occur in response to the local ordering and strain variations but, following the arguments set out here, they need not be regarded as the primary driving force for stabilising the e structure. In this context, it is worth pointing out that f reflections have not been observed in diffraction patterns from the incommensurate structure of pure anorthite, in which there can be no compositional effect.

The importance of a strain coupling mechanism may also be highlighted by the failure of K-Ba feldspars to develop an equivalent incommensurate structure. The Al:Si stoichiometry of the $KAlSi_3O_8$-$BaAl_2Si_2O_8$ solid solution matches that of the $NaAlSi_3O_8$-$CaAl_2Si_2O_8$ solid solution. However, the former is monoclinic over most of its range of composition and degree of order, and therefore has $x_6 \equiv 0$.

3.4.3. *Influences of temperature and composition.* The orientation of the incommensurate repeat will depend, at least in part, on the lattice strains accompanying $C\bar{1}$ and $I\bar{1}$ ordering. The orientation with the minimum elastic and strain energy need not be a rational crystallographic plane, because the crystals are triclinic. This preferred orientation varies systematically with the composition of the crystal (see Fig. 1 of Grove, 1977b, and Fig. 5.15 of Smith and Brown, 1988), but whether it varies with temperature is not certain. The apparent dependence of repeat distance on grade of metamorphism reported by H.-R.Wenk (1979b) is suspect because of the incorrect composition assigned to a critical sample (Smith and Wenk, 1983).

Fleet (1981, 1982, 1984) showed that observed variations can be reproduced by using the slab model as if the incommensurate structure is due to decomposition into albite-rich and anorthite-rich lamellae. Data for the elastic constants and a judicious selection of lattice parameters for the two phases are required. However, the lattice parameters are considerably more sensitive to structural state than they are to purely compositional factors (Carpenter, 1991a). The lattice angle γ, for example, varies by ~1% between albite and anorthite belonging to the "high", i.e. relatively disordered, plagioclase series but by ~4% between different ordered states. The important question is therefore, what are the values of $Q_{od}^{C\bar{1}}$ and $Q_{od}^{I\bar{1}}$ in the differently ordered local regions? The effect of $Q_{od}^{I\bar{1}}$ can be seen in the ordering behaviour of pure anorthite under metastable conditions. When the total degree of order increases, as measured by a macroscopic strain coupled to $Q_{od}^{I\bar{1}}$ (Carpenter, 1991a,b), the incommensurate repeat increases and its orientation rotates (Fig. 12f). Since both $Q_{od}^{C\bar{1}}$ and $Q_{od}^{I\bar{1}}$ vary with stoichiometry and are likely also to vary in magnitude with changing equilibration temperature, the expectation is that the orientation and spacing of the incommensurate repeat in natural crystals should vary with temperature.

A lack of clear evidence for some temperature dependence of the incommensurate ordering could be due to one of two factors. Either the variation is too small to be discerned at the level of precision of currently available data for repeat distances in natural crystals, or there is some mechanistic difficulty associated with resetting the incommensurate repeat during cooling of individual crystals. The first possibility requires the collection of better data, such as comparing suites of samples from the same locality using the same measuring technique in each case. For example, a subset of the measured spacings of the e reflections reported by Gay (1956), and Bown and Gay (1958) are for crystals with different compositions from the Skaergaard intrusion; these give a well defined trend with composition. Slimming (1976) produced a second internally consistent trend, using electron diffraction rather than X-ray diffraction, for crystals from a granulite facies metamorphic terrain. The two trends, presumably reflecting the influence of different crystallisation and equilibration temperatures, are shown in Figure 15 and are essentially the same, except in the vicinity of ~An_{55}. The difference is sufficient to warrant further careful comparisons of similar material using a single experimental method.

A second possible explanation of the lack of variation with thermal history might be that, while the equilibrium spacing and orientation should change with falling temperature, there may be kinetic constraints which prevent this from occurring during the cooling of natural crystals. Once a given incommensurate structure is established, rotation of the repeat direction requires reorientation of all the incommensurate antiphase boundaries in the crystal. The energy advantage of this will

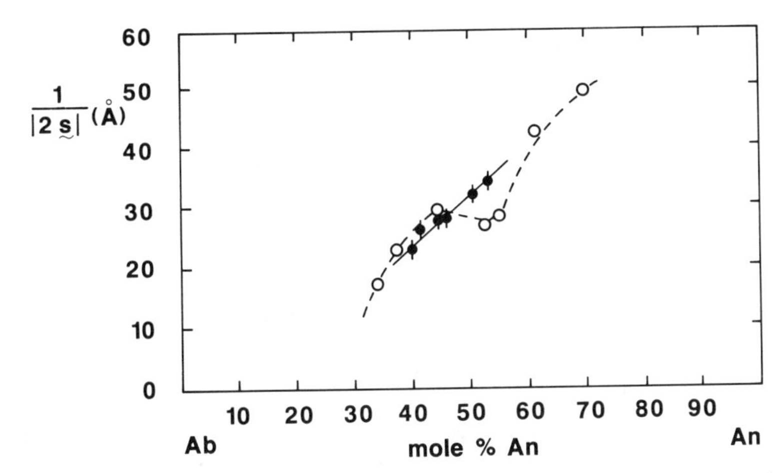

Figure 15. Variation in spacing between type e APB's (1/|2s|) as a function of composition for igneous samples from Skaergaard (filled circles, data from Gay, 1956) and for metamorphic samples from Broken Hill (open circles, data from Slimming, 1976). Note that trends for the two data sets diverge significantly in the region of An_{55}.

probably be small while the degree of structural reorganisation required would be relatively large, and such a reorientation would therefore be slow.

It is interesting to take the separate trends in Figure 15 at face value and speculate on their implications for Bøggild exsolution microstructures. A crystal with composition ~An_{45-50} would begin to exsolve at some temperature below the range of $I\bar{1} \rightleftharpoons e_1$ transition temperatures, and the e_1 and e_2 repeats might correspond to those of the granulite trend - i.e. with both phases across the Bøggild gap having the same repeat. On the other hand, crystals with composition ~An_{53-56} would order on the Skaergaard trend before entering the Bøggild miscibility gap. The exsolved e_2 phase would have a smaller incommensurate repeat. Both crystals would end up with Bøggild lamellae but the An_{45-50} crystal would show no variation of the e spacing with composition while the An_{53-56} crystal would show systematic variations with composition. This corresponds approximately with observations but obviously needs further investigation.

3.4.4. *Distinction between e_1 and e_2 structures.* There have been many discussions of structural breaks approximately at the 50:50 composition in the plagioclase solid solution (e.g. Smith and Gay, 1958; Doman et al., 1965; Bambauer et al., 1967; Nissen, 1969; Slimming, 1976; Carpenter, 1986). The evidence is typically provided by trends in lattice parameters of the type shown in Figure 13, or in trends of the spacing and orientation of e reflections. It can be rationalised in terms of the magnitudes of the local order parameters $Q_{od}^{C\bar{1}}$ and $Q_{od}^{I\bar{1}}$, and the position of the $C\bar{1} \rightleftharpoons I\bar{1}$ transition line in the phase diagram (Fig. 7).

At temperatures appropriate for e ordering there will be a significant free energy driving force for Al/Si ordering on the basis of $I\bar{1}$ symmetry at compositions more anorthite-rich than ~An_{50}. The corresponding driving force will be absent on the $C\bar{1}$ side of the $C\bar{1} \rightleftharpoons I\bar{1}$ line and the difference shows up both as a lower onset temperature for e_2 ordering than for e_1 ordering (Carpenter, 1986), and as a smaller enthalpy of ordering for e_2 relative to that of e_1 structures (Carpenter et al., 1985). With such a fundamental difference in the behaviour of one of the critical order parameters for e ordering, a marked break in the balance of energy contributions due to $Q_{od}^{C\bar{1}}$, $Q_{od}^{I\bar{1}}$, plus coupling terms, might reasonably be expected at the 50:50 composition. There might even be a difference between the Al/Si ordering schemes adopted under local $C\bar{1}$ and $I\bar{1}$ symmetry in the e_1 and e_2 structures (McConnell, personal communication).

In spite of this cumulative evidence for a difference between the two incommensurate structures, the precise relationship between them is not clear. It is commonly claimed that f reflections are only found in crystals with An contents greater than ~An_{50} (e.g. Smith and Brown, 1988), and yet such reflections have been observed in diffraction patterns from crystals within the range ~An_{44} - An_{48} from a granulite facies rock (Carpenter, 1986). Moreover, the albite-rich component of Bøggild exsolution can have type f APB's. Another difference is that diffraction patterns from e_2 crystals are generally found to have diffuse e reflections, though in patterns from crystals with a granulite facies metamorphic history the broadening may only be slight (Carpenter et al., 1985). On the other hand, there does appear to be a significant difference in the distribution of type c reflections. These c reflections have only been observed in diffraction patterns from e_1 crystals, and, although they are diffuse, their intensity varies with An content so as to extrapolate to zero at ~An_{50} (Penzkofer, 1979, in Smith and Brown, 1988). The orientation/spacing relation shown by the incommensurate structure of pure anorthite (Fig. 12f) also shows an as yet unexplained deviation at a point corresponding to the properties of natural crystals with composition ~An_{55}.

3.5. THERMODYNAMIC DATA

Calorimetric data for the enthalpies of ordering as a function of composition give independent evidence for the stability relations between different plagioclase structures. They do not by themselves allow a complete thermodynamic analyis because of the difficulty of assigning values for the entropies, but they must at least provide criteria against which any hypothetical relationships may be tested. Similarly, a proposed phase diagram must be consistent with trends in the activities of components in the solid solution as extracted from ion-exchange data between crystals and fluids. In this section available calorimetric and ion-exchange results are examined in the light of the preceding discussions of order/disorder experiments and microstructural evolution.

3.5.1. *Enthalpies of ordering and mixing.* Using high temperature solution calorimetry, Carpenter et al. (1985) measured enthalpy differences between natural ordered plagioclases and equivalent crystals which had been converted to high structural states experimentally. Broadly, the data showed that: (a) the enthalpy of ordering on the basis of the $I\bar{1}$ structure decreases markedly with increasing albite in solid solution with anorthite; (b) e_1 ordering gives a larger enthalpy of ordering than $I\bar{1}$ ordering at compositions more albite-rich than ~An_{75}; (c) e_1 ordering is associated with larger enthalpy changes than e_2 ordering; (d) the enthalpy of e_1 and e_2 ordering is comparable in magnitude or less than the enthalpy of ordering in pure albite.

A key enthalpy of solution value, that for (equilibrium) disordered anorthite with $C\bar{1}$ symmetry, cannot be obtained directly because anorthite crystals melt before they disorder. Recent calorimetric results specifically on anorthite (Carpenter, 1992b) have provided a better indication of what this value might be, and it is clear that enthalpy changes associated with the $C\bar{1} \rightleftharpoons I\bar{1}$ transition are much greater than had been previously thought. For samples equilibrated in metamorphic rocks the enthalpies of ordering, with respect to relatively disordered $C\bar{1}$ structural states, are ~ -36 $kJ.mole^{-1}$ at An_{100} ($I\bar{1}$), ~ -12 $kJ.mole^{-1}$ at ~An_{65} (e_1), ~ -6 $kJ.mole^{-1}$ at ~An_{40} (e_2) and ~ -13$kJ.mole^{-1}$ at An_0 ($C\bar{1}$, low albite).

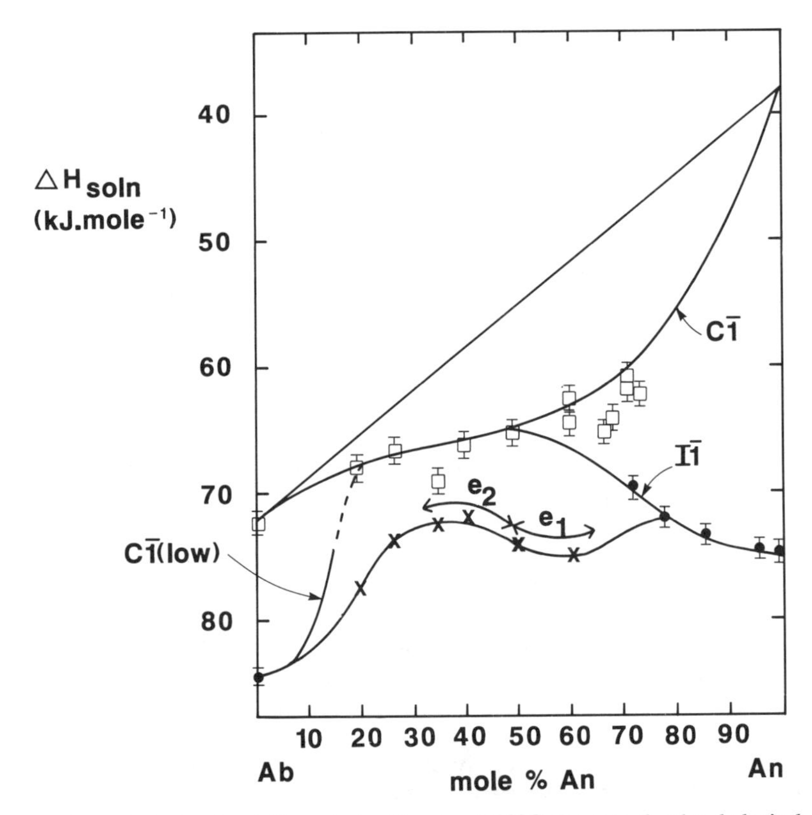

Figure 16. Heat of solution data (ΔH_{soln}, in lead borate at 700°C) for natural ordered plagioclases (filled circles) and experimentally disordered samples (open squares) from Carpenter et al. (1985). Crosses represent values for natural ordered samples plotted by adding the experimentally determined enthalpy of ordering (ΔH_{ord} of Carpenter et al., 1985) to the smoothed data for $C\bar{1}$ crystals. The curve for a $C\bar{1}$ high albite solid solution is taken from Carpenter (1992b). Curves for $I\bar{1}$, e and $C\bar{1}$ solid solutions have been added in an attempt to illustrate possible relationships between the structures. Curves for different structures which merge imply thermodynamically continuous transitions between the structures. Note that as ΔH_{soln} becomes more positive, the enthalpy of a crystal becomes more negative because feldspars dissolve endothermically in lead borate.

Variations in the measured enthalpies of solution must reflect variations in the solid solution mixing properties as well as in the ordering behaviour. Taking into account the enthalpies of ordering deduced for the $C\bar{1} \rightleftharpoons I\bar{1}$ transition by Carpenter (1992b), there appears to be a negative excess enthalpy of mixing in the high structural state $C\bar{1}$ solid solution (Fig. 16). This would imply either local ordering at intermediate compositions in the solid solution or, perhaps, some favourable interactions associated with the substitution $Na^{+} + Si^{4+} \rightleftharpoons Ca^{2+} + Al^{3+}$ due to charge balancing requirements. The ordered structures themselves form separate solid solutions which can be represented by more or less continuous curves between data points for heats of solution of samples with similar ordering schemes. Some scatter arises as a consequence of different impurity and defect contents between samples, and, in an attempt to eliminate this, the data obtained for e crystals have been plotted with reference to the smoothed curve for the $C\bar{1}$ solid solution rather than with reference to their disordered equivalents.

Natural crystals with $I\bar{1}$ symmetry at the calorimeter temperature (700°C) form a well defined trend of decreasing enthalpy of ordering with increasing albite content (Fig. 16). If the $C\bar{1} \rightleftharpoons I\bar{1}$ transition is thermodynamically continuous at intermediate compositions, the curves for $C\bar{1}$ and $I\bar{1}$ solid solutions should meet. The data are not inconsistent with such a confluence. The curve for a solid solution between natural e crystals is less well constrained. If there is a thermodynamically continuous $I\bar{1} \rightleftharpoons e_1$ transition, the $I\bar{1}$ and e_1 curves should also meet. Similarly, if the e_1 and e_2 structures are closely related in a structural sense, the e_1 and e_2 curves might be continuous. Finally, if the $C\bar{1}$ (low) $\rightleftharpoons e_2$ transition is continuous, the $C\bar{1}$ (low) and e_2 curves should also merge. Such a configuration of curves, as shown in Figure 16, is necessarily speculative but is at least not discounted by the data.

With regard to the stability relations, the enthalpy data confirm the effectiveness of incommensurate ordering in the solid solution, relative to either $I\bar{1}$ or $C\bar{1}$ ordering, as a mechanism of stabilising crystals with intermediate compositions. As would be expected, the magnitudes of the enthalpy effects correlate roughly with the values of the equilibrium order/disorder transition temperatures: ~2,000°C in pure anorthite, ~700°C in pure albite, ~800°C for e_1 crystals and < ~700°C for e_2 crystals. The enthalpy of solution values for the most ordered structures are not far from being colinear (Fig. 16), so that only a slightly larger configurational entropy associated with the incommensurate structure than with ordered albite and ordered anorthite would be needed to ensure that crystals with the e structure have a true equilibrium field of stability. The argument here is one of self-consistency with previous discussions, however, and the data do not indicate whether such a field of stability could be extensive either as a function of temperature or of composition.

3.5.2. *Energetics of antiphase boundaries.* Antiphase boundaries are in effect more or less thin zones of disordered ($Q=0$) crystal separating domains of ordered crystal ($Q\neq0$). As defects they are expected to have a positive excess energy, relative to homogeneous ordered material, associated with them. As the energy of ordering increases with falling temperature, so the magnitude of this positive excess energy should also increase. In the case of the APB's resulting from the $C\bar{1} \rightarrow I\bar{1}$ transition in pure anorthite, however, the excess enthalpy per unit area of boundary appears to decrease with falling temperature (Fig. 17; from Carpenter, 1991b). The APB's are evidently being stabilised in some way and one possibility is that ordering on the basis of the local $C\bar{1}$ symmetry occurs in much the same way as suggested earlier for the incommensurate structure.

If the trend really is for increasing stabilisation of the APB's by $C\bar{1}$ ordering as temperature reduces, the boundaries might become stable features of the crystals. This is unlikely to be the case in pure anorthite, but changing the Al:Si stiochiometry by solid solution towards albite would favour such ordering. Once they had a negative excess energy, there would be no energetic advantage associated with domain coarsening. Rather, the domain sizes would reduce to maximise the boundary area until interactions between APB's started to occur. Some degree of self-organisation would take place and preferred alignments and spacings could then be responsible for the development of an incommensurate structure. The APB's are thus likely to be integral to any understanding of the stability of intermediate structures, and the e structure might be understood purely in terms of interactions between them.

Picking up on the model illustrated in Figure 14, the Al/Si ordering variations across a type b APB might appear as shown in Figure 18. The APB would have a finite width, w, separating

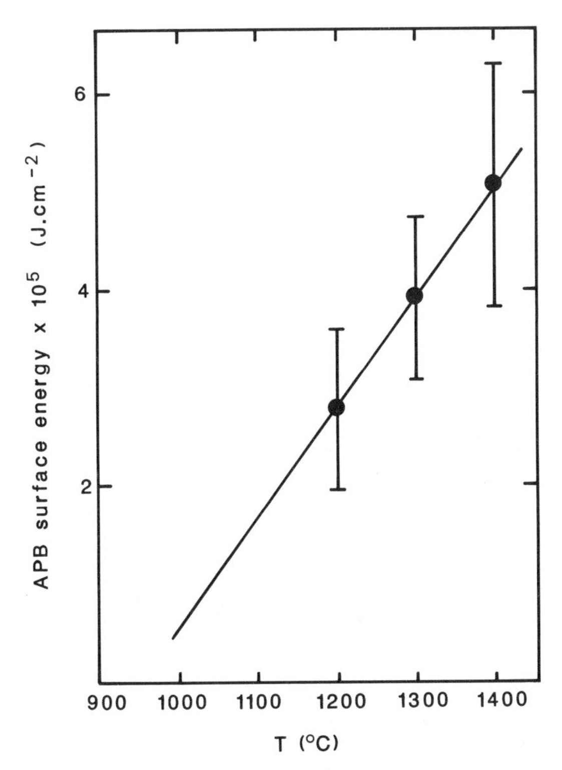

Figure 17. Excess energy of type b APB's in anorthite with respect to homogeneous ordered crystal, expressed as an enthalpy per unit area (from Carpenter, 1991b). The data were obtained by observing the variation of ΔH_{soln} with APD size. Note that the excess energy decreases with falling temperature, contrary to the normal expectation for APB energies. If the energies became negative, the APB's might become stable features in the crystal.

domains with positive and negative values of $Q_{od}^{I\bar{1}}$. Coupling of the gradient in $Q_{od}^{I\bar{1}}$ across the APB would be gradients in $Q_{od}^{C\bar{1}}$ (Fig. 18a). In the limiting case of a favourable energy of formation for such APB's, the crystal would contain so many that they would pile up together (Fig. 18b). The maximum density would correspond to an incommensurate repeat distance of ~2w. If domains with homogeneous states of $I\bar{1}$ order contributed more favourably to the overall stabilisation, the APB's would separate to some equilibrium spacing, l, determined by the relative contributions of all the energy terms (Fig. 18c). A second limiting case for the incommensurate structure is then given by the furthest distance over which APB interactions can occur. Taking the dimensions for the spacing between APB's in the incommensurate structure as being between ~25Å and ~90Å, this implies a typical APB width of $\lesssim$ 25Å, i.e. one or two unit cells, and a maximum interaction distance of ~7 unit cells. High resolution electron micrographs of the incommensurate structure of natural intermediate plagioclases show APB's with apparent widths of ~10-15Å (Nakajima et al., 1977).

According to these arguments the sequence of structural states with falling temperature in crystals of composition ~An_{65} would be, firstly, crystallisation of $C\bar{1}$ crystals from a melt. At the $C\bar{1} \rightleftharpoons I\bar{1}$ transition temperature for the crystals, $I\bar{1}$ ordering would begin. At this stage, and in

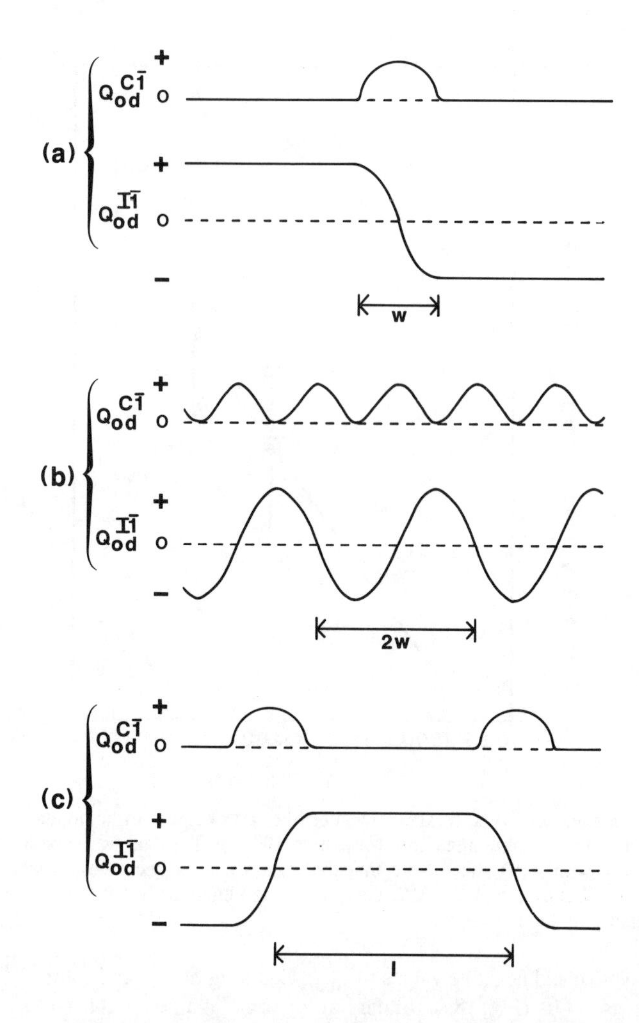

Figure 18. A possible relationship between type b APB's stabilised by local Al/Si ordering on the basis of $C\bar{1}$ symmetry and the incommensurate structure. (a) A single APB with effective width w. As in Figure 14, there is coupling between gradients in $Q_{od}^{C\bar{1}}$ and $Q_{od}^{I\bar{1}}$. (b) APB's pushed together as close as possible to make up the incommensurate structure with the smallest possible repeat distance, 2w. The closest that the APB's can get to each other is equivalent to the effective width of one APB, or half the repeat distance. (c) Separation of APB's by some distance, l. If they still interact over this length scale, the incommensurate repeat will extend to 2l.

some temperature interval below the transition point, the APB's would have either zero, or a small positive excess enthalpy. With further cooling, $C\bar{1}$ ordering would increasingly stabilise the APB's until they became stable equilibrium features of the crystal. At this point the $I\bar{1} \rightarrow e_1$ transition would occur.

3.5.3. *Ion-exchange data.* In principle, experiments involving measurements of the partitioning of Na^+ and Ca^{2+} ions between plagioclase crystals and aqueous solutions can yield quantitative

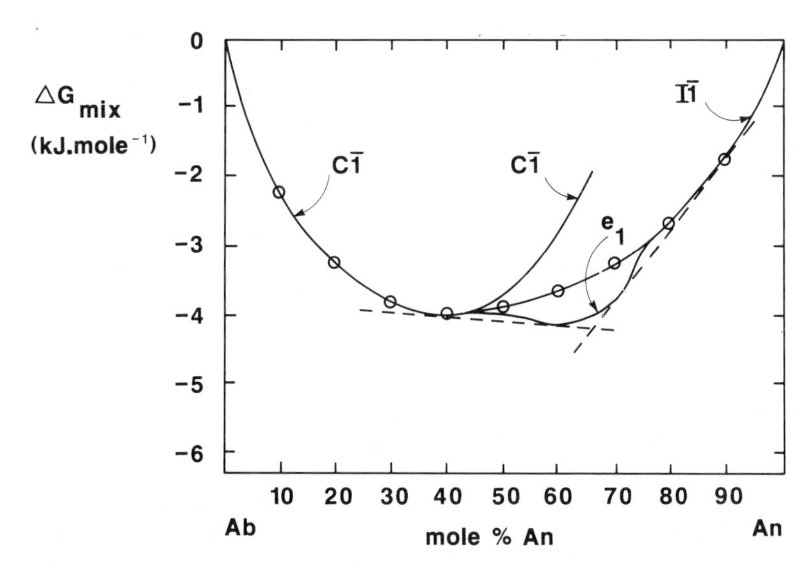

Figure 19. Possible relationship between free energies of mixing for I$\bar{1}$, e and C$\bar{1}$ solid solutions. Circles indicate free energies of mixing derived from the smoothed activity-composition relations obtained by Schliestedt and Johannes (1990) from ion-exchange experiments. If it is assumed that an equilibrium degree of I$\bar{1}$ Al/Si order developed in the plagioclase crystals during the course of the heat treatments, the curves for I$\bar{1}$ and C$\bar{1}$ solid solutions should meet at ~An_{48} (for a thermodynamically continuous transition). The extrapolation of a C$\bar{1}$ curve to more anorthite-rich compositions and the position of the e_1 curve are purely schematic. Possible two phase fields involving coexisting C$\bar{1}$ + e_1 or I$\bar{1}$ + e_1 phases are suggested by common tangents (dashed lines).

activity-composition relations for the albite and anorthite components of the plagioclase solid solution. The results of Orville (1972), Blencoe et al. (1982) and Schliestedt and Johannes (1990) are in qualitative agreement in indicating a small positive excess free energy of mixing at ~700°C. They are also consistent in providing no evidence of any miscibility gaps at this temperature. Unfortunately, in none of the experimental studies have the structural states of crystals equilibrated with fluids been fully characterised. Based on the experience of many hydrothermal experiments (Carpenter, 1986), it is most unlikely that during the run times of up to ~14 days any significant e ordering could have occurred. On the other hand, I$\bar{1}$ ordering in anorthite-rich crystals can occur relatively rapidly, and the experimental data may therefore be most informative about the C$\bar{1}$/I$\bar{1}$ relations.

The free energy of mixing in the plagioclase solid solution, as derived from the smoothed activity-composition relations given by Schliestedt and Johannes (1990), is shown in Figure 19. At this temperature, the C$\bar{1}$ $\rightleftharpoons$ I$\bar{1}$ transition should occur at ~An_{48}, and, taken at face value, the mixing curve is entirely consistent with the transition being thermodynamically continuous. Separate curves for ordered and disordered solid solutions are shown as merging at An_{48} in Figure 19 to signify these relations. The assumption is that equilibrium was achieved with respect to C$\bar{1}$ and I$\bar{1}$ order during the experiments but that no e ordering occurred. If the free energy due to e_1 ordering is added schematically in the composition range ~An_{50} - ~An_{75}, then e_1 + C$\bar{1}$ and e_1 + I$\bar{1}$ two phase fields can be generated (Fig.19). Once again, however, this is a matter for speculation and only provides arguments of self-consistency in support of the phase relations suggested in Figure 6. Some single crystal diffraction observations of the experimental products would be invaluable in assessing the extent to which equilibrium order was attained. Kotel'nikov et al. (1981) reported evidence for a two phase field between ~An_{67} and ~An_{92} in their ion-exchange experiments at 700°C, but, again, without more thorough product characterisation it is not possible to ascertain why they should have obtained such a different result from the other comparable studies.

Implications of these ion-exchange data for thermodynamic mixing models are beyond the scope of the present review. They are discussed at some length by Carpenter and Ferry (1984), Ashworth and Evirgen (1985b), Schliestedt and Johannes (1990), and references therein.

3.6. KINETIC CONSTRAINTS

Most of the uncertainties concerning the plagioclase subsolidus phase relations would probably be resolved if rates of ordering and exsolution were known for geological conditions of pressure, temperature and fluid composition. These would allow tighter constraints to be placed both on the temperature interval over which exsolution can occur and on the cut-off temperatures below which no further equilibration takes place with respect to cation order. Existing experimental data provide two limiting cases, for dry conditions and for high pressures with a hydrous fluid present. The kinetic constraints may also be compared with apparent equilibration temperatures deduced from equilibrium variations in the degree of order with temperature.

In the absence of water, rates of Al/Si ordering become prohibitively slow below ~800°C even on a geological time scale. For example, extrapolating the ordering rate law of Carpenter (1991b) for anorthite implies time scales to achieve equilibrium order under isothermal conditions of ~7,000 years at 800°C, ~10^6 years at 700°C and ~7×10^8 years at 600°C. Similar extrapolation of a dry ordering rate law for albite suggests that the degrees of order observed in natural samples could not be achieved in a geological time scale (Wruck et al., 1991). Increasing pressure appears to enhance the rate of ordering in albite, though at pressures of $\lesssim$ 8kb the effect of adding water is more dramatic (see reviews by Yund, 1983, Smith and Brown, 1988, Goldsmith, 1991). At low pressures, substantial changes in the degree of order of albite can be achieved hydrothermally on a time scale of months down to temperatures of ~600°C (MacKenzie, 1957; McConnell and McKie, 1960). In anorthite, ordering under similar experimental conditions occurs down to at least temperatures of ~800°C (Carpenter, 1992b), and, for experiments on intermediate plagioclase crystals, the comparable cut off temperature is ~750°C (Carpenter, 1986). It follows that the effective cut-off temperatures for ordering in nature will be $\lesssim$ 600-700°C if a hydrous fluid is present, and, hence, that there will be close adherence to equilibrium ordering pathways down to these temperatures. The apparent equilibration temperatures of natural samples suggest even lower kinetic cut-off points. For example, natural microcline must be able to order in its triclinic stability field below ~480°C (Kroll et al., 1991), and natural albites have a degree of order corresponding to a temperature of $\lesssim$ 600°C at low pressures (Goldsmith and Jenkins, 1985). For natural bytownites and anorthites with $I\bar{1}$ Al/Si order, the apparent equilibration temperatures are in the range ~500-800°C, though these values depend on the thermodynamic model used to describe equilibrium ordering (Carpenter, 1992b).

As with ordering rates, the kinetics of interdiffusion in the plagioclase system under dry conditions are too slow to allow exsolution on a realistic geological timescale. Using the standard approximation, $x^2 \approx Dt$, where x is interdiffusion distance, D is the interdiffusion coefficient and t is time, the data of Grove et al. (1984) imply a time scale of ~4×10^6 years at 800°C or ~2×10^9

years at 700°C for Huttenlocher exsolution on a wavelength of ~2,000Å. (The interdiffusion distance has been taken as half the repeat distance of the lamellar texture). At 15kb with a hydrous fluid present, a similar calculation using the data of Liu and Yund (1992) implies time scales of ~30 years, ~2,000 years, ~5 x 10^5 years and ~4 x 10^8 years for Huttenlocher exsolution on a scale of 2,000Å at constant temperatures of 800, 700, 600 and 500°C, respectively. The rate of Na + Si $\rightleftharpoons$ Ca + Al interdiffusion may be less at lower pressures but, again, the implication is that natural plagioclase crystals should carry information relating to the miscibility gaps down to temperatures at least as low as ~600°C. For peristerite exsolution, the extrapolated experimental data of Liu and Yund (1992) may actually underestimate effective interdiffusion rates, since Nord et al. (1978) observed spinodal type exsolution textures with a wavelength of ~150Å in crystals that had experienced a peak metamorphic temperature estimated as ~350°C. The relation $x^2 \approx Dt$ and Liu and Yund's data for homogenisation of natural peristerite textures (P=15kb, H_2O added) gives $t \approx 2 \times 10^9$ and $t \approx 2 \times 10^7$ years for a diffusion distance of 75Å at 350 and 400°C respectively. A time scale of 10^5-10^6 years at 350°C would be more realistic.

In conclusion, the available experimental data for ordering and diffusion rates, combined with observations of the structural states of natural samples, suggest kinetic cut-off temperatures of less than 600°C, and perhaps as low as ~350-400°C, for slowly cooled igneous and metamorphic rocks when a hydrous fluid phase is present. For this reason, the phase relations postulated in Figure 6 have been extended down to ~400°C without the opening of a low albite + anorthite field. If the extrapolated data of Liu and Yund (1992) are approximately correct for determining exsolution rates, it is possible that the Bøggild and Huttenlocher miscibility gaps should peak at slightly lower temperatures than shown in Figure 6.

4. The rudiments of a thermodynamic model

A quantitative thermodynamic model to predict the equilibrium phase relations must define the free energy relationships between all the possible ordered and disordered structural states. Given the close interdependence of ordering and exsolution in the plagioclase system, it is likely that a description of the ordering behaviour will clarify the factors generating the two phase fields. Similarly, the mechanisms by which exsolution can occur in these circumstances are sensitive to the thermodynamic character of the ordering transitions (see, for example, Carpenter, 1981, 1985; Grove et al., 1983). In this section, schematic free energy-composition curves are presented to address these issues. They do not represent a unique set of possible relations but are designed to be consistent with the observational evidence set out above and with the phase diagram proposed in Figure 6. They should provide the basis for a more formal thermodynamic approach, and, in this context, recent developments in the application of Landau theory to minerals show increasing promise for generating equations that are both valid and tractable (see also Salje, 1994). The general forms of some of these equations, as might be applied to plagioclases, are therefore also presented.

4.1. SCHEMATIC G-X CURVES

On the basis of all the available evidence, the C2/m $\rightleftharpoons$ C$\bar{1}$ transition in albite (Kroll et al., 1980; Salje et al., 1985; Carpenter, 1988) and the C$\bar{1}$ $\rightleftharpoons$ I$\bar{1}$ transition in plagioclase crystals with intermediate compositions have been interpreted as being thermodynamically continuous. The free energy curves for C2/m, C$\bar{1}$, and I$\bar{1}$ solid solutions at a temperature of ~1050°C will therefore have

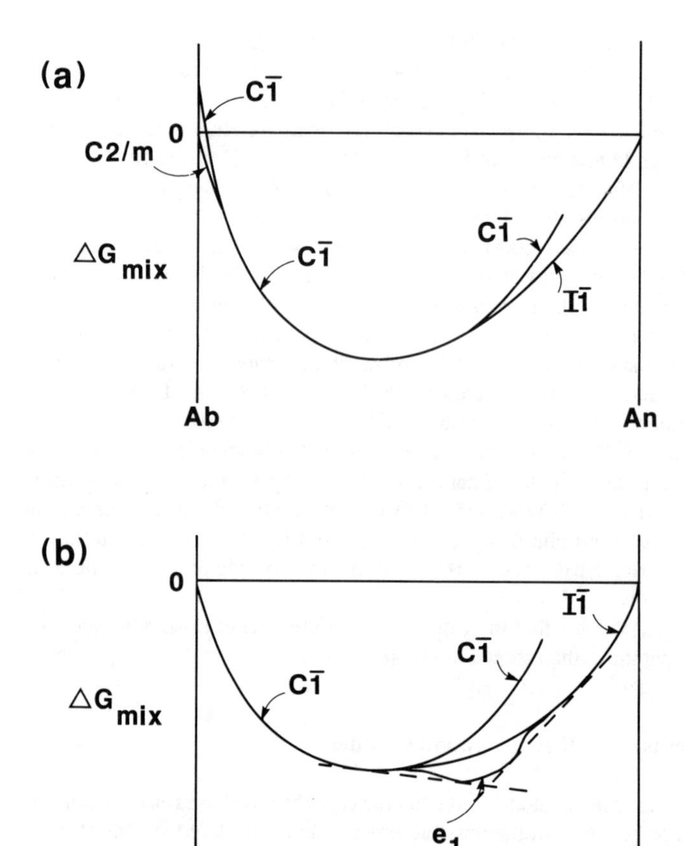

Figure 20. Schematic free energy - composition curves showing possible relationships between different structural states that are consistent with the phase diagram proposed in Figure 6. The structural state of each solid solution is indicated by I$\bar{1}$, e, C$\bar{1}$, etc. Miscibility gaps are indicated by broken lines (common tangents). (a) T ≈ 1050°C. The C2/m ⇌ C$\bar{1}$ and C$\bar{1}$ ⇌ I$\bar{1}$ transitions are shown as being thermodynamically continuous. (b) T ≈ 750°C.

a relationship similar to that shown in Figure 20a. Thermodynamically continuous behaviour as a function of composition, i.e. merging of the free energy curves at the composition of crystals with T_c = 1050°C, follows from thermodynamically continuous behaviour with respect to temperature.

At a temperature of ~700°C, e_1 ordering is interpreted as being stable with respect to I$\bar{1}$ order over a limited composition range, and the Bøggild and Huttenlocher miscibility gaps are believed to then become effective (Fig. 6). If the exsolution mechanisms within these two phase fields are correctly interpreted as involving spinodal decomposition between end members with different states of order, the e_1, C$\bar{1}$ and I$\bar{1}$ free energy curves must also merge in a continuous manner. A possible topology is shown in Figure 20b. Subsumed in this diagram is the implication that the C$\bar{1}$ ⇌ e_1 and I$\bar{1}$ ⇌ e_1 transitions are also thermodynamically continuous. The Bøggild gap then contains a conditional spinodal between C$\bar{1}$ and e_1 end members, and the Huttenlocher gap a conditional spinodal between e_1 and I$\bar{1}$ end members. Figure 20c includes an alternative

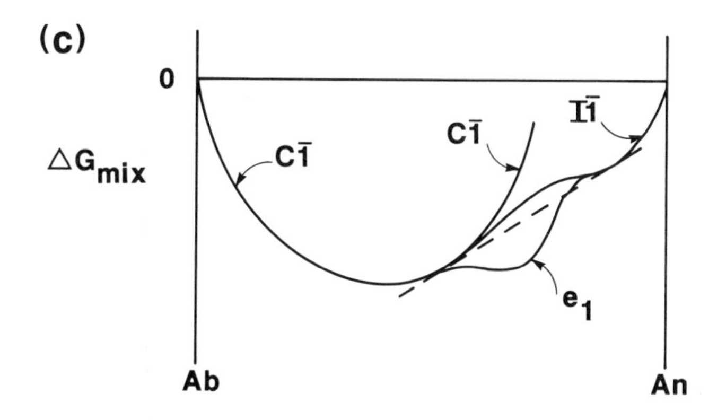

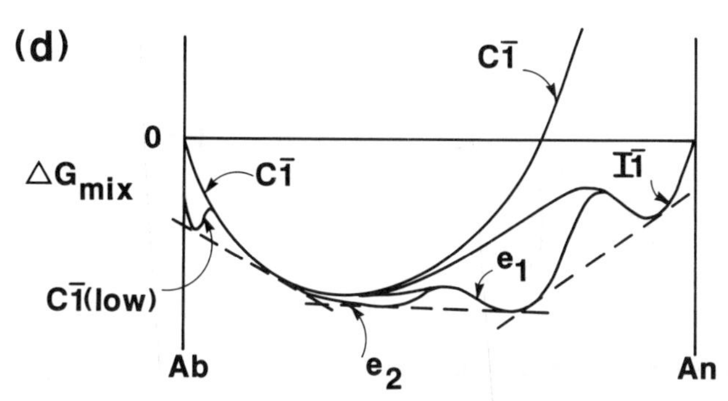

Figure 20 (cont.). (c) T ≈ 750°C. The equilibrium states in this case are the same as in (b) but a metastable miscibility gap with $C\bar{1}$ and $I\bar{1}$ end members is also generated by the topology of the $C\bar{1}$ and $I\bar{1}$ curves. (d) T ≈ 550°C. Three miscibility gaps are now generated, with e_1 and e_2 structures shown as having equilibrium stability ranges. Note that the relationship shown between e_1 and e_2 structures is necessarily speculative.

representation of the $I\bar{1}$ solid solution which is shown as giving rise to a miscibility gap between $C\bar{1}$ and $I\bar{1}$ end members that is metastable with respect both to e_1 ordering and to the two smaller miscibility gaps. This is a distinct possibility, though perhaps occurring at a lower temperature, in view of the observation of coexisting andesine and bytownite crystals in some metamorphic rocks.

Somewhat arbitrarily, the stability field for e_2 ordering has been interpreted in Figure 6 as peaking at ~600-650°C. At ~550°C it is necessary to include the free energy curve for an e_2 solid solution in the schematic G-X relations; in Figure 20d this is shown as merging continuously both with the $C\bar{1}$ and e_1 solid solutions. While it might be reasonable to infer that the $C\bar{1} \rightleftharpoons e_2$ transition is thermodynamically continuous, the relationship between e_1 and e_2 structures is much more problematic. The e_1 and e_2 structures have been represented here as being distinct, to the extent that there should be a discrete $e_1 \rightleftharpoons e_2$ transition at a temperature which varies with composition but which falls within the Bøggild miscibility gap. This topology generates a miscibility gap, together with a conditional spinodal for a thermodynamically continuous transition, between the e_1 and e_2 stability fields. On the other hand, if the e_1 and e_2 structures are essentially the same, the cause of the miscibility gap is less clear. The proposition of McConnell (1974a), that the Bøggild

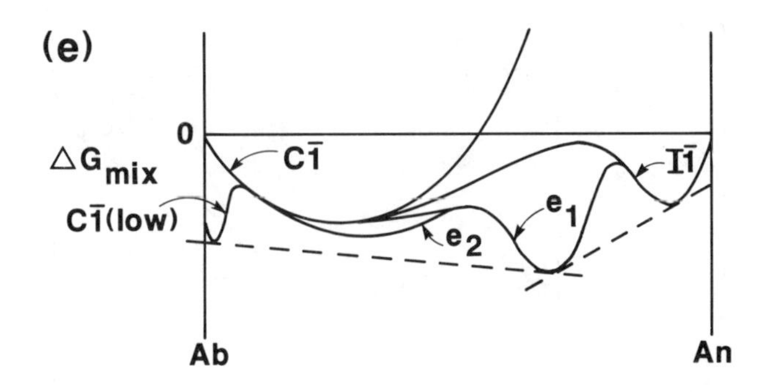

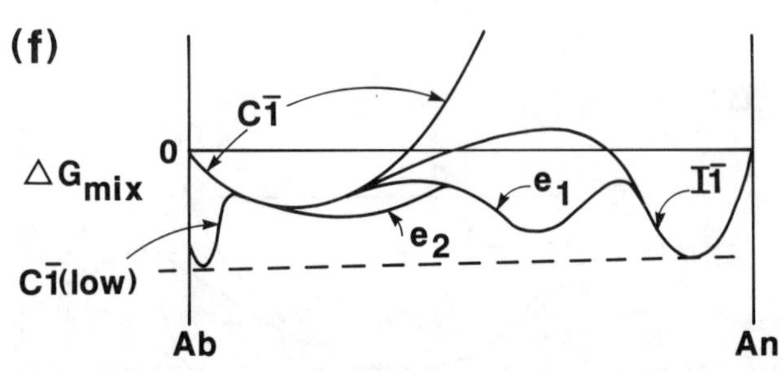

Figure 20 (cont.). (e) T < 400°C. The e_2 field is now replaced by a two phase assemblage of $C\bar{1}$ (low) + e_1. (f) T< 400°C. As an alternative to (e) the assemblage low albite + $I\bar{1}$ anorthite is shown as being the stable assemblage at all intermediate compositions.

gap is a two phase field which closes at low temperature and arises as a consequence only of a difference in the onset temperatures for e_1 and e_2 ordering, remains an alternative possibility. In this case the e_1 and e_2 solid solution curves would still meet continuously below the peak $C\bar{1} \rightleftharpoons e_2$ transition temperature but would not contain a segment with positive $\partial^2G/\partial X^2$.

The topology of the peristerite gap has been discussed in detail elsewhere (see above and review by Smith and Brown, 1988). It is known that increases in the equilibrium degree of Al/Si order in pure albite below ~650-700°C are not due to a discrete phase transition. Rather, they are a reflection of a crossover between the relative influence of largely displacive and largely order/disorder order parameters in determining the thermodynamic evolution. Thus the low albite ($C\bar{1}$, low) and high albite ($C\bar{1}$) curves are necessarily continuous. As shown in Figure 20d, the miscibility gap arises because the low albite ordering scheme is energetically favourable only for a limited compositional range close to the ideal stoichiometry of Al:Si = 1:3.

At progressively lower temperatures, possible forms of the equilibrium phase diagram and related G-X curves become progressively more uncertain. The e_2 structure might become metastable with respect to an assemblage of low albite plus e_1 crystals (Fig. 20e) or crystals with any intermediate composition might become metastable with respect to low albite plus $I\bar{1}$ anorthite (Fig. 20f). Even at ~400°C, the kinetic data for interdiffusion do not rule out unequivocally the possibility of continued development of exsolution microstructures, however. Lamellar exsolution textures with these wider composition limits, as opposed to the apparent breakdown into discrete phases, have not yet been reported.

4.2. LANDAU FREE ENERGY EXPANSIONS

A convenient formal approach for producing quantitative, as opposed to schematic, free energy relations is provided by Landau theory. In pure anorthite, Al/Si ordering variations can be related to the $C\bar{1} \rightleftharpoons I\bar{1}$ transition which would occur at ~2,000°C if melting were suppressed. Only one order parameter, $Q_{od}^{I\bar{1}}$, is required and the excess free energy due to the ordering can be given as:

$$G = \frac{1}{2}a(T-T_c)\left(Q_{od}^{I\bar{1}}\right)^2 + \frac{1}{4}b\left(Q_{od}^{I\bar{1}}\right)^4 + \frac{1}{6}c\left(Q_{od}^{I\bar{1}}\right)^6 \tag{1}$$

where a, b and c are typical Landau coefficients and T_c is the equilibrium temperature for a second order or tricritical transition. The theoretical possiblity of coupling between $Q_{od}^{I\bar{1}}$ and the order parameter, Q, for the largely displacive $C2/m \rightleftharpoons C\bar{1}$ transition appears not to occur to a significant extent in practice (Carpenter, 1992b). Enthalpy and entropy changes can be described by standard manipulations of the free energy expansion (e.g. see Carpenter, 1988, Salje, 1990).

Equation (1) can be extended to describe excess energies accompanying $I\bar{1}$ ordering in the plagioclase solid solution by allowing T_c and the coefficients to vary with albite content. T_c is known to vary approximately linearly with composition (Fig. 7b) and, as a first approximation, linear variations of a, b and c may be assumed. Values for all these coefficients have been determined from calorimetric data for the variation of excess enthalpy with $Q_{od}^{I\bar{1}}$ and from structural data for the variation of $Q_{od}^{I\bar{1}}$ as a function of temperature and composition (Carpenter, 1992b). It appears that the $C\bar{1} \rightleftharpoons I\bar{1}$ transition is first order in pure anorthite but becomes second order with increasing albite content. A tricritical point, in the sense of a Landau tricritical transition with b = 0, is calculated to occur at ~An_{78}, 1455°C.

To describe the thermodynamic evolution of albite it is necessary to account for the variation of two order parameters, Q and $Q_{od}^{C\bar{1}}$, for the largely displacive and largely order/disorder $C2/m \rightleftharpoons C\bar{1}$ transitions respectively. The appropriate Landau expansion has the form (from Salje et al., 1985):

$$G = \frac{1}{2}a(T-T_c)Q^2 + \frac{1}{4}bQ^4 + \frac{1}{6}cQ^6 + \frac{1}{2}a_{od}(T-T_{cod})\left(Q_{od}^{C\bar{1}}\right)^2 + \frac{1}{4}b_{od}\left(Q_{od}^{C\bar{1}}\right)^4 + \frac{1}{6}c_{od}\left(Q_{od}^{C\bar{1}}\right)^6 + \lambda Q Q_{od}^{C\bar{1}} \tag{2}$$

T_c and the coefficients a, b and c relate to Q, while T_{cod} and the coefficients a_{od}, b_{od}, c_{od} relate to $Q_{od}^{C\bar{1}}$. The coupling term with coefficient λ accounts for interactions between the two order parameters and, following symmetry arguments, is linear in Q and $Q_{od}^{C\bar{1}}$ (see Salje, 1985, 1990, or Carpenter, 1988, 1992a). Salje et al. (1985) used structural data for the behaviour of Q as a function both of temperature and $Q_{od}^{C\bar{1}}$, together with calorimetric data for the $C2/m \rightleftharpoons C\bar{1}$ displacive transition, to extract values for the coefficients.

In principle, Equation 2 can also be used to describe the thermodynamic consequences of Al/Si ordering in the $C\bar{1}$ solid solution. At present appropriate experimental data are not available, but all that is required is to determine the composition dependence of the coefficients. Again, linear variations with substitution of anorthite component may provide an adequate first approximation.

For intermediate compositions in the solid solution, it is necessary to produce a description of the e ordering. As has already been discussed, incommensurate structures can be understood in terms of interactions between two order parameters that vary with distance in the crystal. The simplest form of Landau expansion that may be applicable is (e.g. see Carpenter, 1992a, McConnell, 1992, and references therein):

$$G = \frac{1}{2}a(T-T_c)(Q_1^2 + Q_2^2) + \frac{1}{4}b(Q_1^4 + Q_2^4) + \ldots \\ + d[Q_1(\nabla Q_2) - Q_2(\nabla Q_1)] + e[(\nabla Q_1)^2 + (\nabla Q_2)^2] \quad (3)$$

where Q_1 and Q_2 might be $Q_{od}^{C\bar{1}}$ and $Q_{od}^{I\bar{1}}$ in the present case. The gradient coupling terms, with coefficient d, describe interactions between Q_1 and gradients in Q_2, ∇Q_2, and between Q_2 and gradients in Q_1, ∇Q_1. Energies associated with the gradients alone are given by $(\nabla Q_1)^2$ and $(\nabla Q_2)^2$, with the coefficient e. Simple solutions to Equation 3 have Q_1 and Q_2 varying sinusoidally through the structure. This represents a more sophisticated and realistic approach than the simple minded calculations adopted, for example, by Carpenter (1992b). However, as yet no attempts have been made to find values for the coefficients that might generate thermodynamic properties for the e plagioclase structures. An advantage of the approach would be that a detailed understanding of the nature of the structure, or of the precise configuration of Al and Si within it, is not required.

The three Landau expansions set out here give excess energies with respect to some disordered state. In the case of Equation 1 this reference state has C2/m symmetry, while Equations 2 and 3 can be used with respect to disordered crystals that have $C\bar{1}$ symmetry. To complete the thermodynamic analysis some mixing properties for the disordered reference solid solution must also be included. For example, from the analysis of $C\bar{1} \rightleftharpoons I\bar{1}$ ordering, it appears that the disordered $C\bar{1}$ solid solution has a negative enthalpy of mixing (Fig. 16). The addition of this mixing energy, perhaps using a standard regular solution formalism, would generate a complete description of the mixing and ordering behaviour.

5. Conclusion

Some of the discussion and interpretation presented in this review goes beyond what can be fully justified on the basis of evidence from experiments or from observations on natural samples. The intention has been as much to direct questions to critical but unresolved issues as to simply set out what is already known of the plagioclase stability relations. Some important conclusions may be summarised in terms of specific questions that can be addressed using currently available methodological or theoretical expertise.

5.1. Al/Si ORDERING AT An_0 - An_{15}

Thermodynamic changes accompanying the $C2/m \rightleftharpoons C\bar{1}$ transition and Al/Si ordering in albite have been quantified (Salje et al., 1985), and the extent of ordering confirmed experimentally down to ~650°C at high pressures (Goldsmith and Jenkins, 1985). It is postulated that a rather similar pattern of thermal evolution should be followed in crystals with compositions in the range An_0 - ~An_{15}, though with T_c for the displacive transition increasing with increasing An content, and the crossover temperature between Q and $Q_{od}^{C\bar{1}}$ dominated behaviour decreasing. An equivalent approach to that of Salje et al., or Goldsmith and Jenkins on crystals of ~An_5 and ~An_{10} should establish the compositional limits of low albite ordering and clarify the relationship between this ordering and the peristerite gap.

5.2. LOCAL ORDERING IN THE $C\bar{1}$ SOLID SOLUTION

Short range ordering in high albite structures has important consequences for the thermodynamic mixing properties of the complete plagioclase system, since the $C\bar{1}$ solid solution may be used as a reference state in calculations of the long range ordering properties. The nature of this short range order should be identifiable using spectroscopic techniques (vibrational or NMR), or by systematic observations of the diffuse scattering in single crystal diffraction patterns. If the local ordering is related to e_2 ordering, this might have implications for the otherwise inaccessible $C\bar{1} \rightleftharpoons e_2$ transition in crystals with compositions of ~An_{25} - ~An_{50}.

5.3. MICROSTRUCTURAL EVIDENCE FOR SPLIT BEHAVIOUR AT ~An_{52}

Although it appears that many investigations of Bøggild exsolution microstructures have been undertaken, some of the observations seem to be mutually inconsistent. A difference has been proposed here for the structural evolution of crystals with compositions that fall on either side of the peak of the Bøggild miscibility gap. Careful observations of microstructures in crystals with a range of compositions across the Bøggild gap, but with the same thermal history, would confirm or refute this hypothesis. In addition, observations of the orientation and spacing of type e APB's with composition should prove whether the proposed variation with temperature is correct.

5.4. $C\bar{1} \rightleftharpoons e_2$ TRANSITION TEMPERATURES

An important question raised in relation to the Bøggild gap is whether it arises as a consequence of the difference between e_1 and e_2 ordering or as a consequence of the $C\bar{1} \rightleftharpoons e_1$ transition near An_{50} at ~800°C. Experimental data for the ordering sequence $C\bar{1} \rightarrow e_2$ are not available and yet might be obtained for crystals in the composition range ~An_{40}-An_{50} using hydrothermal conditions and high pressures. A possible topology for the intersecting transition lines in the centre part of the phase diagram has been proposed and could be tested given the enhanced ordering kinetics achievable under these experimental conditions.

5.5. THE $I\bar{1} \rightleftharpoons e_1$ ORDERING TRANSITION

The nature of the $I\bar{1} \rightleftharpoons e_1$ transition has implications both for quantitative models of ordering and for the possible mechanisms of exsolution when one end-member has the e_1 structure. What has been proposed is that, in the composition range ~An_{52} to ~An_{72}, APB's due to the $C\bar{1} \rightleftharpoons I\bar{1}$

transition become progressively stabilised by local $C\bar{1}$ ordering until, below the $I\bar{1} \rightleftharpoons e_1$ transition temperature, they become stable features of the crystal. At this temperature they can order and adopt a preferred orientation and spacing. This view of the transition is permissive of second order character, whereas, in the discussion of Carpenter (1986), first order behaviour was proposed. In crystals from volcanic rocks, the $I\bar{1}$ domains may be preserved (e.g. Lake County labradoritc, in Wenk et al., 1980) whereas periodic type e APB's are found in slowly cooled igneous rocks. Such a sequence is contrary to the interpretation of type b domains developing from type e domains by APB recombination, as described by H.-R.Wenk (1979a, b), Wenk et al. (1980), Wenk and Nakajima (1980). Microstructural evidence for the nature of the $I\bar{1} \rightleftharpoons e_1$ transition may be preserved in crystals with intermediate cooling rates that do not appear to have been examined.

5.6. THERMODYNAMICS OF INCOMMENSURATE ORDERING

A viable approach to the calculation of thermodynamic properties for the the complete plagioclase solid solution is to first treat each ordered structure separately. For a quantitative analysis of the e structures, two formal theoretical approaches are available, at least in outline, from the physics literature. A Landau expansion of the form of Equation 3 forms the basis of one possible treatment. Alternatively, the incommensurate phases might be described in terms of interacting APB's. Individual APB's are, in effect, solitons, and their ordered arrangement would be referred to as a soliton lattice (Salje, 1990). Thermodynamic properties would be determined by reference to the energies of the APB's and their interactions. Some of the essential theory is discussed by Salje (1990), and, if a suitable expression for the e plagioclase structure could be generated, experimental data for calibrating the coefficients might be collected. The advantage of this approach is that, by adopting a mesoscopic or macroscopic point of view, the difficult problem of assigning a configurational entropy to the structure on the basis of cation distributions would be avoided. However, some work remains to be done in developing equations with practical applications.

5.7. INCOMMENSURATE STRUCTURES AND DIFFERENCES BETWEEN e_1 AND e_2 ORDERING

Most recent structural studies of the incommensurate plagioclase structure have attempted to produce atomic configurations that can be tested against intensity distributions in single crystal diffraction patterns. Some also include composition as a variable within the periodic structure. It is clear from the metastable behaviour of anorthite crystals produced from glass that type e structures can arise as a consequence of Al/Si ordering alone. This has implications for how the conventional structural studies might be developed. Adopting the coupled order parameter model leads immediately to suggestions as to how density and order modulations are combined without any need for variations in composition. Small variations in strain, proposed to account in large part for the order parameter coupling, are anticipated and these should be evident in the precise shape of type e and f reflections without necessarily requiring a full three dimensional structure refinement. The strain variations in e_1 and e_2 structures may also be quite different if the relative weightings of $Q_{od}^{C\bar{1}}$ and $Q_{od}^{I\bar{1}}$ differ significantly between the two structures.

5.8. PHASE EQUILIBRIUM AND ION-EXCHANGE EXPERIMENTS

Techniques for extracting values of activities for different components in a solid solution from phase equilibrium or ion-exchange experiments are well established in the mineralogical literature (e.g. Windom and Boettcher, 1976, and Orville, 1972, for the plagioclase solid solution). Attempts to extract reliable thermodynamic data relating to the ordering behaviour from such experiments have been frustrated by the lack of adequate sample characterisation, however, (e.g. Carpenter and Ferry, 1984). Single crystal diffraction patterns obtained by transmission electron microscopy and standard lattice parameter determinations would greatly enhance the value of these experiments, even if stable equilibrium states are not obtained. At least the structure type involved would have been identified. It is unlikely that a complete analysis of the plagioclase system can be achieved without more experimental data of this type at a range of temperatures and for a range of structural states. Because they generate numerical activities and free energy contributions, they can provide an independent test of models describing the order/disorder behaviour at different discrete compositions.

Perhaps by the time of the next NATO feldspar conference, these questions will have been resolved, and a quantitative description of the plagioclase phase relations achieved.

Acknowledgements

Financial support from the Nuffield Foundation and the Natural Environment Research Council of Great Britain over a number of years is gratefully acknowledged. G.L.Nord, Jr., J.D.C.McConnell, P.H.Ribbe and J.V.Smith are thanked for their permission to reproduce published diagrams. H.Hashimoto, G.L.Nord, Jr., and H.-R.Wenk are thanked for generously providing electron micrographs. J.D.C.McConnell and J.V.Smith kindly reviewed the manuscript.

References

Ashworth JR, Evirgen MM (1985a) Plagioclase relations in pelites, central Menderes Massif, Turkey. I. The peristerite gap with coexisting kyanite. J Metamorphic Geol 3 : 207-218

Ashworth JR, Evirgen MM (1985b) Plagioclase relations in pelites, central Menderes Massif, Turkey. II. Perturbation of garnet-plagioclase geobarometers. J Metamorphic Geol 3 : 219-229

Bambauer HU, Eberhard E, Viswanathan K (1967) The lattice constants and related parameters of "plagioclases (low)". Schweiz Mineral Petrogr Mitt 47 : 351-364

Blencoe JG, Merkel GA, Seil MK (1982) Thermodynamics of crystal-fluid equilibria, with applications to the system $NaAlSi_3O_8$-$CaAl_2Si_2O_8$-SiO_2-NaCl-$CaCl_2$-H_2O. In: Saxena SK (ed) Advances in Physical Geochemistry, vol 2. Springer, Berlin Heidelberg New York, pp 191-222

Bown MG, Gay P (1958) The reciprocal lattice geometry of the plagioclase feldspar structures. Z Kristallogr 111 : 1-14

Brown WL (1984) Feldspars and Feldspathoids (NATO ASI Series C 137). Reidel, Dordrecht, pp 541

Bruce AD, Cowley RA (1981) Structural Phase Transitions. Taylor and Francis, London, pp 326

Carpenter MA (1981) A "conditional spinodal" within the peristerite miscibility gap of plagioclase feldspars. Am Mineral 66 : 553-560

Carpenter MA (1985) Order-disorder transformations in mineral solid solutions. In: Kieffer SW, Navrotsky A (eds) Microscopic to macroscopic: atomic environments to mineral thermodynamics (Reviews in Mineralogy vol 14). Mineralogical Society of America, Washington DC, pp 187-223

Carpenter MA (1986) Experimental delineation of the "e" $\rightleftharpoons$ $I\bar{1}$ and "e" $\rightleftharpoons$ $C\bar{1}$ transformations in intermediate plagioclase feldspars. Phys Chem Minerals 13 : 119-139

Carpenter MA (1988) Thermochemistry of aluminium/silicon ordering in feldspar minerals. In: Salje EKH (ed) Physical properties and thermodynamic behaviour of minerals (NATO ASI Series C 225). Reidel, Dordrecht, pp 265-323

Carpenter MA (1991a) Mechanisms and kinetics of Al-Si ordering in anorthite: I. Incommensurate structure and domain coarsening. Am Mineral 76 : 1110-1119

Carpenter MA (1991b) Mechanisms and kinetics of Al-Si ordering in anorthite: II. Energetics and a Ginzburg-Landau rate law. Am Mineral 76 : 1120-1133

Carpenter MA (1992a) Thermodynamics of phase transitions in minerals: a macroscopic approach. In: Price GD, Ross NL (eds) The Stability of Minerals. Chapman and Hall, London, pp 172-215

Carpenter MA (1992b) Equilibrium thermodynamics of Al/Si ordering in anorthite. Phys Chem Minerals 19 : 1-24

Carpenter MA, Ferry JM (1984) Constraints on the thermodynamic mixing properties of plagioclase feldspars. Contrib Mineral Petrol 87 : 138-148

Carpenter MA, McConnell JDC (1984) Experimental delineation of the $C\bar{1}$ $\rightleftharpoons$ $I\bar{1}$ transformation in intermediate plagioclase feldspars. Am Mineral 69 : 112-121

Carpenter MA, McConnell JDC, Navrotsky A (1985) Enthalpies of ordering in the plagioclase feldspar solid solution. Geochim Cosmochim Acta 49 : 947-966

Carpenter MA, Putnis A (1985) Cation order and disorder during crystal growth: some implications for natural mineral assemblages. In: Thompson AB, Rubie DC (eds) Metamorphic reactions (Advances in Physical Geochemistry vol 4). Springer, Berlin Heidelberg New York, pp 1-26

Christoffersen RG, Yund RA (1984) An experimental study of the Al,Si distribution in the average structure of andesine. Neues Jahrb Mineral Abh 149 : 91-104

Cooper AF (1972) Progressive metamorphism of metabasic rocks from the Haast schist group of Southern New Zealand. J Petrol 13 : 457-492

Crawford ML (1966) Composition of plagioclase and associated minerals in some schists from Vermont, U.S.A., and South Westland, New Zealand, with inferences about the peristerite solvus. Contrib Mineral Petrol 13 : 269-294

Crawford ML (1972) Plagioclase and other mineral equilibria in a contact metamorphic aureole. Contrib Mineral Petrol 36 : 293-314

Doman RC, Cinnamon CG, Bailey SW (1965) Structural discontinuities in the plagioclase feldspar series. Am Mineral 50 : 724-740

Fleet ME (1981) The intermediate plagioclase structure: an explanation from interface theory. Phys Chem Minerals 7 : 64-70

Fleet ME (1982) Orientation of compositional modulation in plagioclase feldspar. In: Aaronson HI, Laughlin DE, Sekerka RF, Wayman CM (eds) Proceedings of an international conference on solid → solid phase transformations. Metall Soc AIME, Warrendale, Pennsylvania, pp 197-201

Fleet ME (1984) Orientation of feldspar intergrowths: application of lattice misfit theory to cryptoperthites and e-plagioclase. Bull Minéral 107 : 509-519

Garrison JR, Jr (1978) Plagioclase compositions from metabasalts, southeastern Llano uplift: plagioclase unmixing during amphibolite-grade metamorphism. Am Mineral 63 : 143-149

Gay P (1956) The structures of the plagioclase feldspars: VI. Natural intermediate plagioclases. Mineral Mag 31 : 21-40

Gay P (1962) Sub solidus relations in the plagioclase feldspars. Norsk Geol Tidssk 42, 2 : 37-56

Ghose S, Van Tendeloo G, Amelinckx S (1988) Dynamics of a second-order phase transition: $P\bar{1}$ to $I\bar{1}$ phase transition in anorthite, $CaAl_2Si_2O_8$. Science 242 : 1539-1541

Goldsmith JR (1982) Review of the behavior of plagioclase under metamorphic conditions. Am Mineral 67 : 643-652

Goldsmith JR (1991) Pressure-enhanced Al/Si diffusion and oxygen isotope exchange. In: Ganguly J (ed) Diffusion, atomic ordering, and mass transport (Advances in Physical Geochemistry vol 8). Springer, Berlin Heidelberg New York, pp 221-247

Goldsmith JR, Jenkins DM (1985) The high-low albite relations revealed by reversal of degree of order at high pressures. Am Mineral 70 : 911-923

Grapes R, Otsuki M (1983) Peristerite compositions in quartzofeldspathic schists, Franz Josef-Fox Glacier Area, New Zealand. J Metamorphic Geol 1 : 47-61

Grove TL (1976) Exsolution in metamorphic bytownite. In: Wenk H-R (ed) Electron microscopy in mineralogy. Springer, Berlin Heidelberg New York, pp 266-270

Grove TL (1977a) Structural characterization of labradorite-bytownite plagioclase from volcanic, plutonic and metamorphic environments. Contrib Mineral Petrol 64 : 273-302

Grove TL (1977b) A periodic antiphase model for the intermediate plagioclases (An_{33} to An_{75}). Am Mineral 62 : 932-941

Grove TL, Baker MB, Kinzler RJ (1984) Coupled CaAl-NaSi diffusion in plagioclase feldspar: experiments and applications to cooling rate speedometry. Geochim Cosmochim Acta 48 : 2113-2121

Grove TL, Ferry JM, Spear FS (1983) Phase transitions and decomposition relations in calcic plagioclase. Am Mineral 68 : 41-59

Grove TL, Ferry JM, Spear FS (1986) Phase transitions in calcic plagioclase: a correction and further discussion. Am Mineral 71 : 1049-1050

Hashimoto H, Nissen H-U, Ono A, Kumao A, Endoh H, Woensdregt CF (1976) High-resolution electron microscopy of labradorite feldspar. In: Wenk H-R (ed) Electron microscopy in mineralogy. Springer, Berlin Heidelberg New York, pp 332-344

Heine V, McConnell JDC (1984) The origin of incommensurate structures in insulators. J Phys C17 : 1199-1220

Heuer AH, Lally JS, Christie JM, Radcliffe SV (1972) Phase transformations and exsolution in lunar and terrestrial calcic plagioclases. Phil Mag 26 : 465-482

Horst W (1984) Modulated structures in intermediate plagioclases - history, ideas, and perspectives. Fortschr Mineral 62 : 107-154

Horst W, Tagai T, Korekawa M, Jagodzinski H (1981) Modulated structure of plagioclase An_{52}: theory and structure determination. Z Kristallogr 157 : 233-250

Jagodzinski H (1984) Determination of modulated feldspar structures. Bull Minéral 107 : 455-466

Kitamura M, Morimoto N (1984) The modulated structure of the intermediate plagioclases and its change with composition. In: Brown WL (ed) Feldspars and Feldspathoids (NATO ASI Series C 137). Reidel, Dordrecht, pp 95-119

Kotel'nikov AR, Bychkov AM, Chenavina NI (1981) The distribution of calcium between plagioclase and a water-salt fluid at 700°C and P_{fl} = 1000 kg/cm^2. Geochem Int 18 : 61-75

Kroll H, Bambauer H-U (1981) Diffusive and displacive transformation in plagioclase and ternary feldspar series. Am Mineral 66 : 763-769

Kroll H, Bambauer H-U, Schirmer U (1980) Thc high albite-monalbite and analbite-monalbite transitions. Am Mineral 65 : 1192-1211

Kroll H, Müller WF (1980) X-ray and electron-optical investigation of synthetic high-temperature plagioclases. Phys Chem Minerals 5 : 255-277

Kroll H, Krause C, Voll G (1991) Disordering, re-ordering and unmixing in alkali feldspars from contact-metamorphosed quartzites. In: Voll G, Töpel J, Pattison DRM, Seifert F (eds) Equilibrium and kinetics in contact-metamorphism: the Ballachulish igneous complex and its aureole. Springer, Berlin Heidelberg, pp 267-296

Liu M, Yund RA (1992) NaSi-CaAl interdiffusion in plagioclase. Am Mineral 77 : 275-283

MacKenzie WS (1957) The crystalline modifications of $NaAlSi_3O_8$. Am J Sci 255 : 481-516

Maruyama S, Liou JG, Suzuki K (1982) The peristerite gap in low-grade metamorphic rocks. Contrib Mineral Petrol 81 : 268-276

McConnell JDC (1974a) Analysis of the time-temperature-transformation behaviour of the plagioclase feldspars. In: MacKenzie WS, Zussman J (eds) The Feldspars. Manchester University Press, Manchester, pp 460-477

McConnell JDC (1974b) Electron-optical study of the fine structure of a schiller labradorite. In: MacKenzie WS, Zussman J (eds) The Feldspars. Manchester University Press, Manchester, pp 478-490

McConnell JDC (1988) The thermodynamics of short range order. In: Salje EKH (ed) Physical properties and thermodynamic behaviour of minerals (NATO ASI Series C 225). Reidel, Dordrecht, pp 17-48

McConnell JDC (1991) Incommensurate structures. Phil Trans R Soc London A334 : 425-437

McConnell JDC (1992) The stability of modulated structures. In: Price GD, Ross NL (eds) The Stability of Minerals. Chapman and Hall, London, pp 216-242

McConnell JDC, McKie D (1960) The kinetics of the ordering process in triclinic $NaAlSi_3O_8$. Mineral Mag 32 : 436-454

McLaren AC (1974) Transmission electron microscopy of the feldspars. In: MacKenzie WS, Zussman J (eds) The Feldspars. Manchester University Press, Manchester, pp 378-423

McLaren AC, Marshall DB (1974) Transmission electron microscope study of the domain structure associated with b-, c-, d-, e- and f-reflections in plagioclase feldspars. Contrib Mineral Petrol 44 : 237-249

Miura Y (1978) Color zoning in labradorescence. Mineral J 9 : 91-105

Mora CI, Ramseyer K (1992) Cathodoluminescence of coexisting plagioclases, Boehls Butte anorthosite: CL activators and fluid flow paths. Am Mineral 77 : 1258-1265

Nakajima Y, Morimoto N, Kitamura M (1977) The superstructure of plagioclase feldspars. Electron microscopic study of anorthite and labradorite. Phys Chem Minerals 1 : 213-225

Nissen H-U (1968) A study of bytownites in amphibolites of the Ivrea-Zone (Italian Alps) and in anorthosites: a new unmixing gap in the low plagioclases. Schweiz Mineral Petrogr Mitt 48 : 53-55

Nissen H-U (1969) Lattice changes in the low plagioclase series. Schweiz Mineral Petrogr Mitt 49 : 491-508

Nissen H-U (1971) End member compositions of the labradorite exsolution. Naturwiss 58 : 454

Nissen H-U (1974) Exsolution phenomena in bytownite plagioclases. In: MacKenzie WS, Zussman J (eds) The Feldspars. Manchester University Press, Manchester, pp 491-521
Nissen H-U, Eggman H, Laves F (1967) Schiller and submicroscopic lamellae of labradorite. A preliminary report. Schweiz Mineral Petrogr Mitt 47 : 289-302
Nord GL, Jr, Heuer AH, Lally JS (1974) Transmission electron microscopy of substructures in Stillwater bytownites. In: MacKenzie WS, Zussman J (eds) The Feldspars. Manchester University Press, Manchester, pp 522-535
Nord GL, Jr., Hammarstrom J, Zen E-An (1978) Zoned plagioclase and peristerite formation in phyllites from southwestern Massachusetts. Am Mineral 63 : 947-955
Olsen A (1977) Lattice parameter determination of exsolution structures in labradorite feldspars. Acta Crystallogr A33 : 706-712
Orville PM (1972) Plagioclase cation exchange equilibria with aqueous chloride solution: results at 700°C and 2000 bars in the presence of quartz. Am J Sci 272 : 234-272
Penzkofer B (1979) Quantitative Messung der diffusen Streuung an Plagioklasen (An_{50}-An_{75}). Dissertation, Univ Munich
Phillips ER, Chenhall BE, Stone IJ, Pemberton JW (1977) An intergrowth of calcic labradorite in a plagioclase-quartz-biotite gneiss from Broken Hill, New South Wales. Mineral Mag 41 : 469-471
Redfern SAT (1992) The effect of Al/Si disorder on the $I\bar{1}$ - $P\bar{1}$ co-elastic phase transition in Ca-rich plagioclase. Phys Chem Minerals 19 : 246-254
Redfern SAT, Graeme-Barber A, Salje E (1988) Thermodynamics of plagioclase III: spontaneous strain at the $I\bar{1}$ - $P\bar{1}$ phase transition in Ca-rich plagioclase. Phys Chem Minerals 16 : 157-163
Redfern SAT, Salje E (1987) Thermodynamics of plagioclase II: temperature evolution of the spontaneous strain at the $I\bar{1} \rightleftharpoons P\bar{1}$ phase transition in anorthite. Phys Chem Minerals 14 : 189-195
Redfern SAT, Salje E (1992) Microscopic dynamic and macroscopic thermodynamic character of the $I\bar{1}$ - $P\bar{1}$ phase transition in anorthite. Phys Chem Minerals 18 : 526-533
Ribbe PH (1962) Observations on the nature of unmixing in peristerite plagioclases. Norsk Geol Tidssk 42, 2 : 138-151
Ribbe PH (1983a) Feldspar Mineralogy (Reviews in Mineralogy vol 2, 2nd edition). Mineralogical Society of America, Washington DC, pp 362
Ribbe PH (1983b) Exsolution textures in ternary and plagioclase feldspars; interference colours. In: Ribbe PH (ed) Feldspar Mineralogy (Reviews in Mineralogy vol 2, 2nd edition). Mineralogical Society of America, Washington DC, pp 241-270
Salje EKH (1985) Thermodynamics of sodium feldspar I: order parameter treatment and strain induced coupling effects. Phys Chem Minerals 12 : 93-98
Salje EKH (1990) Phase transitions in ferroelastic and co-elastic crystals. Cambridge University Press, Cambridge, pp 366
Salje EKH (1994) Phonon hard mode spectroscopy and phase transitions in feldspars. In: Parsons I (ed) Feldspars and their Reactions (NATO ASI series C), (this volume)
Salje EKH, Kuscholke B, Wruck B, Kroll H (1985) Thermodynamics of sodium feldspar II: experimental results and numerical calculations. Phys Chem Minerals 12 : 99-107
Schliestedt M, Johannes W (1990) Cation exchange equilibria between plagioclase and aqueous chloride solution at 600 to 700°C and 2 to 5 kbar. Eur J Mineral 2 : 283-295
Slimming EH (1976) An electron diffraction study of some intermediate plagioclases. Am Mineral 61 : 54-59

Smith JV (1974) Feldspar minerals. 1. Crystal structure and physical properties. Springer, Berlin Heidelberg New York, pp 627
Smith JV (1983) Phase equilibria of plagioclase. In: Ribbe PH (ed) Feldspar Mineralogy (Reviews in Mineralogy vol 2, 2nd edition). Mineralogical Society of America, Washington DC, pp 223-239
Smith JV (1984) Phase relations of plagioclase feldspars. In: Brown WL (ed) Feldspars and Feldspathoids (NATO ASI Series C 137). Reidel, Dordrecht, pp 55-94
Smith JV, Brown WL (1988) Feldspar minerals. 1. Crystal structures, physical, chemical and microtextural properties. Springer, Berlin Heidelberg New York, pp 828
Smith JV, Gay P (1958) The powder patterns and lattice parameters of plagioclase feldspars. II. Mineral Mag 31, 744-762
Smith JV, Wenk H-R (1983) Reinterpretation of a Verzasca plagioclase: a correction. Am Mineral 68 : 742-743
Spear FS (1980) NaSi $\rightleftharpoons$ CaAl exchange equilibrium between plagioclase and amphibole. An empirical model. Contrib Mineral Petrol 72 : 33-41
Steurer W, Jagodzinski H (1988) The incommensurately modulated structure of an andesine (An_{38}). Acta Crystallogr B44 : 344-351
Voll G (1971) Entmischung von Plagioklasen mit An_{50-55} zu An_{18} + An_{93}. Fortschr Mineral 49 : 61-63
Wenk E (1979) Bevorzugte Zusammensetzung und Variabilität der Plagioklase von Gesteinsserein der Verzasca. Neues Jahrb Mineral Monatsh 1979 : 525-541
Wenk E (1983a) Kristalloptik und Zusammensetzung von Bytownit-Drillingen und -Vierlingen, sowie Verwachsungen von Bytownit mit Andesin in Kalksilicatfels von Bagni Masino (Prov. Sondrio, Italien). Schweiz Mineral Petrogr Mitt 63 : 181-186
Wenk E (1983b) Mikroskopische kristallographische Verwachsungen von Oligoklas mit Andesin. Schweiz Mineral Petrogr Mitt 63 : 177-179
Wenk E, Schwander H, Wenk H-R (1991) Microprobe analyses of plagioclases from metamorphic carbonate rocks of the Central Alps. Eur J Mineral 3 : 181-191
Wenk E, Wenk H-R (1977) An-variation and intergrowths of plagioclases in banded metamorphic rocks from Val Carecchio (Central Alps). Schweiz Mineral Petrogr Mitt 57 : 41-57
Wenk E, Wenk H-R (1984) Distribution of plagioclase in carbonate rocks from the Tertiary metamorphic belt in the Central Alps. Bull Minéral 107 : 357-368
Wenk E, Wenk H-R, Glauser A, Schwander H (1975) Intergrowth of andesine and labradorite in marbles of the central Alps. Contrib Mineral Petrol 53 : 311-326
Wenk H-R (1979a) An albite-anorthite assemblage in low-grade amphibolite facies rocks. Am Mineral 64 : 1294-1299
Wenk H-R (1979b) Superstructure variation in metamorphic intermediate plagioclase. Am Mineral 64 : 71-76
Wenk H-R, Joswig W, Tagai T, Korekawa M, Smith BK (1980) The average structure of An 62-66 labradorite. Am Mineral 65 : 81-95
Wenk H-R, Nakajima Y (1980) Structure, formation, and decomposition of APB's in calcic plagioclase. Phys Chem Minerals 6 : 169-186
Windom KE, Boettcher AL (1976) The effect of reduced activity of anorthite on the reaction grossular + quartz = anorthite + wollastonite: a model for plagioclase in the earth's lower crust and upper mantle. Am Mineral 61 : 889-896
Wruck B, Salje EKH, Graeme-Barber A (1991) Kinetic rate laws derived from order parameter theory IV: kinetics of Al,Si disordering in Na-feldspars. Phys Chem Minerals 17 : 700-710

Yund RA (1983) Diffusion in feldspars. In: Ribbe PH (ed) Feldspar Mineralogy (Reviews in Mineralogy vol 2, 2nd edition). Mineralogical Society of America, Washington DC, pp 203-222

FELDSPARS AT HIGH PRESSURE

R. J. ANGEL
Department of Geological Sciences
University College London
Gower St
London WC1E 6BT
England

ABSTRACT. The thermodynamic stabilities at high pressures of all three rock-forming feldspar end-members are strongly dependent upon the bulk chemistry of the system under consideration, but even in the absence of other components they break down at pressures below 4 GPa to denser minerals. Despite these limits on the thermodynamic stabilities of feldspars they can still be studied to pressures in excess of 5 GPa at low to moderate temperatures under which conditions the feldspars are kinetically stable. The behaviour of feldspars upon compression is reviewed and the available bulk moduli data is summarised. These data are used to demonstrate that the major compressional mechanism of all feldspars is one of T-O-T bond bending. Recent single-crystal diffraction studies have also revealed that some feldspars undergo non-quenchable displacive phase transitions at high pressures, and the data for transitions in anorthite, microcline and sanidine is reviewed.

1. Introduction

This review of the structures and properties of feldspars at high pressure is intended primarily to provide an overview of the advances in the field made since the Third NATO Advanced Summer School on feldspars held in 1983, and its attendant volume (Brown, 1984). Only a brief review is therefore given of the *reactions* and *phase transformations* to other phases that feldspars undergo at high pressures and temperatures as very new little data have been obtained in this field over the past ten years. By contrast, the development of ultra-high-pressure techniques, in particular single-crystal X-ray diffraction in diamond-anvil pressure cells, has resulted in much work being carried out on the "crystallographic" behaviour of feldspars at pressures often far in excess of their normal stability fields. This is, of course, possible because the rates of transformation and reaction to other phases are essentially zero at ambient temperature, and the feldspars are therefore kinetically stable. Although criticisms can be levelled at the pursuit of such studies on the grounds that they are therefore not relevant to the behaviour of feldspars within the Earth, these experiments do provide much needed data on thermodynamic properties such as bulk moduli. Spectroscopic studies in the diamond-anvil cell could, if pursued systematically, provide valuable data on heat capacities and vibrational entropy variations in feldspars at high pressure. Furthermore, the discovery of non-quenchable phase transitions at high pressure that limit the stabilities of the room temperature, pressure phases are of immediate relevance if one wants to extrapolate room P,T properties to geological conditions.

I. Parson (ed.), Feldspars and Their Reactions, 271–312.

Lastly, the rich diversity of phase transition behaviour that is now apparent in feldspars subjected to even modest pressures of up to 5 GPa (50 Kbar) provides a stringent test of our understanding of the principles that govern phase transition behaviour in complex structures.

For these reasons, the main part of this review is concerned with just two aspects of the behaviour of feldspars at high pressure. In section 3 the response of single-phase feldspars to hydrostatic compression is described, including a historical review of early attempts to provide bulk moduli data, and a comparison of results from those experiments with modern determinations of the variation of bulk moduli with pressure, composition, and structural state. This is followed by an attempt to determine the underlying mechanisms of feldspar compression from an analysis of unit-cell parameter variation with pressure, and so to understand these observed variations in bulk moduli. Section 4 addresses a related topic, the phase transition behaviour of feldspars subjected to hydrostatic stress, which can be seen as a response of the feldspar structure to reaching the limit of one particular mode of compression.

2. Stability of Feldspars at High Pressure

The pressure range over which feldspars are thermodynamically stable is extremely limited due to their low densities, although this is partially offset by their high compressibilities (section 3 below). Their stabilities are also strongly dependent upon the bulk chemistry of the system under consideration; in general they are stable to the highest pressures in the absence of other components, this maximum pressure being reduced in the presence of excess water, and further restricted by reactions with other phases in more complex chemical systems. In this section a brief overview of the pure phase stabilities of the geologically important end-members is presented, together with an indication of the reactions that limit their stabilities in the presence of major chemical components. This overview is in no way intended to be complete but is provided as a background to the discussion of the behaviour of feldspars as single phases that constitutes the remainder of this chapter. For more details the reader is referred initially to the experimental papers referenced here, to the various contributions in Bailey (1984) for the reactions involving micas, and to the work of Holland and Powell (1990) for the internally consistent thermodynamic dataset required to calculate the multitude of feldspar breakdown reactions.

2.1. ALBITE

The phase relations and reactions of albite at high pressures are probably the simplest of the three common feldspar end-members. At the pure albite composition, the stability of the feldspar is limited at high pressures by the breakdown to jadeite + quartz, a reaction studied by many workers. For a summary and presentation of the definitive data on this reaction the reader is referred to Holland (1980) who gives a boundary

$$P = 0.00265\ T\ (^{\circ}C) + 0.035\ (GPa)$$

with a quoted uncertainty of ± 0.05 GPa over the temperature range 600-1200°C. At higher pressures the jadeite + quartz assemblage passes through the transitions in silica to coesite and then stishovite. Laser heated diamond-anvil cell experiments (Liu, 1978) suggested that the stishovite + jadeite assemblage reacts to form a single phase hollandite with the albite

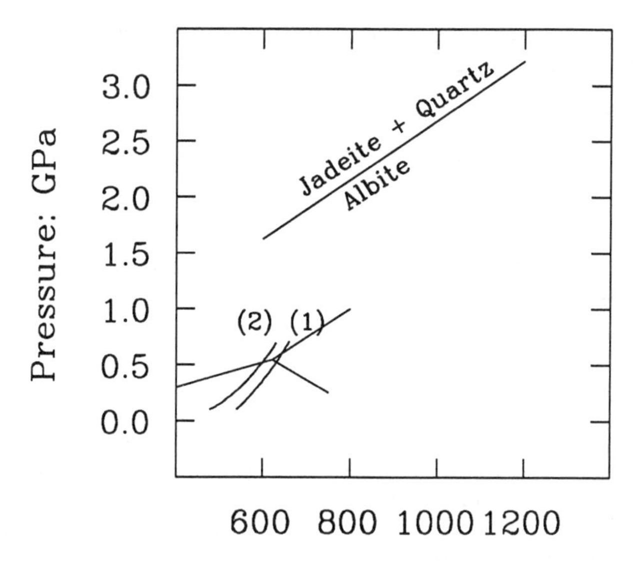

Figure 1. Breakdown reactions of albite overlain upon the kyanite-andalusite-sillimanite phase relations. Reaction (1) is Paragonite = Albite + Corundum + H_2O, and reaction (2) is Paragonite = Albite + Al_2SiO_5 + H_2O. Sources for the positions of the reactions are given in the text.

composition between 21 and 24 GPa at an estimated temperature of 1000°C. Above 24 GPa Liu (1978) produced $NaAlSiO_4$ in a calcium ferrite structure plus free silica as stishovite, the same assemblage found to be stable above 21 GPa at the jadeite composition (Gasparik, 1992). These proposed phase relations are consistent with a recent re-interpretation of shock-wave data by Sekine and Ahrens (1992), although it should be noted that such experiments are carried out far from equilibrium, and that the reduced equation of state data that is obtained can be interpreted in many ways.

Unlike the other feldspars, the addition of water alone to the albite system does not appear to result in decreased stability of albite; water is used to catalyse the breakdown reaction to jadeite + quartz at pressures of up to 3.5 GPa.The further addition of either alumina, or in the case of the silica saturated system aluminosilicate, allows the reaction of albite to the sodium mica paragonite:

$$\text{Albite} + \text{Corundum} + H_2O \rightarrow \text{Paragonite}$$
$$\text{Albite} + Al_2SiO_5 + H_2O \rightarrow \text{Paragonite} + \text{Quartz}$$

These reactions have been bracketed by Chatterjee (1970, 1971) from 0.1 to 0.7 GPa, and lie ~40°C apart at 0.1 GPa (Figure 1). Reaction of albite with other chemical components does not appear to have been explored by experimental reversals, although reactions involving, for example, the addition of an Fe^{+3} component to albite to produce a jadeite-acmite solid solution pyroxene, can be calculated from thermodynamic data. Phase equilibria experiments on the enstatite-jadeite join (Gasparik, 1992) suggest that up to 20 mol% enstatite component is accommodated within jadeitic pyroxene at 2.5 GPa and 900°C. The presence of an $MgSiO_3$ component would therefore displace the equilibrium breakdown of albite to lower pressures as a result of the lowering of the jadeite activity by enstatite.

2.2. K-FELDSPAR

Kinomura et al. (1975) were the first to demonstrate that, in the absence of water, K-feldspar breaks down at pressures between 4 and 10 GPa:

$$\underset{\text{K-feldspar}}{2KAlSi_3O_8} = \underset{\text{Wadeite-type}}{K_2Si_4O_9} + \underset{\text{Kyanite}}{Al_2SiO_5} + \underset{\text{Coesite}}{SiO_2}$$

The structure of the $K_2Si_4O_9$ phase is of the wadeite type and was of particular crystal chemical interest as, at the time of its discovery, it was the first structure known to contain linked SiO_4 tetrahedra and SiO_6 octahedra. The structure determination by Kinomura et al. (1975) was confirmed by Swanson and Prewitt (1983), and a high pressure study to 4 GPa was performed by Ross, Swanson and Prewitt (unpublished). The thermodynamic properties of the wadeite-type $K_2Si_4O_9$ were measured by Geisinger et al. (1987). The equilibrium breakdown curve of K-feldspar calculated by Geisinger et al. (1987) is consistent with the two available synthesis points reported by Kinomura et al. (1975) and Seki and Kennedy (1964).

At higher pressures, in excess of 10 GPa, a hollandite-type phase is formed with the $KAlSi_3O_8$ composition (Ringwood et al., 1967). Preliminary structures of this phase, based upon powder diffraction, were reported by Ringwood et al. (1967) and Yamada et al. (1984). A single-crystal study has recently been performed at various pressures to 4.5 GPa by Zhang et al. (1993). The structure is interesting in that its tetragonal symmetry requires that all of the Al and Si occupy a single octahedral site and are therefore disordered. Zhang et al. (1993) also report an equation of state for this phase with K_T = 180(3) GPa for K' fixed at 4.

The stability of K-feldspar is limited to lower pressures in the presence of water. Seki & Kennedy (1964) reported the reaction of sanidine with excess water to form a sanidine hydrate. A mixture of synthesis runs (from sanidine composition glass) and reversals constrained the boundary to be:

$$P = 0.00345\ T\ (C) + 0.37\ \text{GPa}$$

The structure of the sanidine hydrate phase does not appear to have been studied in detail, but Seki and Kennedy (1964) suggested that, on the basis of density measurements, it should have the composition $KAlSi_3O_8.H_2O$. They noted that the powder diffraction pattern of this phase was very similar to that of the hydrated barium feldspar cymrite, and in a further experiment were able to synthesise a homogeneous phase with the same powder pattern at a composition half-way between the K and Ba end-members. They concluded that a solid solution may exist across the $KAlSi_3O_8.H_2O$ - $BaAlSi_3O_8(OH)$ join based upon the exchange

$$Ba^{+2} = K^{+} + H^{+}$$

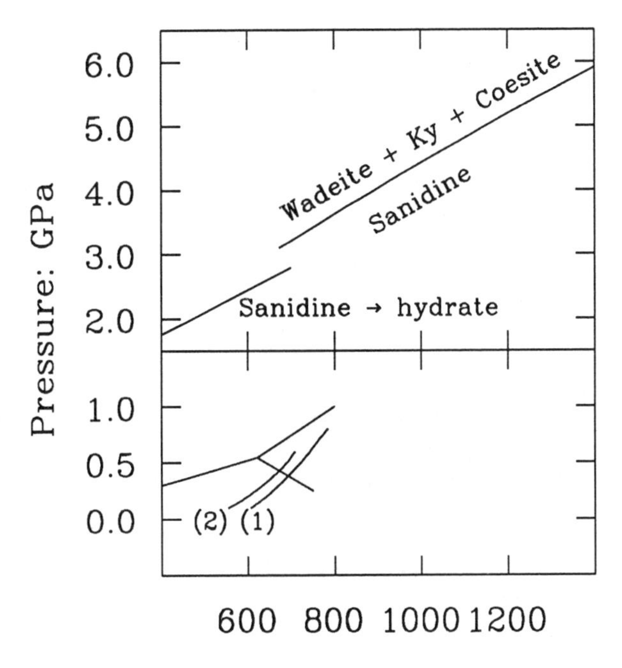

Figure 2. Breakdown reactions of K-feldspar overlain upon the kyanite-andalusite-sillimanite phase relations. Reaction (1) is Muscovite = Sanidine + Corundum + H_2O, and reaction (2) is Muscovite + Quartz = Sanidine + Al_2SiO_5 + H_2O. Note the change in the pressure scale. Sources for the various reactions are to be found in the text.

However, it was subsequently demonstrated that the correct composition for cymrite is $BaAl_2Si_2O_8.H_2O$ (eg. Essene, 1967), so this join may in fact be based upon the coupled exchange

$$Ba^{+2} + Al^{+3} = K^{+} + Si^{+4}$$

at constant water content. Experimental studies to determine the dehydration reaction of cymrite are reported by Graham et al. (1992), Viswanathan et al. (1992) and as an abstract by Hsu (1993).

In most natural systems the high-pressure stability of K-feldspar is limited, like albite, by its breakdown to sheet silicates at pressures below the andalusite to kyanite inversion in Al_2SiO_5. The actual reaction, and the position of the P-T boundary, depends upon the chemistry of the other phases present. Chatterjee and Johannes (1974) report the experimental reversal of two possible end-member breakdown reactions of K-feldspar that produce muscovite. In the presence of excess alumina the reaction is:

$$\text{K-feldspar} + \text{Corundum} + H_2O \rightarrow \text{Muscovite}$$

and with lower alumina contents:

$$\text{K-feldspar} + Al_2SiO_5 + H_2O \rightarrow \text{Muscovite} + \text{Quartz}$$

This second reaction lies only some 30°C below the first (Figure 2). Combined with the similar observation for albite, this indicates that the degree of silica saturation is relatively unimportant in determining the stability of all alkali feldspars with respect to sheet silicates. Other rock chemistries produce other micas; higher Mg,Fe contents, reflected in the presence of orthopyroxene, result in the reaction of opx and K-feldspar to produce biotite at high pressures, while reaction with pyrope-almandine garnets produces siderophyllite. It should be noted, however, that it is more frequent to observe the *breakdown* of these micas to form feldspar during metamorphism, rather than the breakdown of the feldspar which requires high pressures and low temperatures. If the chemistry is suitable it is possible for K-feldspar to be stable at much higher pressures. For example, Smyth and Hatton (1977) described a sanidine-grospydite from a kimberlite containing high sanidine and free silica (in the form of coesite), plus clinopyroxene, kyanite and garnet, for which equilibration conditions were estimated to be >900°C and >3 GPa.

Seki and Kennedy (1964) also performed some experiments on the breakdown, in the presence of excess water, of a natural orthoclase with approximate composition $Or_{63}Ab_{37}$. They found the series of reactions with increasing pressure:

$$\begin{aligned} \text{Orthoclase} \quad &\rightarrow \text{sanidine} + \text{jadeite} + \text{quartz} \\ &\rightarrow \text{sanidine} + \text{jadeite} + \text{coesite} \\ &\rightarrow KAlSi_3O_8.H_2O + \text{jadeite} + \text{coesite} \end{aligned}$$

Thus intermediate alkali feldspars appear to disproportionate and follow the breakdown reactions of the end-members albite and sanidine. This is no doubt the result of the limited solubility of K within the jadeitic clinopyroxene.

2.3. ANORTHITE

The high-pressure breakdown of anorthite to grossular + kyanite + quartz was first observed by Boyd and England (1961), and was subsequently studied experimentally by Hays (1966), Hariya and Kennedy (1968), and Goldsmith (1980). The most reliable reversals (Wood and Holloway, 1984) of this reaction, which is the simplest expression of the important boundary between granulites and eclogites, are those of Goldsmith (1980). His brackets define a boundary (between 1100°C and 1350°C) described by the equation:

$$P = 0.00232T\ (°C) - 0.21\ (GPa)$$

Since the thermodynamic properties of all of the phases involved in this reaction have been measured, this reaction provides an excellent opportunity to assess the entropic effect of a small degree of Al,Si disorder within the anorthite. Wood and Holloway (1984) provided an interesting discussion of this comparison and concluded that 1-2 % Al,Si disorder, sufficient to add between 2 and 4 $J.K^{-1}.mol^{-1}$ to the entropy of the anorthite, was sufficient to bring the calculated boundary of the breakdown reaction into good agreement with the experimental determination. The X-ray structure determinations by Angel et al. (1990) suggested instead that the minimum degree of disorder found in anorthite is about 4%, which would imply a configurational entropy of 5.6 $J.K^{-1}.mol^{-1}$. This figure produces excellent agreement with the reversals of Goldsmith (1980), and is consistent with the estimates of 3.75% to 4.2% disorder obtained by Gasparik (1984) from several anorthite reactions.

Goldsmith (1978,1981) also determined the wet stability limit of anorthite and anorthite-rich plagioclase at high pressures. Below 725°C anorthite reacts with water:

$$\text{Anorthite} + H_2O \rightarrow \text{Zoisite} + \text{Kyanite} + \text{Quartz} + \text{Vapour}$$

The reversals of Goldsmith (1981) bracket a phase boundary with the equation:

$$P = 0.00204\ T\ (^{\circ}C) - 0.459\ \text{GPa}$$

This boundary is valid up to a triple point at approximately 1.02 GPa and 725°C at which point melt appears. Goldsmith (1978) reported experiments on this same breakdown reaction in plagioclases which, although they did not constitute reversals, suggested that the temperature at 1.0 GPa of this breakdown reaction seems to be invariant with composition from An_{100} to An_{40}. From An_{40} to An_{10} the temperature of the boundary appears to drop rapidly to ~450°C.

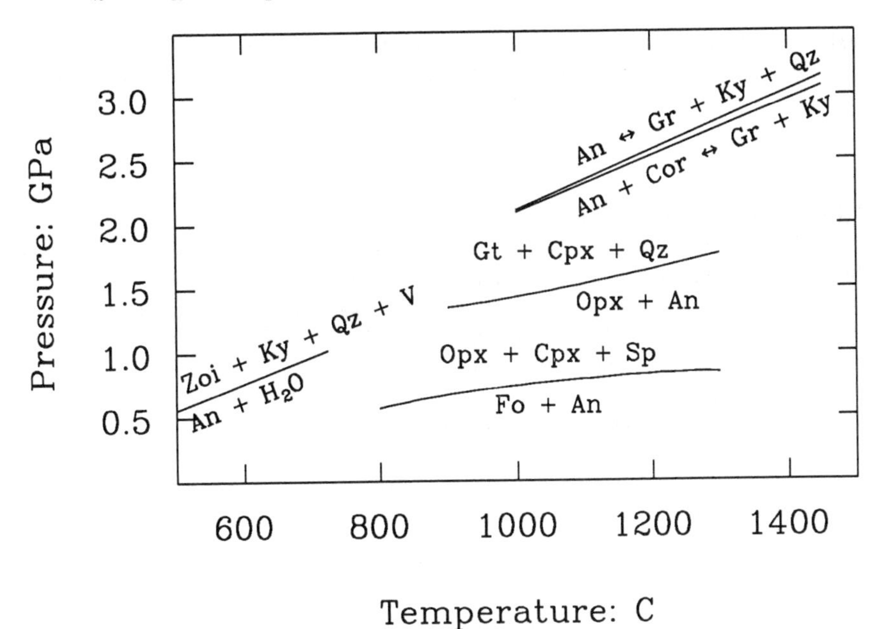

Figure 3. Breakdown reactions of anorthite. The hydrous breakdown Anorthite + H_2O = Zoisite + Kyanite + Quartz + Vapour is terminated at 725°C by the appearance of melt. Sources for the various reactions are to be found in the text.

Experimental reversals by Gasparik (1984) showed that in the CAS system the reaction

$$\text{Anorthite} + \text{Corundum} \rightarrow \text{Grossular} + \text{Kyanite}$$

closely parallels (Figure 3) the silica saturated dry breakdown reaction described above, with very similar slope but offset to lower pressures by < 0.1 GPa:

$$P = 0.00218\ T\ (^{\circ}C) - 0.095\ \text{GPa}$$

As for the alkali feldspars, the degree of silica saturation is therefore relatively unimportant in determining the stability limit of anorthite.

The stability of anorthite can, however, become severely limited by reactions with other phases, and even within the relatively simple CAS and CMAS systems such reactions introduce significant additional theoretical and experimental difficulties due to solid solution and non-stoichiometry, in particular of the pyroxene phases. Of the reactions in which anorthite participates perhaps the best constrained is

$$\text{Forsterite} + \text{Anorthite} \rightarrow \text{Orthopyroxene} + \text{Clinopyroxene} + \text{Spinel}$$

determined experimentally by Kushiro & Yoder (1966) and Herzberg (1976), which limits anorthite stability to below 0.8 GPa in the temperature interval 1000-1400°C (Figure 3). The reaction

$$\text{Orthopyroxene} + \text{Anorthite} \rightarrow \text{Garnet} + \text{Clinopyroxene} + \text{Quartz}$$

in the CMAS system occurs at higher pressures (Figure 3) but the experimental data are not entirely consistent. The brackets of Hansen (1981) can be reconciled with the thermodynamic data of the phases involved provided that the clinopyroxene contains some 12% of the defect Ca-Eskola ($Ca_{0.5}AlSi_2O_6$) component (Wood & Holloway, 1984). There is some doubt as to the validity of all of the reversals of this reaction reported by Perkins (1983), but if they are all correct they would require increasing Ca-Eskola content (from ~3% to ~12%) over the temperature interval from 1000-1300°C. Gasparik (1984) addresses in more detail the reactions of anorthite involving both this pyroxene component and CaTs ($CaAl_2SiO_6$) in the CAS system.

Gasparik (1985) extended his study into the NCAS system (ie. addition of Na_2O component), and his results provide tight constraints upon the breakdown of plagioclase at high pressures. Anorthite rich plagioclase initially disproportionates:

$$\text{Plagioclase} \rightarrow \text{Na-rich Plagioclase} + \text{Garnet} + \text{Kyanite} + \text{Quartz}$$

and at higher pressures (3.0 GPa at 1200°C) the plagioclase gives way to pyroxene. At albite-rich compositions the reaction sequence is dominated by the behaviour of the albite component:

$$\begin{aligned}\text{Plagioclase} &\rightarrow \text{Pyroxene} + \text{Plagioclase} + SiO_2\\ &\rightarrow \text{Pyroxene} + SiO_2\end{aligned}$$

The pressure interval over which the intermediate phase assemblage is stable is limited to less than 0.1 GPa at all temperatures.

More recently there have been several exploratory studies of the behaviour of anorthite at mantle temperatures and pressures undertaken with diamond-anvil pressure cells. A series of laser heating experiments at pressure have been performed by Madon and co-workers. These experiments generally involve the initial pressurisation of the sample, heating by rastering the laser beam across the sample, followed by pressure release and examination of the run products. Madon et al. (1989) reported the recovery of an assemblage consisting of amorphous $CaSiO_3$ (presumed to be the quench product of $CaSiO_3$ perovskite) plus an Al_2SiO_5 phase (with the V_2O_5 structure, Ahmed-Zaid and Madon, 1991) from an anorthite composition glass pressurised to 50 GPa and heated to ~2300°C. Gautron and Madon (1993) reported that when crystalline (synthetic) anorthite is heated to ~1800°C at pressures of 12.5-17.5 GPa, anorthite is recovered. At higher pressures (to 22.5 GPa) heating results in a recovered assemblage of Ca-Al hollandite + Al_2SiO_5 (kyanite) + amorphous $(Ca,Al)(Si,Al)O_3$. These results have been interpreted (Gautron and Madon, 1993) as showing that at these high temperatures anorthite remains stable up to 18 GPa at which point it inverts to produce hollandite + Ca-Al perovskite which is the source of the amorphous phase. However, it should be noted that although Williams and Jeanloz (1989) reported the amorphisation of anorthite above 33 GPa at room temperature, work by Daniel et al. (1993) showed that amorphisation can commence at pressures as low as 14 GPa, and can be complete by 18 GPa. It is quite possible, therefore, that the heating stages of the experiments of Gautron and Madon (1993) were performed on highly metastable anorthite glass. Furthermore, the anorthite that Gautron and Madon (1993) recovered from these experiments has highly anomalous cell parameters, and this may indicate that their results may be severely affected by problems of non-equilibrium and quench effects.

3. Compressional Behaviour

3.1. BULK MODULI

Early determinations of the bulk moduli of feldspars were carried out by a variety of press techniques, wherein the change in length of a cylinder of the material of interest was measured as a function of pressure. These experiments were fraught with difficulties arising principally from the small changes in length that occur on modest compression, the need to maintain hydrostatic (or quasi-hydrostatic) conditions upon the sample, and the need to remove all void space from the sample assembly and to correct for friction in the apparatus. As a result, significant uncertainties are associated with the several determinations of the compressibilities of various feldspars made with these techniques. In addition, the measurements were often made relative to some standard material and the value of the compressibilities of such materials were frequently revised as experimental techniques improved. For example, the value for the bulk modulus of $Ab_{52}An_{48}$ labradorite determined to be 65 GPa by Adams and Williamson (1923) was subsequently revised to 68 GPa by Adams and Gibson (1929) for this last reason. Some indication of the range of the data for two limited compositional ranges is provided by Figure 4.

Hackwell (1993) has re-analysed, as far as is possible, the volume-pressure data available for feldspars from the sources listed in Table 1. Some general trends are quite apparent in these data: Plagioclase feldspars are far less compressible than alkali feldspars, and the isothermal bulk modulus K_T ($= -V (\delta P/\delta V)_T$) increases significantly (Yoder & Weir, 1951, estimated by ~40%)

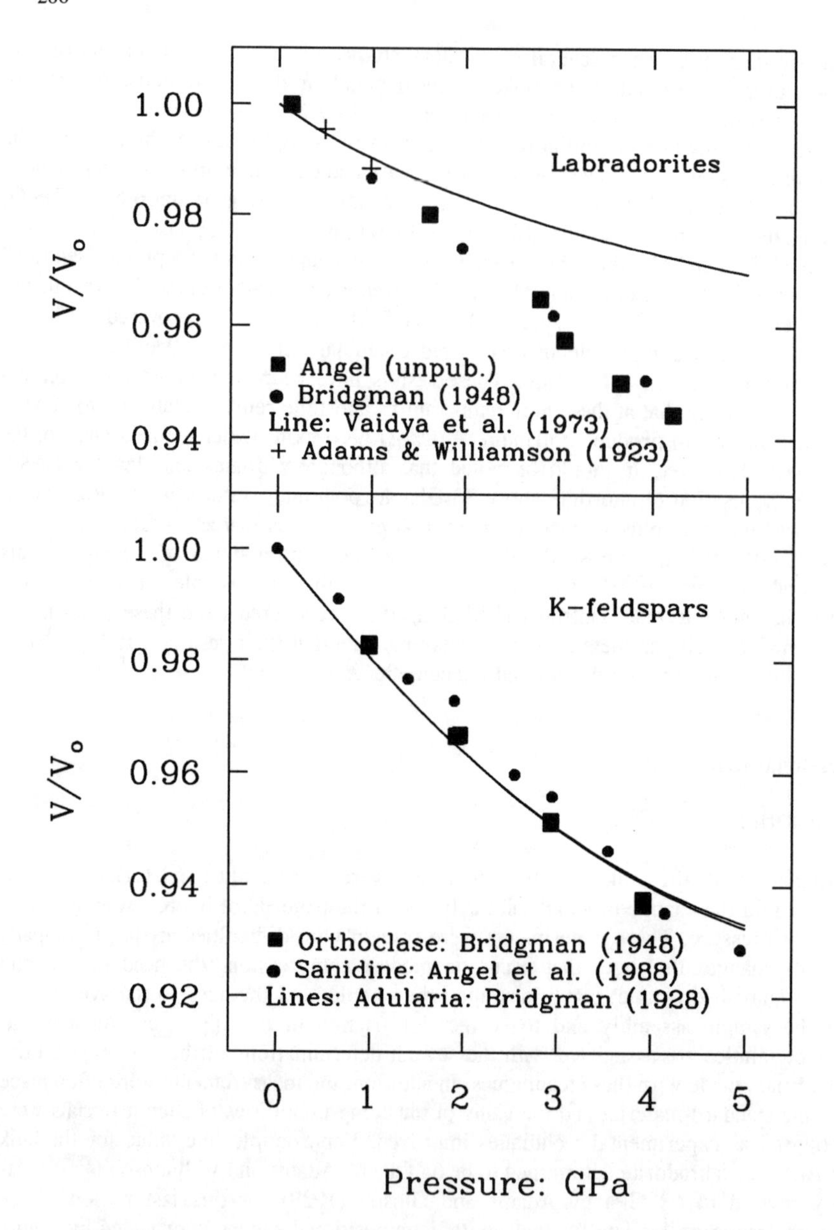

Figure 4. The variation with pressure of the volume of a number of labradorites and K-feldspars determined by press techniques compared to recent single-crystal X-ray determinations.

TABLE 1. Bulk moduli determinations by press techniques.

Reference	Material
Adams & Williamson (1923)	Oligoclase (Ab78Ab22) Labradorite (Ab48An52) Microcline (Or91Ab9)
Bridgman (1928)	Adularia
Adams & Gibson (1929)	Labradorite (Ab48An52)
Bridgman (1948)	Orthoclase (x2) Labradorite
Yoder & Weir (1951)	Albite Oligoclase (Ab78Ab22)
Vaidya et al. (1973)	Labradorite

from albite to anorthite. Beyond these generalisations, and the fact that felds pars are generally stiffer than other framework silicates (cf. quartz, K_T = 41.4 GPa; Glinnemann et al., 1992) but softer than most other low-pressure minerals (eg. pyroxenes K_T = 100-120 GPa, forsterite K_T = 123 GPa; Kudoh and Takeuchi, 1985), no details of systematic variation of bulk moduli composition can be obtained from these data.

The individual elastic constants of some 15 feldspars were measured by ultrasonic techniques by Alexandrov and Ryzhova (1962), Ryzhova (1964) and Ryzhova and Alexandrov (1965). The values of the individual elastic constants so determined are subject to considerable uncertainty (an assessment is given by Alexandrov and Ryzhova, 1962) not least because of the necessary use of twinned crystals and the assumption of monoclinic symmetry for all of the feldspars studied. These uncertainties are reflected in the scatter of the adiabatic bulk moduli, $K_S = -V.(\delta P/\delta V)_S$, obtained by averaging the elastic constants - a complete set of values is tabulated in Simon and Wang (1971). Nevertheless, the increase in bulk modulus (and, indeed, all of the individual elastic constants) with anorthite content across the plagioclase join is very apparent (Figure 5) in the data of Ryzhova (1964). The data for alkali feldspars (Ryzhova and Alexandrov, 1965) exhibit no systematic trends.

X-ray diffraction measurements at high pressures provide unit-cell parameters, and hence unit-cell volume, at pressure from which isothermal bulk moduli can be obtained through the fitting of a suitable equation of state. Such experiments avoid many of the problems associated with the press techniques described above; samples can be small (~ 100μm), and pressure can be measured directly by the ruby fluorescence technique thereby avoiding problems associated with friction corrections and pre-compaction of samples. A hydrostatic pressure environment can be maintained to 10 GPa with alcohol pressure media, and to much higher pressures by using solidified rare gases. Typical precision in the measurement of unit-cell volumes by single-crystal diffraction is 1 part in 5000, and pressure can be determined to a precision of $\pm$0.03 GPa (Angel et al., 1992). Hazen (1976) and Hazen and Prewitt (1977) were the first to study the behaviour of feldspars at high pressures by this technique. Unfortunately, both studies used a secondary pressure standard so the data obtained cannot be used to determine the bulk moduli of the sanidine and albite samples used. The results of Hazen (1976) on possible high-pressure

TABLE 2. Bulk moduli of feldspars determined by X-ray Diffraction.

Composition	K_T:GPa	P:GPa	Reference
An_{100}, Q_{OD}=0.92, $P\bar{1}$	83.3(1.5)	0.0-2.5	Hackwell & Angel (1992)
An_{100}, Q_{OD}=0.92, $I\bar{1}$*	88.6(3.7)	2.6-5.0	Angel (1992)
An_{100}, Q_{OD}=0.87, $P\bar{1}$	88.0(4.2)	0.0-2.4	Hackwell & Angel (1992)
An_{100}, Q_{OD}=0.87, $I\bar{1}$*	77.1(1.7)	3.0-4.9	Angel (1992)
An_{100}, Q_{OD}=0.85, $P\bar{1}$	83.1(1.1)	0.0-2.1	Hackwell & Angel (1992)
An_{100}, Q_{OD}=0.85, $I\bar{1}$*	74.7(2.4)	3.1-4.9	Angel (1992)
An_{100}, Q_{OD}=0.82, $P\bar{1}$	81.0(0.8)	0.0-4.0	Hackwell & Angel (1992)
An_{100}, Q_{OD}=0.82, $I\bar{1}$*	66.3(2.7)	5.1-7.5	Angel (1992)
An_{100}, Q_{OD}=0.78, $P\bar{1}$	79.8(0.3)	0.0-4.0	Hackwell & Angel (1992)
An_{100}, Q_{OD}=0.78, $I\bar{1}$*	66.3(1.7)	5.3-7.7	Angel (1992)
$An_{98}Ab_2$, $P\bar{1}$	83.7(8)	0.0-2.6	Angel (unpublished data)
$An_{98}Ab_2$, $I\bar{1}$*	68.8(2.0)	0.0-3.9	Angel (unpublished data)
$An_{89}Ab_{11}$, $I\bar{1}$	74.1(1.7)	0.0-3.9	Angel (unpublished data)
$An_{72}Ab_{28}$, $I\bar{1}$	66.3(1.6)	0.0-4.2	Angel (unpublished data)
$Ab_{98}An_1Or_1$	57.6(2.0)	0.0-4.9	Angel et al. (1988)
$Or_{82}Ab_{18}$ Sanidine	52.3(0.9)	0.0-5.2	Hackwell (1993)
$Or_{98}Ab_2$ Sanidine	57.2(1.0)	0.0-4.9	Angel et al. (1988)
Or_{87} Microcline	63.3(2.2)	0.0-3.3	Hackwell (1993)
Danburite	113.6(2.9)	4.6	Hackwell & Angel (1992)
Reedmergnerite	68.7(0.7)†	4.9	Hackwell & Angel (1992)

Notes: All bulk moduli determined by least-squares fitting of K_T and V_o of a Murnaghan equation of state to volume-pressure data, with K' constrained to 4.

*High pressure phases.

†Esd incorrectly quoted as 7.0 GPa in Hackwell & Angel (1992).

phase transitions in sanidine are discussed further in Section 4.1.

The available bulk moduli data for feldspars determined by single crystal diffraction experiments are listed in Table 2. The values for isothermal bulk moduli were obtained by fitting K_T and V_o of a Murnaghan equation of state -

$$V=V_o\left(1+\frac{K'}{K_o}P\right)^{-\frac{1}{K'}}$$

- to the volume-pressure data points. Much of the data is limited in pressure range because both anorthite-rich plagioclases and microcline undergo displacive phase transitions at high pressure (Section 4). Since both volume changes and distortions are associated with these transitions it is neither valid nor meaningful to fit a single equation of state to volume-pressure data from both phases together. In these cases reliable estimates of K' (essentially the curvature of the volume-pressure graph) could not be obtained, so its value was fixed at 4 for all samples to provide a proper basis for comparison. In the following discussion it must therefore be recalled that values of, and variations in, the bulk moduli may be a partial artifact of fixing K'; changing K' from 2 to 5 usually results in a $\sim$5% decrease in K_T (Hackwell and Angel, 1992). It should also be noted that the bulk modulus K_T obtained by fitting the Murnaghan equation of state is not equal to the inverse of the linear volume compressibility β, so the K_T values provided in Table 2 *must not* be used to estimate volume as $V(P) = V_o(1 - \beta.P)$.

The data in Table 2 confirm the general trends apparent from the earlier work on feldspars

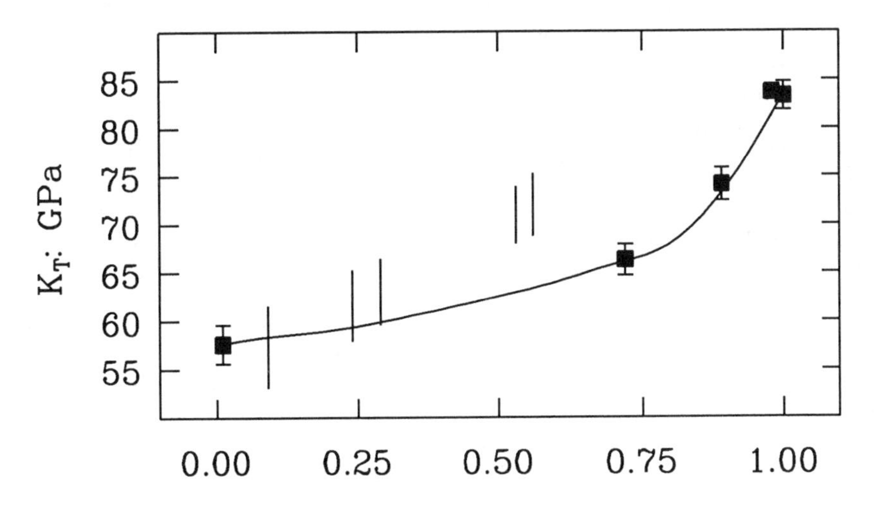

Figure 5. The variation with composition of the bulk moduli of plagioclase feldspars. The square symbols are values of K_T of well-ordered crystals taken from Table 2. Vertical lines represent K_S values calculated from the data of Ryzhova & Aleksandrov (1965). The upper ends of the lines at each composition represent the Voight average, the lower ends the Reuss average.

(Table 1). The bulk moduli of well-ordered plagioclase feldspars increase steeply with increasing anorthite content, anorthite being some 45% stiffer than albite. Figure 5 shows that the trend is not linear in composition, but that K_T increases more rapidly at the anorthite end of the join. More data, especially at albite-rich compositions is required to define the precise form of this trend. The data for alkali feldspars (Table 2) is suggestive of two trends that affect their bulk moduli. First, the bulk moduli of feldspars in both ordered (i.e. albite and microcline) and disordered (i.e. sanidines) structural states increase slightly with increase in K content. Second, the bulk modulus increases with increased Al,Si order from sanidine to microcline.

The bulk moduli data now available for feldspars allows at least a preliminary assessment of the causes of the variations observed in terms of the chemistry, structural state, and the underlying atomic scale compression mechanisms.The structure determinations of anorthite at high pressure (Angel, 1988, and Section 4.2.2) confirmed that, like other silicate minerals, feldspars respond to pressure by bending of T-O-T linkages, while the internal dimensions of the AlO_4 and SiO_4 tetrahedra remain essentially unchanged. The framework of a feldspar therefore behaves under compression as if it were comprised of rigid tetrahedra which can tilt with respect to one another around their shared apices, much the same as their behaviour at high temperatures (eg. Megaw 1974, Prewitt et al., 1976). As the tetrahedra tilt under pressure, the oxygen atoms will be pushed out into the cages containing the extra-framework cations (M) thereby reducing the M-O cation distances. The bulk moduli of feldspars should therefore be controlled by two components - the stiffnesses of the T-O-T bond angles and of the M-O bonds. It is difficult to separate these two factors completely, but the influence of the T-O-T linkages at constant M cation chemistry can be assessed. Increasing disorder of the Al and Si, induced by high-temperature heat treatments of end-member anorthite, results in a small decrease in bulk modulus - the largest decrease being 3.5 GPa for a decrease in Q_{OD} (Angel et al., 1990) from 0.92 (natural material) to 0.78. This decrease must be due to the increase in the number of Al-O-Al and Si-O-Si linkages in anorthite at the expense of Si-O-Al linkages that results from disordering. This result is in agreement with molecular orbital calculations of $H_6T_2O_7$ and $H_7T_2O_7$ molecules which suggest that Al-O-Al and Si-O-Si linkages are slightly stiffer than Al-O-Si linkages (Geisinger et al., 1985). The same trend appears to occur in the alkali feldspars where microcline (the ordered variant) is stiffer than the disordered sanidines (Table 2). Further molecular orbital calculations by Geisinger et al. (1985) also suggested that while these differences between the responses of various (Al,Si)-O-(Al,Si) linkages to bending forces were small, Si-O-B and B-O-B linkages (found in B-feldspars) are significantly stiffer. This is borne out by the observation that the replacement of Al in albite to produce reedmergnerite (with Si-O-B linkages) results in a 19% increase in the bulk modulus. Similarly, the boron analogue of anorthite, danburite (with Si-O-B, Si-O-Si and B-O-B linkages), is more than 30% stiffer than anorthite. It should be noted, however, that the framework topology of danburite differs from that of plagioclase (Phillips et al., 1974) and this may also contribute to the difference in bulk modulus.

One might be tempted to conclude from the fore-going discussion that the T-O-T linkages of the frameworks of feldspars play the dominant role in determining their bulk moduli. Indeed, if the trend of decreasing K_T with decreasing Q_{OD} in anorthite is assumed to be linear then one would estimate the bulk modulus of a fully disordered anorthite to be ~60 GPa, a value remarkably similar to that of the alkali feldspars. The variation in bulk modulus across the plagioclase join would then be attributed entirely to increasing disorder in the anorthite structure as the albite content is increased (Angel et al., 1990), with the implication that M-O repulsions play only a passive role during compression of feldspars. However, the fact that the bulk modulus of Sr-anorthite (a preliminary estimate of K_T = 200±25 GPa can be obtained from the data of McGuinn and Redfern, 1993) is more than twice that of Ca-anorthite clearly shows that this simple conclusion is not valid. Suggestions that feldspar bulk moduli are simple functions of cation properties are equally invalidated by the data now available. Angel et al. (1988) suggested, following Hazen and Finger (1982), that the charge of the extra-framework cation is a dominant factor. Although this might account for the difference between alkali and plagioclase feldspars, it does not account for the bulk modulus of Sr-anorthite. Neither does the size of these extra-framework cations explain the observed variations, as is apparent from the fact that the alkali feldspars have very similar bulk moduli (radius of Na = 1.18Å, K = 1.51Å) whereas anorthites containing Ca (ionic radius 1.12Å) are much stiffer, and the Sr cation (ionic radius 1.25Å) is intermediate between Na and K, yet produces a bulk modulus almost three times greater. The true state of affairs is much more complex, and the bulk moduli variation of feldspars must be due not only to the intrinsic stiffness of the linkages comprising the framework, which are influenced by the framework chemistry and the state of order, but also by the framework configuration which is in itself a partial function of the extra-framework cation. That this is true is demonstrated by the large decrease in the bulk moduli of anorthites (Table 2) that accompanies the $P\bar{1}$ to $I\bar{1}$ phase transition (Section 4.2) that occurs at high pressures, through which there is no change in either the cation chemistry or state of order, merely a change in framework conformation (Angel, 1988). Some further insights into these issues can be provided by examination of the detailed changes in unit-cell parameters of feldspars under hydrostatic compression.

3.2. UNIT CELL PARAMETER VARIATION WITH PRESSURE

Much use has been made in the past of unit-cell parameter variations of feldspars to determine compositions and structural states, and in the case of the survey of Brown et al. (1984, updated in Smith and Brown, 1988) to infer some of the mechanisms of structural response to changes in temperature, pressure, composition and structural state. Brown et al. (1984) showed that the unit-cell parameters of most feldspars follow one of three trends. The two most common, Trend I and Trend II, are illustrated in Figure 6. Trend I, on increasing temperature, consists of a coupled expansion of *a, b* and *c* cell edges together with a decrease in the α and β unit cell angles, while γ remains almost constant. This trend is followed by the alkali feldspars upon heating, and by appropriate chemical substitutions of the M cations, such as K for Na, at constant framework composition and structural state. Beyond the triclinic to monoclinic transition, or its equivalent crossover in ordered alkali feldspars, a slightly different behaviour is apparent (Trend IB) with *b* and *c* increasing at a decreased rate. Trend II occurs beyond the expansion limit of Trend I, and upon T cation substitution at constant M cation composition. The two trends form the edges of quadrilaterals on a *b-c* and α-γ plots whose corners are defined by the cell parameters of low albite, analbite, sanidine, and low-microcline (Figures 6, 7).

Early work on various sanidines (Hazen, 1976) and low albite (Hazen and Prewitt, 1977) suggested that these feldspars follow the reverse of Trend I upon compression. The more recent single-crystal work which has also produced the bulk moduli data in Table 2 shows that this "inverse" behaviour is not generally the case. Most feldspars follow a trajectory on the *b-c* plot (Figure 6) upon compression that has more than twice the slope of Trend I (0.35). Albite is the closest to Trend I with a slope of 0.7, and the general trend of all of the end-member anorthites studied has a slope of 0.8. The natural crystal also shows a small vertical step on the *b-c* plot associated with the phase transition at ~2.55 GPa (Section 4.2), but this step is hardly apparent

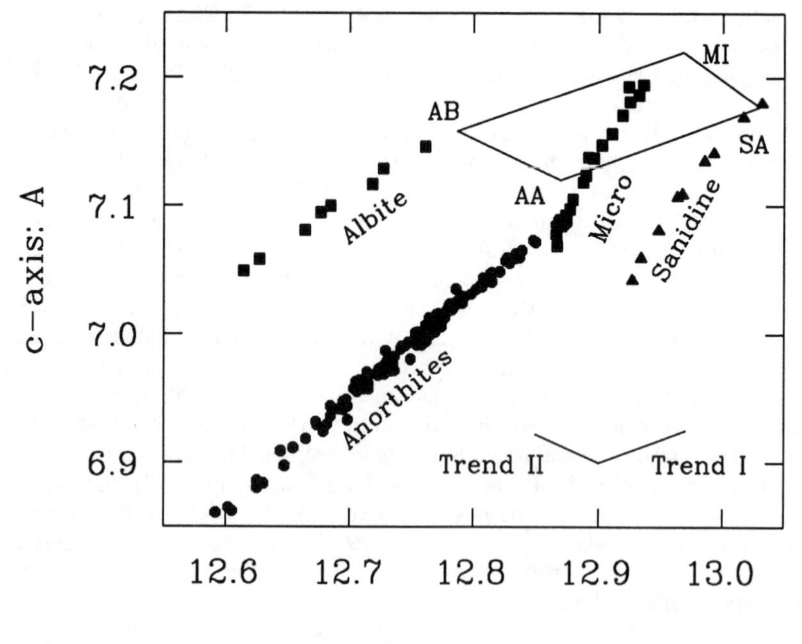
MI
7.2
AB
SA
c−axis: A
7.1
AA
Albite
Micro
Sanidine
7.0
Anorthites
6.9
Trend II
Trend I
12.6
12.7
12.8
12.9
13.0

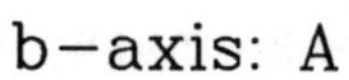
b−axis: A

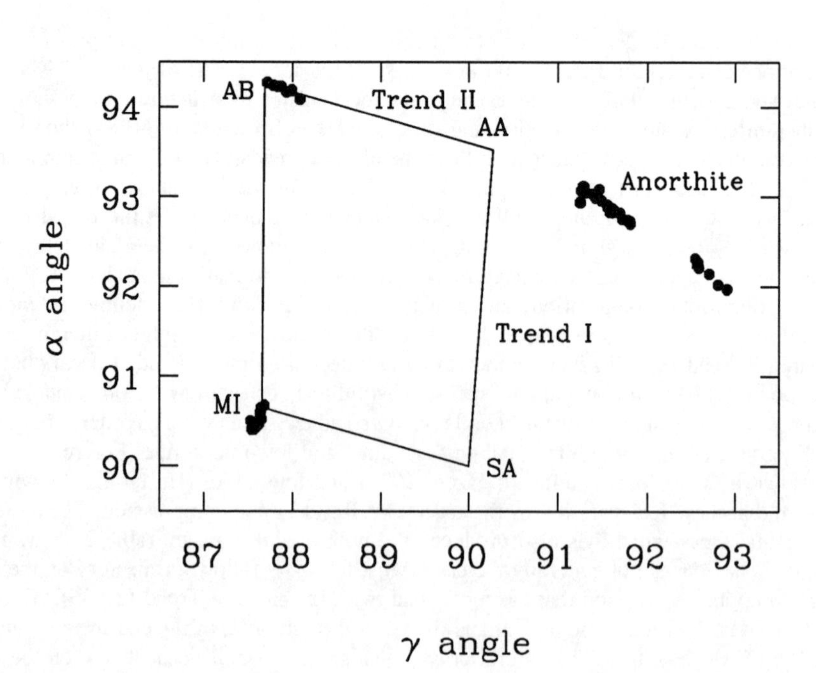
AB
Trend II
AA
94
93
Anorthite
92
α angle
Trend I
91
MI
90
SA
87
88
89
90
91
92
93
γ angle

in the partially disordered crystals. Potassium-rich feldspars display much steeper trends on the *b-c* plot, with slopes of 1.4 for monoclinic sanidine and 1.9 for microcline. The evolution of the unit cell angles with pressure are more variable. Microcline appears to be close to Trend I on an α-γ plot (Figure 7), whereas albite and anorthite are sub-parallel to the reverse of Trend II. It is therefore clear from consideration of the unit cell parameters alone that the effect of pressure upon the feldspar structure is not just the simple opposite of increasing temperature, for which the behaviour of triclinic alkali feldspars is dominated by the elimination of the displacive deviations from monoclinic symmetry (Kroll, 1984). This idea of "analogous variables" originally developed by Hazen (1976) had already been limited in the case of feldspars to those that follow Trend I on heating and compositional change (Brown et al., 1984). These high pressure data show that this concept has even more limited utility as it is not applicable to the behaviour of feldspars upon compression (but see Ribbe, this volume).

3.3. STRAIN TENSORS

As most feldspars are triclinic the choice of unit-cell is, in principle, quite arbitrary and there is nothing significant as far as compression is concerned in any particular setting. Further analysis of the response of the feldspar framework to the application of hydrostatic pressure therefore has to be made through the use of the concept of strain. The change in the unit-cell parameters which occurs upon application of a hydrostatic pressure is a strain, and this physical phenomenon is represented by a second-rank tensor (Nye, 1957). In the following discussion we choose to retain the room pressure unit cell as our reference state, and the strain tensor is calculated as the change in cell parameters induced by application of pressure. An alternative approach (equally valid, but subject to larger uncertainties) would be to calculate the strain due to each increment of pressure. The three principal coefficients of the strain, ϵ_1, ϵ_2, and ϵ_3, obtained by diagonalisation of the strain tensor, represent the fractional change in length along three mutually perpendicular directions denoted the "principal axes" of the strain tensor. Note that the principal strains are all negative for compression, and we use the convention that $|\epsilon_1| > |\epsilon_2| > |\epsilon_3|$. The principal axis along which the strain is ϵ_1 is therefore the most compressible direction in the crystal structure, while that corresponding to ϵ_3 is the stiffest direction. The isothermal *linear* coefficient of volume compressibility, β, is given by -

$$\beta = \frac{V_o - V_P}{V_o} = -(\varepsilon_1 + \varepsilon_2 + \varepsilon_3)$$

Figure 6 (opposite, top). Plot of the unit cell parameters *b* and *c* of feldspars under compression obtained from X-ray diffraction measurements. The quadrilateral at the top right is defined by the corners representing the room pressure cell parameters of albite, microcline, sanidine, and analbite. Its edges define Trends I and II. Note that the compression of most feldspars deviates considerably from Trend I.

Figure 7 (opposite, bottom). An α-γ plot of feldspars under compression. The quadrilateral is defined as Figure 6. Under compression albite unit-cell angles follow Trend II approximately, but those of microcline and anorthite deviate considerably from both trends. The gap in the anorthite data is due to the phase transition discussed in Section 4.2.

which is a different value from the compressibility obtained as the inverse of the bulk modulus, K_T, whose value is given by $-V.(\delta P/\delta V)_T$.

As with other second-rank tensors the strain can be visualised as a representation surface which is in general an ellipsoid with principal axes of lengths $(1+\epsilon_1)$, $(1+\epsilon_2)$, and $(1+\epsilon_3)$. This ellipsoid represents the shape that an initially spherical object would acquire if it were to undergo the same deformation as the crystal structure.The orientation of the strain is best represented visually in terms of the directions of the principal axes, as plotted on a stereogram along with the unit-cell vectors of the reference unit cell. More detailed discussion of the application of the concept of strain to feldspars, and in particular to the thermal expansion behaviour of alkali feldspars, can be found in Kroll (1984). Full expansions of the individual coefficients of the strain tensor for various choices of orthogonal axial sets can be found in Schlenker et al. (1978) and Redfern and Salje (1987).

In monoclinic feldspars the three principal strains are unequal, but one of the principal axes is required by the monoclinic symmetry to lie parallel to the diad along the *b*-axis. This in turn means that the other two principal axes must lie within the (010) plane. The strain ellipsoid calculated from the cell parameters (Angel et al., 1988) of an Or_{98} sanidine (Figure 8) has the least compressible direction parallel to the diad axis, and ϵ_2 almost exactly parallel to the *c*-axis. This means that the most compressible direction of the sanidine structure is parallel to the (100)

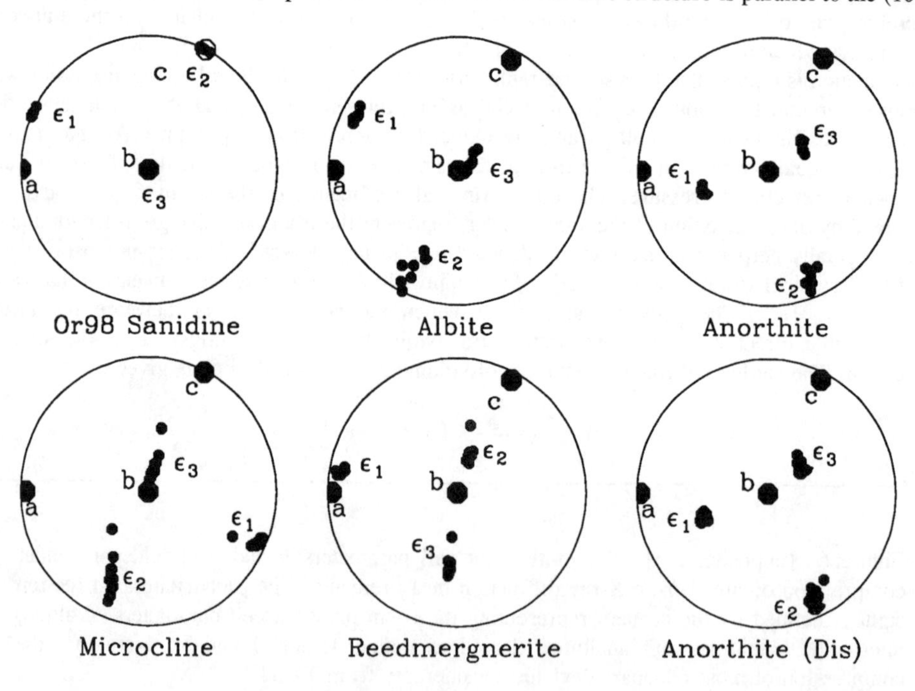

Figure 8. The orientation of the principal axes of the strain ellipsoids describing the behaviour of feldspars upon compression. All axes are plotted in the upper hemisphere, except for the cell axes *a* and *c*. The maximum pressure of the data for anorthite is 2.5 GPa, that of the disordered anorthite (Q_{OD} = 0.78) is 8 GPa, the remainder is 5.0 GPa.

plane normal (approximately the [201] direction). The same orientation is observed in albite and, more approximately, in microcline and reedmergnerite (Figure 8) although it is not required by the symmetry of these crystals. Note that, unlike the other alkali feldspars, the strain ellipsoid of microcline rotates significantly during compression towards the orientation found in sanidine. This rotation, and the magnitude of the strains, is unaffected by the transition at ~3.7 GPa (Section 4.1). Examination of the magnitudes of the principal strains reveals that some 60-70% of the volume compression of the alkali feldspars is accommodated by the linear compression along the (100) plane normal (Table 3). This extreme anisotropy, and its orientation, is almost identical to (but obviously opposite in sign to) the strains accompanying thermal expansion (Williame et al., 1974, Kroll, 1984). In this sense it would be true to say that alkali feldspars do, to a certain extent, exhibit "inverse behaviour", but it must be noted that in the triclinic alkali feldspars the orientations of ϵ_2 and ϵ_3 arising from thermal expansion are very different from those arising from compression.

Analysis of the data for all of the anorthites studied at pressure (Table 2) shows that their strain ellipsoids are very similar to one another. Therefore the overall pattern of deformation upon compression in anorthites is invariant with structural state and composition and is apparently unaffected in orientation by the $P\bar{1} \leftrightarrow I\bar{1}$ transition at high pressure. This orientation is quite different from alkali feldspars (Figure 8) in that the maximum compression occurs in, or close to, the (001) plane as a result of the large increase in the γ unit-cell angle with increasing pressure, and this orientation is also very different from that of the strain induced by thermal expansion. Nevertheless, the same degree of anisotropy as a result of compression is observed as that found in alkali feldspars (Table 3).

TABLE 3. Principal unit strains of feldspars at 4-5 GPa.

Sample	**P: GPa**	ϵ_1	ϵ_2	ϵ_3	$\epsilon_1/(\epsilon_1+\epsilon_2+\epsilon_3)$
An_{100}	5.0	71	24	7	70%
An_{100} (Q_{OD}=0.78)	4.7	71	31	12	62%
An_{98}	4.5	80	30	12	66%
An_{89}	4.6	77	29	14	64%
An_{72}	4.2	68	40	27	50%
Albite	4.9	86	32	25	60%
Reedmergnerite	4.6	77	29	19	62%
Sanidine (Or_{98})	4.9	92	39	16	63%
Microcline (Or_{87})	4.7	91	39	14	63%

Note: All principal strains are in units of 10^4 GPa^{-1}.

We may now turn to the question of whether there is any obvious causal relationship between structural elements of the feldspar structures and the orientation and anisotropy of the strain ellipsoids. The most obvious comparison to make is between the strain ellipsoid and the directions of various inter-atomic vectors within the structures. Figure 9 is a set of stereograms showing the interatomic vectors in sanidine and $P\bar{1}$ anorthite. Stereograms for triclinic alkali feldspars are similar to those of sanidine except that each sanidine vector is replaced by a pair of vectors in the triclinic structures. The first point of note is that the *shortest* M-O bond in sanidine, M-Oa2, is almost exactly parallel to the *most compressible* direction of the structure. This could imply one of two things. Either the M-O bonds (or at least the M-Oa2 bond) has no effective strength, or there is an extremely weak link in line with the M-Oa2 bond which is extremely compressible, so that the M-Oa2 bond itself is only allowed a passive role in determining the compressibility. The stereograms of Figure 9 show that there is great similarity between the directional distribution of the M-O bonds in both alkali and plagioclase structures, whereas the T-T vectors (across shared oxygen atoms) and T-O bonds do show significant differences, thus suggesting again that it is the tetrahedral framework that is controlling the compressibility of feldspars.

In conclusion, it will be apparent from the fore-going arguments that although much is known regarding the compressional behaviour of feldspars and its variation with composition and

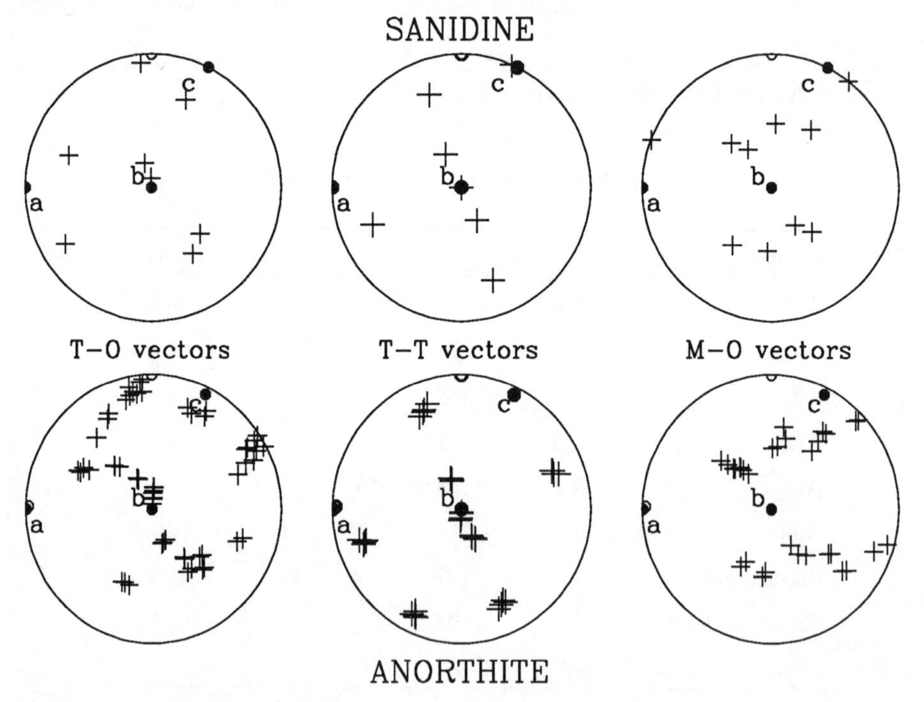

Figure 9. The orientation of interatomic vectors in sanidine and anorthite. All vectors are plotted in the upper hemisphere. The angular distribution of bonds in triclinic alkali feldspars is similar to that in sanidine, except that each single vector in sanidine is replaced by a closely spaced pair of vectors.

structural state, very little is known about the underlying atomic scale mechanisms of compression. The complexity of the feldspar structure makes it difficult to isolate one particular mechanism, as is often done in simpler structures (eg. the work on pyroxenes by Weidner and Vaughan, 1982) where compressibility along a given direction can be identified as being the response of a single bond or set of bonds. Quite clearly the question of compression mechanisms in feldspars can only be resolved by high-quality structure determinations at pressure, which so far have been reported only for the *average* structure of anorthite (Angel, 1988).

4. Structural Phase Transitions

Structural phase transitions in feldspars are, in general, non-quenchable and therefore require *in-situ* studies at high temperatures and/or pressures for their detection and characterisation. Those that have been detected generally involve small (typically 0.1%) or no volume change, and remained undetected during the era of piston-cylinder type measurements of bulk moduli described in Section 3.1. Apart from their relevance to the complete thermodynamic description of feldspars at high temperatures and pressures, the study of such transitions at high pressures in the diamond anvil cell provides an opportunity to study the effect of differing degrees of quenched-in Al,Si order on structural phase transitions in these materials without the dangers of re-ordering associated with experiments at high temperatures.

4.1. PHASE TRANSITIONS IN ALKALI FELDSPARS

The compressional study of albite to ~5 GPa by Angel et al. (1988) demonstrated that it remains as the $C\bar{1}$ phase over this pressure interval at room temperature. By contrast, Hazen (1976) reported unit-cell parameter discontinuities in intermediate alkali feldspars with the monoclinic sanidine structural state. His results, based upon single-crystal precession diffraction techniques suggested that the monoclinic symmetry of sanidine is broken by the application of pressure, resulting in a crystal consisting of $C\bar{1}$ domains twinned on the albite law. Hazen (1976) reported approximate transition pressures of 1.2 GPa for a sample with composition of Or_{67} and 1.8 GPa for a sample of ~Or_{82}. After a failure to detect any such transition in an Or_{98} high sanidine compressed to ~5 GPa (Angel et al., 1988), an attempt was made to confirm with single-crystal diffractometry the results of Hazen (1976) on crystals from the same samples that he had used. Compression of the Or_{82} sample to 5 GPa in a hydrostatic alcohol pressure medium resulted in no anomalous behaviour at all, with all the diffraction peaks remaining sharp and the unit-cell conforming to monoclinic symmetry within the small *esd*'s derived from the experiments (Hackwell, 1993). Similar behaviour was observed in a third sample of Or_{62} composition, which retained monoclinic symmetry to 4.0 GPa. If the assumption is made that the P-T-X surface of the $C\bar{1} \Leftrightarrow C2/m$ displacive phase transition is planar, then these new observations require $\delta T/\delta P$ of the transition surface to be less than ~190°C.GPa^{-1}, a value consistent with the constraint $\delta T/\delta P < 170$°C.GPa^{-1} that can be derived from the microstructures of alkali feldspars observed in high-grade metamorphic rocks (Evangelakakis et al., 1993). By contrast, the results of Hazen (1976) require $\delta T/\delta P$ for the transition surface to be ~ 700°C.GPa^{-1}. To test whether this discrepancy was due to the use by Hazen (1976) of a potentially non-hydrostatic pressure medium, a duplicate compressional study of the Or_{82} was undertaken with glycerine as a non-hydrostatic pressure fluid. This did cause anomalous broadening of all diffraction peaks by 1.9 GPa, suggesting that the effects observed by Hazen (1976) were probably the result of stresses applied to the crystal by a non-hydrostatic pressure medium. It is not clear whether the broadening of the diffraction peaks arises directly from the strain, or whether the non-hydrostatic stresses drive a monoclinic to triclinic transition at pressures far less than the equilibrium transition pressure.

By contrast, recent single-crystal diffraction measurements on microcline (Hackwell, 1993) indicate that well-ordered K-rich feldspars exhibit a change in compressional behaviour at ~3.7 GPa. In order to obtain reasonable single crystals despite the cross-hatched twinning found in microclines, several samples of amazonite, which contain < 1wt % of Pb but tend to have mm-scale twins, were used in these experiments. Unit-cell parameters indicate that these crystals are strained (Kroll and Ribbe, 1983) due to the twinning, but they display sharp, unsplit X-ray diffraction peaks unlike many microcline crystals. Cell parameters indicate that the compositions of the crystals are ~Or_{87}, and almost completely ordered (t_{1o} = 0.96, t_{1m} = 0.04 from equations in Kroll and Ribbe, 1983). On compression to ~3.3 GPa the unit-cell edges and the α unit-cell angle decrease linearly with pressure, but the β and γ angles display significant curvature (Figure 10). Above ~3.8 GPa the rate of change of the unit-cell edges with pressure is unchanged, but the α and β angles show increases with pressure which represent significant deviations from

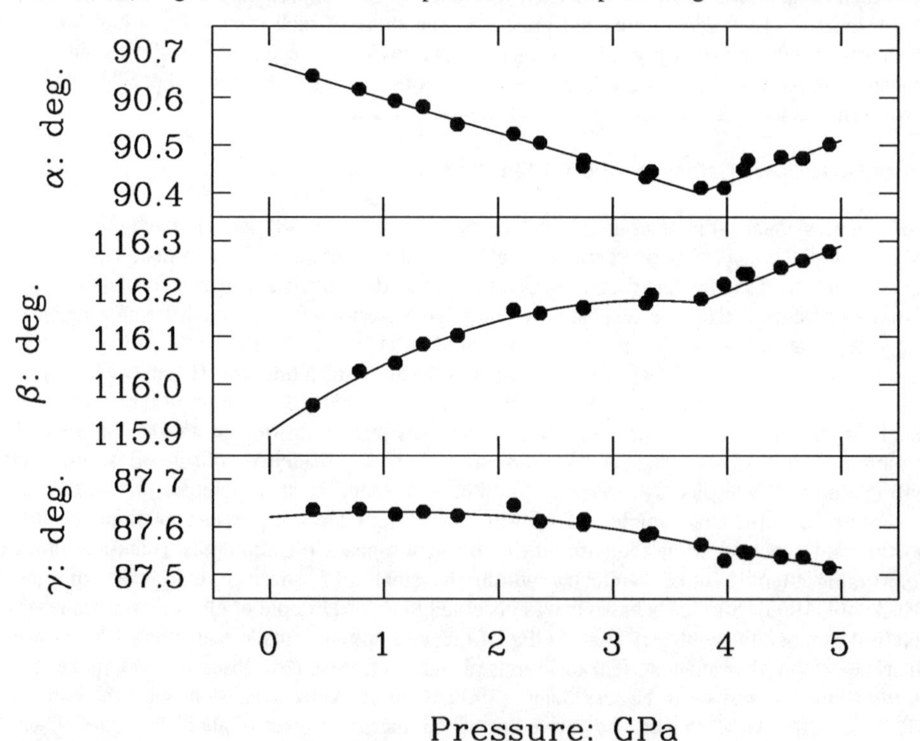

Figure 10. The variation with pressure of the unit-cell angles of an Or_{87} amazonite (microcline with trace Pb) determined by single-crystal X-ray diffraction. The inversion in the trends at ~3.8 GPa may be due to a phase transition or a simple change in compression mechanism.

the low-pressure trends (Figure 10). The volume change accompanying this change in behaviour is so small it is not measurable, and the difference between the unit cell volumes measured above 3.8 GPa and those estimated by extrapolation from the volume-pressure curve measured below 3.8 GPa are ~0.2%. Initial studies of the microcline above 3.8 GPa have not revealed anysymmetry change, but further work to characterise this behaviour is in progress at the time of writing.

4.2. $P\bar{1}$ TO $I\bar{1}$ TRANSITION IN ANORTHITE

The phase transition from the room pressure $P\bar{1}$ phase of anorthite to a high temperature $I\bar{1}$ phase was first detected at ambient pressure by Brown et al. (1963). It has been subsequently studied in great detail using diffraction techniques by Adlhart et al. (1980a,b), Redfern & Salje (1987), Redfern et al. (1988), and Redfern (1992), amongst others. Although there has been some dispute about the detailed nature of the transition at temperature (see Salje, this volume), it is clear from the temperature at which it occurs (~240°C) and its non-quenchable nature that the structural changes associated with it do not involve T-O bond breaking. The persistence of *c* and *d* reflections above the transition has been interpreted (Czank, 1973; Ghose et al., 1988) to suggest that the high-temperature $I\bar{1}$ phase is dynamic in character, with the tetrahedral framework fluctuating between the two configurations it could take up under $P\bar{1}$ symmetry. However, NMR studies (Staehli and Brinkmann, 1974; Phillips, 1990) and infra-red spectra (Redfern and Salje, 1992) point to any structural fluctuations being restricted to within 100K of the transition itself, and probably to the calcium atoms alone.

A phase transition with the same $P\bar{1}$ to $I\bar{1}$ symmetry change also occurs in anorthite at high pressures. This transition has been explored as a function of Al,Si order and composition by a series of single-crystal diffraction experiments, the results of which are reviewed in the following sub-sections. These experiments provide the variation with pressure of unit-cell parameters of the crystals. As anorthite is triclinic the choice of unit-cell is arbitrary, in the sense that it has no symmetrical significance, and changes in specific unit-cell parameters at the $P\bar{1} \leftrightarrow I\bar{1}$ transition are therefore without thermodynamic significance in themselves. However, the changes in cell parameters as a result of the transition can be re-formulated as a spontaneous strain (Aizu, 1970) which is a second-rank tensor. Since the phase transition from the $I\bar{1}$ phase to the P1 phase does not involve a change in the point-group symmetry the transition is properly termed "non-ferroic" (e.g. Wadhawan, 1983), and there are no symmetry-breaking components of the spontaneous strain associated with the transition. All components of the spontaneous strain therefore transform as the identity representation, and these components, ϵ_{ij}, should follow the order parameter Q_o of the transition as $\epsilon_{ij} \propto Q_o^2$. In practice, the propagation of the experimental uncertainties associated with the unit-cell determinations at high pressure results in considerable uncertainties in the values of the individual tensor coefficients. However, since there is considerable correlation between at least some of the *esd's* of the individual unit-cell parameters, the uncertainty in the scalar strain obtained from the diagonalised strain tensor coefficients ϵ_{ii}, $\epsilon_s = \sqrt{\sum_i \epsilon_{ii}}$ (Aizu, 1970; Redfern and Salje, 1987), is proportionately less than that of individual ϵ_{ij}. This scalar strain provides a measure of the magnitude of the distortion ($\epsilon_s \propto Q_o^2$) accompanying the $I\bar{1}$ to $P\bar{1}$ symmetry change, and was used to analyze the results of the high-pressure experiments. The cell parameters of the high-symmetry phase were linearly extrapolated to the pressures at which the $P\bar{1}$ unit cells were determined and the strain tensor obtained from the difference between the two. This approach was shown by Redfern and Salje (1987) to be suitable for the characterisation of the $I\bar{1}$ to $P\bar{1}$ transition in ordered Val Pasmeda

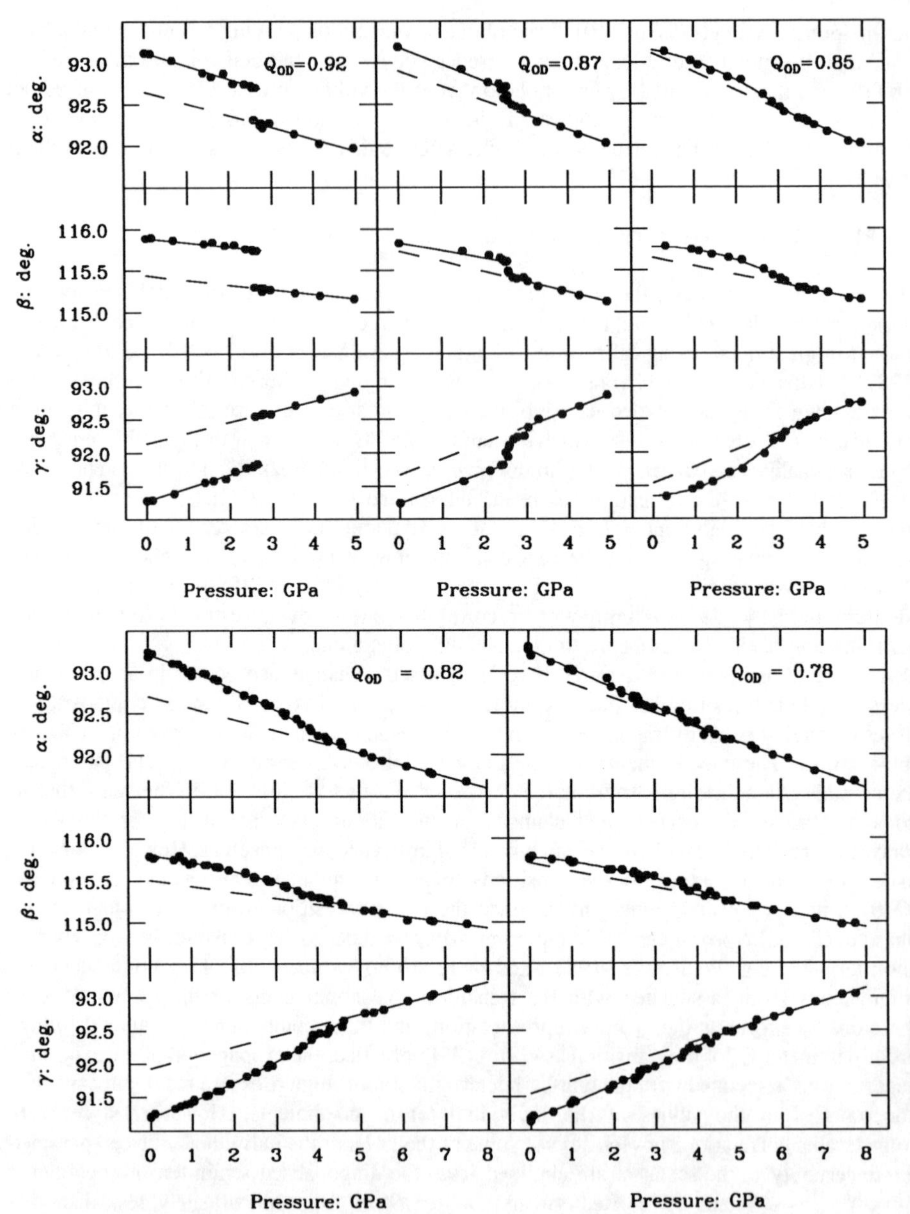

Figure 11. The variation with pressure of the unit-cell angles of five An_{100} anorthites with differing degrees of Al,Si order (indicated by Q_{OD}) determined by X-ray diffraction (Angel, 1992). The solid lines drawn through the data from the $I\bar{1}$ phases and continued to lower pressures as dashed lines were used to extrapolate the $I\bar{1}$ cell parameters for the calculation of the spontaneous strains shown in Figure 15.

anorthite at elevated temperature, and was applied subsequently to studies of the same transition in both anorthite-rich plagioclases (Redfern et al., 1988) and partially disordered end-member anorthite (Redfern, 1992) at high temperatures.

4.2.1. *$P\bar{1} \leftrightarrow I\bar{1}$ in End-Member Anorthite.* A phase transition in end-member (An_{100}) anorthite at high pressure was first detected by Angel et al. (1988) by single-crystal X-ray diffraction through large discontinuities in the unit-cell angles at a pressure between 2.55 and 2.95 GPa. Subsequent work (Angel and Ross, 1988) bracketed the transition by reversals at 2.55 GPa and 2.74 GPa, and showed that the α and β unit-cell angles decrease by $\sim 0.4°$ while γ increases by $\sim 0.7°$ at the transition to the high-pressure phase (Figure 11). There are also small decreases in the unit-cell edges which together result in a volume change at the transition of $\sim 0.15\%$. Together with the demonstrable hysteresis of at least 0.05 GPa at the transition, these results showed that the transition is first order in character. Nonetheless the transition is non-quenchable in the sense that the high-pressure phase reverts to the low pressure $P\bar{1}$ structure once the boundary is overstepped by ~ 0.05 GPa.

Initial investigations (Angel et al., 1988) incorrectly suggested that all four classes (*a, b, c,* and *d*) of reflections were present in the diffraction pattern of the high-pressure phase, leading to the suggestion that the phase transition was either associated with a decrease in symmetry (to $P\bar{1}$) or involved no symmetry change at all. A full structural study of An100 (Val Pasmeda) anorthite just below the phase transition (at 2.5 GPa) and above the phase transition (at 3.1 GPa) was subsequently carried out by Angel (1988) to determine the true symmetry change. All four classes of reflections indicative of a unit-cell with $P\bar{1}$ symmetry were present in the diffraction pattern of the crystal at 2.5 GPa, with approximately the same relative intensities as at ambient pressure ($I_a > I_c > I_b > I_d$). However, owing to the large number of structural parameters and the low numbers of observed *b, c,* and *d* reflections it did not prove possible to completely refine the $P\bar{1}$ structure of anorthite at this pressure, and an "average" structure with $I\bar{1}$ symmetry was instead refined to the intensities of the *a* and *b* reflections alone. This average structure is essentially the same as that at atmospheric pressure, although the poorer data quality resulted in much larger uncertainties in the refined parameters. There are no significant changes in tetrahedral bond lengths and internal (O-T-O) angles between 0.1 MPa and 2.5 GPa indicating that the tetrahedra are essentially rigid towards compression. However, the longer Ca-O bonds do show significant compression (of up to 0.1Å), although the amount of splitting of the pairs of Ca sites (related by a vector of approximately 1/2,1/2,1/2 in the $P\bar{1}$ structure) remains about the same. Thus, between room pressure and 2.5 GPa, compression within the anorthite structure is accommodated by the reduction in Ca-O distances arising from small flexings of the T-O-T bond angles between essentially rigid tetrahedra.

At 3.1 GPa there is a substantial change in the relative intensities of the reflections with $I_a > I_b$, and the *c* and *d* reflections absent. The structure of the high pressure phase therefore has true $I\bar{1}$ symmetry. In particular, the scattering density at the Ca sites within the structure at 3.1 GPa shows a single maximum at each site (Figure 12), in contrast to the split sites observed in the structures at 0.1 MPa and 2.5 GPa, indicating that no averaging is being performed by the refinement in $I\bar{1}$ symmetry. The internal dimensions of the tetrahedra show no significant changes from those below the phase transition, but there are significant changes in the T-O-T bond angles. The largest changes (up to 17°) are those at Obmo, Obmz, Odoo and Odmz, which are the sites showing the largest deviations from $I\bar{1}$ symmetry in the low-pressure $P\bar{1}$ phase. The phase transition therefore arises from a tilting of essentially rigid tetrahedra so as to bring the pairs of tetrahedra related by the pseudo-I symmetry of the $P\bar{1}$ phase into true equivalence. This

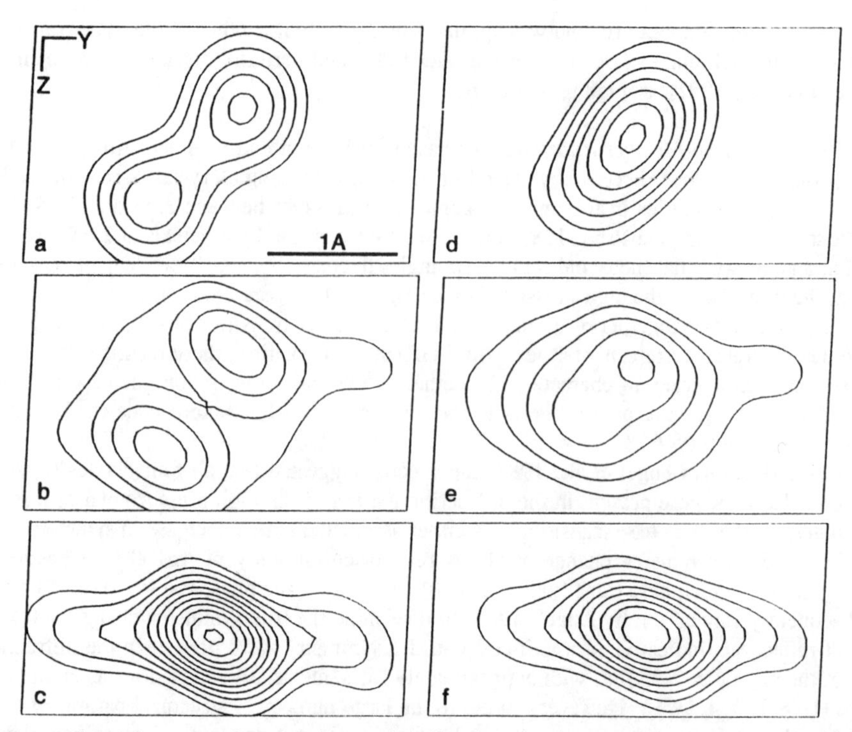

Figure 12. The scattering density at the Ca sites in the anorthite structure at various pressures. Maps on the left are sections through the Mooo site at (a) room pressure, average P$\bar{1}$ structure (b) 2.5 GPa, average P$\bar{1}$ structure (c) 3.1 GPa, I$\bar{1}$ structure. Parts (d), (e) and (f) are the Mzoo sites under the same conditions. Note the distinct change from split sites in the P$\bar{1}$ average structure to single sites in the I$\bar{1}$ structure.

tilting reconfigures the cages in which the calcium atoms reside, making them slightly smaller than in the low-pressure structure and, more significantly, this forces the Ca atoms to occupy a single minima in a more sharply defined potential well.

Well-ordered, end-member (An_{100}) anorthite therefore undergoes the same symmetry change from P$\bar{1}$ to I$\bar{1}$ at high pressure as that at high temperature, but with significant differences between the high-pressure and the high-temperature structures, and also between the characters of the phase transition in these two regimes. While both the high-temperature and high-pressure structures are believed to have a tetrahedral framework with true, static I$\bar{1}$ symmetry, the Ca atoms appear to be disordered within their cavities at high temperature but occupy single sites at high pressure. The transition is strongly first-order in character at high pressure, but is continuous at temperature (Redfern and Salje, 1987). In addition, unit cell parameters extrapolated to room pressure and temperature from the I$\bar{1}$ phases at high pressure and at high temperature are not the same, whereas there ought to be a single, well-defined I$\bar{1}$ reference state at any pressure and temperature below the phase transition. These observations would suggest that the high temperature and high pressure phases are distinct, whereas the measured ΔV and estimates of ΔS (Wruck, 1986) of the transition suggested that the phase boundary should have

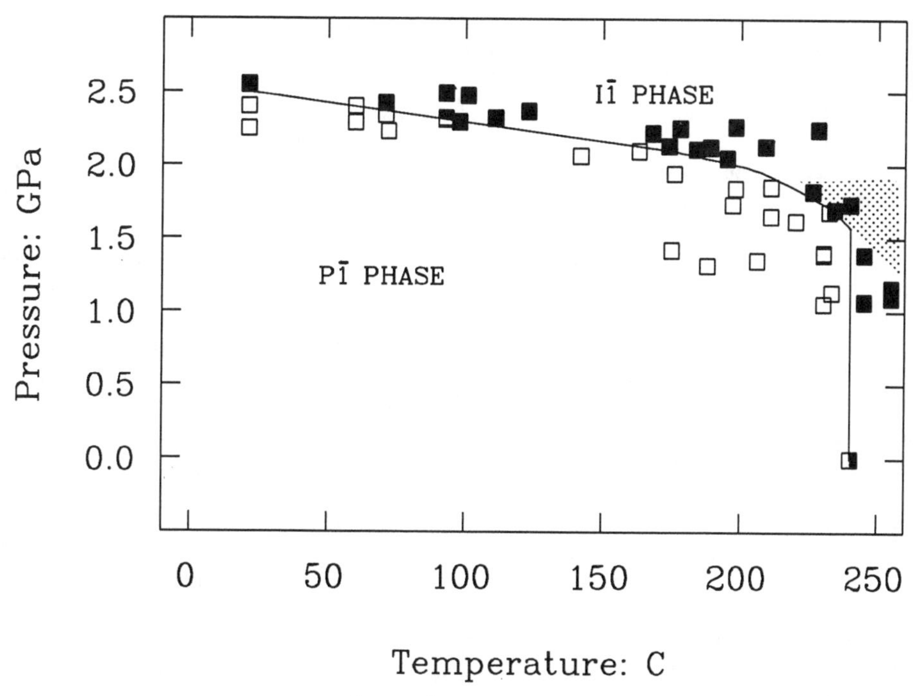

Figure 13. Reversals of the P$\bar{1}$ ↔ I $\bar{1}$ phase boundary in natural An_{100} anorthite as determined by in-situ high-pressure, high temperature single-crystal X-ray diffraction (Hackwell, 1993). Filled symbols are in the I$\bar{1}$ field, open symbols are in the P$\bar{1}$ field. The line is drawn in as a guide to the position of the phase boundary, but its precise form in the region of the crossover in the I$\bar{1}$ phase (indicated by the stipple) is not well constrained.

a negative $\delta P/\delta T$ slope and that the two structures could therefore be the same phase (Angel and Ross, 1988).

In order to determine which is the case Hackwell (1993) undertook a series of experiments using an externally heated diamond-anvil pressure cell to reverse the P$\bar{1}$ to I$\bar{1}$ phase boundary in P-T space. The phase transition was detected by the changes in unit-cell parameters determined in-situ by single-crystal X-ray diffraction, and additionally monitored by following the intensities of several strong *c* type reflections. An experiment performed at atmospheric pressure (ie. without pressure fluid in the cell) gave results in good agreement with those of Redfern and Salje (1987), thus indicating that the phase transition behaviour of the anorthite was not modified by any aspect of the experimental procedure. The phase boundary derived from the reversal experiments is shown in Figure 13. From 2.5 GPa and room temperature the phase boundary has a small negative slope, $\delta P/\delta T \sim -0.0025$ GPa.K^{-1}, up to a temperature of ~165° C. From 165°C to ~230°C the boundary becomes steeper, passing through a well-constrained bracket centred at 230°C and 1.7 GPa. From this pressure to ambient the boundary is almost vertical in P-T space, with $\delta P/\delta T \sim -0.2$ GPa.K^{-1}.

Along the high-pressure limb of the phase boundary the transition is clearly marked by large changes in unit-cell parameters similar to those observed at room temperature, and by the disappearance of the *c* and *d* reflections. The transition is first-order in character. Over the near-vertical limb of the phase boundary at high temperature there are no sharp changes in unit-cell angles at the transition, rather they appear to follow the same sort of variation observed at room pressure. The precision in these measurements is not good however, due to the difficulty in maintaining stable temperatures in the diamond cell and the uncertainties in the estimates of pressure and temperature. Within the limitations of the experimental data it appears that the transition is continuous (or very weakly first-order) across this limb of the phase boundary, with a very small, or zero, volume change, consistent with the very steep slope of the boundary ($\delta P/\delta T = \Delta S/\Delta V$). It is also notable that sharp *c* reflections persist, in a somewhat weakened state, for several tens of degrees above the phase boundary, as is observed at room pressure.

This set of reversals clearly demonstrates that the high-pressure and high-temperature phases are related, but are not identical in thermodynamic properties. The sharp curvature in the phase boundary between 1.7 GPa and 2.2 GPa is clearly related to the change in ΔV, and possibly ΔS, of the transition. Given that the cell parameters of the $P\bar{1}$ phase do not change very much in this pressure-temperature interval while those of the $I\bar{1}$ phase do (Figure 14) it is likely that it is changes in the structure and properties of the $I\bar{1}$ phase that is responsible for the curvature of the phase boundary. One would speculate that the dynamic nature of the $I\bar{1}$ phase within $\sim 100°$ of the phase boundary at ambient pressure (Redfern and Salje, 1992) is suppressed at pressures above $\sim$ 2 GPa and that the $I\bar{1}$ phase undergoes a "crossover" transition to a *static* $I\bar{1}$ structure with increasing pressure. This is certainly supported by the observation that the high-pressure $I\bar{1}$ structure has single Ca sites (Figure 12) whereas the high-temperature structure appears to have split sites between which the Ca atoms "hop" (Czank, 1973). It also explains the discrepancy, noted above, between the cell parameters obtained at room pressure and temperature by extrapolation of the $I\bar{1}$ cell parameters from the high pressure and high temperature fields.

4.2.2 *Effect of Al,Si Disorder on* $P\bar{1} \leftrightarrow I\bar{1}$. Angel (1992) has also reported the results of high-pressure single-crystal diffraction studies of a series of Val Pasmeda anorthite ($\sim An_{100}$) crystals with differing degrees of Al,Si disorder. Full details of the high-temperature annealing used to induce partial Al,Si disorder in these samples, and the methods used to determine the structural states of these crystals, are provided by Angel et al. (1990) and Carpenter et al. (1990). It is sufficient to note that Al,Si order-disorder is quantified by an order parameter Q_{OD} which would have the value of 1 in a completely ordered crystal, or 0 in a completely disordered one. The natural Val Pasmeda anorthite described in Section 4.2.1 has $Q_{OD} = 0.92$ (Angel et al., 1990).

In the sample with the least amount of induced Al,Si disorder ($Q_{OD} = 0.87$) evidence for the same $P\bar{1} \leftrightarrow I\bar{1}$ transition is provided by rapid changes with pressure in the unit-cell angles between 2.5 GPa and 3.0 GPa. The magnitudes of the differences between the unit-cell angles of the low pressure phase and those extrapolated from the high-pressure phase to just below the transition are much smaller than in the natural material: $\Delta\alpha \approx \Delta\beta \approx 0.2°$, $\Delta\gamma \approx 0.3°$. In addition it is clear that instead of a sharp discontinuity at a well-defined transition pressure, these changes are "smeared-out" over the pressure range of 2.5 to 3.0 GPa (Figure 11). These changes in transition character are continued in the next most ordered sample studied ($Q_{OD} = 0.85$), with similar differences between the unit-cell angles of the $P\bar{1}$ and $I\bar{1}$ phases (Figure 11), but with an increased width of transition smear, to 3.3 GPa. By contrast, the unit-cell angles of the two most disordered samples, ($Q_{OD} = 0.82$) and ($Q_{OD} = 0.78$) vary non-linearly with pressure

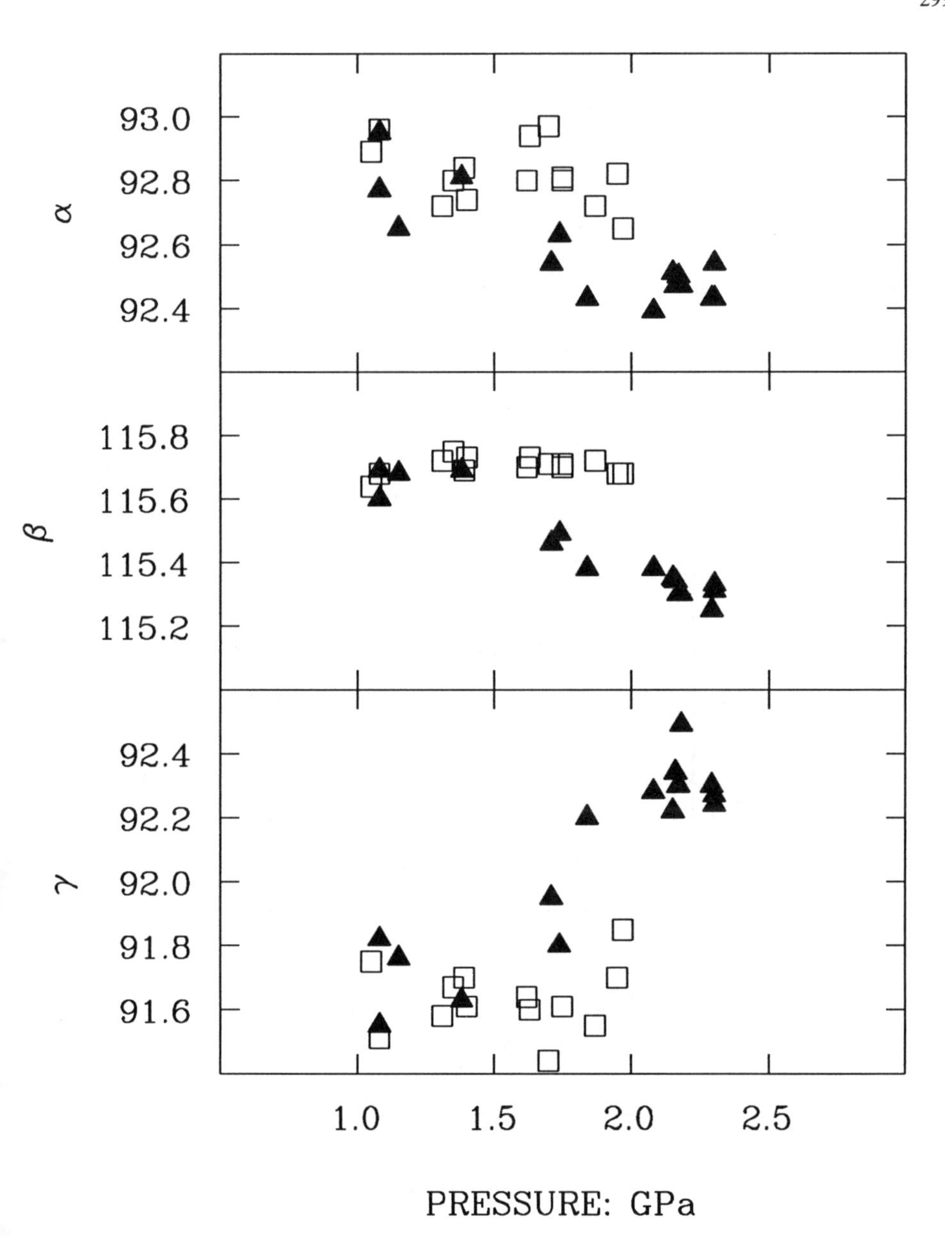

Figure 14. Unit-cell angles of the $I\bar{1}$ (solid symbols) and $P\bar{1}$ (open symbols) of natural An_{100} anorthite between 175° C and 260°C at various pressures. The change in $I\bar{1}$ unit-cell angles at ~1.5 GPa is indicative of a "crossover" transition in this phase.

up to ~4.8 GPa where changes to linear trends are accompanied by distinct changes in slope, but without any detectable discontinuity. Neither is there any detectable step in either the unit-cell edges or the volumes of these two samples, clearly indicating that the transition has become continuous in character.

The scalar strain of the natural sample with Q_{OD}=0.92 displays a sharp discontinuity just above 2.5 GPa, confirming the first-order nature of the transition. With increasing disorder the decrease in the size of the steps in the unit-cell angles at the transition are reflected in the decreased magnitudes of the steps in the scalar strain (Figure 15) which, although they too begin at 2.5 GPa, are smeared out to higher pressures. The two most disordered samples, with Q_{OD} = 0.82 and 0.78, exhibit a continuous decrease in ϵ_s with increasing pressure within the P1

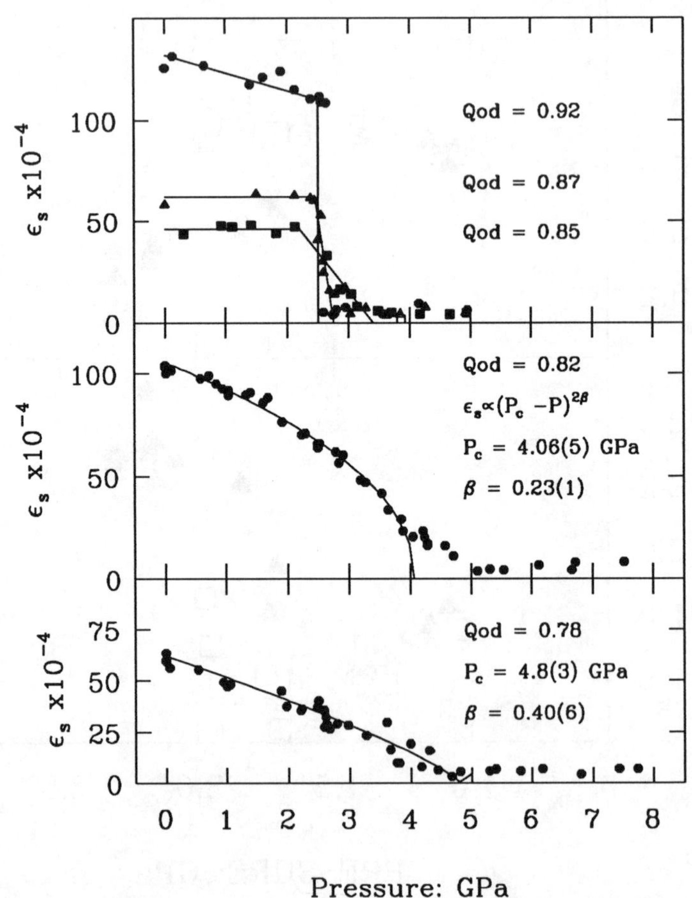

Figure 15. The variation of the scalar strain, ϵ_s, that arises from the I$\bar{1}$ to P$\bar{1}$ transition in An_{100} anorthites with differing degrees of Al,Si order. Data points at pressures above the transitions give an indication of the experimental uncertainties in ϵ_s as these values should ideally be zero. Lines in the top part are guides to the eye, lines in the lower two parts are power law fits to the data, with the parameters given.

stability field, which allows an estimate of the transition pressure, P_{tr}, to be obtained through fitting the ϵ_s data with a power-law equation of the form: $\epsilon_s \propto (P_{tr} - P)^{2\beta}$. While the transition pressure obtained for the most disordered sample (Q_{OD} = 0.78) is 4.8(3) GPa, in good agreement with the pressure at which the variation in unit-cell angles with pressure becomes linear (Figure 11), the transition pressure of the sample with Q_{OD} = 0.82 obtained by fitting the ϵ_s data, 4.06(5) GPa, is some 0.7 GPa below that point. The scalar strain between 4.0 and 4.7 GPa is also quite clearly non-zero in this sample. This smearing of the transition is similar to that displayed by the less disordered samples (Q_{OD}= 0.87, 0.85), and can be interpreted as arising from inhomogeneity in the state of Al, Si order within all three samples. The critical exponents, β, obtained from fitting the ϵ_s data are 0.23(1) for Q_{OD} = 0.82 and 0.40(6) for Q_{OD} = 0.78. These values indicate that the former exhibits close to tricritical behaviour (for which β would be 1/4), while the latter is intermediate between tricritical and second-order (β = 1/2).

These experimental results demonstrate that, as for the $P\bar{1} \leftrightarrow I\bar{1}$ transition at high temperature (Redfern, 1992), the state of Al,Si order within the anorthite has a large effect on both the position of the transition and its character. It may therefore be said that there is strong coupling between the order parameters Q_O and Q_{OD} which describe these two processes.

4.2.3 *Effect of Albite Substitution on* $P\bar{1} \leftrightarrow I\bar{1}$ *Transition.* The influence on the $P\bar{1}$ to $I\bar{1}$ transition at pressure of the coupled substitution of Na+Si for Ca+Al (ie. albite component) was followed by Angel et al. (1989) in a series of single-crystal diamond-anvil cell experiments on well-ordered natural feldspars. Including the natural Val Pasmeda crystal described above, the study used five natural, well-ordered crystals ranging in composition from An_{100} to $An_{88\text{-}90}Ab_{12\text{-}10}$. The details of the structural states of these samples are given in Angel et al. (1990) and Carpenter et al. (1990).

The evolution with pressure of the cell parameters of the Monte Somma anorthite ($An_{97\text{-}100}$) was followed in detail. The unit-cell angles show a sharp change around 2.6 GPa; the transition is actually bracketed by reversals at 2.5 GPa and 2.7 GPa. These cell parameter changes result in a small volume discontinuity which, together with the evolution of the spontaneous strain clearly demonstrates that the transition is first-order in character, but with a smaller step size at the transition than is seen in the Val Pasmeda anorthite. High-pressure experiments on the remaining three samples were restricted to obtaining reversals of the $P\bar{1}$ to $I\bar{1}$ phase transition and estimates (from the cell parameters of the two phases close to the transition) of the scalar strain associated with the transition. Two important results emerge from this study (Table 4). First, the transition pressure increases with increasing albite content (Figure 16). The solid line in Figure 16 is a linear phase boundary obtained by a linear programming analysis of the experimental brackets, and has the equation:

$$P_{tr} = 1.04 N_{Ab} + 2.53 \, (GPa)$$

where N_{Ab} is the mole percent of albite in solid solution. The extreme values allowed by the experimental brackets for the slope and intercept of a linear phase boundary are (0.86, 2.66), (0.95, 2.66), (1.07, 2.41) and (1.28, 2.40) which are shown as the dashed lines in Figure 16. The second result apparent from these experiments is that the size of the step in the cell parameters at the $P\bar{1} \leftrightarrow I\bar{1}$ transition decreases with increasing albite content, a continuation of the trend apparent from the more detailed data collected on the An_{100} (Val Pasmeda) and An_{97}-An_{100} (Monte Somma) samples. The reduced magnitude of the steps in the unit-cell angles is reflected in decreasing scalar strain at the transition with increased albite content (Table 4).

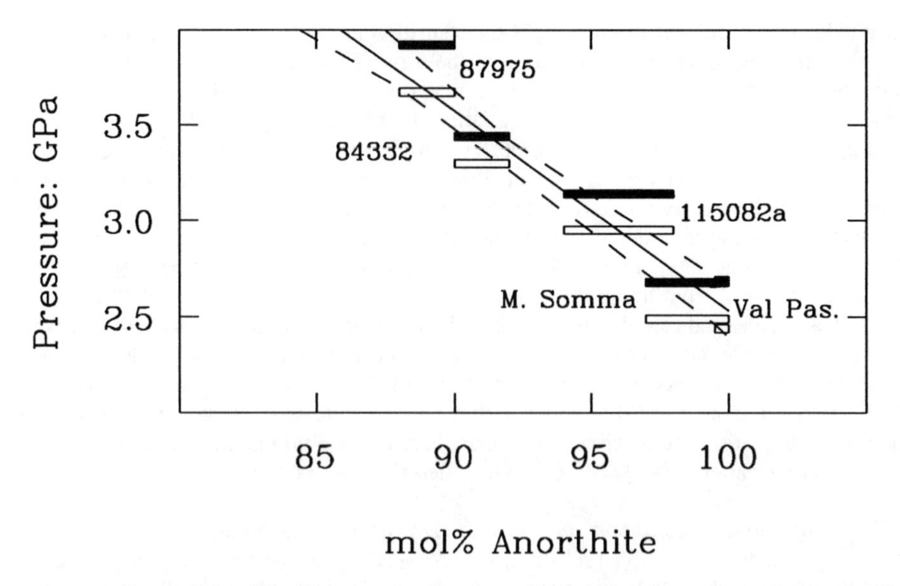

Figure 16.The $P\bar{1} \leftrightarrow I\bar{1}$ phase boundary as a function of albite content in well-ordered natural crystals. The filled symbols represent reversals in the $I\bar{1}$ field, open symbols reversals in the $P\bar{1}$ field. The width of the symbols represents the uncertainties in the compositions of the samples. The solid line is the best linear estimate of the phase boundary, the dashed lines indicate the limits of the range of positions of a straight-line boundary consistent with the data.

TABLE 4. Parameters of the $P\bar{1}$ to $I\bar{1}$ phase transition in anorthite-rich feldspars.

Sample	Composition	Q_{OD}	Transition Bracket (GPa)	ϵ_s
VP	An100	0.92	2.46-2.67	0.0106
MS	An97-An100	0.91	2.51-2.66	0.0095
115082	An94-An98	0.88	2.97-3.12	0.0073
84332	An90-An92	n.d.	3.32-3.42	0.0049
87975	An88-An90	0.85	3.69-3.90	0.0030

Note: Sample numbers refer to the Harker collection of the University of Cambridge. Compositions are the range for these samples reported by Carpenter et al. (1985). Individual crystals used in the high-pressure experiments were not probed.

Extrapolation of this trend with composition (Figure 17) suggests that the first order step would disappear in samples with $N_{Ab} > 15\%$, at which point the transition should become continuous in character.

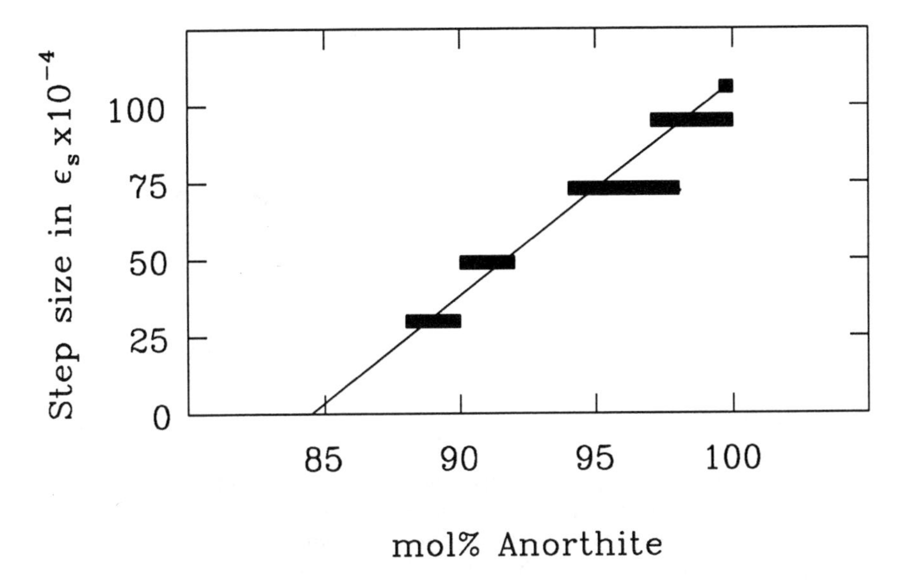

Figure 17. The variation with composition of the step size in the scalar strain ϵ_s at the $I\bar{1} \rightarrow P\bar{1}$ transition in well-ordered anorthites (Table 4). The width of the symbols represents uncertainties in the compositions of the samples.

4.2.4. *Phase Diagram for Anorthite $P\bar{1} \leftrightarrow I\bar{1}$ Transition.* The $P\bar{1} \leftrightarrow I\bar{1}$ phase boundary determined by the experiments described in the preceding sections is shown schematically in Figure 18. The $I\bar{1}$ structure is both the high-temperature and the high-pressure phase, but these two regimes are separated by a cross-over region which also causes the phase boundary to be sharply curved in P-T space. An increase of Al,Si disorder at constant composition stabilises the $P\bar{1}$ phase to higher pressures (Section 4.2.2), in contrast to the reduction in the stability of the $P\bar{1}$ phase at temperature (Redfern, 1992). Note that with increasing disorder, therefore, the *average* slope of the $P\bar{1}$ to $I\bar{1}$ phase boundary is increased, although the precise form of the boundary in P-T space remains to be determined. The actual effect of the addition of albite component (Section 4.2.3) on the stability of the $P\bar{1}$ phase is more difficult to determine because the substitution of Na for Ca is coupled with the substitution of Si for Al, which results in further Al,Si disorder. Redfern (1992) was able to determine that, at atmospheric pressure, the substitution of albite content at constant Q_{OD} results in an *increase* in the phase transition temperature, but that for well-ordered natural materials this effect is over-shadowed by the decrease in Q_{OD}. The high pressure data seem to suggest that an increase in N_{Ab} also leads to stabilisation of the $P\bar{1}$ phase *over and above* that induced by the decrease in Q_{OD}, but much more data need to be collected before the effects of Q_{OD}, albite substitution, and inhomogeneity in Al,Si order can be properly separated.

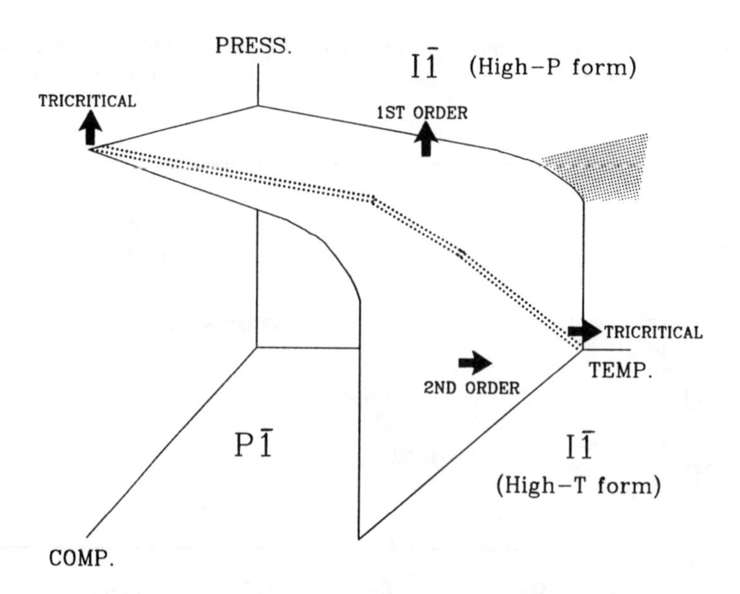

Figure 18. A schematic P-T-X_{Ab} phase diagram for natural, well-ordered anorthite-rich plagioclases. The phase boundary surface separates a $P\bar{1}$ stability field at low pressures, temperatures and albite content from the $I\bar{1}$ phase field which has two distinct regimes separated by a "crossover" transition (indicated at An_{100} by the stippled area). The form of the P-T boundary at compositions other than An_{100} is unknown, and the point of curvature may move to lower temperatures or disappear altogether. The phase boundary surface must, however, be crossed by a line of tri-critical points (dotted line) separating first-order from second-order character, but its precise trajectory is unknown.

4.2.5. *Thermodynamic Analysis.* The thermodynamic analysis of anorthite-rich feldspars developed by Salje (1987) has been remarkably successful in explaining their behaviour at high temperatures. Its basis is that the various processes, such as displacive phase transitions and Al,Si disordering, operating within the feldspar give rise to structural strain. These processes therefore affect one another through this common strain - this is described in terms of the *coupling* of the order parameters describing the processes to the strain ϵ. Here the applicability of Salje's analysis to the changes in transition behaviour at high pressure that have been described in the preceding sections is assessed. The excess free energy of the P-1 phase relative to a (disordered) $C\bar{1}$ reference state can be expressed in terms of the two order parameters Q_o and Q_{od}:

$$G_{ex}=\frac{1}{2}a_o(T-T_{co})Q_o^2+\frac{1}{4}b_oQ_o^4+\frac{1}{6}c_oQ_o^6$$
$$+\frac{1}{2}a_d(T-T_{cd})Q_{OD}^2+\frac{1}{4}b_dQ_{OD}^4+\frac{1}{6}c_dQ_{OD}^6$$
$$+e_o\varepsilon Q_o^2+e_d\varepsilon Q_{OD}^2+f\varepsilon^2$$

where the T_{co} and T_{cd} are the transition temperatures for the $I\bar{1} \leftrightarrow P\bar{1}$ and $C\bar{1} \leftrightarrow I\bar{1}$ transitions in isolation, the *a, b*, and *c* are constants, the e_o and e_d are the coupling coefficients of the two order parameters Q_o and Q_{OD} to the strain, and the coefficient *f* is a function of the elastic constants of the material. By applying the conditions that the crystal is stress free when in equilibrium, and that in a low temperature experiment the state of Al,Si order stays fixed (i.e. Q_{OD} is constant), the excess free energy, G'_{ex}, of the P-1 phase with respect to the I-1 phase may be derived:

$$G'_{ex}=\frac{1}{2}a_o\left(T-T_{co}-\frac{e_o e_d Q_{OD}^2}{a_o}f\right)Q_o^2+\frac{1}{2}\left(b_o-\frac{e_o^2}{f}\right)Q_o^4+\frac{1}{6}c_o Q_o^6$$

By making the assumption that the excess volume associated with the I-1 to P-1 transition, V_{ex}, is proportional to Q_o^2 in the same way that the excess entropy, S_{ex}, is assumed to vary with Q_o^2, the equation can be re-arranged in terms of pressure (Carpenter, 1992) as:

$$G'_{ex}=\frac{1}{2}a_{vo}(P-P_{tr})\,Q_o^2+\frac{1}{4}\left(b_o-\frac{e_o^2}{f}\right)Q_o^4+\frac{1}{6}c_o Q_o^6$$

where the transition pressure, P_{tr}, is a function of both the experimental temperature, T_{exp}, and Q_{OD}:

$$P_{tr}=\frac{a_o}{a_{vo}}\left(T_{co}-T_{exp}+\frac{e_o e_d Q_{OD}^2}{a_o f}\right)$$

This last equation defines the $P\bar{1} \leftrightarrow I\bar{1}$ phase boundary in P-T-Q_{OD} space, and some conclusions can be drawn about the signs and P-T dependencies of the coefficients appearing in this and the preceding equations. At constant Q_{OD} the phase boundary is defined by P_{tr} and T_{exp}. The observation that $\delta P/\delta T < 0$ implies that $a_o/a_{vo} > 0$. The observation that the $P\bar{1}$ phase is destabilised, that is the pressure of the $P\bar{1} \leftrightarrow I\bar{1}$ transition is decreased, by *increasing* Q_{OD} indicates that the coupling between these two order parameters, represented by $e_o e_d/a_{vo}f$ in the above equations, is negative. This negative sign arises from the fact that the strain arising from ordering opposes that due to the $I\bar{1}$ to $P\bar{1}$ transition (Angel, 1992), *decreasing* the stability of the $P\bar{1}$ phase. As the state of order is decreased, however, the magnitude of the strain arising from the $C\bar{1}$ to $I\bar{1}$ transition is decreased, so it becomes easier for the structure to deform as a result of the $I\bar{1}$ to $P\bar{1}$ transition and the stability field of the $P\bar{1}$ phase is thereby expanded. This behaviour is in direct contrast with that of albite and albite-rich alkali feldspars (Salje, 1985; Salje et al., 1985) in which the displacive and ordering transitions couple in such a way that the low symmetry phase is stabilised by *increasing* Al,Si order.

There are however, a number of experimental observations that cannot be explained by the simple application of these free energy expansions. They predict a linear phase boundary in P-T space if the coupling coefficients remain constant, contrary to the observed sharp curvature. The character of the transition is defined by the sign of the Q_o^4 term. The observation that the transition changes character from tricritical at room pressure to first order at high pressure requires a change in the coefficient of Q_o^4 from zero at room pressure to a negative value at high pressure. Changes in the elastic constants of anorthite, which contribute to the *f* term, cannot be responsible because they will lead to an increase in *f*, and more positive values of the Q_o^4 coefficient, as the structure stiffens with the application of pressure. Several other possibilities may be examined to explain the observed behaviour. First, Q_o may be coupled with the order parameter of another displacive phase transition, such as a monoclinic to triclinic inversion. Such a symmetry change, from $C\bar{1}$ to C2/m, has been predicted (Carpenter and Ferry, 1984) to occur at temperatures well above the melting point of anorthite at room pressure. At low temperatures and high pressures the state of Al,Si order would remain unchanged, and the triclinic to monoclinic inversion would result in a symmetry change from $I\bar{1}$ to I2/c. Coupling between such a transition and the $I\bar{1}$ to $P\bar{1}$ transition, allowed by symmetry to be linear-quadratic in form, could be responsible for the changing transition character, although it should be noted that McGuinn and Redfern (1993) report that the application of pressure to I2/c Sr-anorthite results in a decrease in symmetry to $I\bar{1}$. This might be taken as evidence against the possibility of the opposite transformation occurring in anorthite at ultra-high pressures. It is possible that V_{ex} may not be proportional to Q_o^2, but this is unlikely given that V_{ex} is so small, about 0.2% of the total volume. Finally, the changing character of the Ca distribution in the $I\bar{1}$ structure (Section 4.2.1) ought to be reflected by the inclusion of an additional term in the free-energy expansions, presumably in the form of a field. At this point further expansion of this analysis of the free energy adds little to our understanding of the phase transition, especially as more data are needed to determine, for example, the effect of pressure, temperature and Al,Si disorder on the elastic constants of anorthite which contribute to the f term in the equations.

The thermodynamic approach given above can, however, be extended to explain the changes in transition character at pressure that accompany disordering of Al and Si. While consideration of the equations above shows that changes in a homogeneous Q_{OD} cannot be responsible for changing transition character, in his analysis of the high-temperature behaviour of anorthite, Salje (1987) was able to show that smearing of the transition should occur when the state of Al,Si order becomes inhomogeneous on the length scale of a few hundred Å. If each structural block of homogeneous order were completely isolated from the rest of the crystal it would undergo the $P\bar{1} \leftrightarrow I\bar{1}$ transition at a specific pressure determined by its local Q_{OD}. In reality, the structural blocks are all part of the same tetrahedral framework, so the behaviour of each is modified by the strain field of its neighbours. The macroscopic behaviour of the crystal, as reflected in the unit-cell parameters and the derived ϵ_s, results from the coupling together, through the strain, of all of these blocks. The analysis of Salje (1987) also suggests that such a change in the thermodynamic character of the $P\bar{1} \leftrightarrow I\bar{1}$ transition could be induced by a further reduction in the correlation length of the state of Al,Si order within the crystals below that which gives rise to the smearing of the transition. Such a process may contribute to the change to continuous transition behaviour observed in the two samples annealed at the highest temperatures.

4.3. PHASE TRANSITIONS IN OTHER FELDSPARS

Structural phase transitions at pressure have only been reported for one feldspar composition other than those discussed above. In a high-pressure powder-diffraction study with the energy dispersive method McGuinn & Redfern (1993) report a monoclinic to triclinic transition in "Sr-anorthite", $SrAl_2Si_2O_8$, between 2.76 and 3.48 GPa at room temperature. The transition is marked by a change in the β unit-cell angle, significant deviations of α and γ from 90°, and a symmetry reduction from I2/c to $I\bar{1}$. Evaluation of the spontaneous strain associated with the phase transition shows that it is first order in character, but ΔV is smaller than the uncertainties associated with the experimental determination of the cell parameters (of the order of 0.1%).

4.4. CONCLUSIONS

Much remains to be elucidated about the known structural phase transitions in feldspars at high pressure, but several general conclusions can be reached. First, there is no obvious overall trend by which such phase transitions can be predicted to occur, either at high pressure or at elevated pressure and temperature together. Neither can the direction of symmetry change be safely predicted - we have noted that while Ca-anorthite increases symmetry with increasing pressure, Sr-anorthite decreases in symmetry. This complex behaviour must arise from a subtle balance of forces between the response to pressure of the tetrahedral framework and the larger cations residing within the cavities of the framework. As pressure increases the interaction of these two structural features must grow stronger, but whether this will produce a greater or lesser diversity of structural phase transitions at higher pressures remains to be seen. Some common features do emerge from the results obtained so far. All of the phase transitions are displacive in that they involve no bond-breaking within the tetrahedral framework, but may involve reconfiguration of the M-O bonds as a result of a change in framework conformation. Because of these structural changes the transitions display very small or zero ΔV's, and are in general non-quenchable, although some small amount of hysteresis of the order of 0.1 GPa can be demonstrated in some cases. Because of these characteristics *in-situ* single-crystal or high-resolution powder diffraction remain the techniques of choice to explore the complex behaviour of feldspars at high pressures, and their application to other feldspars will no doubt continue to yield a fascinating variety of transition behaviour.

5. Acknowledgements

I would like to express my thanks to Bob Hazen for first suggesting that I study feldspars at high pressures, to Larry Finger and Charlie Prewitt for providing the facilities for the initial studies, and to Michael Carpenter for providing most of the samples used in my experiments. The interpretation of my results has benefitted from discussions with many people, but I would especially acknowledge the help and comments of W.L. Brown, Ekhard Salje, Michael Carpenter and Simon Redfern over the last five years. Most importantly the discussions, advice and support of Dr. Nancy Ross have been instrumental in the progress that I have made in high-pressure research. I would also like to acknowledge the receipt of a Research Fellowship from the Carnegie Institution of Washington, The Royal Society for their support in the form of a 1983 University Research Fellowship, and NERC, the University of London Central Research Fund, and the Mineralogical Society of America for research funds.

6. References

Adams LH, Gibson RE (1929) The elastic properties of certain basic rocks and of their constituent minerals. Proc Nat Acad Sci 15:713-724

Adams LH, Williamson ED (1923) On the compressibility of minerals and rocks at high pressures. J Franklin Inst 195:475-529

Adlhart W, Frey F, Jagodzinski H (1980a) X-ray and neutron investigation of the $P\bar{1}$-$I\bar{1}$ transition in pure anorthite. Acta Crystallogr A36:450-460

Adlhart W, Frey F, Jagodzinski H (1980b) X-ray and neutron investigation of the $P\bar{1}$-$I\bar{1}$ transition in anorthite with low albite content. Acta Crystallogr A36:461-470

Ahmed-Zaid I, Madon M (1991) A high-pressure form of Al_2SiO_5 as a possible host of aluminium in the lower mantle. Nature 353:426-428

Aizu K (1970) Determination of the state parameters and formulation of spontaneous strain for ferroelastics. J Phys Soc Japan 28:706-716

Alexandrov KS, Ryzhova TV (1962) Elastic properties of rock-forming minerals: III Feldspars. Bull Acad Sci USSR, Geophysics Series, 129-131

Angel RJ (1988) High pressure structure of anorthite. Amer Mineral 73:1114-1119

Angel RJ (1992) Order-disorder and the high-pressure $P\bar{1}$-$I\bar{1}$ transition in anorthite. Amer Mineral 77:923-929

Angel RJ, Carpenter MA, Finger, LW (1990) Structural variation associated with compositional variation and order-disorder behaviour in anorthite-rich feldspars. Amer Mineral 75:150-162

Angel RJ, Hazen RM, McCormick TC, Prewitt CT, Smyth JA (1988) Compressibility of end-member feldspars. Phys Chem Minerals 15:313-318

Angel RJ, Redfern SAT, Ross NL (1989) Spontaneous strain below the I-1-P-1 transition in anorthite at pressure. Phys Chem Minerals 16:539-544

Angel RJ, Ross NL (1988) The I-1 to P-1 transition in anorthite-rich feldspars. Ann Rep Director, Geophysical Lab, Carnegie Inst Washington 91-95

Angel RJ, Ross NL, Wood IG, Woods PA (1992) Single-crystal X-ray diffraction at high pressures with diamond-anvil cells. Phase Trans 39:13-32

Bailey SW (1984) Micas (Reviews in Mineralogy, Vol. 13). Mineralogical Society of America, Washington DC

Boyd FR, England JL (1961) Melting of silicates at high pressures. Carn Inst Washington Yearb 60:113-125

Bridgman PW (1928) The linear compressibility of thirteen natural crystals. Amer J Sci 15:287-296

Bridgman PW (1948) Rough compressions of 177 substances to 40,000 Kg/cm^2. Proc Amer Acad Arts Sci 76:71-87

Brown WL (1984) Feldspars and Feldspathoids. Reidel Publishing Co. Dordrecht

Brown WL, Hoffman W, Laves F (1963) Über kontinuierliche und reversible Transformation des Anorthits ($CaAl_2Si_2O_8$) zwischen 25 und 350° C. Naturwissenschaften 50:221

Brown WL, Openshaw RE, McMillan PC, Henderson CMB (1984) A review of the expansion behaviour of alkali feldspars: coupled variations in cell parameters and possible phase transitions. Amer Mineral 69:1058-1071

Carpenter MA (1992) Thermodynamics of phase transitions in minerals: A macroscopic approach. In G.D. Price and N.L. Ross (Eds), Stability of Minerals. Chapman and Hall, London.

Carpenter MA, Angel RJ, LW Finger (1990) Calibration of Al/Si order variations in anorthite. Contrib Mineral Petrol 104:471-480

Carpenter MA, Ferry JM (1984) Constraints on the thermodynamic mixing properties of plagioclase feldspars. Contrib Mineral Petrol 87:138-148

Carpenter MA, McConnell JDC, Navrotsky A. (1985) Enthalpies of ordering in the plagioclase feldspar solid solution. Geochim Cosmochim Acta 49:947-966

Chatterjee ND (1970) Synthesis and upper stability of paragonite. Contrib Mineral Petrol 27:244-257

Chatterjee ND (1971) The upper stability limit of the assemblage paragonite+quartz and its natural occurrences. Contrib Mineral Petrol 34:288-303

Chatterjee ND, Johannes W (1974) Thermal stability and standard thermodynamic properties of synthetic 2M1-muscovite, $KAl_2AlSi_3O_{10}(OH)_2$. Contribs Mineral Petrol 48:89-114

Czank M (1973) Strukturuntersuchungen von Anorthit im Temperaturberiech von 20° C bis 1430°C. Dissertation, ETH Zürich

Daniel I, Gillet P, McMillan P (1993) A Raman spectroscopic study of $CaAl_2Si_2O_8$ at high pressure and high-temperature: polymorphism, melting and amorphization. NATO ASI "Feldspars and their Reactions" abstract volume.

Essene EJ (1967) An occurrence of cymrite in the Franciscan formation, California. Amer Mineral 52:1885-1890

Evangelakakis C, Kroll H, Voll G, Wenk H-R, Meishing H, Köpcke J (1993) Low-temperature cherent exsolution in alkali feldspars from high-grade metamorphic rocks of Sri Lanka. Contrib Mineral Petrol (in press)

Gasparik T (1984) Experimental study of subsolidus phase relations and mixing properties of pyroxene in the system CaO-Al_2O_3-SiO_2. Geochim Cosmochim Acta 48:2537-2545

Gasparik T (1985) Experimental study of subsolidus phase relations and mixing properties of pyroxene and plagioclase in the system Na_2O-CaO-Al_2O_3-SiO_2. Contrib Mineral Petrol 89:346-357

Gasparik T (1992) Enstatite-jadeite join and its role in the Earth's mantle. Contrib Mineral Petrol 111:283-298

Gautron L, Madon M (1993) A study of the stability of anorthite in the (P,T) conditions of the transition zone. Earth Planet Sci Lett, submitted.

Geisinger KL, Gibbs GV, Navrotsky A (1985) A molecular orbital study of bond length and angle variations in framework silicates. Phys Chem Minerals 11:266-283

Geisinger KL, Ross NL, McMillan P, Navrotsky A (1987) $K_2Si_4O_9$: Energetics and vibrational spectra of glass, sheet, and wadeite-type phases. Amer Mineral 72:984-994

Ghose S, Van Tendeloo G, Amelinckx S (1988) Dynamics of a second order phase transition: $P\bar{1}$ to $I\bar{1}$ phase transition in anorthite, $CaAl_2Si_2O_8$. Science 242:1539-1541

Glinnemann J, King HE, Schulz H, Hahn Th, La Place SJ, Dacol F (1992) Crystal structures of the low-temperature quartz-type phases of SiO_2 and GeO_2 at elevated pressure. Zeitschr Kristallogr 198:177-212

Goldsmith JR (1978) Experimental plagioclase-zoisite phase relations. EOS 59:402

Goldsmith JR (1980) The melting and breakdown reactions of anorthite at high pressures and temperatures. Amer Mineral 65:272-284

Goldsmith JR (1981) The join $CaAl_2Si_2O_8$-H_2O (anorthite-water) at elevated pressures and temperatures Amer Mineral 66:1183-1188

Graham CM, Tareen JAK, McMillan PF, Lowe BM (1992) An experimental and thermodynamic study of cymrite and celsian stability in the system BaO-Al_2O_3-SiO_2-H_2O. Euro J Mineral 4:251-259

Hackwell TP (1993) Feldspars at high pressures and temperatures. PhD Thesis, University of London.

Hackwell TP, Angel RJ (1992) Compressibilities of boron-feldspars and their aluminium analogues. Euro J Mineral 4:1221-1227

Hansen B (1981) The transformation from pyroxene granulite facies to garnet clinopyroxene granulite facies. Experiments in the system CaO-MgO-Al_2O_3-SiO_2. Contrib Mineral Petrol 76:234-242

Hariya Y, Kennedy GC (1968) Equilibrium study of anorthite under high pressure and high temperature. Amer J Sci 266:193-203

Hays JF (1966) Lime-alumina-silica. Carnegie Inst Wash Yearb 65:234-239

Hazen RM (1976) Sanidine: Predicted and observed monoclinic-to-triclinic reversible transformations at high pressure. Science 194105-107

Hazen RM, Finger LW (1982) Comparative Crystal Chemistry. John Wiley and Sons, Boston

Hazen RM, Prewitt CT (1977)Linear compressibilities of low albite: high-pressure structural implications. Amer Mineral 62:554-558

Herzberg CT (1976) The plagioclase spinel-lherzolite facies boundary; its bearing on corona structure formation and tectonic history of the Norwegian caledonides. In: Progress in Experimental Petrology, 3, 233-235. NERC Publications, London.

Holland TJB (1980) The reaction albite = jadeite +quartz determined experimentally in the range 600-1200°C. Am Mineral 65:129-134

Holland TJB, Powell R (1990) An enlarged and updated internally consistent thermodynamic dataset with uncertainties and correlations: the system K_2O-Na_2O-CaO-MgO-MnO-FeO-Fe_2O_3-Al_2O_3-TiO_2-SiO_2-C-H_2-O_2. J Met Geol 8:89-124

Hsu LC (1993) Cymrite: Its occurrence and stability. EOS 74:167

Kinomura N, Kume S, Koizumi M (1975) Synthesis of $K_2SiSi_3O_9$ with silicon in 4- and 6-coordination. Mineral Mag 40:401-404

Kroll H (1984) Thermal expansion of alkali feldspars. In: Brown WL (ed) Feldspars and Feldspathoids: Reidel Publishing Co., Dordrecht, pp 163-205

Kroll H, Ribbe PH (1983) Lattice parameters, composition, and Al/Si order in alkali feldspars. In Ribbe PH (ed) Reviews in Mineralogy Vol. 2 Feldspars: Mineral Soc Am, pp 57-99

Kudoh Y, Takéuchi Y (1985) The crystal structure of forsterite Mg_2SiO_4 under high pressure up to 149 kb. Zeit für Kristallogr 171:291-302

Kushiro I, Yoder HS (1966) Anorthite-forsterite and anorthite-enstatite reactions and their bearing on the basalt-eclogite transformation. J Petrol 7:337-362

Liu L (1978) High-pressure phase transformations of albite, jadeite and nepheline. Earth Planet Sci Lett 37:438-444

Madon M, Castex J, Peyronneau J (1989) A new aluminocalcic high-pressure phase as a possible host of calcium and aluminium in the lower mantle. Nature 342:422-425

McGuinn MD, Redfern SAT (1993) Ferroelastic phase transition in $SrAl_2Si_2O_8$ feldspar at elevated pressure. Mineral Mag (in press)

Megaw HD (1974) Tilts and tetrahedra in feldspars. In: Mackenzie WS, Zussman J (eds) The Feldspars: Manchester University Press, pp 87-113

Nye JF (1957) Physical Properties of Crystals. Clarendon Press, Oxford

Perkins D (1983) The stability of Mg-rich garnet in the system CaO-MgO-Al_2O_3-SiO_2 at 1000-1300°C and high pressure. Amer Mineral 68:355-364

Phillips BL (1990) Investigation of structural phase transitions in minerals and analogue systems by high-temperature magic-angle-spinning nuclear magnetic resonance spectroscopy. PhD thesis, University of Illinois at Urbana-Champaign, Urbana, Illinois.

Phillips MW, Gibbs GV, Ribbe PH (1974) The crystal structure of danburite: A comparison with anorthite, albite, and reedmergnerite. Amer Mineral 59:79-85

Prewitt CT, Sueno S, Papike J (1976) The crystal structures of high albite and monalbite at high temperatures. Amer Mineral 61:1213-1225

Redfern SAT (1992) The effect of Al/Si disorder on the $I\bar{1}$-$P\bar{1}$ co-elastic phase transition in Ca-rich plagioclase. Phys Chem Minerals 19:246-255

Redfern SAT, Graeme-Barber A, Salje E. (1988) Thermodynamics of plagioclase III: Spontaneous strain at the $I\bar{1}$-$P\bar{1}$ phase transition in Ca-rich plagioclase. Phys Chem Minerals 16:157-163

Redfern SAT, Salje E (1987) Thermodynamics of plagioclase II: Temperature evolution of the spontaneous strain at the $I\bar{1}$-$P\bar{1}$ phase transition in anorthite. Phys Chem Minerals 14:189-195

Redfern SAT, Salje E (1992) Microscopic dynamic and macroscopic thermodynamic character of the $I\bar{1}$-$P\bar{1}$ phase transition in anorthite. Phys Chem Minerals 18:526-533

Ringwood AE, Reid AF, Wadsley AD (1967) High-pressure $KAlSi_3O_8$, an aluminosilicate with sixfold coordination. Acta Crystallogr 23:1093-1095

Ryzhova TV (1964) Elastic properties of plagioclase. Bull Acad Sci USSR, Geophysics Series 633-635

Ryzhova TV, Alexandrov KS (1965) The elastic properties of potassium-sodium feldspars. Bull Acad Sci USSR, Earth Physics Series 53-56.

Salje E (1985) Thermodynamics of sodium feldspar I: Order parameter treatment and strain induced coupling effects. Phys Chem Minerals 12:93-98

Salje E (1987) Thermodynamics of plagioclase I: Theory of the $I\bar{1}$-$P\bar{1}$ phase transition in anorthite and calcium-rich plagioclases. Phys Chem Minerals 14:181-188

Salje E, Kuscholke B, Wruck B, Kroll H (1985) Thermodynamics of sodium feldspar II: Experimental results and numerical calculations. Phys Chem Minerals 12:99-107

Seki Y, Kennedy GC (1964) The breakdown of potassium feldspar, $KAlSi_3O_8$, at high temperatures and high pressures. Amer Mineral 49:1688-1706

Sekine T, Ahrens TJ (1992) Shock-induced transformations in the system $NaAlSiO_4$-SiO_2: A new interpretation. Phys Chem Minerals 18:359-364

Schlenker JL, Gibbs JV, Boisen MB (1978) Strain tensor components expressed in terms of lattice parameters. Acta Crystallogr A34:52-54

Simons G, Wang H (1971) Single crystal elastic constants and calculated aggregate properties: A handbook. M.I.T. Press, Cambridge, Mass.

Smith JV, Brown WL (1988) Feldspar Minerals, vol 1. Crystal Structures, Physical, Chemical, and Microstructural Properties. Springer-Verlag, Berlin

Smyth JR, Hatton CJ (1977) A coesite-sanidine grospydite from the Roberts-Victor kimberlite. Earth Planet Sci Lett 34:284-290

Staehli JL, Brinkmann D (1974) A nuclear magnetic resonance study of the phase transition in anorthite, $CaAl_2Si_2O_8$. Zeitschr Kristallogr 140:360-373

Swanson DK, Prewitt CT (1983) The crystal structure of $K_2Si^{VI}Si_3^{IV}O_9$. Amer Mineral 68:581-585

Vaidya SN, Bailey S, Pasternack T, Kennedy GC (1973) Compressibility of fifteen minerals to 45 kilobars. J Geophys Res 78:6893-6898

Viswanathan K, Harneit O, Epple M (1992) Hydrated barium aluminosilicates, $BaAl_2Si_2O_8.nH_2O$, and their relations to cymrite and hexacelsian. Euro J Mineral 4:271-278

Wadhawan VK (1983) Ferroelasticity and related properties of crystals. Phase Transitions 3:3-103

Weidner DJ, Vaughan M (1982) Elasticity of pyroxenes: effects of composition and crystal structure. J Geophys Res 87:9349-9354

Williame C, Brown WL, Perucaud MC (1974) On the orientation of the thermal and compositional strain ellipsoids in feldspars. Amer Mineral 59:457-464

Williams Q, Jeanloz R (1989) Static amorphisation of anorthite at 300K and comparison with diaplectic glass. Nature 338:413-415

Wood BJ , Holloway JR (1984) A thermodynamic model for subsolidus equilibria in the system $CaO-MgO-Al_2O_3-SiO_2$. Geochim Cosmochim Acta 48:159-176

Wruck B (1986) Einfluß des Na-Gehaltes und der Al,Si Fehlordnung auf das thermodynamische Verhalten de Phasenumwandlung $P\bar{1}-I\bar{1}$ in Anorthit. Dissertation, Universität Hannover

Yamada H, Matsui Y, Ito E (1984) Crystal-chemical characterisation of $KAlSi_3O_8$ with the hollandite structure. Mineral J 12:29-34

Yoder HS, Weir CE (1951) Change of free energy with pressure of the reaction nepheline + albite = 2 jadeite. Amer J Sci 249:683-694

Zhang J, Ko J, Hazen RM, Prewitt CT (1993) High-pressure crystal chemistry of $KAlSi_3O_8$ hollandite. Amer Mineral 78:493-499

RECENT WORK ON OSCILLATORY ZONING IN PLAGIOCLASE

T. H. PEARCE
Department of Geological Sciences
Queen's University
Kingston, Ontario
Canada, K7L 3N6

ABSTRACT. Growth of crystals from magma is one of the fundamental processes of geology. Changes in composition during growth produce a zonation resembling "tree rings" which is interpreted in a somewhat similar manner (although the periods of growth resulting in zones are not necessarily annual). In principle, zoned crystals have immense potential as tracers of magmatic processes. Major and trace elements have, at least in theory, the potential to record different aspects of the growth phenomenon. Complications result from the fact that crystallization is inherently a non-equilibrium process and the chemistry of the precipitating phase may therefore be influenced by kinetic factors. This zoning is present in a wide variety of rocks and has been studied for well over 100 years, yet it remains a challenging and perplexing subject.

Modern methods of microanalysis and imaging techniques are providing a new view of zoning. The application of interference imaging including Nomarski Differential Interference Contrast (NDIC) has provided hither-to-unknown information regarding the extreme complexity of magmatic zoning. Dissolution and reaction are extremely common features in magmatic phenocrysts. The complexity of typical zoning appears to indicate that phenocrysts move around a magma chamber and may alternatively experience (cycle through) different liquids, experiencing different temperatures and pressures during growth - as was posited by Homma in his classic work of 1932.

It is generally accepted, *a priori*, that crystal growth is a deterministic process and that zoning results from a complex interaction of growth and diffusion processes. New work indicates that, in many instances, zonation is further complicated by dissolution and reaction phenomena. In spite of the complexity of the processes, it should be possible to model plagioclase growth using continuous differential equations; yet, oscillatory zoning of magmatic plagioclase has, so far, defied detailed analysis or explanation. Published models at the present time typically assume a relatively simple growth mechanism and do not yield realistic zoning patterns or profiles. A possible exception is the constitutional

I. Parson (ed.), Feldspars and Their Reactions, 313–349.

supercooling model, but one key assumption of this model - that composition is more important than temperature in determining the initial supercooled crystal composition - has yet to be proven.

Utilizing new techniques recently developed for the analysis of dynamical systems ("chaos" theory), along with computer experiments, it is possible to extract previously unrecognizable kinetic information contained in zoning patterns. Detailed analysis of oscillatory zoning in magmatic plagioclase from several volcanoes indicates that the plagioclase- magma system is a deterministic nonlinear system whose attractor has a relatively low dimension. Patterns of zonation are not consistent with random or stochastic processes.

The ultimate objective of current studies of plagioclase is to find a method of decoding the zonation so that we may read the life history of a crystal treating the zones like "pages" of a book.

1. The Problem

Oscillatory zoning of magmatic plagioclase has, so far, defied explanation, in spite of efforts for more than 100 years (Smith and Brown 1988; Pearce and Kolisnik 1990). The former reference is comprehensive. This present work, in order to avoid duplication, is a specialized review of selected current topics and presents some new findings. It concentrates on empirical observations and the resulting theoretical implications.

Although there seems to be a consensus forming that the phenomenon of oscillatory zoning is a reaction- diffusion problem due to combined growth and diffusion involving a time lag, opinions are by no means unanimous on the exact mechanisms. In this work I review current knowledge and publish new observations of oscillatory zoning in plagioclase using interference imaging of etched polished surfaces. Utilizing new techniques recently developed for the analysis of dynamical systems (Higman and Pearce, 1993a, b), along with computer experiments, it can be shown that random or stochastic processes are not consistent with detailed observations of oscillatory zoning in magmatic plagioclase. The analysis of these new observations suggests that the typical plagioclase-magma system is a deterministic nonlinear system whose attractor has a relatively low dimension. The system may, therefore, be described by a small number of variables (say, 3 to 6).

The composition of plagioclase growing from a melt depends on intensive variables such as magma composition (including water fugacity), pressure, and temperature. In principle, therefore, zoned crystals have immense potential as tracers of magmatic processes (Smith and Brown 1988; Pearce and Kolisnik 1990; Shimizu 1990). Complications result from the fact that crystallization is inherently a non-equilibrium process and the chemistry of the precipitating phase may therefore be influenced by kinetic factors.

While the plagioclase system exhibits oscillatory zoning in a striking way easily seen in the petrographic microscope, the exact shapes of zoning profiles and the incredible richness of the zoning textures have only recently been appreciated (Pearce 1984a, c; Anderson 1983, 1984; Nixon and Pearce 1987). Plagioclase is by no means the only mineral to exhibit oscillatory zoning. The present writer has seen similar zoning in olivine, orthopyroxene, clinopyroxene, sanidine, nepheline, hornblende, and biotite. Thus the problems addressed by this research are of fundamental importance and general interest.

2. Empirical Observations of Zoning

2.1 HYPOTHESES AND ASSUMPTIONS

Most of the assumptions in this work are stated explicitly in Pearce and Kolisnik (1990). A common assumption to make is that a single euhedral horizon traceable completely around a crystal is a time horizon, analogous to a tree ring. Unfortunately, there is no reason to make this assumption, especially if different faces are able to grow with independent growth rates. Another assumption which is conveniently made is that thicker zones grow faster than thinner zones. Indeed, this might be considered a corollary of the previous assumption with regard to time horizons. However, a moment's consideration shows that, if faces are able to grow independently or, if growth alternately starts and stops, then these assumptions are not necessarily true.

Let us consider a face growing very slowly. Growth will be diffusion- controlled and will continue until all the nutrients in the local reservoir from which the crystal is growing are used up. The local reservoir may extend as much as half the distance to the next growing crystal; therefore a very thick single zone may result from such a growth episode. In contrast, let us consider a face growing extremely rapidly - say, the maximum possible growth rate. Such a face will rapidly use all of the nutrients in the layer of liquid immediately in contact with the crystal and growth must then stop resulting in an extremely thin zone. Thus, we have the rather paradoxical result that fast growth may under some circumstances produce thin zones and slow growth, thicker zones. No one-dimensional model can solve this problem. Unforunately, all of the models which are reviewed in this work are one-dimensional, and therefore do not address this intriguing and fundamental problem.

The experimental work of Lofgren (1974a, b; 1980; Lofgren and Norris, 1981) and Tsuchiyama (1985) is used wherever appropriate to interpret textures and zoning patterns. The latter work, which confirms earlier work of Lofgren, has been crucial for recognizing smooth rounded interfaces as due to solution outside of the plagioclase stability field. Rough interfaces and mantles occur when a crystal is more sodic than the equilibrium plagioclase composition and reacts with the calcic melt. Reaction between crystal and melt produces "fritted" or "dusty" zones. Coarse crystal-liquid intergrowths usually described as "wormy", "spongy", or "sieve-like" are thought to be due to partial dissolution - a reaction between the crystal and liquid in which the proportion of solid decreases. Without this experimental work as a guide, many of these textures would still be of problematical origin. It is tacitly assumed by the writer that typical plagioclase growth is inherently a disequilibrium phenomenon that generally takes place by layer spreading or lateral growth facilitated by screw dislocations or growth steps. In cases of a high degree of disequilibrium, layer spreading becomes a less likely mechanism.

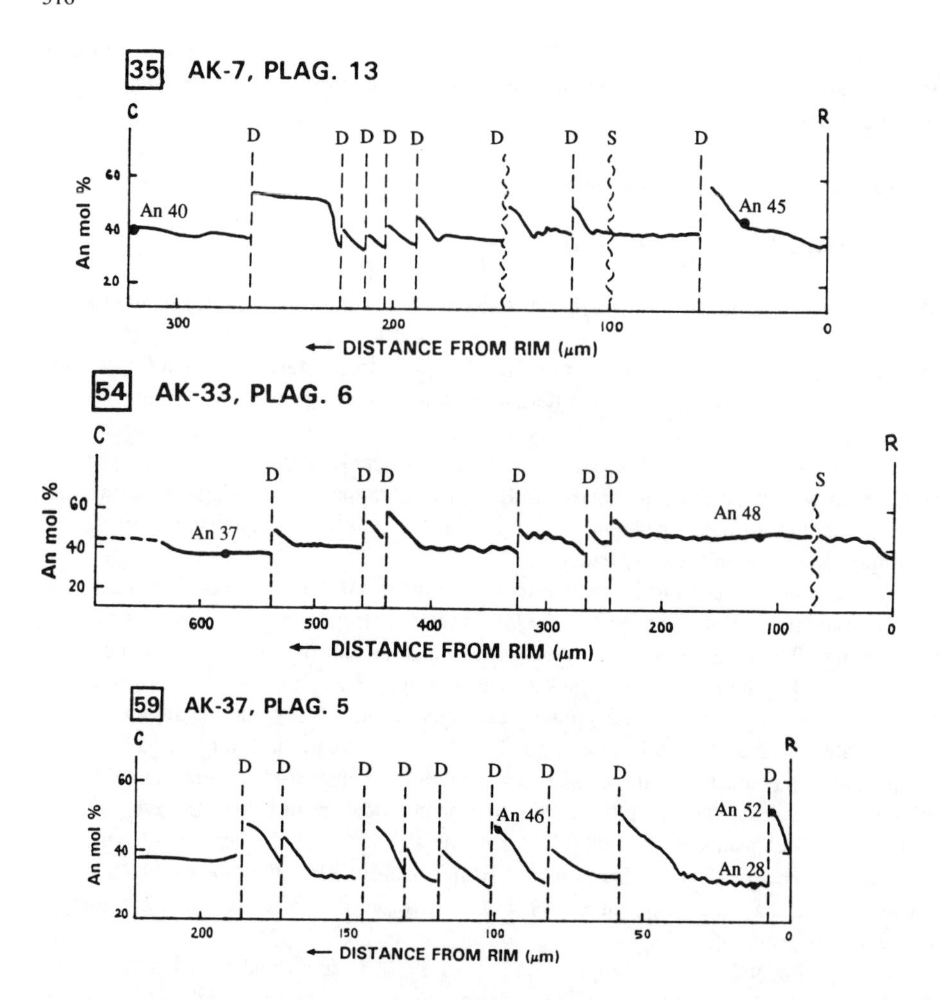

Figure 1. Illustration of actual zoning profiles of plagioclase microphenocrysts traced from Laser interferograms (small Type I oscillations are not shown). Samples are from andesite lavas of Popocatepetl, Mexico (Kolisnik 1990). Dissolution surfaces are indicated by "D". Compare with the model profiles in Fig 7.

2.1.1 *Crystal Growth.* In a supernatant fluid (such as a silicate or hydrous fluid) a crystal may grow, dissolve, react, or co-exist stably or metastably. These processes are so dissimilar that their identification might be thought relatively easy. Such is not the case. Different processes appear capable of yielding similar results, making it difficult to identify the specific genetic processes from the resulting textural pattern (see Tsuchiyama 1985; Pearce and Kolisnik 1990). Consider a crystal (such as that of Plate 1C) as having its "life history" of growth encoded in its pattern of zonation. Can we find a method to decode the zonation so that we may read the crystal's history? Current results indicate that with careful work (using a wide variety of information, including microprobe analyses) it is possible to interpret (or perhaps more appropriately, decode) crystal zoning patterns.

Zones (and interfaces) produced by specific processes (even if problematical) are useful as "marker horizons" indicating particular stages of "growth" (see Plates 1C, 2A, 2B, 3A, F, G). Readers unfamiliar with the methods of magmatic stratigraphy (Wiebe 1968) may wish to consult Pearce and Kolisnik (1990), especially pages 3 to 9, "Methods of Interpretation".

The zoning baseline, in particular, (see Sibley et al 1976; Pearce and Kolisnik 1990) is extremely important for distinguishing permanent changes in the local environment (such as those due to magma mixing of dissimilar liquids) from temporary changes (such as those due to temperature changes and/or diffusion lag in the boundary layer). See section 2.4.2 on dissolution.

2.1.2 *Analysis of Zoning Patterns.* Interference imaging produces two different views of zoning (Pearce 1984c; Anderson 1983, 1984; Clark et al 1986). Narrow fringe laser interferometry produces accurate zoning profiles by measuring changes in refractive indices. The instrument combines a Mach Zehnder interferometer with an optical microscope using an Argon Ion continuous wave laser as a light source. It produces interferograms whose fringes may be treated as “instant” graphs of An% (Pearce 1984a, c) - see Plate 1B. Profiles such as those of Fig. 1 are produced by tracing interferograms and calibrating the profiles with electron probe analyses. Resolution of this method is 1-2 μm spatial resolution and 1-2 % ΔAn (Pearce, 1984c).

Nomarski Differential Interference Contrast (Nomarski and Weill 1954; Nomarski 1955; Anderson 1983) produces a picture or plan view showing zone thicknesses and intricate textures such as cross-cutting features (Plate 1C, 2A, 2B, 3F, G). Details as small as 0.3 μm can be imaged making this the method of choice for observing fine details of oscillatory zoning. Sequences of zone thickness from the interior to exterior of the crystal may be treated as a stratigraphic record of growth incidents (called a time series in dynamical studies). It is important to recognize that such a record is not necessarily a continuous time series. Estimating the time represented by any hiatus in growth between zones is very difficult- it might be insignificant or quite large. We can, however, estimate the chemical effect corresponding to the hiatus. For example, a discontinuity cross-cutting pre-existing zones with an angular unconformity (Plate 1A, 2B, 3C) is likely to be due to a more severe process than a conformable euhedral discontinuity with a small Δ An% measured across the surface (Plate 2A, 2C).

While interpretation of the data must be done with great attention to detail, collection of this "growth sequence" data is relatively straightforward (measuring thickness on a photograph or digitizing a video image). This pseudo-time series data may then be analyzed using standard techniques, assuming it retains the geometry of the phase space (see section 3.1, below). Fractal dimensions are readily measured as well (Fowler et al., 1989).

What other types of information can we obtain from accurate zoning patterns of minerals? If we assume that the crystal grows from a local reservoir from which it obtains its nutrients for growth, then the local system consists of the growing crystal plus its local reservoir. In many cases we may assume the local reservoir is a boundary layer surrounding the crystal. Then, as the crystal grows, each stage of growth of the mineral records the reaction between the crystal and its local reservoir as a single zone. In an earlier study, which will not be discussed in detail in this present work (Pearce et al., 1987b), it was shown that it is possible to obtain information about the size of the local system as well as its chemistry by analyzing the shape of the zoning pattern for individual zones. For plagioclase the variation in An% or An/Ab ratio gives indirect information about the composition of the local reservoir. This is relatively straightforward and (along with information about the liquid composition in the local system) places constraints on the effective distribution coefficient for the plagioclase-melt system. If we can assume a constant Kd over a limited range of growth conditions, then calculations are fairly easy to perform - see, for example, the relationship between mineral and liquid compositions and fractionation curves in Pearce (1978).

This type of empirically-based analysis has theoretical implications but does not require commitment to any particular world-view, e.g. stable thermodynamic, disequilibrium thermodynamic, non-linear dynamical, etc.

2.2 TRACE ELEMENTS AND DISEQUILIBRIUM

Blundy and Shimizu (1991) have published one of the few up-to-date studies involving trace element analyses of zoned plagioclase. Secondary-ion mass spectroscopy (SIMS) was used to determine the trace element composition. The ions analyzed: ^{28}Si, ^{23}Na, ^{24}Mg, ^{39}K, ^{40}Ca, ^{47}Ti, ^{56}Fe, ^{88}Sr, ^{138}Ba and $^{27}Al/^{28}Si$ were used to calculate An%. Only An, Ba, and Sr were reported. The general Sr and Ba zoning patterns were similar for all analyzed crystals - from core to rim there is an increase in Ba (the incompatible or magmaphile element) and a decrease in Sr (the compatible element) - as might be expected. The analytical spot size was 5-8 μm. Typical spacing of analytical spots was 13 μm (30 analyses in 400 μm) although figured oscillatory zones are approximately 2 μm thick. Interestingly, one figured crystal has zones apparently 100 μm thick; unfortunately, the zoning profile is still too coarse to determine the exact shape. Clearly, as none of the figured zoning profiles show enough detail to determine the exact shape of the profile, these results can only show broad trends. Nevertheless, Blundy and Shimizu were able to draw some conclusions which reinforce some inferences regarding recycling of plagioclase within magma chambers undergoing relatively vigorous convection.

In previous work by Shimizu (1990), an antithetic variation was noted between compatible and incompatible elements in finely zoned augite phenocrysts. It was suggested that reaction-transport kinetics was involved and similarities were noted with oscillatory zoning in plagioclase.

Perhaps the most important conclusion we can draw from a review of the literature is that the usefulness of trace elements has barely been explored owing to instrumental limitations (Smith and Brown 1988).

2.3 DISCUSSION OF PLATES

2.3.1 *Plate* 1.

1A. Crossed nichols photomicrograph of half of a plagioclase crystal (No. 6) from sample AK-21 (see Kolisnik 1990). This sample comes from a dacite plug from the Nexpayantla stage of volcano Popocatepetl, southeast of Mexico City, Mexico. The calcic core of this phenocryst is surrounded by a set of about 14 zones of sodic composition (An30) truncated by a strong dissolution surface (D). A similar set of sodic zones occurs immediately next to the last zone of the crystal and is similarly truncated by a massive dissolution surface (D).

1B. A narrow fringed laser interferogram of the same crystal as in Plate 1A. This interferogram (obtained using green light λ= 514.5 nm of an argon ion laser) may be interpreted as a graph of An content across the crystal. On the left there is a calcic core, the dark line (indicating An%) then drops down to a baseline of An_{30} for the 100 µm thick set of sodic zones. There is then a jump of 15 An % at a calcic plateau approximately 250 µm thick. Another layer of sodic zones overlies this calcic plateau and is in turn cut by a dissolution surface upon which is deposited the last "calcic spike". The approximate composition of the calcic plateau is An_{44}.

1C. A Nomarski differential interference contrast (DIC) image of a 230 µm long plagioclase microphenocryst from a basaltic lava (MX-85-16, near Cuernavaca, Mexico). This image shows the intricate detail which can be seen in reflected light Nomarski DIC images. The thinnest visible zone is 0.75 µm thick, while the thickest zone at the edge of the crystal is 20 µm thick. The fine set of well preserved zones under the middle arrow average 1.25 µm thick and the crystal as a whole has more than 60 zones preserved. Interesting features of this crystal are the dissolution surfaces indicated by the arrows. This crystal has more than four of these prominent dissolution surfaces. The individual angular unconformities formed by the surfaces as they intersect earlier formed zones cut through as many as 20 preexisting layers suggesting massive dissolution events.

2.3.2 *Plate 2 "Corners"*

2A. Nomarski DIC image of an etched polished surface in reflected light. The sample is from site 834 ODP Leg 137. This interesting sample shows a rare feature in which the corners of the individual zones appear to be incomplete. Note the curved arrows which

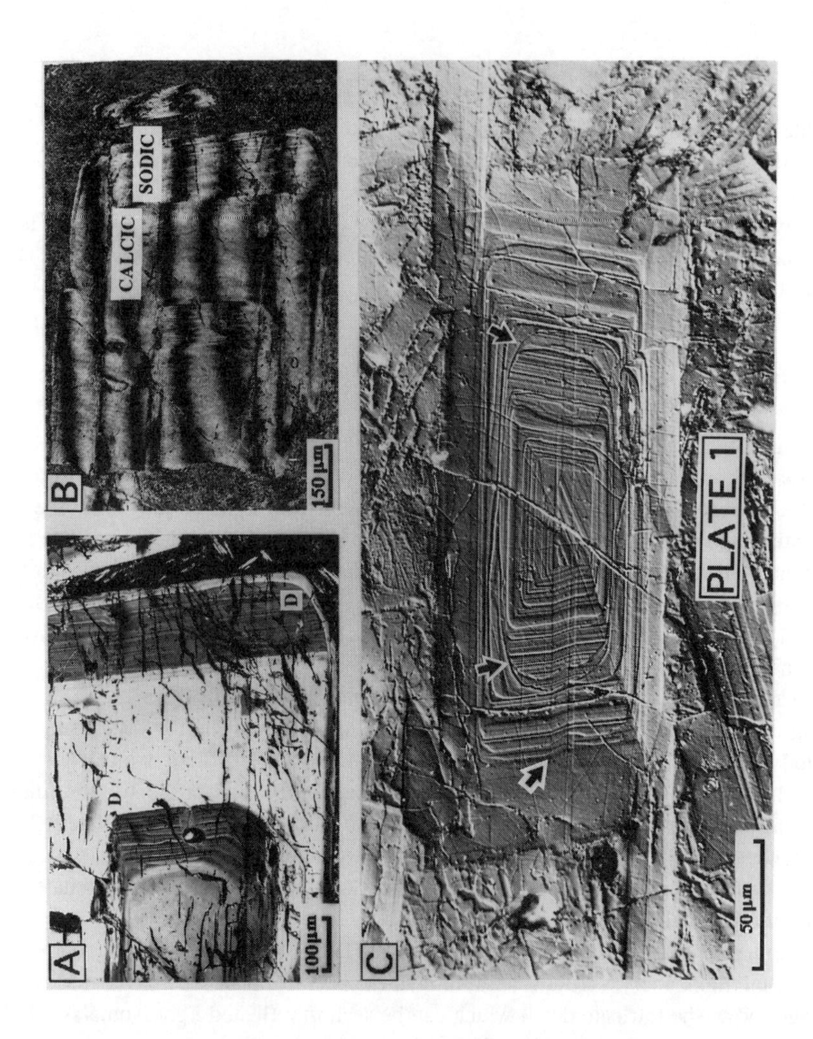
A
B
C
D
CALCIC
SODIC
100 µm
150 µm
50 µm
PLATE 1

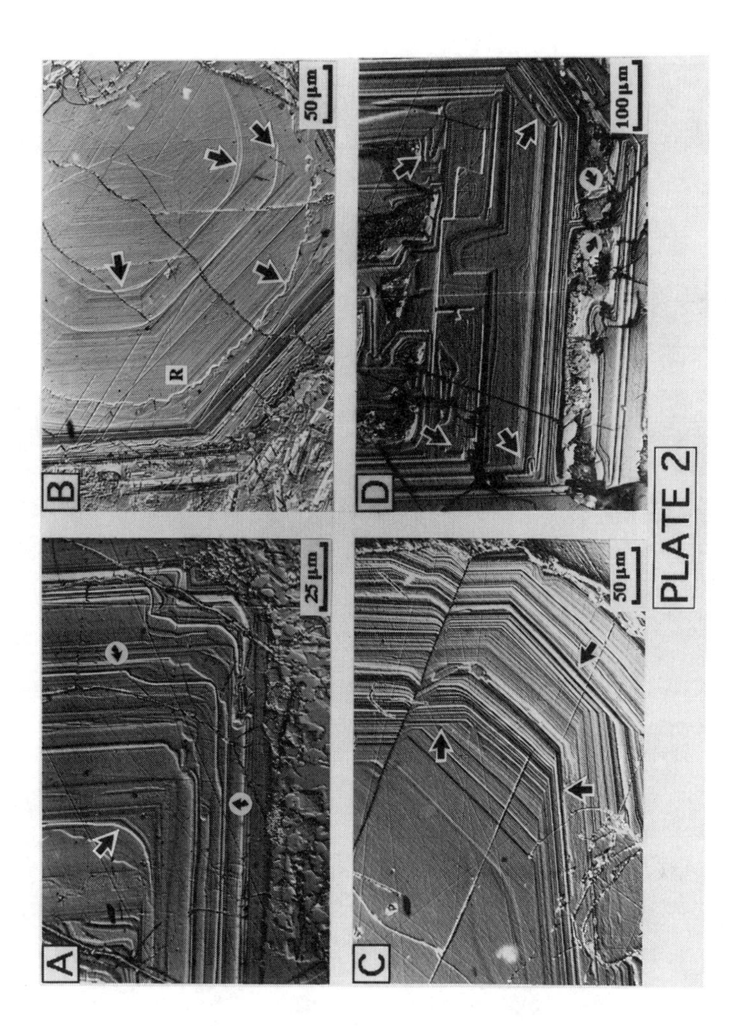
A
25 μm
B
R
50 μm
C
50 μm
D
100 μm
PLATE 2

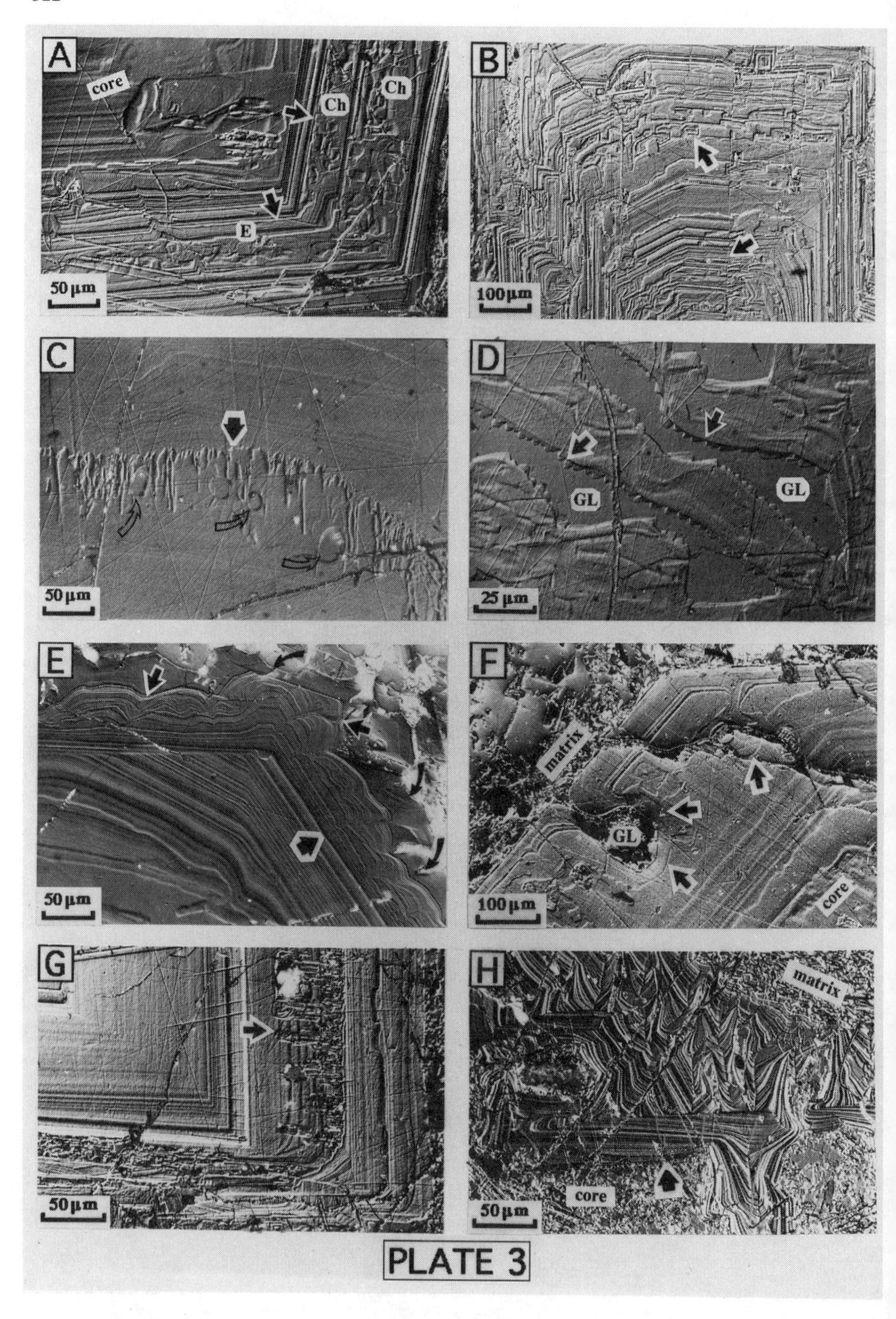
A
core
Ch
Ch
E
50 µm
B
100 µm
C
50 µm
D
GL
GL
25 µm
E
50 µm
F
matrix
GL
core
100 µm
G
50 µm
H
matrix
core
50 µm
PLATE 3

indicate the initiation of a cuspate concave form to the corner in contrast to the normal euhedral shape of the crystal faces. The arrow in the upper left hand corner of the image points to an earlier rounded convex solution surface confirming that the concave features along that sector are growth features rather than due to dissolution. This is a very uncommon feature as plagioclase shows a very strong ability to produce euhedral corners and planar faces. Preservation of features like this are consistent with lateral spreading mechanisms of growth.

2B. Nomarski DIC image of etched polished surface of a plagioclase phenocryst illustrating two different types of dissolution surfaces. This sample, AK-6, comes from the margin of a blocky andesite flow just south of the village of San Baltazar and is part of a young fissure eruption from the east flank of Popocatepetl, Mexico. Two smooth interfaces probably due to simple solution outside of the stability field of plagioclase (Tsuchiyama 1985) are illustrated by the arrows. Note that the inner surface is in fact a doublet of two dissolution surfaces, one on top of the other. Extremely sharp corners have formed from euhedral surfaces immediately above the rounded corners. A rough crosscutting surface is indicated by R and an arrow near the margin of the crystal. This surface has removed a considerable amount of preexisting zoned plagioclase and the cuspate rough character is probable indicative of reaction as well as solution along this surface.

2C. Nomarski DIC image of an etched polished surface of a plagioclase showing well defined euhedral zoning. The sample is andesite number 2-9 from Fuego Volcano, Guatemala and was kindly loaned to the author by A.T. Anderson. Anderson (1983) has already referred to missing zones in some of the crystals that he has described. In this sample, we see that the three faces that are visible in the image all contain missing zones as indicated by the arrows; that is to say, two or more zones on one face suddenly become one or fewer when traced around to an adjacent face. Anderson took this to be evidence of lateral spread of faces.

2D. Nomarski DIC image of an etched polished surface of plagioclase from hole number 834b Leg 137 ODP. This unusual crystal shows hopper growth patterns, many of which have been filled in completely by plagioclase. Note the incomplete corner which must have been hollow and then subsequently filled by plagioclase as indicated by the arrows. The curved arrows indicate a small neck of plagioclase which connects the bulk of the crystal to a zoned rim which encloses glass and also contains some features similar to those of the incomplete corner but on a larger scale. It appears that this crystal grew as a porous skeletal crystal and that some of the voids were subsequently filled by plagioclase and others retained their glass infilling. Hopper-like features such as these are extremely rare in most of the volcanic rocks which the writer has studied.

2.3.2 *Plate 3 "Interfaces"*

3A. Nomarski DIC image of an etched polished surface. This plagioclase crystal 3-2, from a Guatemalan andesite, was kindly provided to the writer by A.T. Anderson. This unusual crystal shows a repeated change from euhedral to chaotic growth (patchy zoning of Anderson 1983, and Vance 1965). Within the regions “Ch” in the figure there is conformable but highly contorted zoning patterns and inclusions of glass. It appears that during (rapid?) growth the planar surface of the crystal became de-stabilized and grew in a highly contorted fashion. The presence of portions of liquid now preserved as glass suggest rapid growth. The change over from euhedral to chaotic zoning is abrupt, occurring within one cycle; on the other hand, the change from chaotic to euhedral takes place somewhat more gradually, usually within two or three cycles (as noted by Anderson, 1984). This zoning pattern gives the impression that the crystal was capable of flipping from one mode of growth to another and back again. This type of behavior is a characteristic feature of nonlinear dynamical systems and is permissive evidence of the dynamical nature of plagioclase zonation (section 3, this work). Of particular interest in this crystal is the fact that the same zone marked “E” is euhedral on one face and on a second adjacent face is chaotic. During the next incident of growth, chaotic zoning was present on both of the faces. This feature proves conclusively that this type of zoning is an inherent feature of the crystal and its local boundary layer and is not due to global changes within the magma chamber, as these would have to effect all faces of the crystal simultaneously. This assumes of course that at least some of the zones in the regions marked “E” and “Ch” are contemporaneous - a reasonable assumption given that when the next event of chaotic zoning took place, it appears to have occurred completely around the crystal without any sectors.

3B. Nomarski DIC image of an etched polished surface of plagioclase from an andesite, St. Kitts Lesser Antilles (GG1). This sample shows a remarkable persistence of intricate euhedral growth reminiscent of hieroglyphic texture (Pearce and Kolisnik, 1990) and is similar in many aspects to the hopper shape seen in Plate 2D. The scarcity of cross-cutting features indicates that this is probably a growth feature.

3C. Nomarski DIC image of an etched polished surface in reflected light of an andesite from Martinique, Lesser Antilles (nuée ardente, St. Pierre). This image shows a reaction phenomena which took place on a rounded dissolution surface marked by the large arrow. Note the fingers of plagioclase which penetrate the substrate. Rounded glass inclusions indicated by the curved open arrows are present within the reaction zone. This phenomenon is a coarser version of what is commonly known as fritted texture.

3D. Nomarski DIC image of an etched polished surface of plagioclase from an andesite (St. Kitts, Lesser Antilles, sample GG1). This porous crystal contains large embayments of glass (GL) of problematical origin as there is not sufficient evidence to indicate whether this is due to growth or solution. The interesting feature of this is that the plagioclase grew a tooth-like euhedral layer surrounding the enclosed liquid. This growth event is probably related to a temperature or pressure drop prior to eruption. This feature emphasizes the tendency for plagioclase to grow euhedral forms even under restricted circumstances.

3E. Nomarski DIC image of an etched polished surface of plagioclase from an abyssal lava sample 841B from leg 137 ODP (see Bryan and Pearce 1993). This unusual crystal shows both euhedral and convolute conformable zoning undoubtedly due to growth. Within the outer most 50 to 100 μm there are unusual lines which appear to perturb the zoning pattern. These have the characteristics of "dislocations" . They average about 50 μm apart and there are more than 30 of them located around this corner of the crystal. As none of them appear to penetrate the euhedral zone indicated by the large arrow and, since, several are seen to begin at that surface, it appears that they may all have begun at that horizon and have propagated upwards throughout the subsequent zoning. There are in excess of 20 contorted layers and, interestingly, this feature is not healed over by subsequent growth, but appears to be perpetuated by effecting each of the zones as they were laid down. The outermost edge of the crystal is indicated by the curved black arrows. The last layer to grow was homogeneous and has a highly irregular form.

3F. Nomarski DIC image of an etched polished surface of plagioclase from a lava (IZ-75-100, Iztaccíhuatl volcano, Mexico, kindly provided by G. T. Nixon). Note the large embayments (arrows) whose cross-cutting features indicate they were caused by solution. One of the embayments has been completely healed over forming an euhedral surface. A second one containing glass (GL) still has a thin neck of glass connecting it to the matrix. Features like this show the tendency for plagioclase to heal over highly contorted surfaces, filling them in and covering them over with an euhedral surface.

3G. Nomarski DIC image of an etched polished surface of plagioclase from an andesite (St. Pierre, Martinique, sample Mar 4A/B). This sample from a clast in a nuée ardente deposit shows remarkable euhedral zoning in the core with very faintly defined sectors. The arrow points to a fritted texture due to magmatic reaction.

3H. Nomarski DIC image of an etched polished surface of plagioclase from an Aegean andesite (A-1, courtesy of K. St.-Seymour). This unusual sample shows highly contorted and euhedral growth on an irregular substrate indicated by the arrow. The entire core of this crystal suffered massive dissolution and reaction and is completely fritted. The zoned plagioclase which grew on top of this has an unusual texture and structure somewhat reminiscent of that of Plate 3E, but with a tendency for more euhedral forms. The propagation of lines of apparent "dislocation" is the same as in Plate 3E.

2.4 CONCLUSION FROM EMPIRICAL OBSERVATIONS

2.4.1 *The Consequences of Saw-toothed Patterns.* Is growth continuous or intermittent? I believe the textural evidence presented here and elsewhere (Pearce and Kolisnik 1990) is unequivocal and indicates that intermittent growth is the rule rather than the exception during oscillatory growth. Consider a typical saw-toothed pattern, say with ΔAn 5% and ΔX 5 µm (a pattern such as that of Figs. 4A, or simulated in 5A). The only way to avoid an interpretation which has the crystal stop growing (presumably at equilibrium) is to assume a late stage dissolution event as an integral part of each and every growth cycle. In this case, problems with an asymptotic approach to equilibrium are avoided, but the crystal must still stop growing (since there is usually no evidence of magmatic reaction).

There is one possible way in which perfectly euhedral zones might grow continuously but have apparent optical-scale discontinuities - if the An% changed almost instantaneously within a sub-microscopic (nanometer scale) layer. As there is no evidence for this feature, the present writer considers this rather unlikely (see discussion in section 4.2.2).

During destabilized chaotic growth (the so-called patchy zoning of Anderson 1983,

1984) the crystal may not stop growing, i.e. the growth rate may not approach zero because a planar base has not formed under equilibrium conditions. In other words, a convolute surface even under equilibrium conditions may still have the possibility of many growth steps at the surface and therefore the possibility of continued growth.

2.4.2 *Significance of Dissolution.* Cross-cutting evidence of solution is exceedingly common in zoned plagioclase. In fact it is quite difficult to find more than a few contiguous euhedral zones which are not separated by a dissolution surface. Trial calculations testing the effect of solution on subsequent plagioclase growth have been made by the author and Kolisnik (1991) using the plagioclase saturation surface of Langmuir (1980); and, Langmuir and Hansen (1985). They show that changing the local boundary layer liquid composition by dissolution of plagioclase (say, due to a temperature increase) has the effect of abruptly increasing the An% of the next layer of plagioclase to precipitate. This mechanism is much more efficacious than magma mixing at abruptly increasing the An% of plagioclase. It has the added advantage that the increase in An% is temporary (as the baseline composition is not affected). Magma mixing of disparate liquids, in contrast, would be expected to change the zoning baseline to a permanently higher An%.

2.4.3 *Constraints on Growth Mechanisms and Laws.* Empirical observations are entirely consistent with growth involving dislocations (Plate 2A, C, D, 3A, B, E), and an implied growth step mechanism (see infilling of solution pits, Plate 3F). Boundary layers have been reported in glass (Bottinga et al 1966) and the laser interferometric zoning profiles of Figure 1 would seem to imply a similar phenomenon to account for the repeated calcic jumps in An%. Layers of bubbles (Plate 3 A, C) may be bubble showers (Tiller 1977) and the exact textural and chemical features of their occurrence are noteworthy.

3. **Spatiotemporal Dynamics in Oscillatory Zoning**

3.1 INTRODUCTION

Deterministic systems influenced by non-linear dynamics (popularly called chaos) exhibit behavior varying from regular repetitive, to pseudo-regular, to chaotic, depending on values of significant parameters (May 1976, Shaw 1987, Middleton 1990). Chaotic behavior is patterned and repetitive - yet unpredictable - and extremely sensitive to initial conditions (Ruelle 1979). Chaos is not a stochastic process - it is deterministic although it may appear random. The weather, for example, is the archetypal non-linear system. We might note in passing that patterns of plagioclase zoning do somewhat resemble graphs of weather data.

Given a data set, can we determine if it was generated by a chaotic or stochastic process? The answer appears to be a qualified yes but this is a non-trivial problem. Unambiguous interpretation of data as due to noise or chaos is difficult (Glass and Mackey 1988).

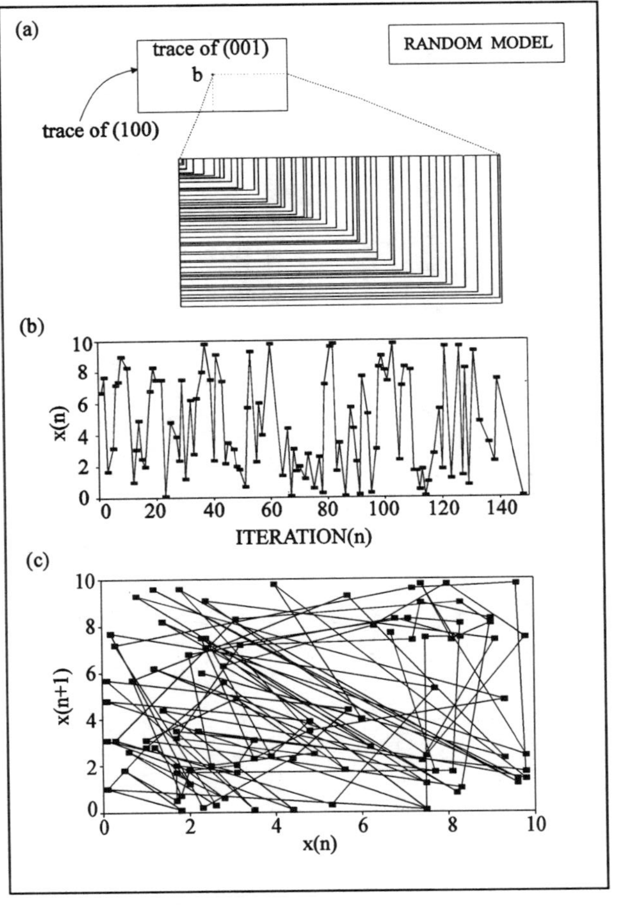

Figure 2. Computer- generated randomly zoned crystal.

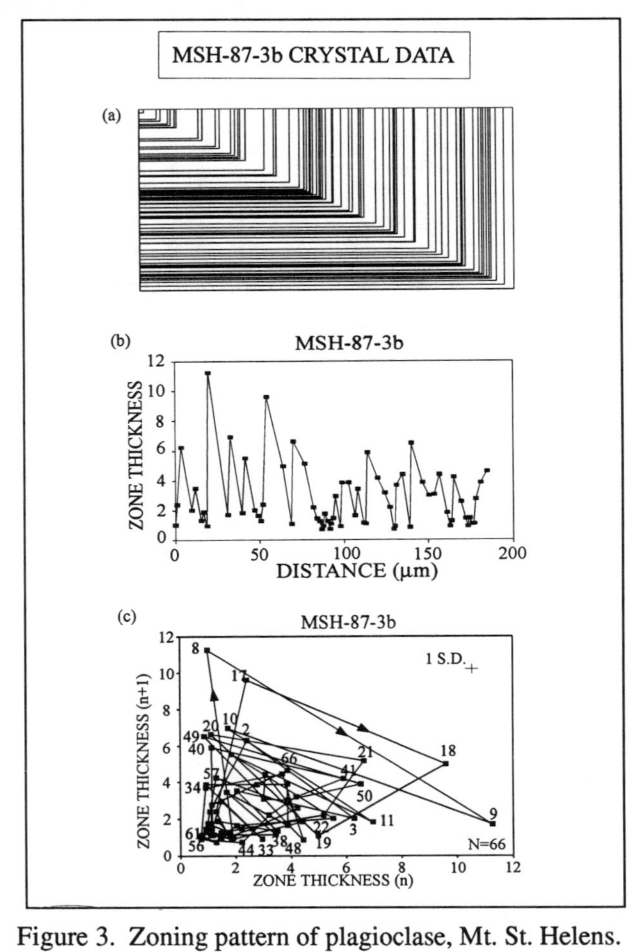

Figure 3. Zoning pattern of plagioclase, Mt. St. Helens.

Assuming accurate data, we can discriminate between the two possibilities by plotting the data in phase space. In this context, phase space is the mathematical space of the dynamical variables of the system (Baker and Goolub 1990). A point in phase space completely describes the corresponding state of the system. In phase space, chaotic data define an attractor while stochastic data fills the available space. This is the primary means of distinguishing the two types of data sets.

There are several cogent reasons, both empirical and theoretical, for suspecting that non-linear dynamics or chaos is involved in the production of oscillatory zoning and other repetitive effects in magmatic plagioclase. This reasoning is summarized below.

3.2 THEORETICAL RATIONALE

Several workers (Haase et al 1980; Simakin 1983; Provost 1985; Higman and Pearce 1992, 1993a, b) have proposed models of oscillatory zoning in plagioclase based on non-linear equations. Indeed, repetitive growth of plagioclase from solution seems so much like a non-linear process that one is tempted to assume non-linearity, a priori. Abrupt changes in zoning style such as those imaged in Plate 3A would certainly lend credibility to this assumption. However, the theoretical models developed at the time of writing of this work produce zoning profiles which do not match those which are observed using interference imaging (Pearce and Kolisnik 1990). This mismatch in zoning profiles between theory and observation does not mean that growth was not chaotic. Most of the theoretical workers have not had access to the precise profiles generated in the author's laboratory. Consequently, their equations and/or parameters are, at the very least, in need of adjustment.

Reasons that growth of plagioclase is considered to be governed by non-linear equations and might become chaotic are as follows (Haase et al 1980; Ortoleva et al 1987):
i) the system is out of equilibrium in order to obtain either growth or solution, see Sibley et al (1976), Tiller (1977). The system may, in fact, be "far from equilibrium" (as assumed by Haase et al 1980)
ii) there are feedback and coupled reactions in the crystal/liquid system, both thermal and chemical (see discussion in Swinney and Roux 1985)
iii) there are at least 3 variables (in the simplest model, albite, anorthite and solvent liquid).

In addition the following features are suggestive of chaotic dynamic systems:
iv) there are two possible dynamic steady-states of the system (growth and solution). Chaotic or otherwise interesting behavior tends to occur when a system is supercooled (or superheated) at the boundary between the two steady-states. This is exactly what the present writer would postulate for the plagioclase-liquid system.
v) patterns of zoning tend to show clustering (short-range similarity) but are unpredictable in the long range and never repeat exactly.

3.3 EMPIRICAL TEXTURAL EVIDENCE

Empirical observations from interference imaging which tend to support the possibility of

nonlinear dynamics during growth have been figured in Pearce and Kolisnik (1990) and are further illustrated in this present work:
i) repetitive features: growth, solution, reaction (see Plate 1C, 2, 3)
ii) similar shape of zones (see Plates 1C and especially 2B, C and 3E, H)
iii) patterns never repeat exactly - like weather, the classic chaotic system (see Plate 2D, 3B).
iv) abrupt changes from ordered to disordered growth and back again (Plate 3A) suggest that a parameter value of the system may have passed through successive bifurcation values.
v) repeated groups of zones are evidence of clustering, a common feature in nonlinear systems. Apparent clustering of Type I (fine-scale zones) in groups of 4 to 12 separated by irregular dissolution surfaces is not figured in this present work (but see Pearce and Kolisnik 1990 Plate II D).
vi) S. Higman and the writer have discovered spiral patterns in phase space for zone thicknesses (Higman and Pearce 1991a, b, 1992, 1993a, b; Higman 1992) and section 3.4, of this work.

3.4 IDENTIFICATION OF CHAOTIC GROWTH

3.4.1 *General Comments.* Identification of chaotic processes is a current "hot" research topic (see references in Middleton 1991). Distinguishing chaotic from stochastic processes is not an easy task. A considerable amount of effort has been put into developing techniques for identifying chaotic dynamics and these have been used with varying degrees of success. The technique used in this work is the return map or Santa Cruz method for reconstructing phase space from a single variable time series (Packard et al 1980), and the prediction of the time series from the reconstructed phase space (Farmer and Sidorowich 1987, 1988; Sugihara and May 1990). The return map (Figs 2C, 3C) is a plot of a single variable time series of the type X_n vs X_{n+1} where n varies from 1 to m (the total number in the series). Other time series methods, such as the "simplex" method and its simplification the "zeroth order method" are techniques for predicting times series data. The original simplex method of Farmer and Sidorowich (1987) requires much more data (thousands of points) than is likely to be available in zoned crystal studies (60 measurable zones is a large number for a zoned crystal). At the present time, a simple return map is the method of choice.

If we consider crystal growth to be reflected by a growth index such as zone thickness measured in μm, it is theoretically possible to determine the minimum number of variables (i.e. dimensions or equations) needed to describe the crystal growth. This is done by measuring the correlation dimension and Lyapunov exponents (Wolf et al 1985) of the zone thickness- time series. Note that in this case time is relative - not absolute. The attractor dimension is commonly calculated according to the Grassberger-Procaccia algorithm (1983). For completeness, we must also consider errors associated with estimating the attractor dimension from empirical data.

The above approach which determines the dimensionality of a system from an empirical data set, constrains the number of variables needed for the equations in the system and is an exciting and much needed practical technique. The ability to constrain the number of variables is fundamental to scientific inquiry and recalls the "hidden variable" problem of Einstein,and, Bell's Theorem (for a non-mathematical discussion of this problem see Peat 1990).

3.4.2. *Random Zoning Patterns.* Figure 2a plots data of a simulated zoned crystal 'grown" by a random growth law (each zone can have any value from 0 to 10 units thickness selected at random). The crystal is assumed to be oriented ⊥ b and the enlargement of the lower right-hand section of the crystal shows the random zoning pattern produced by the model. In this model very thick and very thin zones can occur in contact. This feature of the model can be seen by inspection of the resulting pattern and is very rarely observed in natural data. Figure 2b shows the time series resulting from measuring the thickness of successive zones along the upper edge of the crystal as shown in Fig. 2a. Although there is no coherent pattern to the time series data, it is difficult for the eye to detect this lack of coherence. Using a return map (plotting X_n vs X_{n+1}) as in Fig. 2c, the eye can more readily detect the lack of coherent pattern.

The random pattern of Fig. 2c is said to be space filling and this feature, along with the spatial juxtaposition of thick and thin layers, is a characteristic of random data.

3.4.3. *Real Crystal Zoning Patterns.* Higman 1992, and Higman and Pearce 1993a, b, have presented zoning thickness data on crystals from four volcanoes (Mt Misery, St. Kitts, Fuego, Guatemala, Popocatepetl, Mexico, and Mt. St. Helens, Washington). Crystals from all these volcanoes showed broadly similar patterns so we will examine one typical crystal here. Sample MSH-87-3b is a blast dacite from the May 18, 1980 eruption of Mt. St. Helens, Washington. Figure 3a represents a series of 66 zone thicknesses measured in a euhedral microphenocryst (the Type II crystal of Pearce et al 1987a). To the eye, the pattern is not dissimilar from that of the random simulation in Fig. 2a (note that the straight sector is an artifact of the method of presenting a single "time" series). The only feature recognizable to the eye in a pattern presented like this is the slight apparent tendency for thin zones of similar thickness to occur together in groups. Even so, this is an "arguable" feature and we should look at the data in a more quantitative way.

Figure 3b present the data as a time series. There is more apparent order to this graph than to the comparable Fig. 2b but describing the pattern is challenging. The return map of Figure 3c illustrates the order which is difficult to see in other depictions of this data set. Following the trajectory of points produces a spiral pattern and triangular segments are quite common. Following Takens reconstruction theorem (Takens 1980, Packard et al 1980), we assume that the geometry of the attractor in multidimensional phase space is consistent with the geometry displayed in the return map phase portrait plotted using a single variable. The triangular pattern (which is very common in other crystals studied by the above authors) is produced in a return map by two thin zones deposited on top of a thicker zone. This pattern is so common that any zoning model which realistically describes zonation must allow for (or better still, require) this geometric configuration of zones. There is also a suggestion of a limit cycle (the oval concentration of data centered about the apparent fixed point 3,3).

The suggestion from the above work is that the plagioclase- magma system is deterministic and has a low dimension attractor. A possible limit cycle and fixed point are consistent with a Hopf bifurcation. The common occurrence of triangles in the return map suggests to the writer that the correct phenomenological equations may be discrete differential equations rather than continuous differential equations typically used in modelling.

4. Attempts at Modelling Zonation

4.1 THEORETICAL AND EMPIRICAL MODELS

Most models assume that oscillatory zoning is a reaction-diffusion problem (Garcia-Ruiz 1987). A variety of methods (from analytical mathematical to computer simulations) have been used to treat it. A common feature of many (but not all) theoretical models is that thermal heat transfer rates are higher than chemical diffusion so that the process is essentially isothermal. Bearing in mind that differential equations have no general solution, a common problem of the analytical approach is that simplifying assumptions must be made to solve or approximate a solution; for example, variables such as growth rate, boundary layer thickness, etc., are held constant. In addition, the exact form of equations is generally not known so that further assumptions of the form of the equations and other ad hoc assumptions are necessary (such as which parameters and variables are important). Some of these difficulties are detailed below. Almost all theories use as a starting point a set of equations describing a reaction-diffusion process of the following form:

$$\partial C/\partial t = V\, \partial C/\partial x + D\partial^2 C/\partial x^2 \qquad [1]$$

$$V = f(C_b, Y, \alpha, \gamma, \lambda ...) \qquad [2]$$

where t is time, C is concentration, x is distance, D is the diffusion coefficient, V is crystal growth velocity, C_b, is the composition at the boundary immediately in contact with the crystal, Y is a reduced or nominal interface growth rate, and α, γ, λ are various parameters thought to affect crystal growth rates. In almost all theories outlined here, the Fickian diffusion part of [1] is either approximated with a discretized form or solved explicitly. Equation [2] represents the major differences between the various theories and is very poorly constrained. In fact, various workers have felt free to give it almost any form necessary to provide the desired results.

Most quantitative models (Allègre et al, Lasaga, Ghiorso, Pearce) involve a coordinate system whose origin is fixed at the crystal-liquid interface, thus avoiding the Stefan moving boundary problem familiar to engineers. In many of the models below, the concept of a boundary layer is used. There are two types of boundary layers. One is a mechanical or hydrodynamic boundary layer, which, as used by Lasaga (1982), is reckoned to be 10-100 μm thick based on experimental work. Other models use a diffusion boundary layer which has a maximum thickness equal to the hydrodynamic boundary layer but, it is clear, under certain circumstances may have a value much less than the mechanical boundary layer. The present writer believes that the thickness of the boundary layer is not a valid parameter for a model, since it is a function of the growth rate which, itself, is one of the most important variables in any model. A satisfactory model will have a diffusion-limited boundary layer which is allowed to develop in step with the growth rate of the plagioclase. It is evident from the observations presented in this work that the growth rate itself is a variable which may have values from zero (or even negative if dissolution is considered) to some maximum value. No model in which oscillatory zoning is explained by a constant growth rate is worthy of consideration at this point in the development of the theory of zoning.

4.2 DETAILS OF MODELS

4.2.1 *Sibley et al (1976).* Sibley and co-workers have produced a much quoted qualitative model based on the concept of constitutional supercooling borrowed from metallurgy (see also Tiller 1977). The model is isothermal because thermal diffusion is assumed to be much faster than chemical diffusion. The critical assumptions of this model are that growth ceases at equilibrium and that, having stopped, growth can not begin again until a certain degree of supercooling is re-established. Modern work on crystallization theory appears to substantiate the idea that nucleation difficulties may be involved in cyclic growth (see Brandeis and Jaupart 1987, for example). In addition, the composition of the liquid is assumed to be more important than the temperature in determining the initial composition of plagioclase that grows in each cycle (accounting for the jump in An% at the beginning of each cycle). Sibley's model involves a constant ΔT but variable ΔC hence the term constitutional supercooling.

The model is based on a simple interpretation of supercooled crystallization at a constant T2 in the plagioclase system below (Figure 6). Their model is as follows. The magma cools down past the liquidus temperature T_1 to a supercooled temperature T_2. Crystals of composition An1 nucleate and grow from liquid L_1 (the bulk composition). As growth proceeds, the crystals become more sodic (zoning to An2) while the liquid (boundary layer) in contact with the crystals changes composition to L_2. This represents equilibrium at T_2 therefore crystallization stops. The crystal is surrounded by a boundary layer L_2 around which there is a large reservoir of composition L_1. As diffusion begins to wipe out the boundary layer and move its composition back towards L_1, the crystal remains in a quiescent phase - not growing because it does not have the required amount of constitutional supercooling to initiate nucleation. When diffusion has eliminated the boundary layer, the system has returned to the initial state and the cycle can begin again. Repetition of this process will produce a saw-tooth pattern in the zoning profile. Sibley et al drew attention to both the form of the expected zoning pattern and the fact that the lower baseline of composition represented equilibrium and therefore should be "constant", claiming that this latter observation would be a critical method to validate the model.

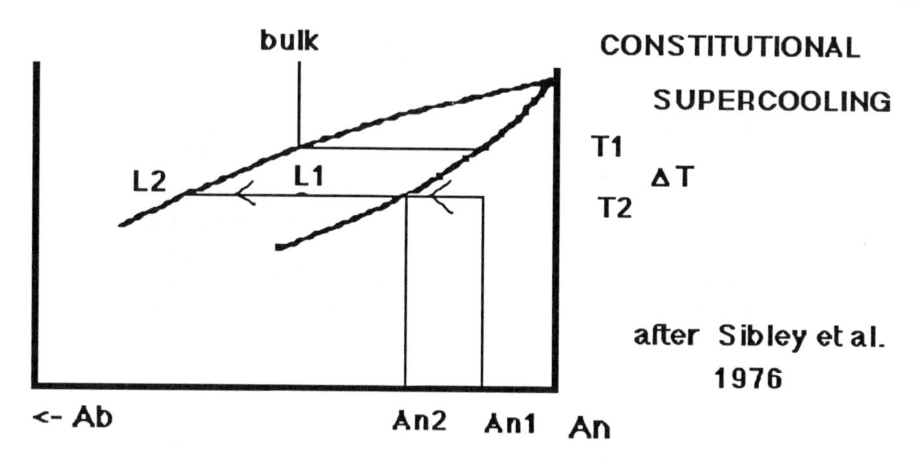

Figure 6. Constitutional supercooling model after Sibley et al 1976.

The major advantage of this model is that it is the only model proposed to date that appears to predict the correct form of oscillatory zoning (of the Type I). A major difficulty (as with all qualitative models) is that it is not possible to determine if the prediction is exactly correct or merely in the right direction. In addition, the model predicts exactly repetitive zoning which is not what is observed (as we have seen in section 3).

Although Phemister (1934), suggested the possibility of plagioclase ceasing to grow (and even dissolving!) during cycles of deposition, Sibley et al seem to have been the first to relate the cessation of growth to attainment of equilibrium conditions. This concept appears to presage new work on nonlinear dynamical systems in which equilibrium is merely a stable point in a phase portrait rather than a state towards which the system is inexorably driven and which it must attain.

4.2.2 *Haase et al* (1980). Haase et al (1980) have produced one of the most sophisticated approaches to oscillatory zoning in plagioclase feldspar published to date. They formulated the problem as a Stefan moving-boundary problem and assumed some phenomenological laws for plagioclase growth from solution. Their approach is the first work which treats oscillatory zoning as an example of a structure that is far from equilibrium and results from an interplay between nonlinearity and non-equilibrium conditions.

Details of their model are as follows. Their model assumes that plagioclase grows from a melt by a layer-filling mechanism and that the layers begin at the base of kinks or other growth sites. They simplify the melt to consist of Na^{+}, Ca^{++}, Al^{3+}, SiO^{4-}. However, the key assumption of their model is that the rate constants for precipitation of Ab and An units have a positive feedback dependence such that An growth is favored by An-rich surfaces

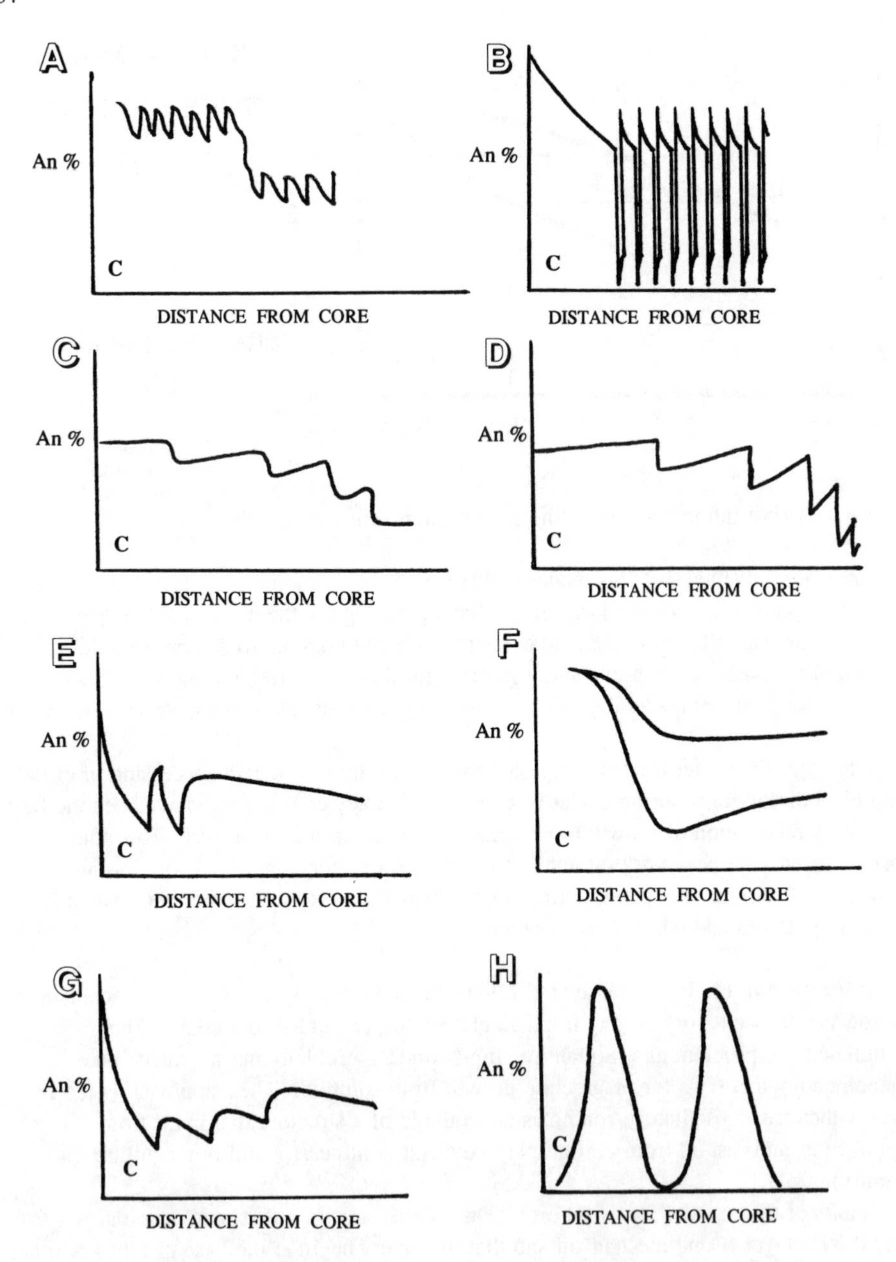

Figure 4. Zoning profiles depicted in the geological literature.

and Ab growth by Ab-rich surfaces. This assumption is required for autocatalytic oscillatory behavior, and is presented without justification. Since there is simply not enough known about the plagioclase-magma system, they were forced to make assumptions concerning the phenomenological functions for growth rates. They found several forms with a functional dependence which lead to oscillatory behavior and they discussed only those with a rather complicated form, for example for the Ab growth rate:

$$B(f) = B(0)[\beta + (1 - \beta)(1 - f)^2] \qquad [3]$$

where B(0) is the growth constant for Ab, f is the surface fraction of An, and ß is one of three parameters which control the intensity of the surface compositional feedback effect. Their equations are highly nonlinear and produce an auto-catalytic mechanism which, in turn, produces oscillations when the various parameters have the correct values. The validity of the phenomenological function [3] is not made clear by their exposition. One result of this model is that while Haase et al show a hysteresis effect in parameter space, there are no corresponding discontinuities in the zoning profile. Even though their profile (see Figure 4B) shows abrupt changes in An percent, growth is essentially continuous.

Ortoleva has argued (personal communication. 1989) that rapid changes in a continuous zoning profile might be sub-microscopic (nanometer scale) and appear, therefore, to the optical microscope as discontinuities. If so, then the zoning profile of Figure 4B would more closely resemble the natural zoning profile (as in Figs. 1, 4A), although the sense of the curvature of the zone is opposite to what is normally observed. In addition, evidence of solution, which is exceedingly common in Type I zoning profiles, requires that there be discontinuities in the zoning profile; therefore, Ortoleva's explanation, although plausible, is probably not generally correct. Because of the autocatalytic nature of their model, phase portraits (i.e. sections of state space for their model) show a limit cycle which is similar to that inferred from the observations presented in Section 3 of this work. Unfortunately, the results produced by this particular dynamical simulation do not resemble naturally occurring zoning profiles. It remains to be seen, whether fine tuning the parameters of Haase's work can produce more realistic zoning profiles.

This approach has recently been successful in many biological, chemical and physical systems and the present writer considers that much more work should be done treating geological problems such as oscillatory zoning in a nonlinear dynamical fashion. The work of Haase et al, therefore, would repay more detailed consideration which is beyond the scope of this present work.

4.2.3 *Allègre et al* (1981). In this work, Allègre et al develop a model which oscillates like simple harmonic motion ($\sin\omega t$). They conclude that oscillatory zoning is rare ("pathological" in their terminology) and requires quite specific conditions. Their model is one-dimensional with two component Fickian diffusion. Only one component (Ab) is independent and this precludes the "poisoning" effect of extra components alluded to by Simakin (1983) - see section 4.2.6.

In common with other models, they begin with equation [1] above. They then postulate a coupling between the growth rate and the initial boundary layer composition. They further postulate that the form of this coupling is a differential equation rather than algebraic i.e.:

$$\tau \cdot dV/dt + V = V(C) \quad [4]$$

The parameter "τ" is a characteristic time lag which they postulate is necessary for the surface to "rearrange" itself for growth. After more assumptions, they arrive at an equation with a sine term ($\sin\omega t$) remarkably similar to the equations of simple harmonic motion. The results of these equations certainly produce oscillations. This work is flawed by an excess of assumptions and the belief that oscillatory zoning is "the exception rather than the rule" and that the shape of the zoning is a skewed sine-curve. Their ad hoc assumptions were designed to produce a sine curve of zoning rather than the discontinuous saw-tooth shape which is actually observed using modern interference imaging (Pearce and Kolisnik 1990, etc.). The correct shape of zoning patterns was figured long ago by Harloff (1927) and Homma (1932) and more recently by Bottinga et al 1966; see Fig. 4A. The model of Allègre et al requires quite specific conditions for oscillations and they suggest that the plagioclase-melt system is one of those "rare" systems which fulfill certain critical requirements. They went so far as to refer to oscillatory zoning as "pathological", emphasizing their opinion that it was an unusual phenomenon. In fact, oscillatory zoning has been observed in many different mineralogical systems - it is hardly unusual.

4.2.4 *Lasaga* (1982). Lasaga, in his work on a "master equation" for crystal growth has developed a general model which assumes local exchange equilibrium between the melt and the surface of the crystal. Crystal growth rate either varies explicitly with time or is a general function of concentration (as for example during interface controlled growth). Lasaga generated zoning patterns calculated using his "master equation" (see Figs.. 4E, 4G) but assumptions such as a multivalued growth function or a step function are apparently necessary to produce oscillations. Lasaga admits (p. 1284) the analytic form of his equation V(c) never allows oscillations to take place. Lasaga measures the flux, J as follows:

$$J = -D\, \partial c_L/\partial x - V(t)c_L \quad [5]$$

where t is time, D is the diffusion coefficient, c_L is liquid composition at distance x. As a boundary condition at x = 0, the crystal growth is treated as an external flux so that J = 0; hence, for a trace element with distribution coefficient k:

$$D\, \partial c_L/\partial x\, |_{x=0} + (1-k)V(t)c(0,t) = 0 \quad [6]$$

An important result of his work is that oscillations cannot arise from growth rates that are analytic functions of the melt composition and in a later work, Ghiorso (1987), concurs with this finding. In order to obtain oscillations it is necessary to make assumptions such as a multi-valued growth assumption or a step-function and these results are displayed in Figs. 4E, G. In Lasaga's model, the reference frame is fixed at the crystal boundary and the liquid is assumed to move towards the boundary producing an external flux. The most important part of the model, the form of the growth term in equation [6] above, is left general, thus necessitating the assumptions necessary to yield results which are time dependent.

4.2.5 *Loomis* (1979, 1982). Loomis has produced a rather unusual model, a computer simulation of plagioclase growth in which ΔT varies, unlike most of the other models discussed here. His model also includes water and his approximate calculations suggest that water accumulates at the surface of growing crystals promoting local convection which might wipe out diffusion gradients. In spite of this however, his model does not produce oscillations (Figures 4D, F). In order to make his model oscillate, an assumption of stepwise decreasing temperature is necessary. This constraint is similar to the experimental method of Lofgren and indeed the results do resemble Lofgren's experimental work (see Figure 4C). It appears therefore that Loomis has produced a good computer simulation of Lofgren's experiments rather than naturally occurring plagioclase. Loomis himself admits that the scarcity of kinetic data and the modelling assumptions made for the purposes of tractability limit the accuracy of the simulations. A key part of Loomis' model is the following equation from Kirkpatrick et al 1979:

$$\upsilon\eta/\Delta T = (1.8\text{x}10^{-6}\,\Delta T) - (9.9\text{x}10^{-9}\,\Delta T) \qquad [7]$$

where υ = velocity (cm/s) and η = viscosity (poise) and ΔT is the undercooling. The above equation is nonlinear in ΔT, yet even this nonlinearity is not sufficient to make the system of equations produce oscillations.

4.2.6 *Simakin* (1983). Simakin, has developed a phenomenological model which describes rhythmic zoning in magmatic minerals accompanied by periodic impurity accumulation and dispersal. As with other models, Simakin's model is a one-dimensional approximation which solves the diffusion term by discrete differentiation. The model assumes thermal heat transfer rates are higher than chemical diffusion so that the process is essentially isothermal. The growth rate is single-valued and depends on composition. Simakin emphasizes minor component concentration in the boundary layer as a method of producing supersaturation as well as "poisoning" the crystal surface. The minor component retards growth by poisoning the surface and reducing the supercooling $\Delta C = C_L - C_b$ (concentration at liquidus - concentration in boundary layer).

Simakin treats growth in an interesting fashion. He postulates that a parameter λ describes the surface state of the growing crystal and that the growth velocity is a function of both the parameter λ and the supercooling as measured by $\Delta C = C_L - C_b$. Although he does not explain or define this λ parameter, it is this critical parameter which allows the system to oscillate. Simakin further assumes that diffusion takes place only within a boundary layer of width δ, and neglects the motion of the boundary layer in the Stefan moving boundary problem by assuming that the velocity is very small. His model involves periodic reductions in growth rate "proceeding as far as halting". His equations are very similar to the standard equations [1] and [2].

$$\partial(\int C dx)/\partial t = VkC_b - D\partial C/\partial x \quad [8]$$

from 0 to δ

$$V \approx \lambda \cdot \Delta T \approx \lambda \cdot \Delta C \quad [9]$$

or $P \cdot \delta \cdot dC_b/dt = VkC_b - D\Delta C_{b,0}/\delta$, where δ is the defined width of the diffusion boundary layer, P is growth rate (scaled in units of D/δ), D is the diffusion coefficient; and $\Delta C_{b,0} = (C_b - C_0)$ boundary composition - starting composition and $\Delta C = (C_L - C_b)$ liquidus - boundary composition

Having set up the standard equations, Simakin performs a linear stability analysis on a system of first-order differential equations, one of which,[11] has an unknown form.

$$dC_b/dt = (\lambda k)/P.\ C_b(C_e - C_b)/\delta - D/P\ .\ (C_b - C_o)/\delta^2 \quad [10]$$

$$d\lambda/dt = F(\lambda, C) \quad [11]$$

Simakin could proceed no further with such a general equation [11] unless he made some assumptions. The key assumptions which he made are that the system possesses an unstable fixed point in λ-C phase space. Furthermore, it appears that he expects to see a Hopf bifurcation (which must possess a fixed point and a limit cycle) in the system of equations. He therefore assumed a polynomial form for $F(\lambda, C)$ and determined the values of the parameters which will yield a fixed point corresponding to an unstable steady-state. It is interesting to note, as an aside, that such a point could not be physically observed as the system cannot occupy it for any finite period of time. In λ-C phase space, the trajectory would move rapidly towards the limit cycle.

In summary, it appears that Simakin has merely assumed that there is a Hopf bifurcation and an unstable fixed point in λ -C phase space and determined what sort of assumptions and polynomial equations are necessary to yield this desired result. This comes perilously close to assuming what you have to prove. That said, the present writer believes that linear stability analysis is a valid and useful approach and it should be encouraged, but with better empirical data to constrain theorizing. It is interesting to note that as with most of the models in this work the zoning patterns that Simakin produces with his models do not resemble natural plagioclase zoning profile (compare Fig. 4H with 4A). Again, we must wonder if by suitably tuning the parameters or exploring different forms of the $F(\lambda, C)$ equation, a more realistic zoning pattern could be produced. Although the lack of a discontinuities in these continuous zoning profiles is held to be a serious flaw by the present writer this approach (which comes from the study of dynamical systems) should be continued as it represents the best opportunity for success in modelling and understanding complex natural systems.

One other point, not touched on by Simakin, and not generally considered by other workers is confounding of the variables. Because the parameters k, δ and λ are tied closely together (usually by multiplication) in the fundamental equations, an error in estimating k, for example, can be compensated by an error in λ, giving a "realistic" model profile whose parameters may be in error perhaps by a very large amount. The fact that a realistic profile has been produced by modelling, therefore, does not mean that the magmatic parameters are known. There are, however, natural constraints on the values which the parameters may take in natural processes and these must guide theoreticians in their work.

4.2.7 *Ghiorso* (1987). Ghiorso, in his treatment of chemical mass transfer in magmatic processes devoted a considerable amount of space to plagioclase growth. He was concerned with modelling interface growth kinetics in naturally occurring silicate liquids and focussed on chemical affinity and the contribution of equilibrium description of melt behavior to problems of this sort. That is to say, he explored time-dependent chemical composition crystallization of minerals from a supersaturated, silicate melt. He showed that the geometry of the multi-component phase loop has a profound effect on equilibrium compositional trends with the result that projections made from the simple Ab-An binary system may not be correct when applied to the multi-component natural system. Ghiorso's model involved an assumption of a fixed amount of undercooling close to equilibrium. His basic equation is similar to equation [1] above. The flux vector is

$$\mathbf{J} = - \mathbf{D}\partial \mathbf{C}_L/\partial x - Y\nu \mathbf{C}_L \qquad [12]$$

His form of equation [2] is a reduced growth rate, Y, multiplied by a distribution coefficient vector, ν, multiplied by a vector, $\mathbf{C}_L$, describing the composition of the liquid in contact with the solid. Ghiorso reports that the composition of the precipitated plagioclase is always more anorthitic than that which would be produced at equilibrium, and this is in direct contrast to the results obtained for the two component phase loop as reported by Lasaga (1982). Even so, the disequilibrium plagioclase is more anorthitic than the eventual equilibrium composition. Hence the crystals should be normally zoned to a more sodic composition as crystallization continues.

The interesting feature of Ghiorso's work is that the initial growth rate is dominated by interface kinetics which rapidly depletes the interface meltin anorthite content. As growth continues, the crystal becomes more albite-rich and the growth rate begins to be controlled more by chemical diffusion than by interface kinetics. Curiously, although describing zoning in crystal growth, Ghiorso displays only profiles of chemical composition within the liquid and shows no profiles of zoning patterns in the crystals. It is therefore difficult to assess the accuracy of his model. In common with all quantitative models, Ghiorso makes necessary assumptions. Nevertheless, his description of zoning profiles within the liquid as the crystal undergoes a change from interface-controlled to diffusion-controlled kinetics greatly resembles the results of the finite-element model in this current work - see section 4.2.8.

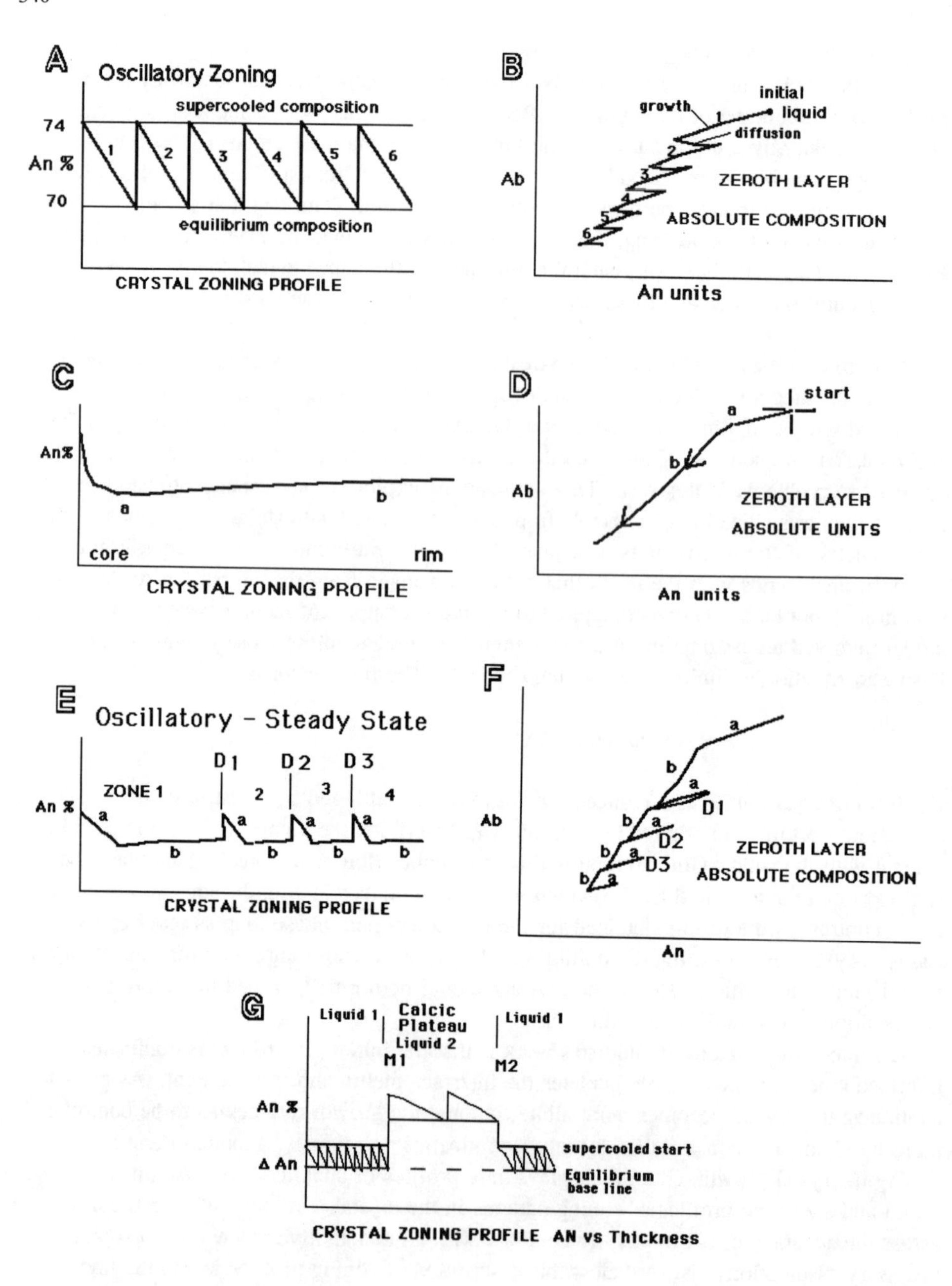

Figure 5. Zoning profiles produced by a finite element model (section 4.2.8).

Ghiorso's work is an interesting attempt at applying what is basically an equilibrium thermodynamic approach to a non-equilibrium problem. The overall thrust of the work appears to be more directed at calculating zoning profiles within the liquid during crystal growth rather than modelling the zoning profiles of the crystals.

4.2.8 *Pearce* (1993). The computer model developed by Pearce (1993) and expanded in this present work is a one-dimensional finite-element model with three or more chemical components and m (>2) cells in the liquid. The model consists of solid crystal, the zeroth layer (the liquid immediately in contact with a solid and from which the solid grows) and the local reservoir consisting of cells 1, 2... m. The coordinate system has its origin fixed at the crystal-liquid interface and the crystal is assumed to be extracted from the liquid part of the system. Diffusion between adjacent liquid cells is controlled by Fick's laws of diffusion. The model, therefore, uses equations [1] and [2] above in discretized form as shown below.

Scaled for a unit distance ($\Delta x = 1$), for any component, C, Y_1 is the nominal growth rate for a unit of undercooling; Y_2 is an interface growth mechanism not tied to ΔC; $\Delta C_{0,1}$ is the difference between concentrations in the first and zeroth cells; k is a partitioning coefficient. The parameter α models nucleation difficulties by having the value 1 (growth) or 0 (equilibrium). It acts as a switch by turning growth on or off. The variable C_L is the concentration at the liquidus (i.e. equilibrium), C_n is the concentration at the nth layer of the model and C_{-1} is the first layer of the zoned crystal, and Jc is the flux of C. The system is initialized to unit distances between cell centers and unit volumes for the cells.

The discretized rate of change of concentration of the component in the zeroth layer of liquid (i.e. the layer in contact with the crystal) is given by:

$$Jc_{0,-1} = -\alpha \cdot k \cdot (\Delta C_{L,0} \cdot Y_1 + Y_2) \quad + \quad D \cdot (\Delta C_{0,1}/\Delta X_{0,1}) \qquad [13]$$

This model is a numerical refinement and extension of a constitutional supercooling model (Sibley et al 1976). The key part of this model is that growth stops at equilibrium and does not restart again until a certain minimum amount of constitutional supercooling ($\Delta C_{L,0}$) is re-established by diffusion. The model has a hysteresis effect and produces a saw-toothed zoning pattern with a discrete jump in An% following the hiatus when there is no growth. In common with many of the models described in this section, the process is essentially isothermal (thermal diffusion is assumed to be much faster than chemical diffusion). Turbulent convection within the boundary layer (<100 μms) is assumed to be negligible. All parameters except concentration are held constant for a given computer run and the boundary layer caused by the interplay of growth and diffusion is allowed to assume its characteristic value as the growth process proceeds. Even this simple model can give some insight into the complexity of plagioclase growth. The model is capable of either producing regular oscillations, as in Fig. 5A, or a steady-state zoning profile as in Fig. 5C. Trial runs indicate that the system oscillates only when growth (removal of An and Ab units from the zeroth layer) is much faster (as much as ten times) than diffusion of these constituents into the layer. If growth is slow then a steady-state results. The change in composition (An and Ab units) within the zeroth layer is shown in Figs. 5B, 5D.

An unexpected feature of this model is that, in the intermediate case where growth is not quite slow enough to produce a steady-state, an intermediate type of zoning is produced as in Figure 5E (liquid path shown in Fig. 5F). This type of zoning, which for lack of another term I will refer to as "mixed mode zoning", is remarkably similar to certain laser profiles of zoning which have been determined by Kolisnik (1990) and shown in Figure 1.

The calculated results from this simple model resemble natural plagioclase patterns with one major exception - they are much more regular. Major changes in the base line are readily modelled by an abrupt change in the composition of the liquid reservoir (as due to magma mixing, for example). Compare Fig. 5G with Plate 1B of this work. A feature of this model which closely resembles observations of natural crystals is shown in Fig. 5G. In the model, an abrupt change is made in the composition of the liquid of the local reservoir. Liquid #1 (say, dacite) grows narrowly spaced oscillations of a low An and when the reservoir liquid is abruptly changed to a second liquid #2 (say, andesite) which precipitates a more calcic plagioclase, the thickness of the zones increases without any other change in the parameter other than the change in the composition of the liquid. This phenomenon is observed in some actual plagioclase phenocrysts. Zoning in the calcic plateau such as that exhibited in Fig. 5G, if it is present at all, tends to be thicker than the zones in the adjacent sodic areas.

One other feature, which serves to emphasize the complexity of results obtainable from a simple model, is that at least two growth mechanisms seem to be necessary. One (Y_1) is proportional to the degree of supercooling, the second (Y_2) is capable of operating near or perhaps even at equilibrium (a process perhaps similar to Ostwald ripening of precipitates). This second mechanism appears to be necessary in order to drive a system to equilibrium. Otherwise the system will never get to equilibrium since the growth rate diminishes exponentially as the system nears equilibrium. This raises the question of whether equilibrium is ever achieved even with the long time span available in geological systems. While this appears to be a philosophical question, it is not something that can be completely ignored when dealing with dynamical systems.

This simple finite-element deterministic model is capable of complex behavior on the assumption that there is a hysteresis effect in the growth rate due to nucleation problems (see Brandeis and Jaupart 1987). The calculated results resemble natural plagioclase patterns but they are not complicated enough to be chaotic. As we have seen from the previous section, even if the features of this model were to be basically correct, a one-dimensional model must still be regarded as an oversimplification. Nevertheless, the finite-element approach appears worth pursuing perhaps in a 2- or 3-dimensional extension of the technique.

4.2.9 *Wang and Merino* (1993). Recently, these authors produced a model which predicts oscillatory behavior when two minerals such as biotite and plagioclase grow simultaneously in a melt - thus competing for the same chemical species (Al^{3+}). The model was developed to explain repetitive igneous layering rather than oscillatory zoning per se; however, their approach appears sound. They set up some equations (with assumptions concerning kinetics, advection, and coupled chemical reactions), then did a linear stability analysis around the steady-state solution of the equations. They concluded that their system of equations has oscillatory solutions under certain conditions: very restricted composition of liquids, and the excluded liquid species must have a higher diffusivity than the compatible elements. Although not strictly about the present subject, this work points the way toward future studies.

5. Discussion and Conclusions

5.1 GENERAL COMMENTS

1) There are two main zoning types (Pearce and Kolisnik 1990): Type I, small scale, diffusion controlled (both interface and diffusion controlled using the nomenclature of Ghiorso 1987, and Lasaga 1982) and Type II, larger scale, liquid composition controlled. The local reservoir is a packet of liquid picked up by the crystal during magma overturns or mixing events and modified by dissolution of the entrained plagioclase crystals.

2) A satisfactory model must involve repetitive growth and cessation of growth. The possibility of repetitive growth and solution (as envisaged as long ago as the 1930s by Homma, and Phemister) is consistent with empirical observations.

3) In plagioclase-melt systems, oscillatory zoning is the most common form of growth and is observed in systems which vary in viscosity (and possibly growth rates) by as much as 100,000 (volcanic basalts vs plutonic granites). There is also a wide variation in temperature (700° - 1200°C) and pressure (1 atmosphere to >3 kb) indicated by the geology of the rock types.

4) Oscillatory behavior is observed in other minerals as well as plagioclase so that a general solution must be found.

5) Earlier workers appeared to have been mesmerized by the concept of "equilibrium crystallization" and were reluctant to consider that growth must stop at equilibrium. Indeed, one of the best definitions of equilibrium is that nothing happens to the system at equilibrium no matter how long it is observed. A system at equilibrium cannot, by definition, be changed by a time-dependant process. Phases are of homogeneous composition and their proportions do not change with time. Since, in the present case, we are dealing with zoned, i.e., heterogeneous crystals which grow, dissolve or react with time, it should be axiomatic that equilibrium has very little if anything to do with zoned crystals.

6) In several attempts (Lasaga, Allègre, Ghiorso), analytical, continuous equations have failed to produce oscillations without an imposed external template (step function or variable). In other attempts (notably Simakin, Haase et al, Wang and Merino) linear stability analysis has revealed oscillatory behavior for systems of continuous equations. Unfortunately, the oscillatory behavior yields the wrong shape to the zoning patterns and apparently can not account for repetitive cessation of growth or solution. I take this to mean that the correct equations which describe the natural phenomenon may be discrete

differential equations rather than continuous. The assumption that continuous equations are adequate to describe nature is not merely a philosophical question and needs to be addressed.

5.2 PREFERRED MODEL

A model must take into account: 1) the ubiquitous repetitive occurrence of dissolution surfaces; 2) the inevitable interaction between the dissolving plagioclase and the adjacent boundary layer; 3) the complications in growth dynamics due to 2) and 3). No formally presented model is capable of handling all these constraints and producing realistic zoning patterns. The present writer favours a model with the following features. Plagioclase cycles around a turbulently convecting magma chamber as envisioned by Homma (1932). As it cycles around, it grows in the cooler parts and dissolves (but may also react) in the hotter parts. The only crystals we see at the surface are "survivors" whose total growth exceeded their dissolution. Repetitive dissolution removes large quantities of existing solid plagioclase as detailed in Fig. 7 and recycles it through the magma via the boundary layer. What this does to the trace element composition of the magma is problematical. The removal by dissolution of so much pre-existing solid means that the surviving crystals have truncated profiles as they go through various stages of growth and dissolution (4 stages in Fig. 7).

The enrichment of the hydrodynamic boundary in plagioclase components after each solution event increases the initial An% by 10-20% at the next growth stage. The key observation which determines if the liquid reservoir is the same after the solution event (simple convection) or has a different composition (magma mixing) is the baseline steady-state composition. Most profiles (as in Fig. 1), if they record a steady-state composition, do not require a major change in composition of the magma reservoir. This suggests that serious cases of magma mixing (as in basalt + dacite = andesite) may not be all that common. This type of zoning (Type II of Pearce and Kolisnik 1990) is not self-organized oscillatory because the oscillations are externally induced by turbulence and changes in the local reservoir. However, for many petrologists, this is the most spectacular form of oscillatory zoning. Several models (Lasaga, Ghiorso, Pearce) can model steady-state zoning like this although no published model appears to handle both growth and solution.

Having developed a plausible explanation for the overall pattern of zonation in Fig. 1, we are still left with the smaller Type I oscillatory zoning which ornament the larger scale zones. This small scale zoning which is the object of the modelling by Haase et al, Allègre, and Pearce is best developed when a steady-state or quiescent conditions have been established. It does not occur, for example, immediately after a massive dissolution event. This observation is consistent with the diffusion-reaction assumptions in most models. We must conclude that, while the dynamical models appear very interesting, no model yet proposed explains the detailed observations of this type of zoning. Oscillatory zoning thus remains a current topic of research.

ZONING PROFILES DUE TO GROWTH & SOLUTION

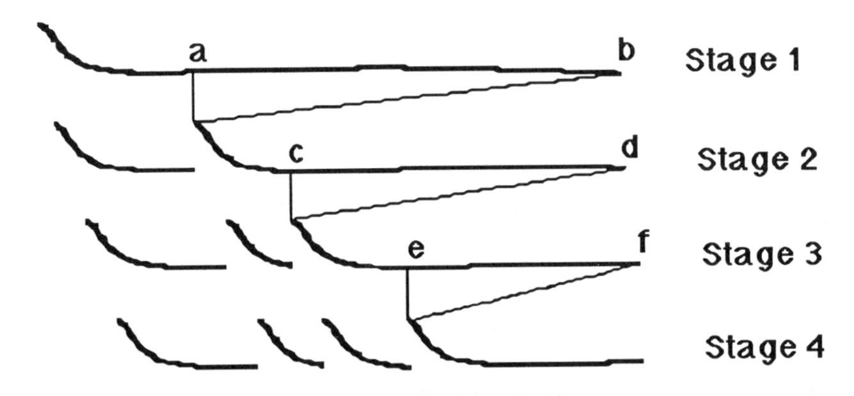

Figure 7. Hypothetical zoning profiles due to repeated growth and solution An% vs distance from core. Parts of the crystal a-b, c-d, and e-f, are successively grown on the existing crystal then removed by solution leaving the profile at Stage 4. Compare Fig. 1.

5.3 SUGGESTIONS FOR FURTHER WORK

One of the conclusions resulting from this work is that the potential use of trace elements has barely begun. It is clear to the writer that detailed accurate analyses of both compatible and incompatible trace elements within individual zones are necessary in order to constrain dynamical models of zoning. It is observed that in volcanic and sub-volcanic rocks, the outermost zone corresponding to the last episode of growth in plagioclase is frequently much larger, sometimes an order of magnitude larger than the zones which are present in the centre of the crystal. It is not unusual to find such zones which measure 50 or even 100 μms in thickness. Given the difficulties in spatial resolution with most microprobe techniques, it makes sense to the writer to obtain detailed probe traverses of such thick zones rather than incomplete traverses of much thinner zones within the crystal. It would be worthwhile having detailed analytical data from such zones for additional reasons: first, it is sometimes possible to estimate the length of time taken for the growth of this thick zone because it is the last growth episode in the crystal; second, unlike other zones within the crystal, in the case of the last zone we can usually estimate the composition of a liquid from which the zone grew, as it must be the liquid with which the zone is presently in contact.

6. Acknowledgements

This study is, in part, a result of 10 years work in the Laser Lab, Department of Geological Sciences, Queens University at Kingston. It was supported by National Sciences and Engineering Research Council of Canada Grants to the writer. I thank Ian Parsons and W.

L. Brown for helpful reviews, S. Chapman and A. Grammatikopoulos for assistance with the manuscript, and, Nancy and Rodney for their continued support.

7. References

Allegre CJ, Provost A, Jaupart C (1981) Oscillatory zoning: a pathological case of crystal growth. Nature 294: 223-294

Anderson AT Jr (1983) Oscillatory zoning of plagioclase: Nomarski interference contrast microscopy of etched sections. Amer Mineral 68: 125-29

Anderson AT Jr (1984) Probable relations between plagioclase zoning and magma dynamics, Fuego Volcano, Guatemala. Amer Mineral 69: 660-676

Baker G L, Goolub JP (1990) Chaotic Dynamics: An Introduction, Cambridge University Press.

Blundy J D, Shimizu N (1991) Trace element evidence for plagioclase recycling in calc-alkaline magmas. Earth Planet Sci Letters 102: 178-197

Bottinga Y, Kudo A, Weill D (1966) Some observations on oscillatory zoning and crystallization of magmatic plagioclase. Amer Mineral 51: 792-806

Brandeis G, Jaupart C (1987) Crystal sizes in intrusions of different dimensions: Constraints on the cooling regime and the crystallization kinetics, Magmatic Processes: Physicalchemical Principles. The Geochemical Society, Special Publication No 1, Ed. B. O. Mysen

Bryan WB, Pearce TH (1993) Plagioclase zoning in selected lavas from holes 834B, 839B, and 841B. O. D. P. Leg 137 volume, Journ Geophys Res (accepted for publication)

Clark A H, Pearce TH, Roeder PL, Wolfson I (1986) Oscillatory zoning and other micro-structures in magmatic olivine and augite: Nomarski interference contrast observations on etched polished surfaces. Amer Mineral 71: 734-741

Farmer JD, Sidorovich JJ (1987) Predicting chaotic time series. Physical Review Letters 59: no. 8, 845-848

Farmer JD, Sidorowich JJ (1988) Exploiting chaos to predict the future and reduce noise. Unpublished manuscript, version 1.2

Fowler, AD, Stanley HE, Daccord G (1989) Disequilibrium silicate mineral textures; fractal and non-fractal features. Nature 341 (6238): 134-138

Garcia-Ruiz JM, Santos A, Alfaro EJ (1987) Oscillatory growth rates in single crystals growing under diffusional control. J G R 84: 555-558

Ghiorso MS (1987) Chemical mass transfer in magmatic processes III. Crystal growth, chemical diffusion and thermal diffusion in multicomponent silicate melts. Contrib Mineral Petrol 96: 291-313

Glass L, Mackey MC (1988) From Clocks to Chaos: The Rhythms of Life. Princeton University Press: pp 248

Grassberger P, Procaccia I (1983) Measuring the strangeness of strange attractors. Physica 9D: 189-208

Haase CS, Chadam J, Feinn D, Ortoleva P (1980) Oscillatory zoning in plagioclase

feldspar. Science 209: 272-274
Harloff C (1927) Zonal structure in plagioclases. Leidsche Geol Med 2: 99-114
Higman S, Pearce TH (1991a) Preliminary applications of chaos theory to plagioclase zoning. GAC-MAC, Annual Meeting Toronto
Higman S, Pearce TH (1991b) Applications of chaos theory to crystal morphology and zoning in magmatic plagioclase. GSA, Annual Meeting, San Diego
Higman S, Pearce TH (1992) Nonlinear dynamics and crystal growth from the morphology of oscillatory zoned plagioclase. A.G.U. Annual Meeting, Montreal, EOS, April 7 1992, 353
Higman S, Pearce TH (1993a) A nonlinear dynamical approach to the growth of oscillatory zoned magmatic plagioclase. Geol Assoc Canada, abstracts, 18, A-44
Higman S, Pearce TH (1993b) Spatiotemporal dynamics in oscillatory zoned magmatic plagioclase. Geophys Res Letters (accepted for publication June 1993)
Higman S (1992) An application of nonlinear dynamics to the theory of oscillatory zoned magmatic plagioclase: an empirical approach. Unpubl. M.Sc thesis, Queen's University 185
Homma F (1932) Über das Ergebnis von Messungen an zonaren Plagioklasen aus Andesitien mit Hilfe des Universaldrehtisches. Sweizerische Mineralogische und Petrographische Mitteilungen 12: 345-352
Kirkpatrick RJ, Klein L, Uhlmann DR, Hays JF (1979) Rates and processes of crystal growth in the system anorthite-albite. J Geophys Res 84: 3671-3676
Kolisnik AM, Pearce TH (1991) Magma-mixing textures from intermediate rocks of Volcan Popocatepetl, Mexico. GAC-MAC Annual Meeting, Toronto
Kolisnik AM (1990) Phenocryst zoning and heterogeneity in andesites and dacites of Volcan Popocatepetl, Mexico. Unpubl M.Sc., Queen' University, 219p
Langmuir CH (1980) A major and trace element approach to basalts. Ph.D. thesis, S.U.N.Y, Stony Brook, 351 pp
Langmuir CH, Hansen GN (1985) Plagioclase saturation surface. Unpublished manuscript
Lasaga AC (1982) Toward a master equation in crystal growth. Amer J Sci 282: 1264-1288.
Lofgren G (1974a) An experimental study of plagioclase crystal morphology: isothermal crystallization. Amer Jour Sci 274: 243-273.
Lofgren G (1974b) Temperature induced zoning in synthetic plagioclase feldspar. In: The Feldspars, Proceedings of the NATO Advanced Study Institute, edited by W.S. MacKenzie and J. Zussman. Manchester University Press, Manchester, 362-375
Lofgren G (1980) Experimental studies on the dynamic crystallization of silicate melts. Chapter 11 in Physics of Magmatic Processes, Hargraves RB, (ed). Princeton University Press
Lofgren GE, Norris PN (1981) Experimental duplication of plagioclase sieve and overgrowth textures. Geol Soc Am, Abstr Prog
Loomis TP (1979) An empirical model for plagioclase equilibrium in hydrous melts. Geochem Cosmochim Acta 43: 1753-1759
Loomis TP (1982) Numerical simulations of crystallization processes of plagioclase in

complex melts: the origin of major and oscillatory zoning in plagioclase. Contrib Mineral Petrol 81: 219-229

May RM (1976) Simple mathematical models with very complicated dynamics. Nature 261: 461-467

Middleton G (1990) Non-linear dynamics and chaos: Potential applications in the earth sciences. Geoscience Canada 17: no. 1, 3-11

Middleton G (1991) editor. Non-linear dynamics, chaos, and fractals. Short Course Notes volume 9, Geological Association of Canada 235 pp

Nixon GT, Pearce TH (1987) Laser interferometry study of oscillatory zoning in plagioclase: The record of magma mixing and phenocryst recycling in calc-alkaline magma chambers, Iztaccihuatl volcano, Mexico. Amer Mineral 72: 1144-1162

Nomarski G (1955) Microinterferometrie differentielle a ondes polarisees. J de Physique et la Radium 16: 9-13

Nomarski G, Weill AR (1954) Sur l'observation des figures de croissance des cristaux par les methodes interferentielles a deux ondes, Bulletin de la Societe Francaise de Minearalogie et de Cristallographie 77: 840-868

Ortoleva P, Chadam J, Merino E, Sen A (1987) Geochemical self-organization, I. Reaction-transport feedbacks and modelling approach. Amer Jour Sci 287: 979-1007

Packard NH, Crutchfield JP, Farmer JD, Shaw RS (1980) Geometry from a time series. Physical Review Letters 45: 712-716

Pearce TH (1982a) Laser interferometric observations of plagioclase from Mt St Helens. Geol Assoc Can - Mineral Assoc Can, Abstr. Program 7: 72

___________ (1982b) Multiple frequency laser interference microscopy: a new technique for mineralogy. Geol Assoc Can - Mineral Assoc Can, Abstr. Program 7: 72

___________ (1982c) Observations of plagioclase phenocrysts from andesites using multiple frequency laser interference microscopy. Geol Soc Amer, Abstr. Program 14: 585

Pearce TH (1978) Olivine fractionation equations for basaltic and ultrabasic liquids. Nature 276: 771-774

Pearce TH (1984a) Optical dispersion and zoning in magmatic plagioclase: Laser interference observations. Can Mineral 22: 383-390

Pearce TH (1984b) The analysis of zoning in magmatic crystals with emphasis on olivine. Contrib Mineral Petrol 86: 149-154

Pearce TH (1984c) Multiple Frequency Laser Interference Microscopy: A new technique. The Microscope 32: no. 2, 69-81 (with colour cover photo)

Pearce TH (1993) A simple deterministic model of oscillatory zoning in magmatic plagioclase Geol Assoc Canada Abstracts, 18: A-81

Pearce TH, Clark AH (1989) Nomarski interference contrast observations of textural details in volcanic rocks. Geology 17: 757-759

Pearce TH Kolisnik AM (1990) Observations of plagioclase zoning using interference imaging. Earth Science Reviews 19: 9-26

Pearce TH Russell JK, Wolfson I (1987a) Laser-interference and Nomarski interference imaging of zoning profiles in plagioclase phenocrysts from the May 18, 1980 eruption

of Mount St. Helens, Washington. Amer Mineral 72: 1131-1143
Pearce TH Griffin MP, Kolisnik AM (1987b) Magmatic crystal stratigraphy and constraints on magma chamber dynamics: Laser interference results on individual phenocrysts. Jour of Geophys Res 92: pt. 13, 13,745-13,752
Peat FD (1990) Einstein's Moon - Bell's Theorem and the Curious Quest for Quantum Reality. Contemporary Books, Inc, Chicago
Phemister J (1934) Zoning in plagioclase feldspar. Min Mag 23: 541-555
Provost A (1985) Oscillatory zoning in plagioclase: a logical issue after rapid ascent. EOS 66: 362
Ruelle D (1979) Sensitive dependence on initial conditions and turbulent behavior of dynamical systems. Annals N.Y. Acad Sci 316: 408-416
Shimizu N (1990) The oscillatory trace element zoning of augite phenocrysts. Earth-Science Reviews 29: 27-37
Sibley DF, Vogel TA, Walker BM, Byerly G (1976) The origin of oscillatory zoning in plagioclase: A diffusion and growth controlled model. Amer Jour Sci 276: 275-284
Simakin AG (1983) A simple quantitative model for rhythmic zoning in crystals.
Geokhimiya 12: 1720-1729. (English translation by Scripta Publishing 1984).
Shaw HR (1987) The periodic structure of the natural record, and nonlinear dynamics. EOS 68: 1651-1665
Shimizu N (1990) The oscillatory trace element zoning of augite phenocrysts. Earth Science Reviews 29: 27-37
Smith JV, Brown WL (1988) Feldspar Minerals.Volume 1 Crystal Structures, Physical, Chemical and Microtextural Properties.Second revised and extended version. Springer Verlag, Berlin, 828 pp
Sugihara G, May RM (1990) Nonlinear forecasting as a way of distinguishing from measurement error in time series. Nature 344: 734-741
Swinney HL, Roux JC (1985) Chemical chaos. In Non-equilibrium Dynamics in Chemical Systems.Vidal C, Pacault A, (eds) Springer-Verlag
Takens F (1980) Detecting strange attractors in turbulence. In: Lecture Notes in Mathematics, Springel-Verlag, New York, 366-381
Tiller WA (1977) On the cross-pollenation of crystallization ideas between metallurgy and geology. Phys Chem Minerals 2: 125-151.
Tsuchiyama A (1985) Dissolution kinetics of plagioclase in the melt of the system diopside-albite-anorthite and origin of dusty plagioclase in andesites. Contrib Mineral Petrol 89: 1-16
Vance JA (1965) Zoning in igneous plagioclase: patchy zoning. J Geol 73: 636-651
Wang Y, Merino E (1993) Oscillatory magma crystallization by feedback between the concentrations of the reactant species and mineral growth rates. Journ Petrol 34: 369-382
Wiebe AR (1968) Plagioclase stratigraphy: a record of magmatic conditions and events in a granite stock. Amer Jour Sci 266: 690-703
Wolf A, Swift JB, Swinney HL, Vastano JA (1985) Determining Lyapunov exponents from a time series. Physica D 16: 285-31

ISOTOPIC EQUILIBRIUM/DISEQUILIBRIUM AND DIFFUSION KINETICS IN FELDSPARS

BRUNO J. GILETTI
Department of Geological Sciences
Brown University
Providence, R.I. 02912, U.S.A.

ABSTRACT. The use of isotopes as tracers in diffusion kinetics measurements on feldspars is discussed with particular emphasis on oxygen and cations. The effect of feldspar composition, temperature, fugacity of water, fugacity of oxygen, anisotropy of diffusion, and sub–solidus phases are treated, with a view toward understanding the diffusion mechanisms and to predict diffusion coefficients where measured data are lacking. The strong dependence of oxygen diffusion on f_{H2O} is suggested to derive from having H^+ act to aid in breaking the Si–O bonds, and H_2O aid in the transport through the crystal. The lack of such dependence by cations, together with their systematic behavior as a function of cation charge, and, secondarily, on ionic radius has led to a diffusion model utilizing a vacancy mechanism for most of the cations in feldspars, with Na transport being by an interstitial mechanism. The ionic porosity prediction model can predict oxygen diffusion under hydrothermal conditions, but fails for cations.

Either oxygen or Sr isotopes can be used in feldspars, together with the other minerals in the rock, to determine cooling histories in rocks. The thermal history of a metamorphic event can also be determined using Sr diffusion kinetics and Rb–Sr systematics. Examples are given.

1. Introduction

Various stable and radioactive nuclides have been used to study feldspars, either from the perspective of understanding the physics and chemistry of the feldspars themselves, or to determine aspects of the geological history of a particular feldspar specimen. An approach that has proven useful in both types of study has been the measurement of the diffusion kinetics of different elemental species in feldspars and the application of the results to determine thermal histories. This paper will focus on diffusion kinetics and isotopes in the feldspars and the geological applications.

The study of the properties of feldspars has involved both the naturally occurring isotopic abundances and artificially introduced, enriched isotopic compositions. A few comments may serve to remind the reader of concepts that practitioners in this area take for granted in their papers.

1.1 ISOTOPE FRACTIONATION

While isotopes of an element behave almost identically, a difference of one mass unit will have a larger percent difference in mass for light elements than for the heavy ones. The subtle differences in vibrational and other thermal energy modes of the different isotopes of an element result in subtle differences in relative isotopic concentrations in coexisting phases containing that element, even when the phases are in equilibrium. This isotopic fractionation occurs for all isotopes of all elements to some extent, but the effect is greater, and so more easily measurable, for the light elements. Fractionation may also occur in non–equilibrium processes, such as instances where the geological process occurs too rapidly for equilibrium to be attained.

Two procedures tend to obscure or ignore the fractionation concept in practice. The first is the case where highly enriched isotopes are used in an experiment. A highly enriched isotope of an element that is normally in low abundance purchased and used as a readily identified tracer. In this case,

I. Parson (ed.), Feldspars and Their Reactions, 351–382.

the fractionation effects, which are often on the scale of 1% or less, are totally overshadowed by the actual differences in isotopic composition of the laboratory starting materials, and can be ignored.

The second procedure relates to a few elements, one of which is Sr. In measuring Sr isotopic compositions of natural materials with a mass spectrometer, the variations sought are caused by ^{87}Rb decay to ^{87}Sr. These variations are often very small, and thus it is necessary to correct for fractionation produced by the mass spectrometer itself, in order that a reported $^{87}Sr/^{86}Sr$ reflects only the natural variations as produced by the Rb decay. Essentially all Sr isotope compositions are measured and then the $^{87}Sr/^{86}Sr$ ratios are corrected by normallizing the $^{86}Sr/^{88}Sr$ to 0.1194. By doing this, all fractionation effects, whether induced by nature or produced in the mass spectrometer, are eliminated. It is not possible, therefore, to use literature data on $^{87}Sr/^{86}Sr$ to infer fractionation effects.

1.2 DIFFUSION MEASUREMENT

The goal of most diffusion measurements is to determine how rapidly atoms of a given species migrate in a crystal and how this varies with temperature. The most commonly expressed way to describe this motion is through the Arrhenius relation:

$$D = D_0 \exp(-Q / RT)$$

where D is the diffusion coefficient (expressed in m^2/sec); D_0 is the pre–exponential factor; Q is the activation energy, or the temperature dependence (kjoules/g–atom); R is the gas constant; and T is temperature in K. This relation is usually plotted as log D versus (1/T), which yields a straight line over some temperature interval that depends on the mineral.

Diffusion kinetics measurements have been made using a variety of methods and experimental conditions. Criteria for judging the validity of a reported diffusion measurement depend, in part, on the experiment itself, and on the way the effect is analyzed. For example, the sample must not undergo changes in chemistry, crystal structure, or consist of more than one phase, during the course of the experiment, unless that specific phenomenon is being studied (Giletti, 1974).

Broadly speaking, diffusion rates are determined in two ways: the integrating method and the profile method. Both are "cook and look" experiments, employing "solid" (crystalline or glass) material and an external reservoir. The integrating method starts with a collection of mineral particles, all of the same size, exposes them to a reservoir (often a fluid) that contains some species of diffusing atom, at some temperature and pressure for a controlled length of time. The collection is then separated from the reservoir, and the total (integrated) concentration of the diffusing species is measured. The increase of diffusing species in the solid (or decrease in the fluid) is then used as a measure of the amount transferred from the reservoir. Provided that equilibrium has not been attained, the fractional approach to equilibrium during the time of the experiment is a measure of the diffusion coefficient.

In the second type of experiment, a planar surface of the mineral is exposed to the external reservoir. The experiment is run as in the previous case. The run product is then analysed by measuring the concentration of the diffusing species from the surface inward. Normally, the profile is measured deep enough to reach crystal that was unaffected (no increase in concentration of the diffusing species) by the experiment.

Variations on this profiling method exist. One is to accelerate ions (monoenergetically) of the diffusing species at the surface of the crystal. These will bury themselves in the crystal at some depth, largely determined by the ion energy. The experiment is then conducted to see how the ions diffuse away from this 'planar' layer of ions that started out near, and parallel to, the surface (see: Cherniak et al., 1991). A second method is to deposit a new crystal layer continuous with the existing solid, where the additional material differs in the concentration or isotopic composition of the diffusant. The resulting solid is then run in the diffusion experiment (see Elphick and Graham, 1986, also Giletti et al., 1976).

Computation of diffusion coefficients from the experimental data is strongly dependent on both the boundary conditions of the experiment and the method of analysis. Many solutions to the diffusion equation, specifically designed to solve real world problems, may be found in Crank (1967), but care must be exercised to see that his solution is applicable in all details to the experiment and analysis. Other useful texts on diffusion include Manning (1968), Shewmon (1963), Girifalco (1964), and Barrer (1951). A compilation of diffusion data for geological materials was published by Freer (1981). A new compilation by Brady (in press) will include literature to 1993.

The slow diffusion rates in minerals mean that laboratory experiments must be run for long times in order to attain sufficient transport into the crystal for it to be measureable quantitatively. This becomes a major problem for experiments at low temperatures, regardless of the method used to measure the amount of transport. On the other hand, there frequently is rapid interaction between the external reservoir and the crystal surface. This may extend into the crystal for one, or a few, unit cells and is not diffusion–controlled. It poses a problem for those cases of very slow diffusion rates, where the total length of a profile may be much less than 1 μm (1 micrometer), even if the experiment were run for a long time. The enhanced surface adsorption and reaction may yield apparent diffusional transport that is much higher than diffusion alone would produce. This will affect both the integrating and profiling modes of experimental design. Profiles that are too short are often what prevents measureents to lower temperatures.

Experiments must be run for sufficiently different durations to ensure that the calculated D (diffusion coefficient) is correct. Computed D values that decrease with run duration may be the result of the effect just described. Increasing run durations may be necessary to obtain diffusion data that do not depend on run duration. A good control on this phenomenon is to vary the duration enough to increase the profile length or amount of exchange by a factor of three. The time needed for this, however, is the square of the increased length factor, or nine times the time of the short time run. A six month run that produced a very short profile, for example, would then require almost five years for a reliable test.

For the integrating method, short profiles (or small degrees of approach to equilibrium) are particularly critical if the particle size is small, because this means the sample presents a very large surface area per unit weight. A modest amount of surface adsorption may contribute a significant amount of diffusant to the solid, making the apparent uptake by diffusion much larger than it actually is.

This discussion is limited to small amounts of interaction at the surface. Cases where new phases are formed that coat the surface for thicknesses on the order of microns, require special treatment that deals with the multiple processes that occurred and are beyond the scope of this paper (see Brady, 1975). Such effects are often readily apparent when the surface is examined with SEM (scanning electron microscopy) (for example, Elphick et al., 1986).

No fixed rule can be stated for the uncertainties in diffusion data, but a general assumption of an uncertainty of a factor of two would describe much of the published data. The more difficult experiments may well have uncertainties that are much larger. Until the reasons for the uncertainties are better understood, this limitation will persist. It should be noted that in application to thermal histories of real minerals, this limitation is likely to be less of a problem than others, such as the actual shape and size of the mineral particle, whether the entire particle was measured, and the degree to which the system may have been open to external fluids.

1.3 TERMINOLOGY

Each field has its own vocabulary. In this paper, particle is taken to mean the crystal as seen by the naked eye, or in an optical microscope. The term grain is used to mean the effective grain size for diffusion in the crystal. Sometimes the two are the same. Sometimes, owing to the existence of micro–domains or to minute fissures, there are bounding surfaces that exist within the particle. Such surfaces usually permit much more rapid diffusion (Nagy and Giletti, 1986, Farver and Yund, 1992) than does the solid crystal. Exchange with the particle can be much faster if these bounding surfaces are numerous. The result is that the effective grain size for diffusion is smaller than the particle size.

It is also important to distinguish between diffusion and net transport by diffusion. Net transport will occur if the crystal and its surroundings are not in equilibrium. Once equilibrium is attained (the activity of the diffusing species is everywhere the same), however, diffusion in the crystal will continue at the same rate, but no net transport will occur. Diffusional transport is a random process.

Diffusion occurs because the thermal motion of the atoms will provide some small fraction of the atoms with sufficient energy to jump from one place in a crystal to another. In a perfect crystal, this process is temperature dependent, and the process is called intrinsic diffusion. If a crystal has impurity atoms (particularly of different valence), or if it has been deformed so as to produce defects, these 'imperfections' may enhance the diffusion rate because they provide vacancies or other anomalous sites that permit more rapid transport. In such cases, the diffusion rate is controlled partly by the abundances of these defects, which usually add to the normal thermally dependent diffusion. This mode of diffusion is called extrinsic (because they are externally introduced anomalies that are not an essential property of the mineral) and will normally have a different temperature dependence, Q from that for intrinsic transport. A special case of extrinsic diffusion occurs when the effect of water on diffusion rates is considered.

1.4 APPLICATIONS

The use of isotopes and of diffusion kinetics in tectono–thermal applications generally involves many minerals in addition to the feldspars, although the latter are often of major importance. Because of the pronounced dependence of diffusion on temperature, the applications are generally related to the temperature history of the mineral.

Here, two illustrations of the approaches which can be used to examine thermal histories will be treated. In one, the diffusion kinetics themselves help determine the temperature and time sequence because the diffusion rates are temperature dependent. In the other, the kinetics are used along with a separate time–dependent process, exemplified by some geochronological system.

In the first case, the cooling rate of a rock is determined. The rate of diffusional exchange between a crystal and an external reservoir depends on the mineral, its size, and the temperature. In addition, the coolong rate is important. If the other parameters are known, then the extent of exchange (fractional approach to equilibrium) depends on time. Theoretically, it should be possible, with enough minerals behaving differently, to determine all the parameters and then find the cooling rate from the diffusion kinetics.

The second approach examines systems that have an independent time dependence, such as the Rb–Sr system, where the ^{87}Rb decay rate is temperature independent. In this case, the amount of exchange will be time and temperature dependent, but the rate of ^{87}Sr production depends only on the [Rb]. In this case, cooling rates can be measured, but details of metamorphic events can also be determined.

2. Part 1. Experiments and Theory

The diffusion kinetics of different atomic species in the feldspars depend on the same basic physical and chemical properties of the constituent atoms as do the lattice structure and the other properties. Study of the diffusion kinetics, therefore, provides insights into these other properties, as well as having application to tectonic problems. In the decade since the last NATO Advanced Study Institute, significant progress has been made in understanding diffusion in feldspars of: oxygen, argon, and cations. Other authors in this volume will discuss the studies on argon diffusion and its application to thermal histories (Foland; Brown and Parsons, this vol). The paper by Graham and Elphick (this volume) gives a more detailed discussion of oxygen diffusion with an emphasis on hydrogen ion

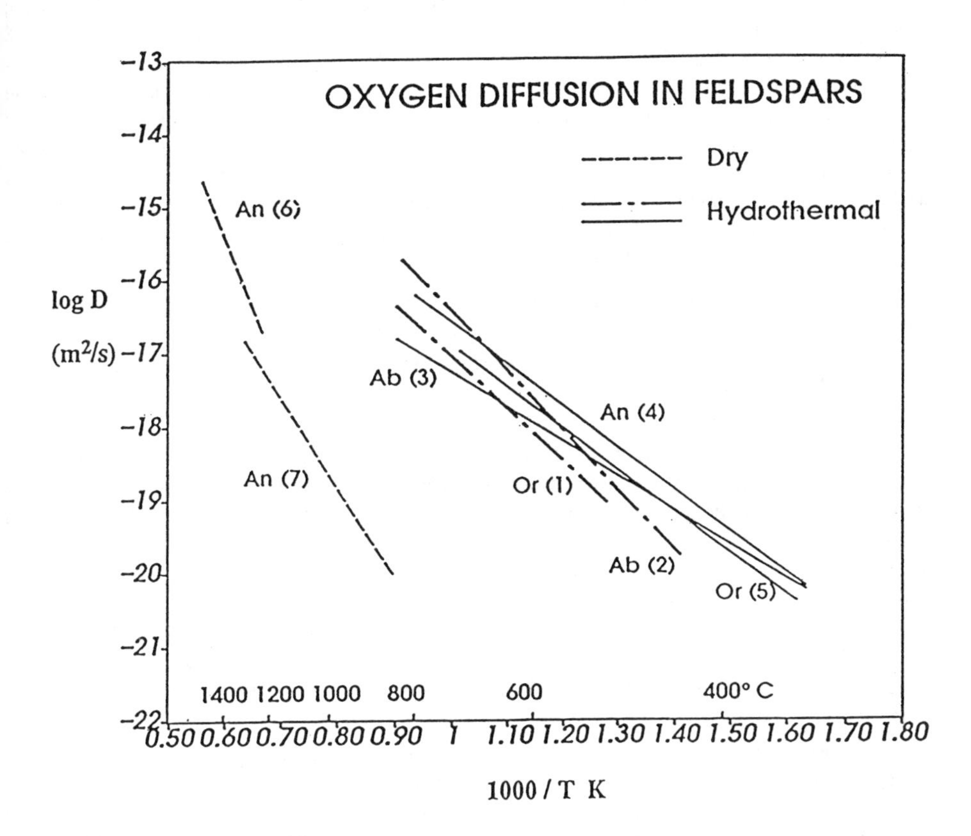

Figure 1. Arrhenius relations comparing 'wet' versus 'dry' oxygen diffusion in feldspars. Numbers are for references and activation energies in kj/g–atom O: (1) and (2), Merigoux (1968), 134 and 155, respecively; (3), (4), and (5), Giletti et al. (1978), 89, 110, and 107, respectively; (6) Muehlenbachs and Kushiro (1974), 377; and (7) Elphick et al. (1988), 236. Lines span approximate range of temperatures reported. Note the much higher activation energies for the 'dry' experiments. Labels Ab, An, and Or for albite, anorthite, and K–feldspar, respectively, denote approximate end–member composition.

activity, while that of Behrens (this volume) examines the role of water as well as hydrogen. Wondratschek (this volume) considers the order–disorder of Si–Al in sanidine, and the factors affecting the rate of the transformation. This part of the paper will consider the diffusion kinetics of oxygen diffusion, partly from an historical approach, but principally to discuss the current understanding of this process.

2.1 OXYGEN DIFFUSION KINETICS

Hydrothermal Experiments. The first Arrhenius relation for measurements of oxygen diffusion was reported by Merigoux (1968). He used the integrating method, where sized alkali feldspar powders were heated in ^{18}O–enriched water, in vessels that were sealed, so that the water in the vessel, upon being heated, would increase in pressure (see Graham and Elphick, this volume, for details on the early history). The change in the ^{18}O concentration in the feldspar was the measure of the amount of net transport that occurred between water and crystal. His experimental technique meant that the pressure differed for each temperature (Figure 1 lines (1) and (2)). Later work (Giletti et al., 1978 Figure 1 lines (3),(4), and (5); Yund and Anderson, 1974; Kasper, 1975) served to confirm those results, except for a slight discrepancy in temperature dependence. In 1978, Yund and Anderson showed that there is a strong dependence of the oxygen diffusion rate on water pressure. When this effect is used to correct the Merigoux (1968) data to a single pressure, his results are in good agreement with the later data. All these experiments were carried out with feldspar and water charges sealed in gold tubing, and run in a water medium to support the pressure increases when the charges were heated. The integrating method using a mass spectrometer was used in the earlier years. Later, much of the work was done with the profiling method and ion microprobes.

The results of all this work were to show that all the feldspars exchanged oxygen readily by diffusion in hydrothermal media. A crude envelope spanning less than one order of magnitude could encompass the diffusion coefficients for all the feldspar compositions over the range of geologically interesting temperatures.

Dry Experiments In 1974, Muehlenbachs and Kushiro reported experiments of oxygen diffusion in plagioclase, where the exchange medium was ^{18}O–enriched oxygen as CO_2 or O_2. These were dry experiments. The Arrhenius relation (Figure 1 line (6)) was totally incompatible with the hydrothermal data. The dry experiments had to be run at much higher temperatures, in order to obtain sufficient exchange to be measured, because the rates were far slower. Further, the activation energies were much greater. The observed difference in behavior was confirmed by Elphick et al. (1988) (Figure 1 line (7)). The Yund and Anderson (1978) data resolved the source of the differences when they showed the strong water pressure dependence of the diffusion rates in the feldspars. This served to focus attention on the diffusion mechanisms that operate, and how they differ between wet and dry experiments.

The importance of water pressure in the physics of mineral behavior was of great interest because of the recent discovery of the 'water–weakening' phenomenon in quartz (Griggs, 1967, 1974). Quartz was more easily deformed under 'wet' conditions at high pressure, relative to dry. Models that could explain the deformation in quartz might involve the more rapid diffusion of oxygen (in the course of breaking Si–O bonds) at high fluid pressure. Discussion continues today.

Numerous species, OH^-, H^+, H_2O, etc., have been suggested as the 'facilitators' in the diffusion process involving water. The common thread has been the idea that Si–O bonds would be easier to break in the presence of species that would link to the Si or O, thereby helping to weaken the Si–O bond (Goldsmith, 1988, Graham and Elphick, 1991).

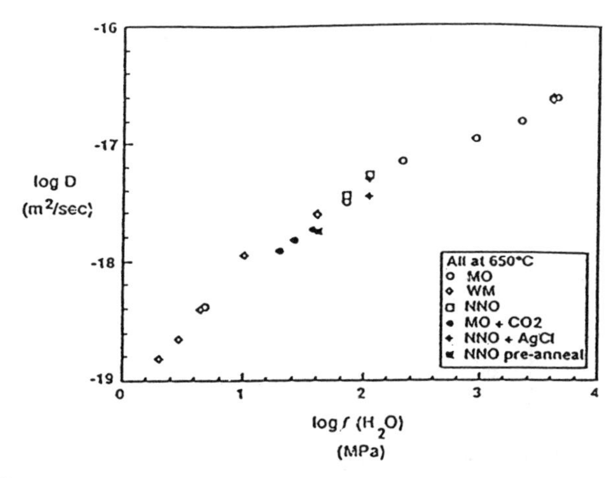

Fig. 2 The oxygen diffusion coefficient as a function of the water fugacity for adularia at 650°C. The symbols distinguish different buffering assemblages: open circles = Mn_2O_3 + Mn_3O_4, open diamonds = wustite + magnetite, open squares = Ni + NiO, filled circles = Mn_2O_3 + Mn_3O_4 with CO_2, filled diamonds = Ni + NiO and Ag + AgCl, and filled squares = Ni + NiO pre-annealed (see text for details). Uncertainties on the data are approximately the size of the symbols.

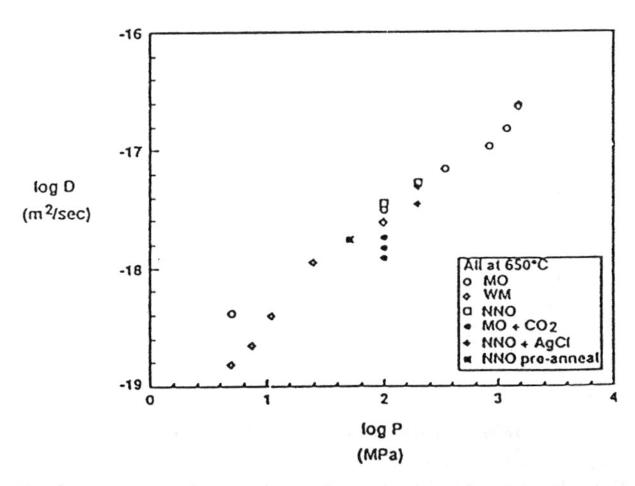

Fig. 3 The oxygen diffusion coefficient as a function of the confining pressure for adularia at 650°C. Symbols are as described in Fig. 2.

Figures 2 and 3. Taken from Farver and Yund (1990)

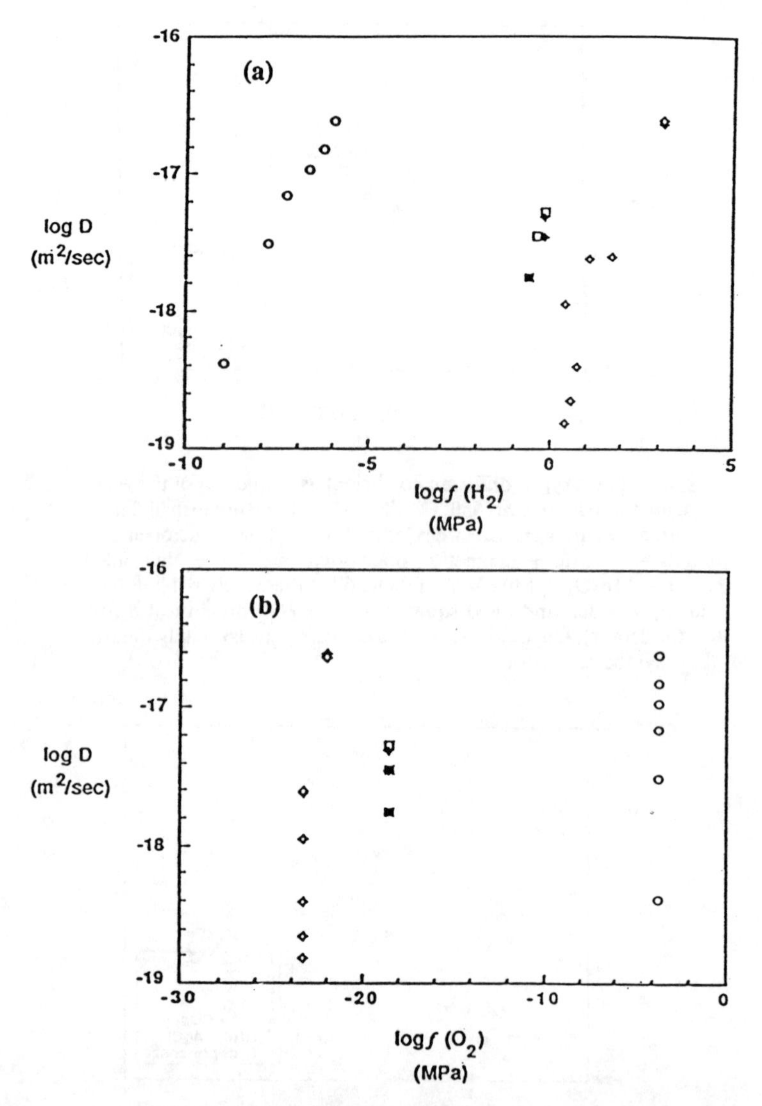

(a) The oxygen diffusion coefficient as a function of the hydrogen fugacity for adularia at 650°C.

(b) The oxygen diffusion coefficient as a function of the oxygen fugacity for adularia at 650°C. Symbols are as described in Fig. 2.

Figure 4. Taken from Farver and Yund (1990)

Much of the uncertainty about the externally controlling parameters was dispelled by Farver and Yund (1990). In a set of experiments designed to contrast the dependence of the diffusion coefficient on the different possible species, they reported the series of results depicted in the following figures. Figure 2 shows the dependence of D on water fugacity to be a simple continuous function. Figure 3 plots the log D vs log P (total pressure), and the relation again seems simple until the data at log P =2 are examined. The set of points are all for a total pressure of 100 MPa, but they differ considerably, and the difference is directly related to the water fugacity, which was varied by making the fluid a mixture of CO_2 and H_2O. Lower f_{H2O} yields lower D, even at the same total P. The same is seen for the lowest pressure data at 5 MPa. Figure 4 shows the dependence of D on the H_2 and O_2 fugacities, and there is clearly an enormous spread of values for a given fugacity. Lastly, Figure 5 shows the dependence on a_{H+}. Of particular importance are the filled diamond–shaped points (using the Ag + AgCl hydrogen buffer) at the highest a_{H+}, which fall approximately an order of magnitude below the continuation of the trend of the data in D. On the other hand, these points agree well with the open square at $a_{H+} = -5.52$ (three orders of magnitude lower activity), but with the same water fugacity. This suggests that the deplendence of D is on the f_{H2O} rather than on a_{H+}. When the values are normallized to a constant water fugacity, there is no dependence on a_{H+} (see lower part of Figure 5). This area is still a matter of active discussion, however, and different interpretations are still being advocated (see Graham and Elphick, this volume).

The suggestion has been made (Farver and Yund, 1990; Zhang et al., 1991) that the neutral water molecule may be able to travel rapidly, relative to a dry system. This could be the critical oxygen transport agent, which would be consistent with the data. We shall return to this suggestion later. The known very rapid transport rate for H^+ in minerals (see Graham and Elphick, this volume), provided a means for bringing a (+) charge to an oxygen ion so that the oxygen's Si–O bond might be broken more easily.

It may be that the qualitative resolution of the H^+ versus H_2O debate will be that the proton is indeed the agent that provides a ready mechanism for release of the O from the Si–O bond, and that the relatively rapid transport of the oxygens through the structure depends on their moving as the neutral water molecule. In this sense, both species play a role. The ready dissociation of H_2O at the temperatures of these experiments ensures that increased water pressure means increased H+. More experimental data are sure to come, perhaps guided by ab initio calculations, but the different views may be convergent.

2.2 CATION DIFFUSION KINETICS

Cation diffusion studies in the feldspars have a more complex history, as there are different ways to get at this problem. Volume diffusion has long been recognized as playing a role in processes of order–disorder in the Al–Si tetrahedral position, zoning of plagioclases, and exsolution of perthites and anti–perthites (most of this volume and previous volumes in this series could be cited here). This discussion will not deal with most of that work, though it is important, and gave clues as to which cations migrate faster or slower. Instead, the focus will be on isotope studies.

Early attempts to measure cation diffusion in feldspars yielded a variety of sometimes contradictory data (Jensen, 1952; Sippel, 1963; Bailey, 1971; Petrovic, 1972, 1974) as discussed in the last ASI volume (Yund, 1984). Foland (1974) made the first successful measurement of Na, K, and Rb diffusion in orthoclase feldspar. This was followed by a study of Na and K diffusion in albite (Kasper, 1975). Both studies used the integrating method, where hydrothermal solutions containing alkali chlorides were used along with the feldspar powder. The Na/K in the solutions was prepared to be in chemical equilibrium with the feldspar, but the diffusing species in solution was artificially

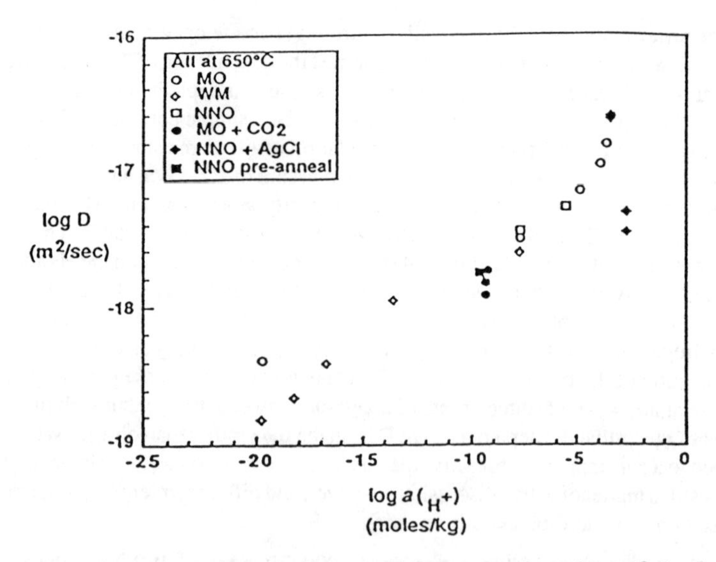

The oxygen diffusion coefficient as a function of the hydrogen ion activity for adularia at 650°C. Symbols are as described in Fig. 2.

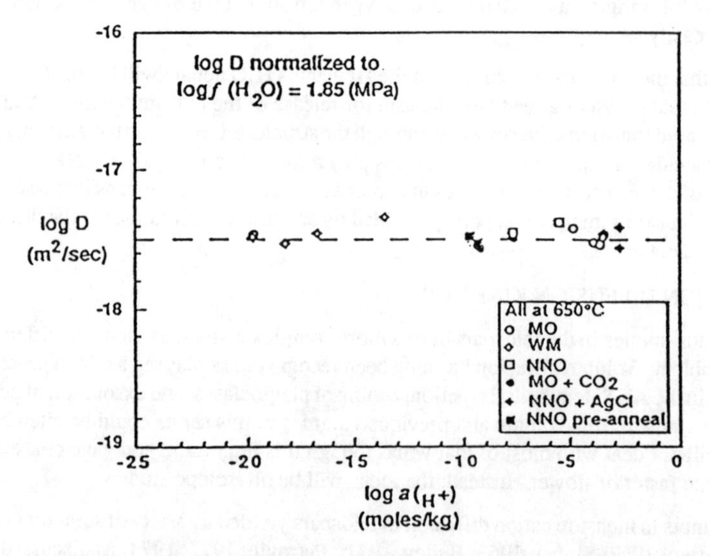

The oxygen diffusion coefficient normalized to a constant water fugacity of 71 MPa (the f_{H_2O} fixed by the MO buffer at 100 MPa confining pressure) as a function of the hydrogen ion activity for adularia at 650°C. Symbols are as described in Fig. 2.

Figure 5. Taken from Farver and Yund (1990)

enriched in some isotope.

Strontium Diffusion. Diffusion of Sr in minerals is of particular importance because of its relevance to Rb–Sr dating, specifically, to Sr isotope exchange between minerals, or between minerals and other phases such as fluids. Early reported Sr diffusion data in K–feldspar (Misra and Venkatasubramanian, 1977) yielded high values inconsistent with Rb–Sr observations in field settings.

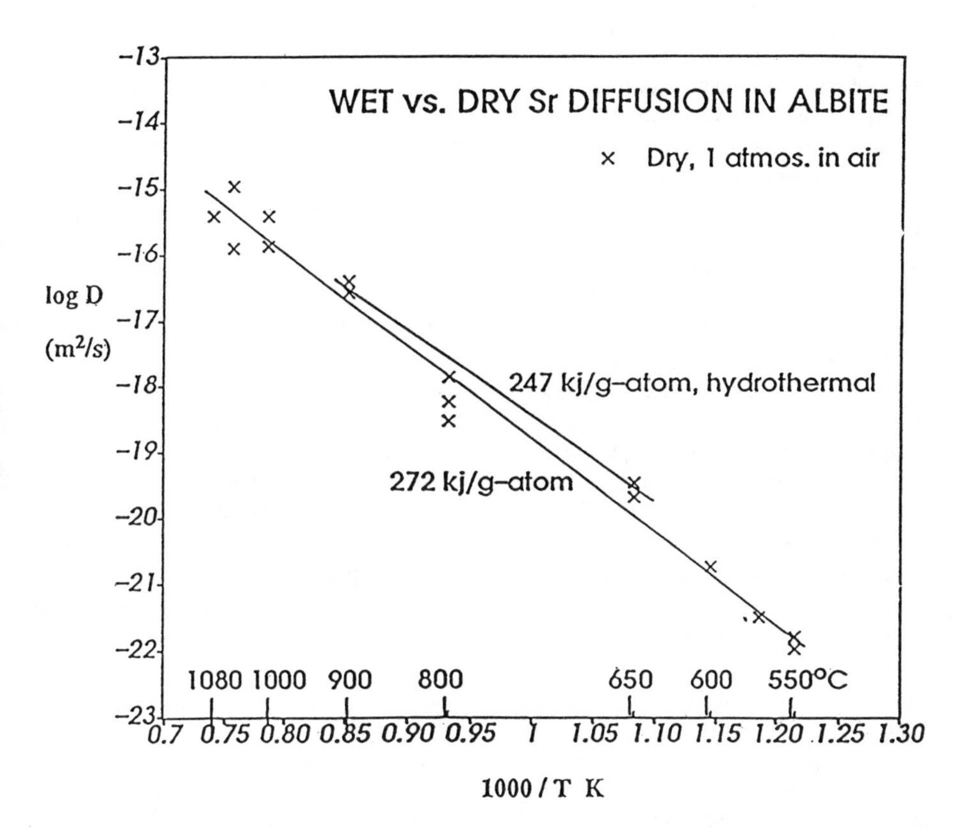

Figure 6. Sr diffusion in albite. Data points are for the dry experiments, while the hydrothermal experiments are summarized by the shorter line. Runs at 1050 and 1080° C used the step scanning analytical technique on the ion microprobe.

Interest in this problem has been renewed recently, and Sr diffusion in albite and orthoclase, as well as Rb diffusion in orthoclase have been reported (Giletti, 1991). Calcium diffusion in labradorite was reported (Behrens et al., 1990). In addition, Giletti and Casserly (submitted) have measured Sr diffusion in albite under dry conditions (in air) and found the same Sr kinetics, within the uncertainty, as in the hydrothermal experiments of Giletti (1991) (Figure 6). Measurements of diffusional transport parallel, and normal to the albite twin lamellae yielded the same D values within the uncertainty.

The results of Giletti and Casserly (submitted) show that the diffusion mechanism for Sr is independent of water prèssure (and fugacity) between approximately 0.04 bar and 1000 bars. The diffusion kinetics at 1000 bars were measured for experiments buffered by the Rene metal of the vessels (at approximately the Ni–NiO oxygen fugacity buffer), while the 1 bar experiment was run in air at 0.2 bars of O_2, demonstrating that the diffusion rate was not affected by f_{O2}. An experiment at one bar, sealed with air and graphite present yielded the same value as those in air. The sealed charge had an oxygen fugacity approximately 10^{-19} that in air, so that the independence from oxygen fugacity is not related to the presence of water in the charge. Further, this means that diffusion kinetics can be measured much more easily, requiring only a simple furnace.

All four of the specimens used were chosen to be high quality crystals of normal compostion, which means low Fe and Mn contents. The absence of an f_{O2} dependence of the Sr diffusion kinetics may simply be a confirmation of the absence of cations that will change valence readily. For the Fe-rich K–feldspars, varying the f_{O2} should change the Fe^{2+}/Fe^{3+} ratio, in the crystal, which should change the abundance of vacancies in the structure. This could have a significant effect on the kinetics of the vacancy diffusion mechanism.

Sr diffusion as a function of plagioclase An content was measured by Giletti and Casserly (submitted), and found to be highly dependent on An (Figure 7). The D values for Sr in albite ($Ab_{98}Or_2An_{0.1}$), oligoclase (An_{30}), labradorite (An_{60}), and anorthite (An_{96}) differed systematically, with the end members differing by a factor of 10^4. The Sr diffuses fastest in albite and is slowest in anorthite. The activation energies have slightly higher values at higher An content, but the difference is within the uncertainty.

No significant diffusional anisotropy, with regard to crystallographic directions, for Sr was found in albite, labradorite, or anorthite, within the measurement uncertainties. It is possible that differences of a factor of two or less exist.

The large difference in D values is surprising in that the structures in the plagioclase series are quite similar. The numerous sub–solidus crystallographic structures in the plagioclase series are quite subtle. In fact, part of the challenge has been to sort out these subtleties. The exsolution of different phases, often as a result of Al–Si ordering, however, may produce phase boundaries that, where non–coherent, provide rapid avenues for transport. Diffusion of oxgyen, however, appears to have very similar kinetics regardless of the feldspar composition (Giletti et al., 1978).

The compositions of the plagioclases used for the Sr measurements were chosen to avoid some of the complications such as the peristerites, Bφggilds, and Huttenlochers (see Carpenter, this volume). While these solid state effects occur at temperatures generally well below those for which the diffusion data are reported, by avoiding the 'solvi' as much as possible, the experiments avoid the liklihood that the low temperature structures might persist during even part of the diffusion experiments. Pre–annealing the sample before doing the experiment is sometimes helpful.

At lower temperatures than those of the diffusion experiments, however, the different exsolved phases may have an influence on the Sr diffusion rate. Consequently, while the four Arrhenius

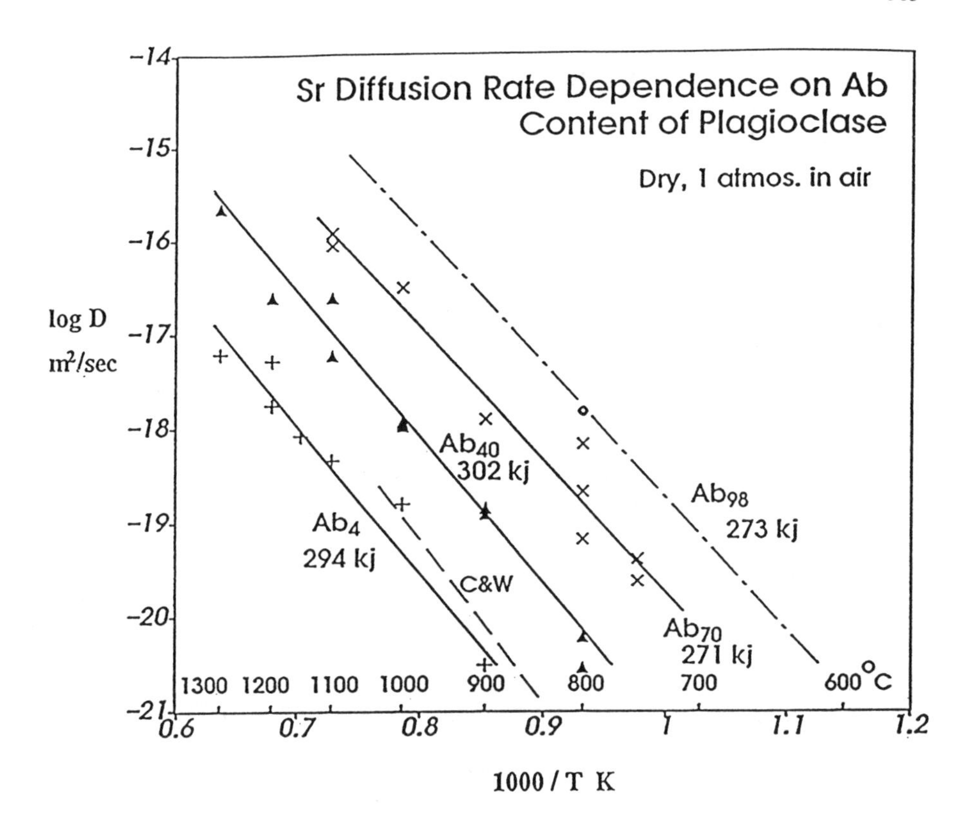

Figure 7 Four Sr tracer diffusion Arrhenius curves, with activation energies in kjoules/g–atom Sr. Data for three are shown, while the Ab_{98} data were shown in Figure 6. The solid point at 800°C on the Ab_{98} curve is for the graphite buffer experiment on the albite (see text). The line labelled C & W is for Sr interdiffusion data of Cherniak and Watson (1992).

relations may be assumed to be subject to the usual risks when extrapolated below the lowest measured temperatures, guesses as to diffusion behavior for compositions between those reported, may become even more highly suspect at temperatures below the respective sub–solidus solvi.

The principal cause for the large difference in Sr diffusion rate between Ab and An is probably the charge difference between the major host cations, Na^+ versus Ca^{2+}. If diffusion of Sr is by a vacancy mechanism, then a vacancy must be created near a Sr ion. Either a sodium or calcium will be the ion that creates the vacancy. The next nearest neighbor Sr ion to a Sr will be tens of atomic diameters away, and irrelevant. Creation of a vacancy in a structure that already exists requires the 'vacating' ion to enter an interstitial position. This position was electrically neutral prior to the jump. The 2+ charge on the Ca, therefore, makes this jump much more difficult than for the 1+ Na. Statistically, this means fewer vacancies produced by Ca, and so fewer vacancies available to the Sr.

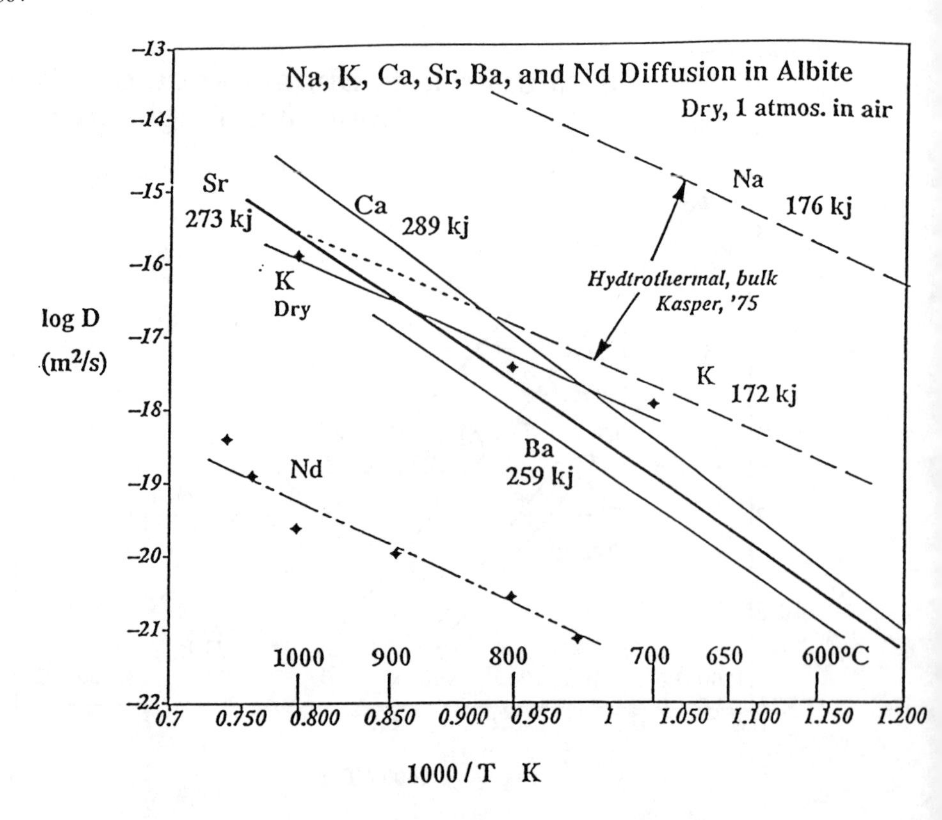

Figure 8. Cation diffusion in albite for one atmosphere in air experiments on K^{1+}, Ca^{2+}, Sr^{2+}, Ba^{2+} and Nd^{3+} (Giletti and Casserly, submitted; Giletti, in prep.). Hydrothermal Na^{1+} and K^{1+} diffuion data (Kasper, 1975) shown for comparison.

In albite, on the other hand, substitution of a Sr^{2+} in place of a Na^{1+} would create a nearby vacancy at a Na^{1+} site to maintain charge balance (an extrinsic diffusional effect). Thus, a vacancy would be built–in for the Sr in an albite, while having Sr^{2+} substitute for a Ca^{2+} in an anorthite would require no charge balancing process. However, an alternative scenario for albite would be an Al^{3+} substitution for Si^{4+}, when the Sr went in place of a Na. In that case, the nearby Na sites could still all be occupied by Na, and no vacancy is required.

Other Cations. The importance of ionic radius and charge will next be considered from the perspective of the diffusing species. The suggestion that the charge on the host Na and Ca ions plays a significant role can be examined by varying the size and charge of the diffusing species.

Comparison of Ca, Sr, and Ba ions diffusing in albite shows that they differ only very slightly from each other (Figure 8) (Giletti, in prep). The activation energies are quite similar, and the Arrhenius curves differ by factors of 2 or 3. Ca diffusion in anorthite is similar to Sr and, of course, very different from Ca in albite. Differences in ionic radius from Ca^{2+} to Ba^{2+} of 0.120 to 0.150 nm (see Table

Table 1. Ionic radii for cations relevant to text. Data from Whittaker and Muntus (1970)

Element	Charge	IONIC RADIUS (nm) Coordination Number of Ion		
	(+)	IV	VI	VIII
Na	1	0.107	0.110	0.124
K	1		0.146	0.159
Rb	1		0.157	0.168
Cs	1		0.178	0.182
Mg	2	0.066	0.080	0.097
Ca	2		0.108	0.120
Sr	2		0.121	0.133
Ba	2		0.144	0.150
Ga	3	0.055	0.070	—
La	3		0.113	0.126
Nd	3		0.106	0.120
Er	3		0.097	0.108
Lu	3		0.094	0.105

1)(Whittaker and Muntus, 1970), therefore play a minor role in the diffusion kinetics of the alkaline earths.

Figure 8 also shows the behavior of Na, K, and the rare earth element Nd in albite. The Na and K data for the same albite specimen (Kasper, 1975) were from hydrothermal experiments using the integrating method. The other K curve is from dry experiments (Giletti, in prep.) Note that the two sets of K diffusion data differ only slightly, and are nearly parallel. The difference may result from the irregular surfaces and shapes of the albite particles in the integrating method experiments. The diffusion model used by Kasper was a sphere. It is not surprising that the apparent D value is somewhat higher, as more exchange is possible with the real crystals, which are irregular. The comparison also suggests that the potassium kinetics do not depend on water pressure.

The similarity in the D values for the profile method data and those of Kasper for the integrating method suggests that the effective grain size for diffusion in the Kasper (1975) data is nearly the same as the particle size, and not some smaller domain. This supports the suggestion made earlier, that the sub–solidus complexities of the plagioclases were avoided. Further, preliminary data (Giletti, in

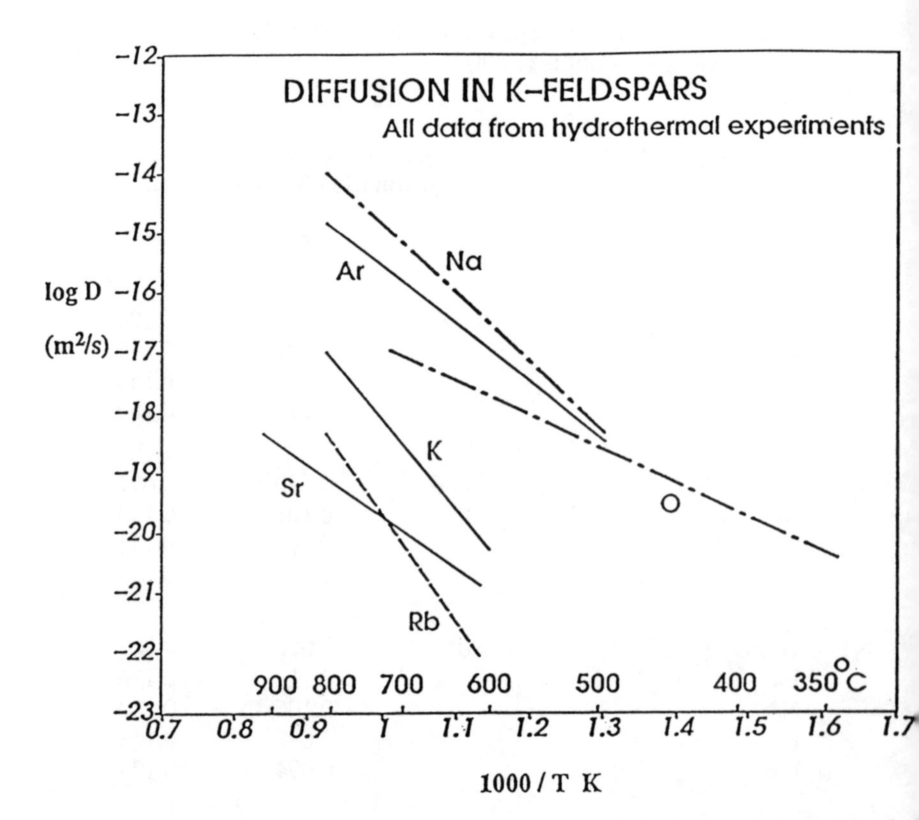

Figure 9. Arrhenius relations for diffusion in K–feldspars. Na and K (Foland, 1974a); O (Giletti et al., 1978); Sr and Rb (Giletti, 1991)

prep.) using peristeritic albite–oligoclases show little, if any, difference from what interpolation of the albite and oligoclase data would predict. It is important to note that this is for cation behavior, and neutral atoms, such as Ar, may behave differently.

Returning now to the comparison based on charge, the Na and K have lower activation energies than the alkaline earths, but in the range of interest to geologists, they have higher, to signifiantly higher, D values. The Nd^{3+}, however, has a much lower D value than the alkaline earths. Eight–fold coordinated Na^{+}, Ca^{2+}, and Nd^{3+} ions have essentially the same ionic radii (0.124, 0.120, and 0.120 nm, respectively; see Table 1). Clearly, ionic radius makes for modest differences in the diffusion kinetics, while the charge makes differences of orders of magnitude. The one exception to this is the difference between Na^{+} and K^{+} ions. This difference may be due to the ability of the Na^{+} to migrate as an interstitial ion, while the K^{+} may be too large to migrate easily as an interstitial.

K–feldspars. There are probably more diffusion data for the K–feldspars than all the other feldspars combined, but most of these data are for Ar diffusion (see papers by Foland, and by Parsons and Brown, this volume). The variety of other elements for which there are data is small.

Figure 9 shows some of these. As with albite, the highest diffusion rates are for Na^{1+}, and O, with Ar also diffusing rapidly. The diffusion rate for K is approximately three orders of magnitude slower than for Na.

Systematics of Cation behavior. In looking beyond the actual measurements of diffusion for specific elements, a key starting point is the large difference in rate between K^{1+} and Na^{1+}, which suggests two types of diffusion mechanism. Although modest differences in ionic radius have little effect on D in the case of the the alkaline earths, similar differences in the alkalis have a large effect. There may be some threshhold radius, specific to the particular crystal structure, which might be the result of different mechanisms of transport (such as interstitial versus vacancy) becoming dominant. Small ions, such as Li^{1+} or Be^{2+} may diffuse rapidly if they can travel by an interstitial mechanism. The large difference in diffusivities between Na and K suggests that the critical ionic radius may lie between them.

The large increase in D in going from 3+ to 1+, in the plagioclase, suggests that neutral atoms may travel rapidly. This seems to be the case for Ar. The discussion on the relative effects of H^+ and H_2O on oxygen diffusion is relevant here, because the prediction would be for the neutral water molecule to diffuse rapidly. The very rapid transport of the H^+ to aid in the release of the O from an Si–O bond, coupled with the observed fairly rapidly O transport, supports the idea that both H^+ and H_2O may be important in aiding rapid diffusion under wet conditions.

The role of fluorine in the kinetics of cation diffusion has been examined by Snow and Kidman (1991) who found that interdiffusion of Na and K, between albite and K–feldspar is enhanced by four orders of magnitude in the presence of F_2, relative to a fluorine–absent system (Christoffersen et al., 1983). In addition, Snow and Sherman (personal comm. and in prep.) find enhanced kinetics for Al–Si order–disorder in alkali feldspars in the presence of F_2.

2.3 PREDICTION OF D VALUES

Given the large number of minerals, it is desirable to have a means of predicting D values on the basis of some crystallographic parameters, rather than measuring all the elements in all the minerals. Systematic behavior has been sought for many years. One approach was the diffusion compensation model (Winchell, 1969), elaborated on by Hart (1981). Inherent in this model is the assumption that there is some single temperature, for any given mineral, where the D values of all the different diffusing species are the same (the Arrhenius relations for all the diffusing species in the mineral intersect at one temperature). As a consequence, a plot of the Arrhenius parameters as log D_o versus Q should give a straight line (Figure 10). The solid dots in the figure, together with the lines, are taken from Hart (1981).

New data that have been determined since the publication of the Hart (1981) compensation plot, have been added to Figure 10 as the open symbols, and include the new Sr data for plagioclase, points 18 through 21 (Giletti, submitted) the Sr and Rb data for orthoclase, points 22 and 23 (Giletti, 1991) and the O data for anorthite, point 24, of Muehlenbachs and Kushiro, (1974). In the same vicinity as the new Sr data would be those for Ca and Ba in albite and Ca in anorthite (Giletti, in prep.). Point 23 (Rb in orthoclase, Giletti, 1991) falls relatively near to the point 16, which is from Foland (1974). The latter was measured using the integrating method, and has a larger uncertainty owing to the very low D values encountered, and the small number of points available to define the Arrhenius relation. In addtion, the points numbered 9 and 10, which do fall on the Hart line, are for Sr diffusion in K–feldspar (Misra and Venkatasubramanian,, 1977), but are four orders of magnitude higher than the Giletti

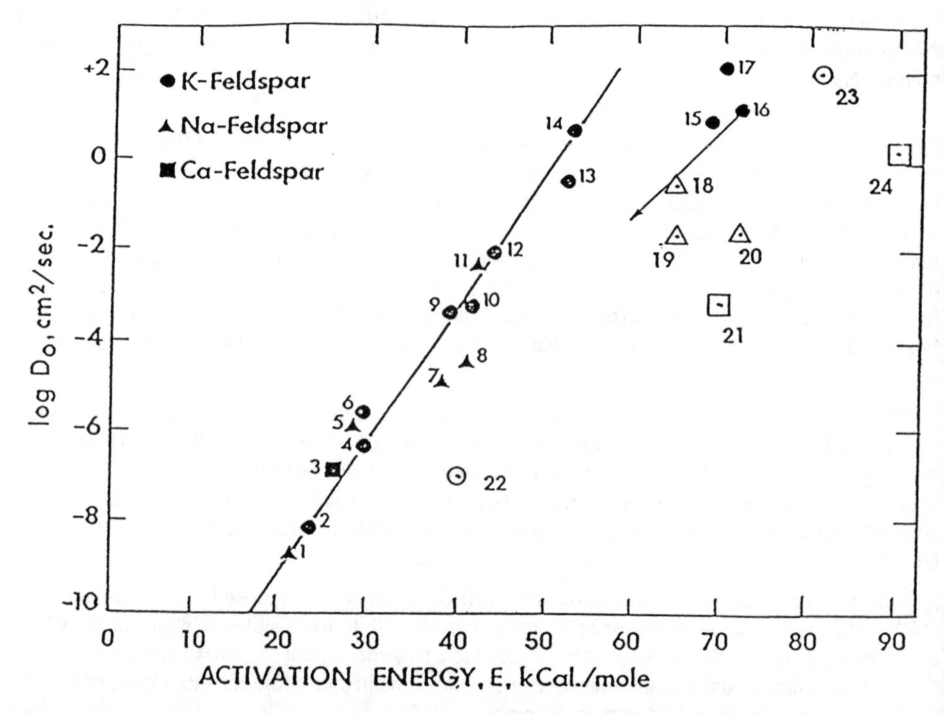

Figure 10. Compensation model plot of log D_0 vs activation energy, from Hart (1981). Filled symbols and lines are from the Hart paper, and open symbols are: 18, 19,20, and 21 = Sr diffusion in albite, oligoclase, labradorite, and anorthite (Giletti and Casserly, submitted); 22 and 23 = Sr and Rb in orthoclase (Giletti, 1991); 24 = O in anorthite (Muehlenbachs and Kushiro, 1974). Note: activation energies as reported in Hart (1981) were in kCal/mole, and other data plotted were converted to these units.

(1991) data, point 22. It has become increasingly clear, in the years since Hart (1981) published his paper, that the Misra and Venkatasubramanian data predict very ready exchange of Sr isotopes in geological settings where this is known not to occur. Unfortunately, complete details of those older experiments have never been published. What is noteworthy here, however, is that, although probably is serious error, these data do fall on the Compensation plot line.

While the model cannot predict any D values for a diffusing species in the mineral, except near the compensation temperature, where all D values are assumed to be the same, it has been used informally as a test of the validity of new diffusion data for a mineral. No theoretical physical basis has been established for this compensation relation. The assumption that there is a compensation temperature, even for diffusing species using the same diffusion mechanism, has not been shown to apply generally. The new data (Figure 10) show that there are substantial numbers of cases which fall far from the line. The plagioclase data, in particular, which consist of four parallel lines spaced over four orders of magnitude, demonstrate the difficulty of a model that requires a common point of intersection. At present, the compensation "law" is principally of historical interest.

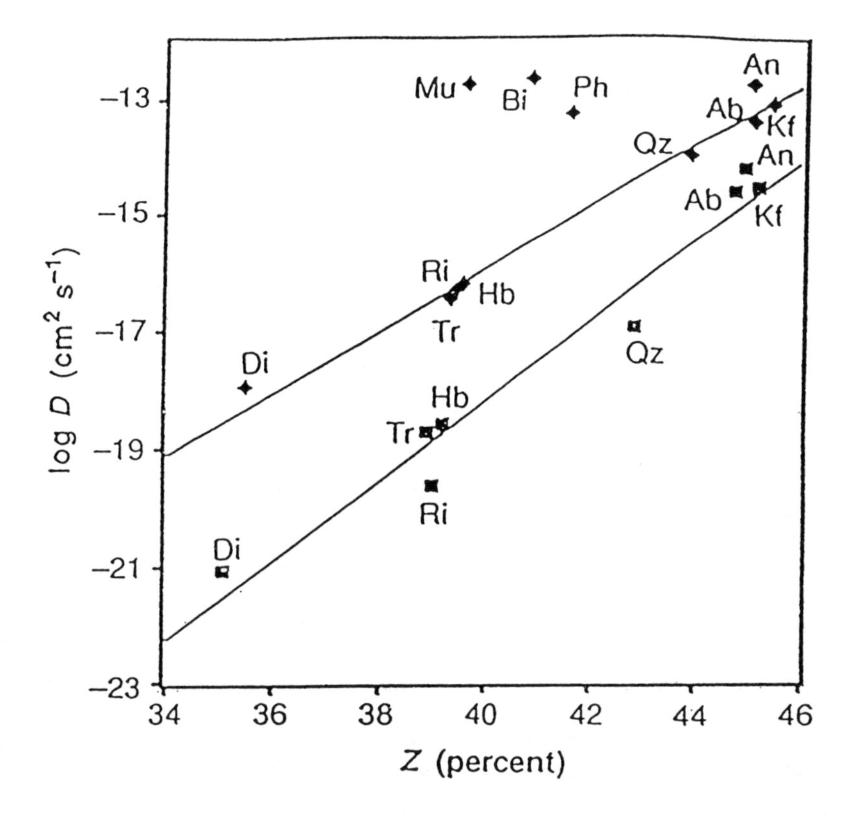

Figure 11. Ionic porosity model graph of Z, ionic porosity, (percent void space in unit cell) vs. log D for oxygen diffusion at 500°C = squares, and 700°C = diamonds, for different minerals (from Fortier and Giletti, 1989). Minerals are: Di = diopside; Ri = richterite; Tr = tremolite; Hb = hornblende; Qz = quartz; Ab = albite; Kf = K–feldspar; An = anorthite; Mu = muscovite; Bi = biotite; Ph = phlogopite.

In a different effort to predict diffusion coefficients, Dowty (1980) suggested the concept of anion porosity in the crystal lattice as a measure of ease of diffusional transport. The anion porosity is defined as the fraction of the unit cell not occupied by the anions. The anions are taken as hard spheres of the appropriate ionic radius. As he had very little data, he made this suggestion as something to be tested. With the acquisition of significantly more data, it became possible to explore the model. Fortier and Giletti, (1990) found that a plot of log D vs total ionic porosity for a given temperature gave a straight line (Figure 11) for a large number of minerals. The total ionic porosity is the fraction of the unit cell not occupied by either anions or cations, again assuming hard spheres. Only the micas failed to fall on the line. A single equation, with parameters of ionic porosity and temperature, will yield the oxygen diffusion coefficient to within approximately a factor of three. The ionic porosity depends only on the size of the unit cell and the ionic radii of the constituent atoms. The model predicted oxygen diffusion rates for minerals that had not been studied at the time, but have been since, to within a factor of approximately three for any given temperature. The reason why the micas still do not fit remains an open question.

Although data for oxygen transport in hydrothermal systems fit the model, dry feldspar oxygen diffusion data do not fit. A plot of log D versus ionic porosity for a given temperature does not yield a linear relation for the dry data alone, even one with different coefficients to the equation. An additional problem, however, is that the dry data do not agree amongst themselves, so that it is not clear where the problem lies.

Data for cation diffusion in the feldspars (Giletti and Casserly, submitted; Giletti, in prep.) do not fit an ionic porosity model. The ionic porosity is very similar for all the plagioclases, but the Sr and Ca diffusion coefficients differ by four orders of magnitude. This empirical approach using porosity may only be valid for neutral atoms and molecules.

3. Part 2. Applications of Diffusion Kinetics to Tectono-Thermal Histories

Before discussing applications of diffusion kinetics to the solution of tectonic problems, the concept of closure temperature must be examined. Dodson (1973) defined the closure temperature, T_c, in the context of a particular process, where a rock cooled from some high temperature to a temperature where all net exchange of the diffusing species between a solid phase and its surroundings has ceased to be significant. Because of the exponential dependence of D on temperature, there is a narrow range of temperatures that starts when the central part of the solid phase can no longer maintain equilibrium with the surroundings outside of the phase (as the rock cools, the diffusion rate is no longer rapid enough to maintain this equilibrium). The other end of this narrow range of temperatures is when the rock has cooled to the point that no further significant exchange occurs between the solid phase and the surroundings. The range of temperatures between start and end is relatively small, and a single value chosen for the closure temperature serves to represent the range.

In his derivation of the equation to calculate the T_c, Dodson (1973) assumed exchange with an infinite, well mixed, external reservoir. Consequently, the surface of the solid is always at the same composition. The Dodson equation is:

$$T_c = \frac{Q/R}{\ln\left[\dfrac{A\,R\,T_c^2\,D_0/a^2}{Q\,(CR)}\right]}$$

where: Q and D_0 are the activation energy and pre–exponential factor for diffusion of the diffusing species; R = gas constant; A = an anisotropy factor; a = a dimension of the crystal (mineral grain size); and CR = the cooling rate. The equation is solved by iteration of the value of T_c. The anisotropy term A is 55, 27, or 8.7, depending on whether the diffusion in the crystal is, respectively, isotropic, occurs in one plane, or occurs only in one direction. Examples that are good approximations of each of the three can be found in different minerals.

The T_c is clearly not a single value for a diffusing species in a given mineral, because it depends on the mineral grain size and the cooling rate. The larger the grain size, or the higher the cooling rate, the higher will be the value of T_c. For this reason, it is incorrect to speak of the closure temperature of a diffusing species in a mineral. Depending on cooling rate and grain size, the T_c may differ by as much as 200°C.

For those cases where exchange with another phase results in the diffusing species having an equilibrium chemical concentration, or isotopic composition, that is a function of temperature, the decreasing temperature will result in different equilibrium concentrations in the mineral relative to the external reservoir. When the mineral has closed, it will have a variable composition, with a gradient in concentration of the diffusing species from the surface inwards (Dodson, 1986), because the interior portions stopped equilibrating with the reservoir earlier, when it was hotter.

An important aspect of the Dodson model is that the reservoir be infinite and well–mixed. When a rock cools below the solidus, the mineral in question is exchanging with a number of other minerals, all of which will have their own diffusion rates, and differing activation energies. As a consequence, the external reservoir is neither infinite nor well mixed. Even if it is assumed that no fluid phase is present, the system is too complex for the Dodson (1973, 1986) treatments. More elaborate models, involving finite element computations, have been developed by Eiler et al. (1992) and Jenkin et al. (1993, in press) to determine cooling rates using oxygen isotope compositions for rocks with several minerals.

It is important to distinguish the closure temperature, T_c, from exchange occurring when the temperature is not dropping steadily. In the case of metamorphic events, for example, a high temperature may be maintained for a long time relative to the rate of exchange permitted by diffusion. In such cases, the idea of a T_c is not relevant. For example, if the temperature is held constant at some temperature, even below the estimated T_c determined by assuming some cooling rate, there may be significant diffusional exchange, particularly if this isothermal condition persists for a sufficiently long time.

3.1 OXYGEN ISOTOPES AND THE RATES OF COOLING FROM HIGH TEMPERATURES

The first application combines the known temperature dependence of equilibrium oxygen isotope fractionation between minerals with oxygen diffusion kinetics in different minerals. The objective is to determine the rate at which a rock cooled from high temperature (Giletti, 1986). This model is based on the assumption that the hand specimen acted throughout as a closed system to oxygen. The premise is that different minerals in a rock at high temperature exchange oxygen isotopes freely, maintaining the differences in $\delta^{18}O$ between minerals that are the appropriate equilibrium values for those minerals, but changing to maintain isotopic equilibrium as the temperature decreases (Figure 12).

When the temperature of the rock decreases to the closure temperature of the first mineral to close, the isotope composition of that mineral, A, becomes fixed. As the temperature continues to decrease, the other minerals continue to exchange isotopes, until each of their T_c's are reached. Finally, all exchange ceases. The record of this process left in the rock is a set of mineral $\delta^{18}O$ values. Because the different minerals closed at different temperatures, they give a jumble of apparent equilibrium temperatures, or, depending on which pair of minerals is considered, $\delta^{18}O$ values that could not have been in equilibrium at any temperature.

The diffusion kinetics data provide a means to estimate the closure temperature, T_c, for each mineral. This closure temperature is a function of the grain size of the mineral, the diffusion kinetics of the mineral, and the cooling rate. It is possible to compute the T_c for different cooling rates, and determine what the mineral $\delta^{18}O$ would be for each mineral, based on the whole rock $\delta^{18}O$, the mode of the rock, and the mineral grain sizes. Each cooling rate produces a different set of mineral isotope compositions. It is then possible to compare these with the observed mineral $\delta^{18}O$'s. The rock cooling rate was, presumably, the one where the model fits the observation.

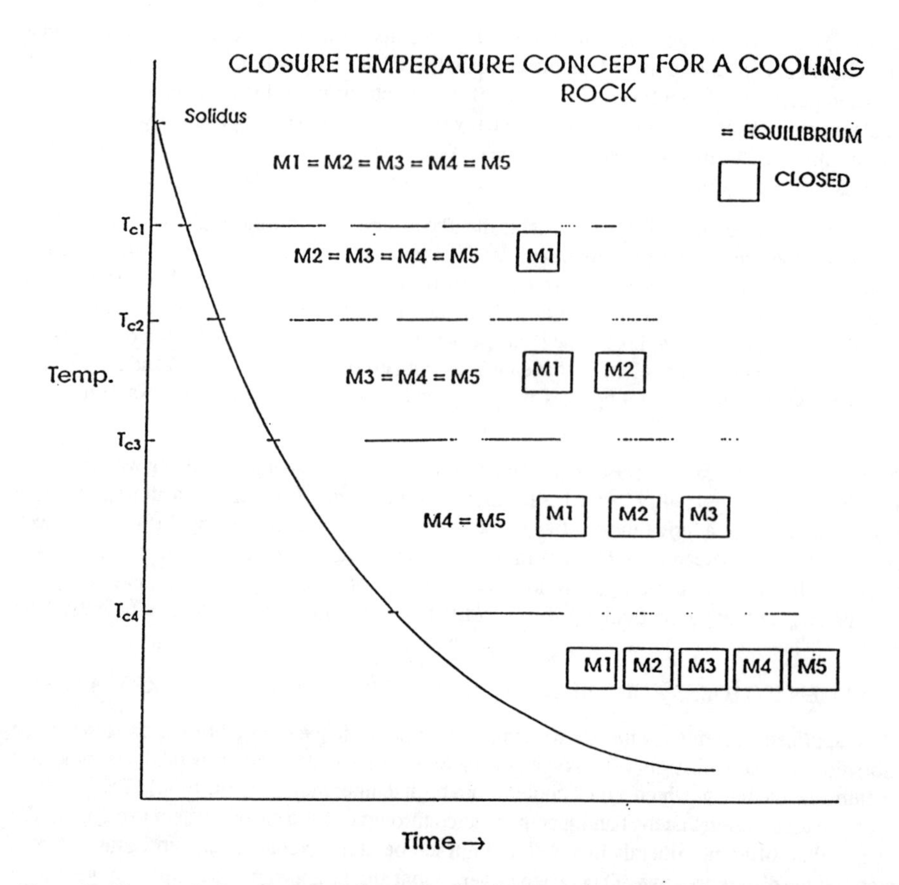

Figure 12. Hypothetical cooling curve for a rock, showing five minerals that are in isotopic equilibrium initially. As temperature decreases, closure temperatures, T_c, are reached which result in isolation of the different minerals in sequence, while the remaining minerals continue to exchange. When T_{c4} is reached, all minerals are closed because whole rock specimen is closed, and mineral 5 has no other minerals with which to exchange.

Three examples of this sort of approach are shown in Figure 13. In each case, the plot is of $\delta^{18}O$ versus cooling rate. The horizontal lines are the observed data, and the sloping lines are the model values. Ideally, all the intersections of the respective mineral data should occur at the same cooling rate. In fact, the intersections scatter. Despite this, it is possible to conclude that the three examples are showing values of cooling rate that are approximately 20, 200, and 2000°C/m.y., with an uncertainty of a factor of approximately four.

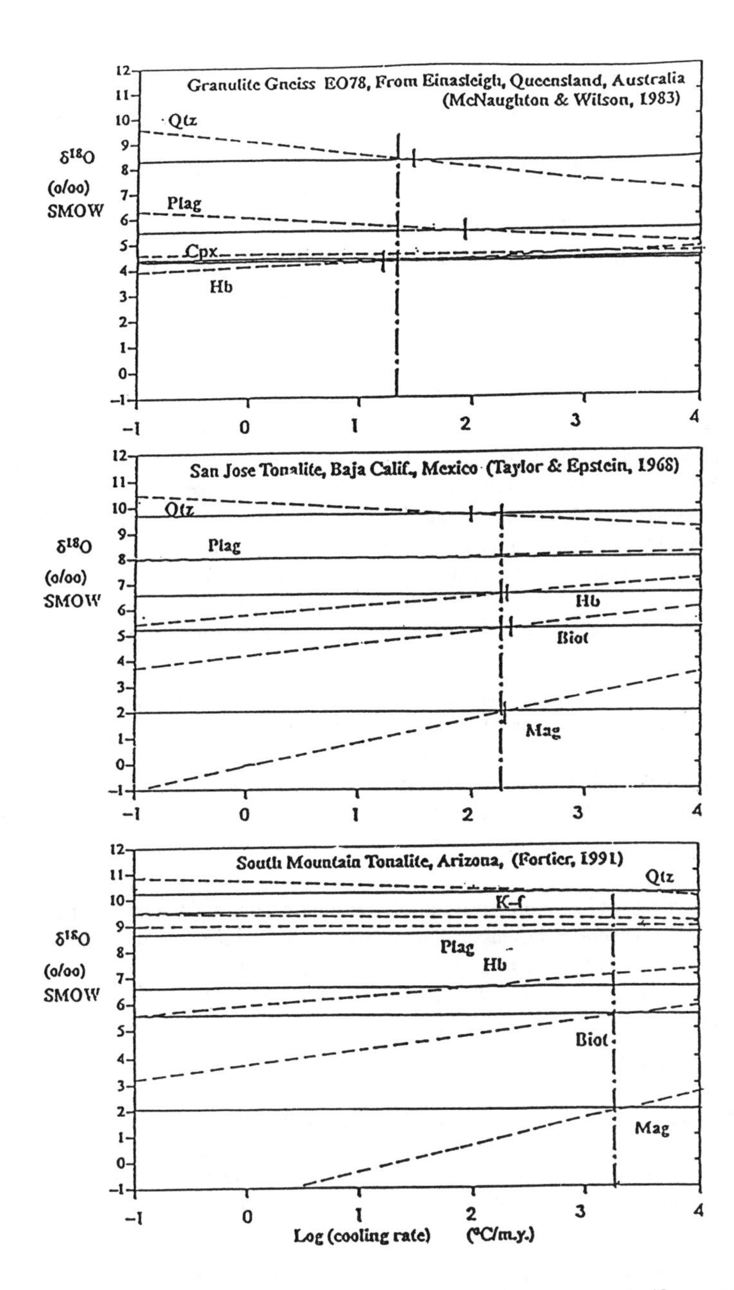

Figure 13. Oxygen isotope Coolrate model examples, showing measured $\delta^{18}O$ (solid lines) and model–predicted $\delta^{18}O$ (dashed lines) from Farver (1989), Giletti (1986), and Fortier (1991) (top to bottom). Estimated coolling rates are dash–dot lines, at approximately 20, 200, and 2000°C/m.y., respectively.

There are several sources of uncertainty in this cooling rate approach, and improvements are possible. Examples of these sources include: (1) some oxygen isotope fractionation factors are not known well, such as those for biotite mica; (2) the model assumes that the hand specimen was closed to fluids migrating through it, and thus to oxygen isotope exchange throughout the cooling and subsequently; and (3) the model is imprecise in its method of calculation, and Eiler et al. (1992) and Jenkin (in press, 1993) have made significant improvements in this regard.

Given the relatively small amount of variation in the oxygen isotope fractionation as a function of temperature, one or a few permil, the differences in predicted $\delta^{18}O$ as a function of cooling rate will always be quite small. Consequently, cooling rates may not be known to better than a factor of 2 or 3, even with refinements.

3.2 Rb-Sr SYSTEMATICS AND COOLING RATES FROM HIGH TEMPERATURE

A similar approach to the oxygen isotope cooling rate model may be taken using Rb–Sr geochronological systematics to determine cooling rate (Giletti, 1991). In this case, isotope fractionation is not measured (indeed, the usual nature of the Sr isotope measurement makes it impossible to know the rock isotope fractionation, as this is normallized out of the results). A re–examination of Figure 12, however, shows that a cooling curve intersects successive T_c's (this time for Sr diffusion) at different times. The whole point of the Rb–Sr systematics is that ^{87}Rb decays to ^{87}Sr at a rate determined by the half–life (see Faure, 1986, for the basics of Rb–Sr geochronology). Consequently, growth of the ^{87}Sr, as measured from the growth of the $^{87}Sr/^{86}Sr$ ratio, will occur during the cooling process, as well as subsequently.

If a granite crystallizes and cools slowly, the time between reaching the solidus and the first mineral T_c (closure to Sr diffusion in this case) may be a few million years. This is long enough to permit ^{87}Rb decay to affect the $^{87}Sr/^{86}Sr$ ratios of the minerals. Because the different minerals have different Rb/Sr chemical ratios, the $^{87}Sr/^{86}Sr$ ratios of the different minerals will increase at different rates from their common value (when they had reached the solidus). Diffusional exchange of Sr isotopes between minerals in a hand specimen, however, will tend to equalize the $^{87}Sr/^{86}Sr$ ratios of the different minerals.

Consider the exchange of Sr isotopes in terms of a strontium isochron diagram (Figure 14), where a granite is assumed to behave in the classical whole rock (WR) manner. That is, upon reaching the solidus of the granite, all WR specimens act like closed systems, and all the WR specimen points start out exactly colinear and with the same $^{87}Sr/^{86}Sr$ ratio, and the line rotates from horizontal as a function of time, with all the WR points remaining exactlly colinear.

While the WR isochron behaves in the classical sense, inside each WR specimen the minerals exchange Sr isotopes at high temperature. If all diffusion rates are rapid, the different minerals will exchange Sr isotopes completely, and all minerals will have the same $^{87}Sr/^{86}Sr$ ratio, which is also that of the particular WR specimen. This means that mineral points on the isochron diagram will fall off the WR isochron. Instead, they will lie on a horizontal line through the WR specimen point. Low Rb–Sr minerals (relative to the WR) will, therefore, fall above the WR isochron, and high Rb/Sr minerals will occur below the WR isochron.

This exchange amongst the different minerals will continue until the first mineral to close reaches its T_c. If this first T_c mineral is plagioclase, then from this time on, the plagioclase is a closed system, and the $^{87}Sr/^{86}Sr$ ratio will continue to grow, but at a slower rate, governed by the Rb/Sr of the plagioclase instead of the larger value of the WR specimen. In the example chosen, the solidus was attained at 750°C and, for plagioclase at a cooling rate of 20°C/m.y., closure occurred 10 m.y. later, at 549°.

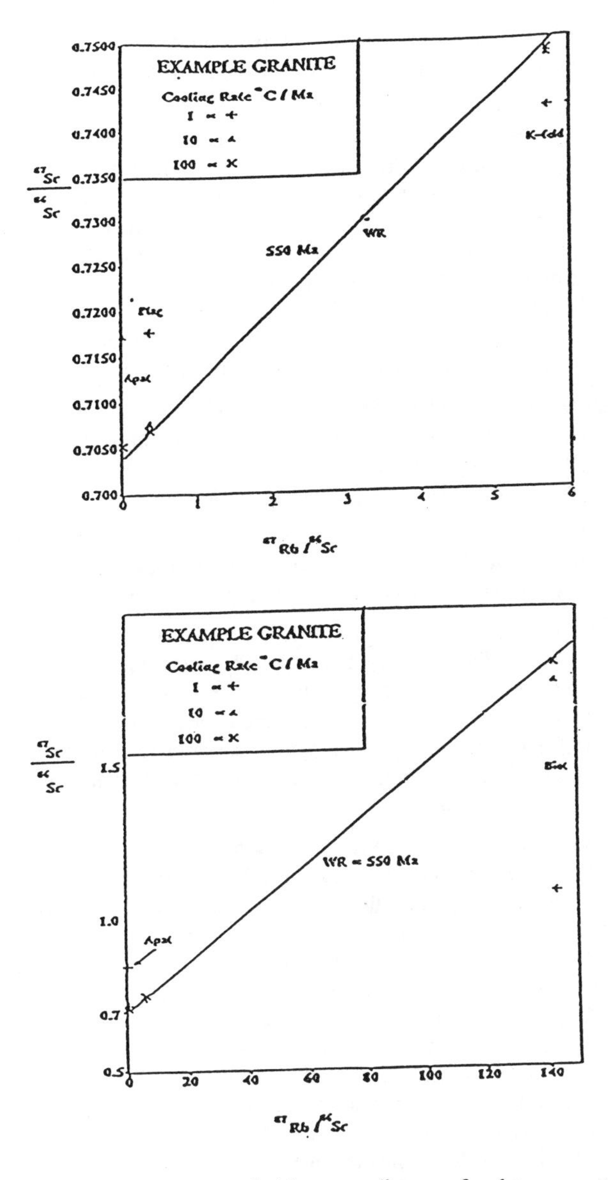

Figure 14. Example of Rb–Sr system development diagram for data computed by assuming ideal behavior of a hypothetical intrusive granite which cooled from the solidus at 1, 10, or 100°C/m.y. WR = whole rock; and line is WR isochron in both figures. Figures from Giletti (1991).

The plagioclase $^{87}Sr/^{86}Sr$ will grow at the rate of the WR point until 549°, and then will grow more slowly, at a rate governed by its own Rb/Sr ratio, but it will maintain its distance above the WR line.

The remaining minerals will continue to exchange until the second mineral closes. With each closure, the Rb/Sr of the residual exchanging mineral set will change, but the principle will be the same. As in the case with the oxygen isotopes, the T_c's can be computed if the radii of the minerals and the cooling rate are known (or assumed). If the mode of the rock and the concentrations of the Rb and Sr are known for each mineral, a simple material balance can be computed. Further, the growth of radiogenic Sr (or the $^{87}Sr/^{86}Sr$) can be computed. This approach will permit the computation, for any cooling rate, of the values of the $^{87}Sr/^{86}Sr$ ratio today in each mineral. This value, however, will depend on how long the minerals exchanged with each other, i.e. the cooling rate and when closure occurred. The forward calculation can be performed for several cooling rates, and the one that most closely matches the actual mineral data should be the actual rate at which the rock cooled.

This type of calculation was used for the example granite, and yielded the distribution of mineral values shown in Figure 14. This shows that the cooling rate is best determined if it is very slow. For rates in excess of 100°C/m.y. the points are likely to fall so near to the WR isochron that they will not yield rates that are distinguishable from a rapid quench of the rock.

The simple–minded model just described will yield rough values of the cooling rate, but a more elaborate calculation, using a finite element model, needs to be developed to treat this problem in a more rigorous way. This is currently being pursued (Jenkin, pers. comm.).

An altogether different approach can be made with the case of a regional metamorphism subsequent to the emplacement and cooling of an igneous rock. This meta (metamorphism) model employs the Sr diffusion kinetics and all the trappings of the previous model, except that it cannot use the T_c concept. The process includes a rise in temperature, then constant metamorphic temperature, then a drop in temperature, all of which is very different from the simple cooling from high temperature implicit in the T_c concept. The meta model has been applied to the data of Frey et al. (1976) for the Monte Rosa granite and gneiss in the western Alps with modest success (Figure 15).

The parameters needed for the meta model are: mode of the rock, mineral particle sizes for all minerals that are significant Rb or Sr reservoirs, concentration of Rb and Sr in each mineral, and $^{87}Sr/^{86}Sr$ for each mineral particle. Because many minerals will have gradients in $^{87}Sr/^{86}Sr$, the whole particle must be extracted and analyzed. The assumptions to be made for the forward model are the age of the rock, the time the metamorphism began, the duration of the metamorphism, and the maximum temperature of the metamorphism. The rock age can be obtained from the Rb–Sr whole rock isochron, which must have the points be colinear. The model assumes an instantaneous temperature rise at the start of the metamorphism, constant temperature for the duration of the metamorphism, and an instantaneous drop in temperature at the end of the metamorphism. The forward calculation will then predict the present–day mineral $^{87}Sr/^{86}Sr$ values for a rock that experienced such a heating. The closeness of fit between the measured $^{87}Sr/^{86}Sr$ mineral values today, and those predicted by the forward calculations will determine which combination of age, duration of heating, and temperature of metamorphism is most likely to be correct.

The model suggests that the best fit of a forward model calculation would be to have these Hercynian age intrusive rocks be heated starting at approximately 30 m.y. ago, for approximately 4 m.y. duration. Two different localities, that were heated to different temperatures (see Frey et al., 1976), are found to have been heated at 600–700° and 350–400°C respectively, which is consistent with the geology. The point is not to note the exact values, which may be wrong, but to see that the model can give potentially useful results.

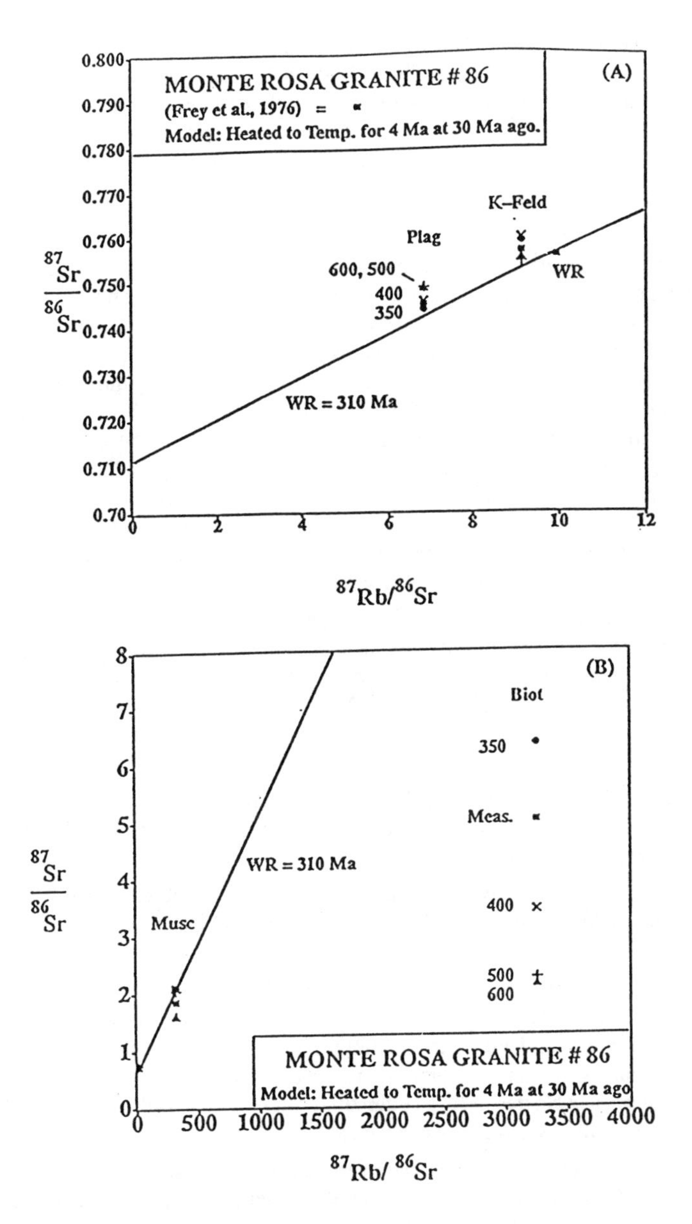

Figure 15. Comparison of Meta model Rb–Sr data with measured data of Frey et al. (1976) for different metamorphic temperatures. WR shows whole rock isochron line, but with only this hand specimen point shown. Values for plagioclase, K–feldspar, and WR of Figure 15A appear as a cluster at lower left of WR line in 15B owing to change in scale. Metamorphic temperature of 350 to 400°C fits reported data best.

While the older data of Frey et al. (1976) did yield interesting results, more careful measurement of the mineral sizes, particularly for those actually analyzed, is needed, the analysis of the entire mineral particle must be done, and better analytical precision with more modern mass spectrometry is needed. A more rigorous computational method would also be desirable. A more refined determination, based on new analyses, is being attempted (Giletti, in prep.).

4. Conclusions

In the decade since the last Advanced Study Institute, significant progress has been made in the measurement and understanding of diffusion kinetics in the feldspars. The differences in the diffusion kinetics of oxygen as a function of water fugacity have been explored, and largely resolved. Hydrogen appears to be the key agent required to help release oxygen ions from their Si–O bonds, and seems capable of aiding oxygen exchanges because it can migrate extremely rapidly through the structure. Although there is still some debate, it appears that the mode of transport of the oxygen through the structure is via the neutral water molecule, rather than OH^{1-}. There is general agreement as to the diffusion rate of oxygen under hydrothermal conditions, given that those conditions are specified, although some discrepancies still persist for transport under dry conditions. There is general agreement that the transport is isotropic. For a given set of conditions, the diffusion rate for oxygen falls well within one order of magnitude for any plagioclase or K–feldspar thus far reported (with the exception of some data where different reports yield different values on the same mineral when run dry).

In contrast, cation diffusion (other than Si and Al) appears to occur independent of the fugacity of water. Further, variation of oxygen fugacity by a factor of approximately 10^{19} has no measureable effect on Sr diffusion rates in albite. It remains to be seen if this holds true for transport is such cases as the Fe–rich K–feldspars, where varying the Fe^{2+}/Fe^{3+} ratio by changing the oxygen fugacity should change the vacancy concentrations significantly, thereby affecting the vacancy transport mechanism.

As with oxygen diffusion, cation diffusion is isotropic within the uncertainty of the measurement (approximately a factor of 2). There is no difference in diffusion rate for Sr measured parallel to the albite twin lamellae versus normal to the lamellae.

A major departure from the oxygen behavior, however, is the large difference in diffusion rate of Sr and Ca as a function of An/Ab content of the plagioclase. Sr and Ca diffuse faster in albite by a factor of 10^4 relative to anorthite, at the same temperature.

Diffusion rates for cations in plagioclases are strongly dependent on charge, with 1+ ions diffusing faster than 2+, which are faster than 3+. The differences are several orders of magnitude between the valence states. On the other hand, ionic radius differences between ions of the same charge cause differences in diffusion rates of are only factors of two or three. The exception is Na, which diffuses very rapidly. This suggests that Na travels by an interstitial mechanism of diffusion, whilst the other ions are transported principally by a vacancy mechanism.

Oxygen isotope compositions of minerals have been used to reconstruct thermal histories of cooled igneous and metamorphic rocks by application of oxygen diffusion kinetics. Cooling rates were determined with a precision of approximately a factor of four, and agree with expectations based on the geological setting.

Rb–Sr dating systematics can also be combined with Sr diffusion kinetics to obtain cooling rates for rocks, as well as the on timing and temperatures of metamorphic episodes. The metamorphic episode model shows promise of being useful in the Monte Rosa nappe region of the western Alps. The cooling rate model is currently being tested for the first time.

Acknowledgements

The author is very grateful to Drs. K.A. Foland and C. M. Graham for useful comments on an early version of this manuscript, and to Prof. J.A. Tullis for a remarkably thorough, critical review of the entire manuscript in its late stages. Support for much of the author's work referred to in this paper derives from grants (the latest of which is EAR–9018622) from the U.S. National Science Foundation, which is acknowledged with thanks. The author also thanks Pergamon Press for permission to reproduce figures from the paper by Farver and Yund, 1990, and from Giletti (1991), in Geochimica et Cosmochimica Acta; and the American Association for the Advancement of Science for permission to reproduce a figure from Fortier and Giletti (1989)

References

Bailey, A. (1971) Comparison of low–temperature with high–temperature diffusion of sodium in albite. Geochim. Cosmochim. Acta, v. 35, p. 1073–1081.

Barrer, R.M. (1951) *Diffusion in and Through Solids* Cambride Univ. Press, 464 pp.

Behrens, H., Johannes, W. and Schmalzried, H. (1990) On the mechanisms of cation diffusion processes in ternary feldspars. *Phys. Chem. Minerals* **17**, 62–78.

Brady, J.B. (1975) Reference frames and diffusion coefficients. Amer. Jour. Sci., v. 275, p. 954–983.

Brady, J. (in press) Diffusion data for silicate minerals, glasses, and liquids. in *AGU Handbook of Physical Constants*, Ahrens, T.H. ed.

Christoffersen, R., Yund, R.A., and Tullis, J.A. (1983) Inter–diffusion of K and Na in alkali feldspars. Amer. Min., v. 68, p. 1126–1133.

Cherniak, D.K., Lanford, W.A., and Ryerson, R.J. (1991) Lead diffusion in apatite and zircon using ion implantation and Rutherford backscattering techniques. Geochim. Cosmochim. Acta, v. 55, p. 1663–1673.

Cherniak, D.K. and Watson, E.B. (1992) A study of strontium diffusion in K–feldspar, Na–K feldspar and anorthite using Rutherford backscattering sprctroscopy. Earth Planet. Sci. Lett., v. 411–425.

Crank, J. (1967) *The Mathematics of Diffusion*, Oxford Univ. Press, London, 347 pp.

Dodson, M.H. (1973) Closure temperature in cooling geochronological and petrological systems. Contrib. Mineral. Petrol. v. 40, 259–274.

Dodson, M.H. (1986) Closure profiles in cooling systems. Materials Sci. Forum, v. 7, p. 145–154.

Dowty, E. (1980) Crystal–chemical factors affecting the modbility of ions in minerals. Amer. Min., v. 65, p. 174–182.

Eiler, J.M., Baumgartner, L.P., and Valley, J.W., (1992) Intercrystalline stable isotope diffsion: a fast grain boundary model. Contrib. Mineral. Petrol., v. 112, p. 543–547.

Elphick S.C., Dennis, P.F., and Graham, C.M. (1986) An experimental study of the diffusion of oxygen in albite using an overgrowth technique. Contrib. Mineral. Petrol. v. 92, p. 322–330.

Elphick S.C. and Graham, C.M. (1988) The effect of hydrogen on oxygen diffusion in quartz: evidence for fast proton transients? Nature v. 335, p. 243–245.

Elphick, S.C., Graham, C.M., and Dennis, P.F. (1988) An ion microprobe study of anhydrous oxygen diffusion in anorthite: a comparison of hydrothrmal data and some geological implications. Contrib. Mineral. Petrol., v. 100, 490–495.

Farver, J.R. (1989) Oxygen self–diffusion in diopside with application to cooling rate determinations. Earth Planet. Sci. Lett. **92**, 386–396.

Farver, J.R. and Yund, R.A. (1992) Oxygen diffusion in a fine–grained quartz aggregate with wetted and nonwetted microstructures. Jour. Geophys. Res. v. 97, p. 14017–14029.

Farver, J.R. and Yund, R.A. (1990) The effect of hydrogen, oxygen, and water fugacity on oxygen diffusion in alkali feldspar. *Geochim. Cosmochim. Acta* **54,** 2953–2964.

Faure, G. (1986) *Principles of Isotope Geology,* 2nd ed., John Wiley & Sons, New York, p. 117–ff.

Foland, K.A., (1974a) Alkali diffusion in orthoclase. in *Geochemical Transport and Kinetics,* Hofmann, A.W., Giletti, B.J., Yoder, H.S., and Yund, R.A., eds., Carnaegie Inst. Washington, pub., p. 77–98.

Foland, K.A. (1974b) Ar^{40} diffusion in homogeneous orthoclase and an interpretation of Ar diffusion in K–feldspars. Geochim. Cosmochim. Acta **38,** 151–166.

Fortier, S.R. (1991) Empirical models for predicting diffusion kinetics in silicate minerals and the thermal history of the South Mountains metamorphic core complex, Arizona, derived from oxygen isotope and diffusion data. PhD Thesis, Brown Univ.,.

Fortier, S. R. and Giletti, B.J. (1989) An empirical model for predicting diffusion coefficients in silicate minerals. Science, v. 245, p. 1481–1484.

Freer, R. (1981) Diffusion in silicate minerals and glasses: a data digest and guide to the literature. Contrib. Mineral. Petrol., v. 76, p. 440–454.

Frey, M., Hunziker, J.C., O'Neil, J.R., and Schwander, H.W., (1976) Equilibrium–disequilibrium relations in the Monte Roxa granite, western Alps: petrological, Rb–Sr and stable isotope data. Contrib. Mineral. and Petrol. v. 55, p. 147–179.

Giletti, B.J. (1974) Diffusion related to geochronology. in *Geochemical Transport and Kinetics,* Hofmann, A.W., Giletti, B.J., Yoder, H.S., and Yund, R.A., eds., Carnegie Inst. Washington, Publ. 634, 61–76.

Giletti, B.J. (1986) Diffusion effects on oxygen isotope temperatures of slowly cooled igneous and metamorphic rocks. Earth Planet. Sci. Lett. **77,** 218–228.

Giletti, B. J. (1991) Rb and Sr diffusion in alkali feldspars, with implications for cooling histories of rocks. *Geochim. Cosmochim. Acta* **55,** 1331–1343.

Giletti, B.J. (in prep.) Systematics and mechanisms of cation diffusion in feldspars.

Giletti, B.J. and Casserly, J.E.D. (submitted) Sr diffusion kinetics in plagioclase feldspars. Geochim. Cosmochim. Acta.

Giletti, B.J., Semet, M.P., and Yund, R.A. (1978) Studies in diffusion –III: an ion microprobe determination. Geochim. Cosmochim. Acta, v. 42, p. 45–57.

Giletti, B.J., Yund, R.A., and Semet, M. (1976) Silicon diffusion in quartz. Geol. Soc. Amer., Abstracts with Programs, v. 8, p. 883–884.

Girifalco, L.A. (1964) Atomic migration in crystals. Blaisdell Publ. Co., N.Y. 162 pp.

Goldsmith, J.R. (1988) Enhanced Al/Si diffusion in $KAlSi_3O_8$ at high pressures: the effect of hydrogen. Jour. Geol. v. 96, p. 109–124.

Graham, C.M. and Elphick S.C. (1991) Some experimental constratints on the role of hydrogen in oxygen and hydrogen diffusion and Al–Si interdiffusion i silicates. in *Diffusion, Atomic Ordering, and Mass Transport: Selected Problems in Geochemistry.* Ganguly, J. ed., Springer–Verlag, New York, p. 248–285.

Griggs, D.T. (1967) Hydrolytic weakening of quartz and other silicates. Geophys. Jour. Roy. Astron. Soc., v. 14, p. 19–32.

Griggs, D.T. (1974) A model of hydrolytic weakening in quartz. Jour. Geophys. Res., v. 79, p. 1653–1661.

Hart, S.R. (1981) Diffusion compensation in natural silicates. Geochim. Cosmochim. Acta, v. 45, p. 279–291.

Jenkin, G. (in press) Oxygen isotope exchange in cooling rocks and oxygen isotope closure temperatures.

Jensen, M.L. (1952) Solid diffusion of radioactive sodium in perthite. Amer. Jour. Sci. v. 250, p. 808–821.

Kasper, R.B. (1975) Cation and oxygen diffusion in albite. PhD Thesis, Brown University.

Manning, J.R. (1968) *Diffusion Kinetics for Atoms in Crystals.* Van Norstrand, Princeton, N.J., 257 pp.

McNaughton, N.J. and Wilson, A.F. (1980) Problems in oxygen isotope geohermometry in mafic granulite facies rocks from near Einasleigh, northern Queensland. Precambrian Res. v. 13, p. 77–86.

Merigoux, H., 1968, Etude de la mobilite de l'oxygene dans les feldspaths alcalins, Bull. Soc. Francaise Mineral. Crystallzgr., v. 91, p. 51–64.

Misra, N.K. and Venkatasubramanian V.S. (1977) Strontium diffusion in feldspars– a laboratory study. Geochim. Cosmochim. Acta **41**, 837–838.

Muehlenbachs and Kushiro (1974) Oxygen isotope exchange and equilibrium of silicates with CO_2 or O_2. Carnegie Inst. Washington Ybk., v. 73, p. 232–236.

Nagy, K.L. and Giletti, B.J. (1986) Grain boundary diffusion of oxygen in a macroperthitic feldspar. Geochim. Cosmochim. Acta v. 50, p.1151–1158.

Petrovic, R. (1972) Alkali ion diffusion in alkali feldspars. PhD Thesis, Yale University, 131 pp.

Petrovic, R. (1974) Diffusion of alkali ions in alkali feldpsars. in *The Feldspars.* NATO Advanced Study Institute, Mackenzie, W.S. and Zussman, J. eds., Manchester Univ. Press, p. 174–182.

Shewmon, P.G. (1963) *Diffusion in Solids.* McGraw–Hill, New York, 202 pp.

Sipple, R.F. (1963) Sodium self–diffusion in natural minerals. Geochim. Cosmochim. Acta, v. 27, p. 107–120.

Snow, E. and Kidman, S. (1991) Effect of fluorine on solid–state alkali interdiffusion rates in feldspar. Nature, v. 349. p. 231–233.

Snow, E. and Sherman, S. (in prep.) The effect of fluorine on cation diffusion in feldspar: implications for microstructural development.

Taylor, H.P. and Epstein, S. (1962) Relationship between O^{18}/O^{16} ratios in coexisting minerals of igneous and metamorphic rocks. Part I: Principles and experimental results. Bull. Geol. Soc. Amer. v. 73, p. 461–480.

Whittaker, E.J.W. and Muntus, R. (1970) Ionic radii for use in geochemistry. Geochim. Cosmochim. Acta, v. 34, p. 945–956.

Winchell, P. (1969) The compensation law for diffusion in silicates. High Temp. Sci. v. 1, p. 200–215.

Yund, R.A. (1984) Alkali feldspar exsolution: kinetics and dependence on alkali interdiffusion. in *Feldspars and Feldspathoids, Structures, Properties and Occurrences*. Brown, W.L., ed., NATO ASI Series, p. 281–315 D. Reidel Publ. Co., Dordrecht/Boston/Lancaster.

Yund, R.A. and Anderson, T.F. (1974) Oxygen isotope exchange between potassium feldspar and KCl solution. in *Geochemical Transport and Kinetics*, Hofmann A.W., Giletti, B.J., Yoder, H.S., and Yund, R.A., eds., Carnegie Inst. Washington Publ. 634, p. 99–105.

Yund, R.A. and Anderson, T.F. (1978) The effect of fluid pressure on oxygen isotope exchange between feldspar and water. Geochim. Cosmochim. Acta **42**, 235–239.

Zhang, Y., Stolper, E.M., and Wasserburg, G.J. (1991) Diffusion of a multi–species component and its role in oxygen and water transport in silicates. Earth Planet. Sci. Lett. v. 103, p. 228–240.

HYDROGEN IN FELDSPARS AND RELATED SILICATES

COLIN M. GRAHAM *and* STEPHEN C. ELPHICK
Department of Geology and Geophysics
University of Edinburgh
Edinburgh EH9 3JW
Scotland

ABSTRACT. The importance of trace quantities of hydrogen in feldspars in determining the rates of intra-crystalline transport processes was suggested by several early experimental studies of Al-Si order-disorder in feldspars and oxygen isotope exchange between feldspars and water. With the aid of modern experimental and analytical techniques, it has been possible to distinguish solid-state from solution-controlled processes in hydrothermal experiments, and to confirm the conclusions of earlier studies. The identities of the hydrogen-bearing species which enhances diffusion processes and of the oxygen-transporting species have however remained controversial.

By careful control of the activities of relevant hydrogen-bearing species in high P-T experiments, it has been possible to demonstrate the importance of protons in enhancing the rates of Al-Si order-disorder in alkali feldspars. The enhancement of oxygen diffusion in feldspar and quartz also appears to be activated by protons, but the identity of the oxygen-transporting species is more controversial. Some recent experimental studies have pointed to molecular water as the oxygen-transporting species, but ab initio calculations provide contrary evidence that interstitial water has much higher energy in the α-quartz structure than the hydrogarnet defect. It may not be possible to unambiguously constrain the identity of the oxygen-transporting species using existing experimental techniques.

The effects of $P(H_2O)$ and T on the solubility of hydrogen in feldspars are unknown. At high $P(H_2O)$ and T, K and Ba feldspars break down to pseudo-hexagonal hydrous phases which can accommodate molecular water in varying amounts depending on $P(H_2O)$ and T.

1. Introduction

Hydrogen may occur in feldspars in major element abundance. The ammonium feldspar buddingtonite, ($NH_4AlSi_3O_8$), was synthesised and characterised by Voncken et al (1988) as an anhydrous feldspar, although natural examples of this mineral may also contain some structural molecular water (eg Erd et al 1964). Hydrogen feldspar ($HAlSi_3O_8$) was prepared by cation exchange of Al-Si disordered sanidine and reaction of the resulting albite with concentrated H_2SO_4 at 320°C (Muller 1988), and characterised by Deubener et al (1991). These feldspars, containing hydrogen as a cation substituent, are from a geological point of view largely mineralogical curiosities, and will not be considered further in this review.

Here we are concerned primarily with the trace amounts of hydrogen which occur in natural and synthetic common alkali and plagioclase feldspars, which are of the greatest importance in determining their structural and mechanical properties (eg Griggs 1967; Tullis 1983; Gandais and

I. Parson (ed.), Feldspars and Their Reactions, 383–413.

Williame 1984), their colour (eg Hofmeister and Rossman 1983, 1985a,b), and in particular the rates of intra-crystalline solid-state diffusion and order-disorder properties (eg Yund 1983, 1984). This review will focus largely on the role and speciation of hydrogen involved in enhancing intra-crystalline transport properties. These properties, being sensitive to the presence of hydrogen-bearing impurities within the feldspar lattice, provide evidence on the identity of the impurities and the mechanisms by which they modify intracrystalline processes. Because the transport properties of quartz have been more intensively studied than those of feldspar, particularly in relation to the role of hydrogen and the diffusion of oxygen, we can exploit information and data on quartz where appropriate as a framework silicate analogue for feldspar.

2. Abundances and Substitution Mechanisms of Structural "Water"

2.1 CHEMICAL ANALYSIS

In common with many nominally anhydrous rock-forming silicates, feldspars have been found to contain small amounts of water, revealed in the first instance by chemical analysis (Deer et al 1963; Martin and Donnay 1972). The latter authors present a compilation of analyses showing feldspars with a range of water contents up to over 0.8 wt% (Fig 1; see also Wilkins and Sabine 1973), and point out that inclusion of OH with O in recalculating structural formulae often improves the stoichiometric formulae, lending some credence to the presence of truly structural "water" rather than simply fluid inclusions or impurities. Water in feldspars at the abundances suggested by Martin and Donnay (1972) could constitute a significant crustal reservoir. Furthermore, if its incorporation into feldspars is systematically dependent on temperature and/or water pressure, then the cooling and uplift of high P-T feldspar-rich rocks could well exsolve significant quantities of water, providing an "internal" source of fluid to explain the common porosity and turbidity of crustal feldspars (eg Worden et al 1990; Walker 1990). Nothing is currently known about the influence of pressure and temperature on the incorporation of water into feldspars, which remains an important but complex experimental task.

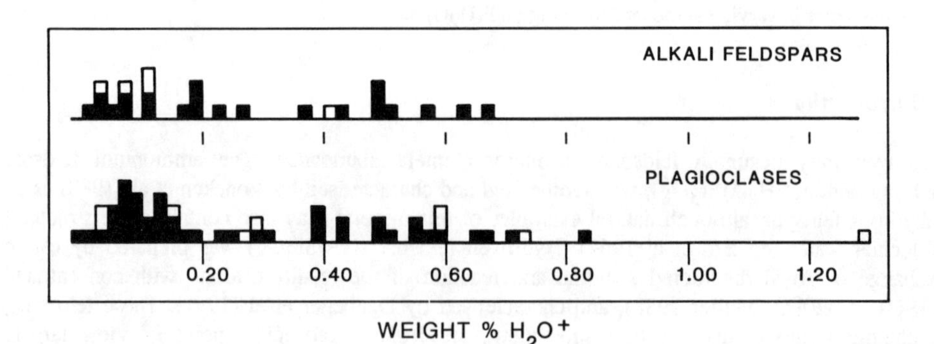

Figure 1. Water contents (H_2O+) of alkali and plagioclase feldspars, after Martin and Donnay (1972). Black squares - igneous feldspars; open squares - metamorphic feldspars.

TABLE 1. Water content and hydrogen speciation in feldspars determined by infra-red spectroscopy

Sample	Composition	H_2O(wt%)	Hydrogen Speciation
sanidine	$Or_{86}Ab_{14}$	0.017*	H_2O
		0.013	
orthoclase	$Or_{90}Ab_{10}$	<0.002	OH
microcline	$Or_{92}Ab_8$	0.14*	H_2O
amazonite	$Or_{91}Ab_9$	0.001	H_2O
orthoclase	$Or_{92}Ab_8$	0.004	H_2O
amazonite	$Or_{96}Ab_4$	0.051*	H_2O
		0.001	
amazonite	$Or_{96}Ab_4$	0.09*	H_2O
		0.001	
oligoclase	An_{17}	0.020	OH
andesine	An_{30}	0.052	OH
labradorite	An_{59}	0.030	OH
labradorite	An_{66}	0.013	OH

*Denotes water content determined by hydrogen manometry; other water contents determined by infra-red spectroscopy. Data from Hofmeister and Rossman (1985b), Beran (1986, 1987).

2.2 SPECTROSCOPIC CONSTRAINTS

Infrared spectroscopy provides a powerful modern method of characterizing the abundance, speciation and structure of water or other H-bearing species in feldspars. While an in-depth account of the spectroscopy of feldspars is beyond the scope of this review and the expertise of the authors, a summary of the principal conclusions of recent key studies is helpful in constraining abundances and speciation of structural "water" and possible mechanisms of hydrogen substitution.

Abundances of structural "water" in feldspars determined by infra-red spectroscopy, using well-characterised standards with carefully analysed water contents (eg by hydrogen manometry) tend to be significantly lower than those suggested by earlier compilations of analyses (eg Martin and Donnay 1972). A tabulation of water in alkali and plagioclase feldspars from several recent spectroscopic studies (Hofmeister and Rossman 1985b; Beran 1986, 1987; Rossman and Smyth 1990) indicates water contents generally less than 0.2 wt% (Table 1). Of particular importance in these studies is the identification of *structural molecular* H_2O, and its allocation to the *M cation site* (eg Hofmeister and Rossman 1985a,b; Beran 1986), where it is presumably charge-balanced by replacement of divalent for monovalent cations or Si for Al (ie $[Si_{Al}^{\cdot}]$). Molecular water has been identified as the dominant hydrogen-bearing species in potassium feldspars, while plagioclases commonly contain structural OH (eg Hofmeister and Rossman 1985a,b).

Other possible substitution mechanisms for hydrogen in feldspars have been suggested, or may be inferred by analogy with other framework silicates, especially quartz. These mechanisms relate to the formation of defects in the structure, occurring as vacancies (eg $V_{Si}^{''''}$, $V_{O}^{\cdot\cdot}$), interstitials (eg $Na_i^{\cdot}$, $H_i^{\cdot}$) or substitutions (eg $Al_{Si}^{'}$). Possible mechanisms include for example:

(i) $[H_i^{\cdot}] = [Al_{Si}^{'}]$

(eg Dennis 1984a)

(ii) Si-O-Si + H_2O = Si-OH:HO-Si

(the "Griggs" defect; eg Griggs and Blacic 1965; Hobbs 1984)

(iii) $(4H)_{Si}$; $[(3H)_{Si}^{'}] = [Ca_{Na}^{\cdot}]$

(the "hydrogarnet" defect; eg Hobbs 1984). Recent ab initio calculations (Purton et al 1992, 1993; Price, pers comm) have shown that the hydrogarnet defect in α-quartz is stable while the Griggs defect is not.

Measurements of the quantity and speciation of hydrogen or "water" in feldspars or "analogue" quartz reported in the literature have been made on natural samples or the quenched products of high P-T annealing experiments. When we are considering a hydrogen species of potentially high mobility (Elphick and Graham 1988), there is no assurance that the abundance or speciation relate to those pertaining at high pressures and temperatures.

3. Early Studies and Models

3.1 OXYGEN ISOTOPE EXCHANGE EXPERIMENTS

There is nothing novel about the use of oxygen isotopes in hydrothermal experiments as tracers of the mechanisms of oxygen exchange in feldspars and other silicates, and of the role of hydrogen in enhancing the exchange process. Similarly, the concept that rapid transport of some hydrogen-bearing impurity through the feldspar lattice might enhance the rates of intra-crystalline processes has been current for over thirty years. A number of French studies in the 1950's not only used this technique, but addressed the same questions which we are now able to approach in more sophisticated and definitive ways with the application of modern analytical and experimental techniques. The mechanisms proposed to account for the volume diffusion of oxygen in silicates and the role of hydrogen are those that remain the subject of present-day controversy.

Wyart et al (1959, 1961) studied the kinetics of exchange of oxygen isotopes between silicates (microcline, quartz, granite) and water under hydrothermal conditions at pressures of 150 to 1800 bars and temperatures of 360 to 800°C. The experimental method involved bulk exchange between sized mineral powders and water of known of $^{18}O/^{16}O$. 25μm powders of microcline were found to have exchanged 37% of their oxygen in 24 hours at 400 bars, 690°C, and the exchanged fraction of oxygen in 60μm quartz ranged from 10% at 170 bars, 360°C to 16% at 350 bars, 610°C. These authors proposed a solid-state replacement mechanism for oxygen exchange involving transient breaking and reforming of Si-O-Si and Al-O-Si bonds (Si-O-Si + H_2O = Si-OH:HO-Si; eg Griggs and Blacic 1965), in which H^+ and OH^- ions were identified as playing an important role. Of particular interest are the published discussion questions (Wyart et al 1961), which remain topical thirty years later:

Q: "How large was the OH^-content in the crystal after the experiment and does this content depend upon pressure?"

A: "The water content was less than 0.2%. We have no results on the variation as a function of pressure". [Donnay et al (1959) report less than 0.1 wt % H_2O.]

Q: "As the unmolten samples in your experiment are apparently unchanged, there may be a solution and redeposition from the gas phase. ..there is a problem of the diffusion rate and activation energy. "

A: " We do not have a definitive experimental argument for eliminating the hypothesis of exchange by solution and reprecipitation, consistent with diffusion in the solid. We are carrying out experiments with the purpose of proving the simultaneous diffusion of hydrogen and oxygen."

There is general agreement that the rate of oxygen exchange between feldspar and water is greatly enhanced both by the *presence* of water and by the *pressure* of water, but differing views have arisen as to the *mechanisms* of this exchange. There are three possible mechanisms, operating singly or in combination (eg Giletti 1985): (i) **solution-reprecipitation,** in which the crystal continuously dissolves at the solid-fluid interface, then recrystallises again, either on existing surfaces or as new crystals, as other parts of the crystal go into solution; new crystals will form initially in isotopic equilibrium with the fluid. Since all minerals have a finite solubility in water, this process will occur to some degree in any hydrothermal experiment, and will thus complicate hydrothermal experiments designed to measure solid-state diffusion (eg Elphick et al 1986a); (ii) **chemical reaction,** in which the crystal breaks down to some degree, probably initially at crystal surfaces, with the formation of new crystals of different composition (thus distinguishing this case from (i)), probably in isotopic equilibrium with the fluid; (iii) **diffusion**, in which the crystal remains a single solid reservoir of constant shape and size, and isotopic exchange occurs by volume diffusional transport across the faces and through the body of the crystal (volume diffusion). In most experimental and natural situations in which solids and aqueous fluids coexist, some combination of two or more of these processes may occur to some degree. Some approaches to resolving this experimental problem are summarised below.

The study by Yund and Anderson (1978) on the effect of $P(H_2O)$ on oxygen isotope exchange between adularia and aqueous solution (2M KCl) illustrates the difficulty in identifying which of the above mechanisms operates in kinetic studies of oxygen isotope exchange between feldspar and water. Using a similar bulk exchange technique to earlier studies (the "integrating technique" of Giletti (1985 and this volume)), with sized adularia powders and solutions of known $^{18}O/^{16}O$, they measured the extent of oxygen isotope exchange as a basis for calculating the diffusivity of oxygen over the pressure range 125-4000 bars. They found an approximately linear relationship between diffusivity and $P(H_2O)$, and pointed out that a similar relationship could be demonstrated with $\sqrt{P(H_2O)}$. Following Donnay et al (1959) they proposed that "water" present as molecular H_2O, OH^-, or H^+ in the feldspar structure increases the rate of oxygen diffusion. However, Norton and Taylor (1979) observed that the linear relationship between oxygen diffusivity and water pressure might be interpreted alternatively in terms of a surface hydrolysis (dissolution) reaction in which H^+ is involved in the rate-determining step, consistent with a $\sqrt{f(H_2O)}$ relationship, rather than volume diffusion of oxygen (Fig. 2). This ambiguity serves to emphasise the need to distinguish surface reaction processes from solid-state diffusion processes

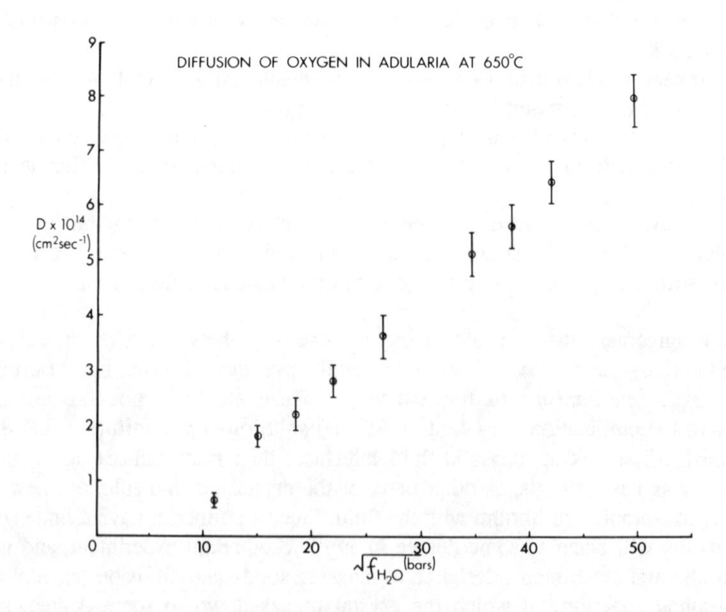

Figure 2. Effect of water fugacity on the diffusivity of oxygen in adularia feldspar at 650°C, using data of Yund and Anderson (1978). After Graham and Elphick (1991).

before attempting to draw conclusions about the role and mechanism of hydrogen in modifying the rates of intra-crystalline processes, and the identity of the oxygen-transporting species. It is not possible to distinguish unambiguously surface-controlled and diffusion-controlled mechanisms in bulk exchange (integrating) experiments.

The questions of water contents, solution-reprecipitation vs solid-state diffusion, diffusion rates and mechanisms, and diffusing species, have all been addressed in detail by experimental studies over the past twenty years, and we are now able to provide definitive answers to some, if not all, of these questions with the aid of modern experimental and analytical techniques.

3.2 Al-Si ORDER-DISORDER AND INTERDIFFUSION EXPERIMENTS

The idea that a hydrogen species plays a key role in the rates and mechanisms of order-disorder and Al-Si interdiffusion in feldspars is again of long standing. MacKenzie (1957) undertook a detailed experimental study of the structural state of albite synthesised and/or annealed under a wide range of T, $P(H_2O)$ conditions. The most important general conclusions of this study were: (1) the lattice parameters of crystalline albite (which are dependent on the state of Al-Si disorder) depend upon the duration of the experiment; (2) no changes in Al-Si order could be produced in laboratory time-scales unless water vapour was present. For example, at 700°C, 2 kbars, albite glass crystallises within 30 minutes, but the lattice parameters continue to change with time in longer duration experiments. Amelia albite annealed at 800°C and 2 kbars for 90 hours was found to have disordered but to have retained its shape, and the outer zones of crystals disordered before the interiors. MacKenzie therefore proposed that water is able to pass through the feldspar lattice to effect the breaking and reforming of Al-O and Si-O bonds.

A kinetic analysis of MacKenzie's hydrothermal data (McConnell and McKie 1960) showed that the *rate* of Al-Si ordering was proportional to the *square root* of the water pressure in the experiment, that is water influences the *rate* of Al-Si interdiffusion but not the *equilibrium* ordering state. McConnell and McKie (1960) concluded that this kinetic relationship was consistent with "the direct participation of H^+ in the transformation [disordering] process, since the activity of H^+ in the vapour phase bears a corresponding relationship to the total water-vapour pressure". As in the case of oxygen transport in adularia (above), we find a relationship between the rate of a presumed intra-crystalline process (Al-Si interdiffusion) and $\sqrt{(fH_2O)}$, indicating the involvement of some dissociated species (?H^+) in the rate-enhancing mechanism, with textural evidence supporting the case for a solid-state mechanism rather than a surface reaction enhanced mechanism in this instance.

In considering the effect of water on the structural state of feldspars, Donnay et al (1959) concluded that "the mobility of the Si and Al ions rests on the catalytic action of protons and hydroxyl ions", this action involving the attack of bridging oxygens by protons, which precede the hydroxyl ions when water diffuses in from the surface of the crystal. Water is thus viewed as the diffusing species. Donnay et al (1959) concluded their paper by stating: "Our explanation of the mobility of the Si and Al ions rests on the catalytic action of protons and hydroxyl ions. We must therefore ascertain the presence of these ions wherever Si and Al are found to rearrange themselves in the solid state. The amount of water to be looked for may be extremely small and it may be firmly held between the mosaic blocks and at the surface of the crystal. Methods other than chemical analysis and infrared spectroscopy may have to be used to detect it." As will emerge below, this situation is close to that which emerges from a distillation of much of the recent experimental literature!

4. The Question of Solution-Reprecipitation vs Solid-State Diffusion

Models of proton, hydroxyl or molecular water involvement in mechanisms of oxygen exchange or Al-Si ordering in feldspars depend upon the unsubstantiated assumption that the progress of these processes during hydrothermal experiments was not the result of dissolution and reprecipitation of the feldspar crystals. Following bulk exchange (integrating) experiments using mineral powders of known grain size, the exchanged mineral powder is analysed in bulk to determine the extent of exchange. The analysis provides a measure of the extent and kinetics of exchange, but is not capable of distinguishing diffusion from solution-reprecipitation or chemical reaction (or some combination of these) as the rate-determining step in the measured exchange. In reality, all minerals have a finite solubility in aqueous solution, and the amount of material dissolved will increase with the total surface area (or reciprocal grain size) of the sample. How can we distiguish these processes in order to establish whether the exchange kinetics provide a measure of volume diffusion?

In attempting to measure the equilibrium fractionation of oxygen isotopes between feldspars and water under hydrothermal conditions, O'Neil and Taylor (1967) enhanced the kinetics of isotope exchange by inducing simultaneous cation and oxygen exchange between feldspar and aqueous chloride solutions (eg albite + KCl + H_2O). For example, natural Amelia albite underwent only 15% oxygen exchange with *pure* water in 62 hours at 650°C, compared to 92% exchange with KCl solution in only 2 hours at the same temperature. A direct correlation was found between cation and oxygen exchange, consistent with solution-reprecipitation as the dominant oxygen exchange mechanism. O'Neil and Taylor envisaged a reaction front sweeping through feldspar crystals exchanging oxygen and (where relevant) cations by a micro-solution-reprecipitation mechanism, rather than by the solid-state diffusion of oxygen and hydrogen bearing species envisaged by earlier workers. The pure water and alkali halide experiments are thus envisaged to differ only in degree rather than in kind. By contrast, Merigoux (1968) used the extent of oxygen exchange between alkali feldspars and ^{18}O-enriched water to provide a measure of the diffusivity of oxygen in the feldspar. Their results are comparable to diffusivities measured in more recent studies where volume diffusion is demonstrably the dominant mechanism. More rapid exchange between feldspar and alkali chloride solution is attributed by Merigoux to a quite distinct *reaction* mechanism.

Yund and Anderson (1974) demonstrated that oxygen exchange between feldspar and fluid does indeed proceed dominantly by volume diffusion of oxygen in the absence of cation disequilibrium when the integrating method is used. They showed that experimentally exchanged microcline and adularia feldspars had not changed their Al-Si order state and that K diffusivity under identical experimental conditions was independent of oxygen diffusivity. If solution-reprecipitation or reaction was the dominant mechanism of exchange, these elements should have yielded the same apparent diffusivities at all temperatures. This point was re-emphasised by Giletti (1985) in comparing the experimentally-determined diffusivities of Na, Ar, O, K and Rb in K-feldspar (Fig. 3).

Two modern analytical techniques have subsequently been used to support these conclusions. Matthews et al (1983) examined the mechanisms and kinetics of oxygen isotope exchange between feldspar and water with the aid of scanning electron microscopy (SEM). Unlike quartz-water experiments, substantial recrystallisation did not occur during oxygen exchange between albite, anorthite and water at 500-600°C, 2-15 kbars; degradation into smaller fragments was dominantly mechanical, with oxygen isotope exchange proceeding by diffusion-controlled mechanisms. The specific surface area of the feldspars did not change with time or fractional

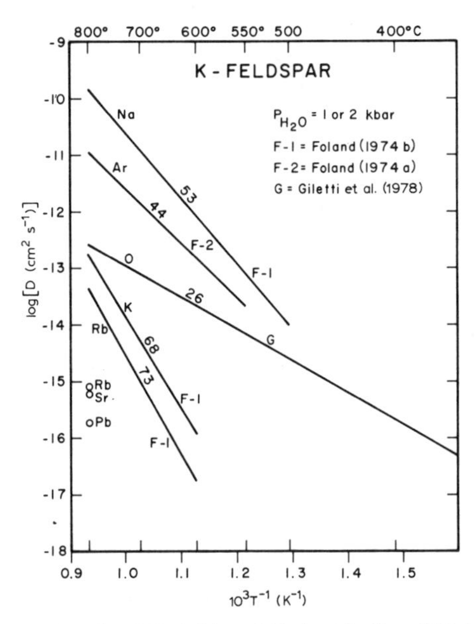

Figure 3. Arrhenius plot comparing the diffusivities of O, Ar, Na, K and Rb in K-feldspar. From Giletti (1985).

exchange after initial degradation, while in quartz the specific surface area decreased with time as grain growth (Ostwald ripening) occurred.

Ion imaging by ion microprobe (SIMS) provides a novel method of identifying mechanisms of fluid-mineral interaction during hydrothermal experiments. The technique of scanning ion imaging, described and illustrated by Elphick et al (1991), has been used to distinguish the relative importance of volume diffusion and solution-reprecipitation during oxygen isotope exchange and Al-Si disordering of albite. Albite crystals were exchanged with $H_2{}^{18}O$-labelled saturated NaCl solution at 850°C, 1 and 17 kbars, and the annealed run product crystals mounted in epoxy and polished to reveal grain cross-sections (Graham and Elphick 1990). The mounts were examined by scanning ion imaging using a Cameca ims-4f ion microprobe, the ion beam being raster-scanned to give a 256x256 pixel ion image of the distribution of ^{18}O. The resulting ^{18}O maps (Fig. 4) clearly distinguish areas of solution and reprecipitation characterised by overgrowths of high ^{18}O on certain crystal faces and edges, narrow zones of ^{18}O enrichment on other crystal faces where diffusional exchange is dominant, and large unmodified cores of original low-^{18}O albite. These images demonstrate unequivocally that solution-reprecipitation occurs *but is very limited in extent*, supporting the conclusions of Giletti (1985), Matthews et al (1983), Yund and Anderson (1974) and earlier French studies. This is an important step in the logic of understanding the mechanistic role of hydrogen in intra-crystalline processes in feldspars; in the case of the albite studied by Graham and Elphick (1990), the feldspar could be shown to have undergone extensive Al-Si disordering by dominantly solid-state mechanisms, with a minor contribution from solution processes. It should be noted that both SEM and SIMS techniques are capable of identifying new phases which might result from chemical reaction or incongruent dissolution during mineral-fluid exchange.

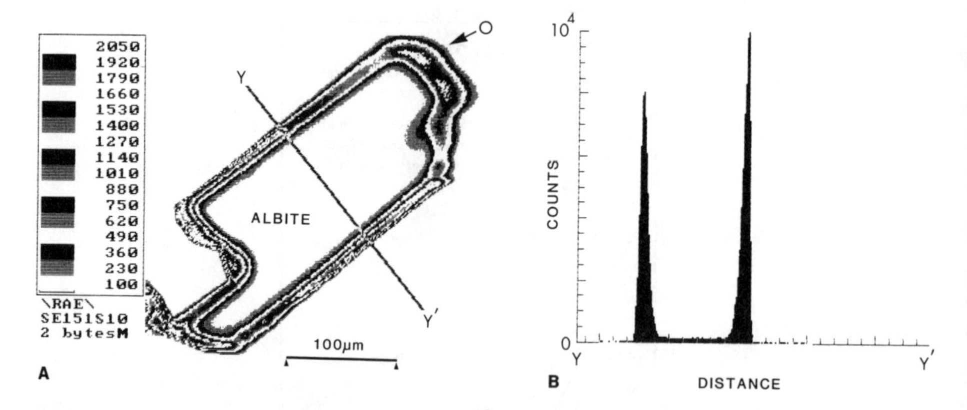

Figure 4. Ion microprobe scanning ion image of ^{18}O distribution in Amelia albite heated at 17 kbar, 850°C for 94 hrs in NaCl-saturated $H_2{}^{18}O$ (99%). **A.** Section across cleavage fragment showing diffusion of ^{18}O into long cleavage faces, and precipitation of ^{18}O labelled material on other faces. Inset scale bar shows ^{18}O counts per pixel. Note clean facets on areas of overgrowth (O). **B.** Concentration vs distance plot of ^{18}O distribution along a 10-pixel width strip at position Y-Y' in A. Note constant diffusion profiles for both faces. From Graham and Elphick (1990).

5. Al-Si Order-Disorder and Interdiffusion in Feldspars as a Guide to the Role of Hydrogen

The rate-enhancing effect of hydrogen-bearing species on Al-Si order-disorder and interdiffusion in feldspars was proposed in the early studies summarised above. The demonstration that the role of solution-reprecipitation in these early experiments was indeed minor confirms the validity of the experimental interpretations, but raises questions about the identity of the hydrogen-bearing species and the precise mechanism involved. These questions have been addressed in recent experimental studies, in which careful consideration has been given to the modification of the experimental configuration and chemical environment in order to control the activity of relevant hydrogen-bearing species.

Experimental study of the kinetics of Al-Si disordering in alkali feldspars at high $P(H_2O)$ is of course limited by the intervention of melting at the temperatures at which Al-Si interdiffusion becomes experimentally measureable. However, Goldsmith and Jenkins (1985) examined the ordering and disordering of albite powders annealed in the absence of water in sealed Pt capsules run in NaCl pressure media in the piston cylinder apparatus at 15-20 kbars. They found that the rate of Al-Si interdiffusion in albite is enhanced by orders of magnitude at these high pressures in the absence of water, relative to 1-2 kbar hydrothermal experiments, and used this discovery to bracket the order-disorder relations in albite by reversal experiments. They measured the degree of order from the $\Delta 131$ (= $2\theta(131) - 2\theta(1\bar{3}1)$) value derived from X-ray powder diffraction data. The effects of temperature and pressure on the order-disorder relations of albite were documented, and the process was shown to be continuous. It was concluded that enhanced Al-Si interdiffusion rates were dependent on either confining pressure alone or on mediation by some hydrogen-bearing species present in the NaCl pressure medium in the piston-cylinder apparatus and capable of diffusing through the platinum container material. This hydrogen may have originated in the NaCl pressure cells, from which it is hard to eliminate traces of H_2O. In the absence of water in direct contact with the albite, the process of Al-Si interdiffusion must have proceeded by a solid-state mechanism. The effect of pressure in *enhancing* interdiffusion is the opposite of the expected *inhibiting* effect of increased pressure on solid-state diffusion. This study stimulated further experimental effort to constrain the mechanism and hydrogen species involved.

The role of hydrogen in Al-Si disordering was more clearly identified in a series of experiments (Goldsmith 1986) in which the experimental environment around sealed capsules containing dry albite run in the piston-cylinder apparatus was modified either to inhibit production of hydrogen-bearing species or to absorb these species. The run conditions (850-900^oC, 10-16 kbars) were sufficient to extensively disorder the ordered albite starting powders. The incorporation of soft glass into the NaCl pressure cell around the Pt sample capsule strongly suppressed the disordering of albite relative to normal NaCl, presumably due to absorption by the glass of free water in the experimental environment. The packing of loose hematite around the capsules had a similar rate-inhibiting effect by an H_2-absorbing reaction such as: $3Fe_2O_3 + H_2 = 2Fe_3O_4 + H_2O$, implicating H_2 in the rate-enhancing mechanism. Hydrogen could have formed by the water-gas reaction between water and the graphite furnace: $H_2O + C = H_2 + CO$. Hydrogen diffusing through the Pt capsule wall must clearly be incorporated into the feldspar lattice to enhance the solid-state disordering process. The importance of these novel experiments is that they demonstrate unequivocally the previously speculative role of a hydrogen species within the feldspar lattice in the rate-enhancing process, in an experimental environment in which dissolution mechanisms are eliminated.

The effect of pressure on the rate of Al-Si disorder in albite was investigated systematically by Goldsmith (1987). Ordered albite sealed dry in Pt capsules and run in NaCl pressure media in the piston-cylinder was disordered at 800-950^oC, 6-24 kbars, providing a measure of the degree of Al-Si disorder relative to the equilibrium state of disorder at any temperature as determined by Goldsmith and Jenkins (1985). The time required to reach equilibrium Al-Si disorder was found to decrease dramatically in the pressure range 6-9 kbars, giving a non-linear relationship between pressure and Al-Si disordering rate (Fig. 5). A similar relationship was demonstrated for potassium feldspar (Goldsmith 1988), but the dramatic decrease in Al-Si disordering rate was found to occur at higher pressures of 10-12 kbars. If the relationship between disordering rate and pressure is controlled by the activity of hydrogen in the experimental environment, then this hydrogen must increasingly partition itself into the feldspar lattice with increasing pressure. The fundamental relationship is that between the activity of the relevant hydrogen-bearing species in the experimental environment and the rate of Al-Si disorder or interdiffusion. Such a relationship is very plausible since it is well known that aqueous solutions become increasingly *dissociated* with increasing pressure at constant temperature (eg Eugster 1981, 1986; Fig. 6). In the salt cell surrounding the Pt capsule any H_2O will be strongly dissociated, and the activity of H^+ will increase with increasing pressure. The form of this relationship can be approximated by the dissociation constant for HCl (K_{HCl}), whose variation with P and T may be estimated from the experimental data of Frantz and Marshall (1984) and provides an empirical monitor of [H^+] in the experimental environment (the numerous assumptions and approximations involved in this approach are stated in Graham and Elphick 1990). Graham and Elphick (1990) found that a first-order rate constant for the disordering of Al and Si in albite was linearly proportional to [H^+] as calculated from K_{HCl} (Fig. 7), thus lending support to the proton enhancement model. There is nothing unique about the NaCl pressure medium as a source of protons; the critical parameter is the increased ionic dissociation of water in any water-bearing pressure medium (eg talc) with increasing pressure.

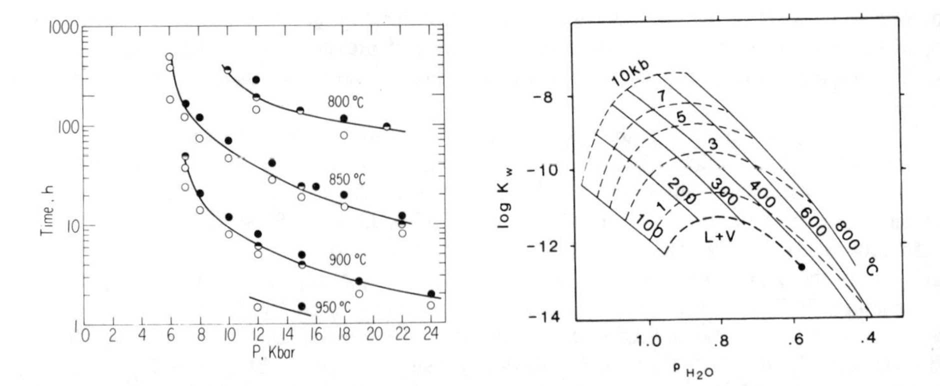

Figure 5 (left). Effect of pressure on the time necessary to develop the equilibrium degree of Al-Si disorder in albite at 800-950°C, using the piston-cylinder apparatus with a NaCl pressure medium. Solid symbols indicate attainment of equilibrium state of disorder. Open symbols indicate unequilibrated runs. Modified after Goldsmith (1987).

Figure 6 (right). Dissociation constant for water vapour (K_W) as a function of water density. Solid contour lines are isotherms (°C) and dashed contour lines are isobars (kbars). Lower heavy dashed line indicates liquid + vapour curve. From Eugster (1986).

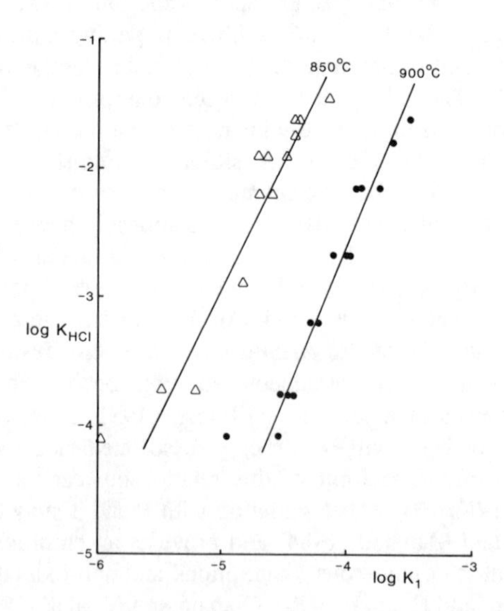

Figure 7. Plot of K_{HCl} (dissociation constant for HCl) against rate constant (K_1) for rate of Al-Si disordering in albite at 850-900°C, calculated from experimental data of Goldsmith (1987). Data fitted by least-squares regression. From Graham and Elphick (1990).

In order to constrain experimentally the precise species of hydrogen involved in activating diffusion in the feldspar lattice, it is necessary to design experiments in which the activities of H^+ and H_2 are held at extreme values, or else varied continuously in a systematic way. High fH_2 is readily achieved in presence of a tantalum getter which reacts rapidly with water to generate H_2 according to the reaction: $2Ta + 5H_2O = Ta_2O_5 + 5H_2$. When Ta metal powder is mixed with the NaCl pressure medium, the Al-Si disordering reaction is strongly inhibited (Graham and Elphick 1990), demonstrating that H_2 cannot be the activating species.

The possibility exists, however, that Ta may also act as an H_2 getter (Yntema and Percy 1954). In order to circumvent this possibility, albite was exposed to an H_2-rich, H_2O-poor fluid in a further experiment by the authors. Dried Amelia albite was placed in a small graphite capsule fitted with a graphite lid, and sealed in a closed, second Pt capsule together with stearic acid and CaC_2 to excess. This double capsule was then embedded in a salt cell, and run at 16 kbars, 850°C for 24 hours. Under these P-T conditions in the NaCl pressure cell, albite would normally reach its maximal state of disorder. During heating, the following reactions occur:

$$CH_3(CH_2)_{16}.COOH + 2CaC_2 = 22C + 18H_2 + 2CaO$$

$$C + 2H_2 = CH_4.$$

The presence of an inner carbon capsule avoids kinetic problems with graphite precipitation. Under these run conditions, $f(H_2O) = 0.6E\text{-}4$ bars; $fH_2 = 0.8E4$ bars, $f(CH_4) = 0.2E7$ bars, $f(C_2H_6) = 0.4E6$ bars. After the anneal, the charge was carefully examined for the presence of unreacted CaC_2 and the albite checked by XRD. The Al-Si disorder state of the albite was found to be close to that of the starting albite, confirming the negative effect of H_2 in Al-Si interdiffusion rate enhancement.

The effect of fluid chemistry on the extent of disordering of albite was investigated by performing hydrothermal anneals at 1 kbar, 850°C, for 48 hours, in the presence of aqueous NaCl solutions of varying molarities (Graham and Elphick 1990). The greatest disordering was shown by the albite annealed in the most saline fluid, in which $[H^+]$ should be highest (Eugster 1986), providing positive evidence for the rate-enhancing role of protons in the Al-Si interdiffusion process. The absence of major solution-reprecipitation under these experimental conditions has already been demonstrated by ion imaging (Fig. 4), confirming that Al-Si interdiffusion enhancement occurred by a solid-state process without the necessity for high pressures.

The experiments described in this section have been designed to generate extreme conditions of high aH^+ or fH_2 in order to attempt to isolate the possible hydrogen-bearing species responsible for intracrystalline rate enhancement, and thus gain some insight into the interdiffusion mechanism. These and similar techniques have been used in an attempt to gain a similar insight into the mechanisms of oxygen diffusion in feldspars and quartz.

6. The Role of Hydrogen in Oxygen Diffusion in Feldspars

6.1 INTRODUCTION

The framework of feldspar comprises Si, Al and O. Co-operative Si-Al interdiffusion has been convincingly shown to be enhanced in a high $[H^+]$ environment. The diffusivity of oxygen in feldspars and quartz provides another avenue of investigation into the role of hydrogen in intracrystalline transport processes.

6.2 FIRST ORDER EFFECTS: WET VERSUS DRY OXYGEN DIFFUSION

Yund and Anderson (1978), Norton and Taylor (1979), and Freer and Dennis (1982) all drew attention to the high diffusivities and low activation energies (typically 100-150 kJ mol^{-1}) for oxygen diffusion in feldspars under wet (hydrothermal) conditions compared to the low diffusivities and high activation energies (300-400 kJ mol^{-1}) associated with oxygen diffusion under dry conditions. The wet and dry diffusivities were determined by different experimental and analytical methods. The "wet" data pertained to experiments in which diffusion profiles were measured in single crystals of feldspar and quartz previously annealed in the presence of $H_2{}^{18}O$, using an ion microprobe in depth-profile mode (eg Giletti et al 1978; Freer and Dennis 1982; Dennis 1984a; Giletti and Yund 1984) to measure the short (<1.5μm) diffusion profiles in the surfaces of the crystals directly (the "profiling method" of Giletti (1985 and this volume)). Oxygen diffusion profiles in quartz under dry conditions were also measured by ion microprobe depth-profiling, but exchange was effected using dry $^{18}O_2$ gas at 1 atmosphere (Dennis 1984b). Dry diffusion data for anorthite feldspar were initially obtained in integrating-type experiments in which powdered mineral samples were exchanged with dry $^{18}O_2$ or $C^{18}O_2$ gas at 1 atm and high (>1000°C) temperatures, followed by mass spectrometric analysis of gas for $^{18}O/^{16}O$ after exchange (eg Muehlenbachs and Kushiro 1974). In the latter type of experiment there is inherent uncertainty about whether volume diffusion is the process being measured. Oxygen diffusion in anorthite under dry conditions was recently remeasured by the profiling method, using single crystals exchanged with dry $^{18}O_2$ gas down to 850°C (Elphick et al 1988). These experiments provide direct comparison with hydrothermal experiments at similar temperatures, and direct ion microprobe depth-profile analysis of the resulting short $^{18}O/^{16}O$ diffusion profiles (Fig. 8).

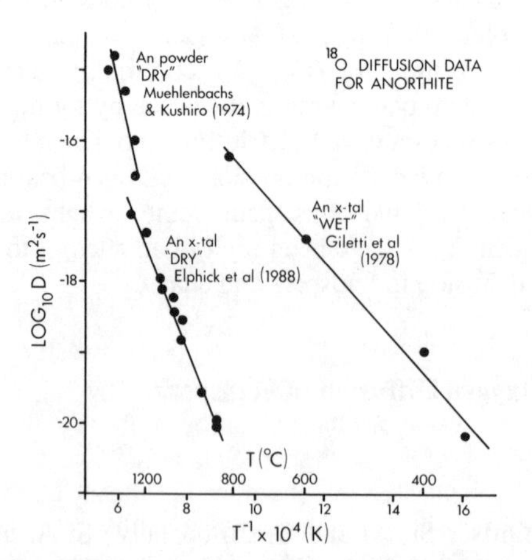

Figure 8. Arrhenius plot comparing diffusivity of oxygen in anorthite under hydrothermal conditions (data of Giletti et al 1978), and dry conditions (data of Muehlenbachs and Kushiro 1974; Elphick et al 1988). From Elphick et al (1988).

The unequivocal conclusion of these studies is that a mobile hydrogen-bearing species mediates in the oxygen jump mechanism in feldspar and quartz when water is present, in such a way as to enhance diffusion rates by orders of magnitude relative to dry conditions. The dry data may perhaps represent *intrinsic* diffusion, and the hydrothermal data *extrinsic* diffusion (Foland, Giletti, this volume). The identity of this rate-enhancing hydrogen species may in principle be ascertained in experiments in which the diffusivity of oxygen is measured under extreme fugacities of relevant hydrogen species, as has been done to aid identification of the rate-enhancing hydrogen species for Al-Si interdiffusion. Here we describe these experiments as "first order", in comparison to the "second order" experiments reviewed below in which the activities of relevant hydrogen-bearing species were varied *continuously* over some more limited range.

Experiments of the first-order type were conducted for quartz (Elphick and Graham 1988), in the presence in turn of H_2O and H_2. An ^{18}O diffusion profile was established in a polished quartz crystal annealed hydrothermally in $H_2{}^{18}O$ in a sealed Pt capsule at 700^oC, 1 kbar, and the profile measured by ion microprobe. A portion of the crystal was annealed again in direct contact with dry, pressurised H_2 in an internally heated pressure vessel at 800^oC, 1.5 kbars. The ^{18}O diffusion profile was re-measured, and found to be identical within error to the first. Had H_2 been capable of providing the enhancing hydrogen impurity to the quartz, the second anneal would have produced a much longer diffusion profile. It did not. Elphick and Graham (1988) proposed that oxygen diffusion in framework silicates is enhanced by high $[H^+]$ but inhibited by high $f(H_2)$ in an analogous manner to Al-Si interdiffusion in feldspars. They further proposed that the mediating proton species is a fast transient species which is highly mobile within the quartz lattice and may therefore be unquenchable and differ from other hydrogen species (charge compensating point-defect hydrogen or molecular water) present in quenched quartz samples. Although the pertinent experiments have not yet been conducted on feldspars, quartz probably represents a good working analogue for the likely behaviour of oxygen in feldspar under similar experimental conditions.

Enhancement of Al-Si interdiffusion in feldspars sealed in Pt capsules under fluid-absent conditions in the solid media apparatus using NaCl and related pressure media has been shown above to be the result of greatly enhanced proton activities in the experimental environment (Goldsmith 1987, 1991; Elphick and Graham 1990). In this environment a high aH^+ is achieved but H_2O is effectively eliminated. This technique has been exploited to measure the diffusivity of oxygen in α-quartz in comparison with the results with those for oxygen diffusion at low pressures under hydrothermal conditions.

Polished crystals of quartz were embedded in dried $CaC^{18}O_3$ in sealed Pt capsules (Elphick and Graham 1992), or an ^{18}O signature was introduced into the quartz as a surface overgrowth during hydrothermal pre-annealing at 700^oC, 1 kbar, prior annealing in sealed Pt capsules in an NaCl pressure medium at 16 kbars, 500-850^oC (Graham and Elphick 1991). Resultant ^{18}O diffusion profiles were determined by ion microprobe. Oxygen diffusivities were found to be in excellent agreement with 1 kbar hydrothermal data of Elphick et al (1986b), where an overgrowth technique had been used to prevent surface dissolution during annealing (Fig. 9). The main conclusion is that the crystal behaved *as if in contact with water* although water was absent. Crystal surfaces were intact, ensuring that oxygen transport was by volume diffusion. Inasmuch as the principal variables aH^+ and $f(H_2O)$ were separated by eliminating water from contact with the quartz, enhancement of oxygen diffusivity at high pressures is the result of high proton activity and not high $P(H_2O)$. In addition the agreement between 1 kbar and 16 kbar data demonstrate conclusively that there is no lattice or isostatic pressure effect. If it was argued that traces of water were still in fact present in either the $CaCO_3$ or the quartz sufficient to generate free H_2O at 16 kbars despite precautions taken to dry the samples before experiments, then the water fugacities were coincidentally equivalent to fugacities at 1 kbar in each experiment, which seems unlikely. On the other hand if the free water was present and subject to the same pressure as the experimental or confining pressure, then there is no $P(H_2O)$ effect on oxygen diffusivity. Parallel experiments with feldspar are currently being undertaken to demonstrate the same result as was achieved with quartz, but at present we must argue that quartz can be used as a framework silicate analogue for feldspar.

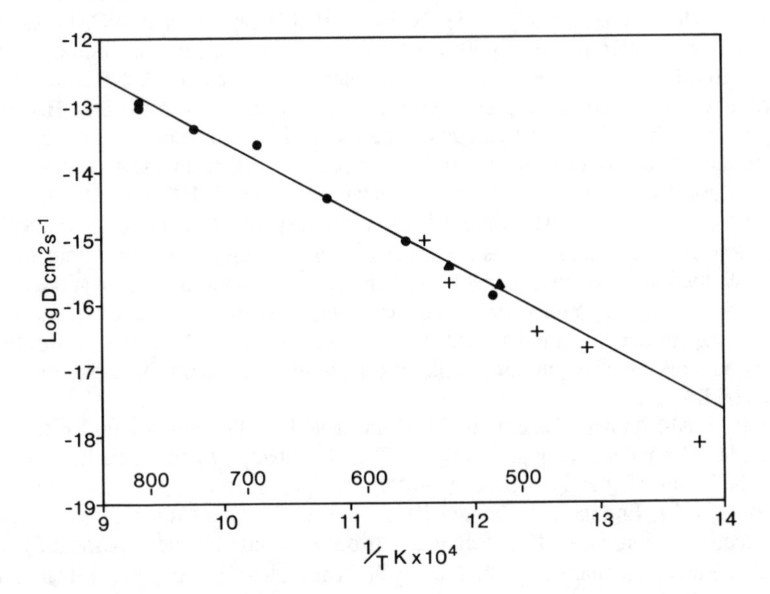

Figure 9. Arrhenius plot comparing diffusivity of oxygen in α-quartz determined at 16 kbars under "dry" conditions in piston cylinder using a NaCl pressure medium (circles)(Elphick and Graham 1992), and at 1 kbar under hydrothermal conditions (triangles)(Elphick et al 1986). Line is least-squares regression through these data. Also shown for comparison (crosses) are results of Farver and Yund (1991) for hydrothermal conditions.

6.3 SECOND ORDER EFFECTS: $P(H_2O)$?

The dramatic first-order difference between oxygen diffusion under "wet" versus "dry" conditions for feldspars is unequivocally a difference between hydrogen-present and hydrogen-absent diffusion, indicating that a mobile hydrogen species mediates in the oxygen jump mechanism. By analogy with quartz, H^+ is the likely mediating species.

The second-order effect of $P(H_2O)$ on the magnitude of oxygen diffusivity in feldspars under hydrothermal conditions, while presumably the result of the same hydrogen-bearing impurity in the feldspar lattice, has long been controversial, with different studies reaching opposed conclusions. These studies are referred to here as "second order" studies, in which the activities of relevant hydrogen-bearing species in OH-bearing fluids have been varied continuously over a more restricted range of values than in the first-order experiments described above. Here we review and discuss these studies and their mechanistic implications, and attempt to identify possible sources of the discrepancies between the results of these studies.

Yund and Anderson (1978) found that oxygen diffusivity in adularia feldspar increased in proportion to water pressure up to 4 kilobars at 650°C in integrating-method exchange experiments between sized adularia powders and ^{18}O-labelled KCl solution. However, the linear relationship between D and $\sqrt{f(H_2O)}$ (Yund and Anderson 1978; Norton and Taylor 1979; Graham and Elphick 1991; Fig. 2) implies the involvement of a species derived from the dissociation of water (eg $H_2O = H^+ + OH^-$) in the exchange mechanism, whether this mechanism involves solution-reprecipitation, surface hydrolysis or H^+-enhanced diffusion.

In conflict with the results of Yund and Anderson (1978), Freer and Dennis (1982) failed to find any pressure dependence for the diffusivity of oxygen in albite under hydrothermal conditions between 1 and 8 kilobars at 600°C, in experiments conducted in internally-heated pressure vessels, using an argon pressure medium. They noted that the surface concentration of ^{18}O in the albite crystals increased with increasing $P(H_2O)$, and suggested that it is the surface exchange mechanism and not the oxygen diffusivity which changes with pressure (Fig. 10). Ewald (1985) interpreted the pressure dependence of oxygen diffusion in feldspar kinetically in terms of a two stage process, in which the increase of rate with pressure is the result of increased surface concentration of ^{18}O due to increasing density of the fluid phase, while the volume diffusion stage is pressure-independent. Dennis (1984a) found that oxygen diffusivity in quartz between 100 bars and 1 kilobar $P(H_2O)$ was independent of water pressure, whereas Giletti and Yund (1984) report a strong dependence of oxygen diffusivity in quartz on $f(H_2O)$ between 0.25 and 3.5 kilobars, reflecting the increase in some hydrogen-bearing impurity such as OH^-, H^+, H_2O, or H_3O^+ with increasing $P(H_2O)$.

Recently, Farver and Yund (1990, 1991) have also reported strong positive correlations between oxygen diffusivity and $P(H_2O)$ for adularia, albite and quartz. Oxygen diffusion in single-crystal albite and adularia was studied hydrothermally from 50 bars to 15 kilobars using a range of oxygen buffers, a hydrogen ion buffer, and variable mole fractions of H_2O. The purpose of this study was to test the influence of the fugacities of H_2O, H_2 and O_2, the activity of H^+, and the total confining pressure on oxygen diffusivity. Solid oxygen buffers spanned the range of wustite-magnetite, nickel-nickel oxide, and Mn_3O_4-Mn_2O_3, and hydrogen ion activities were buffered by the Ag-AgCl buffer in two experiments. Water fugacity was also varied by introducing CO_2 from the breakdown of $Ag_2C_2O_4$ in some runs. By varying the hydrogen, oxygen and water fugacities, confining pressure and hydrogen ion activity, the effect of each of these parameters on oxygen diffusivity was assessed.

Farver and Yund found a strong but non-linear positive correlation of log D with log $f(H_2O)$ for adularia (Fig.11) and quartz, and concluded from this that the diffusing oxygen-bearing species in these framework silicates was molecular water. Although a strong positive correlation between confining pressure and diffusivity is observed, this relation breaks down at very low pressures when confining pressure and water pressure are varied independently by varying fO_2 and $X(CO_2)$. No correlation of diffusivity with fO_2 or fH_2 was found, and the strong correlation between diffusivity and hydrogen ion activity (Fig.12) was considered to be simply an artifact of

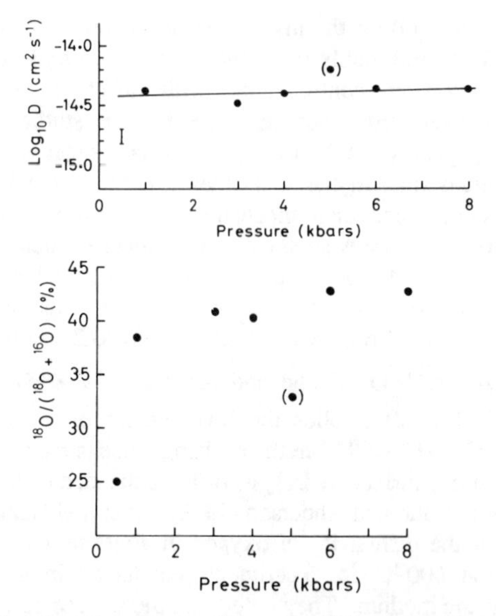

Figure 10. (above). Diffusivity of oxygen in albite at 600°C under hydrothermal conditions determined as a function of fluid pressure. Line shows least-squares fit of data excluding 0.5 and 5 kbar data. (below). Surface oxygen isotope composition of albite as a function of fluid pressure after hydrothermal exchange at 600°C. From Freer and Dennis (1982).

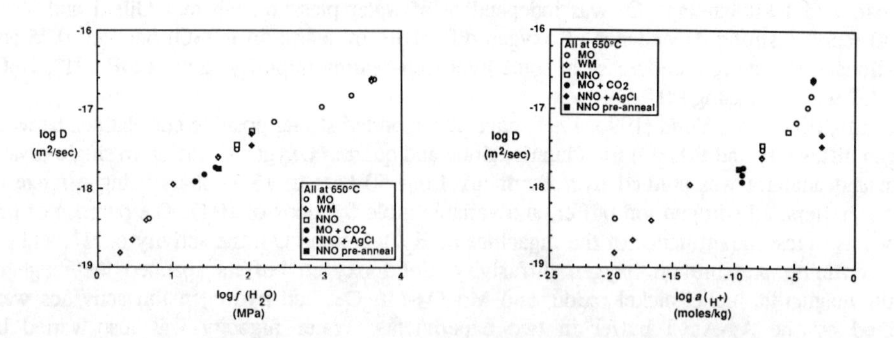

Figures 11 and 12. The diffusivity of oxygen in adularia feldspar at 650°C as a function of water fugacity (left) and hydrogen ion activity (right). Symbols refer to different buffering assemblages, as shown. MO = Mn_2O_3-Mn_3O_4; WM = wustite + magnetite; NNO = nickel + nickel oxide; AgCl = Ag + AgCl. From Farver and Yund (1990).

the primary correlation between $f(H_2O)$ and aH^+. In support of the absence of an aH^+ effect, experiments in which aH^+ was increased by about three orders of magnitude relative to water-only experiments in the presence of Ag-AgCl buffer did not show any proportional increase in oxygen diffusivity.

The resolution of the disagreement amongst different second-order studies of $P(H_2O)$ effects on oxygen diffusivity is of some importance for the identification of the mechanism by which water enhances oxygen diffusivity, and for the identity of both the hydrogen-bearing rate-enhancing species and the oxygen-bearing species involved in oxygen transport in the mineral lattice. The first-order experiments designed to identify the rate-enhancing species, which achieved a rigorous separation of the key variables ($f(H_2O)$, aH^+, $f(H_2)$) summarised above, have demonstrated convincingly that a high concentration of protons in the mineral lattice achieves this first-order effect of rate enhancement, but these experiments do not unambiguously constrain the oxygen transporting species. Current discrepancies in results and interpretations of experiments reviewed here which attempted to constrain these second order $P(H_2O)$ effects are now considered under several headings.

6.3.1. *Experimental Problems.* Central to the interpretation of data from any experimental diffusion study is the demonstration that the process being measured is volume diffusion rather than surface reaction, as discussed above. To summarise, the oxygen isotope exchange experiments with adularia and aqueous solution used by Yund and Anderson (1978) to demonstrate a positive correlation between diffusivity and $P(H_2O)$ might be re-interpreted to be the result of surface hydrolysis in which the measured extent of isotopic exchange shows a dependence on water fugacity, while Ewald (1985) proposed that the measured pressure effect relates to the increase in surface concentration of oxygen with pressure resulting from the increasing density of aqueous solution. Positive correlation between diffusivity and water fugacity in oxygen diffusion studies is therefore subject to various possible interpretations. Hydrothermal isotope exchange experiments using powdered mineral samples with a large solid surface area are particularly susceptible to surface reaction, and the method cannot by its nature distinguish diffusion from solution-reprecipitation (Giletti 1985).

The ion microprobe provides a potentially ideal means of ready and direct measurement of short diffusion profiles in polished or cleaved single crystal surfaces exchanged hydrothermally (the profiling method), yielding both direct measurement of crystallographic controls on diffusional anisotropy (eg Dennis 1984a; Giletti and Yund 1984) and information on surface solution and reaction processes through direct measurement of surface isotopic and chemical compositions. However, when surface reaction is a problem the length-scales of diffusion (0.1 to 5 μm) are sufficiently small that physically or chemically induced surface reaction may affect a significant proportion of the near-surface region of a crystal in which diffusion profiles are measured. Surface processes may include solution-reprecipitation, recrystallisation of highly strained surface layers generated by mechanical polishing, or dislocation or damage enhanced diffusion. An overgrowth/interdiffusion technique to circumvent this problem was proposed by Elphick et al (1986), and dry encapsulation in NaCl pressure cells in the piston-cylinder at high pressures (Goldsmith 1987; Elphick and Graham 1992; see above) offers a novel way of simulating hydrous environments.

Dennis (1984a) attributed time-dependent changes in the diffusivity of oxygen in quartz in Giletti and Yund's (1984) study to possible foreshortening of profiles by progressive surface dissolution and resultant small differences in diffusivity. Elphick et al (1986b) also suggested such a possibility to explain small differences in activation energy for oxygen diffusion in β-

quartz between their interdiffusion experiments in which dissolution of the polished crystal surface was prevented by an overgrowth of ^{18}O-enriched quartz on the surface of the crystal, and the exchange experiments of Dennis (1984a) and Giletti and Yund (1984) in which the polished crystal surface was in direct contact with the aqueous fluid. These small differences at the experimental temperatures nonetheless may potentially lead to larger differences in diffusivities when the data are extrapolated down-temperature to geologically relevant conditions.

The magnitude of the $f(H_2O)$ effect measured by Farver and Yund (1990) for adularia spans over two orders of magnitude in diffusion coefficient over the pressure range investigated, and would appear to be too large to be explained by surface reaction processes. In addition, Farver and Yund (1990) report no variation in measured surface concentrations of ^{18}O, nor any step functions in the depth profiles which might indicate surface reaction. Thus the discrepancy between the results of different studies and laboratories regarding the reality of a water pressure effect on diffusivity cannot at present be ascribed to any obvious experimental artifact.

6.3.2. *Interdependence of Variables in H_2O-bearing Systems*. Farver and Yund (1990, 1991) were persuaded by the correlation of log D and log $f(H_2O)$ for oxygen diffusion in adularia and quartz to conclude that molecular water was the oxygen-transporting species in oxygen diffusion in framework silicates. Attribution of the relationship between log D and log (aH^+) to the strong (but non-linear) correlation of $f(H_2O)$ with aH^+ for hydrous fluids, and normalisation of D values to a constant $f(H_2O)$, depends heavily upon the results of two experiments in which aH^+ was increased by the presence of an Ag-AgCl buffer with no concomitant increase in oxygen diffusivity.

Experiments in which water-rich fluid is always present do not permit a rigorous separation and independent consideration of the key experimental parameters $f(H_2O)$, $f(H_2)$, $f(O_2)$ and aH^+, and are therefore dependent upon the identification and inter-comparison of trends. Log-log plots enhance apparent correlations, and inevitably obscure power-law relationships. The *non*-linearity of the plots of log D against aH^+ and $f(H_2O)$ (Figs. 11, 12) may reflect the operation of more than one mechanism in the rate-determining step. It is likely that the enhancement of oxygen diffusion by protons is a non-linear process to the extent that there may be a critical but low concentration of protons required to enhance the oxygen diffusion process (Chacko and Goldsmith 1988; Farver and Yund 1990; Graham and Elphick 1990). Such a threshold in proton concentration would explain why increased proton concentrations in experiments in which proton activity was buffered to high values by Ag-AgCl did not enhance oxygen diffusivity relative to unbuffered runs (Farver and Yund 1990). As protons are progressively incorporated onto the feldspar (or quartz) lattice, the first protons to be incorporated may have a much larger effect in enhancing the oxygen jump mechanism than subsequent protons. Such a process will depend on the solubility of hydrogen on the lattice and/or the mobility of hydrogen through the lattice.

Other power-law relationships could also be derived from the experimental data of Farver and Yund (1990). For example, when the data for oxygen diffusion in adularia are replotted against $\sqrt{f(H_2O)}$, a significant improvement in correlation is achieved compared to the equivalent $f(H_2O)$ plot (Fig. 13), implying the involvement of some species derived from the dissociation of water in the activating mechanism. Such a relationship could be interpreted in terms of hydrolysis of crystal surfaces, solution-reprecipitation or a proton-enhanced oxygen transport mechanism. We are not arguing here that any specific mechanism can be identified from these data, but simply make the important point that $f(H_2O)$ effects deduced from trends in plots of diffusivity against $f(H_2O)$ may plausibly be re-interpreted in other ways, implying quite different

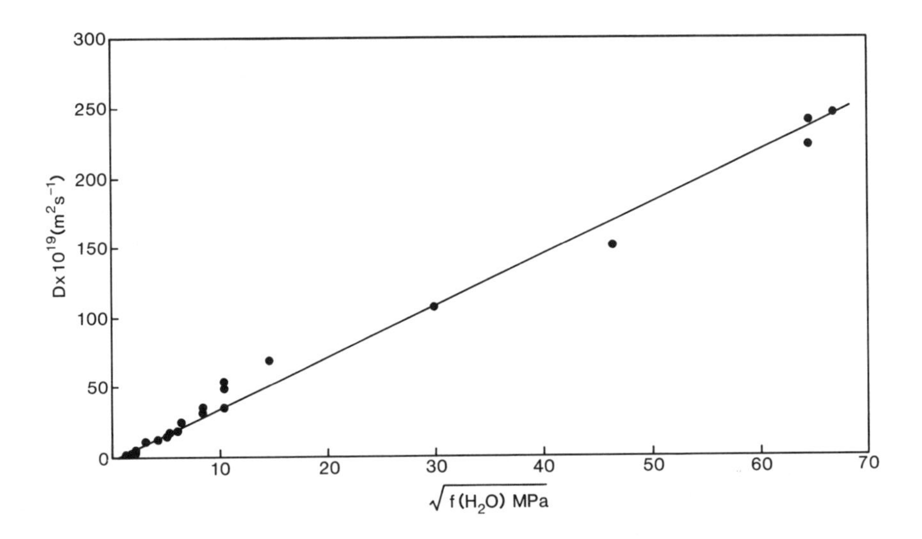

Figure 13. Replot of data of Farver and Yund (1990) for diffusivity of oxygen in adularia at 650°C (from Fig. 11) plotted against the square root of the water fugacity.

mechanisms or transport species. These second-order experiments either provide negative evidence, which is nevertheless important, or are ambiguous.

In conclusion, it is not possible to deduce the rate-determining oxygen transport mechanism or the oxygen-transporting species *unambiguously* from the second-order experiments, although the conclusion that H_2O is the oxygen-transporting species is plausible (see also Zhang et al 1991). There is no a priori reason to expect that the fugacity of the hydrogen-bearing impurity which enhances oxygen diffusion should bear a linear relationship to the oxygen diffusivity.

6.3.3. *Discussion.* In considering the mechanism of oxygen diffusion enhancement it is important to distinguish the rate-enhancing species from the oxygen-bearing species being transported. In this connection, comparative data on the mobility of oxygen and hydrogen in feldspars provide important constraints. Kronenberg and Yund (1988) and Kronenberg et al (1989) have annealed natural crystals of adularia containing molecular water for varying lengths of time at 600-800°C in order to eliminate the hydrogen-bearing impurity. Diffusivities calculated from the rate of loss of hydrogen are three to four times faster than diffusivities of oxygen in feldspar, and are comparable to those of proton transport in quartz (Kats et al 1962). The introduction of this defect into feldspars under high $f(H_2O)$ or high aH^+ conditions will inevitably be very much faster than the transport rate of the oxygen-transporting species. These data imply that enhancement of oxygen diffusion in feldspars by hydrogen-bearing impurities is likely to be a two-stage process of the type proposed by Donnay et al (1959), with initial rapid introduction of the mobile hydrogen defect into the feldspar followed by slower transport of the oxygen-bearing species.

The transport of protons through silicates has been discussed by Ernsberger (1983). The bare proton has no independent existence in a silicate mineral, but rather the proton will reside within the outer electron shell of the oxygen atoms. Framework silicates have an extended and continuous three-dimensional oxygen sub-lattice in which we can envisage that protons are readily transported, with potentially high mobility. The progressive absorption of protons by an oxygen ion to successively produce OH^- , H_2O and H_3O^+ is accompanied by a progressive *decrease* in the effective volume of the oxygen. The presence of one or more protons in the outer electron shell of an oxygen in a silicate could then modify the jump frequency of the oxygens without modifying the diffusion mechanism; in other words, in terms of the Arrhenius relationship:

$$D = D_o \exp(-Q/RT)$$

Q (the activation energy) will be modified by the availability of protons, while D_o will be unmodified. We note from a comparison of the diffusivity of oxygen in feldspar and quartz under wet (proton present) and dry (proton absent) conditions (Table 2) that it is indeed Q that is modified, with the presence of hydrogen impurities decreasing the activation energy (increasing the jump frequency) as predicted, while D_o is unchanged.

TABLE 2. Comparison of oxygen diffusivity in feldspar and quartz under dry and hydrothermal conditions

	hydrothermal		dry	
	D_o	Q	D_o	Q
anorthite	$2\text{x}10^{-11}$	109	$9\text{x}10^{-10}$	234
	(1 kbar, 400-600°C)		(1 bar, 800-1300°C)	
quartz	$2\text{x}10^{-11}$	138	$3\text{x}10^{-11}$	222
	(1 kbar, 600-850°C)		(1 bar, 600-850°C)	

Data from Giletti et al (1978) and Elphick et al (1988) for anorthite, and Dennis (1984a,b) for quartz. Table after Elphick and Graham (1988). Units D_o (m^2s^{-1}), Q ($kJ\ mol^{-1}$).

Regarding the identity of the rate-enhancing species, consideration of the evidence of the first-order experiments indicates that protons are the relevant mobile species. Regarding the oxygen-transporting species involved in the diffusion process, one can be less certain. Molecular water is one possible species, which would be compatible with recent infra-red spectroscopic studies showing the presence of molecular H_2O in cation sites and with the recent oxygen diffusion studies of Farver and Yund (1990). However, this conclusion is not supported by ab initio calculations of the stabilities of hydrogen-bearing species in α-quartz (McConnell, pers comm; Lin et al, in prep), which indicate unambiguously that interstitial molecular water has very much higher energy than the hydrogarnet defect. It is not obvious that one unique species need be transported intact through the feldspar lattice during the oxygen diffusion process. More likely is the possibility that the mobile oxygen transiently exists in more than one form or species during the progress of oxygen diffusion.

7. The Effect of Water on Cation Diffusion

The effect of hydrogen-bearing impurities on the diffusion of cations in feldspars has been investigated by undertaking homogenisation experiments on perthitic alkali feldspars under dry and hydrothermal conditions (Yund 1983, 1986; Brady and Yund 1983; Hokanson and Yund 1986). No evidence for enhancement of interdiffusion has been observed. The enhancement effects induced by hydrogen-bearing impurities described above relate to O, Al and Si - the framework-forming elements. There is therefore no a priori reason to anticipate a similar effect of hydrogen on cation diffusion.

By contrast, interdiffusion of the Ca^{2+} and Na^+ cations in plagioclases is coupled to Al-Si interdiffusion in order to maintain charge balance. Interdiffusion of NaSi and CaAl is likely to be limited by the rate of Al-Si interdiffusion as the rate determining step, and therefore some dependence on hydrogen-bearing impurities would be expected, by analogy with experimental evidence described above. Again, NaSi-CaAl interdiffusion rates have been determined experimentally by homogenisation of fine intergrowths or lamellae in plagioclase feldspars, and examination by TEM. A comparison of the results of Grove et al (1984) for the homogenisation of Huttenlocher intergrowths (An_{70-90}) in air, and those of Yund (1986), who homogenised peristerite (An_{0-26}) intergrowths with water at 15 kbars, indicates that the difference of about five orders of magnitude (Fig. 14) is attributable either to confining pressure or to the activity of some hydrogen-bearing defect or impurity.

Yund and Snow (1989) examined the effects of confining pressure and hydrogen fugacity at 1000°C by repeating the homogenisation experiments on peristerite lamellae with three types of experiment: annealing experiments in H_2 atmospheres, unbuffered annealing experiments with small amounts of added water at 2.5-15 kbars, and experiments at three different oxygen buffers (Mn_3O_4-Mn_2O_3; hematite-magnetite and wustite-magnetite). The latter experiments generated a wide range of fO_2 and fH_2 at fairly constant fH_2O. The interdiffusion rate is reported to increase with increasing fH_2 and with confining pressure (Fig. 15), and enhancement by some hydrogen-bearing defect is proposed. This analysis, however, ignores the possibility that the small amounts of added water may enhance the interdiffusion rate at high pressures via a proton enhancement mechanism. Accordingly, Snow (1989) has re-interpreted these experiments in terms of a proton enhancement mechanism, and has reported that diffusivity is a linear function of aH^+ above 7 kbars for the unbuffered experiments, reflecting the increased ionic dissociation of water with increasing pressure. These results are mechanistically consistent with those of Goldsmith (1987) and Graham and Elphick (1990) for Al-Si interdiffusion in albite.

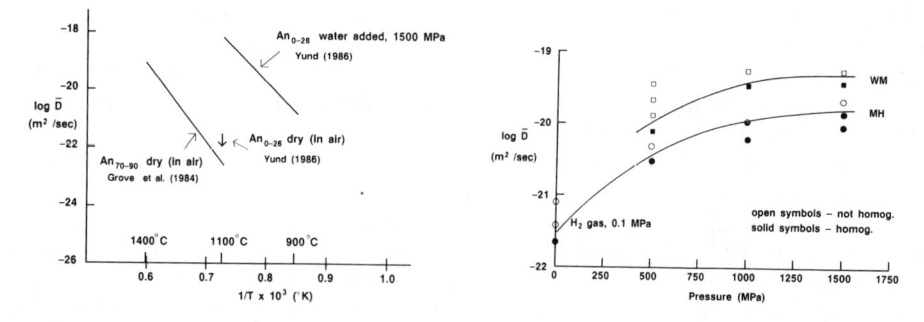

Figure 14 (left). Arrhenius plot comparing NaSi-CaAl interdiffusion in An_{70-90} under dry, low pressure conditions (Grove et al 1984) and in An_{0-26} under water added conditions at 15 kbars (Yund 1986), using the lamellar homogenisation method. From Yund and Snow (1989).

Figure 15 (right). Plot of NaSi-CaAl interdiffusion versus confining pressure for hydrogen fugacities defined by wustite-magnetite (WM) and hematite-magnetite (HM). 0.1MPa H_2 gas data are at fH_2 close to MH. From Yund and Snow (1989).

8. Properties of Eifel Sanidine: A Natural Example of the Catalytic Role of Hydrogen in Feldspars?

No review of the role of hydrogen in the transport properties of feldspars would be complete without some mention of unusual order-disorder properties of Eifel sanidine. These sanidines, occurring as large (kg-size) megacrysts from the Volkesfeld area in the Eifel region of Germany, have been found to show unusually rapid Al-Si order-disorder kinetics in annealing experiments at temperatures as low as 750°C, based on optic axial angle measurements and lattice parameters (Bernotat-Wulf et al 1988). They also have particularly perfect structures with exceptionally low dislocation density. Beran (1986) has identified the presence of molecular water occupying the cation sites (about 0.046 wt%) from infra-red spectroscopic data.

Slabs of sanidine were annealed at 850-1050°C and times from 0.5 to 1600 hours, and 2V measured at various distances from the surface of slabs cut parallel to (010) (Bernotat-Wulf et al 1988). The near-surface regions were found to disorder much more slowly than the centres of slabs (Fig. 16). When the data are fitted to a simple first-order rate law of the type found empirically to describe Al-Si disorder of albite (Graham and Elphick 1990), the rate-constant is observed to be time dependent, with a different time-dependence in the near-surface region (<0.05mm from surface) than in the central regions (>0.05mm from surface)(Fig. 17). Powdered samples of Eifel sanidine show the same sluggish transformation behaviour as the outer regions of the slabs.

The time-dependent rate-constant for this material during annealing implies that the defect chemistry and structure of the material is actively changing during annealing, and it is tempting to suggest by analogy with Al-Si disordering experiments on albite that a hydrogen impurity is being lost through the surfaces of sanidine slabs during the progress of the anneal. Confirmation of the identity of this impurity remains elusive at the present time however.

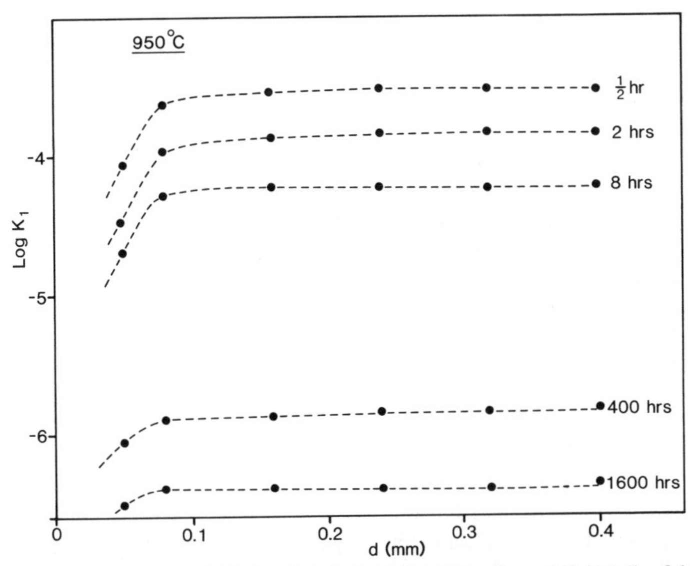

Figure 16. Plot of rate constant (K_1) for disordering of sanidine from Eifel/Volkesfeld (as measured by optic axial angle $2V_x$) as a function of distance d from the sample surface of slabs cut parallel to (010) and annealed at 950°C for various times as shown. Data from Bernotat-Wulf et al (1988).

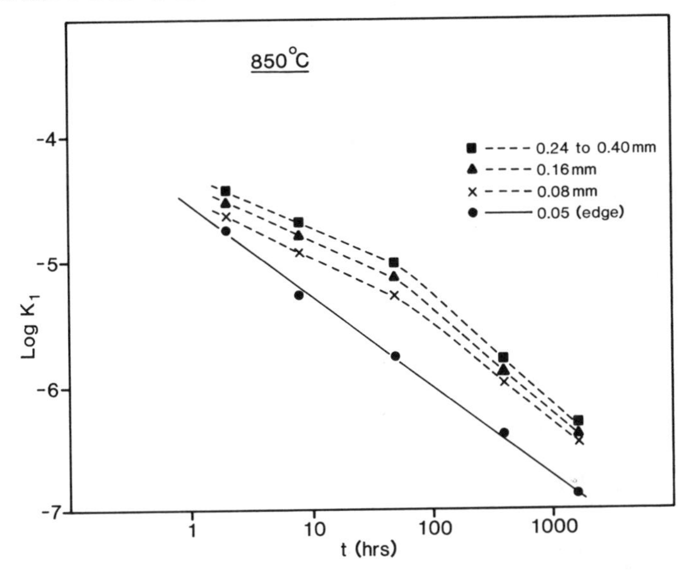

Figure 17. Plot of the variation of rate constant (K_1) with time (t hrs) for disordering of Eifel/Volkesfeld sanidine as a function of distance (d) from the surface of slabs cut parallel to (010) and annealed at 850°C for various times. Note the different kinetic and disordering behaviour of the surface of the slab compared to the interior. Data from Bernotat-Wulf et al (1988).

9. Hydrogen in Feldspars: Conclusions and Prospects for Future Progress

Classical experimental techniques employed to date have successfully provided quantitative constraints on oxygen diffusivities and Al-Si interdiffusion kinetics which are vital for the calculation of rates of geological processes and the interpretation of stable isotope data from feldspar-bearing crustal rocks (eg Giletti 1986). These diverse experiments have also provided a useful guide to possible hydrogen-bearing activating species in the enhancement of diffusion processes, but are by their nature incapable of unambiguously identifying diffusing species and diffusion mechanisms. Thus the controversies initiated by Donnay et al (1959) and other early workers remain unresolved. Further progress in constraining the identity of the hydrogen and mobile oxygen species involved in these intra-crystalline transport mechanisms is likely to come from in-situ spectroscopic or electrical conductivity measurements conducted at high P and T, combined with ab initio or molecular dynamic calculations of mineral and defect structure.

10. Footnote: Water in Feldspars at High P(H_2O), and Feldspar Hydrates

The solubility of water in feldspar as a function of P(H_2O) and T is not known. Farver and Yund (1990) have suggested that water concentrations might correlate with water fugacity, as has been reported for quartz (Cordier and Doukhan 1989). At high water pressures, the large-cation feldspars, sanidine and celsian, break down to a pseudo-hexagonal mineral with a double-layer

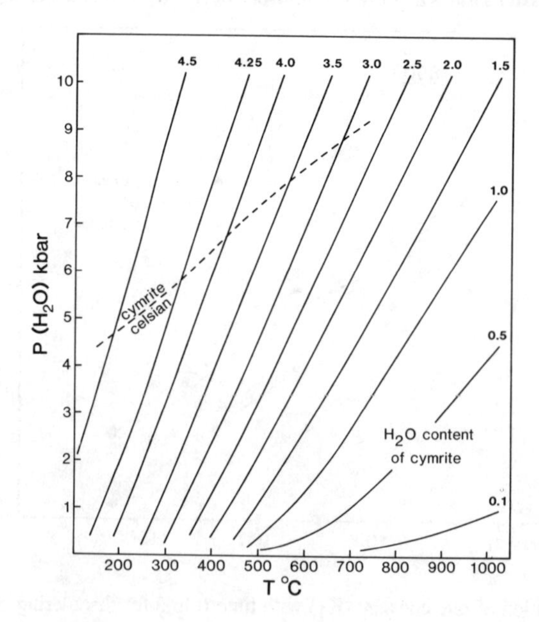

Figure 18. Variation of H_2O content of cymrite ($BaAl_2Si_2O_8.nH_20$; where 0<n<1) as a function of P(H_2O) and T. Contours in wt% H_2O. Dashed line shows the position of the equilibrium: $BaAl_2Si_2O_8.nH_2O = BaAl_2Si_2O_8 + nH_2O$. After Graham et al (1992).

structure (Drits et al 1975; Viswanathan et al 1992), represented by the minerals sanidine hydrate, $KAlSi_3O_8.H_2O$ (Seki and Kennedy 1964) and cymrite, $BaAl_2Si_2O_8.nH_2O$ ($0<n<1$) (Graham et al 1992; Viswanathan et al 1992). The latter mineral has been shown to correspond to hexacelsian in structure (Viswanathan et al 1992) and to have a water content which varies systematically with pressure and temperature (Fig. 18). Little is known about the structure and chemistry of "sanidine hydrate", although it has the same X-ray diffraction pattern as cymrite and hexacelsian. There is no reason to doubt that it may have the same variable water content. The upper $P(H_2O)$ stability limit of the feldspar structure increases from less than 10 kbars for the barium end-member (Fig. 18) to over 20 kbars for the potassium end-member (Seki and Kennedy 1964).

Acknowledgements

We firstly thank Ian Parsons and his organising committee for organising a stimulating meeting. The first draft of this chapter benefitted from helpful reviews from Bruno Giletti and Ken Foland. Des McConnell and David Price drew our attention to the importance of ab initio and molecular dynamic calculations in constraining the stabilities of hydrogen-bearing impurities in minerals. Research undertaken in the Ion Microprobe and Experimental Petrology Laboratories at the University of Edinburgh is supported by NERC. We thank John Craven for his assistance with depth-profile analysis using the ion microprobe, and Bob Brown for support in the experimental petrology laboratories.

References

Beran A (1986) A model of water allocation in alkali feldspar, derived from infrared spectroscopic investigations. Phys Chem Minerals 13:306-310

Beran A (1987) OH groups in nominally anhydrous framework structures: an infrared spectroscopic investigation of danburite and labradorite. Phys Chem Minerals 14:441-445

Bernotat-Wulf H, Bertelmann D, Wondratschek H (1988) The annealing behaviour of Eifel sanidine (Volkesfeld) III. The influence of the sample surface and sample size on the order-disorder transformation rate. N Jb Miner Mh 11:503-515

Brady JB, Yund RA (1983) Interdiffusion of K and Na in alkali feldspars: homogenization experiments. Am Mineral 68:95-105

Chacko T, Goldsmith JR (1988) The effect of pressure and protons on Si/Al diffusion and oxygen exchange in anhydrous systems. EOS, Trans Am Geophys Union 69:1518

Cordier P, Doukhan JC (1989) Water solubility in quartz and its influence on ductility. Eur J Mineral 1:221-237

Deer WA, Howie RA, Zussman J (1963) Rock-forming Minerals. Vol. 4, Framework Silicates. Longman: London

Dennis PF (1984a) Oxygen self-diffusion in quartz under hydrothermal conditions. J Geophys Res 89:4047-4057

Dennis PF (1984b) Oxygen self-diffusion in quartz. Progress in Experimental Petrol. NERC (UK) Publ Series D, 25:260-265

Deubener J, Sternitzke M, Muller G (1991) Feldspar $MAlSi_3O_8$ (M = H, Li, Ag) synthesized by low-temperature ion exchange. Am Mineral 76:1620-1627

Donnay G, Wyart J, Sabatier G (1959) Structural mechanism of thermal and compositional transformations in silicates. Zeitschrift fur Krist 112:161-168

Drits VA, Kashaev AA, Sokolova GV (1975) Crystal structure of cymrite. Kristallografiya 20:280-286

Elphick SC, Dennis PF, Graham CM (1986a) Oxygen diffusion studies in minerals: the effect of surface reaction processes in hydrothermal exchange experiments. Materials Science Forum 7:235-242

Elphick SC, Dennis PF, Graham CM (1986b) An experimental study of the diffusion of oxygen in quartz and albite using an overgrowth technique. Contrib Mineral Petrol 92:322-330

Elphick SC, Graham CM (1988) The effect of hydrogen on oxygen diffusion in quartz: evidence for fast proton transients? Nature 335:243-245

Elphick SC, Graham CM, Dennis PF (1988) An ion microprobe study of anhydrous oxygen diffusion in anorthite: a comparison with hydrothermal data and some geological implications. Contrib Mineral Petrol 100:690-695

Elphick SC, Graham CM (1992) Measurement of oxygen diffusivity in minerals under high pressure, high aH^+ conditions. EOS 73:373

Elphick SC, Graham CM, Walker FDL, Holness MB (1991) The application of SIMS ion imaging techniques in the experimental study of fluid-mineral interactions. Mineral Mag 55:347-356

Erd RC, White DE, Fahey JJ, Lee DE (1964) Buddingtonite, an ammonium feldspar with zeolitic water. Am Mineral 49:831-857

Ernsberger FM (1983) The nonconformist ion. J Amer Ceram Soc 66:747-750

Eugster HP (1981) Metamorphic solutions and reactions. In: Wickman FE, Rickard DT (eds) Chemistry and Geochemistry of Solutions at High Temperatures and Pressures. Physics and Chemistry of the Earth 13/14. Pergamon, New York, pp 461-507

Eugster HP (1986) Minerals in hot water. Am Mineral 71:655-673

Ewald AH (1985) The effect of pressure on oxygen isotope exchange in silicates. Chem Geol 49:179-185

Farver JR, Yund RA (1990) The effect of hydrogen, oxygen and water fugacity on oxygen diffusion in alkali feldspar. Geochim Cosmochim Acta 54:2953-2964

Farver JR, Yund RA (1991) Oxygen diffusion in quartz: Dependence on temperature and water fugacity. Chem Geol 90:55-70

Foland KA (1974a) Ar^{40} diffusion in homogeneous orthoclase and an interpretation of Ar diffusion in K-feldspars. Geochim Cosmochim Acta 38:151-166

Foland KA (1974b) Alkali diffusion in orthoclase. In: Hoffman AW, Giletti BJ, Yoder HS, Yund RA (eds), Geochemical Transport and Kinetics. Academic Press, New York, pp 77-98

Frantz JD, Marshall WL (1984) Electrical conductances and ionization constants of salts, acids and bases in supercritical aqueous fluids. I. Hydrochloric acid from 100^oC to 700^oC and at pressures to 4000 bars. Am J Sci 284:651-667

Freer R, Dennis PF (1982) Oxygen diffusion studies I. A preliminary ion microprobe investigation of oxygen diffusion in some rock-forming minerals. Miner Mag 45:179-192

Gandais M, Williame C (1984) Mechanical properties of feldspars. In: Brown WL (ed) Feldspars and Feldspathoids, NATO ASI Series C 137, D Reidel Publ Co, Dordrecht, pp207-246

Giletti BJ (1985) The nature of oxygen transport within minerals in the presence of hydrothermal water and the role of diffusion. Chem Geol 53:197-206

Giletti BJ (1986) Diffusion effects on oxygen isotope temperatures of slowly cooled igneous and metamorphic rocks. Earth Planet Sci Letters 77:218-228

Giletti BJ, Yund RA (1984) Oxygen diffusion in quartz. J Geophys Res 89:4039-4046
Giletti BJ, Semet MP, Yund RA (1978) Studies in diffusion, III oxygen in feldspars: An ion microprobe determination. Geochim Cosmochim Acta 42:45-57
Goldsmith JR (1986) The role of hydrogen in promoting Al-Si interdiffusion in albite at high pressures. Earth Planet Sci Letters 80:135-138
Goldsmith JR (1987) Al/Si interdiffusion in albite: effect of pressure and the role of hydrogen. Contrib Mineral Petrol 95:311-321
Goldsmith JR (1988) Enhanced Al/Si interdiffusion in $KAlSi_3O_8$ at high pressures. J Geol 96:109-124
Goldsmith JR (1991) Pressure enhanced Al-Si diffusion and oxygen isotope exchange. In: Ganguly, J (ed) Diffusion, Atomic Ordering and Mass Transport. Advances in Physical Geochemistry vol 8, Springer-Verlag, New York, pp 221-247
Goldsmith JR, Jenkins DM (1985) The high-low albite relations revealed by reversal of degree of order at high pressures. Am Mineral 70:911-923
Graham CM, Elphick SC (1990) A re-examination of the role of hydrogen in Al-Si interdiffusion in feldspars. Contrib Mineral Petrol 104:481-491
Graham CM, Elphick SC (1991) Some experimental constraints on the role of hydrogen in oxygen and hydrogen diffusion and Al-Si interdiffusion in minerals. In: Ganguly, J (ed) Diffusion, Atomic Ordering and Mass Transport, Advances in Physical Geochemistry vol 8, Springer-Verlag, New York, pp 248-285
Graham CM, Tareen JAK, McMillan PF, Lowe BM (1992) An experimental and thermodynamic study of cymrite and celsian stability in the system $BaO-Al_2O_3-SiO_2-H_2O$. Eur J Mineral 4:251-269
Griggs DT (1967) Hydrolytic weakening of quartz and other silicates. Geophys J Roy Astron Soc 14:19-31
Griggs DT, Blacic JD (1965) Quartz: anomalous weakness of synthetic crystals. Science 147:292-295
Grove TL, Baker MB, Kinzler RJ (1984) Coupled CaAl-NaSi diffusion in plagioclase feldspar: experiments and applications to cooling rate speedometry. Geochim Cosmochim Acta 48:2113-2121
Hobbs B (1984) Point defect chemistry of minerals under a hydrothermal environment. J Geophys Res 89:4026-4038
Hofmeister AM, Rossman GR (1983) Color in feldspars. In: Ribbe PH (ed) Feldspar Mineralogy, Mineral Soc Am, Reviews in Mineralogy 2:271-280
Hofmeister AM, Rossman GR (1985a) A spectroscopic study of irradiation coloring of amazonite: structurally hydrous, Pb-bearing feldspar. Am Mineral 70:794-804
Hofmeister AM, Rossman GR (1985b) A model for the irradiative coloration of smoky feldspar and the inhibiting influence of water. Phys Chem Minerals 12:324-332
Hokanson SA, Yund RA (1986) Comparison of alkali interdiffusion rates for cryptoperthites. Am Mineral 71:1409-1414
Kats A, Haven Y, Stevels JM (1962) Hydroxyl groups in α-quartz. Phys Chem Glasses 3:69-75
Kronenberg AK, Yund RA (1988) Diffusion of hydrogen-related species in feldspar. EOS, Trans Am Geophys Union 69:478
Kronenberg AK, Rossman GR, Yund RA, Huffman AR (1989) Stationary and mobile hydrogen defects in potassium feldspar. EOS, Trans Am Geophys Union 70:1406
MacKenzie WS (1957) The crystalline modifications of $NaAlSi_3O_8$. Am J Sci 255:481-516
Martin RF, Donnay G (1972) Hydroxyl in the mantle. Am Mineral 57:554-570

Matthews A, Goldsmith JR, Clayton RN (1983) On the mechanisms and kinetics of oxygen isotope exchange in quartz and feldspars at elevated temperatures and pressures. Geol Soc Amer Bull 94:396-412

McConnell JDC, McKie D (1960) The kinetics of the ordering process in triclinic $NaAlSi_3O_8$. Mineral Mag 32:436-457

Merigoux H (1968) Etude de la mobilite de l'oxygene dans les feldspaths alcalins. Bull Soc franc Mineral Cristallogr 91:51-64

Muller G (1988) Preparation of hydrogen and lithium feldspar by ion exchange. Nature 332:435-436

Muehlenbachs K, Kushiro I (1974) Oxygen isotope exchange and equilibrium of silicates with CO_2 or O_2. Carnegie Instn Washington Yb 74:232-240

Norton D, Taylor HP (1979) Quantitative simulation of the hydrothermal systems of crystallizing magmas on the basis of transport theory and oxygen isotope data: an analysis of the Skaergaard Intrusion. J Petrol 20:421-486

O'Neil JR, Taylor HP (1967) The oxygen isotope and cation exchange chemistry of feldspars. Am Mineral 52:1414-1436

Purton J, Jones R, Heggie M, Oberg S, Catlow CRA (1992) LDF pseudopotential calculations of the α-quartz structure and hydrogarnet defect. Phys Chem Minerals 18:389-392

Purton J, Jones R, Catlow CRA, Leslie M (1993) Ab inition potentials for the calculation of the dynamical and elastic properties of α-quartz. Phys Chem Minerals 19:392-400

Rossman GR, Smyth JR (1990) Hydroxyl content of accessory minerals in mantle eclogites and related rocks. Am Mineral 75:775-780

Seki Y, Kennedy GC (1964) The breakdown of potassium feldspar, $KAlSi_3O_8$, at high temperatures and pressures. Am Mineral 49:1688-1706

Snow E (1989) The effect of aH^+ on Al/Si interdiffusion in plagioclase feldspar at high pressure and temperature: a new look at old data. EOS, Trans Am Geophys Union 70:1405

Tullis J (1983) Deformation of feldspars. In: Ribbe PH (ed) Feldspar Mineralogy (2nd edn), Mineral Soc Amer, Reviews in Mineralogy 2:297-323

Viswanathan K, Harneit O, Epple M (1992) Hydrated barium aluminosilicates, $BaAl_2Si_2O_8.nH_2O$, and their relations to cymrite and celsian. Eur J Mineral 4:271-278

Voncken JHL, Konings RJM, Jansen JBH, Woensdregt CF (1988) Hydrothermally grown buddingtonite, an anhydrous ammonium feldspar ($NH_4AlSi_3O_8$). Phys Chem Minerals 15:323-328

Walker FDL (1990) Ion microprobe study of intragrain micropermeability in alkali feldspars. Contrib Mineral Petrol 106:124-128

Wilkins RWT, Sabine W (1973) Water content of some nominally anhydrous silicates. Am Mineral 58:508-516

Worden RH, Walker FDL, Parsons I, Brown WL (1990) Development of microporosity, diffusion channels and deuteric coarsening in perthitic alkali feldspars. Contrib Mineral Petrol 104:507-515

Wyart J, Sabatier G, Curien H, Ducheylard G, Severin M (1959) Echanges isotopiques des atomes d'oxygene dans les silicates. Bull Soc franc Miner Cristallogr 82:387-389

Wyart J, Curien H, Sabatier G (1961) Echanges isotopiques des atomes d'oxygene dans les silicates et mecanisme d'interaction eau-silicate. Instituto Luca Mallada C.S.I.C (Espana), Cursillos Y Conferencias 8:27-29

Yntema LF, Percy AL (1954) Tantalum and columbium. In: Hampel CA (ed) Rare Metals Handbook. Reinhold, New York, pp 389-404

Yund RA (1983) Diffusion in feldspars In: Ribbe PH (ed) Feldspar Mineralogy (2nd edn), Mineral Soc Amer, Reviews in Mineralogy 2:203-222

Yund RA (1984) Alkali feldspar exsolution: kinetics and dependence on alkali interdiffusion. In: Brown WL (ed) Feldspars and Feldspathoids, NATO ASI Series C 137, D Reidel Publ Co, Dordrecht, pp281-315

Yund RA (1986) Interdiffusion of NaSi-CaAl in peristerite. Phys Chem Minerals 13:11-16

Yund RA, Anderson TF (1974) Oxygen isotope exchange between potassium feldspar and KCl solution. In: Hoffman AW, Giletti BJ, Yoder HS, Yund RA (eds) Geochemical Transport and Kinetics. Academic Press, New York, pp 99-105

Yund RA, Anderson TF (1978) The effect of fluid pressure on oxygen isotope exchange between feldspar and water. Geochim Cosmochim Acta 42:235-239

Yund RA, Snow E (1989) Effects of hydrogen fugacity and confining pressure on the interdiffusion rate of NaSi-CaAl in plagioclase. J Geophys Res 94:10662-10668

Zhang Y, Stolper EM, Wasserburg GJ (1991) Diffusion of a multi-species component and its role in oxygen and water transport in silicates. Earth Planet Sci Letters 103:228-240

ARGON DIFFUSION IN FELDSPARS

K. A. FOLAND
Department of Geological Sciences
Ohio State University
Columbus, Ohio 43210
U.S.A.

ABSTRACT. Diffusion of argon in feldspar is important because of its significance to K-Ar geochronology and becomes especially so with increasingly widespread use of both plagioclase and alkali feldspar for dating by the $^{40}Ar/^{39}Ar$ technique. Despite study and debate for more than 30 years, the basic feldspar Ar transport phenomena are subject to considerable uncertainty. Many studies report release kinetics but the results show a very large range in apparent fundamental parameters such as activation energy. While these measurements document the reactivity and structural complications of feldspars, few can be interpreted unambiguously. The most definitive description of Arrhenius parameters is provided by measurements on the Benson Mines orthoclase which is devoid of the internal features that commonly provide pathways for rapid Ar transport and for which physical grain size is the effective diffusion dimension. Long-term isothermal experiments at 500 to 800°C show Ar loss that conforms to theoretical behavior producing a linear Arrhenius relation. For the spherical model of isotopic Ar transport, diffusion coefficients are described by a frequency factor of 0.00982 cm^2/sec and an activation energy of 43.8 kcal/mol. The behavior of Ar in this specimen leads to several generalizations. Diffusion of Ar appears to follow kinetics predicted by basic theory with a single mechanism although some trapping appears at high temperatures and degrees of outgassing. Kinetics of transport are not affected by the presence of external atmospheres of elevated water activity or atmospheric gases. Although neutron irradiation will have a great effect on defect concentration and may enhance or retard migration, it does not apparently alter diffusion rates at temperatures in excess of about 600°C. However, recoil redistribution of ^{39}Ar will be important for many feldspars. There are several reasons to expect that Arrhenius parameters given above describe intrinsic Ar volume diffusion. They may be used for modeling and, acknowledging uncertainty about the operative mechanisms, extrapolated to lower temperatures. At the same time, behavior during incremental-heating degassing shows important differences. There are serious departures from a simple linear Arrhenius relation at higher temperatures (in excess of about 925°C) and large fractional losses; this serious effect appears to be due to Ar trapping with heating at high temperature for long times. In addition, rates of Ar release are enhanced during step-heating experiments due to reduction in grain size as a result of cracking and spalling. Both trapping and grain size reduction have profound implications for alkali feldspar laboratory incremental-heating release kinetics and their use for modeling of Ar loss in nature.

1. Introduction

Argon in feldspars and other minerals has been examined for many years. Its diffusive transport has been studied for the last several decades which have been marked by periods of differing interest and intensity. Currently, the level of interest is high and characterized by disagreement, controversy, and ambiguity particularly about Ar behavior in feldspars. Uncertainties surround laboratory measurements, their interpretations, and natural behavior.

I. Parson (ed.), Feldspars and Their Reactions, 415–447.

1.1. PERSPECTIVE

Understanding Ar diffusive transport is of interest primarily because of its importance to K-Ar geochronology. This method of measuring geologic time is the predominate one whereby the thermal history of the crust is deduced. The common feldspars, moreover, are important minerals for dating and this type of geochronology. The closure theory developed for radiogenic isotopes in minerals undergoing cooling links temperature and time by way of fundamental kinetics parameters. Application of this theory permits determination, or estimation, of both temperature and time information insofar as key assumptions are valid and the relevant transport parameters are known. Using the $^{40}Ar/^{39}Ar$ incremental-heating technique as is commonly practiced, it is potentially possible to obtain appropriate kinetic parameters and age information simultaneously from a single set of measurements. An implicit assumption in applying such laboratory data to nature is that the same transport parameters apparent from the dating experiment are valid and apply to behavior in nature. This is clearly not the case for common K-bearing minerals which are used for Ar dating, e.g., micas and amphiboles. These minerals are typically hydrous and undergo serious reaction and physical degradation during the vacuum Ar extraction that is essential for the age measurement. As a result, the observed kinetics of Ar release during such dating experiments do not characterize the rates of diffusion of Ar in nature. The use of feldspars offers promise because they do not pose the same level of basic laboratory instability during heating. They do, however, have other characteristics and properties, for example, microstrucural features, which pose serious problems or at least complications.

The wide distribution of feldspars in the crust, coupled with their potential for providing both geologic ages and thermal history information, underscores the necessity of defining their Ar transport phenomena. While it is important to understand both natural and laboratory behavior, there are at the present several unresolved issues about both dimensions.

1.2. SCOPE

This report addresses Ar diffusion in feldspar in terms of basic transport characteristics, mechanisms, and phenomena. With the intent to focus upon fundamental behavior, there is no attempt to cover the many feldspar geochronological applications and results. The treatment is not intended to be of uniform coverage. Many references to published reports are given for the purpose of providing sources rather than complete details. On several aspects that are not well covered or widely considered in the literature, more extensive discussion is provided.

Some useful background material is provided in the next section. The purpose is to provide a brief introduction to feldspar Ar geochronology and diffusion in order to provide the framework for the subsequent discussion. Next a brief survey of previous results relative to Ar transport is presented. Then the discussion turns to a consideration of basic transport phenomena in terms of the current understanding provided mainly by selected laboratory data and concludes with consideration of general diffusion characteristics.

2. Background

2.1. $^{40}Ar/^{39}Ar$ TECHNIQUE

2.1.1. *Some Basics.* This variant of the well-known K-Ar method of dating has seen increasingly widespread use since its introduction (Merrihue, 1965; Merrihue and Turner, 1966) largely because of the ability to examine heterogeneities in the ratio of K to ^{40}Ar. The basic principles of K-Ar geochronology are available from general treatises such as Faure (1986); many practical and theoretical aspects of the ^{40}Ar/^{39}Ar technique are presented by McDougall and Harrison (1988) which is an excellent source for details that cannot be provided here. The terrestrial geochemistry of noble gases is reviewed by Ozima and Podosek (1983).

The minor isotope (0.01167 atom %) of natural potassium, ^{40}K, undergoes branching decay (with a total half-life of 1.25 billion years) via electron capture (approximately 11%) to ^{40}Ar or negatron emission (89%) to ^{40}Ca . As a non-reactive gas, argon is readily analyzed to ultra-trace concentration levels. In combination with the wide distribution of K, this produces a geochronometer of great utility. A specimen's "age" is a function of the concentrations of parent and daughter nuclides. An age measures the time since the specimen became a closed system with respect to gain or loss of K and Ar assuming insignificant or no initial ^{40}Ar.

With the ^{40}Ar/^{39}Ar technique, specimens are subjected to neutron irradiation to produce ^{39}Ar, via the reaction ^{39}K(*n, p*)^{39}Ar with neutrons of energy in excess of 0.22 MeV. This serves as an indirect measure of ^{40}K which is then measured along with ^{40}Ar. The general procedure is to: (a) irradiate the sample obtaining a total dose on the order of 10^{18} to 10^{19} neutrons/cm^2; (b) "calibrate" (or determine the J parameter for) the irradiation using a "monitor" mineral of known age; (c) extract Ar from a sample by heating under high vacuum, purify it, and determine its isotopic composition; and, (d) determine an "age" from,

$$T = (1/\lambda) \times \ln [(^{40}Ar^*/^{39}Ar) \times J + 1], \qquad [1]$$

where: T is "age"; λ is the total decay constant of ^{40}K; J is an empirically determined irradiation parameter; and, the ratio ^{40}Ar*/^{39}Ar represents the ^{40}Ar from decay and the ^{39}Ar from ^{39}K which serves as a measure of ^{40}K. This ratio is corrected for interferences from other nuclear reactions that also produce Ar isotopes of masses 36 through 40.

It is useful to distinguish several types of argon: radiogenic, ^{40}Ar formed from *in situ* decay of ^{40}K; excess, ^{40}Ar (in a specimen) which is in excess of that formed by *in situ* decay over the life of the specimen; atmospheric, Ar having the (unique) isotopic composition of modern air (^{40}Ar/^{36}Ar = 295.5 and ^{38}Ar/^{36}Ar = 0.188) of which it makes up almost 1%; and, neutron-induced, Ar produced during neutron irradiation (e.g., ^{39}Ar, ^{37}Ar). All of these play a role in the dating experiment and the interpretation of the apparent age. Of course, the radiogenic Ar is the measure of age. A sample containing excess Ar, either present initially or introduced, will produce an age that is too old. Excess Ar is sometimes a complication for feldspars, particularly plutonic types. All samples contain at least small amounts of atmospheric Ar "contamination"; it is quite successfully taken into account using ^{36}Ar to monitor this component. All irradiated specimens also contain neutron-induced Ar isotopes in addition to ^{39}Ar which need to be monitored for interference but which also contribute important chemical information. Corrections are applied using the amounts of ^{37}Ar and ^{38}Ar which are produced mainly from Ca and Cl, respectively. These two Ar isotopes provide data about the concentrations of Cl and Ca so that mass analysis of Ar liberated from the specimen provides information on K/Ca and K/Cl in addition to apparent age. Both of these chemical signatures are quite important for feldspars because of compositional variations involving Ca and the common abundance of Cl in crustal fluids which commonly affect feldspars.

Since the apparent age and these ratios can be determined from the isotopic composition of liberated Ar, it is possible to perform serial extraction and thereby obtain a series of ages on a given specimen. This procedure, incremental or step heating, adds enormous power to the analysis. A series of apparent ages is obtained for a heating schedule, typically where heating temperature is progressively increased. Isothermal heating steps or even cycled temperatures have been used, albeit infrequently, and are extremely informative about kinetics. The great advantage of the incremental or step-heating approach is the ability to examine heterogeneities in the ratio of ^{40}Ar to K, Ca, and Cl. Such heterogeneities may reflect several geologic processes including partial loss of ^{40}Ar during a heating event, introduction or incorporation of excess Ar, and the presence of components having different ages.

Incremental-heating data are typically presented graphically in a diagram of apparent age versus cumulative fraction of released ^{39}Ar. Such a diagram is called an age spectrum and a simple example is illustrated in Figure 1. A concordant spectrum, where apparent ages are within analytical uncertainty, will be produced when the ratio of $^{40}Ar^{*}/^{39}Ar$ is uniform, whereas a discordant one reflects heterogeneities in this ratio. A concordant or nearly concordant spectrum produces an age plateau which is a broad region of released Ar that indicates the same age within uncertainties. Such plateau ages are generally interpreted to have geological significance. Coupled with concomitant variation in chemical signatures, incremental heating is a powerful approach for revealing the different regions that are releasing Ar.

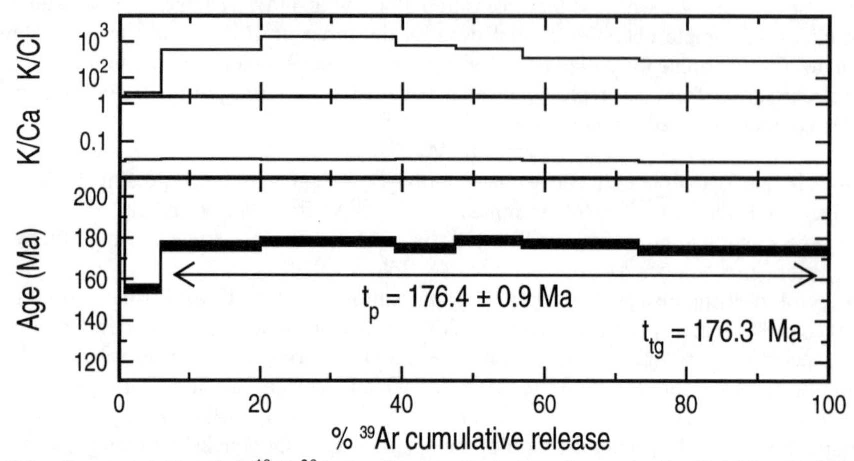

Figure 1. Incremental-heating $^{40}Ar/^{39}Ar$ results for plagioclase from sample 81-13-3 of Kirkpatrick basalt of north Victoria Land, Antarctica (from Heimann et al., in press). It illustrates a simple, concordant spectrum where the plateau age "t_p" is the same as the total-gas or integrated apparent age ("t_{tg}") and defines the time of eruption and cooling. Note the uniformity in K/Ca ratio.

2.1.2. *Dating of Feldspars.* The advent and refinement of the $^{40}Ar/^{39}Ar$ incremental-heating technique has resulted in a broad expansion in the geochronological potential of feldspars of all types compared with the conventional K-Ar approach. Indeed, geochronologic study of feldspars now figures importantly for studies with a variety of geologic objectives. The $^{40}Ar/^{39}Ar$ approach is almost invariably advantageous and frequently fully required for feldspars.

Many examples of the application of the $^{40}Ar/^{39}Ar$ technique to feldspars are illustrated by McDougall and Harrison (1988) as well as later in this report. Two specific applications which have recently seen wide exploitation are for determining times of volcanism and for constraining thermal histories. Plagioclase has been increasingly and widely used for addressing the ages of basaltic rocks of various ages (see, e.g., Renne and Basu, 1991; Pringle et al., 1992; Foland et al., 1993). Feldspar has been used to: determine the ages of young rocks; constrain the interval of volcanism and examine the correlation between volcanism and global extinction events; and, date rocks which have suffered extensive low temperature alteration and Ar loss. Plutonic alkali feldspars are now being used to address crustal cooling histories for a variety of tectonic regimes (e.g., Lovera et al., 1989; Richter et al., 1991).

Several incremental-heating spectra are shown as Figures 1 in 5. Rather than describe the details of results, these serve to illustrate the important general types of $^{40}Ar/^{39}Ar$ behavior.

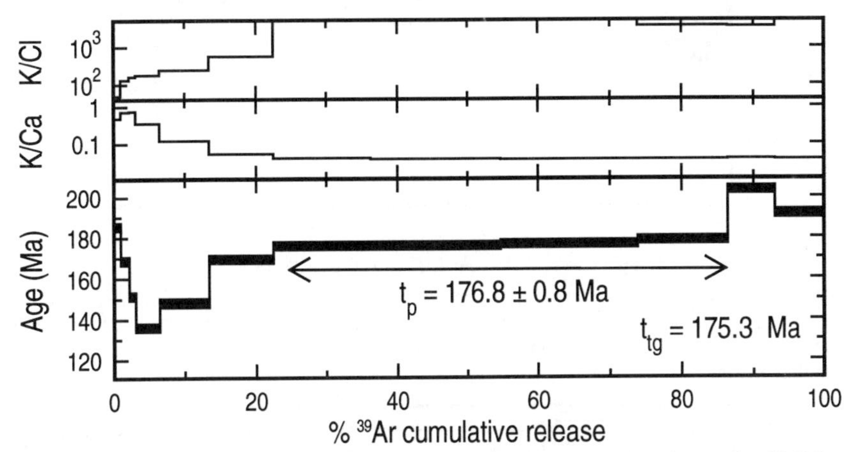

Figure 2. Discordant incremental-heating spectrum for plagioclase A of sample 81-7-2, also a Kirkpatrick basalt from north Victoria Land, Antarctica (from Foland et al., 1993). This sample has the same Jurassic age as that shown in Figure 1 as defined by the age plateau. In this case, discordance is due to minor, inseparable basalt matrix which adheres to feldspar grains. The correlative variations in K/Ca, K/Cl, and apparent age indicate matrix impurities with higher K and Cl and lower Ca and age (reflecting Cretaceous alteration) that release Ar at laboratory temperatures lower than plagioclase. Spectra of this type where discordance is due to impurities or mixed phases are common.

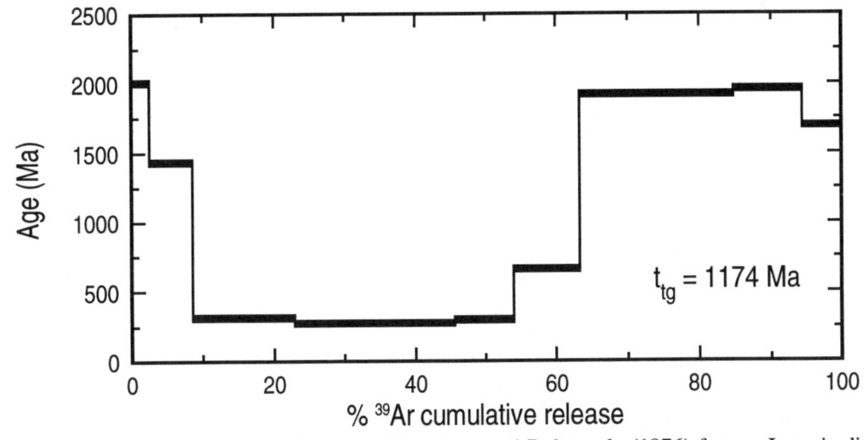

Figure 3. Age spectrum for plagioclase L2 of Lanphere and Dalrymple (1976) from a Jurassic diabase dike in Liberia. This sample contains excess ^{40}Ar with all gas increments giving dates older than age of emplacement. This type of age spectrum, called "saddle-shaped," has generally been regarded as indicating the presence of excess Ar and has been observed for a number of both alkali and plagioclase feldspars.

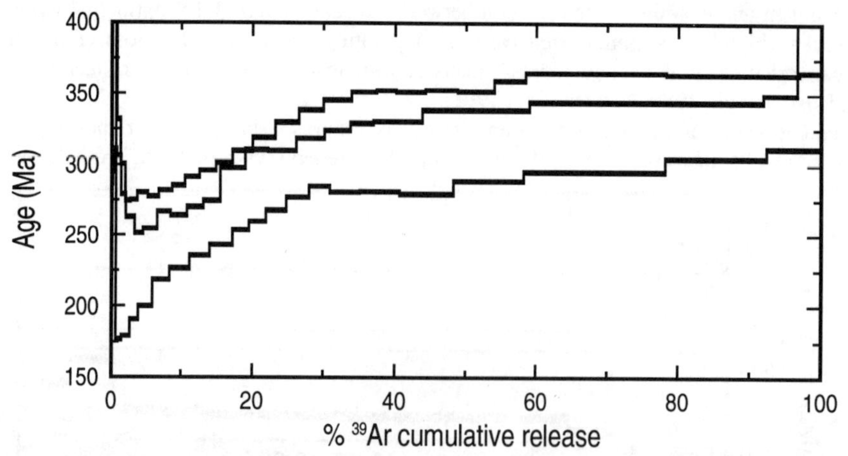

Figure 4. Age spectra for microcline and orthoclase specimens MH-10, MH-8, and MH-42 of Heizler et al. (1988) from the Chain of Ponds pluton in Maine. Lovera, Richter, and Harrison (1989) interpret the discordant patterns to reflect slow cooling where there is a wide variation in the sizes of effective diffusion dimensions.

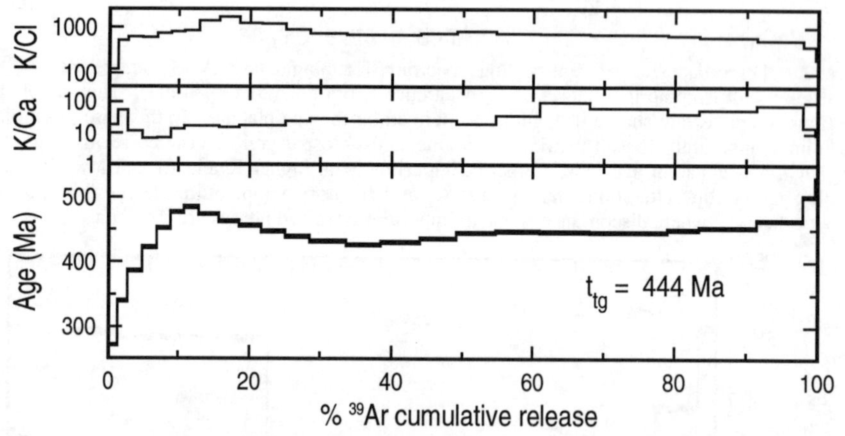

Figure 5. Age spectrum for authigenic adularia, sample R-39, described by Mensing and Faure (1983) from the top of the basement of Ohio (Heimann and Foland, unpublished data). This feldspar formed at low temperature at 500 Ma (Heimann et al., 1991). The K/Ca pattern indicates that the discordance is due, at least in part, to the presence of plagioclase lamellae probably via recoil redistribution of ^{39}Ar. Complicated patterns of discordance of this type are relatively common for feldspar.

2.1.3. *Effects of Neutron Irradiation.* The nuclear reactions and collisions attending neutron irradiation with the very large integrated doses of energetic neutrons associated with the ^{40}Ar/^{39}Ar technique produce several types of effects. The consequences of this neutron bombardment are very important to Ar behavior but often ignored. Both normal lattice atoms and newly produced nuclides suffer displacements from their original positions. Such effects of irradiation are significant with regard to crystal defect concentrations and to the possible redistribution of neutron-induced Ar.

Lattice atom displacement in crystalline material subjected to neutron beams is to be anticipated from head-on collisions, when a neutron directly hits a target, and collision cascades, when secondary displacement is induced by primary target atoms. With the energetic neutrons necessary in order to produce ^{39}Ar, the energy transmitted to an atom will be in the keV range which far exceeds the few eV typically required to overcome atom binding energy and displace it. With respect to $^{40}Ar/^{39}Ar$ dating, Horn et al. (1975) specifically consider the importance of this effect. For a total fluence of 10^{19} n_f/cm^2, they estimate that 30% of the lattice atoms are displaced and conclude that an order of magnitude of several percent are displaced. In addition, Horn et al. (1975) suggest that radiation damage and displacements due to reactor γ-rays may be important for some reactors. They further propose that some of these generated defects may combine and act as potential traps for gas atoms.

The issue has also been considered by Xu (1990) who independently arrived at a similar result. He considered direct neutron impact and cascades and estimated that 1.7% of the total atoms would be displaced for a total fluence of 10^{18} n_f/cm^2. Since the fraction of atoms displaced is proportional to the total neutron dose, these estimates of lattice disruption, albeit rough ones, are consistent. They indicate that for a typical neutron irradiation for feldspar dating, the fraction of atoms displaced is at a level on the order of percent. The end result must be a very large number of point defects and perhaps higher-order ones.

In addition to these direct *n*-induced displacements, atoms undergoing nuclear reaction undergo recoil in order to conserve momentum. For those in which nucleons with >~ 10 keV enter or leave the nucleus, the residual nucleus will have kinetic energy that greatly exceeds bond energy. For transmutations involving *n*-induced Ar, such recoil is large as discussed below. However, there are many other more numerous (n, γ) reactions on the feldspar major and trace elements.

Recoil is significant even for γ emission which may be associated with capture or with a decay of generated unstable nuclides. The recoil energy following γ emission, as required for conservation of momentum (see Friedlander et al., 1981), is given by,

$$E_r = 0.537\ (E_\gamma)^2 / M, \qquad [2]$$

where: E_r, is the recoil energy (in keV); E_γ is the γ energy (in MeV); and, M is the atomic mass (in u). For nuclides of masses 20 to 40 and E_γ of 1 to 2 MeV, E_r will be ~10 to 100 eV. This is large compared to the bond strength and will result in displacement. To illustrate with a relevant example, ^{38}Cl, produced via the (n, γ) reaction on ^{37}Cl, undergoes β decay accompanied by γ emission of 1.6 (or 2.2) MeV. The ^{38}Ar daughter will thus have recoil energy of roughly 36 (or 68) eV. It follows that the probability is very high that neutron-induced ^{38}Ar will not occupy the same precise position as its Cl progenitor. Similar effects occur for the essential feldspar elements particularly Si, Al, Na, and K for which one part in 10^6 (within a factor of about 10) will undergo reaction. Not only will there be displacement and lattice damage, but also the production of different nuclides will result.

These considerations and expression [2] also apply to natural decay of ^{40}K. The 1.46 MeV γ emitted following electron capture imparts 29 eV recoil energy to the ^{40}Ar daughter. As pointed out long ago by Brandt and Voronovskiy (1967), this would be sufficient to displace ^{40}Ar produced in nature from the normal alkali position. The specific structural position of radiogenic ^{40}Ar in an untreated sample is a matter of conjecture. Long periods, frequently at temperatures above those at the surface, are available for structural readjustment or annealing of damage produced by ^{40}K decay. It is illustrative to recall that subtle small-scale features can be affected at relatively low temperatures over geologic periods. An example is the "fading" or "annealing" of fission tracks in minerals in nature which is well known to occur in some minerals at temperatures as low as ~100°C. In sum, one cannot be certain of the position now occupied by natural radiogenic ^{40}Ar. Some of it may even reside in major imperfections.

Recoil is especially important for neutron-induced Ar because it can result in redistribution of these isotopes or even in loss of Ar. In these reactions the recoil energies are relatively large, comparable in magnitude to daughter recoil accompanying alpha decay which is known to produce extensive lattice damage in silicates. The affected Ar is the substance of prime concern. Redistribution of Ar due to recoil will be especially significant for feldspars that can have large variations in K and Ca concentrations over very short distances.

The recoil energy of neutron-induced Ar is substantial and there are several methods for estimating recoil range for a given energy. Ranges may be estimated from the work of Schiott (1970), which for recoil energies of approximately 25 to 500 keV and the appropriate density and atomic masses of feldspar, leads to,

$$R_r \cong 7.5 \times 10^{-4} (E_r), \qquad [3]$$

where: R_r is the recoil range (in μm) and E_r is the energy of the recoiling atom (in keV).

For ^{39}Ar produced via the *(n, p)* reaction on ^{39}K, the recoil energy varies because of the spectrum of neutron energies and the angular distribution of proton emission. Turner and Cadogan (1974) provide the first and most thorough evaluation of the recoil energetics. They show that the recoil energies of ^{39}Ar produced in a typical $^{40}Ar/^{39}Ar$ irradiation range up to about 500 keV with a median of about 150 keV. Using expression [3], this translates to recoil ranges of up to approximately 0.40 μm with a median distance of ~ 0.11 μm.

The impact of ^{39}Ar recoil displacements on the order of ~ 0.1 μm in terms of loss from grains or regions and the reimplantation into neighboring ones is stressed theoretically by Turner and Cadogan (1974) and experimentally by Huneke and Smith (1976). Turner and Cadogan (1974) model the recoil effects empirically to describe the recoil loss of ^{39}Ar near a grain surface; the depletion layer near a surface may be described by,

$$C(x) = 1 - 0.5 \exp(-x/x_0), \qquad [4]$$

where: C(x) is the relative concentration of ^{39}Ar at a depth x measured from the grain surface and x_0 is 0.082 μm. This expression, which should hold for a typical irradiation in various reactors, may be evaluated for simple shapes (see, e.g., Foland and Xu, 1990) to estimate the fraction of ^{39}Ar which recoils out of a given volume. Such an exercise shows that only minor losses of ^{39}Ar are expected from physical grains or grain regions with diameters of several tens to hundreds of μm. However, it also indicates that losses will be significant from grains or regions which are physically small, for example, on the order of 1 μm or less.

Recoil loss has been addressed and measured by several workers including Foland et al. (1984), Hess and Lippolt (1986), and Smith et al. (1993); the results converge to show that it is an important effect especially for small grains. Possible losses in feldspar have been briefly considered by Hess and Lippolt (1986) and Foland and Xu (1990). Hess and Lippolt (1986) report small yet significant (~ 1% including that released with heating to about 250 °C) recoil loss of ^{39}Ar from 100 to 500 μm sanidine grains.

Considering the typical recoil range of ^{39}Ar in the context of the scale of compositional variations and other microstructures common in feldspars, it is clear that recoil of neutron-induced Ar will be important in redistributing Ar within a specimen. The net transfer will be from regions of higher to lower K. Obviously, the redistribution will affect the spatial distribution of ^{39}Ar which will aberrations in the $^{40}Ar/^{39}Ar$ ratio and apparent age. It will also produce some variations in the distribution of ^{39}Ar in the context of modeling mass transport by diffusion. Some ^{39}Ar may also recoil into micropores since it appears that not all that recoils out is actually reimplanted within adjacent regions (see York et al., 1992; Smith et al., 1993). This has also been suggested by Turner and Wang (1992).

Another consequence of recoil is a damaged region which is potentially up to about 0.4 μm long; moreover, the region should be associated with a neutron-induced Ar atom. Such a region, which is probably tube like, potentially offers a fast diffusion path for a short distance much like dislocation assistance addressed by Yund et al. (1981). Thus, the result may be enhanced transport rates which may be very important in cases of small grain dimensions.

In summary, radiation damage is an inevitable consequence of the technique and has important implications. The discoloration of feldspar subjected to typical neutron doses serves as a dramatic reminder of this. Radiation damage has long been recognized as capable of enhancing mass transport rates in solids (see, e.g., Borg and Dienes, 1988; Dienes and Vineyard, 1957). The recoil energy associated with the production of ^{39}Ar and radiogenic ^{40}Ar differs by a factor of about 10^4 and this difference may manifest itself in contrasting lattice positions with very different local characteristics.

2.2. DIFFUSION OF ARGON

2.2.1. *General.* In elementary terms, diffusion may be regarded as the process of the relative motion of atoms or molecules that occurs spontaneously in response to gradients in concentration or chemical potential. The fundamental material property for mass transport is the diffusion coefficient or diffusivity, which is the constant of proportionality between mass flux and concentration gradient. It is defined from Fick's first law,

$$J = -D_{ii}\, C, \qquad [5]$$

and generally operationalized with Fick's second law,

$$\partial C/\partial t = D_{ii}\, \nabla^2 C, \qquad [6]$$

where: J is the flux or the number of particles of a given species passing a plane of unit area per unit time; C is the concentration of the species of concern; t is time; and, the D_{ii} are its diffusion coefficients along the principal crystallographic directions. These diffusion coefficients take on the dimensions of area/time; most commonly the units cm^2/sec have been used and this convention is followed here. As defined, the diffusion coefficient is assumed to be independent of concentration which should be valid for Ar in minerals because the concentrations are typically at the ppm level or lower.

Of principal interest for Ar will be the volume (sometimes also referred to as lattice, solid-state, or atomic) diffusion coefficient which relates transport through a crystal within the normal atomic framework. Expressions [5] and [6] also describe transport by non-volume diffusion such as along grain boundaries or high diffusivity paths. The diffusion coefficients for these processes are expected to be several orders of magnitude higher in general (see, e.g., Shewmon, 1963; Borg and Dienes, 1988) and in feldspar (see, Yund, 1983; Giletti, this volume). Therefore, the general assumption is that volume diffusion within the crystal lattice is the rate limiting step for the process of Ar loss from grains. Thus, the remaining discussion addresses this process.

The distance of migration by a given species in a given period of time may be estimated using a convenient approximation,

$$\overline{x}^2 \cong 2\, D\, t, \qquad [7]$$

where $\overline{x}$ is the average distance traveled over a time interval, t. For example, the average Ar atom will travel 1 μm in 10 m.y. for a diffusion coefficient of $\sim 1 \times 10^{-21}$ cm^2/sec.

2.2.2. *Temperature Dependence.* The temperature dependence of diffusion is described by the Arrhenius relation which applies generally to kinetics phenomena. It has the form,

$$D = D_0\, e^{-E/RT}, \qquad [8]$$

where: D_0 is the pre-exponential or frequency factor which is a constant independent of temperature; E is the activation energy; R is the gas constant; and, T is absolute temperature.

For diffusion by a single mechanism, such as one of the classical ones (e.g., vacancy or interstitial mechanism), diffusivity may be described by the two parameters D_0 and E. In an Arrhenius diagram, a plot of logarithm D against reciprocal temperature as shown in Figure 6, transport by one mechanism is expressed as a "linear" Arrhenius relation. The values for D_0 and E may be obtained from the infinite T intercept and slope, respectively. A linear relationship results from motion involving thermally activated defects inherent to the material. A break or "knee" in the Arrhenius plot for a given set of measured diffusion coefficients, i.e., change in activation energy, implies one of two alternatives for a material that does not react or undergo a phase transition within the range of temperatures. One explanation is that it indicates a change in diffusive transport mechanism. Although there may be simultaneously several feasible mechanisms, the fastest one will predominate. A second alternative is that it indicates the importance of defects, for example, substitutional impurities or vacancies, whose concentrations are not thermally controlled by a customary Boltzmann distribution. The region controlled by thermally generated defects, the intrinsic region, will have the higher activation energy and be at higher temperature whereas that characterized by temperature-independent defects, the extrinsic region, will dominate at lower temperatures, as illustrated schematically in Figure 6.

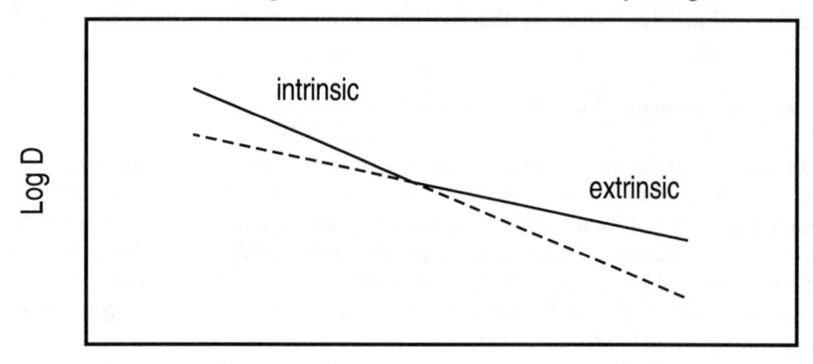

Figure 6. Schematic Arrhenius diagram illustrating a break in slope or activation energy. Such a break or "knee" will occur for a shift from diffusion controlled by intrinsic defects at higher temperature to that dominated at lower temperature by extrinsic defects such as those related to impurities, natural imperfections, or radiation-induced defects.

The transition from extrinsic to intrinsic control at low temperatures is important for the application of Ar diffusion parameters, and indeed virtually all diffusion data in geological materials, to lower temperatures by linear extrapolation of the Arrhenius relation. The relative temperature of a laboratory Ar diffusion measurement is much higher than those of attendant transport in nature because the typical heating time is a factor of about 10^8 less than that available in nature. As a result, long extrapolations of Arrhenius relations are required so that it is possible that different mechanisms or extrinsic control may prevail at low temperature.

Extrinsic control is pertinent relative to the very large numbers of defects expected from irradiation. Whether or not such defects produce extrinsic behavior depends upon the rates of their annealing relative to the thermal production of intrinsic ones. Drawing upon studies of irradiation effects in metals, it appears that the defects may lead to dramatic enhancement in diffusion rates (by a factor on the order of 10^{10}) at low temperatures (< 150°C) with a transition from extrinsic to intrinsic probably occurring prior to about 600°C (Xu, 1990). The transition will depend upon the annealing characteristics which are poorly constrained.

2.2.3. *Measurements and Models.* Both the measurement of diffusion coefficients and their application to specific circumstances require solutions to equation [7] for given initial and boundary conditions and geometries. Solutions to models, all of which involve some simplifications of the more complex feldspar, are available from a number of sources (e.g., Crank, 1975; Mussett, 1969; McDougall and Harrison, 1988; Lovera et al., 1989).

The measurement of Ar diffusion in feldspar involves degassing Ar with experiments for a given heating temperature and duration. Direct profiling of concentrations, such as is performed for cations and oxygen (see Giletti, this volume), has not been possible for Ar in feldspar. Instead, observed losses from bulk materials are used to calculate diffusion coefficients by assuming a simple geometric transport model, i.e., either the sphere, infinite cylinder, or plane sheet solutions. For a specific fractional loss, the solutions yield a value for the dimensionless time parameter of the form (Dt/a^2) where t is heating time and a is the characteristic or effective diffusion dimension. For the above models, this is the radius for the sphere and infinite cylinder or the halfwidth for the plane sheet. Knowledge of the effective dimension for Ar diffusion is required to calculate D, the fundamental transport quantity. Obviously, the application of these geometric models involves great simplifications in terms of anisotropy and particle shapes.

Diffusion experiments for feldspar have basically used one of two approaches: isothermal or step heating. Isothermal heating involves holding grains at a constant elevated temperature for periods of days or longer. The fractional losses of Ar during this diffusion experiment, which are determined from the difference between the initial (measured on another portion of sample) and final concentrations, are typically several percent or more. With the incremental-heating approach, a sample is heated serially for various times and temperatures until eventually all Ar is liberated which is typically at melting. This is essentially a dating experiment (if the specimen has been irradiated) and is performed in a high-vacuum furnace attached to a mass spectrometer. The fractional losses of Ar liberated are determined from the individual Ar increments and the summation of all steps where the mass spectrometer is used as a manometer to measure signal intensities. The total value of (Dt/a^2) is evaluated at the end of each step and the values for each increment obtained from the summation. Comparing the two approaches, incremental heating has much shorter heating times (minutes to a few hours) with smaller losses and at higher temperatures (> 950°C) which are eventually required to liberate most of the Ar within an acceptable interval; it also readily permits many more measurements on the same sample portion. Either the fractional losses of ^{40}Ar or ^{39}Ar may be used to calculate diffusivities. Authors have typically used ^{39}Ar fractional losses because the apparent ages suggest that radiogenic ^{40}Ar is heterogeneously distributed.

The accurate measurement of heating temperature is of obvious importance but not always a trivial matter. Accurate control is not usually a problem for long-term isothermal experiments, which are typically heated in a tube furnace, but can be for step-heating degassing which is performed under demanding vacuum conditions (see McDougall and Harrison, 1988). The indicated or estimated temperatures of older heating arrangements are notoriously questionable. However, modern furnaces can be controlled precisely and are capable of very rapid temperature changes. Regardless, it is important to assess carefully the accuracy of the delivered temperatures because large deviations and gradients are possible.

Using a chosen mathematical solution, the diffusion coefficient is obtained from the heating time and observed loss. The determined value is a model diffusivity inasmuch as it implicitly involves a number of assumptions. These include: (1) sample stability in the sense of retaining integrity and not undergoing reaction either promoting or retarding Ar loss (namely, transport by volume diffusion and uninfluenced by other processes); (2) a given transport anisotropy and grain geometry; (3) effective diffusion dimensions; (4) uniform or known concentration of Ar (e.g., ^{39}Ar or ^{40}Ar) initially ($t = 0$); (5) zero concentration of Ar at grain edges ($r = a$) for $t > 0$; (6) zero concentrations (or total loss) for heating to $t = \infty$; and, (7) transport unaltered by other laboratory treatment or preparation.

Failure of experiments on feldspar to satisfy these many conditions has been discussed by many authors, in particular Giletti (1974a), Foland (1974a), McDougall and Harrison (1988) and Harrison (1983; 1990). All the above experimental assumptions are potential problems but the issue of dimensions, number (3), presents particular complications because of feldspar microstructures. The stability of the sample is also key inasmuch as exsolution structures may homogenize or the Si-Al ordering may change during heating. Potentially both processes affect degassing kinetics but in unknown ways.

The failure to meet one or more of the above requirements has meant that many measurements of D are flawed. The failure of most specimens to satisfy the demanding assumptions above is the general explanation for non-linear Arrhenius plots that characterize feldspars. Recognition of an incorrect assumption (for example, an effective diffusion dimension that is smaller than physical particle size) is critical in assessing results. It seems appropriate that the model diffusivity be qualified as an "apparent" value where it is known or suspected that the assumptions may not obtain.

2.3 DIFFUSION AND THE $^{40}Ar/^{39}Ar$ TECHNIQUE

2.3.1. *Closure Temperature.* It has long been recognized that K-Ar dates, with a few exceptions, are "cooling ages" that reflect cooling below a "closure temperature" at a time when radiogenic Ar is quantitatively retained. At high temperatures, Ar will be lost as rapidly as it is produced but with cooling it will begin to accumulate in the crystal and eventually will be totally retained. Although Ar may be lost by non-diffusive mechanisms (e.g., mineral reaction), diffusion is assumed to be the rate limiting process.

The formalism used for closure temperature, which may be defined as the temperature of a system (e.g., mineral sample) at the time given by its age, is provided by Dodson (1973; 1979). Assuming volume diffusion as a function of temperature according to the Arrhenius relation and cooling to be linear with 1/T, Dodson (1973) solved the equations for Ar loss by volume diffusion and production by decay to obtain,

$$T_c = E/\{R \ln [A\,\tau\,(D_0/a^2)]\}, \qquad [9]$$

where: $\tau = -\{[R\,T_c^2]/[E\,(dT/dt)]\}$; T_c is closure temperature; E, R, T, D_0, and a are as defined previously; (dT/dt) is the rate of cooling; and A is a numeric factor depending on geometry, equal to 55, 27, or 8.7 for the sphere, cylinder or plane sheet models, respectively.

Expression [9] has been used widely to calculate closure temperatures for given diffusion parameters and cooling rates. An example for specific parameters appropriate to alkali feldspar is shown in Figure 7 which demonstrates a wide range of closure temperatures. The closure temperature is not dramatically dependent upon cooling rate but is strongly controlled by the effective diffusion dimension for given Arrhenius parameters. Domains or grains of divergent diffusion distances, therefore, will have different values of T_c and age.

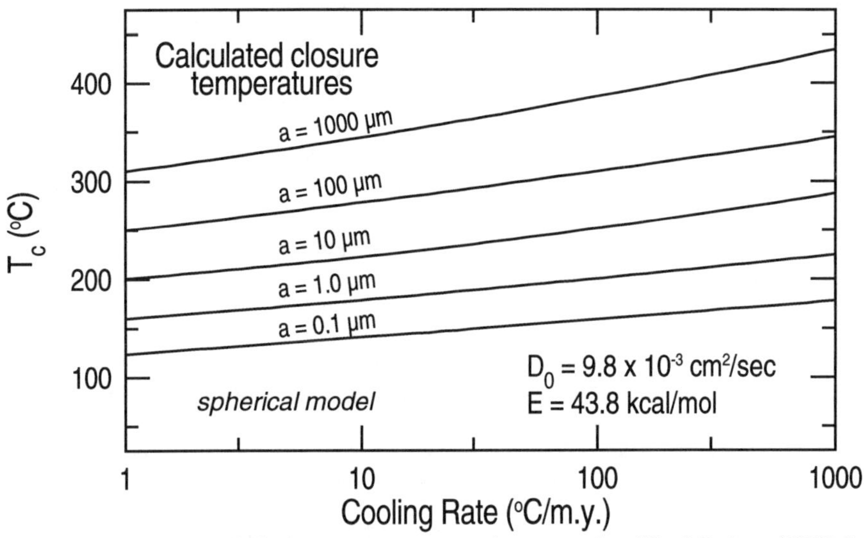

Figure 7. Calculated Ar method closure temperatures using expression [9] of Dodson (1973) for the parameters shown and for various effective diffusion dimensions (a). These closure temperatures are appropriate for alkali feldspar.

To estimate closure temperature, expression [9] requires values for D_0/a^2 (rather than needing both D_0 and a) and E. These quantities are potentially derived from an Arrhenius plot of kinetics data from a dating experiment without knowledge of the actual effective diffusion dimension. They can then be used to calculate a closure temperature. This approach has some essential major assumptions, namely: (1) Ar loss in the laboratory is via volume diffusion with all the assumptions above for the diffusion measurement; (2) the Arrhenius parameters are valid; and, (3) extrapolation of the Arrhenius relation to the much lower effective transport temperatures in nature is valid. The loss of Ar in the laboratory must follow the same mechanisms as in nature and with the same temperature dependence.

The approach has been utilized many times since its application by Berger and York (1981). Unfortunately, such experiments rarely yield linear Arrhenius relations, certainly in part due to violation of the assumptions of the diffusion measurement. It is now well established (see Gaber et al., 1988, Wartho et al., 1991; Lee, 1993 and references therein) that Ar degassing under vacuum heating of hydrous materials (micas, amphiboles) leads to modification and reaction. As a result, Ar release kinetics of such experiments are influenced by these changes and thus do not reflect volume diffusion. This is shown by comparison hydrothermal and vacuum-heating experiments. Feldspar clearly does not pose the same level of instability as the hydrous minerals. However, non-linear and complex Arrhenius plots are invariably obtained. The meaning of this behavior has been a matter of debate. Some (e.g., Harrison et al., 1986; Zeitler, 1987; Heizler et al., 1988) have suggested that the lower temperature apparent D/a^2 values reflect volume diffusion with higher temperature deviation due to reaction, for example, homogenization. Others (e.g., Parsons et al., 1988) suggest only higher temperature release reflects volume diffusion. Yet another interpretation is that deviations reflect the presence of effective diffusion domains of different dimensions (e.g., Zeitler, 1987; Lovera, et al., 1989). This is clearly a vital issue surrounding the application of the approach.

2.3.2. *Argon Gradients.* Variations in the ratio of $^{40}Ar/^{40}K$ within a sample are generally assumed to reflect either Ar loss or introduction. The loss may occur episodically or with protracted cooling within the transition region of partial Ar accumulation. Variations may also be present reflecting different diffusion dimensions. An expectation of the $^{40}Ar/^{39}Ar$ step-heating approach is that it will reveal spatial variations in radiogenic ^{40}Ar (or more precisely $^{40}Ar/^{39}K$ ratios) which reflect the gradients developed in nature.

Models to describe these gradients or variations are typically constructed assuming volume diffusion transport both in nature and the laboratory. A model relating episodic loss of ^{40}Ar in nature to the age spectrum with diffusion as the operative mechanism was presented by Turner (1968); subsequently, a number of authors have modeled more complex scenarios in terms of geometry, number and types of events, and size distributions (e.g., see McDougall and Harrison, 1988; Lovera et al., 1991; Xu, 1993). The simplest such model behavior is illustrated in terms of Ar gradients in Figure 8 and derivative age spectra in Figure 9. Here, isotropic diffusion in a spherical grain (or collection of uniform grains) experiencing a recent episode of radiogenic Ar loss is assumed.

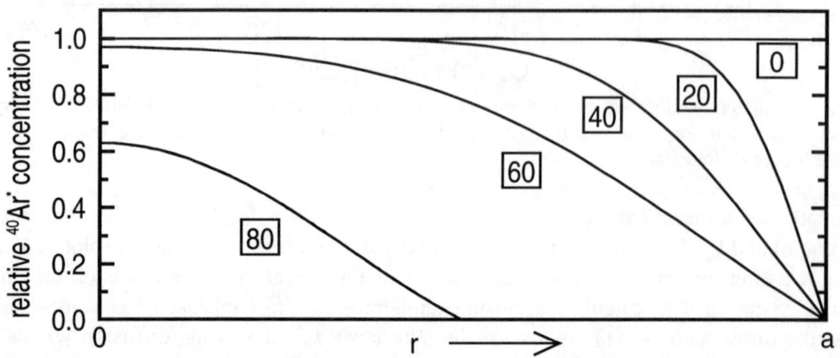

Figure 8. Concentration gradients for a grain which has experienced recent diffusive loss of radiogenic Ar shown as the relative concentration as a function of radial position, center to edge. Model curves are for isotropic transport in a sphere with an initially uniform concentration. Numbers in boxes indicate the model values in percent (e.g., 20, 40) of the initial ^{40}Ar that was lost.

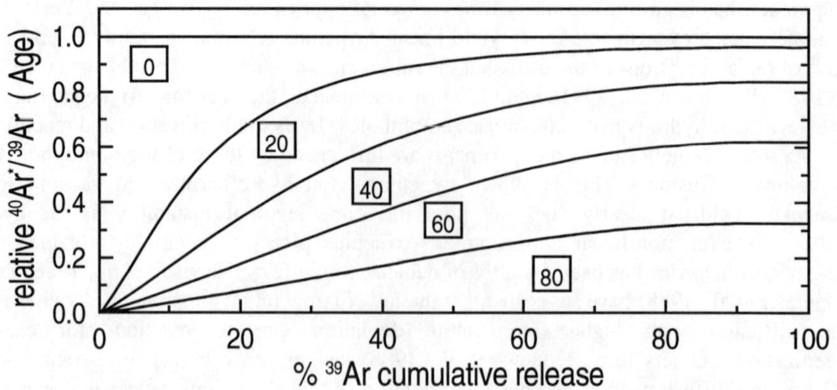

Figure 9. Ideal incremental-heating release pattern in terms of normalized $^{40}Ar^{*}/^{39}Ar$ ratio (or age) for the grain, or a uniform collection of grains, having the distribution shown in Figure 8 and releasing Ar via diffusion in the dating experiment. Assumptions as in Figure 8.

Models of this nature, although often more complex than shown above, have been applied to explain age spectra of various sorts. Some of these models are quite successful at fitting the relationships. However, there are no unambiguous experimental data that verify that gradients can be accurately revealed by step heating of any material. Again, it is now clear that release of Ar from hydrous minerals and the accurate reflection of Ar gradients are compromised by sample instability (e.g., Lee, 1993).

3. Brief Review of Selected Feldspar Argon Studies

3.1. OVERVIEW

It is useful to review some aspects of Ar research on feldspars in order to set issues into perspective. This survey is only a brief chronological summary of selected studies dealing with Ar behavior of natural samples since it is impossible to treat all related work here. The focus is upon alkali feldspars because this is timely and the area of most current interest. While it is not possible to mention many significant papers, the intention is to capture the progression of ideas and the points of new ones or significant departures from earlier work in the area.

3.2. EARLY DIFFUSION STUDIES

The history of Ar diffusion studies of feldspar and other minerals is punctuated by periods of doom and gloom, unbridled optimism, and intense interest and controversy. It is perhaps appropriate to start here with the pessimistic, yet insightful review by Mussett (1969) who summarized early measurements made mostly during the 1960's. Mussett (1969) found that Ar "diffusion" rates in various minerals showed remarkable spread and, in some cases, prohibitive apparent behavior. At that time in the context of development of Ar geochronology, the K-Ar ages of potassium feldspars were acknowledged as "unreliable" in terms of giving the times for formation of rocks. Indeed, it had been widely appreciated for more than 15 years that plutonic varieties, even those from regions which had no history of thermal disturbance, frequently gave ages which were "too young." This became immediately apparent during some of the pioneering work in geochronology including attempts to use alkali feldspar for making a geologic determination the of ^{40}K branching decay ratio (Wetherill et al., 1955; Wasserburg et al., 1956). By this time it was also recognized that volcanic feldspars, both plagioclase and sanidine, usually provided "reliable" ages (see Lanphere and Dalrymple, 1969).

Mussett (1969) compiled the extant Ar diffusion data for feldspars, shown here as Figure 10, emphasizing that there was an enormous apparent spread in D at one temperature. Almost all samples show distinctly non-linear Arrhenius relations that frequently are concave up or even have inflections. Mussett (1969) concluded that the loss of Ar from minerals in laboratory experiments differs from natural loss in several ways: (a) grains may be damaged during extraction from a rock; (b) separated grains lose Ar into a vacuum or other atmosphere but not, as in nature, into the rest of the rock; and, (c) lab measurements are made at much higher temperatures. He further pointed out the importance of radiation damage and its "probable" effect on defects and rates.

A new and much more encouraging perspective on the kinetics of diffusion measurements emerged in the early 1970's, largely from the work of B. J. Giletti and associates (Hofmann and Giletti, 1970; Giletti, 1974a, b; Foland, 1974a). The problems of many previous measurements were recognized as largely twofold (Giletti 1974a): sample stability and breakdown of mineral phases; and, multiphase specimens. The first factor is illustrated by the very rapid loss of Ar by phlogopite heated in vacuum compared to hydrothermal conditions where the former is dominated by reaction rather than diffusion (Giletti, 1974b). The second factor is well illustrated by perthitic alkali feldspars for which assumed boundary conditions, for example, the effective diffusion dimension, were incorrect.

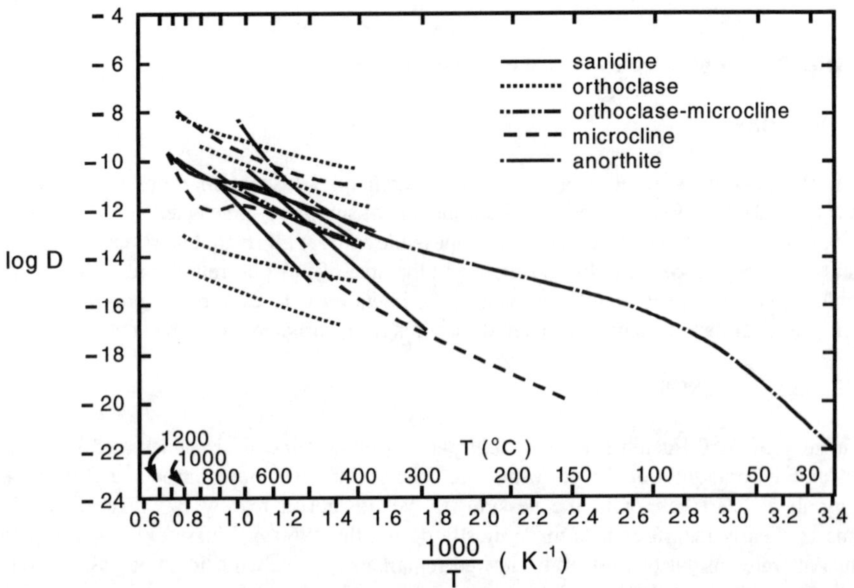

Figure 10. Arrhenius plots for feldspars from Mussett (1969), showing the wide variations in apparent diffusion coefficients.

Recognizing the potential pitfalls outlined above, Foland (1974a) examined the diffusion of Ar in a sample of orthoclase and found a linear Arrhenius relation with an activation energy of about 44 kcal/mol based upon long-term (days to months) isothermal heating experiments. The material studied was termed "homogeneous," meaning non-perthitic and uniform when subjected to examination by microscopic, microprobe, and X-ray methods. The basic conclusions were that Ar transport is rapid relative to the large cations which also have higher activation energies. Moreover, the Ar diffusion rate in feldspar compared to that in mica (e.g., phlogopite) was more rapid as implied by the typical pattern of relative ages. However, the diffusion rate at low temperature, according to linear extrapolation of the Arrhenius relation, was not inherently so rapid that Ar should be lost readily for typical grain sizes. The offered explanation was that the actual transport distance was much smaller than actual grain size. That is, Ar need only travel a small distance by volume diffusion before encountering a grain boundary discontinuity that was characterized by rapid transport. Foland (1974a) suggested that the boundaries of perthite lamellae. which were regarded to be discontinuous, provide such pathways. This was not entirely new as others had suggested a relationship between K-Ar age and 'perthitization'. The large range for D previously observed was attributed to a number of factors, including heterogeneities in Ar distribution, incorrect boundary conditions, incorrect value for the effective dimension, loosely held Ar in grain-boundary regions, and sample reaction.

3.3. $^{40}Ar/^{39}Ar$ STUDIES

Most more recent studies addressing Ar diffusion in feldspar have used the $^{40}Ar/^{39}Ar$ technique with apparent diffusion coefficients calculated from step-heating measurements in a high vacuum system. A large number of studies over the last 15 years report Ar release kinetics and apparent diffusivities. The spread of behavior is huge and may be attributed to the above factors. It is clear that one of the key assumptions fails to hold in the context of measuring basic material properties. Indeed, the discordant age patterns which are more common for feldspar, especially slowly cooled ones, imply serious complications. While the significance of many apparent kinetic data cannot be evaluated unambiguously, a number of studies, some of which are mentioned below, provide important suggestions or information.

Harrison and McDougall (1981) reported calculated diffusion coefficients for plagioclases with excess Ar and proposed three different sites of Ar giving rise to separate apparent Arrhenius relations. They suggested that excess and radiogenic Ar occupy different lattice positions, with anion vacancies more retentive than cation sites, that can lead to the saddle-shaped spectra (see Figure 3), well known since the work of Lanphere and Dalrymple (1976). In another study, Harrison and McDougall (1982), publishing results of measurements for microclines and anorthoclases, concluded that age spectra for some microclines were consistent with ^{40}Ar gradients developed during slow cooling. They also suggested that Ar diffusion in K-feldspar was rather strongly dependent upon structural state with activation energy for microclines at about 30 kcal/mol and much less than the roughly 45 kcal/mol for monoclinic K-feldspars. With the scatter in diffusivities and with some uncertainties in the actual temperatures of these experiments (McDougall and Harrison, 1988), the significance of this observation is not clear. Harrison and McDougall (1982) proposed a new explanation for U or saddle-shaped spectra, namely, that they were the result of inherited excess Ar superimposed upon a ^{40}Ar loss gradient. Along with Berger and York (1981) this is among the early attempts to use feldspar age spectrum step-heating data to define geological closure temperatures.

Zeitler and Fitz Gerald (1986) studied potassium feldspars containing excess Ar by $^{40}Ar/^{39}Ar$ and TEM methods. They noted that microstructural pathways were abundant and reduced the effective grain size and would promote uptake of Ar. Their samples showed the saddle-shaped type of spectra where excess Ar dominated at both low and high temperatures. Drawing upon the earlier suggestion by Harrison and McDougall (1981) and the correlation of excess ^{40}Ar and Cl-derived Ar noted by Claesson and Roddick (1983), Zeitler and Fitz Gerald (1986) proposed to account for this by having two different sites of Ar, cation- and anion-sited, which were liberated differentially. According to this proposal, the cation-sites contain both radiogenic and excess Ar that is released mainly at lower temperature whereas the anion sites contain excess Ar that is released predominantly at higher temperature.

Zeitler (1987) published results on important experiments in which several feldspars were heated to induce Ar loss and then subjected to $^{40}Ar/^{39}Ar$ analysis. The resulting spectra failed to conform to that predicted for simple diffusive loss from uniform grains. Zeitler (1987) suggested that the patterns could be explained if the samples had multiple grain sizes or characteristic diffusion distances. The consequences of having such variations had been considered by several previous workers, in particular, Turner (1968), and Gillespie et al. (1982). Zeitler (1987) heated a sample with excess Ar at 400°C both under 0.1 GPa water pressure and in air and noticed a relatively small but significant difference between the loss of the apparent excess component for hydrothermal heating. This was interpreted as further support for anion vacancy siting in view of the very different oxygen diffusion rates under wet and dry conditions. Finally, Zeitler (1987) concluded that heating changed Arrhenius parameters for the lower temperatures of Ar release. Both of these latter points were challenged by Villa (1988; see also, Zeitler, 1988), especially the modified Arrhenius trajectories which are not only ambiguous in terms of sample modifications but also have major implications for the validity of extrapolating release data to nature.

A new and different view of Ar loss was presented by Parsons et al. (1988) who studied $^{40}Ar/^{39}Ar$ and microstructural relations in alkali feldspars from the Klokken syenite in Greenland. These authors found a variety of spectra with the non-turbid cryptoperthites giving the flattest spectra that show an age in agreement with the time of intrusion. The turbid and more coarsely perthitic samples were more discordant and had lower ages. Turbidity was the only feature systematically correlated with Ar loss. The interpretation was that submicron "micropores" or channels causing the turbidity were responsible for Ar loss. This "hole" model was advocated with the repudiation of previously proposed "perthitization" or "microclinization." The coherent boundaries in perthites were rejected as pathways for fast transport. Parsons et al. (1988) also suggested that the step-heating data were compromised by recoil in cryptoperthites and by lab readjustment at about 1000°C. The rate controlling factor for laboratory degassing was taken to be volume diffusion at high temperature but not low. Previously, most workers adopted just the opposite view; for example, suspicions about homogenization and disordering led to the notion that low temperature diffusivities (e.g., Harrison et al., 1986; Heizler et al., 1988) reflected volume diffusion. Parsons et al. (1988) concluded that degassing in the laboratory is not analogous with natural loss.

The microtextures in the Klokken feldspars were further characterized by Worden et al. (1990) who document the development of micropore "diffusion channels" that are proposed to enhance reactivity. Laser probe and *in vacuo* crushing of Klokken samples containing excess Ar by Burgess et al. (1992) demonstrated that crushing liberated much of this anomalous component as well as ^{38}Ar derived from Cl. This study made a convincing case for the origin of excess Ar from ^{40}Ar- and Cl-rich fluids and its presence in fluid inclusions that were liberated in crushing experiments. Burgess et al. (1992) proposed inclusion-hosted excess Ar as an alternative to previous siting propositions. Turner and Wang (1992) also demonstrate the correlation of excess Ar and Cl in crushing experiments and emphasize the importance of micropores. They suggest that pores also contain both radiogenic ^{40}Ar due to diffusion and ^{39}Ar due to recoil.

A different but not completely incompatible view of K-feldspar spectra for slowly cooled samples was presented by Lovera, Richter, and Harrison (1989) who proposed that discordance be explained as reflecting the presence of diffusion domains of different sizes (and thus ages). Subsequently, this hypothesis, the multi-domain diffusion model, has been explored and defended in a series of papers by T. M. Harrison and coworkers (Lovera et al., 1991, 1993; Harrison et al., 1991; Richter et al., 1991; Fitz Gerald and Harrison, 1993). The model posits a distribution of discrete diffusion domain sizes to account for the apparent diffusivities of Ar and the non-linear Arrhenius behavior and the age spectrum. The basic idea is that the smaller domains release gas at lower temperatures in the laboratory and also have younger ages due to lower closure temperatures. Both the kinetic and age data are forward modeled for a given domain distribution to obtain a self-consistent fit. For example, the spectra illustrated in Figure 4 have been modeled with this approach to obtain realistic time-temperature cooling scenarios.

Lovera and coworkers have used as few as three and as many as seven discrete domain sizes in modeling. In effect, the kinetics of Ar release during a step-heating experiment are interpreted to reflect volume diffusion described by an activation energy and different values of (D/a^2). Their assumed sizes span more than three orders of magnitude in one specimen, MH-10. With TEM study, Fitz Gerald and Harrison (1993) were able to identify qualitatively candidates for the larger and smaller domains but not a clear one for the intermediate size. Harrison et al. (1991) have further extended the original model proposing that various domains have different activation energies. A multiple activation energy model produced improved fits although there is no obvious reason for variable activation energy. If this proposition is valid, then the age spectrum obtained will depend upon the details of the heating schedule.

Of course, the determination of a cooling history which follows from quantitative modeling in this manner presents great promise for constraining geologic processes. However, this modeling involves a number of critical assumptions. The most significant is that the release of Ar in the laboratory occurs by the same mechanisms and with the same temperature dependence as that over much longer times and lower temperatures in nature. This is a stringent requirement in view of feldspar microstructures and reactivity and the long extrapolation. At the same time, the cooling rates that have been obtained in these studies appear to make good geological sense and the compatibility of the age spectrum and kinetics of Ar release provide powerful support for the basic model.

Several workers have challenged applicability of the multi-domain model. Girard and Onstott (1991) found that such a model was inconsistent with kinetics of laboratory degassing from detrital and authigenic K-feldspars. They indicate that more complex mechanisms must be involved and suggest Ar transport by volume diffusion through the lattice and short-circuit diffusion along pathways of rapid migration. Lee (1993) has described a model of this type. Parsons et al. (1991) contend that major changes in microtextural diffusion pathways occur at low temperatures in nature. Villa (1991) questioned volume diffusion as the operative process and proposed that the physics of Ar loss is different in experiments and in nature. Xu and Foland (1990) and Foland et al. (1991) note that step-heating kinetic data are mediated by processes that modify the normal diffusive losses from feldspars at high temperatures.

4. Defining the Parameters of Ar Diffusive Transport in Feldspar

In view of the many interpretations of feldspar Ar behavior, it seems critically important to define basic transport phenomena. This is even more imperative if models, either simple or elaborate, are constructed to explain complicated patterns and define thermal histories. There are a number of outstanding questions some of which might be formulated as follows: Is Ar transport affected by irradiation and, if so, does this lead to an enhancement of mass transport rates or even trapping? Are there different normal crystallographic sites that hold Ar and which manifest themselves in dramatically different kinetics? Is the loss of Ar by feldspar with normal laboratory heating controlled by volume diffusion and Fick's law? If so, do different Ar components and isotopes migrate by the same mechanism and at the same rate? Is transport intrinsically or extrinsically controlled in the laboratory and in nature? Does transport during step-heating analysis conform to the rates and behavior in nature or even that predicted from longer-term experiments?

Although not all the answers to these and related questions are obvious, some can be addressed satisfactorily while still others may be constrained. The most profitable approach for doing this is to build upon simple behavior rather than deal with observations that can be explained only by higher order modeling. The simplest behavior and most definitive description of Ar diffusion in feldspar are for the Benson Mines orthoclase as presented by Foland (1974a) and in subsequent studies (Foland and Xu, 1990; Xu, 1990; Xu and Foland, 1991; Foland et al., 1992). These results are summarized in the next section.

4.1. BEHAVIOR IN BENSON MINES ORTHOCLASE

4.1.1. *Specimen.* The characteristics of the material and its behavior during experimentation are obviously all-important to interpretation of degassing kinetics. The Benson Mines orthoclase comes from a pegmatite in the Adirondack Mountains of New York State and is described by Foland (1974a). It is a clear and colorless monoclinic K-feldspar which might be described as "gem quality." It has the composition $K_{0.915}Na_{0.067}Al_{1.06}Si_{2.96}O_8$, corresponding to Or_{94}, with only approximately 150 ppm Ca. By examination using visible-light microscopy, electron probe microanalysis, the scanning electron microscope, and X-ray powder and precession techniques, it appears to be a single phase sample without any alteration, zoning, twinning, measurable major element variations, or separate albite phases. Heating to temperatures sufficient to homogenize perthite lamellae does not change the positions of compositional-dependent reflections. R.A. Yund has performed detailed TEM examination of the Benson Mines orthoclase and found the sample to be essentially an ideal material for diffusion studies, viz., it is very uniform with no alteration or second phase impurity, no cracks, no dislocations, etc. It has the "tweed" microstructure showing a pattern with a wavelength of modulation on the order of 5 nm. This characterization is consistent with its known behavior.

With K and radiogenic ^{40}Ar concentrations of 3.23×10^{-3} mol/g (12.6 wt % K) and 2.40×10^{-8} mol/g, respectively, it has a K/Ar age of 856 Ma. Its incremental-heating spectrum is illustrated in Figure 11. The pattern shows minor discordance with a range of only ~ 2% in apparent age with a consistent trend. The K is uniform (based upon microanalysis) and the implication of the spectrum is that radiogenic ^{40}Ar is also uniform or sufficiently so for the purpose of diffusion modeling. When annealed at 800°C for 10 days, the feldspar showed no indication of Al-Si disordering or other changes in powder pattern which might indicate reaction.

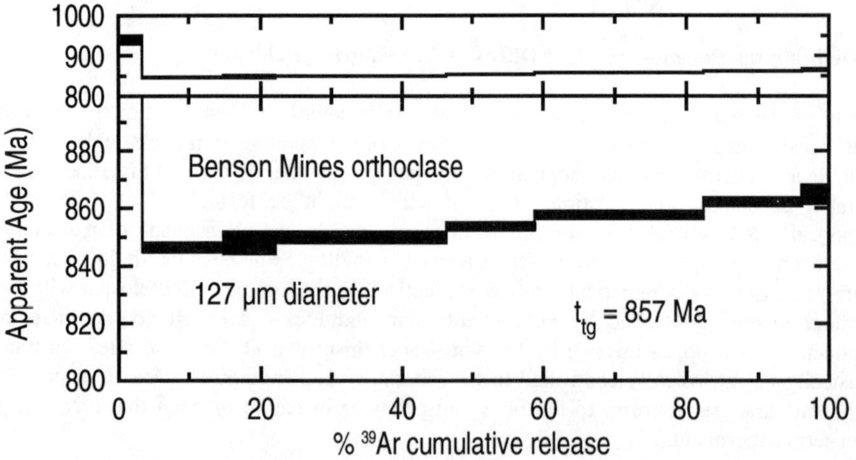

Figure 11. Incremental-heating spectrum for the Benson Mines orthoclase (from Foland and Xu, 1990).

The grains used for diffusion study were prepared from larger pieces by hand grinding and repeated sieving. They have a variety of shapes but are typically square to rectangular in cross section. As is inevitable, there are many irregularities such as stepped cleavage surfaces. The use of different size fractions permits examination of the effective dimension of diffusion. Three carefully sized fractions with diameters averaging 127 ± 6, 233 ± 10, and 480 ± 17 μm have been studied. Using the same sizes Foland (1974b) showed that the physical grain dimension was the effective dimension of diffusion for cation transport. Ion probe profiling experiments of even larger grains subjected to hydrothermal exchange with isotopically enriched solutions for large cations (see Giletti, this volume) and oxygen (Foland and Giletti, unpublished data) are consistent with volume diffusion transport controlled by physical grain dimensions both in terms of values of the diffusion coefficients and the shape of depth profiles.

4.1.2. *Behavior of Ar in Natural Material.* The diffusivity of Ar is best described by the initial isothermal experiments at 500 to 800°C for 2 days to several months using unirradiated material. A series of runs at 700°C for varying lengths of time (up to about one month) addresses the nature of the loss and the dimension controlling it. Shown in Figure 12 in the form of the well-known test of diffusion control, Ar losses yield a single value for the D which is consistent with a volume diffusion mechanism with a single effective diffusion dimension. The values are consistent with a single (D/a^2) without significant time dependence.

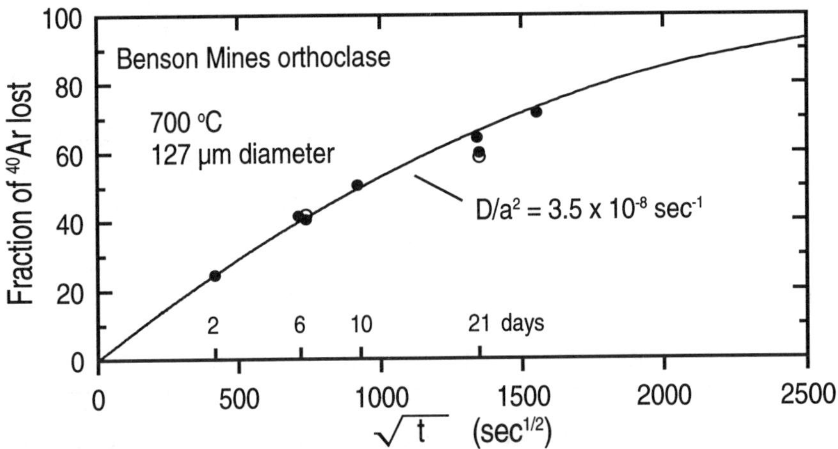

Figure 12. Fractional ^{40}Ar loss as a function of square root of heating time for isothermal experiments at 700°C using 127 µm grains (data from Foland 1974a and Foland and Xu, 1990). Open symbols represent irradiated samples, closed untreated. Observed losses conform to loss by diffusion with a single diffusivity for all data up to the highest loss of about 70%. There is virtually no time dependence in measured D value.

Diffusion coefficients for all sizes of grains and for experiments of various conditions are shown in an Arrhenius diagram in Figure 13 using the spherical model. These D values are calculated using the measured grain sizes that vary by a factor of four and the (D/a^2) values derived from the solution for the spherical model. To emphasize the relationships of various D values, they are also shown in Figure 14 normalized to the value from the Arrhenius relation. The diffusion coefficients show close agreement for different sizes although the (D/a^2) values show substantial and expected differences which reflect the size disparity. This indicates that the actual grain size is the effective transport dimension. If this were not the case, the D values would be disparate and correlated with grain size. The physical size-diffusion dimension relationship is precisely the same as that found for other elements.

The experiments in Figures 13 and 14 are for several types of heating conditions: sealed in a vacuum capsule, open to the air, or hydrothermally at 0.2 GPa water pressure with either distilled water or alkali chloride solutions. The transport rate does not depend upon these conditions. It is also notable that a sample heated in air with an Ar partial pressure of about 10^3 Pa does not absorb noticeable quantities of atmospheric Ar; this observation is consistent with the very strong partitioning of Ar to a gas phase relative to the crystal lattice.

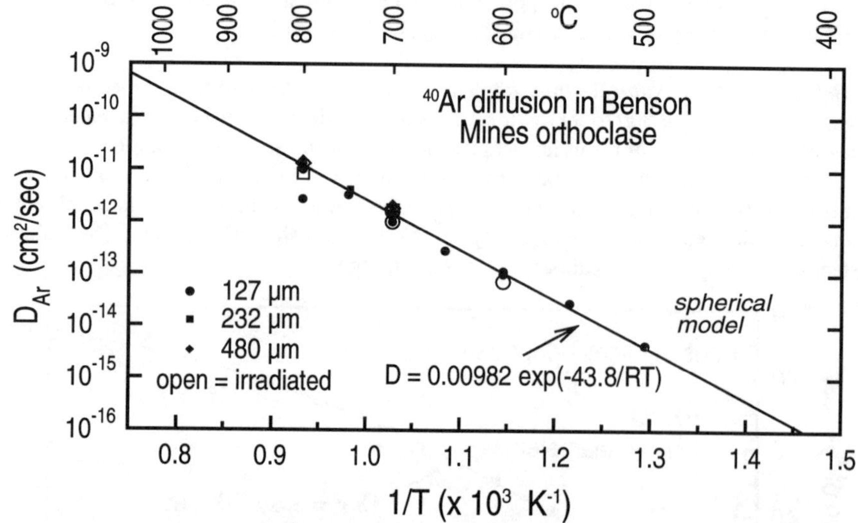

Figure 13. Arrhenius diagram for ^{40}Ar diffusion in Benson Mines orthoclase (data from Foland et al., 1974a; Foland and Xu, 1990) calculated using a spherical model. Shown are all data for the three different grain diameters and for both unirradiated and irradiated samples.

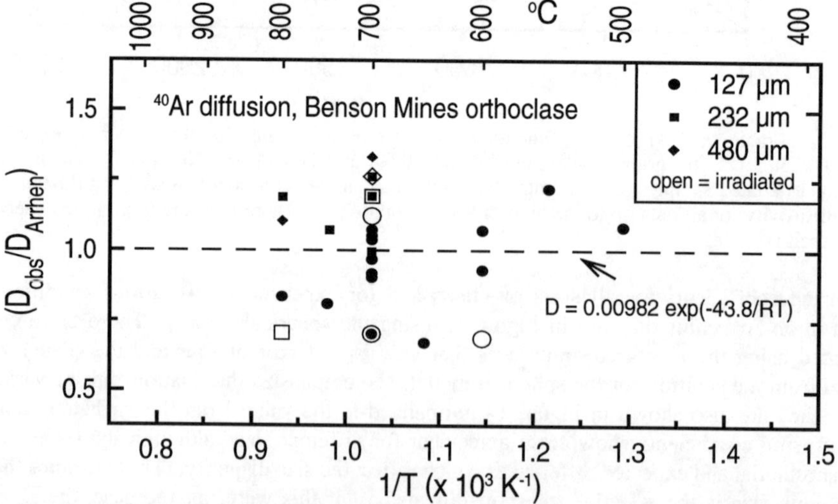

Figure 14. Comparison of individual observed D values with those predicted by the Arrhenius relation shown. The individual determinations deviate from the Arrhenius value only within about ± 30% or less. N.B. Different size fractions and also irradiated specimens give the same D value when the measured grain size is used as the effective diffusion dimension. There is no apparent systematic difference in these values.

The diffusion coefficients from the isothermal heating data above define a linear Arrhenius relation with D_0 = 0.00982 (with one-sigma limits of + 0.00660 and -0.00371) cm^2/sec and E = 43.8 (± 1.0) kcal/mol for the spherical model. The same data could be modeled with alternative geometries and Foland (1974b) did so for the infinite cylinder. For other models, cylinder or plane sheet, the results will be similar but with a higher D_0 by factors of about 2.0 and 6.3 consistent with the ratios of the values for A in expression [9]. Simply using the same activation energy and these values to obtain a frequency factor is sufficient for virtually any modeling.

4.1.3. *Behavior of Ar in Irradiated Material.* Foland and Xu (1990) reported on results for irradiated Benson Mines feldspar. Isothermal heating experiments at 600 to 800°C for periods of 4 days to 3 weeks yielded diffusion coefficients which were indistinguishable from those for unirradiated grains (see Figures 13 and 14). Under these conditions, the massive doses of neutrons appear to have little or no effect on the kinetics.

The nearly flat spectra demonstrate that the diffusion coefficients of ^{39}Ar and ^{40}Ar cannot be very different because incremental heating is a sensitive measure of relative diffusivities. Only very minor differences (a few percent or less) in D_0 or E are permissible because of a flat age spectrum as shown in Figure 11. It follows that both radiogenic and neutron-induced Ar have the same transport mechanism in spite of the different effects associated with their production (see Foland and Xu, 1990).

Small yet significant variations in the age spectrum are observed. The elevated apparent age observed for the lowest temperature is a minor effect probably due to recoil. The small yet steady increase of apparent age with increasing degassing implies either initial variations in ^{40}Ar or slightly faster transport of ^{39}Ar. The first explanation is possible if there are $^{40}Ar/^{39}K$ heterogeneities, for example, the result of a domain distribution under slow cooling. Foland and Xu (1990) discount this explanation for several reasons and suggest small differences in behavior of ^{39}Ar and ^{40}Ar. These could reflect either a kinetic isotope effect in diffusivity or enhancement of ^{39}Ar diffusion that may reflect recoil damage or retardation of ^{40}Ar transport.

A kinetic isotope effect of full magnitude predicts the diffusivity of ^{39}Ar to be 1.3% greater than ^{40}Ar which would correspond to the inverse ratio of the square roots of their masses. This possibility was further investigated by Xu (1990) for a specimen irradiated in the more conventional way for $^{40}Ar/^{39}Ar$ dating and also with 14 MeV neutrons that produce appreciable ^{38}Ar from ^{39}K (Foland et al., 1987). The step-heating results failed to confirm the effect. With increasing Ar release the $^{39}Ar/^{38}Ar$ ratios were nearly constant or even decreased slightly which is the opposite of the trend predicted for a kinetic isotope effect where ^{38}Ar diffusivity would be slightly greater. Unfortunately, this result is not fully definitive because measurement uncertainties are relatively large compared to the magnitude of expected variations. Moreover, there are possible recoil effects which will be different for ^{38}Ar compared to ^{39}Ar.

4.1.4. *Behavior of Ar in Incremental-Heating Experiments.* Subsequent step-heating analyses on unirradiated, irradiated, and previously outgassed and then irradiated material have been reported by Xu and Foland (1991) and Foland et al. (1992). While the full details have not yet appeared in the literature, the basic findings are important enough to mention briefly. Figures 15 and 16 illustrate these by showing the results of two experiments in terms of both the patterns of the age spectra and the kinetics of outgassing.

One experiment is for 127 μm diameter grains that were heated for three weeks at 700°C to induce ^{40}Ar loss and then were irradiated in the normal manner. The feldspar was subjected to incremental-heating analysis in a normal manner over about 100 hours where the furnace temperature was carefully calibrated and the grains wrapped in Pt foil to guard against any possible reaction with the substrate materials. The resulting age spectrum and the apparent diffusion coefficients for calculated ^{39}Ar are shown in Figures 15 and 16, respectively. Another experiment shown in Figure 15 is for an irradiated but otherwise untreated sample of the same grain size that was subjected to the same experimental protocol and heating schedule.

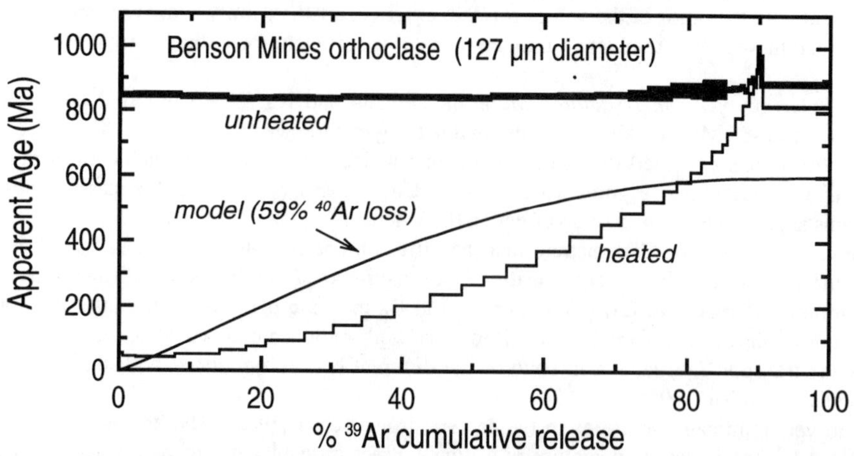

Figure 15. Incremental-heating diagram for Benson Mines feldspar which was heated (700°C) prior to irradiation compared to an unheated aliquot. Both experiments follow the same heating schedule which lasted for a total of about 100 hours with the last 10% of ^{39}Ar given off in a single fraction at very high temperature. Both show the same abrupt rise in apparent age near 90% release. The heated sample lost 59% of ^{40}Ar prior to irradiation. Shown also is a model spectrum for this degree of loss assuming isotropic transport in spherical grains as is illustrated in Figures 8 and 9.

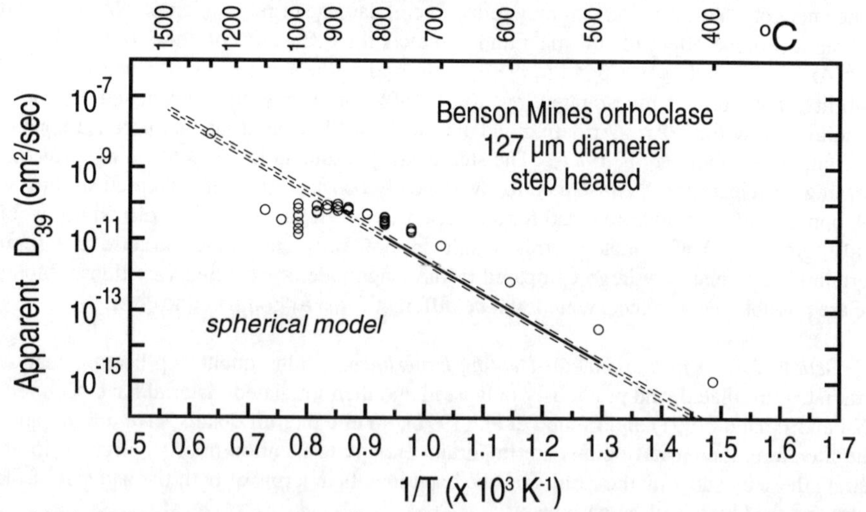

Figure 16. Arrhenius diagram of apparent ^{39}Ar diffusion coefficients calculated for the sample shown in Figure 15. The solid line is the Arrhenius relation determined from long-term experiments as shown in Figure 13 and is bounded by dashed lines at ± 30%. Apparent diffusion coefficients for ^{39}Ar are calculated using the spherical model for uniform grains having the original measured radius.

The step-heating data show important deviations from those predicted on the basis of the behavior already described. All experiments show three common anomalous features: Ar is liberated more rapidly than expected at lower temperatures but much more slowly at higher ones especially those in excess of about 950°C; with extended heating at high temperature, the apparent age rises steeply to extremely high values, continually doing so until the last several percent of Ar are released in a final very high temperature step; and, the age spectra for previously outgassed samples fail to conform to theoretical models. All of these features appear to reflect two processes: cracking and spalling of grains during heating; and, trapping of Ar at high temperature. The first process can be verified directly by visual inspection. The second, however, is by inference since trapping is the only plausible way to account for incredibly low outgassing rates.

Several aspects of the step-heating data are similar to those observed for many alkali feldspars and have recently been attributed solely to variations in domain size. However, in this case such an explanation is untenable because of the effective diffusion dimension is known to be the physical size. This explanation would require repudiation of a large quantity of other diffusion data and relationships established for oxygen and several cations as well as the data for Ar itself at temperatures of 800°and below (see Figure 12). The aberrations simply need to be accounted for within the processes of the step-heating experiments themselves. Obviously, new grains or domains that have small characteristic transport distances are generated by the experiment itself.

A process which affects the release rates is grain fracturing in which larger grains crack apart and small pieces spall off. This is clearly and easily documented with simple examination of grains after heating in a vacuum furnace. The amount and size of fines produced has not yet been quantified nor has the possibility that larger and still intact grains might contain induced fractures. These factors are probably at least somewhat dependent on heating schedule. Cracking is observed for both irradiated and natural grains. It presumably reflects rapid temperature changes which characterize incremental-heating equipment and probably continues with spalling accompanying successive temperature changes. An obvious consequence of the size reduction is enhanced rates of Ar loss which will be reflected in anomalously high apparent diffusion coefficients as seen in Figure 16.

Grain cracking will not only affect the release kinetics but also impact the shape of the $^{40}Ar/^{39}Ar$ profile of previously degassed samples. A consequence of the process will be to expose regions having higher $^{40}Ar/^{39}Ar$ ratios that were initially interior to larger grains. As a result, somewhat elevated apparent ages will be observed for the low fractional losses whereas lower ages will be seen for intermediate releases. The spectra for such samples show this behavior as illustrated in Figure 15.

A different effect is observed at high temperatures where the release rates decrease substantially and yield apparent diffusion coefficients that are much too low. The dramatic decrease in calculated D results in inflection of points on Ahrrenius diagrams as shown in Figure 16. This is similar to the behavior seen in virtually all feldspars at temperatures of roughly 950°C and is consistent with trapping of Ar. The exact cause of this trapping is not obvious but it conceivably reflects either the complete annealing of defects that are key to transport or the capture and retention of Ar in some imperfection or trap. There are several types of a potential traps including vacancy clusters, recoil tracks, or healed or partially healed cracks. It is likely that the number of traps is temperature dependent and increases at high temperature. Argon may become trapped simply by random motion during the normal transport process. After a jump into a trap, the atom has difficulty escaping because the activation energy needed is quite high. An analogy would be the formation of bubbles that trap gas because the gas has a much lower fugacity in the void space than in the surrounding crystal. Such a preference for occupying void space is extremely strong for a noble gas atom as is clear from distribution coefficients between crystalline solids and gas phases (see Ozima and Podosek, 1983).

The trapping phenomenon has a number of implications and affects release spectra because those atoms with the largest required transport distance have the highest probability of being trapped. Thus, trapping will preferentially retain Ar present in the deep interiors of grains. As a result, the pattern of $^{40}Ar/^{39}Ar$ release will be influenced and it is not surprising that observed spectra for outgassed samples depart from the model ones. This is illustrated in Figure 15 where the apparent ages rise considerably above those predicted from Figure 9 but are consistent with the ratios for Ar in the deeper interiors of grains as expected from the spatial distributions illustrated in Figure 8. The anomalous and rapid increase in apparent age observed at high temperature and at about 80% or more of released Ar implies that ^{40}Ar and ^{39}Ar are trapped differentially.

The spectra observed for material outgassed prior to irradiation (e.g., Figure 15) are broadly similar to those found by Zeitler (1987) although he attributed his results to different sizes of diffusion domains. Trapping provides another method of producing this behavior and provides an explanation for the very large domain sizes suggested by results for some natural samples.

There are some interesting ramifications of trapping Ar. The larger the required transport distance, the higher the probability of trapping and the higher the temperature required to liberate Ar within an acceptably short measurement interval. The higher temperature suggests a greater abundance of imperfections capable of serving as traps. In effect, the process will tend to accentuate differences that may already be present in a sample containing a spectrum of sizes.

Overall, the incremental-heating experiments show important differences from the kinetics of Ar loss observed in long-term experiments below 800°. The differences have obvious significance to the interpretation of data for complex samples and the application of incremental-heating data for the calculation of closure temperatures. The processes responsible, cracking and trapping, greatly complicate both the kinetics and the resolution of initial Ar gradients.

4.2. GENERALIZATIONS

The relationships for the Benson Mines orthoclase can be used to stimulate a more global consideration of Ar behavior in feldspars. This section discusses several basic phenomena, draws conclusions where possible, or points out areas for additional research. In summarizing, it is clear that much remains to be learned about the basic processes and noble gases in minerals.

4.2.1. *Diffusion following "Fickian" behavior.* There are few, if any, sets of Ar kinetics data that are simpler than those for the long-term experiments on Benson Mines orthoclase. These data make a compelling case for Ar laboratory transport by volume diffusion as described by Fick's Laws for temperatures below 800°C. The trapping observed in step-heating experiments could be a second-order effect important only at the higher temperatures of laboratory study. With this exception, it seems reasonable to regard transport through the "normal" crystal lattice as volume diffusion for feldspars in general. There are, however, a few caveats, e.g., the nature and effect of microstructures and the experimental assumptions emphasized above.

4.2.2. *Anisotropy of Ar Transport.* The directional dependence of Ar diffusion is essentially unknown. Without definitive data or obvious structural indication, it seems most reasonable to infer anisotropy based upon that observed for large cations and oxygen. The variations in diffusion coefficients for these, as determined by profiling experiments in different crystallographic directions, are relatively small (Giletti, this volume). The suggestion that diffusion in feldspar be regarded as approximately isotropic by Yund (1983) continues to seem appropriate.

An issue related to anisotropy in terms of diffusion modeling is the geometry of effective diffusion domains. As pointed out by Fitz Gerald and Harrison (1993), the complex shapes of microstructural features and their relation to one another complicate the task of modeling. The impact of the necessary simplifications concomitant with the choice of models is not obvious.

4.2.3. *Effect of external atmospheres.* The possible effects of water or a related species on kinetics have important implications especially for those concerning the applicability of laboratory rates of vacuum degassing. Even though the data set is limited, there is clear evidence is that the presence of water or an oxidizing atmosphere has no effect on Ar kinetics. Experiments by Zeitler (1987) showed an effect on excess Ar content for a sample heated hydrothermally. However, this observation seems more simply explained as due to effects of water on complex microstructures, for example, small amounts of dissolution that open micropores containing Ar.

The presence of aqueous fluids is obviously important in Ar behavior given their roles in such processes as dissolution and precipitation of new phases or the capacity for bearing excess ^{40}Ar. Aqueous species have a dramatic catalytic effect upon oxygen isotope exchange in feldspar (see Graham and Elphick, this volume). However, there appears to be no similar effect on Ar. This means that this phenomenon does not present a problem for the application of laboratory data to natural settings.

4.2.4. *Effects of neutron irradiation.* The effects of irradiation may manifest themselves in a number of ways mentioned in section 2.1.3., but there are few data that address them. The imperfections attendant upon irradiation do not produce a dramatic effect on the transport rates in experiments at temperatures of 600 to 800°C over period of days. Apparently, annealing of the damage is sufficiently rapid under these experimental conditions. There are possible effects for shorter times and at significantly lower temperatures which need evaluation.

Horn et al. (1975) found that plagioclase irradiated with a high total fast neutron fluence of some 10^{19} n/cm^2 was more retentive of Ar and suggested that this reflected trapping due to coalescing of induced defects. Their neutron dose was about an order of magnitude higher than that used by Foland and Xu (1990) and typically for terrestrial samples which probably accounts for the difference in results. If the Horn et al. (1975) observation is correct, some effects albeit probably not as cannot be precluded for lower doses.

The energetic recoil of ^{39}Ar (and other nuclides) may affect transport rates because of the production of damaged regions on the order of 0.1 μm. These regions may provide kinetic enhancement much like the dislocation-assisted diffusion of oxygen in albite (Yund et al., 1981). By analogy, this effect would be expected to be relatively small considering total transport through bulk material. Assuming that ^{39}Ar is associated with a region of damage, it may experience rapid transport along this area. Compared with a total transport distance of 100 μm for example, the initial rapid migration will be small and contribute only negligibly to the overall rate. If the transport distances are short, however, this effect could become significant.

4.2.5. *Diffusion Mechanisms.* Although several workers have commented upon it, the mechanism for Ar volume diffusion through the bulk feldspar structure is unknown. With a radius of about 1.58 Å, argon is a large in terms of the structure. Petrovic (1974) concluded that Ar can be accommodated at (0,0,½) interstitial sites and possibly others; however, direct interstitial migration is prohibited by its large size and would be prevented by the large cations. Petrovic (1974) concluded that migration of Ar from one interstitial site to an adjacent one requires a cation vacancy between the two sites. This scenario suggests further the possibility of an impurity-vacancy complex participating in transport. Foland (1974b) speculated that each Ar atom may be associated with a characteristic defect related to structural distortion from the impurity or ^{40}K radiation damage. A kinetic isotope effect for Ar transport in feldspar suggested by Foland and Xu (1990) would be significant for the mechanism but this effect has not been confirmed by subsequent work.

In fact, there is no certainty that Ar transport is by point defects as opposed to more complex ones. For example, divacancy diffusion is believed to be important for rare gases in some alkali halides. Higher-order features (e.g., micropores) and pipe diffusion are certainly important to non-volume diffusion in common microstructurally complex feldspars.

4.2.6. *Sites of Argon.* A related matter concerns "siting," the possibility of different "phases" or "components" of Ar, and different transport mechanisms. Harrison and McDougall (1981), Zeitler (1987), and others have suggested that excess and radiogenic Ar occupy different defect sites, with the former in anion vacancies, and that these components have different transport rates and mechanisms. One line of supporting evidence is a correlation of Cl-derived ^{38}Ar and excess ^{40}Ar. Considering the formation of various isotopes with attendant recoil, it seems probable that all Ar will be displaced during generation from the specific site occupied by its predecessor. This includes ^{38}Ar which would no longer be associated with Cl or the postulated anion vacancy. Moreover, a very large number of defects, including anion vacancies, are produced during irradiation. The experimental evidence argues against such an effect because irradiation does not greatly modify the diffusivity of either ^{40}Ar or ^{39}Ar. Local damage will be very different for radiogenic ^{40}Ar and ^{39}Ar so if the initial point defect were important, profound differences in their diffusivities should result. It follows that the occurrence of different Ar components displaying dramatically different transport properties within the normal lattice is unlikely to obtain.

At the same time, it is quite reasonable to expect that there may be different locations of Ar in a microstructurally complex feldspar. Grain boundaries, healed cracks, fluid inclusions, etc. are all good locations for Ar. A convincing case for such features in feldspar is the fluid inclusion work of Burgess et al. (1992) which accounts for the correlation of Cl-derived and excess Ar. In fact, features of this type will be good traps for Ar if they are not interconnected to the surface or have very high tortuosity. It is likely that several types of such features may also trap Ar.

The proposition of Ar holding traps has some similarities with earlier ideas of siting but differs in important ways. The trap is a major imperfection in which Ar may be quite content to remain because it has a lower fugacity than that in structural point defects. Trapping does not imply anything about actual diffusion mechanism.

4.2.7. *Effect of Twin and Lamellar Boundaries.* Several lines of evidence suggest that coherent boundaries or twin planes do not provide rapid pathways for Ar transport. Parsons et al. (1988) make an effective argument by noting that the finest cryptoperthites with coherent boundaries are more retentive. Using the Arrhenius relation for Benson Mines orthoclase and expression [7], the average Ar atom will travel in only 1 m.y. approximately 1 nm at 100°, 10 nm at 125°, or 40 nm at 150°C. For many volcanic feldspars which are coherent cryptoperthites, a significant portion of their Ar is therefore within reach of boundary; yet, such samples are invariably retentive. Finally, cation diffusion data are consistent with this conclusion as Giletti (this volume) finds no difference in kinetics for directions parallel or perpendicular to twin planes in albite.

4.2.8. *Intrinsic versus Extrinsic Control.* Based upon available data, it seems that a strong case may be made for intrinsic control of Ar diffusion over the range 500 to perhaps as high as 900° C. There are three main reasons for this conclusion. First, there is no appreciable enhancement of rate with irradiation which seems likely to have had an important impact on defects. Second, the data for cation and especially oxygen diffusion (Giletti, this volume) in different samples of a given feldspar type show only limited range in diffusion coefficients. This suggests that these elements are intrinsically controlled. And finally, two other well defined Ar diffusion studies yield diffusion rates very similar to the Benson Mines orthoclase. These samples are sanidines studied by Baadsgaard (1961) and Newland (1963); for both the samples the effective dimension for diffusion appears to be the physical grain size. The results for these three samples are very similar as pointed out by Foland (1974a). Significantly, both of these sanidines show a dip in the diffusion coefficient below the linear Arrhenius relation somewhere above 900°C.

As pointed out by Parsons et al. (1988), the differences in cell parameters among the potassium-rich alkali feldspars of various structural state are very small and should not therefore make a significant difference in diffusion rates. It seems appropriate to regard the Arrhenius parameters for Benson Mines orthoclase feldspar as appropriate to monoclinic and triclinic potassium feldspar.

4.2.9. *Argon Trapping.* The concept of trapping of noble gases in solids has been discussed for many years (see Ozima and Podosek, 1983) but has been largely ignored in connection with Ar diffusion processes. It seems likely that the trapping effects outlined above are going to be important for feldspars. Microstructural studies show a wealth of potential features (e.g., embedded inclusions, sealed micropores) that may serve as traps during laboratory treatment and in nature. Considering the distances expected for Ar migration at low temperatures over many millions of years, a significant amount of radiogenic Ar could end up in a trap depending on the availability of appropriate structures. Many of these features may be so small that they are not easily revealed, for example, by breaching during crushing.

4.2.10. *Applicability to Nature.* A basic question is whether step-heating experiments faithfully reflect the same processes that control Ar transport in nature. Hence, it is important to consider whether laboratory experiments meet the requirements and, if so, whether a linear extrapolations of the Arrhenius relations are appropriate. Various aspects of these questions have been entertained throughout this review and do not need to be repeated here.

It is clear that incremental-heating results described above raise serious questions about the validity of the diffusion parameters derived from such experiments. At the very minimum, the reduction in grain size during heating poses problems for determining both diffusivities and Ar gradients that characterize the initial sample. It is unlikely that the phenomenon is restricted only to this specimen. Indeed, Fitz Gerald and Harrison (1993) note some cracking and T.C. Onstott (electronic communication, 1993) reports finding abundant cracks in samples heated above 900°C. These effects and the other complications posed by natural feldspars present challenges for understanding the basic processes so that the age and additional information recorded by this commonest of crustal minerals can be fully utilized.

5. References

Baadsgaard H, Lipson J, Folinsbee RE (1961) The leakage of radiogenic argon from sanidine. Geochim Cosmochim Acta 25:147-157

Berger GW, York D (1981) Geothermometry from $^{40}Ar/^{39}Ar$ dating experiments. Geochim Cosmochim Acta 45:795-811

Borg RJ, Dienes GJ (1988) An Introduction to Solid State Diffusion. Academic Press, New York

Brandt SB, Voronovskiy SN (1967) Dehydration and diffusion of radiogenic argon in micas. Internat Geol Rev 9:1504-1507

Burgess R, Kelley SP, Parsons I, Walker FDL, Worden RH (1992) $^{39}Ar/^{40}Ar$ analysis of perthite microtextures and fluid inclusions in alkali feldspars from the Klokken syenite, South Greenland. Earth Planet Sci Lett 109:147-167

Claesson S, Roddick JC (1983) $^{40}Ar/^{39}Ar$ data on the age and metamorphism of the Ottfjälet dolerites, Särv Nappe, Swedish Caledonites. Lithos 16:61-73

Crank J (1975) The Mathematics of Diffusion, 2nd edn. Oxford University Press, Oxford

Dalrymple GB, Lanphere MA (1969 Potassium-Argon Dating. Freeman , San Francisco

Dienes GJ, Vineyard GH (1957) Radiation Effects in Solids. Interscience Publishers, New York

Dodson MH (1973) Closure temperature in cooling geochronological and petrological systems. Contrib Mineral Petrol 40:259-274

Dodson MH (1979) Theory of cooling ages. In: Jäger E, Hunziker JC (eds) Lectures in Isotope Geology. Springer Verlag, Berlin, pp 194-202

Faure G (1986) Principles of Isotope Geology, 2nd edn. John Wiley and Sons, New York

Fitz Gerald JD, Harrison TM (1993) Argon diffusion domains in K-feldspar I: microstructures in MH-10. Contrib Mineral Petrol 113:367-380

Foland KA (1974a) ^{40}Ar diffusion in homogeneous orthoclase and an interpretation of Ar diffusion in K-feldspar. Geochim Cosmochim Acta 38:151-166

Foland KA (1974b) Alkali diffusion in orthoclase. In: Hofmann AW, Giletti BJ, Yoder HS, Yund RA (eds) Geochemical Transport and Kinetics. Academic Press, New York, pp 77-98

Foland KA, Borg RJ, Mustafa MG (1987) The production of ^{38}Ar and ^{39}Ar by 14 MeV neutrons on ^{39}K. Nuc Sci Engin 95:128-134

Foland KA, Fleming TH, Heimann A, Elliot DH (1993) Potassium-argon dating of fine-grained basalts with massive Ar loss: Application of the $^{40}Ar/^{39}Ar$ technique to plagioclase and glass from the Kirkpatrick Basalt, Antarctica. Chem Geol Isot Geosci Sect 107:173-190

Foland KA, Linder JS, Laskowski TE, Grant NK (1984) $^{40}Ar/^{39}Ar$ dating of glauconites: Measured ^{39}Ar recoil loss from well-crystallized specimens. Isot Geosci 2:241-264

Foland KA, Xu Y-p (1990) Diffusion of ^{40}Ar and ^{39}Ar in irradiated orthoclase. Geochim Cosmochim Acta 54:3147-3158

Foland KA, Xu Y-p, Hubacher FA (1992) Reality of "Ar loss" $^{40}Ar/^{39}Ar$ step-heating spectrum: Confirmation and complications for K-feldspars. Eos Trans. Am. Geophys Union, Spring Meeting Suppl 73 (14):363

Friedlander G, Kennedy JW, Macias ES, Miller JM (1981) Nuclear and Radiochemistry, 3rd edn. John Wiley and Sons, New York

Gaber LJ, Foland KA, Corbató CE (1988) On the significance of argon release from biotite and amphibole during $^{40}Ar/^{39}Ar$ vacuum heating. Geochim Cosmochim Acta 52:2457-2465

Giletti (1974a) Diffusion Related to Geochronology. In: Hofmann AW, Giletti BJ, Yoder HS, Yund RA (eds) Geochemical Transport and Kinetics. Academic Press, New York, pp 61-76

Giletti BJ (1974b) Studies in diffusion I: Ar in phlogopite mica. In: Hofmann AW, Giletti BJ,

Yoder HS, Yund RA (eds) Geochemical Transport and Kinetics. Academic Press, New York, pp 107-115

Gillespie AR, Huneke JC, Wasserburg GJ (1982) An assessment of ^{40}Ar-^{39}Ar analysis of partially degassed xenoliths. J Geophys Res 87:9247-9257

Harrison TM (1983) Some observations on the interpretation of ^{40}Ar/^{39}Ar age spectra. Isot Geosci 1:319-338

Harrison TM (1990) Some observations on the interpretation of feldspar ^{40}Ar/^{39}Ar results. Chem Geol Isot Geosci Sect 80:219-229

Harrison TM, McDougall I (1981) Excess ^{40}Ar in metamorphic rocks from Broken Hill, New south Wales: Implications for ^{40}Ar/^{39}Ar age spectra and the thermal history of the region. Earth Planet Sci Lett 55:123-149

Harrison TM, McDougall I (1982) The thermal significance of potassium feldspar K-Ar ages inferred from ^{40}Ar/^{39}Ar age spectrum results. Geochim Cosmochim Acta 46:1811-1820

Harrison TM, Morgan P, Blackwell DD (1986) Constraints on the age of heating at the Fenton Hills Site, Valles Caldera, New Mexico. J Geophys Res 91:1899-1908

Harrison TM, Lovera OM, Heizler MT (1991) ^{40}Ar/^{39}Ar results for multi-domain samples with varying activation energy. Geochim Cosmochim Acta 55:1435-1448

Heimann A, Foland KA, Linder JS, Gaber LJ (1991) Simultaneous application of the K-Ca and K-Ar methods: Early results for potassium feldspars and selected micas. Geol Soc Am Abs with Prog 23:A261

Heimann A, Fleming TH, Elliot DM, Foland KA (in press) A short interval of Jurassic continental flood basalt volcanism in Antarctica as demonstrated by ^{40}Ar/^{39}Ar geochronology. Earth Planet Sci Lett

Heizler MT, Lux DR, Decker ER (1988) The age and cooling history of the Chain of Ponds and Big Island Pond Plutons and the Spider Lake granite, West-Central Maine and Quebec. Am J Sci 288:925-952

Hess JC, Lippolt HJ (1986) Kinetics of Ar isotopes during neutron irradiation:^{39}Ar loss from minerals as a source of error in ^{40}Ar/^{39}Ar dating. Chem Geol Isot Geosci Sect 59:223-236

Hofmann AW, Giletti BJ (1970) Diffusion of geochronologically important minerals under hydrothermal conditions. Eclogae Geol Helv 63:141-150

Horn P, Jessberger EK, Kirsten T, Richter H (1975) ^{40}Ar-^{39}Ar dating of Lunar rocks: Effects of grain size and neutron irradiation. Proc 6th Lunar Sci Conf, Geochim Cosmochim Acta Suppl 6:1563-1591

Huneke JC, Smith SP (1976) The realities of recoil: ^{39}Ar recoil out of small grains and anomalous patterns in ^{40}Ar-^{39}Ar dating. Proc 7th Lunar Sci Conf, Geochim Cosmochim Acta Suppl 7:1987-2008

Lanphere MA, Dalrymple GG (1976) Identification of excess ^{40}Ar by ^{40}Ar/^{39}Ar spectrum techniques. Earth Planet Sci Lett 32:141-148

Lee JKW (1993) The argon release mechanisms of hornblende in vacuo. Chem Geol Isot Geosci Sect 106:133-170

Lovera OM, Richter FM, Harrison TM (1989) The ^{40}Ar/^{39}Ar thermochronometry for slowly cooled samples having a distribution of diffusion domain sizes. J Geophys Res 94:17917-17935

Lovera OM, Richter FM, Harrison TM (1991) Diffusion domains determined by ^{39}Ar released during step heating. J Geophys Res 96:2057-2069

Lovera OM, Heizler MT, Harrison TM (1993) Argon diffusion domains in K-feldspar II: Kinetic properties of MH-10. Contrib Mineral Petrol 113:381-393

McDougall I, Harrison TM (1988) Geochronology and Thermochronology by the $^{40}Ar/^{39}Ar$ Method. Oxford University Press, Oxford

Mensing TM, Faure G (1983) Identification and age of neoformed Paleozoic feldspar (adularia) in a Precambrian basement core from Scioto County, Ohio, U.S.A. Contrib Mineral Petrol 82:327-333

Merrihue C (1965) Trace-element determinations and potassium-argon dating by mass spectroscopy of neutron-irradiated samples. Trans Am Geophys Union 46:125

Merrihue C, Turner G (1966) Potassium-argon dating by activation with fast neutrons. J Geophys Res 71:2852-2857

Mussett AE (1969) Diffusion measurements an the potassium-argon method of dating. Geophys J R astr Soc 18:257-303

Newland BT (1963) On the diffusion of radiogenic argon from potassium feldspar. M Sc Thesis, Univ of Alberta, Alberta.

Ozima N, Podosek FA (1983) Noble Gas Geochemistry. Cambridge University Press, New York

Parsons I, Rex DC, Guise P, Halliday AN (1988) Argon-loss by alkali feldspars. Geochim Cosmochim Acta 52:1097-1112

Parsons I, Waldron KA, Walker FDL (1991) Microtextural controls of ^{40}Ar loss and ^{18}O exchange in alkali feldspars. Eos Trans Am Geophys Union, Spring Meeting Suppl 72 (17):290

Petrovic R (1974) Diffusion of alkali ions in alkali feldspars. In: MacKenzie WS, Zussman J (eds) The Feldspars. Manchester University Press, Manchester, pp 174-182

Pringle MS, McWilliams M, Houghton BF, Lanphere MA, Wilson CJN (1992) $^{40}Ar/^{39}Ar$ dating of Quaternary feldspar: Examples from the Taupo volcanic zone, New Zealand. Geology 20:531-534

Renne PR, Basu AR (1991) Rapid eruption of the Siberian Traps flood basalts and the Permo-Triassic boundary. Science 253:176-179

Richter FM, Lovera OM, Harrison TM, Copeland PC (1991) Tibetan tectonics from a single feldspar sample: An application of the $^{40}Ar/^{39}Ar$ method. Earth Planet Sci Lett 105:266-276

Schiott HE (1970) Approximation and interpolation rules for ranges and range stragglings. Rad Effects 6:107-113

Shewmon PG (1963) Diffusion in Solids. McGraw-Hill Book Company, New York

Smith PE, Evenson NM, York D (1993) First successful ^{40}Ar-^{39}Ar dating of glauconites: Argon recoil in single grains of cryptocrystalline material. Geology 21:41-44

Turner G (1968) The distribution of potassium and argon in chondrites. In: Ahrens LH (ed) Origin and Distribution of the Elements. Pergamon Press, Oxford, pp 387-398

Turner G, Cadogan PH (1974) Possible effects of ^{39}Ar recoil in $^{40}Ar/^{39}Ar$ dating. Proc 5th Lunar Sci Conf, Geochim Cosmochim Acta Suppl 5:1601-1615

Turner G, Wang S (1992) Excess argon, crustal fluids and apparent isochrons from crushing K-feldspar. Earth Planet Sci Lett 110:193-211

Villa IM (1988) Ar diffusion in partially outgassed feldspars: insights from $^{40}Ar/^{39}Ar$ analysis - comments. Chem Geol Isot Geosci Sect 73:265-269

Villa IM (1991) Ar loss in nature and the lab - is it really volume diffusion. Eos Trans Am Geophys Union, Spring Meeting Suppl 72 (17):291

Wartho J, Dodson MH, Rex DC, Guise PG, Knipe RJ (1991) Mechanisms of Ar release from Himalayan metamorphic hornblende. Am Mineral 76:1446-1448.

Wasserburg GJ, Hayden RJ, Jensen KJ (1956) A^{40}-K^{40} dating of igneous rocks and sediments. Geochim Cosmochim Acta 10:153-165

Wetherill GW, Aldrich LT, Davis, GL (1955) Ar^{40}-K^{40} ratios of feldspars and micas from the same rock. Geochim Cosmochim Acta 8:171-172

Worden RH, Walker DL, Parsons I, Brown WL (1990) Development of microporosity, diffusion channels and deuteric coarsening in perthitic alkali feldspars. Contrib Mineral Petrol 90:507-515

Xu Y-p (1990) Theoretical and experimental study of $^{40}Ar/^{39}Ar$ diffusion-related phenomena in K-feldspar. M Sc Thesis, Ohio State University, Columbus, Ohio.

Xu Y-p (1993) ARLOSS: A fortran program for modeling the effects of initial ^{40}Ar losses on $^{40}Ar/^{39}Ar$ dating. Comp Geosci 19:533-546

Xu Y-p, Foland KA (1991) Diffusion of Ar isotopes in irradiated minerals during $^{40}Ar/^{39}Ar$ incremental heating. Eos Trans Am Geophys Union, Spring Meeting Suppl 72 (17):291

York D, Evenson NM, Smith PE (1992) Recoil. Eos Trans Am Geophys Union, Spring Meeting Suppl 73 (14):363

Yund RA (1983) Diffusion in feldspars. In: Ribbe PH (ed) Reviews of Mineralogy, vol 2, 2nd edn. Mineralogical Society of America, Washington, pp 203-222

Yund RA, Smith BM, Tullis J (1981) Dislocation-assisted diffusion of oxygen in albite. Phys Chem Minerals 7:185-189

Zeitler PK (1987) Ar diffusion in partially outgassed feldspars: insights from $^{40}Ar/^{39}Ar$ analysis. Chem Geol Isot Geosci Sect 65:167-181

Zeitler PK (1988) Ar diffusion in partially outgassed feldspars: insights from $^{40}Ar/^{39}Ar$ analysis - reply. Chem Geol Isot Geosci Sect 73:268-269

Zeitler PK, Fitz Gerald JD (1986) Saddle shaped $^{40}Ar/^{39}Ar$ age spectra from young, microstructurally complex potassium feldspars. Geochim Cosmochim Acta 50:1185-1199

FELDSPARS IN IGNEOUS ROCKS

WILLIAM L. BROWN[1] and IAN PARSONS[2]
[1]*CRPG-CNRS, BP 20,*
54501 Vandœuvre-lès-Nancy Cedex, France

[2]*Department of Geology and Geophysics,*
Grant Institute, West Mains Road,
Edinburgh EH9 3JW, Scotland

ABSTRACT. Feldspars are the most abundant minerals in almost all crustal igneous rocks, those without feldspars being rarities. Feldspars define better than any other mineral group the three stages in the evolution of igneous rocks, the *magmatic growth*, the *subsolidus transformation* and the *deuteric alteration* stages. Feldspars in igneous rocks show the largest variation in chemical composition, form the basis for the classification of these rocks and have played a dominant role in the development of ideas on differentiation. Plagioclase is the only mineral whose composition can be estimated in a few seconds using a method as simple as measuring the extinction angle. Because of their low symmetry the plagioclases show zoning and growth textures better and because of very slow Si,Al interdiffusion they preserve them better than any other minerals, so that they are important in the fields of crystal growth and zoning, and in magma mixing and dynamics. Plagioclase feldspars also play an important role in ideas on the origin of igneous layering and lamination, and of how and where 'cumulus' crystals nucleate in layered rocks, because of controversy over whether plagioclase can sink in basic liquids. The subsolidus phase-transformation and exsolution microtextures in the plagioclase and the alkali feldspars are very different, and are more complex and informative than in any other mineral group. Alkali feldspars store up to 2-4 $kJmol^{-1}$ of elastic energy in cryptoperthites or orthoclase, accounting for their reactivity at low temperature. It may be released during the deuteric alteration stage resulting in cloudiness, caused by fluid-filled micropores, and in an important permeability, which make extensive chemical and isotopic exchange with fluids easy.

1. Introduction

Igneous rocks crystallize from a magma (Sec. 2.1) and after emplacement, three stages in the history of the rocks may be recognized, a *magmatic growth*, a *subsolidus transformation* and a *deuteric alteration* stage (Parsons and Brown 1984; Brown and Parsons 1993); these stages are shown better by feldspars than any other mineral group. For this reason it is impossible to deal adequately with so vast a subject in a single chapter, but for the convenience of readers an extensive bibliography is given. The *magmatic growth* stage occurs in all except fully glassy rocks, whereas the other stages may be more or less well expressed, depending on cooling rate and the abundance of fluids. The first stage involves nucleation and growth (Sec. 2.2) of feldspars with high Si,Al disorder. Depending on emplacement conditions, crystallization may occur at different degrees of undercooling and under conditions in which magma dynamics play an important role. If magma mixing occurs, crystals may be in contact with different liquids during their formation, so that both growth and partial dissolution may be involved. Because of Ca-Al

I. Parson (ed.), Feldspars and Their Reactions, 449–499.

and (Na,K)-Si coupling and very slow Si,Al diffusion in plagioclase, crystals are often zoned (Sec. 2.3); zoning paths are well preserved and may approach those of ideal fractionation (Sec. 2.5; Brown 1993). A major factor is the effect of the water content of the liquid on the lowering of liquidus temperatures and the way in which it varies during crystallization---low water contents lead to hypersolvus crystallization and often strongly ternary compositions, whereas high water contents, especially under water saturation, lead ultimately to subsolvus crystallization of two feldspars (Tuttle and Bowen 1958). Feldspars in pegmatites are discussed by Černý (this volume). Depending on cooling rate, further, largely intracrystalline changes may occur during the *subsolidus transformation* stage, involving phase-transformation and exsolution microtextures (Sec. 2.6). The scale and extent of development of the microtextures depend inversely on cooling rate, because they usually involve Si,Al ordering or interdiffusion, or both. The microtextures are generally submicroscopic in scale and for this reason they are are often not given adequate attention by petrologists and geochemists. Because of lattice coherency and misfit they may store very large amounts of elastic energy, especially in alkali feldspars. This stored elastic energy greatly increases alkali feldspar reactivity and may be released during the *deuteric alteration* stage (Secs. 2.7 and 6) leading to visible modification of the textures and the production of an important porosity and permeability (Brown and Parsons 1993). The term 'deuteric', meaning 'secondary', was introduced before the submicroscopic subsolidus stage was discovered. The term should be restricted to alteration which does not involve large quantities of water introduced from outside the rock, for which the term hydrothermal should be used. Deuteric alteration produces only small to negligible changes in the bulk composition of the rock, whereas hydrothermal alteration can produce drastic changes leading to complete replacement of feldspar by other minerals. Finally, igneous rocks must in most cases be uplifted and eroded to reach the surface in order to be accessible for study by mineralogists and petrologists (Sec. 2.8). Feldspars in weathering and diagenesis are treated by Blum (this volume).

The commonly used classification schemes for igneous rocks (Sec. 3) are based essentially on only two criteria "(1) the ratio of [modal] alkali feldspar to plagioclase and (2) some measure of silica saturation e.g. the presence of quartz or feldspathoids" (Le Maitre 1976a). Thus, of the two limbs of Bowen's reaction series (Bowen 1928), only the one dealing with the tectosilicates is used for classification, the minerals in the other being used largely only as qualifiers. Insofar as it affects classification, the differentiation of igneous rocks depends mainly on the fractionation of plagioclase; rocks in which early fractionation of Or-rich feldspar is important form only a small fraction of all igneous rocks. Nevertheless, small differences in potassium content in parental magmas may make a great difference in the later stages of the differentiation. It is essential, however, that any classification scheme take feldspar phase relationships into account, otherwise grave difficulties will arise. Because of the primary role of cooling rate in determining rock textures and feldspar microtextures, feldspars in volcanic rocks (Sec. 4) are dealt with separately from those in plutonic rocks (Sec. 5). Finally, the importance of deuteric and hydrothermal alteration which gives rise to feldspar turbidity, micropores, enhanced rock permeability and to chemical and isotopic exchange is stressed (Sec. 6).

2. Conditions of Crystallization and Alteration of Feldspars in Rocks

2.1 THE FELDSPATHIC COMPONENT OF MAGMAS

Igneous rocks crystallize from magmas which by definition may comprise liquid and crystals. The feldspathic component of a magma may come from the melting of feldspar or other minerals in the source, or by the incorporation of unmelted feldspar crystals from the source or during ascent before emplacement. The depth of origin of the magma is a guide to whether feldspars are stable in the source and whether they may be entrained or not. The stability fields of feldspars (Angel this volume) are limited towards high pressures (P) by breakdown reactions to give denser phases and towards high temperatures (T) by melting. At geologically reasonable T of 1100-1300°C for the formation of basic melts in the mantle, albite breaks down in the absence of water at P of ~3.0-3.5 GPa (Holland 1980) and anorthite at slightly lower P of ~2.4-2.8 GPa (Goldsmith 1980); intermediate plagioclases probably break down at similar P under the same conditions. In the presence of small amounts of water the breakdown reactions occur at lower P (*cf.* Goldsmith 1982). Sanidine is stable in the absence of water to higher P (Liu 1978, 1987), but may react in the presence of water to give 'sanidine hydrate' (Seki and Kennedy 1964; Graham this volume). Thus, in the absence of water, plagioclase is unlikely to be intrinsically stable to depths >~75-90 km, whereas sanidine may extend to greater depths.

Feldspar stability in the upper mantle is, however, restricted to shallower depths by reaction with other minerals and by melting. The upper mantle is composed predominantly of peridotite, varying in composition from lherzolite to harzburgite with minor amounts of more differentiated rocks. Plagioclase peridotite occurs only at relatively shallow depths of up to about 30-35 km and is replaced by spinel peridotite at intermediate depths and garnet peridotite at greater depths (Ringwood 1975). Plagioclase may be abundant in the differentiated rocks occurring as dykes or sills within the peridotite, as inferred from ophiolite complexes (Nicolas 1989). Melting in the upper mantle occurs at lower T than that of pure minerals, because of simultaneous reaction of other phases and the stabilizing effect of water, where present, on the melt. Magmas of deep origin (DL Anderson 1987) may thus form and collect under conditions in which plagioclase is not stable, but they percolate along grain boundaries depending on wetting angle (Toramura and Fujii 1986; Riley and Kohlstedt 1990) or rise up by hydrofracturing (Turcotte 1987; Nicolas 1989, 1990) through plagioclase-bearing rocks in the upper mantle and crust. Plagioclase can fractionate from liquids only at depths where it is stable. During ascent of the *primary* magma either directly towards the surface or during residence in intermediate chambers at the base of or within the crust, complex differentiation involving fractional crystallization, crystal settling, filter pressing, assimilation of igneous or other rocks or their minerals, mixing with other magmas and degassing may occur. The composition of the source, the degree of partial melting, the abundance and nature of volatiles and the extent of reaction during ascent strongly affect the composition of the resulting *parental* magmas (in the sense of Carmichael et al. 1974, p 44-46).

The processes during ascent, along with those involved during emplacement and crystallization, account for the chemical diversity of igneous rocks, the complexity of their structures and textures, and of the internal textures of the feldspars. In order to elucidate liquid lines of descent (Bowen 1928), it is important to determine to what extent the compositions of igneous rocks may be considered to be representative of liquids, because of the possible entrainment of crystals. In the case of non-porphyritic volcanic rocks it is generally admitted that their compositions represent those of liquids, especially for common rocks. Liquid compositions can also be calculated for porphyritic volcanic rocks by subtracting the phenocrysts, on the assumption that they grew and were concentrated elsewhere, and were then entrained by the liquid. Many intrusive rocks, especially those whose textures and compositions are remarkably homogeneous, such as many granitoids, may also be considered to represent liquid compositions or to be close thereto. This is, however, not true for layered basic rocks, because the modal variation may be extreme (nearly monomineralic dunite, pyroxenite or anorthosite).

2.2 THE EFFECT OF EMPLACEMENT CONDITIONS ON NUCLEATION AND GROWTH

2.2.1 *Nucleation.* Because magmas may be formed in one environment and crystallize on their way to and in another, feldspars in igneous rocks must have nucleated and/or undergone complex phases of growth and dissolution. Feldspar crystals may have been absent from the magma, either because it rose from a depth where feldspars are absent, or because any crystals present were removed during percolation, by filter pressing, by settling, by melting either through mixing with another magma or through heating by a hotter magma. If feldspar crystals are absent but feldspar is stable, nucleation may occur and this requires a certain degree of undercooling relative to the stable liquidus *T*. Feldspar nucleation rates and densities depend on the composition of the liquid, especially on the volatile content, on undercooling and on the nucleation mechanism, whether homogeneous, occurring within the liquid, or heterogeneous occurring on previously existing crystals (Smith and Brown 1988, Sec. 17.1). Heterogeneous nucleation can also occur on already existing feldspar crystals, if for some reason growth has stopped, and is called self-nucleation. Homogeneous nucleation requires greater undercooling than heterogeneous. Nucleation of plagioclase occurs early in most igneous rocks, even though it is more difficult to nucleate tectosilicates than less-polymerized minerals, such as olivine or pyroxene, in weakly polymerized basic or ultrabasic liquids. This is because of the very large primary phase volume of plagioclase in typical basic liquids. Even though plagioclase is stable, it does not occur in some unusual rapidly cooled, partly glassy rocks, such as boninites, although the residual glasses are very rich in normative plagioclase. The absence of plagioclase in boninites was explained by the abundance of Cr in the initial liquid, which favours early nucleation of chrome spinel (Libourel 1991), and of water in the residual liquid, which inhibits nucleation of plagioclase (Ohnenstetter and Brown 1992). During and after emplacement through intrusion or extrusion, undercooling in a liquid varies with time and position (Brandeis and Jaupart 1987a,b). After emplacement it immediately increases close to the margins because of the large *T* difference between the magma and the wall rocks. Some time after emplacement a steady state is probably approached, because of the release of enthalpy of crystallization, when the undercooling becomes very small in the centres of large slowly cooled basic intrusions, moderate in intermediate-sized igneous bodies, and higher in small bodies or flows. If convection occurs in large basic magma chambers, the degree of undercooling varies regularly with height and time because of adiabatic cooling and heating of the liquid during upward and downward motion and this may affect nucleation and zoning (Sec. 2.3).

Alternatively, depending on magma viscosity and speed of flow, feldspars may have been *entrained* with the liquid from depth and may give rise to preferred orientation of crystals defining flow structures. After emplacement the crystals may grow or partly dissolve at first, before cooling produces increased undercooling. Depending on cooling rate, there will be competition between homogeneous nucleation in the liquid and heterogeneous nucleation on the entrained crystals followed by overgrowth. If cooling is rapid, undercooling will be high enough for homogeneous nucleation and overgrowth to occur, the rock containing two sizes or 'generations' of feldspar, as in lavas with small groundmass feldspars and large phenocrysts. If cooling is slow, undercooling should always be insufficient for homogeneous nucleation to occur in the liquid, the entrained crystals being simply overgrown heterogeneously *in situ*, unless diffusion is very slow as in acid liquids. Heterogeneous nucleation and self-nucleation should be frequent in areas close to contacts, such as walls, roofs and floors. It is the dominant mechanism for plagioclase crystals growing downwards from the rapidly cooled crust at the roofs of two lava lakes in Hawaii, the calculated heterogeneous nucleation rates on different minerals decreasing in the order plagioclase (self-nucleation), pyroxene and olivine (Kirkpatrick 1977; see also Cashman 1993); no information is available about conditions of nucleation below the crust in the still-liquid part. Heterogeneous nucleation is also thought to be the cause of crescumulate or comb layering on the walls of many intrusions (Wager and Brown 1968; Moore and Lockwood 1973; Lofgren and Donaldson 1975). It may be inferred to have occurred in glomeroporphyritic and 'snowflake' plagioclase in many lavas or close to contacts in intrusions (Berg 1980; *cf.* MacKenzie et al. 1982; Brown et al. 1989).

If heterogeneous nucleation were the only probable process, large basic intrusions, where undercooling is low, would be dominated by roof, wall and bottom growth, the latter giving rise most probably to crescumulate-like textures. Compaction would not be expected to occur without breaking the crystals. Igneous lamination, most frequently defined by plagioclase, should be absent. In spite of the greater ease of heterogeneous nucleation, the textures of many porphyritic lavas strongly suggest that the phenocrysts grew independently of neighbouring crystals until they impinged on other crystals at the end of crystallization. This implies that plagioclase and other minerals may nucleate homogeneously in the liquid in magma chambers at intermediate depths before extrusion. Furthermore, a study of phenocryst frequencies in a series of porphyritic lavas from the Deccan traps and elsewhere strongly suggests that dense crystals may have been removed by settling before eruption, and plagioclase crystals entrained by the liquid (Cox and Mitchell 1988; MC Jackson 1993). We can see no compelling reason why homogeneous nucleation should be common in liquids which were subsequently extruded, but rare or impossible in the supernatant liquid in large basic intrusions which crystallized *in situ* (*cf.* Langmuir 1989). Thus it seems probable that that homogeneous nucleation may occur and crystals be separated from the liquid in intrusions, for example by settling to the bottom of the chamber, as invoked by Wager and Brown (1968) to explain layering, or concentrated by floating of light plagioclase as suggested by the existence of massif anorthosites. Although many workers on layered intrusions (ED Jackson 1961; Campbell 1978; McBirney and Noyes 1979; Langmuir 1989) have questioned the possibility of both homogeneous nucleation and settling of plagioclase, especially in non-Newtonian or more evolved Fe-rich liquids in which it should float and not sink, the wind may have turned and crystal settling may once more be an acceptable hypothesis under certain circumstances (papers in Parsons 1987; Morse 1988; Hort et al. 1993; Sparks et al. 1993).

2.2.2 *Growth.* If cooling is extremely rapid, strongly disequilibrium growth will occur at large undercoolings and compositions and zoning paths are then largely unpredictable in detail (Baker and Cahn 1971; Smith and Brown, 1988, Sec. 17.2.4); they may approach, but not reach, non-fractional (Bowen 1928, p 99-100) or partitionless (Cahn 1980) crystallization, giving crystals which are closer to the bulk composition of the liquid than expected and which have strongly skeletal to spherulitic morphology, the latter having been compared to fractal geometry (Fowler 1990). If cooling is very slow, equilibrium crystallization will be approached giving euhedral crystals. Complete equilibrium crystallization requires continuous interdiffusion within the crystal and exchange with the liquid at the interface so that the growing crystal always remains fully homogeneous and of the correct composition. It is probable that complete equilibrium crystallization of plagioclase never occurs because Si,Al interdiffusion is so extremely slow relative to growth rates. Growth of plagioclase may thus occur at best under conditions approaching ideal fractional crystallization (Sec. 2.5), equilibrium being maintained only at the interface and not within the crystal which becomes compositionally zoned. Where the crystal was in contact with different liquids or with the same liquid under very different conditions of P or activity of some of the components, it may show quite complex textures and zoning paths and may have suffered periods of dissolution (Pearce this volume).

In volcanic rocks it is likely that the zoning observed in phenocryst overgrowths and in groundmass feldspars closely approaches ideal fractional crystallization, at least during periods of growth. The situation is more complex in plutonic rocks because of slow cooling and magma dynamics. The relative homogeneity of cumulus crystals (Wager et al. 1960) in large layered intrusions is evidence in favour of unconstrained growth (Tiller 1977) in a large volume of well mixed magma, although very slow cooling may have partly blurred any μm-scale oscillatory, but not large-scale, zoning through Si,Al interdiffusion (Grove et al. 1984; Smith and Brown 1988, Sec. 17.3.1). Constrained growth (Tiller 1977) occurs on the walls, roof and floor of intrusions. If, however, crystals grow in a permeable mush on the floors of magma chambers, efficient compositional convection (Tait et al. 1984; Langmuir 1989) may allow zoning paths to approach those of unconstrained growth, giving adcumulus textures. The late stages of crystallization of all magmatic rocks probably involve *in situ* growth from trapped liquids, as the porosity and permeability decrease, giving rise to overgrowths on all the minerals which define local fractionation paths. Whatever the mechanism of large-scale fractionation, it is clear that it does occur, as shown by the existence of suites of volcanic rocks related by low-P fractionation, and it can only do so (see Sec. 2.2.1) either by separating crystals from liquid (crystal settling, Wager and Brown 1968), liquid from crystals (Tait et al. 1984; Langmuir 1989) or by interaction between crystals and a percolating liquid in a T gradient (Lesher and Walker 1988). There must, moreover, be a fundamental constancy in the process in a given geological context, as shown by the repeated production of batches of magma which follow the same liquid line (Cox et al. 1979). The differences between suites, however, may arise through melting of different sources, through different degrees of melting of the source or through contamination.

2.3 INTRACRYSTALLINE TEXTURES AND CHEMICAL ZONING

Because the conditions of formation of feldspars in igneous rocks may be inferred from their textures and internal structures, they should be carefully studied by conventional optical methods, and also by Nomarski interference-contrast and laser-interference microscopy (AT Anderson 1983; Pearce 1984; Pearce and Kolisnik 1990; Pearce this volume), by cathodoluminescence (Smith and Stenstrom 1965; Ramseyer et al. 1989) or by back-scattered electron imaging using SEM. Xenocrysts, phenocrysts, glomerocrysts and cumulus crystals should be identified, and intergrowth, overgrowth and dissolution textures, and complex reverse, oscillatory and discontinuous zoning distinguished. Compositional zoning has been intensively studied in plagioclase, because it is well preserved and well seen optically. The recent use of Nomarski interference-contrast and laser-interference microscopy has led to a renewal of interest in zoning in plagioclase, especially in volcanic rocks. Measurements should be carried out with care, however, the whole compositional range established and results given in terms of all three components and not only as An/(Ab+An). Even if rocks with complex crystallization conditions are avoided, the zoning paths may still differ to a certain extent from ideal fractional crystallization (Sec. 2.5) and crystals may have suffered modification after growth. Compositional variation may be difficult to determine by electron microprobe in the latest overgrowth stages or in fine-grained volcanic rocks because of the size of the area excited by the beam.

The complexity of zoning in plagioclase is at present not fully understood. This is especially so for 'simple' oscillatory zoning, which is almost ubiquitous in plagioclase but occurs occasionally in other minerals, such as olivine (Gerlach and Grove 1982) and pyroxene (Tsukui 1985; Ohnenstetter and Brown 1992) in rapidly cooled rocks. Oscillatory zoning may depend (1) on fluctuations in external variables, such as total P, gas pressure or partial pressure, or T, related to cyclic movement of magma, (2) to coupling in the diffusion of several species (Ghiorso 1987; Wang and Merino 1993) or (3) possibly to growth mechanisms at the interface (Smith and Brown 1988, Sec. 17.3.5). Progress has, however, be made in our understanding of sieve and other textures thought to be produced by magma mixing, through experimental study of melting and dissolution (Smith and Brown 1988, Sec. 17.2.5; Johannes and Holtz 1992; Johannes this volume). Using the recently introduced methods which enable the 'stratigraphy' of zoning, especially in volcanic plagioclase, to be studied, the relative speed of growth of different faces, the mechanisms of growth and interruptions in growth, as well as dissolution, can be seen and related to changes in composition. Readers should refer to detailed descriptions of intracrystalline textures in Smith and Brown (1988, Sec. 17.3) and in papers cited in Pearce and Kolisnik (1990) and in Singer et al. (1993) and to the excellent illustrations of intercrystalline textures in MacKenzie et al. (1982).

2.4 EQUILIBRIUM PHASE RELATIONSHIPS

Because of the effect of water on lowering liquidus and solidus, but not solvus, T, phase relationships in feldspars are best discussed in terms of P_{H_2O}, or P and a_{H_2O}. The effects of components other than SiO_2 on phase relationships projected into the An--Ab--Or plane are not well known, but they are probably small judging from the similarity between predicted and observed zoning paths, at least in oversaturated and mildly undersaturated rocks (Secs. 2.5 and

4.2). Fig. 1 is a simplified equilibrium phase diagram in terms of T and the three end members An, Ab and Or at constant but moderate P_{H_2O} of ~0.3 GPa. Two liquidus surfaces exist, a large, steep plagioclase one, MABEC, and a much smaller flatter sanidine one, MOEC, which give the compositions of liquids coexisting with crystals, whose compositions lie on the two solidus surfaces, a steep plagioclase one, MABIPK$_E$, and a flatter sanidine one, MOJSK$_E$; Or thus has a drastic effect on lowering liquidus and solidus T in the feldspar system. The two liquidi intersect along a boundary line, EC, which stops before reaching the Ab--Or join on which a minimum M occurs. The two solidi intersect the ternary feldspar solvus along a line, IPK$_E$SJ. The plagioclase solidus is flat near AM and the two solidi coalesce between C and M. KK$_E$ is the critical solution line on the solvus along which the two feldspars at a given T become *identical*; it is metastable at T > K$_E$.

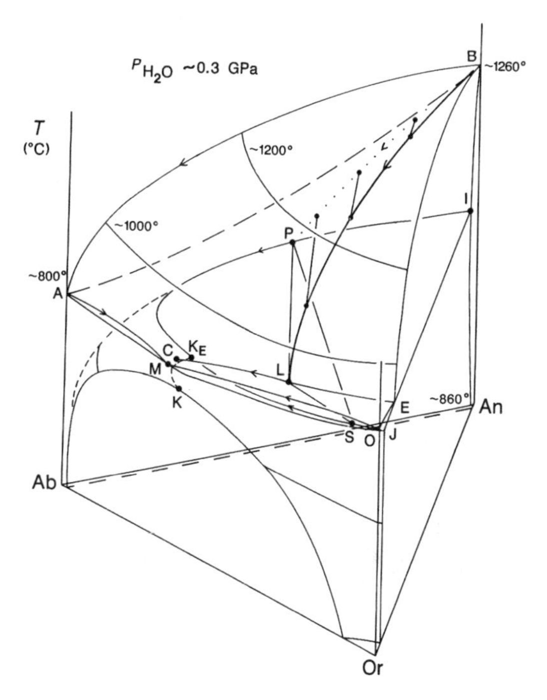

Figure 1. Equilibrium phase relationships at intermediate P_{H_2O} in the ternary feldspar system. T are approximate. From Brown (1993, Fig. 1)

Liquids on the liquidi coexist with one feldspar on the corresponding solidi, whereas those on the boundary line coexist with two feldspars, plagioclase, P, and sanidine, S, on the solvus; the liquid at C coexists with the critical feldspar at K$_E$. The boundary line starts at a eutectic, E, on the An--Or join close to Or and falls slightly in T towards Ab, always lying close to the Ab--Or join. The position and nature of the termination of the boundary line depend on P_{H_2O}, or P and a_{H_2O}. An increase of P_{H_2O}, or a_{H_2O} at constant or increasing P, greatly lowers liquidus and solidus T leading to a larger intersection with the solvus---compare the effect of P_{H_2O} on phase relationships in the Ab--Or system (Bowen and Tuttle 1950). Where K$_E$ is stable, as at low and moderate P_{H_2O} or a_{H_2O}, the boundary line does not reach the Ab--Or join and ends at C, changing from cotectic (even) to peritectic (odd) at a neutral point, N, before reaching the end point C (Ricci 1966). At high P_{H_2O}, the solidi intersect the solvus on the Ab--Or join, the critical solution line becoming entirely metastable; the boundary line also reaches the join and is cotectic along its entire length. Details of the progressive effects of increasing P_{H_2O} (or a_{H_2O} and/or a_{SiO_2}) are discussed by Brown (1993).

The maximum extent of feldspar solubility under different conditions is best shown in T projections, the one-feldspar field(s) being separated from the two-feldspar field by the line of the solvus/solidi intersection. If part of the critical solution line is stable, the area occupied by single-feldspar compositions is very large and continuous (Fig. 2), but it may be divided into two parts, the plagioclase field meeting the sanidine field along the projection of the KK_E line where the two crystalline phases become identical; this corresponds in salic rocks to the well known hypersolvus crystallization of Tuttle and Bowen (1958). If on the other hand the solidi intersect the solvus on the Ab--Or join (*i.e.* the KK_E line is entirely metastable), the one-feldspar field is divided into two parts separated by a gap (Fig. 3): this corresponds in salic rocks to the well known subsolvus crystallization of Tuttle and Bowen (1958).

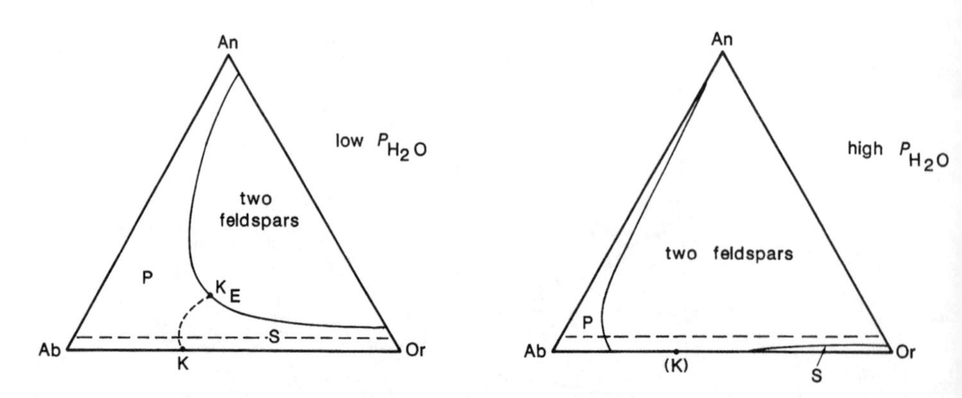

Figure 2. Projection (*left*) at low P_{H_2O}, or a_{H_2O} and P, showing the solvus/solidi intersection with close to maximum mutual solubility in the ternary feldspar system. The critical solution line on the solvus is metastable at T greater than that of K_E. Three fields exist, a two-feldspar one inside the solvus (*cf.* Fig. 14), a plagioclase field, P, and a sanidine one, S, the latter two being separated by the critical solution line along which pairs of coexisting feldspars become identical. The dashed line at 5 mol% An corresponds to the limit between 'plagioclase' and 'alkali feldspar' in the IUGS classification scheme (see Sec. 3.1 and Figs. 8, 9)

Figure 3. Projection (*right*) at high P_{H_2O}, or a_{H_2O} and P, showing the solvus/solidi intersections with limited solubility in the ternary feldspar system. Three fields exist, a very large two-feldspar one inside the solvus (*cf.* Fig. 15), a plagioclase field, P, and a sanidine one, S, the latter two being separated by a wide gap. Dashed line as in Fig. 2

2.5 EQUILIBRIUM AND FRACTIONAL CRYSTALLIZATION

During *equilibrium crystallization* both the growing crystal and the residual liquid remain homogeneous throughout the whole process, but their compositions change continuously so as to comply with mass-balance considerations--*i.e.* the tie lines always pass through the original bulk composition. In order to achieve this, the elements involved must diffuse freely through the liquid, those required for growth diffusing towards the advancing interface and those not required diffusing away; this must also happen within the growing crystal. Because of the complex geometry of the ternary feldspar phase relationships, it is possible, even during equilibrium crystallization of two feldspars, for one of them to undergo partial or complete dissolution (Nekvasil 1990, this volume).

Whereas diffusion rates in liquids are generally able to keep up with growth, this probably cannot happen in plagioclase because of very slow Si,Al interdiffusion (see Smith and Brown 1988, Chaps. 16 and 17). If the inside of a growing plagioclase crystal does not adjust its composition continuously during the growth of successive layers, it is effectively removed from the system (fractionated), the bulk composition of the ever-smaller residual system changing progressively during growth; this process is described as fractional crystallization. During *ideal fractional crystallization*, which can occur, if at all, only at intermediate cooling rates to allow for interdiffusion in the liquid, residual liquid compositions *follow* liquidus fractionation *lines*, which start at maxima in T and evolve towards minima on the liquidus surfaces (*cf.* Fig. 12). *Instantaneous crystal compositions* (called π for plagioclase or σ for sanidine, Figs. 4-6) *define* solidus fractionation *paths* which descend from An or Or to meet either the solvus/solidus intersection or M. All components are free to fractionate into the crystals or to remain in the liquid depending on their distribution coefficients. Experiments on feldspar systems in the laboratory are often carried out under conditions in which some of the components may be buffered by the presence of a pure phase such as quartz, *i.e.* quartz-buffered. This is also approximately true for studies in the presence of water at fixed P_{H_2O}, insofar as the vapour phase may be considered to be pure H_2O, or at fixed P and a_{H_2O}, if the feldspars may be considered to be insoluble in the vapour (see, however, Paillat and Brown unpublished data). Fractionation under such conditions can be described approximately as *water buffered.*

An ideal water-buffered liquidus fractionation line, BL, at moderate P_{H_2O} (Fig. 1) starts at An and descends steeply in T towards the boundary line at L. The instantaneous compositions of plagioclase which fractionates from the liquid define a steep solidus path BP which starts at An and descends to meet the solvus at P (=π), coexisting phases being joined by tie lines. From L the liquid follows the boundary line to N (not shown), crystallizing both plagioclase and sanidine whose compositions follow paths from P (=π) towards K_E and from S (=σ) towards K_E, respectively, without reaching it (*cf.* Fig. 5). At any one T the compositions of the three phases lie on a three-phase triangle. The shapes of three-phase triangles close to the low-T end of the boundary line change dramatically with increasing P_{H_2O}, or a_{H_2O} and P. At N the liquid leaves the boundary line, its course depending on the shape of the three-phase triangle at N.

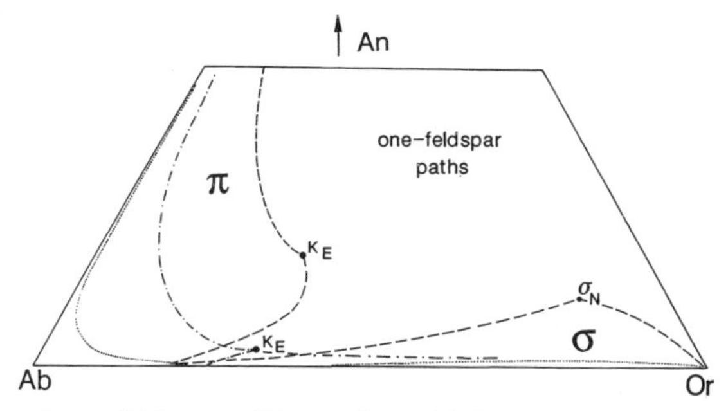

Figure 4. Summary of one-feldspar solidus paths at high (dotted lines), intermediate (dot-dash lines) and low (dashed lines) values of P_{H_2O} (or a_{H_2O} and P) for plagioclase (π) or sanidine (σ). Paths are simple at high P_{H_2O} (or a_{H_2O} and P) and become more complex at successively lower values of P_{H_2O} (or a_{H_2O} and P). Plagioclase paths bend back on themselves at K_E, whereas sanidine paths show an inflection first at K_E and then at σ_N. The paths from K_E are unique for each set of conditions, the phase which fractionates being strictly neither plagioclase nor sanidine but 'feldspar'. From Brown (1993, Fig. 12)

The topology and probable geometry of ideal water-buffered liquidus fractionation lines and crystal paths in ternary feldspars (Brown 1993) were developed as far as possible as a function of P_{H_2O}, or P and a_{H_2O}, on the basis of experimental data and thermodynamic modelling from the literature. Because of the nature of the ternary feldspar phase relationships, it was shown that water-buffered *one-feldspar solidus paths* and *two-feldspar solvus paths* exist. One-feldspar paths comprise plagioclase and sanidine (alkali feldspar s.l.) paths (Fig. 4), whereas two-feldspar paths are of two types, *simultaneous* (Fig. 5) and *sequential* (Fig. 6) paths, the latter not having been previously described. The distinction between the latter two is due to the effect of the cotectic (even) or peritectic (odd) nature of the boundary line on fractionation. During the crystallization of igneous rocks, fractionation probably occurs under water-unbuffered conditions so that the amount of H_2O, and other components, build up in the liquid as fractionation of feldspar proceeds (*cf.* Nekvasil 1992).

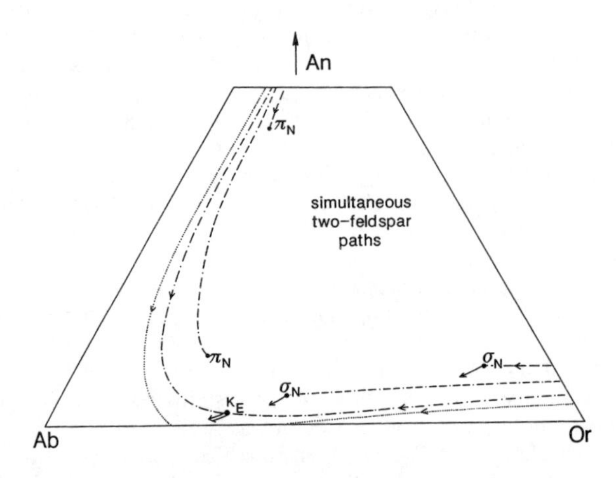

Figure 5. Summary of simultaneous cotectic two-feldspar solvus paths at high (dotted lines), intermediate (dot-dash lines) and low (dashed lines) values of P_{H_2O} (or a_{H_2O} and P) for plagioclase (π) and sanidine (σ). Two-feldspar paths exist only if the liquid meets the boundary line on an even part and cease when it reaches the neutral point N (not shown). Two-feldspar paths are simple at high P_{H_2O} (or a_{H_2O} and P) and follow the solvus to the Ab--Or join. Two-feldspar paths are more complex at lower values of P_{H_2O} (or a_{H_2O} and P), because they stop at particular points on the solvus, fractionation changing from *subsolvus* to *hypersolvus*. The arrows show the start of one-feldspar hypersolvus paths followed from K_E and σ_N (Fig. 4). The critical solution line, or its metastable extension, does not lie between the ends of the paths but between pairs of points such as π_N and σ_N. From Brown (1993, Fig. 13)

2.6 PHASE-TRANSFORMATION AND EXSOLUTION MICROTEXTURES

A subsolidus transformation stage occurs at intermediate T, normally after crystallization of the melt is complete, the extent of which depends inversely on cooling rate, because of the role of Si,Al ordering and interdiffusion which are slow processes. The feldspar crystals normally act essentially as closed systems. The modifications include intracrystalline interdiffusion which blurs zoning (especially in alkali feldspars) or produces phase transformations or exsolution. Microtextures that develop are generally submicroscopic, regular and almost entirely coherent; *i.e.* the Si,Al-O tetrahedral framework is continuous. Depending on composition and cooling rates, high-T alkali feldspars may exsolve and/or undergo two types of C2/m-to-C$\bar{1}$ phase

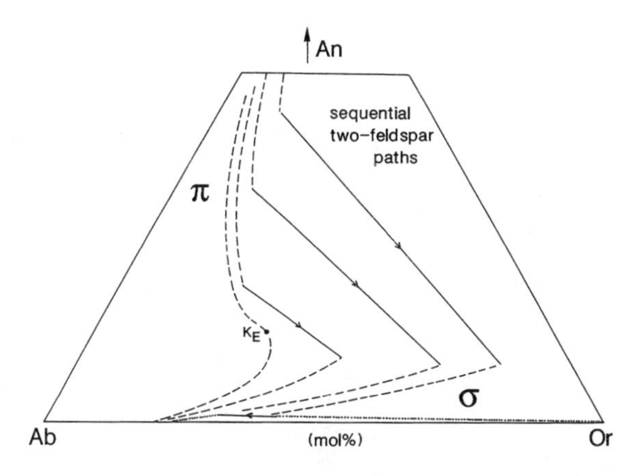

Figure 6. Summary of sequential peritectic two-feldspar solidus paths at high (dotted lines) and low (dashed lines) values of P_{H_2O} (or a_{H_2O} and P) for plagioclase (π) and sanidine (σ). Sequential two-feldspar paths exist only if the liquid meets the boundary line on an odd part. Crystallization is always *hypersolvus* but two different feldspars have fractionated, the change-over of the phase which fractionates occurs by a jump from one side of the solvus to the other (thin straight lines with arrows). At high P_{H_2O} (or a_{H_2O} and P) sanidine fractionates first and is followed by plagioclase, whereas at low P_{H_2O} (or a_{H_2O} and P) plagioclase fractionates first followed by sanidine. If the two phases are close in composition and thus in lattice parameters, the second phase may overgrow the first epitaxially; at high P_{H_2O} (or a_{H_2O} and P) plagioclase may overgrow sanidine, whereas at low P_{H_2O} (or a_{H_2O} and P) sanidine may overgrow plagioclase. The path which passes through K_E is an extreme case where the two feldspars are identical--it is also shown as a one-feldspar path in Fig. 4. From Brown (1993, Fig. 14)

transformations; extreme Ab- or Or-rich compositions transform before exsolving, whereas the reverse is true for intermediate compositions (*cf.* Brown and Parsons 1989). In Ab-rich feldspars the transformation produces anorthoclase, whereas in Or-rich feldspars it results from Si,Al ordering, giving orthoclase. Both transformations produce twins or twin-like microtextures. Exsolution involves phase separation and chemical maturation (Smith and Brown 1988, Chap. 19). Simple microtextures develop at first in the alkali feldspars and, if cooled slowly, they may coarsen and be modified by the transformations in one or both of the phases (Brown and Parsons 1984b). Morphologies of these intergrowths are strikingly regular (Fig. 7; see also Brown et al. 1983; Brown and Parsons 1984b, 1988a) and are controlled by elastic strain energy (Willaime and Brown 1974). Strain energy per mole (Brown and Parsons 1993) increases during cooling, because the compositions, or degree of Si,Al order, of the domains differ more and more, and the feldspars become elastically stiffer. In plagioclase feldspars phase transformations and exsolution occur in three compositional domains (Smith and Brown 1988; Carpenter 1988, this volume). Because exsolution is order-dependent and involves coupled NaSi and CaAl interdiffusion, it is very sluggish, occurring only in very slowly cooled intrusions and being rarely visible optically. The microtextures in plagioclase store very little elastic strain energy, whereas alkali feldspars may store up to 2-4 $kJmol^{-1}$ in cryptoperthites and orthoclase. This is because lattice misfit in exsolution microtextures is small in plagioclases and large in cryptoperthites due to the large difference in ionic size between Na and K. The elastic strain energy is also large in orthoclase, because of the very fine-scale transverse modulations (Brown and Parsons 1993).

Figure 7. Bright-field transmission electron micrograph of a regular coherent braid microperthite (KB 14) from the Klokken intrusion. Beam approximately parallel to *c*. Zig-zag lamellae of low microcline outline lozenges of Albite-twinned low albite. The fine lamellae of microcline define a primary periodicity for the intergrowth. A few thicker bands of microcline, some of which show Albite twinning, define a secondary periodicity. This is visible on the etched surface shown in Fig. 21. (Micrograph taken by Kim Waldron)

Extensive phase separation (unmixing) of exsolved feldspars, which occurs in very slowly cooled rocks, is most widespread in Ab-rich ternary and alkali feldspars in dry rocks and gives rise to coarse sometimes irregular intergrowths (mesoperthites and antiperthites). The initial magmatic bulk compositions of regular intergrowths (but not the zoning patterns) may be obtained by simple areal integration. If large-scale phase separation and Na--K exchange between plagioclase and alkali feldspar has occurred giving rise to irregular intergrowths as in large intrusions (*cf.* Fuhrman et al. 1988; Kolker and Lindsley 1989) or granulite facies rocks (Brown and Parsons 1988b; Kroll et al. 1993), it may be difficult to establish the original compositions.

2.7 DEUTERIC AND HYDROTHERMAL ALTERATION

A deuteric alteration stage (Sec. 6) occurs at lower T (<~450°C) in most igneous rocks, the extent of which depends on the availability of water, on the amount of stored elastic energy and possibly inversely on cooling rate. The crystals act as partly open systems and react with water-rich fluids present in the rock; water and other fluids are also possibly released from solution in the feldspar (Worden et al. 1990). Reactions are of three types: (1) partial replacement by non-feldspar phases, as in the replacement of alkali feldspar by clay minerals; (2) chemical exchange with the fluid accompanied by bulk compositional changes in the feldspar, as in the albitization of plagioclase. The alteration is usually concentrated in the centres of crystals, An-rich zones being more affected than An-poor ones; (3) mainly microtextural, as in exsolution and twin-domain coarsening in alkali feldspar. An up to thousand-fold coarsening of the textures occurs (Parsons 1978; Worden et al. 1990), which we called 'unzipping' (Parsons and Brown 1984; Brown and Parsons 1989), with only small changes in Si,Al order and/or composition of the phases. It also affects grain boundaries giving an unusual intercrystalline texture called 'swapped rims' (Smith and Brown 1988, p 603-604). This drastic coarsening may make it difficult to determine the magmatic textures and compositions of alkali feldspars in many water-rich rocks. It is accompanied by the formation of fluid-filled micropores (Folk 1955; Montgomery and Brace 1975; Ferry 1985; Worden et al. 1990; Walker 1990, 1991; Guthrie and Veblen 1991; Burgess et al. 1992), which cause cloudiness or turbidity in alkali feldspar (Folk 1955). In spite of the ubiquity of the cloudiness, the small size of the micropores (a few µm down to <~10 nm) has lead to them being more or less ignored by petrologists! The porosity is associated with a permeability which may play an important role in chemical and isotopic exchange between the rock and hydrothermal fluids (Walker 1990, 1991). Whereas the driving force for reactions of the first two types is reduction of chemical free energy, this is not true for isochemical microtextural reactions for which elastic strain energy developed in the subsolidus transformation stage provides the driving force (Brown and Parsons 1993).

Hydrothermal alteration, which is outside the scope of this review, may lead to extensive chemical, textural, structural and mineralogical modification of igneous rocks, either locally close to veins (Meyer and Hemley 1967) or on a larger scale. Feldspars may be partly or completely replaced by other feldspars, as in the albitization of plagioclase or alkali feldspar giving rocks dominated by chess-board albite and almost devoid of K or Ca, or replaced by other minerals (Giggenbach 1981), such as clay minerals (kaolinite), white mica, topaz, tourmaline giving extreme rocks such as greisen, and tourmaline or topaz granites. Such chemical changes may be accompanied by extensive isotopic exchange.

2.8 DEFORMATION AND UPLIFT

Intrusive rocks need to be uplifted and their cover eroded in order to reach the surface. Anorogenic intrusive rocks may reach the surface slowly by erosion alone, although they may also be slightly uplifted by movements along the faults which favoured the initial ascent of the magma and were responsible for their spatial distribution. Plutonic rocks associated with an orogenic cycle may have been intruded at different times during the cycle; syn- and late-orogenic intrusives may have been more affected hydrothermally and have suffered more intense deformation than post-orogenic ones. In extreme cases the initial igneous fabric of the rock may have been completely reworked leading to gneisses. Even if the effects are apparently small, they may partially or substantially modify the microtextures developed in feldspars during the subsolidus-transformation and deuteric stages. The modifications induced may potentially be used to elucidate the cooling and uplift history of igneous rocks, for example from argon-loss studies, provided that care is taken to determine the nature of the microtextures and to establish their influence on the isotopic behaviour (Sec. 6).

3. Feldspar in the Classification of Igneous Rocks

3.1 THE MODAL IUGS CLASSIFICATION SCHEME

Igneous rocks are classified on the basis of their modal mineral compositions. The scheme adopted by the International Union of Geological Sciences (IUGS), and based on Streckeisen (1976), was reported by Le Maitre (1989). Where it is not possible to determine the mode because of the presence of glass or because of the very fine-grained nature of the rock, use may be made of chemical classifications such as the total alkalis--silica (TAS) diagram (Le Bas et al. 1986), but the relationship between the two is not fully clear in the published schemes (Sec. 3.2) In the IUGS scheme the primary modal classifications for common plutonic and volcanic rocks are based on the relative volume proportions of four mineral groups: Q (quartz and other silica minerals), A (alkali feldspar, including sanidine, orthoclase, microcline, perthite, anorthoclase, and albite An_{0-5}), P (plagioclase An_{5-100} and scapolite), F (feldspathoids) and M (mafic and related minerals, including ferromagnesian minerals, accessory minerals, melilite, monticellite and primary carbonate), where Q+A+P+F+M=100 vol%. The separation of alkali feldspar from plagioclase at An_5 is arbitrary and does not correspond to usage in the plagioclase series, where albite is divided from oligoclase at An_{10}. The first four groups comprise the felsic minerals, whereas the last group forms the mafic minerals. For rocks with minerals in equilibrium Q and F are mutually exclusive. The colour index M' is defined as M minus muscovite, apatite and primary carbonates, these minerals being considered to be colourless for defining the colour index. For rocks with a colour index less than 90 %, use is made of a QAPF diagram renormalized to 100 % (Streckeisen 1976; Le Maitre 1989) in the form of a double triangle (Figs. 8, 9), the two opposite corners being Q and F and the common base AP. Oversaturated rocks lie in the upper QAP triangle, undersaturated ones in the lower FAP triangle and saturated rocks along the AP line. Because feldspars form more than 50 % of most igneous rocks and are by far the main felsic minerals, most rocks plot close to the AP line in the QAP triangle stretching out from P towards a point on the AQ side about a third of the way from A to Q. Two slightly different versions were proposed, one for plutonic (Fig. 8) and one for volcanic rocks (Fig. 9), the main differences concerning the suppression of plutonic fields close to QP (the fields of granodiorite and tonalite being combined to define dacite and six fields being reduced to one with two names, basalt and andesite) and also close to F. In addition to use for classification, trends in magma series can also be plotted on a QAP diagram as done, for example, for granitoid rocks by Lameyre and Bowden (1982) and Lameyre and Bonin (1991).

No rock classification scheme can ever be perfect and it is easy to draw attention to shortcomings which arise because the classification is not complete. One is that plagioclases with An>5 plot at P (Fig. 8), so that two different plutonic rock names fall in each of eight fields close to P, four in the QAP triangle and four in the FAP triangle (e.g. gabbro and diorite, monzogabbro and monzodiorite and their quartz- and feldspathoid-bearing equivalents) although gabbro and diorite *etc.* are distinguished by the presence of labradorite or andesine, respectively. In addition, these are the only fields in which the rock name depends on minerals other than those used to define the fields, gabbro being defined by the additional presence of Ca-rich clinopyroxene. Furthermore, there is no compelling reason to restrict the names in the field at the corner close to P to gabbro and diorite, along with anorthosite. It would be equally valid to include norite, gabbronorite and troctolite. For volcanic rocks, only the names basalt and andesite are given to rocks plotting close to P (Fig. 9), there being no volcanic equivalents to the other plutonic names. However, basalt and andesite are distinguished not on the composition of the plagioclase, as for gabbro and diorite, but are divided at a silica content of 50 wt%. This is independent of whether the rock is undersaturated or not and also of its colour index, as if all mafic minerals had a silica content of 50 wt%! Although the use of a chemical criterion to distinguish between basalt and andesite in a modal classification scheme is strange, it is put forward because of the frequent presence of phenocrysts of labradorite or bytownite in andesites.

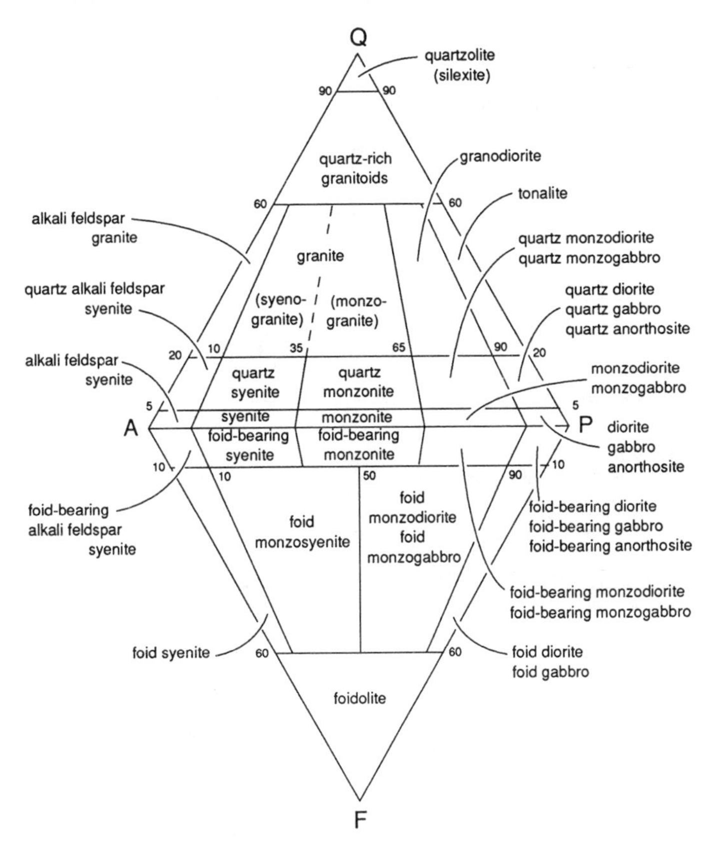

Figure 8. The International Union of Geological Sciences (IUGS) modal mineral classification scheme for plutonic rocks with M<90 vol%, where M includes mafic and related minerals (ferromagnesian minerals, melilite, monticellite and primary carbonate). The diagram is renormalized so that Q+A+P=100 or F+A+P=100 for each rock. From Le Maitre (1989)

In spite of these minor but important imperfections, the classification of holocrystalline igneous rocks using the two QAPF diagrams would be straightforward if feldspars formed two *distinct* mineral groups with no solid solution between them, namely 'alkali feldspar' and 'plagioclase' with An>5. However, feldspar phase relationships do not conform at all to this simple but arbitrary picture (Fig. 1), so that grave problems are encountered with many rocks using the IUGS scheme. These are of two kinds, fundamental and practical. Fundamentally, because of phase continuity, Ab-rich feldspars are plagioclases and should be classified as P and not A. This would lead to total rejection of the IUGS scheme. It is not our aim to make such a radical proposal as it would require a redefinition of all common igneous rock names, a momentous undertaking. Nevertheless, severe practical problems still exist which it would be unwise to neglect completely. These include (1) misnaming of alkali feldspars and hence of rocks, (2) identification of the wrong number of feldspars in strongly zoned crystals, (3) the treatment of coarsely exsolved alkali feldspars and (4) apparent jumps in rock series:

(1) Ternary feldspars exist which are alkali feldspars from their position with respect to the critical solution line, but in which the solubility of An is greater than 5 % (Fig. 2). Such feldspars occur in dry, high-*T* igneous rocks; they would be described optically as alkali feldspar and the rocks as alkali feldspar syenites by most petrologists. However, the feldspars would have to be called P and the rocks diorites in the IUGS sense.

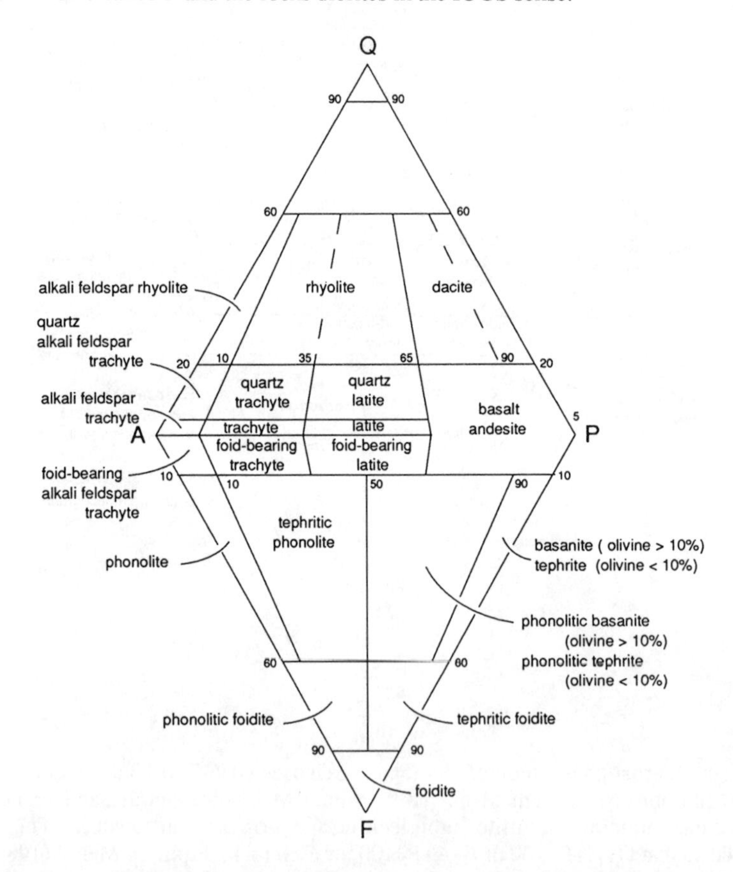

Figure 9. The International Union of Geological Sciences (IUGS) modal mineral classification scheme for volcanic rocks with M<90 vol%, where M includes mafic and related minerals (ferromagnesian minerals, melilite, monticellite and primary carbonate). The diagram is renormalized so that Q+A+P=100 or F+A+P=100 for each rock. Compare with Fig. 8. From Le Maitre (1989)

(2) Igneous rocks with strongly zoned feldspars are common, in which plagioclase cores may be overgrown by weakly ternary rims (Sec. 4.2). Using the IUGS definitions, the cores would be classified as P and the rims as A, even though only one feldspar phase crystallized throughout the history of the rock. The number of feldspar phases which crystallized is fundamental for understanding rock differentiation and any modal classification. If only one feldspar crystallizes it follows a solidus path, whereas if two crystallize they follow solvus paths. In the first case lowering *T* will lead to increased solid solution, whereas the reverse is true on solvus paths. Maximum solubility occurs in hot, dry rocks where the limit of ternary solubility reaches ~$An_{25}Ab_{50}Or_{25}$.

(3) If dry rocks with alkali feldspars or ternary feldspars are cooled slowly, the feldspars undergo extensive exsolution and coarsening so that the exsolved phases become visible under the optical microscope (forming mesoperthites), even though the individual crystals may retain their original form and bulk composition. For An-poor feldspars the IUGS classification makes no distinction between such hypersolvus rocks, and subsolvus rocks in which the two feldspars (albite and potassium feldspar) crystallized as separate phases, because they both plot at A. This is not true for more An-rich feldspars in syenites (see 1 above) or for 'igneous-looking' charnockitic rocks, which are often characterized by the presence of coarsely exsolved mesoperthite, but for which it is recommended to assign the phases "equally between A and P as the amounts of alkali feldspar and plagioclase (usually oligoclase or andesine) components are roughly the same" (Le Maitre 1989). It is further recommended to use the prefix m-, being short for mesoperthite, as in m-charnockite, whereas this recommendation is not made for non-charnockitic mesoperthite-bearing hypersolvus granites and syenites.

(4) Rock series poor in K exist which contain only plagioclase. Rocks in such series would be classified as gabbro/diorite, quartz gabbro/diorite or tonalite plotting along the QP edge (as long as the plagioclase is more basic than An_5) and then as alkali feldspar granite along the AP edge (where the plagioclase is more acid than An_5), although such rocks could be incorrectly called plagiogranite or trondhjemite (Le Maitre 1989, p 106, 124), in which case they would be 'expected' to plot in the tonalite field. Thus the points for the modal compositions would jump from one edge of the diagram to the other even though the rocks may grade continuously in the field; most petrologists would expect a large chemical break between rocks named tonalite and granite (see for example the average compositions for the two rocks in Le Maitre 1976c or Cox et al. 1979).

In spite of the grave shortcomings involved in the treatment of Ab-rich feldspars in rock nomenclature using modal compositions, we recommend use of the present IUGS scheme for want of a better one and to avoid further confusion. Any future revision of the IUGS scheme should make a more serious attempt to take account of feldspar phase relationships.

3.2 THE CHEMICAL TAS DIAGRAM

Where it is difficult (very fine-grained rocks) or impossible (glassy rocks) to determine the mode of volcanic rocks, it is recommended that a total alkalis--silica (TAS) diagram (renormalized on an H_2O- and CO_2-free basis) be used (Le Maitre 1984, 1989; Le Bas et al. 1986). The TAS diagram (Fig. 10; *cf.* Cox et al. 1979), on which are shown the points for the three normative feldspar end members *an*, *ab* and *or* as well as an_{50}, an_5, nepheline (*ne*), leucite (*lc*) and SiO_2 (*Q*) for comparison, should be used only for unaltered or weakly altered rocks, because of the great mobility of alkalis. Most other minerals plot along the base line if alkali-free (such as olivine), or slightly above it if they contain alkalis (such as micas, amphiboles, some pyroxenes or glass). Although there is no direct relationship between the TAS and the QAPF diagrams, the positions of Q, A, F and P have been indicated. Because feldspars are by far the dominant minerals in igneous rocks, rock compositions will plot in or close to the area *an-ab-or*. If low-alkali or alkali-free minerals are present, they will plot below this area, whereas if feldspathoids are present, they will plot above it. As for the QAPF diagram, rock series can be plotted on a TAS diagram (Lameyre and Bonin 1991), as can feldspar zoning paths which plot in the *an-ab-or* triangle.

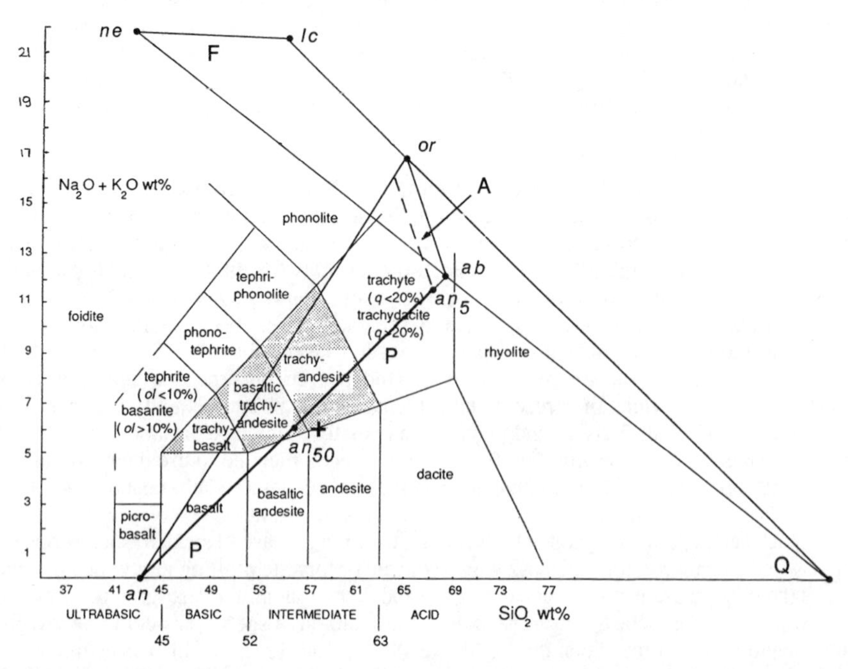

Figure 10. Total alkalis--silica (TAS) diagram (renormalized on a volatile-free basis) for essentially unaltered fine-grained or glassy volcanic rocks which cannot be classified using Fig. 9. The positions of quartz (*Q*), albite (*ab*), orthoclase (*or*), anorthite (*an*), leucite (*lc*) and nepheline (*ne*) are also plotted for comparison; *Q*, *or* and *lc* are collinear as are *Q*, *ab* and *ne*. The average composition of nearly 26,000 rocks is shown by (+) close to an_{50}. Modified from Le Maitre (1989)

There are, however, inconsistencies in the use of names compared with the QAPF diagram. Only nine names are common to both (rhyolite, dacite, andesite, basalt, trachyte, basanite, tephrite, phonolite and foidite) but two more can be considered to be spelling variants (phonotephrite and tephriphonolite compared to phonolitic tephrite and tephritic phonolite). Rock names involving mineral names such as quartz, alkali feldspar and feldspathoid are obviously absent from the TAS diagram. Moreover, there are five composite names (trachydacite, trachyandesite, basaltic trachyandesite, trachybasalt and picrobasalt) which do not occur on the QAPF diagram, some of which (trachyandesite, basaltic trachyandesite, trachybasalt) are further subdivided into sodic and potassic variants. The absence from the TAS diagram of the name *latite*, a primary field in the QAP diagram, defined restrictively in the former as a *potassic* trachyandesite, is remarkable; trachyandesite would have been a better choice than latite for the QAP diagram. As the QAP diagram does not distinguish between rocks with sodic and potassic feldspar, both being included in the general term alkali feldspar, there is no reason why names for potassic rocks should appear on it. A very serious inconsistency is the difference in the definition of basalt and andesite at 50 wt% SiO_2 in the QAPF diagram (see above) and between 52 and 57 wt% SiO_2 in the TAS diagram with basaltic andesite in the gap between. In spite of the many difficulties involved in devising rock classification schemes, it is highly regrettable that the differences between the two should be so great, especially if one wishes students to adopt them.

3.3 NORMATIVE PLAGIOCLASE COMPOSITIONS AND CLASSIFICATION SCHEMES

Just as it is possible to indicate the positions of feldspar compositions on a TAS diagram, one can contour the QAPF diagrams for plagioclase compositions for common plutonic and volcanic rocks. This can rarely be done directly because both modes and plagioclase compositions are often not given in publications; in addition, plagioclases are generally zoned so that it is difficult to determine average compositions. On the other hand published chemical analyses of igneous rocks are extremely common, although often without adequate petrographical description. It is thus tempting to make use of chemical data to classify rocks from calculated norms. Le Maitre (1976c) made a compilation of nearly 26,000 chemical analyses of igneous rocks, the rock names being taken from the publications. He plotted the chemical data on various diagrams including a TAS diagram and a normative feldspar diagram. On a TAS plot (Le Maitre 1976c, Fig. 1, p 599) almost all rocks lie within the fields named on Fig. 10; the vast majority lie on a broad band stretching from ~45 wt% SiO_2, ~0 wt% alkalis to ~75 wt% SiO_2, ~7.5 wt% alkalis, straddling both the line (not shown) separating alkali basalts from tholeiites from Macdonald and Katsura (1964) and the A-P line in Fig. 10. The average composition is close to 58 wt% SiO_2, 6 wt% alkalis.

Although the standard norm calculation does not give the true amounts of feldspar in the rock, which it tends to overestimate (because it neglects Na in pyroxene and amphibole, K in mica and Ca in amphibole), it may have only a small effect on relative proportions (*cf.* Le Maitre 1967a). A further small error arises because the norm involves wt% whereas feldspar compositions are given in mol%; because An and Or have almost the same molecular weight and that of Ab is about 6 % smaller, the effect is to displace points on a wt% diagram slightly away from the Ab corner, the greatest displacement being in the centre. The total normative feldspar contents of average intermediate and acid rocks taken from Cox et al. (1979) show remarkably small variations from ~60-72 wt% (except for syenites, trachytes and anorthosites which are larger), basic rocks having values around 50-55 wt%; ultrabasic and undersaturated alkaline rocks have, of course, much lower values.

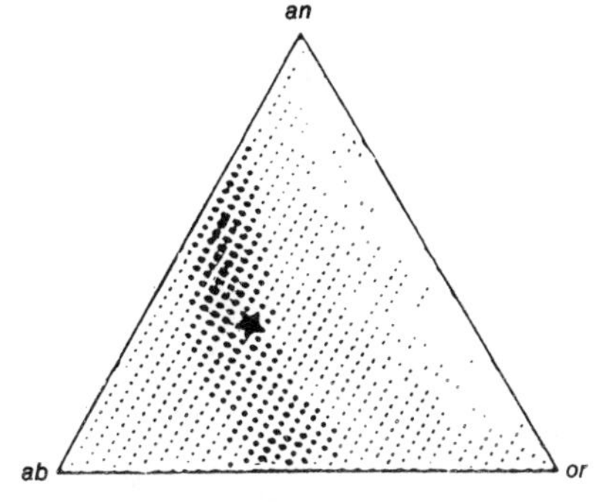

Figure 11. Plot of normative feldspar compositions of oversaturated plutonic and volcanic igneous rocks with more than 20 wt% normative feldspar. The star gives the average composition. A diagram for undersaturated rocks (with *ne* added to *ab*, and *lc* and *kp* added to *or*) is very similar but more dispersed. After Le Maitre (1976c, Fig. 1, p 599).

A normative feldspar plot (where small italic letters are used for normative *an*, *ab* and *or* because norms are in wt%) of the 26,000 rocks of Le Maitre (1976c) shows that most points fall in a rather narrow band (Fig. 11) stretching from close to the *an--ab* side line near an_{70} to close to the *ab--or* side line near $ab_{50}or_{50}$, much of the band being nearly parallel to the *an--or* side line at $ab_{50\pm5}$. The average value falls close to $an_{35}ab_{45}or_{20}$. Moreover, average values for most volcanic rocks also lie on the same band (*cf.* Le Maitre 1976b; also data from Cox et al. 1979). This observation and the fact that the shape of the band is similar to liquidus fractionation lines in the feldspar system (Fig. 12) strongly suggest that it defines a very broad *liquid line of descent.* Thus, other components in the liquid would appear to have a minor effect on the shape of such lines projected into the feldspar system. It is also interesting that average plutonic rocks from Le Maitre (1976b) and from Cox et al. (1979) plot on the same line as the volcanic rocks, possibly suggesting that most of the rocks selected are not significantly cumulative, at least for the feldspars.

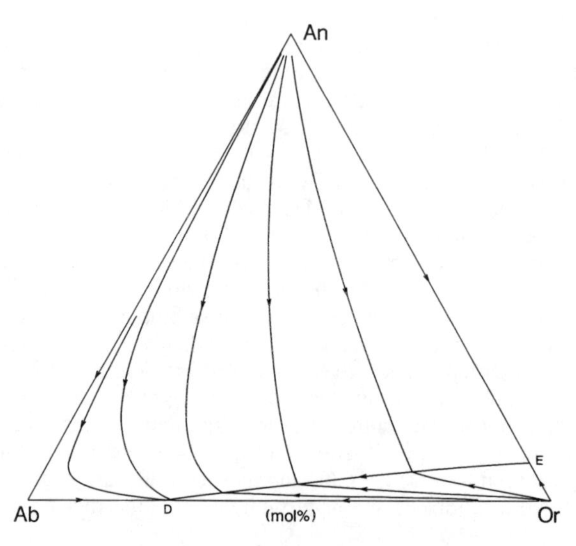

Figure 12. Liquidus fractionation lines at high P_{H_2O} (≈0.5 GPa) diverge from An and Or and converge on the boundary line ED. They are convex to Ab. They descend steeply in T in the plagioclase field and much less steeply in the sanidine field. From Brown (1993, Fig. 6)

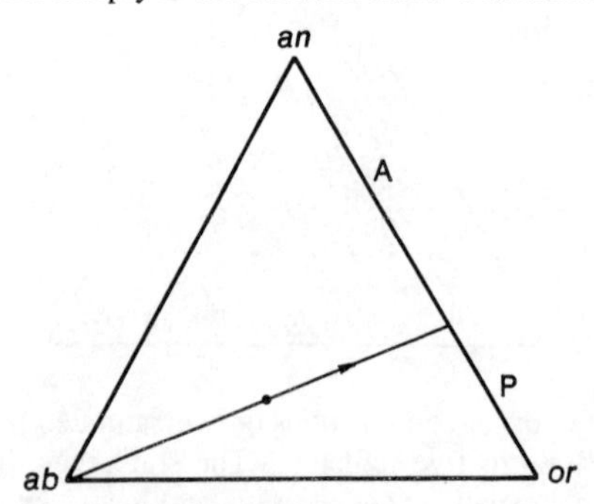

Figure 13. Method of projection from the *ab* apex through the normative feldspar composition (•) in order to estimate A/P ratio from chemical analyses. After Le Maitre (1976a)

Le Maitre (1976a) used normative feldspar compositions, which give the bulk feldspar composition, to estimate modal compositions to classify rocks on a QAP diagram, by partitioning the normative components between the phases. There is no direct way of doing this, although it could be calculated using thermodynamic models, such as that of Lindsley and Nekvasil (1989), if T and P_{H_2O} were known. Le Maitre proposed that the least bad way was to ignore *ab* and calculate the A/P ratio as *an*/*or*. This is equivalent to projecting the normative composition from the *ab* corner onto the *an*--*or* side line (Fig. 13), and would be reasonable only if (1) the solubility of *or* in P and of *an* in A were 0 or nearly so and (2) the tie lines were always parallel to the *an*--*or* side line. As can be seen (Fig. 2) the solubility is significant close to the Ab corner, especially at low P_{H_2O}, the tie lines which are parallel to the An--Or side line swing round to being roughly parallel to the Ab--Or side line at low P_{H_2O} (Fig. 14) and to being strictly parallel thereto at moderate to high P_{H_2O} (Fig. 15). Moreover, Le Maitre's proposal requires one to ignore the IUGS cut-off at An_5 for the distinction between P and A.

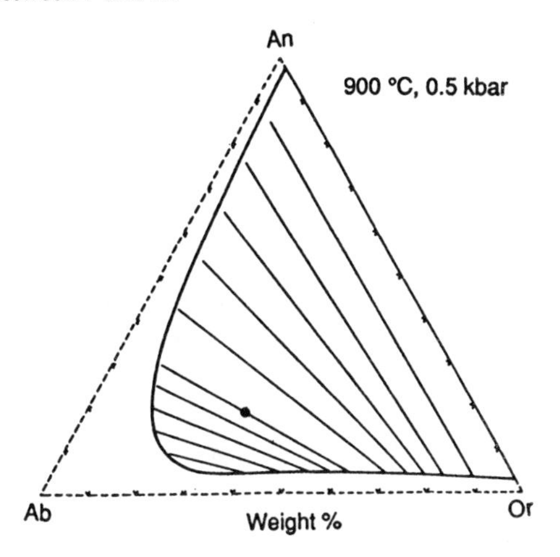

Figure 14. Section at 900°C, P_{H_2O} 0.05 GPa showing solvus isotherm and tie lines joining co-existing feldspar pairs in the two-feldspar field. (•) from Fig. 13. From Nekvasil (1988, Fig. 1c)

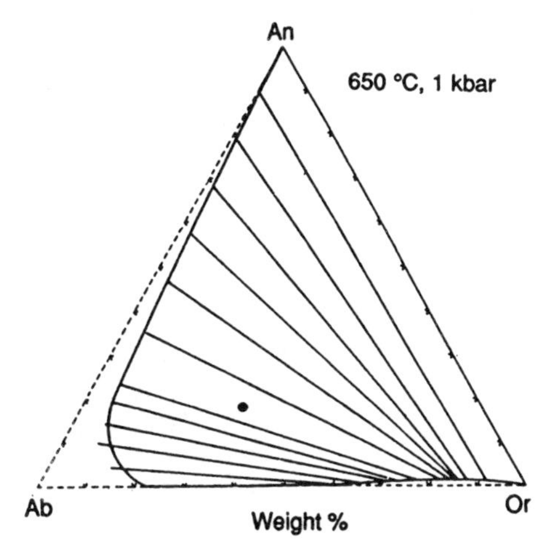

Figure 15. Section at 650°C, P_{H_2O} 0.1 GPa showing solvus isotherm and tie lines joining co-existing feldspar pairs in the two-feldspar field. (•) from Fig. 13. From Nekvasil (1988, Fig. 1a)

One simple graphical way to estimate the A/P ratio would be to plot the bulk normative feldspar composition on a ternary phase diagram at low and high P_{H_2O} (or P and a_{H_2O}) and make use of the tie lines to estimate the proportions of the two phases and the compositions of the plagioclase. This was done for normative plagioclase compositions of average rocks from Cox et al. (1979), and the averages plotted on a QAP diagram (Fig. 16); they differ significantly from those which would be obtained using the method proposed by Le Maitre.

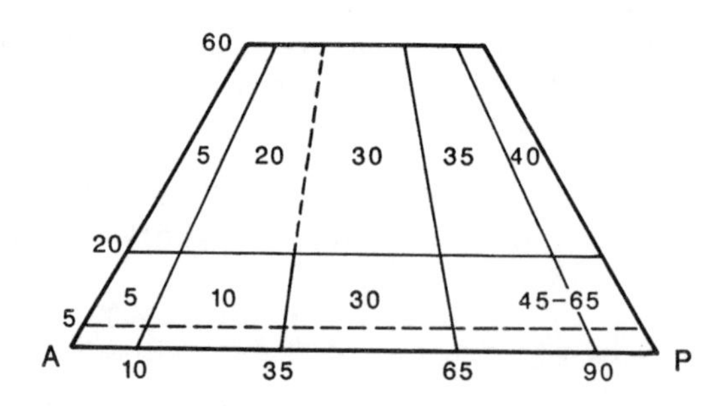

Figure 16. Estimated values of plagioclase compositions (An contents), calculated from average chemical analyses of rocks from Cox et al. (1979), plotted on a QAP diagram in the field corresponding to the rock name

4. Feldspars in Volcanic and Hypabyssal Rocks

4.1 STRUCTURES AND TEXTURES IN VOLCANIC ROCKS

4.1.1 *Structures in Volcanic Rocks.* The distinction between structures and textures in volcanic rocks is largely a question of scale. Large-scale volcanic structures are outwith the subject of this review, and include pillows, banding and heterogeneities found in mixed and brecciated rocks, tuffs and ignimbrites (Fischer and Schmincke 1984). Textures involve the shapes of crystals or the mutual spatial relationships of a few crystals (or of crystals in glass), whereas structures involve many crystals. Most petrologists call the preferential orientation (alignment) of a large number of platy feldspars, a flow structure, although MacKenzie et al. (1982) call it a trachytic texture---they suggest that flow structure or texture should be avoided because of genetic connotation. Many basic and intermediate volcanic rocks contain fragments of refractory foreign rocks of mantle (ultramafic nodules) or deep crustal origin (basic granulites) called *xenoliths*, which may contain feldspars. Of particular interest is the occurrence in vents, tuffs and flows of alkali basalt *s.l.* of suites of *xenocrysts* and *megacrysts* of plagioclase, anorthoclase and sanidine (Irving 1974; Laughlin et al. 1974; Chapman and Powell 1976; Gray and Anderson 1982; Upton et al. 1984; Bertelmann et al. 1985; Barker 1987; Fitton 1987; Schulze 1987; Upton and Emeleus 1987), because they give information about the crystallization of feldspar near the top of the mantle or the base of the crust. Basalts also frequently include feldspar-rich rocks, such as leucogabbro, melagabbro and anorthosite, which could have crystallized from the liquid in a chamber at intermediate depths in the crust, and these may be called *cognate enclaves.* The vol% of enclaves and megacrysts may reach very high values, as in the 'big-feldspar' dykes from the Gardar province, South Greenland, which contain up to 80 vol% of anorthosite and plagioclase crystals (Emeleus and Upton 1976); Bridgwater and Harry (1968) found that the more basic the host rock, the higher the An content of the plagioclase, from which one may infer that they are cognate.

4.1.2 *Volcanic Intercrystalline Textures.* Textures involve crystallinity (*i.e.* the fraction of crystals relative to residual glass) and the size, shape, orientation and mutual relationships of the crystals (MacKenzie et al. 1982). Textures in volcanic and hypabyssal rocks are generally quite different from those in plutonic rocks, and are not simply finer-grained or glassy variants of the latter. The sizes and shapes of feldspars depend on the conditions of nucleation and growth, which are best seen in partly glassy rocks. Feldspars tend to grow late in the crystallization history of ultrabasic and many basic volcanic rocks. Plagioclase is almost always absent from Archæan komatiites, although the residual glass, which is almost invariably altered, may be strongly feldspar-normative (Arndt and Nisbet 1982); fresh glass and plagioclase occur in the Phanerozoic komatiite from Gorgona island, Columbia (Echeverría 1982). Plagioclase is also absent from intermediate boninites, the glasses of which are also strongly plagioclase-normative (Ohnenstetter and Brown in prep.). It crystallizes late in picrites and picritic basalts, which may also be partly glassy. It generally crystallizes at an intermediate to early stage in glassy basalts, and early in andesites, dacites and rhyolites (pitchstones), the composition changing from basic through acid plagioclase to alkali feldspar. Plagioclase frequently shows subhedral, skeletal, dendritic, branching, radiate or spherulitic morphology in such rocks, which may be interpreted as resulting from growth at progressively higher rates under progressively higher undercooling, on the basis of experiments in the laboratory (Lofgren 1974a,b; Lofgren and Donaldson 1975; Smith and Brown 1988, Sec. 17.2). Glasses tend to be divitrified or altered with time, and many tuffs or ignimbrites may now have a cryptocrystalline texture. Other features such as rounded corners and certain embayments may be interpreted as signs of partial dissolution (resorption) in some intermediate glassy rocks (Smith and Brown 1988, Sec. 17.3). Feldspars frequently contain glass inclusions which were incorporated during growth; these range from regularly arranged, often euhedral to subhedral inclusions outlining growth faces, to myriads of small, irregular inclusions giving what is called a sieve texture. This is inferred to have formed by partial dissolution during mixing of different magmas (Tsuchiyama 1985; Smith and Brown 1988, Secs. 17.2 and 17.3).

Feldspars frequently have complex size distributions. Many lavas are porphyritic with large cognate plagioclase or alkali feldspar phenocrysts or smaller microphenocrysts; the rhomb-porphyries from southern Norway, and oceanic basalts, are well known examples. The crystals formed during at least a two-stage cooling history, if not during a two-stage emplacement history. In many cases it can be shown that the feldspars grew in an intermediate magma chamber and were subsequently erupted. Highly porphyritic lavas with 'excess plagioclase' must have concentrated plagioclase (Cox and Mitchell 1988; MC Jackson 1993). In some cases cognate plagioclase megacrysts may reach very high vol% as in the 'big-feldspar' dykes from the Gardar province (Emeleus and Upton 1976). Feldspars may also vary greatly in shape. Where they grew early, they are often large and tabular on (010) and may show strong preferred orientation, defining a flow structure (see above). In more rapidly cooled, glassy rocks their morphology is more complex, as described above.

Where the feldspar crystals grew at the same time as other feldspar crystals or other minerals, their textures are determined by the mutual relationships of the minerals. Where feldspar has self-nucleated on earlier crystals, glomeroporphyritic textures are formed (*cf.* MacKenzie et al. 1982), and these are common in many basalts and andesites; in oceanic basalts plagioclase glomerocrysts may be more common than phenocrysts (Kuo and Kirkpatrick 1982). Intergrowth textures involving plagioclase, such as *poikilitic* inclusion in clinopyroxene, are common. Variants of the poikilitic texture include the *ophitic* and *subophitic* (or doleritic) textures. Two common interstitial textures involving plagioclase laths are the *intersertal* texture, where the spaces between the laths are filled by glass, and the *intergranular* texture where they are filled by one or more pyroxenes. See MacKenzie et al. (1982) for illustrations of these textures.

4.1.3 *Volcanic Intracrystalline Textures.* The internal textures in feldspars are probably among the best developed and most well studied in minerals. Readers should refer to the detailed review of growth and dissolution features, inclusions and chemical zoning in feldspars in Smith and Brown (1988, Secs. 17.2 and 17.3, especially 17.3.3) and in recent papers using Nomarski and laser-interferometry methods (AT Anderson 1983; Pearce 1984; Pearce and Kolisnik 1990; Singer et al. 1993, Pearce this volume).

4.2 ZONING PATHS IN VOLCANIC FELDSPARS

Zoning paths in volcanic feldspars may differ from ideal fractionation paths in the simple hydrous ternary feldspar system (Sec. 2.5) because (1) water is not buffered and (2) other components are present in the liquid. Furthermore, because of the dynamic uprise and eruption of magmas, relative movement of crystals and liquid may occur, so that periods of growth may alternate with periods of dissolution. Even where growth is continuous, it seems improbable, however, that growth conditions approach those of equilibrium crystallization (see Nekvasil this volume). On eruption of lavas decompression occurs, which causes a relative increase in a_{H_2O} (because most magmatic liquids are not water-saturated at depth), and if saturation is reached, a loss of water through vesiculation but at constant a_{H_2O} of 1. Zoning paths in groundmass feldspars thus correspond to those at low P and gradually increasing a_{H_2O}, so that the final compositions are less ternary and are closer to the Ab corner at the end than expected; paths in phenocrysts may consist of two parts, one at moderate P during growth in a magma chamber at intermediate depth and a second as in groundmass feldspars. The paths in minor intrusions such as sills (Sec. 5.2) are formed under conditions intermediate between those of lavas and large intrusions. Paths for silica-saturated rocks such as rhyolites correspond to expectations based on calculations by Nekvasil (1990), being closer to An--Ab and Or--Ab side lines and also to the Ab corner than in the dry or hydrous ternary system. The reason is that silica lowers significantly the T of the liquidi and solidi so that the intersection with the ternary feldspar solvus occurs at lower T on the steep limbs where solubility is much lower. The effects of components other than silica are largely unknown. Zoning paths in normal basic and intermediate volcanic rocks appear to follow those for the simple feldspar system quite well, so that ferromagnesian components do not significantly modify feldspar phase relationships; there are, however, slight systematic differences which correspond with the nature of the magma series. This does not appear to be true for sanidine paths in strongly undersaturated peralkaline rocks, in which there is more Na or K than can be incorporated in feldspar.

4.2.1 *One-Feldspar Paths--Plagioclase.* The zoning paths in plagioclase in volcanic series can be characterized by (1) the most An-rich composition, which depends on how primitive the liquid was, (2) the total range of An in each rock, which depends on how perfect the fractionation was, (3) the Or content at an arbitrary value of An (taken here as 50 mol%), on which depends the magma series to which the liquid belongs (tholeiitic, calc-alkaline, alkaline or peralkaline) and (4) the maximum value in Ab at the curve-round towards Or, Ab_{max}, in paths where plagioclase is overgrown by an Ab-rich alkali feldspar (anorthoclase and sanidine), which depends largely on the water and silica activities.

Calc-alkaline basalts and basaltic andesites from Tonga (Ewart 1976) show a *small range* in An (about 10-20 %); the phenocrysts are richer in An than the groundmass plagioclases and both are very low in Or. Many series show a *moderate range* in An (30-50 % An, anorthite/bytownite to andesine), especially rocks of calc-alkaline affinities with low Or (Arculus 1978; Huijsmans and Barton 1989) or moderate Or (Heming 1977; Wyers and Barton 1989), or alkaline rocks with higher Or (FH Brown and Carmichael 1969). These paths can be distinguished by the Or content at An_{50}: Or rises from <1 % in tholeiitic and calc-alkaline rocks to ~4 % in alkaline rocks. Where phenocrysts and groundmass feldspars have been reported separately (AL Smith and Carmichael 1968; Lowder 1970; Heming 1977), the two often have similar overall compositional ranges in the different rocks, although individual crystals have smaller ranges. In general the cores of phenocrysts in orogenic rock suites are more An-rich than the groundmass plagioclases and may show complex mantles indicative of partial resorption. Where only phenocrysts were analyzed, the points for each rock define paths which differ in An content but fit on a smooth trend for the whole series, even though the individual crystals show complex zoning (Wyers and Barton 1989).

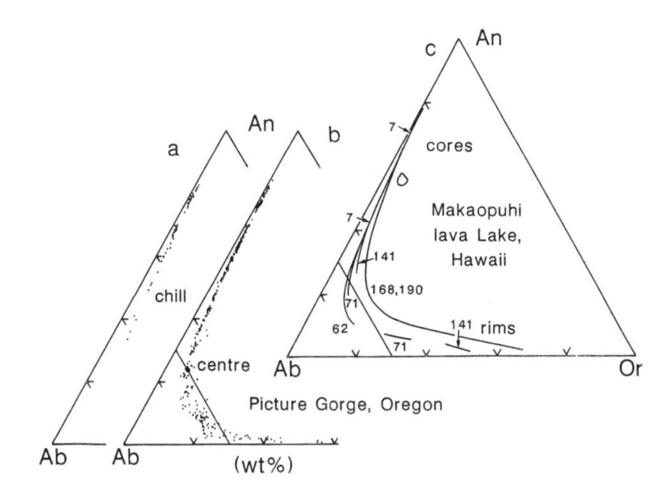

Figure 17. Variation of the composition of phenocrysts and groundmass feldspars from two specimens (a) chill and (b) 'centre' of a thick olivine tholeiite lava flow from Picture Gorge, Oregon (Lindsley and Smith 1971) and (c) in groundmass feldspars from specimens of olivine tholeiite in a lava lake from Hawaii (Evans and Moore 1968). Numbers refer to depths in feet

Where *in situ* differentiation has occurred extensively, the zoning paths may show a *very large range* in An. The outer zones of plagioclase crystals (anorthoclase or Ab-rich sanidine) may show increases first in Ab and then in Or so that the paths curve strongly away from the Ab corner towards the Ab--Or join. Feldspars from olivine tholeiite from a lava flow (Picture Gorge, Oregon Lindsley and Smith 1971) or a lava lake (Hawaii, Evans and Moore 1968) vary in composition with position in the flow or depth in the lake (Fig. 17). Specimens which cooled most quickly, from near the base of the flow or the top of the lake, show a moderate range reaching only ~30-45 % An (bytownite to andesine) with values of Or at An_{50} of ~2-2.2 %. Evans and Moore (p. 99) suggested that the small range near the top was caused by strongly disequilibrium growth due to rapid cooling, following Bowen (1928); where cooling is fast, zoning is also restricted by the presence of glass. A specimen from 45 m above the base of the flow ('centre' in Fig. 17b) and all specimens from deep within the lake have large ranges (bytownite through anorthoclase to Ab-rich sanidine) with greater values of Or at An_{50} of ~2.7 %. The maximum value of An in the cores of plagioclase crystals is greatest at intermediate depths (~An_{78}) and lower at the top and bottom (An_{72} and An_{68}). A strong curve-round away from Ab occurs, reaching *maximum* values Ab of 75-65 % before trending towards the Ab--Or join, this value of Ab_{max} decreasing systematically with depth in the lake. Whereas the path is complete for specimens from inside the flow and analyses are numerous close to the curve-round, this is not so for specimens from the lake at intermediate depths (at 62, 71 and 141 feet), in which the alkali feldspar occurs as distinct rims which were clearly seen by cathodoluminescence (Evans and Moore 1968).

Very similar paths to those from the Picture Gorge olivine tholeiite were observed by Russell and Nicholls (1987) in alkali olivine basalts from Diamond Craters, Oregon and from alkali olivine basalts, trachybasalts and trachytes from Lake Rudolf, Kenya (FH Brown and Carmichael 1971), the value of Ab_{max} for both being ~65-70 %, although individual rocks from Kenya had smaller

An ranges. Henderson and Gibb (1983) also gave very similar zoning paths for plagioclase from mildly alkaline minor intrusions, with a range of values of Ab_{max} of 75-60. The curve-round occurred at lower values of Ab_{max} in alkaline to peralkaline rocks (basalts, trachytes, rhyolites)---close to 65 % from Nandewar volcano, New South Wales (Stolz 1986) and 60 % from Gough Island, South Atlantic (Le Roex 1985). Three distinct An-poor curves were found for alkali feldspars from trachytes from Gough Island and Le Roex explained the curves with lower An as being due to higher water contents. Similar differences in feldspar compositions from intermediate lavas forming two series from Tristan da Cunha were explained, however, as being correlated with different bulk-rock chemistries (Le Roex et al. 1990). In general the smaller the value of Ab_{max}, the larger the value of Or at An_{50}. Similar compositions in the anorthoclase--sanidine range were found in peralkaline trachytes and phonolites from Mount Suswa, Kenya (Nash et al. 1969).

In most of the non-peralkaline basic to intermediate rocks dealt with in this section, the zoning paths appear to extend continuously outwards from the plagioclase cores towards the most Or-rich compositions at the rims of the crystals, although the latter are hard to analyze by electron microprobe, because the volume percentage of each composition decreases with fractionation and the extreme compositions often exist only in small groundmass crystals. As a result the direction of the zoning path in Ab-rich alkali feldspars is not always clear from the descriptions in the literature. However, most rocks with a few exceptions (see Sec. 4.2.4) probably crystallized only one feldspar, but further detailed textural and chemical studies are required to confirm this.

4.2.2 *One-Feldspar Paths--Sanidine.* Volcanic rocks with sanidine only are very rare and belong to ultrapotassic series. Wallace and Carmichael (1989) described groundmass sanidines from minettes and leucitites from Western Mexico (with very high contents of BaO and especially SrO up to 5 wt%) and Ewart (1981) from peralkaline lavas from Australia, but zoning directions were not given. Potassic and ultrapotassic basaltic lavas either with sanidine alone or with two feldspars (Sec. 4.2.4), from Wyoming (Nicholls and Carmichael 1969b) or from the central Sierra Nevada (Van Kooten 1980) have been described. As pointed out for non-peralkaline salic rocks by Nicholls and Carmichael (1969a) "a fractionating feldspar and its liquid are tied together in the sense that both either get more sodic or both get more potassic. In the pantellerites this relationship of feldspar to liquid is upset and becomes opposite in sense." Similar observations have been made for peralkaline salic rocks by Nash et al. (1969), Stolz (1986) and others and explanations given by Bailey and Schairer (1964), Nicholls and Carmichael (1969a) and Nash et al. (1969)---in liquids corresponding to such rocks, more Na and K is available than is required to form Ab and Or, so that crystallization paths are not well represented by projection in an An--Ab--Or diagram.

4.2.3 *Simultaneous Two-Feldspar Paths.* In many volcanic suites only the more differentiated rocks contain two feldspars, earlier members containing only plagioclase; the earlier the simultaneous crystallization of two feldspars occurs, the more alkaline is the suite. Crystallization must first follow solidus paths (hypersolvus conditions) before the liquid meets the boundary line on an even part, when crystallization of two feldspars occurs simultaneously (subsolvus conditions, Fig. 4). In the calc-alkaline lavas of central North Island, New Zealand, only the rhyolites contain two feldspars (Ewart 1969), whereas in the mildly alkaline rocks of the Banks Peninsula, South Island, two feldspars occur in the trachytes and are said to "converge towards one common, more sodic, composition" (Price and Taylor 1980), although no clear textural description was given. Two feldspars coexist in more basic rocks in potassium-rich suites and zoning paths have been given for various alkaline volcanics (Nicholls and Carmichael 1969b; Carmichael et al. 1974; Baldridge et al. 1981). Luhr and Giannetti (1987) and Turbeville (1993) described two-feldspar phenocryst associations in tuff and in pumice and scoria respectively, in which the plagioclase compositions may be as An-rich as 80-95 mol%. Luhr and Giannetti showed that the tie lines for the more Ab-rich pairs were less steep than expected (i.e. more closely parallel to the An--Or sideline, Fig. 14) and that sanidine for Ab-poor pairs may be more Ab-rich than plagioclase. Ferguson and Cundari (1982) suggested that one should not base crystallization relationships on phenocryst compositions because they may not have been in equilibrium with the liquid. Henderson and Gibb (1983) gave two-feldspar paths for rocks from three mildly alkaline minor intrusions in which both feldspars crystallized from an early stage and four others in which the alkali feldspar crystallized as a late-stage discrete phase.

4.2.4 *Sequential Two-Feldspar Paths.* If the liquid meets the boundary line on an odd part, crystallization of the first feldspar will cease and be replaced by the second (Brown 1993). This occurs most frequently at high T in dry rocks in which early crystallization of plagioclase is replaced by that of sanidine (Fig. 5). The second feldspar could nucleate and grow independently of the first, but it is much more likely that it overgrows it. Mantles of sanidine occur on calcic plagioclase in peralkaline rocks from the Roman province and in many shoshonites (Nicholls and Carmichael 1969b; Carmichael et al. 1974; Ferguson and Cundari 1982). {Under equilibrium crystallization this texture would correspond to a reaction relationship between the two feldspars (Bowen 1928). Partial resorption without overgrowth could also occur under equilibrium crystallization on an even part of the boundary line (Nekvasil 1992)}. Ferguson and Cundari (1982) suggested that crystallization of plagioclase and sanidine in rocks from the Roman province and from south-west Uganda is *sequential* and not cotectic. It is for this reason that the term 'sequential' was adopted (Brown 1993). However, sanidine in the groundmass consistently becomes richer in Or towards the rims (Cundari 1979; Ferguson and Cundari 1982).

4.2.5 *Critical Solution Line.* The stable part of the critical solution line, KK_E, is the locus of critical solution points lying on the ternary feldspar solvus at T below the intersection with the solidi. If crystallization is simultaneous, the line must lie between the end points of the zoning paths of the two feldspars. Henderson and Gibb (1983, Fig. 1a) and others (*e.g.* Fig. 17) found apparent one-feldspar paths which extend closer to Or than the position of the critical solution point in the binary alkali feldspar system near $Ab_{63}Or_{37}$. This would require either that crystallization in all such cases was disequilibrium, or that the critical solution line curves strongly away from the Ab corner on entering the ternary system. The position of the critical solution line, calculated by Fuhrman and Lindsley (1988), however, curves *towards* the Ab corner and not away from and it is compatible with that deduced from zoning paths in igneous rocks by Parsons and Brown (1983). Where zoning is apparently complete or nearly so from intermediate plagioclase cores to rims with compositions beyond $Ab_{60}Or_{40}$, it is possible that crystallization might have been sequential at the end with only be a small difference between the composition of the last plagioclase and the overgrowing sanidine. (The sanidine rims would have to zone to more Ab-rich compositions, but a detailed textural and microprobe study would be required to show whether this is true or not.) If crystallization is sequential, determination of the position of the critical solution point will be difficult, because it will not always lie between the ends of the two zoning paths as proposed by Henderson and Gibb (1983). Furthermore, it may be possible that some of the incomplete zoning paths in specimens at intermediate depths (71 and 141 feet) in the lava lake from Hawaii (Fig. 17) may also correspond to sequential crystallization, because distinct rims were seen by cathodoluminescence (Evans and Moore 1968). Evans and Moore suggested that the gradual increase of Ab_{max} with depth was due to a closer approach to equilibrium crystallization, although the opposite would be expected (Smith and Brown 1988, Sec. 17.2.4). Other possibilities are variation in the activity of water or in the composition of the interstitial liquid.

4.3 VOLCANIC SUBSOLIDUS TRANSFORMATION AND EXSOLUTION MICROTEXTURES

Feldspars in igneous rocks grow initially with high Si,Al disorder and mutual solubility. Cooling rates in volcanic and hypabyssal rocks are, however, too fast for significant Si,Al ordering to occur in plagioclase, so that order-dependent exsolution microtextures (Carpenter 1981; Smith and

Brown 1988, Chap. 19) are not found; the only microtextures in basic plagioclases are antiphase domains due to the C$\bar{1}$-to-I$\bar{1}$ and the I$\bar{1}$-to-P$\bar{1}$ transformations (Smith and Brown 1988, Chap. 5). Si,Al ordering and interdiffusion of Na and K in alkali feldspars, on the other hand, are very fast relative to cooling rates so that transformation and exsolution microtextures are the rule even in volcanic rocks.

4.3.1 *Anorthoclase.* Alkali feldspars in volcanic rocks are initially disordered and monoclinic. Two types of C2/m-to-C$\bar{1}$ phase transformation may occur during cooling of high-T alkali feldspars, depending on composition (Brown and Parsons 1989). The transformation in Or-rich feldspars involves Si,Al ordering, and would give a very fine, modulated structure called orthoclase, if sufficient time were available. Because ordering is too slow, orthoclase is absent from lavas and dykes, only high or low sanidine being found. The transformation in Ab-rich feldspars is reversible, diffusionless, unquenchable and due to spontaneous shearing of the framework, and gives anorthoclase. If cooling is slow, partial Si,Al ordering may occur before transformation. Even if cooling is very fast, as in crystal ejecta or very thin dykes or lavas, an Albite- and Pericline-twinned microtexture characteristic of anorthoclase is always found (McLaren 1978; Gray and Anderson 1982; Akizuki 1983; KL Smith et al. 1987; Smith and Brown 1988, Sec. 18.3.1). It is usually sufficiently coarse to be visible and is used to identify anorthoclase in thin section (MacKenzie and Guilford 1980; MacKenzie et al. 1982); twins range in size from submicroscopic to up to a few hundred μm (Gray and Anderson 1982). The microtexture stores very little elastic energy, because the large distances between composition planes lead to almost complete relaxation of shear stresses at the interfaces (Brown and Parsons 1993). Anorthoclase with up to 20 mol% An (Smith and Brown 1988, Fig. 7.27), occurs as phenocrysts in sodic trachyte, rhyolite (pantellerite) and certain phonolite lavas and dykes (De Pieri and Quareni 1973; De Pieri et al. 1974, 1977; Harlow 1982) and as megacrysts in alkali basalts (Irving 1974; Laughlin et al. 1974; Chapman and Powell 1976; Gray and Anderson 1982; Upton et al. 1984; Barker 1987; Fitton 1987; Schulze 1987). It also occurs as overgrowths on strongly zoned plagioclase in mildly alkaline basalts (Sec. 4.2.1).

4.3.2 *Cryptoperthites.* Disordered An-poor alkali feldspars of intermediate composition may exsolve without ordering on fast cooling, or order partially before exsolving on slower cooling. Because Na,K interdiffusion in alkali feldspars is so fast, homogeneous sanidines of intermediate composition should possibly not exist except in crystal ejecta, although Hay (1962) claimed to have described a homogeneous low sanidine of composition $An_3Ab_{56}Or_{41}$ from a pyroclastic tuff and De Pieri et al. (1977) one of similar composition from an alkali rhyolite, although TEM might show that they are exsolved. More extreme compositions, which should exsolve at lower T from their position relative to the alkali-feldspar solvus (Brown and Parsons 1989), such as ejected crystals of anorthoclase (close to $An_{0-20}Ab_{85-60}Or_{15-30}$) and sanidine (close to $Ab_{15-25}Or_{85-75}$) may be homogeneous (Sec. 4.3.1; Bertelmann et al. 1985; Smith and Brown 1988). Feldspars of intermediate composition from lavas, dykes, ignimbrites and tuffs show incipient or well developed exsolution. The earliest stages consist of two monoclinic phases (Jung 1965; Soldatos 1965; Willaime et al. 1973; De Pieri et al. 1977) with slightly anastomosing lamellar shape and poorly defined boundaries (McConnell 1969) close to (60$\bar{1}$) (Willaime and Brown 1974) which on slower cooling coarsen slowly and develop Albite and/or Pericline twins in the Ab-rich high albite phase, whereas the Or-rich phase is high or low sanidine (MacKenzie and Smith 1956; De Pieri et al. 1974, 1977; Christoffersen and Schedl 1980; Yund and Chapple 1980; Snow and Yund 1985; Smith and Brown 1988, Sec. 19.2.2). The finest periodicity is ~10 nm and it may coarsen to 50-200 nm, so that the microtextures are always invisible under the microscope, although they may give a pale blue iridescence; the feldspars are then called moonstones. Because of the very fine periodicity, cryptoperthites store large amounts of elastic energy due to lattice misfit and to twinning of the high albite (Brown and Parsons 1993) as shown by the strained lattice parameters

(Smith and Brown 1988, Fig. 19.8). The rates of coarsening of the lamellar microtextures in disordered alkali feldspars have been measured in the laboratory (Owen and McConnell 1974; Yund and Davidson 1978). The relationship between cooling rate and the development of the microtextures is given by Brown and Parsons (1984b). Cooling in volcanic rocks is, however, too fast for the development of twin-like modulations or twins in the Or-rich phase so that orthoclase and microcline are absent from cryptoperthites in volcanic rocks. Deuteric alteration of volcanic feldspars is generally much less intense than that of plutonic feldspars (Sec. 6), because of the low water content or the low crystallization P which favours loss of water through degassing.

5. Feldspars in Plutonic Rocks

5.1 STRUCTURES AND TEXTURES IN PLUTONIC ROCKS

5.1.1 *Structures in Plutonic Rocks.* Plutonic igneous rocks are often devoid of noticeably well developed structures: i.e. they are to a first approximation homogeneous over much of the intrusion and isotropic or nearly so. This is more frequent in intermediate to acid plutonic rocks, the so-called granitoids, than in basic intrusions. Close to the margins they may be less homogeneous, because of the presence of a border facies or of xenoliths of country rock. They may also contain cognate enclaves (Didier and Barbarin 1991) or show evidence of magma mixing. Modal, grain-size and rhythmic *layering* (Irvine 1987) is frequent in moderate-sized to large basic intrusions such as Bushveld, Kiglapait, Skaergaard, or Stillwater (Hess 1960; Wager and Brown 1968; Morse 1969; McBirney and Noyes 1979; Parsons 1987), although it also occurs in intermediate and acid central intrusions such as those in the Gardar province, South Greenland (Emeleus and Upton 1976; Parsons 1987; Upton and Emeleus 1987). It also occurs in the Tugtutôq giant dykes of the Gardar province (Upton 1987), but is generally absent from sills (*cf.* Sec. 5.1.2). Plutonic rocks frequently show, often in addition to layering, a strong preferred orientation of plagioclase or sanidine laths called *igneous lamination* (Hess 1960; McBirney and Noyes 1979; Parsons and Becker 1987; Brown et al. 1989; Higgins 1991). Such structures have been studied from the point of view of grain shape and grain orientation in order to determine whether the lamination could have been formed by simple compaction (McKenzie 1987) or by shear during flow (Benn and Allard 1989; Higgins 1991). According to Curie's principle the symmetry of the fabric must reflect the strain regime (Paterson and Weiss 1961; Fernandez and Laporte 1991). Two extreme cases of interest are fabrics with orthorhombic symmetry produced by co-axial flattening in a viscous liquid, due for example to simple compaction, and fabrics with monoclinic or lower symmetry produced by non-co-axial flattening, due for example to viscous shear flow. Criteria may be deduced to determine the sense of shear during non-co-axial flattening and hence the sense of magmatic flow (Fernandez and Laporte 1991; Nicolas 1992).

5.1.2 *Plutonic Intercrystalline Textures.* Intercrystalline textures in plutonic rocks are less varied than in volcanic rocks, because of the much smaller range of cooling rates. Plutonic igneous rocks are phanerocrystalline, medium- (1-5 mm) to coarse-grained (>5 mm) with a surprisingly small range of grain sizes compared to volcanic rocks, very slowly cooled rocks (e.g. the Bushveld intrusion) being not particularly coarser-grained than more quickly cooled ones. Crystals are generally equant and equigranular, porphyritic textures being much less common except in some granitoids, than in volcanic rocks. Textures seen in thin section are subhedral or anhedral granular, but little work has been done on studying the three-dimensional shapes or inter-relationships of crystals. In non-cumulative granitoids (Petford et al. 1993; Bryon et al. in press) the early-formed minerals, such as plagioclase, have a tendency to form a framework of crystals with euhedral, Euclidean shapes, whereas later ones, such as alkali feldspar and quartz, tend to fill the interstices with anhedral, possibly 'fractal' geometries. The textures of minerals in granitoids thus depend on the order of crystallization and also on the length and nature of the liquid path in the phase diagram. There should be differences in the textures of rocks whose compositions approach those of liquids close to a low-T eutectic compared to those far from a eutectic, which may either have had a long crystallization history or have entrained significant amounts of crystals.

Many *layers* in basic and ultrabasic layered intrusion have compositions which are, however, far from those of liquids. The relationships between layers allows one in most cases to exclude an origin by separate intrusion, so that some process of local differentiation must be invoked. Moreover, crystals of some of the minerals in the layers have textures which suggest a different origin from the other crystals. Wager et al. (1960) distinguished between the generally larger, chemically homogeneous, euhedral crystals, which they called *cumulus crystals* and by extension cumulus phases, and the usually anhedral minerals filling the interstices which they called *intercumulus crystals* or intercumulus phases. They described three main cumulus textures and defined three types of cumulate rocks, namely orthocumulate, adcumulate and mesocumulate. They explained the origin of the cumulus phases by nucleation and growth of crystals during movement in large-scale convection currents followed by settling to the bottom. The difference between the different types of cumulates involved the way in which the intercumulus phases were produced. For orthocumulates Wager et al. invoked *in situ* crystallization of the interstitial liquid, whereas for adcumulates movement of components between the interstitial liquid and the supernatant liquid was invoked; extreme examples of adcumulates are monomineralic rocks such as anorthosites. In some cases it is possible to show that movement leading to exchange of the interstitial liquid must have occurred (Moreau et al. 1987). Many workers on layered intrusions (ED Jackson 1961; Campbell 1978; McBirney and Noyes 1979; Langmuir 1989) questioned the possibility of settling of plagioclase, especially in non-Newtonian or more evolved Fe-rich liquids in which it should float and not sink, and suggested *in situ* nucleation and growth. However, cumulates have a 'packstone' texture (Bryon et al. in press) and look as if they formed a three-dimensional framework of settled crystals, rather than having a texture suggesting bottom growth (crescumulate texture). Whatever one's views on the possibility of crystal settling and the origin of cumulates (*cf.* papers in Parsons 1987; Morse 1988; Hort et al. 1993; Sparks et al. 1993; Bryon et al. in press), the cumulus classification scheme is a valuable way of describing the rocks (Irvine 1987). It is clear that a serious study of the three-dimensional relationships between cumulus crystals in rocks in a layered intrusion is urgently required to see whether they form a self-supporting framework or not. In slowly cooled, high-T basic and ultrabasic cumulates such as those from Rum, sufficient time may be available for textural adjustment to occur between grains of the same or different minerals leading to textural equilibrium (Hunter 1987). In smaller, more rapidly cooled intrusions such as sills, Gibb and Henderson (1993) suggested that large-scale convection and settling of crystals are absent. This is in agreement with the fact that sills of basaltic composition almost always have doleritic, intersertal textures and that cumulate textures appear to be absent, possibly implying that convection, crystal settling and cumulus textures are closely related.

5.1.3 *Plutonic Intracrystalline Textures*. Internal growth textures are less well-developed in feldspars in plutonic than volcanic rocks. Chemical zoning in plagioclase (see below) is less strong except in some diorites and tonalites. Plagioclase often shows signs of periods of partial dissolution, and may also contain inclusions of glass or fluids (Smith and Brown 1988, Sec. 17.2.6). Intergrowths with other minerals are not uncommon, sometimes fine-scale symplectites being found. Thus in small, shallow granitic intrusions granophyric or micrographic intergrowths of feldspar and quartz are often found. Intergrowths and overgrowths with feldspar are very frequent, the most spectacular overgrowths being perhaps those found in rapakivi granites in which rounded alkali feldspar is overgrown by plagioclase, usually oligoclase (*cf.* Emslie 1991). Numerous hypotheses have been proposed to explain the origin of the texture, one of the main ones being changes in pressure or water content related to magma ascent (*cf.* Nekvasil 1991) or unroofing, or to magma mixing. Stimac and Wark (1992) described rapakivi-like plagioclase mantles on sanidine in silicic lavas from Clear Lake, California and by analogy suggested that true rapakivi textures may arise by magma mixing. An example of complex feldspar intergrowths involving plagioclase and two microperthitic ternary feldspars from the Klokken gabbro was described by Parsons and Brown (1983).

5.2 ZONING PATHS IN PLUTONIC FELDSPARS

Zoning paths in feldspars in plutonic rocks are not as well known as those in volcanic rocks, which have been extensively studied, largely through the sustained efforts over the last twenty-five years of Carmichael and coworkers using the electron microprobe (*cf.* Sec. 4.2). More qualitative studies of the complexities of zoning, similar to those on volcanic feldspars (Pearce this volume), have also not yet been carried out on plutonic feldspars to the best of our knowledge. This may be due partly to the more altered nature of plutonic rocks (Sec. 6). Fine-scale oscillatory zoning is completely absent, either because it never formed or because it was destroyed by interdiffusion during slow cooling (Maaløe 1976; Smith and Brown 1988, Chap. 17), whereas zoning is present on the scale of a few μm to tens of μm. During the crystallization of intrusions water will tend to build up gradually in the liquid at constant P causing a progressive increase in a_{H_2O}, which leads to a lowering of the liquidi, a greater intersection with the solvus, and a reduction of mutual solubility of the feldspars; this occurs to a much greater extent than in volcanic rocks which tend to lose water during ascent and on reaching the Earth's surface. Water saturation may be reached in many cases as shown by the presence of drusy cavities, often filled with secondary minerals. Where the magma was rich in volatiles other than water such as fluorine, fluid saturation does not imply water saturation so that a_{H_2O} may nevertheless be <1.

Examples of one- or two-feldspar zoning paths in a variety of plutonic rocks are given for the Tugtutôq older giant dyke complex by Upton et al. (1985), for a ring complex from Cameroon by Parsons et al. (1986), for cumulus and intercumulus feldspars in a monzo-anorthosite from Aïr by Moreau et al. (1987), for a syenite from the Larimie complex, Wyoming by Fuhrman et al. (1988), for a leucotroctolite--anorthosite complex from Aïr (Brown et al. 1989), for a microdiorite dyke from New Zealand by Brothers and Yokoyama (1991) and for an alkaline carbonatite complex from Brazil by Beccaluva et al. (1992). Perhaps the most detailed study of the compositions, and textural and microtextural relationships of feldspars from a plutonic rock is that involving the Klokken gabbro--syenite (Brown et al. 1993; Parsons and Brown 1983; Parsons and Brown 1988). Plagioclase with normal zoning is overgrown in a complex way by ternary to alkali feldspars, the compositions of which change systematically from a gabbro sample near the rim through the unlaminated syenodiorites and syenites, plagioclase being no longer present in the layered and laminated syenites in the centre (Parsons and Brown 1988). Crystallization could be interpreted as sequential, plagioclase crystallizing alone in the gabbro, then being replaced sequentially and overgrown by ternary feldspar in the umlaminated syenites; finally alkali feldspar crystallizes alone in the layered and laminated syenites.

Zoning paths are considerably shorter in plutonic rocks than in volcanic rocks. For example, Lindsley and Smith (1971) pointed out that the range observed in a *single sample* of olivine tholeiite from Picture Gorge, Oregon is very similar to that observed in the *whole* Kiglapait intrusion (Speer and Ribbe 1973; Morse pers. comm.), although complications arise in interpreting the mesoperthites. This possibly suggests that *in situ* fractionation has not gone as far in plutonic rocks as in volcanic rocks. There are two possible explanations. The first is that because plutonic rocks cool much more slowly than volcanic rocks, crystallization in the former approaches equilibrium crystallization (Nekvasil this volume). Thus exchange through interdiffusion must occur between the centres and rims of the crystals leading, if complete, to homogeneous crystals or, if partial, to ones in which the zoning paths are much shorter. Because of extremely slow interdiffusion rates in plagioclase (see Smith and Brown 1988, Chap. 16 and Fig. 17.25), we believe that true equilibrium crystallization probably does not occur. A second possibility is that relative movement of liquid and crystals may occur much more easily in large, slowly cooled basic intrusions, so that complete fractionation does not occur *locally* but only on the scale of the *whole* intrusion. Thus nearly the whole crystallization history of the Picture Gorge tholeiite can be seen in the zoning of one crystal and its overgrowth into the interstices, whereas this is not true for large intrusions because of the separation in space of the early and late crystallization phases.

5.3 PLUTONIC SUBSOLIDUS TRANSFORMATION AND EXSOLUTION MICROTEXTURES

5.3.1 *Plagioclase*. Cooling rates in plutonic environments are sufficiently slow that order-dependent exsolution microtextures may develop in plagioclase. An excellent discussion of the still-controversial subsolidus phase relationships and a summary of what is known from the study of natural crystals is provided by Carpenter in the present volume. The microstructures present (Carpenter, Sec. 3.2) include peristerite, Bøggild and Huttenlocher intergrowths, the 'e' microtexture and various antiphase-domain microtextures which interact with the foregoing microtextures in complex ways. However, much of our knowledge of plagioclase microtextures is built up from small sample sets or isolated samples which were not collected with a geologically well-defined thermal history in mind. The studies of Skaergaard plagioclases by Gay (1956) and Bown and Gay (1958), and of metamorphic plagioclases from Broken Hill by Slimming (1976), represent two of the very few examples of studies of sets of plagioclases, which cooled under similar conditions, but which have a range of compositions. It is likely that microtextural work on plutonic plagioclases may reveal important information on the cooling history of rocks, as has been done successfully with alkali feldspars, but at the present time this approach has not been attempted systematically. Another valuable approach would be to look at microtextural changes with respect to composition in chemically zoned crystals. Petrologically, perhaps the most important implication of experimental and microtextural work on low plagioclase to date is the non-ideality of mixing below ~800°C implied by Carpenter's Figure 6. This should provide food for thought for readers concerned with thermodynamic modelling of metamorphic reactions.

5.3.2 *Ternary and Alkali Feldspars in Hypersolvus Rocks*. Together with our coworkers we have published several studies of microtextural variation and evolution of feldspars in hypersolvus syenites (Brown et al. 1983; Brown and Parsons 1984a,b; Worden et al. 1990; Waldron and Parsons 1992) and in rock series grading from gabbro into syenite (Parsons and Brown 1983; Brown and Parsons 1983; Parsons and Brown 1988; Brown and Parsons 1988a). The topic has also formed part of several reviews (Parsons and Brown 1984 in the previous NATO feldspar volume; Brown and Parsons 1989; Parsons and Brown 1991; Brown and Parsons 1993). Only the main conclusions since 1984 are summarized here, with the more recent developments stressed. Almost without exception, in our experience, TEM shows that plutonic alkali feldspars (including those in subsolvus rocks, Sec. 5.3.3) have a *dual microtexture*. *Pristine* parts of crystals have 'strain-controlled' exsolution microtextures which are regular, coherent and/or semicoherent, and *turbid*, deuterically altered parts of crystals have much coarser, more irregular microtexture in which the exsolved phases may be semi- or incoherent. This dual texture can occur on an extremely fine scale (Figs. 20 and 21) and provides an explanation for the multiplicity of feldspar types identified even in small cleavage fragments by single-crystal X-ray diffraction (MacKenzie and Smith 1962, in the first NATO feldspar volume). The following subsections deal with strain-controlled microtextures. Deuteric alteration is discussed in Section 6.

The morphology of strain-controlled microtextures is dependent on bulk composition and cooling rate, and the scale is largely dependent on cooling rate. The character of strain-controlled crypto- and microperthites in plutons is shown in Fig. 18. The relationship between texture and cooling rate, as judged from proximity to intrusive contacts, has been demonstrated for the Klokken intrusion (Brown et al. 1983) and the Coldwell intrusion (Waldron and Parsons 1992).

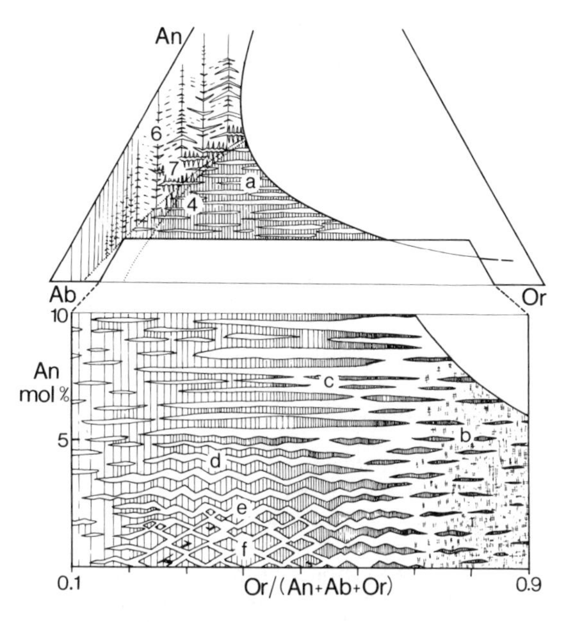

Figure 18. Schematic diagram showing strain-controlled microtextures, viewed down [001], developed as a function of bulk composition in slowly cooled ternary feldspars. The diagram, which is highly schematic, is largely based on microtextures in a series of spatially related feldspars and on variation within chemically zoned crystals in the Klokken intrusion. Vertical hatching: Albite twins in plagioclase. Cross hatching: tweed orthoclase. Unornamented: other Or-rich feldspar, low sanidine in straight lamellar intergrowths, high microcline in wavy intergrowths, and low microcline in zig-zag and lozenge intergrowths. (from Brown and Parsons 1988a)

The stages in the development of the zig-zag and lozenge textures, which lead to optical braid microperthite, were described by Brown and Parsons (1984b), and reviewed in the previous NATO volume (Parsons and Brown 1984). The critical factor which leads to the complex morphology of the intergrowths, in contrast with the nearly straight microtextures found in volcanic alkali feldspars (Sec. 4.3.2), is Si,Al ordering. This is coupled with Na--K interdiffusion and also controls coarsening rates. For compositions around f (Fig. 18), near the critical solution curve on the ternary solvus (Fig. 1) exsolution probably occurs by spinodal decomposition, in a monoclinic feldspar with some framework order (Al concentrated in T_1 sites), and leads to straight lamellae of two monoclinic feldspars. When the Ab-rich phase cools through the C2/m--$C\bar{1}$ transformation, twinning develops to reduce strain energy and transfer of Al from undifferentiated T_1 sites to differentiated T_10 sites begins. The 'primary periodicity' (Fig. 7) of the exsolution microtexture seems to be fixed at this point. Provided cooling is slow enough (see Parsons and Brown 1991, for the most recent discussion of rates) *Y*-ordering begins in the Or-rich feldspar and a wavy interface develops. The waves develop into zigs and zags which represent Albite twins in high microcline. Relatively coarsely twinned microcline develops because of strain induced by the Albite twins in the albite. Eventually albite forms detached rods which are lozenge-shaped when viewed parallel to *c* (Fig. 7). These intergrowths are fully coherent and reduction of coherency strain is achieved entirely by interface rotation from $(\bar{6}01)$ to $(\bar{6}\bar{6}1)$, and by twinning, as explained by Brown and Willaime (1974). A 'secondary periodicity', in the form of broader oblique lamellae of microcline which sometimes show 'M'-twinning, may develop (Figs. 7 and 21).

In the Klokken syenite, both primary and secondary periodicities coarsen downwards from the roof of the intrusion (Brown et al. 1983) but almost all microtextures for feldspars in this compositional range were of braid type, even when extremely fine scale. In the Coldwell syenite (Waldron and Parsons 1992) feldspars near the contact have nearly straight lamellae, but coarsen and become wavy and ultimately lozenge-shaped further into the intrusion. It should be possible to use these textures to estimate relative cooling rate in hypersolvus plutons directly, but because the rate-controlling factor is Si,Al ordering, we do not, at present, have any experimental calibration. Crude estimates were made by Brown and Parsons (1984b) and Parsons and Brown (1984, 1991), using heat-flow calculations.

In the Coldwell syenite, Waldron and Parsons (1992) found an additional textural complication which they called 'ripples'. These are relatively long-period (~10 μm) bulk compositional variations superimposed on the regular, mostly braid, microtextures. The ripples have Ab- and Or-rich bulk compositions, which correspond with an inferred T of ~530°C, but they are themselves linear cryptoperthites and crypto-antiperthites which formed by coherent exsolution at an inferred low T, ~380°C. The ripples possibly record a relatively high-T (~530°C) phase of fluid--rock interaction.

The Klokken intrusion provides a continuum of feldspar compositions ranging from Or-rich andesine in alkali gabbros (above 6 in Fig. 18), through strongly ternary feldspars in syenodiorites (7, 4 and a), to almost An-free braid intergrowths in syenites (f). These are spread through an orderly, ordered series of sidewall cumulate rocks, which become more evolved inwards from the contact of the pluton, and there is also a considerable range of microtextures in chemically zoned crystals (Parsons and Brown 1988, Brown and Parsons 1988a). In the latter we can be sure that the microtextural variation is wholly dependent on composition and not cooling rate. In more Ab-rich and strongly ternary feldspars (4 and a, Fig. 18) the Or-rich areas form lenses in the a*-b* plane, and regularly spaced dislocations develop at the interfaces (Brown and Parsons 1984a). Unlike the fully coherent mesoperthites between c and f (Fig. 18) these antiperthites may be described as *regularly semicoherent*. We suggested that the interfaces in these antiperthites are not able to rotate to the ($\bar{6}61$) position to relieve strain, because of the high An content. Because of CaAl--NaSi coupling, the Si,Al--O framework controls diffusion and the interface becomes blocked in the ($\bar{6}01$) position. Interface and strain energy are reduced by the formation of dislocations. The Or-rich feldspar does not transform to microcline, forming ordered and anti-ordered zigs and zags in the diagonal association, but develops instead the tweed domain texture of orthoclase, with twin-like ordered--anti-ordered domains on the scale of a few unit cells.

Antiperthites nearer the plagioclase join (7 and 6, Fig. 18) are fully coherent. Strain is minimized by the relatively small volume of the Or-rich phase, by its wedge-shaped terminations, and by the development of wedge-shaped Albite twins in the enclosing plagioclase (7). The Or-rich phase may nucleate heterogeneously on pre-existing Albite-twin composition planes in the plagioclase, or homogeneously between them. Bulk compositions on the flanks of the ternary solvus are unlikely to exsolve by spinodal decomposition, because of the relatively low T of the coherent spinodal and the relatively large T interval between the coherent solvus and spinodal. The Or-rich phase in the crypto-antiperthites from the Klokken intrusion is low sanidine.

Feldspar with an Or-rich bulk composition (b on Fig. 18) is uncommon in hypersolvus rocks. It has been found as parts of alkali feldspars with unusually Or-rich bulk compositions from both Klokken and Coldwell, particularly in 'ripples' in the latter (Waldron and Parsons 1992). These are linear, fully coherent, tweed orthoclase cryptoperthites. Or-rich feldspars are common in subsolvus rocks and are described in the next Section.

5.3.3 *Alkali Feldspars in Subsolvus Rocks*. In principle, if stable equilibrium is maintained during cooling, pairs of feldspars crystallized on solvus tie-lines (Fig. 15) should react continuously by reciprocal exchange of the three components An, Ab and Or. Phase compositions will follow *two-feldspar solvus paths* on the strain-free solvus which differ markedly from two-feldspar fractionation paths in the presence of liquid (Sec. 2.5) in that they move steeply down T towards Or and the An--Ab join at the ends of tie lines passing through the bulk composition. There is evidence that in granulite-facies rocks, some intercrystalline bulk compositional exchange does occur (Brown and Parsons 1988b, Evangelakakis et al. 1993), but it is not certain to what extent, if any, such exchange occurs in common granitic rocks. Eventually, in all granitic rocks, the Or-rich feldspar (the alkali feldspar) leaves the strain-free solvus, is no longer in equilibrium with the plagioclase feldspar, and eventually exsolves giving the orthoclase-- and/or microcline--low albite microperthites characteristic of so many granites. More-or-less straight, lamellar microperthites can often, but not always, be seen using the light microscope in the optically clear parts of these feldspars, but patch or film perthites developed in the somewhat turbid parts are usually more conspicuous. The latter are discussed in Section 6.1.

There are many X-ray diffraction studies of the character of the potassium feldspar in granites (e.g. Dietrich 1962; Parsons and Boyd 1971; Stewart and Wright 1974) but, with the exception of a few single-crystal studies (e.g. MacKenzie and Smith 1962), these give no microtextural information. The petrological significance of the 'obliquity' and 'triclinicity' measurements popular thirty years ago (Dietrich 1962) is still obscure, although in granites they very probably reflect variation in the amount and perhaps T range of the deuteric interactions that the sample has experienced. The relative proportions of orthoclase and microcline in samples (Parsons and Boyd 1971) certainly reflect the completeness of 'unzipping' reactions during the deuteric stage (Sec. 6).

TEM studies of microcline and orthoclase were published by McLaren (1984) and the character of orthoclase firmly established by Eggleton and Buseck (1980) but neither study treated the minerals in a petrological context. Bambauer et al. (1989) illustrated microclines from metamorphic rocks and subdivided them into 'regular' and 'irregular' varieties, on the basis of the appearance of the twin microtextures. There are, to the best of our knowledge, no published TEM studies of twin microtextures in Or-rich feldspars from granites which put them in a geological context, and as far as we know no studies at all of strain-controlled exsolution microtextures. Despite their abundance and often, in porphyritic examples, very striking appearance, potassium feldspars in granites are very poorly characterized at a TEM scale.

With our colleagues, Kim Waldron and Martin Lee, we have been looking at potassium feldspars from a variety of granitic rocks using TEM. As the much earlier X-ray diffraction work showed, many crystals are mixtures of orthoclase and microcline. Irregular microcline is extremely common and seems to be related to deuteric recrystallization (Sec. 6). Regular microcline also occurs and is well developed in pristine microcline ovoids in rapakivi granites intruded into granulite-facies rocks in South Greenland. We do not, at present, have a clear view of the factors which lead to the development of these different types of microcline, although the persistence of tweed orthoclase is a good, although not perfect, marker of crystals (or parts of crystals) which have escaped interactions with fluids (Sec. 6). It is not clear whether microcline can develop in crystals which have not interacted with fluids, and the role of external deformation in granites with long cooling histories is also not understood, although localized examples of the role of deformation have long been known (Harker 1954).

We have seen fine micro- or cryptoperthitic coherent or semicoherent exsolution lamellae in pristine parts of all the granitic alkali feldspars so far examined (e.g. frontispiece to this volume). These are mostly from Scottish calc-alkaline intrusions, but are also from rapakivi granite batholiths from South Greenland. The idea, prevalent in the literature, that cryptoperthite is characteristic of relatively rapidly chilled rocks is simply wrong, and this observation is confirmed by the discovery of cryptoperthites in high-grade metamorphic rocks from Sri Lanka (Evangelakakis et al. 1993), in charnockite/mangerite rocks from the Adirondacks (Waldron et al. 1993) and in orthoclase porphyroblasts in regional migmatites from N Scotland (work in progress).

The feldspar from the Shap granite, Northern England, which forms the frontispiece, is characteristic of the pristine part of feldspars from subsolvus granites. The photograph shows a secondary-electron SEM image of a cleavage fragment which has been lightly etched in HF vapour (Waldron et al. submitted), a technique which provides a simple way of obtaining a three-dimensional overview of exsolution textures with excellent resolution (see also Fig. 21). TEM shows the Or-rich phase to be tweed orthoclase. Because the Or-rich phase is dominant (Fig.18, field b), the minor Ab-rich phase exerts insufficient stress to promote microcline development, in the form of long-period twins, and the rotation of the interfaces into the $(\bar{6}61)$ position characteristic of braid perthite does not occur. Ordering in the Or-rich phase leads to tweed structure of very fine ordered and anti-ordered domains which have significant residual strain (see Sec. 6). Strain in the perthite lamellae, which are wing-shaped, extended in a direction close to *c* but forming blunt lenses parallel to *b*, is relaxed either by the presence of Albite- and Pericline-twin corrugations in the albite, by wedge-shaped terminations to the lamellae, or by the presence of spaced dislocations (Waldron et al. submitted). Etch pits visible in the frontispiece have developed at the site of dislocations, which develop only on the larger lamellae. The development of such dislocations was treated by Aberdam (1965), who imaged carbon replicas by TEM, but since then this work have been ignored. The dislocations shown etched in the frontispiece form two sets of lens-shaped loops approximately in the *a-b* and *a-c* planes, and some, but not all, may reach the surface of the cleavage fragment. Dislocations at the interfaces of exsolution lamellae may have a role in the rapid diffusion of radiogenic argon from alkali feldspar crystals (see Sec. 6.2 and Fitz Gerald and Harrison 1993).

At present we do not know the geological factors governing the development of the coherent lamellae in these feldspars, and we are investigating whether they can be related to estimated cooling rates in the same way as they can in hypersolvus rocks (Brown et al. 1983; Waldron and Parsons 1992). It is of interest that the coarser lamellae in the feldspar from the Shap granite, which is a relatively small, high-level calc-alkaline intrusion, are considerably coarser than those in the regional metamorphic rocks listed above.

6. Deuteric Alteration of Feldspars and Low-Temperature Reactivity

6.1 DEUTERIC MICROTEXTURAL CHANGES IN ALKALI FELDSPARS

As introduced in Section 2.7, a deuteric alteration stage occurs at T <450°C in most igneous rocks (hydrothermal alteration is not dealt with here), which can lead to three types of reaction: (1) partial replacement by other phases such as clay minerals; (2) replacement by feldspars of different composition, as in albitization of plagioclase; (3) microtextural changes not involving changes in bulk composition. The first two reactions are also particularly important in weathering and diagenesis (Blum this volume). The present section, however, deals mainly with the essentially isochemical microtextural changes in alkali feldspars, produced by reaction with a local fluid, which include the development of an important microporosity which causes optically visible turbidity in the crystals (Worden et al. 1990; Walker 1990, 1991; Guthrie and Veblen 1991; Burgess et al. 1992). The micropores (a few μm to a few tens of nm in size) develop in conjunction with the breakdown of the submicroscopic strain-controlled microtextures, which are replaced by a much coarser patch or film perthite consisting of a mosaic of semicoherent or incoherent subgrains of Or- and Ab-rich feldspars. Optical micrographs illustrating the relationship between turbidity and perthite coarsening were provided by Parsons (1978), and Finch and Walker (1991) showed that areas of red cathodoluminescence seen in many alkali feldspars correlate with the microporosity. Micrographs at the TEM scale by Worden et al. (1990), Guthrie and Veblen (1991), Waldron and Parsons (1992) and Waldron et al. (1993) allow submicroscopic details to be seen. Almost all plutonic igneous rocks are affected to a greater or lesser extent. Relatively unaffected rocks, such as the decorative Oslo larvikite, with its dark-coloured feldspars, are rarities compared with the pink, white or creamy, translucent crystals which predominate in granites.

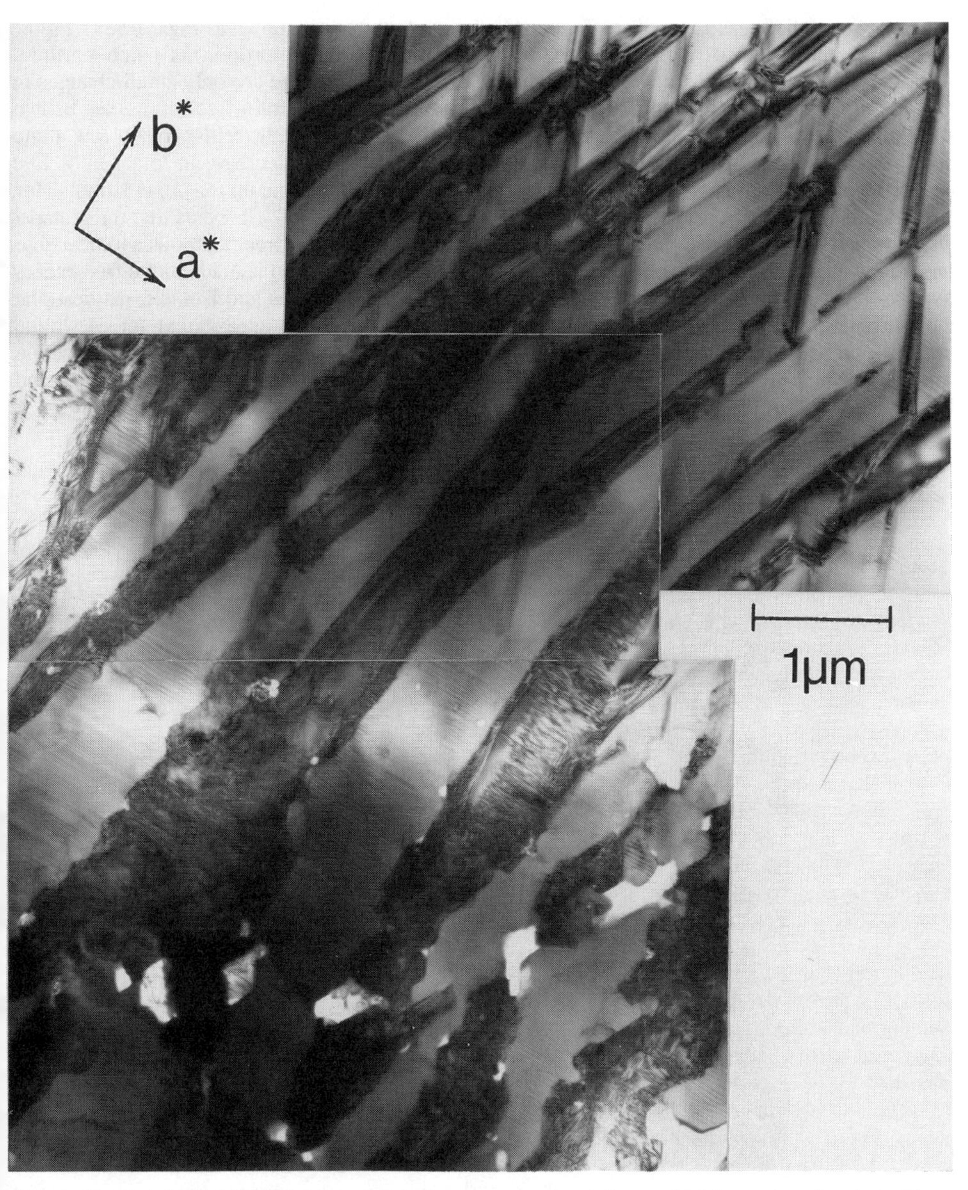

Figure 19. Bright-field TEM micrograph, beam approximately parallel to *c*, showing transition between coherent braid cryptoperthite (*top*, compare Fig. 7) to irregular, microporous, incoherent patch and film perthite (*bottom*), within an alkali feldspar crystal from the Klokken syenite. Note the presence of subgrains in the patch perthite, most easily seen in the Albite-twinned albite, and the fine, irregular twinning in the microcline. The micropores are often defined by {110} planes, as are subgrains in the patch perthite, which sometimes form {110} serrations at the incoherent boundaries of perthite. (From Worden et al. 1990)

Examples of the breakdown of strain-controlled microtextures are shown as Figs. 19-21. Figure 19 shows breakdown of a braid microperthite to an irregular, microporous patch perthite. Although the exsolution texture is coarser and coherency is lost, there are only small changes in phase compositions (as the feldspars move from the coherent to the strain-free solvus, see Brown and Parsons 1989) or degree of order, because in both intergrowths, the feldspars are low albite and low microcline. The driving force for this textural change is decrease in total Gibbs free energy through loss of elastic strain energy, which was calculated to be up to ~2.5-4 $kJmol^{-1}$ for lamellar intergrowths in cryptoperthites by Brown and Parsons (1993). It seems that the feldspar with strained microtexture dissolves locally in an aqueous fluid and reprecipitates as a more stable, unstrained mosaic of the two phases. Although a positive contribution is made to the free energy by introduction of subgrain boundaries, incoherent phase boundaries and isolated dislocations, there is nevertheless an overall reduction in free energy because of the release of strain. A similar process can occur in tweed orthoclase (Fig. 20) when it is converted to unstrained low microcline. In this case we estimated the strain energy in tweed orthoclase to be 2-3 $kJmol^{-1}$ (Brown and Parsons 1993).

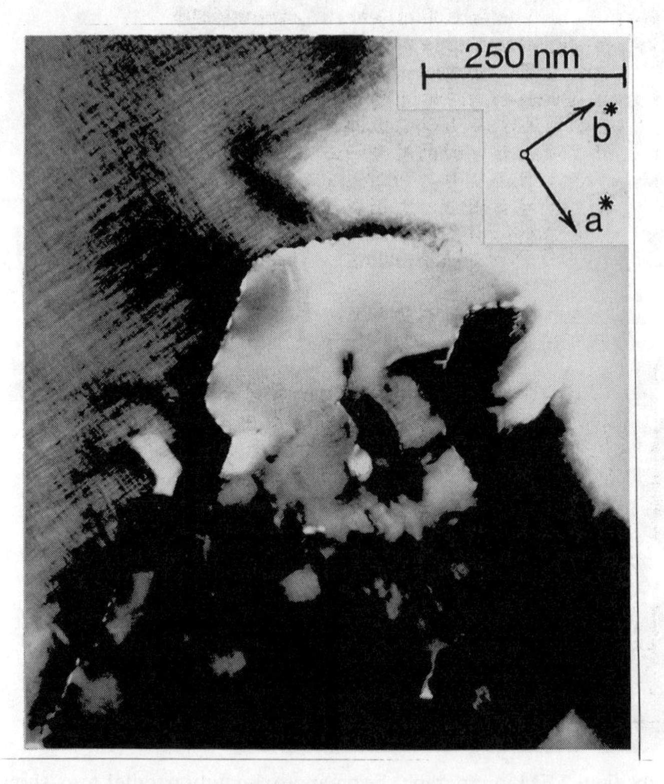

Figure 20. Bright-field TEM micrograph, beam approximately parallel to *c*, showing a discontinuous boundary between tweed orthoclase (*top*) and irregular microcline (*bottom*). The microcline forms suboptical 'seams' in the Or-rich phase of a mesoperthite crystal from a mildly retrogressed granulite from the Adirondacks. Note the {110} habit of the interface, and subgrains in the microcline. (From Waldron et al. 1993)

At the TEM scale, loss of textural regularity is the most striking result of deuteric reactions. Sometimes, mosaics of well-defined subgrains develop (as in Worden et al. 1990, Fig. 7B) which may be of the same phase, or different phases with incoherent phase boundaries (Waldron et al. 1993, Fig. 3). In other cases subgrain walls are discontinuous and boundaries are semicoherent. Many Or-rich feldspars consist of complex mixtures of 'tweed' orthoclase and 'tartan' microcline, on a TEM scale, and this microcline has extremely irregular twin widths, leading to the patchy texture called 'irregular microcline' by Bambauer et al. (1989). Irregular microcline possibly forms from tweed orthoclase by fluid--feldspar reaction. There are many aspects of the deuteric stage which are uncertain and require further work.

Development of micropores always occurs together with deuteric coarsening, and the pores often occupy the spaces between subgrains (Worden et al. 1990; Walker 1991). The etching technique employed (Waldron et al. submitted) to produce the frontispiece is an extremely effective way of illustrating the textural changes that occur. This is much less laborious than TEM and provides a much better overview of the parts of the crystal affected by deuteric reactions. Back-scattered electron imaging (BSE) can also be employed (e.g. Worden et al. 1990; Burgess et al. 1992) but has much poorer resolution and requires that the sample be polished. Figure 21 shows the abrupt cut-out of the regular strain-controlled 'braid' texture and its replacement by relatively featureless subgrains, in this case largely of albite. Etching reveals the subgrain boundaries and shows how micropores occur between them. Similar relationships are seen using TEM (see Worden et al. 1990). The recrystallization shown on Fig. 21 appears to have occurred along a pre-existing fracture, but many alkali feldspar crystals from the Klokken syenite show pervasive development of patch perthite (see BSE micrographs in Worden et al. 1990 and Burgess et al. 1992). In Or-rich feldspars from calc-alkaline granites, similar textures develop (corresponding with the variations in turbidity visible in thin section and hand specimen), and we have also observed suboptical recrystallization of Ab-rich lamellae (Lee et al. in prep.) which does not affect surrounding orthoclase. Suboptical recrystallization of tweed orthoclase to microcline was reported by Waldron et al. (1993), in mildly retrogressed granulite-facies rocks (Fig. 20).

A common feature of the boundaries of subgrains is that they have a {110} 'Adularia' habit (Fig. 20; also Waldron and Parsons 1992, Fig. 8C; Waldron et al. 1993, Figs. 2, 3; Brown and Parsons 1993, Fig. 7). The habit often defines the shape of micropores (Waldron et al. 1990, Figs. 5 and 7C; Waldron and Parsons, 1992, Fig. 8D). The true Adularia habit is characteristic of low-T hydrothermal veins and diagenetic overgrowths in sediments, and this is consistent with the idea that deuteric recrystallization occurs by dissolution--reprecipitation in a fluid film. There is occasionally circumstantial textural evidence (Waldron and Parsons, 1992, Fig. 8D) that pores have advanced through the structure, leaving a trail of new feldspar in which elastic strain has been relaxed and subgrains have formed.

6.2 GEOCHEMICAL IMPLICATIONS AND TEMPERATURES OF DEUTERIC REACTIONS

Because of the pervasive character of the recrystallization which accompanies deuteric reactions, particularly in alkali feldspars, the process has important geochemical implications in stable-isotope and ^{39}Ar-^{40}Ar work, and possibly for trace elements which reside in the feldspar structure or occur in inclusions. The reduction of elastic strain (Brown and Parsons 1993) provides a driving force for recrystallization which can lead to exchange of ^{18}O or ^{40}Ar with an aqueous fluid. As Giletti (1985, this volume) has pointed out, there is no significant driving force for oxygen exchange by volume diffusion at low T as the free energy change represented by a few parts per mil adjustment in ^{18}O is minuscule. Coherent cryptoperthite lamellae and/or tweed orthoclase are extremely common in the alkali feldspars in plutonic igneous rocks, and all such feldspars reach low T with a built-in driving force which permits dissolution and reprecipitation. It is this feature which facilitates ^{18}O exchange and leads to the low 'closure temperature' for oxygen which is characteristic of alkali feldspar. Once deuteric reactions have occurred, the effective dimension for

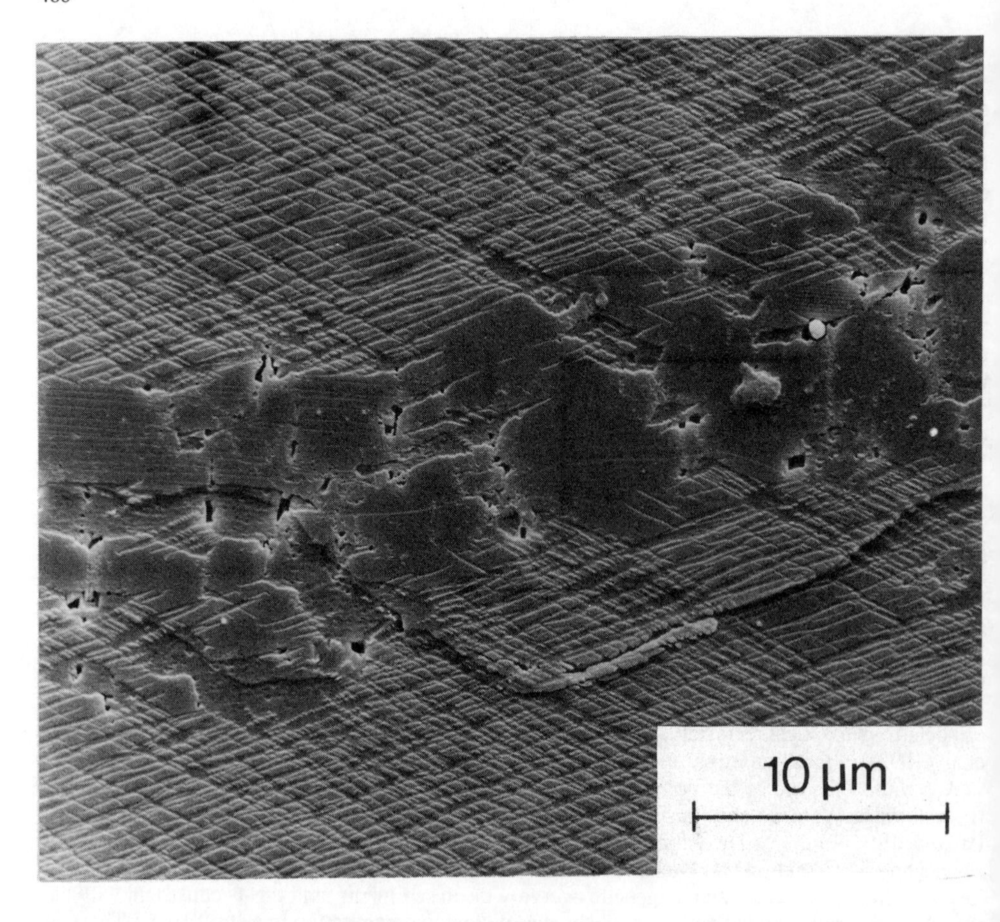

Figure 21. Secondary-electron SEM image of the (001) cleavage surface of an alkali feldspar from the Klokken intrusion (KB 14) lightly etched in HF vapour (see Waldron et al. submitted). Preferential etching of microcline in the top and bottom areas shows the secondary periodicity of the coherent cryptoperthite, which forms the broader bands visible in Fig. 7. A band of deuteric coarsening runs across the centre of the photograph, largely composed of albite subgrains, with a few areas of Or-rich feldspar. The etching has brought out the subgrain boundaries, which are also in places defined by micropores, slightly enlarged by etching. (From Waldron et al. submitted)

volume diffusion may decrease by a factor of at least 10^3, leading to much more rapid ^{18}O exchange at low *T*, and also providing pathways for rapid loss of ^{40}Ar. Walker (1990) used ion-probe imaging of ^{18}O to demonstrate experimentally that microporous feldspar was also permeable at 750°C and $P_{H_2{}^{18}O}$ 0.1 GPa. The relationship between microtexture and argon retention has been demonstrated directly, using laser microprobe extraction, by Burgess et al. (1992), although there are complications brought about by ^{40}Ar contained in fluids in the micropores themselves (see also Burgess and Parsons 1993). Textures which permit relatively rapid loss of Ar to the exterior are implicit in the 'multi-diffusion domain' models of Ar-loss proposed by Lovera et al. (1989, 1991,

1993), and Fitz Gerald and Harrison (1993) have had some success in relating microtextures seen by TEM to Ar-loss behaviour, although there are many uncertainties.

It is clearly of importance to determine the *T*'s at which deuteric reactions begin and down to which they may extend. The {110} habit, which characterizes subgrains and defines the outline of many micropores, is characteristic of diagenetic overgrowths in sediments (e.g. Worden and Rushton 1992). Feldspar grows readily in diagenesis and is one of the first overgrowth minerals to form in clastic sediments during burial. Thus feldspar--fluid interactions involving dissolution and reprecipitation of feldspar may continue to surface *T*. Replacement of one feldspar by another at a buried unconformity was demonstrated by Lidiak and Ceci (1991) and large-scale replacement of feldspar by a different feldspar occurs during albitization in sedimentary basins (Boles 1982). Alkali exchange between feldspars is continuing at measured downhole *T* of 250-350°C in the Salton Sea geothermal field (McDowell 1986). There is reason to accept, therefore, that deuteric reactions may, under some circumstances, continue to very low *T*.

Limits on the *T* at which deuteric reactions begin can be deduced from the microtextures (see below), but the timing of these reactions with respect to the cooling of plutons is not always clear. In the case of simple high-level intrusions such as the Klokken syenite, a case can be made that the effects occurred within a few tens of thousands of years after emplacement (Parsons et al. 1988). Similarly, the retention of Ar by the extremely altered feldspars in the Skye epigranites (Burgess and Parsons 1993) can only be explained if the high-level intrusions developed their deuteric textures rapidly after emplacement and cooled quickly to near-surface *T*. The situation is not so clear for calc-alkaline granites in orogenic environments, for which very slow cooling is often postulated (e.g. Lovera et al. 1989, 1991, 1993), and the *T* range of the reactions may not be reached until millions, or even tens of millions, of years after emplacement. This may be of critical importance in thermal modelling, using the ^{39}Ar-^{40}Ar methods proposed by Lovera et al., because microtextures which control the release of argon during laboratory step heating may not have been present throughout the cooling path being modelled. It is clearly essential that the time-temperature evolution of these microtextures be taken into account, as well as the accumulation of radiogenic argon.

An upper limit for deuteric recrystallization can be deduced from consideration of the *T* evolution of the microtextures (see Brown and Parsons, 1989, Figs. 8 and 9). In the hypersolvus alkali feldspars from the Klokken syenite (Worden et al. 1990) the deuteric textures (Fig. 19) cross-cut low microcline implying a maximum *T* of ~450°C. In the compositionally similar Coldwell syenite (Waldron and Parsons 1992) the textures cut coherent compositional fluctuations ('ripples') which imply maximum *T* of ~380°C. In more Or-rich alkali feldspars from many granites even lower *T* are implied. For example, feldspar MH-10, described by Fitz Gerald and Harrison (1993) for which the authors give a composition of Or_{95}, would not, if An-free, intersect the coherent solvus, leading to the coherent cryptoperthites illustrated, until *T* <200°C. This is a minimum estimate, because An has a large but poorly known effect on coherent-exsolution *T*. The coherent ternary solvus lies inside, and sub-parallel to, the strain-free ternary solvus depicted in Fig. 1. Nevertheless, it is likely that the deuteric textures in MH-10 formed at *T* <300°C. Alkali feldspars in subsolvus granites commonly have bulk compositions near $Ab_{20}Or_{80}$ implying coherent exsolution at *T* <400°C. Our experience is that such feldspars often contain coherent cryptoperthites cut by deuteric textures (e.g. the phenocryst from Shap forming the frontispiece and illustrated by Waldron et al. submitted); it seems that the feldspar--fluid interactions begin, in some cases, at <300°C. It is interesting to note that both coherent exsolution and deuteric coarsening in MH-10 (Fitz Gerald and Harrison 1993) would have taken place at *T* near the low-*T* end of the cooling path calculated on the basis of ^{39}Ar-^{40}Ar step-heating behaviour. In other words, the textures present at the beginning of the laboratory heating experiments were not present for much or most of the cooling history calculated from them. Readers should interpret the results of 'argon thermochronometry' (e.g. Lovera et al. 1989, 1991, 1993) with great care, unless they can be sure of the *T* and time when the microtextures relevant to Ar-loss formed.

Deuteric recrystallization of feldspars produces considerably more profound structural rearrangement than is apparent from optical microscopy. Not only may stable and radiogenic isotopes be affected, but also trace elements (Walker 1991). Petrologists and geochemists should take account of the fact that plutonic igneous rocks, which would meet all normal criteria of 'freshness', have almost all suffered some degree of deuteric interaction and hence potentially have suffered resetting of isotopes and trace elements that reside in feldspar.

7. Acknowledgements

Electron micrographs in this article were taken by our colleagues Richard Worden, Kim Waldron and Martin Lee who were supported by NERC Research Grants GR3/6697 and GR3/8374. Our ideas on the relationships betwen microtexture and argon loss have developed through collaboration with Ray Burgess under NERC grant GR3/7201. David Walker enhanced our understanding of the importance of micropores while holding an NERC Research Studentship. WLB acknowledges research support from CNRS. This is CRPG publication 1019.

8. References

Aberdam D (1965) Utilisation de la microscopie électronique pour l'étude des feldspaths. Observations sur les microperthites. Sciences de la Terre, 6, 76 pp.

Akizuki M (1983) An electron microscopic study of anorthoclase spherulites. Lithos 16:249-254

Anderson AT Jr (1983) Oscillatory zoning of plagioclase: Nomarski interference contrast microscopy of etched polished surfaces. Am Mineral 68: 125-129

Anderson DL (1987) The depths of mantle reservoirs. In: Mysen BO (ed) Magmatic processes: physicochemical principles: Geochem Soc Spec Publ 1, Penn State Univ, University Park, Penn., USA, 3-12

Arculus RJ (1978) Mineralogy and petrology of Grenada, Lesser Antilles Island Arc. Contrib Mineral Petrol 65: 413-424

Arndt NT, Nisbet EG (1982) Komatiites. Allen and Unwin, London, 526 p

Bailey DK, Schairer JF (1964) Feldspar--liquid equilibria in peralkaline liquids---the orthoclase effect. Am J Sci 262: 1198-1206

Baker JC, Cahn JW (1971) Thermodynamics of solidification. In: Solidification, Am Soc Metals, Metals Park, Ohio, 23-58

Baldridge WS, Carmichael ISE, Albee AL (1981) Crystallization paths of leucite-bearing lavas: examples from Italy. Contrib Mineral Petrol 76: 321-335

Bambauer HU, Krause C, Kroll H (1989) TEM-investigation of the sanidine/microcline transition across metamorphic zones: the K-feldspar varieties. Eur J Mineral 1: 47-58

Barker DS (1987) Tertiary alkaline magmatism in Trans-Pecos Texas. In: Fitton JG, Upton BGJ (eds) Alkaline igneous rocks: Blackwell, Oxford, 415-431

Beccaluva L, Barbieri M, Born H, Brotzu P, Coltorti M, Conte A, Garbarino C, Gomes CB, Macciotta G, Morbidelli L, Ruberti E, Siena F, Traversa G (1992) Fractional crystallization and liquid immiscibility processes in the alkaline-carbonatite complex of Juquia (Sao Paulo, Brazil). J Petrol 33: 1371-1404

Benn K, Allard B (1989) Preferred mineral orientation related to magmatic flow in ophiolite layered gabbros. J Petrol 30: 925-946

Berg JH (1980) Snowflake troctolite in the Hettasch intrusion Labrador: evidence for magma-mixing and supercooling in a plutonic environment. Contrib Mineral Petrol 72: 339-351

Bertelmann D, Förtsch E, Wondratschek H (1985) Zum Temperverhalten von Sanidinen: Die Ausnahmerolle der Eifelsanidin-Megakristalle. N Jb Mineral Abh 152: 123-141

Boles JR (1982) Active albitization of plagioclase, Gulf Coast Tertiary. Am J Sci 282: 165-180
Bowen NL (1928) Evolution of the igneous rocks. Princeton Univ Press, 334 p
Bowen NL, Tuttle OF (1950) The system $NaAlSi_3O_8$--$KAlSi_3O_8$--H_2O. J Geol 58: 489-511
Bown MG, Gay P (1958) The reciprocal lattice geometry of the plagioclase feldspar structures. Z Kristallogr 111: 1-14
Brandeis G, Jaupart C (1987a) Characteristic dimensions and times for dynamic crystallization. In: Parsons I (ed) Origins of igneous layering: D Reidel Publ Co, Dordrecht, 613-639
Brandeis G, Jaupart C (1987b) Crystal sizes in intrusions of different dimensions: constraints on the cooling regime and crystallization kinetics. In: Mysen BO (ed) Magmatic processes: physicochemical principles: Geochem Soc Spec Publ 1, Penn State Univ, University Park, Penn., USA, 307-318
Bridgwater D, Harry WT (1968) Anorthosite xenoliths and plagioclase megacrysts in Precambrian intrusions in South Greenland. Bull GrØnlands Geol Unders 77: 1-243
Brothers RN, Yokoyama K (1990) Fe-rich pyroxenes from a microdiorite dike, Whangarei, New Zealand. Am Mineral 75: 620-630
Brown FH, Carmichael ISE (1969) Quaternary volcanoes of the Lake Rudolf region: I. The basanite--tephrite series of the Korath Range. Lithos 2: 239-260
Brown FH, Carmichael ISE (1971) Quaternary volcanoes of the Lake Rudolf region: II. The lavas of North Island, South Island and the Barrier. Lithos 4: 305-323
Brown WL (1993) Fractional crystallization and zoning in igneous feldspars: ideal water-buffered liquid fractionation lines and feldspar zoning paths. Contrib Mineral Petrol 113: 115-125
Brown WL, Parsons I (1983) Nucleation on perthite-perthite boundaries and exsolution mechanisms in alkali feldspars. Phys Chem Min 10: 55-61
Brown WL, Parsons I (1984a) The nature of potassium feldspar, exsolution microtextures and development of dislocations as a function of composition in perthitic alkali feldspars. Contrib Mineral Petrol 86: 335-341
Brown WL, Parsons I (1984b) Exsolution and coarsening mechanisms and kinetics in an ordered cryptoperthite series. Contrib Mineral Petrol 86: 3-18
Brown WL, Parsons I (1988a) Zoned ternary feldspars in the Klokken intrusion: exsolution microtextures and mechanisms. Contrib Mineral Petrol 98: 444-454
Brown WL, Parsons I (1988b) Intra- and intercrystalline exchange and geothermometry in granulite-facies feldspars. Terra cognita 8: 263
Brown WL, Parsons I (1989) Alkali feldspars: ordering rates, phase transformations and behaviour diagrams for igneous rocks. Mineral Mag 53: 25-42
Brown WL, Parsons I (1993) Storage and release of elastic strain energy: the driving force for low-temperature reactivity and alteration of alkali feldspar. In: Boland JN, Fitz Gerald JD (eds) Defects and processes in the solid state: geoscience applications. The McLaren volume: Elsevier, Amsterdam, 267-290
Brown WL, Willaime C (1974) An explanation of exsolution orientation and residual strain in cryptoperthites. In: Mackenzie WS, Zussman J (eds) The Feldspars: Manchester University Press, 440-459
Brown WL, Becker SM, Parsons I (1983) Cryptoperthites and cooling rate in a layered syenite pluton: a chemical and TEM study. Contrib Mineral Petrol 82: 13-25
Brown WL, Moreau C, Demaiffe D (1989) An anorthosite suite in a ring-complex: crystallization and emplacement of an anorogenic type from Abontorok, Niger. J Petrol 30: 1501-1540
Bryon DN, Atherton MP, Hunter RH (1994) The description of the primary textures of 'Cordilleran' graniitic rocks. Contrib Mineral Petrol (in press)
Burgess R, Parsons I (1993) Argon and halogen geochemistry of hydrothermal fluids in the Loch Ainort granite, Isle of Skye, Scotland. Contrib Mineral Petrol (in press)

Burgess R, Kelley SP, Parsons I, Walker FDL, Worden RH (1992) ^{40}Ar-^{39}Ar analysis of perthite microtextures and fluid inclusions in alkali feldspar from the Klokken syenite, South Greenland. Earth Planet Sci Lett 109: 147-167

Cahn JW (1980) Thermodynamics of metastable equilibria. In: Mehrabian R, Kear BH, Cohen M (eds) Rapid solidification processes: vol 2, Claitors Publ Div, Baton Rouge, 24-34

Campbell IH (1978) Some problems with the cumulus theory. Lithos 11: 311-323

Carmichael ISE, Turner FJ, Verhoogen J (1974) Igneous petrology. McGraw-Hill, New York, 739 p

Carpenter MA (1981) A "conditional spinodal" within the peristerite miscibility gap of plagioclase feldspars. Am Mineral 66: 553-560

Carpenter MA (1988) Thermochemistry of aluminium/silicon ordering in feldspar minerals. In: Salje EKH (ed) Physical properties and thermodynamic behaviour of minerals: Reidel, Dordrecht, 265-323

Cashman KV (1993) Relationship between plagioclase crystallization and cooling rate in basaltic melts. Contrib Mineral Petrol 113: 126-142

Chapman NA, Powell R (1976) Origin of anorthoclase megacrysts in alkali basalts. Contrib Mineral Petrol 58: 29-35

Christoffersen R, Schedl A (1980) Microstructural and thermal history of cryptoperthites in a dike from Big Bend, Texas. Am Mineral 65: 444-448

Cox KG, Mitchell C (1988) Importance of crystal settling in the differentiation of the Deccan Trap basaltic magmas. Nature 333: 447-449

Cox KG, Bell JD, Pankhurst RJ (1979) The interpretation of igneous rocks. Allen and Unwin, London 450 p

Cundari A (1979) Petrogenesis of leucite-bearing lavas in the Roman volcanic region, Italy. The Sabatini lavas. Contrib Mineral Petrol 70: 9-21

De Pieri R, Quareni S (1973) The crystal structure of an anorthoclase: an intermediate alkali feldspar. Acta Cryst B29: 1483-1487

De Pieri R, Gregnanin A, Piccirillo EM (1974) I feldspati alcalini delle rocce eruttive dei Colli Euganei. Mem Ist Geol Mineral Univ Padova 31: 1-21

De Pieri R, De Vecchi G, Gregnanin A, Piccirillo EM (1977) Trachyte and rhyolite feldspars in the Euganean Hills (Northern Italy). Mem Ist Geol Mineral Univ Padova 32: 1-22

Didier J, Barbarin B (1991) Enclaves and granite petrology. Elsevier, Amsterdam, 625 p

Dietrich RV (1962) K-feldspar structural states as petrogenetic indicators. Norsk Geol Tidsskrift 42: 394-414

Echeverria LM (1982) Komatiites from Gorgona Island, Columbia. In: Arndt NT, Nisbet EG (eds) Komatiites: Allen and Unwin, London, 199-209

Eggleton RA, Buseck PR (1980) The orthoclase microcline inversion: a high resolution transmission electron microscope study and strain analysis. Contrib Mineral Petrol 74: 123-133

Emeleus CH, Upton BGJ (1976) The Gardar period in southern Greenland. In: Escher A, Watt WS (eds) Geology of Greenland: Geol Surv Greenland, Copenhagen,153-181

Emslie RF (1991) Granitoids of rapakivi--anorthosite and related associations. Precamb Res 51: 173-192

Evangelakakis C, Kroll H, Voll G, Wenk H-R, Meisheng H, Köpcke J (1993) Low-temperature coherent exsolution in alkali feldspars from high-grade metamorphic rocks of Sri Lanka. Contrib Mineral Petrol 114: 519-532

Evans BW, Moore JG (1968) Mineralogy as a function of depth in the prehistoric Makaopuhi tholeiitic lava lake, Hawaii. Contrib Mineral Petrol 17: 85-115

Ewart A (1969) Petrochemistry and feldspar crystallisation in the silicic volcanic rock, central North Island, New Zealand. Lithos 2: 371-388

Ewart A (1976) A petrological study of the younger Tongan andesites and dacites, and the olivine tholeiites of Niua Fo'ou Island, S.W. Pacific. Contrib Mineral Petrol 58: 1-21
Ewart A (1981) The mineralogy and chemistry of the anorogenic Tertiary silicic volcanics of S.E. Queensland and N.E. New South Wales, Australia. J Geophys Res 86: 10242-10256
Ferguson AK, Cundari A (1982) Feldspar crystallization trends in leucite-bearing and related assemblages. Contrib Mineral Petrol 81: 212-218
Fernandez A, Laporte D (1991) Significance of low symmetry fabrics in magmatic rocks. J Struct Geol 13: 337-347
Ferry JM (1985) Hydrothermal alteration of Tertiary igneous rocks from the Isle of Skye, northwest Scotland. II. Granites. Contrib Mineral Petrol 91: 283-304
Finch AA, Walker FDL (1991) Cathodoluminescence and microporosity in alkali feldspars from the Blå Måne SØ perthosite, South Greenland. Mineral Mag 55: 583-589
Fischer RV, Schmincke H-U (1984) Pyroclastic rocks. Springer-Verlag, Berlin, 472 p
Fitton JG (1987) The Cameroon line: a comparison between oceanic and continental alkaline volcanism. In: Fitton JG, Upton BGJ (eds) Alkaline igneous rocks: Blackwell, Oxford, 273-291
Fitz Gerald JD, Harrison TM (1993) Argon diffusion domains in K-feldspar 1: Microstructures in MH-10. Contrib Mineral Petrol 113: 367-380
Folk RL (1955) Note on the significance of "turbid" feldspars. Am Mineral 40: 356-357
Fowler AD (1990) Self-organized mineral textures of igneous rocks: the fractal approach. Earth Sci Rev 29: 47-55
Fuhrman ML, Lindsley DH (1988) Ternary-feldspar modeling and thermometry. Am Mineral 73: 201-215
Fuhrman ML, Frost BR, Lindsley DH (1988) Crystallization conditions of the Sybille monzosyenite, Laramie Anorthosite Complex, Wyoming. J Petrol 29: 699-729
Gay P (1956) The structures of plagioclase felspars: VI. Natural intermediate plagioclases. Mineral Mag 31: 21-40
Gerlach DC, Grove TL (1982) Petrology of Medecine Lake Highland volcanics: characterization of endmembers of magma mixing. Contrib Mineral Petrol 80: 147-159
Ghiorso MS (1987) Chemical mass transfer in magmatic processes III. Crystal growth, chemical diffusion and thermal diffusion in multicomponent silicate melts. Contrib Mineral Petrol 96: 291-313
Gibb FGF, Henderson CMB (1993) Convection and crystal settling in sills. Contrib Mineral Petrol 109: 538-545
Giggenbach WF (1981) Geothermal mineral equilibria. Geochim Cosmochim Acta 45: 393-410
Giletti BJ (1985) The nature of oxygen transport within minerals in the presence of hydrothermal water and the role of diffusion. Chem Geol 53: 197-206
Goldsmith JR (1980) The melting and breakdown reactions of anorthite at high pressures and temperatures. Am Mineral 65: 272-284
Goldsmith JR (1982) Plagioclase stability at elevated temperatures and water pressures. Am Mineral 67: 653-675
Gray NH, Anderson JB (1982) Polysynthetic twin width distributions in anorthoclase. Lithos 15:27-37
Grove TL, Baker MA, Kinzler RJ (1984) Coupled CaAl-NaSi diffusion in plagioclase feldspar: Experiments and applications to cooling rate speedometry. Geochim Cosmochim Acta 48: 2113-2121
Guthrie GD, Veblen DR (1991) Turbid alkali feldspars from the Isle of Skye, Scotland. Contrib Mineral Petrol 108: 398-404
Harker RI (1954) The occurence of orthoclase and microcline in the granite gneisses of the Carn Chuinneag--Inchbae Complex, E Ross-shire. Geol Mag 91: 129-136

Harlow GE (1982) The anorthoclase structures: the effects of temperature and composition. Am Mineral 67: 975-996

Hay RL (1962) Soda-rich sanidine of pyroclastic origin from the John Day formation of Oregon. Am Mineral 47: 968-971

Heming RF (1977) Mineralogy and proposed *P--T* paths of basaltic lavas from Rabaul Caldera, Papua New Guinea. Contrib Mineral Petrol 61: 15-33

Henderson CMB, Gibb FGF (1983) Felsic mineral crystallization trends in differentiating alkaline basic magmas. Contrib Mineral Petrol 84: 355-364

Hess HH (1960) Stillwater igneous complex, Montana. Geol Soc Am Mem 80, 230 p

Higgins MD (1991) The origin of laminated and massive anorthosite, Sept Iles layered intrusion, Québec, Canada. Contrib Mineral Petrol 106: 340-354

Holland TJB (1980) The reaction albite=jadeite+quartz determined experimentally in the range 600-1200°C. Am Mineral 65: 129-134

Hort M, Marsh BD, Spohn T (1993) Igneous layering through oscillatory nucleation and crystal settling in well-mixed magmas. Contrib Mineral Petrol 114: 425-440

Huijsmans JPP, Barton M (1989) Polybaric geochemical evolution of two shield volcanoes from Santorini, Aegean Sea, Greece: evidence for zoned magma chambers from cyclic compositional variations. J Petrol 30: 583-625

Hunter RH (1987) Textural equilibrium in layered igneous rocks. In: Parsons I (ed) Origins of igneous layering: D Reidel Publ Co, Dordrecht, 473-503

Irvine TN (1987) Appendix I. Glossary of terms for layered intrusions and Appendix II. Processes involved in the formation and development of layered igneous rocks. In: Parsons I (ed) Origins of igneous layering: D Reidel Publ Co, Dordrecht, 641-647 and 649-656

Irving AJ (1974) Megacrysts from the newer basalts and other basaltic rocks of south-eastern Australia. Geol Soc Am Bull 85: 1503-1514

Jackson ED (1961) Primary textures and mineral associations in the ultramafic zone of the Stillwater Complex, Montana. US Geol Surv Prof Pap 358: 106 p

Jackson MC (1993) Crystal accumulation and magma mixing in the petrogenesis of tholeiitic andesites from Fukujin seamount, northern Mariana Island arc. J Petrol 34: 259-289

Johannes W, Holtz F (1992) Melting of plagioclase in granite and related systems: composition of coexisting phases and kinetic observations. Trans Roy Soc Edin 83: 417-422

Jung D (1965) Ein natürliches Vorkommen von Monalbit? Z Kristallogr 121: 425-430

Kirkpatrick RJ (1977) Nucleation and growth of plagioclase, Makaopuhi and Alae lava lakes, Kilauea volcano, Hawaii. Bull Geol Soc Am 88: 78-84

Kolker A, Lindsley DH (1989) Geochemical evolution of the Maloin Ranch Pluton, Laramie Anorthosite Complex, Wyoming: Petrology and mixing relations. Am Mineral 74: 307-324

Kroll H, Evangelakakis C, Voll G (1993) Two-feldspar geothermometry: a review and revision for slowly cooled rocks. Contrib Mineral Petrol 114: 510-518

Kuo L-C, Kirkpatrick RJ (1982) Pre-eruption history of phyric basalts from DSDP Legs 45 and 46: evidence from morphology and zoning patterns in plagioclase. Contrib Mineral Petrol 79: 13-27

Lameyre J, Bonin B (1991) Granites in the main plutonic series. In: Didier J, Barbarin B (eds) Enclaves and granite petrology: Elsevier, Amsterdam, 3-17

Lameyre J, Bowden P (1982) Plutonic rock types series: discrimination of various granitoid series and related rocks. J Volc geotherm Res 14: 391-428

Langmuir CH (1989) Geochemical consequences of *in situ* crystallization. Nature 340: 199-205

Laughlin AW, Manzer GK, Carden JR (1974) Feldspar megacrysts in alkali basalts. Bull Geol Soc Am 85: 413-416

Le Bas MJ, Le Maitre RW, Streckeisen A, Zanettin B (1986) A chemical classification of volcanic rocks based on the total alkali--silica diagram. J Petrol 27: 745-750

Le Maitre RW (1976a) Some problems of the projection of chemical data into mineralogical classifications. Contrib Mineral Petrol 56: 181-189

Le Maitre RW (1976b) A new approach to the classification of igneous rocks using the basalt-andesite-dacite-rhyolite suite as an example. Contrib Mineral Petrol 56: 189-203

Le Maitre RW (1976c) The chemical variability of some common igneous rocks. J Petrol 17: 589-637

Le Maitre RW (1984) A proposal by the IUGS Subcommission on the Systematics of Igneous Rocks for a chemical classification of volcanic rocks based on the total alkali silica (TAS) diagram. Austral J Earth Sci 31: 243-255

Le Maitre RW (1989) A classification of igneous rocks and a glossary of terms. Blackwell, Oxford, 193 p

Le Roex AP (1985) Geochemistry, mineralogy and magmatic evolution of the basaltic and trachytic lavas from Gough Island, South Atlantic. J Petrol 26: 149-186

Le Roex AP, Cliff RA, Adair BJI (1990) Tristan da Cunha: geochemistry and petrogenesis of a basanite--phonolite lava series. J Petrol 31: 779-812

Lesher CE, Walker D (1988) Cumulate maturation and melt migration in a temperature gradient. J Geophys Res 93: 10295-10311

Libourel G (1991) The effect of chromium in synthetic basalt-like compositions. Terra Abstr 3: 433

Lidiak EG, Ceci VM (1991) Authigenic K-feldspar in the Precambrian basement of Ohio and its effect on tectonic discrimination of the granitic rocks. Can J Earth Sci 28: 1624-1634

Lindsley DH, Nekvasil H (1989) A ternary feldspar model for all reasons. Eos 70: 506

Lindsley DH, Smith D (1971) Chemical variations in feldspars. Carnegie Inst Washington Yearb 69: 274-278

Liu L (1978) High-pressure phase transitions of kalsilite and related potassium bearing aluminosilicates. Geochem J 12: 275-277

Liu L (1987) High-pressure phase transitions of potassium aluminosilicates with an emphasis on leucite. Contrib Mineral Petrol 95: 1-3

Lofgren G (1974a) An experimental study of plagioclase crystal morphology: isothermal crystallization. Am J Sci 274: 243-273

Lofgren G (1974b) Temperature induced zoning in synthetic plagioclase feldspar. In MacKenzie WS, Zussman J (eds) The feldspars: Manchester Univ Press, 362-375

Lofgren G, Donaldson CH (1975) Curved and branching crystals and differentiation in comb-layered rocks. Contrib Mineral Petrol 49: 309-319

Lovera OM, Richter FM, Harrison TM (1989) ^{40}Ar-^{39}Ar thermochronometry for slowly cooled samples having a distribution of diffusion domain sizes. J Geophys Res 94: 17917-17935

Lovera OM, Richter FM, Harrison TM (1991) Diffusion domains determined by ^{39}Ar released during step heating. J Geophys Res 96: 2057-2069

Lovera OM, Heizler MT, Harrison TM (1993) Argon diffusion domains in K-feldspar II: Kinetic properties MH-10. Contrib. Mineral Petrol 113: 381-393

Lowder GG (1970) The volcanoes and caldera of Talasea, New Britain: mineralogy. Contrib Mineral Petrol 26: 324-340

Luhr JF, Giannetti B (1987) The brown leucitic tuff of Roccamonfina volcano (Roman region, Italy). Contrib Mineral Petrol 95: 420-436

MaalØe S (1976) The zoned feldspars of the Skaergaard intrusion, East Greenland. J Petrol 17: 398-419

Macdonald GA, Katsura T (1964) Chemical composition of Hawaiian lavas. J Petrol 5: 82-133

MacKenzie WS, Guilford C (1980) Atlas of rock-forming minerals in thin section. Longman, Harlow, Essex, 98 p

MacKenzie WS, Smith JV (1956) The alkali feldspars: III An optical and X-ray study of high-temperature feldspars. Am Mineral 41: 405-427

MacKenzie WS, Smith JV (1962) Single crystal X-ray studies of crypto- and microperthites. Norsk Geol Tidsskrift 42: 72-103
MacKenzie WS, Donaldson CH, Guilford C (1982) Atlas of igneous rocks and their textures. Longman, London, 148 p
McBirney AR, Noyes RM (1979) Crystallization and layering of the Skaergaard Intrusion. J Petrol 20: 487-554
McConnell JDC (1969) Electron-optical study of incipient exsolution and inversion phenomena in the system $NaAlSi_3O_8$--$KAlSi_3O_8$. Philos Mag 19: 221-229
McDowell SD (1986) Compositional and structural state of coexisting feldspars, Salton Sea geothermal field. Mineral Mag 50: 75-84
McKenzie DP (1987) The compaction of igneous and sedimentary rocks. J Geol Soc London 144: 299-307
McLaren AC (1978) Defects and microstructures in feldspars. In: Chemical Physics of Solids and their Surfaces: Chemical Soc, London, 1-30
McLaren AC (1984) Transmission electron microscope investigations of the microstructures of microclines. In: Brown WL (ed) Feldspars and feldspathoids: D Reidel Publ Co, Dordrecht, 373-410
Meyer C, Hemley JJ (1967) Wall rock alteration. In: Barnes HL (ed) Geochemistry of hydrothermal ore deposits: Holt, Rinehart and Winston, New York, 166-235
Montgomery CW, Brace WF (1975) Micropores in plagioclase. Contrib Mineral Petrol 52: 17-28
Moore JG, Lockwood JP (1973) Origin of comb layering and orbicular structure, Sierra Nevada batholith, California. Bull Geol Soc Am 84: 1-20
Moreau C, Brown WL, Karche J-P (1987) Monzo-anorthosite from the Tagueï ring-complex, Aïr, Niger: a hybrid rock with cuumulus plagioclase and an infiltrated granitic intercumulus liquid? Contrib Mineral Petrol 95: 32-43
Morse SA (1969) Geology of the Kiglapait layered intrusion, Labrador. Mem Geol Soc Am 112
Morse SA (1988) Motion of crystals, solute, and heat in layered intrusions. Can Mineral 26: 209-224
Nash WP, Carmichael ISE, Johnson RW (1969) The mineralogy and petrology of Mount Suswa, Kenya. J Petrol 10: 409-439
Nekvasil H (1988) Calculation of equilibrium crystallization paths of compositionally simple hydrous felsic melts. Am Mineral 73: 956-965
Nekvasil H (1990) Reaction relations in the granite system: implications for trachytic and syenitic magmas. Am Mineral 75: 560-571
Nekvasil H (1991) Ascent of magmas and formation of rapakivi. Am Mineral 76: 1279-1290
Nekvasil H (1992) Ternary feldspar crystallization in high-temperature felsic magmas. Am Mineral 77: 592-604
Nicholls J, Carmichael ISE (1969a) Peralkaline acid liquids: a petrological study. Contrib Mineral Petrol 20: 268-294
Nicholls J, Carmichael ISE (1969b) A commentary on the absorokite--shoshonite--banakite series of Wyoming, USA. Schweiz Mineral Petr Mitt 49: 47-64
Nicolas A (1989) Structures of ophiolites and dynamics of oceanic lithosphere. Kluwer, Dordrecht, 367 p
Nicolas A (1990) Melt extraction from mantle peridotites: hydrofracturing and porous flow, with consequences for oceanic ridge activity. In: Ryan MP (ed) Magma transport and storage: Wiley, New York, 159-173
Nicolas A (1992) Kinematics in magmatic rocks with special reference to gabbros. J Petrol 33: 891-915
Ohnenstetter D, Brown WL (1992) Overgrowth textures, disequilibrium zoning and cooling history of a four-pyroxene boninite dyke from New Caledonia. J Petrol 33: 231-271

Owen DC, McConnell JDC (1974) Spinodal unmixing in an alkali feldspar. In: WS MacKenzie and J Zussman (Editors), The Feldspars: Manchester University Press, Manchester, pp. 424-439

Parsons I (1978) Feldspars and fluids in cooling plutons. Mineral Mag 42: 1-17

Parsons I (1987) Origins of igneous layering. D Reidel Publ Co, Dordrecht, 666 p

Parsons I, Becker SM (1987) Layering, compaction and post-magmatic processes in the Klokken intrusion. In: Parsons I (ed) Origins of igneous layering: D Reidel Publ Co, Dordrecht, 29-92

Parsons I, Boyd R (1971) Distribution of potassium feldspar polymorphs in intrusive sequences. Mineral Mag 38: 295-311

Parsons I, Brown WL (1983) A TEM and microprobe study of a two-perthite gabbro: implications for the ternary feldspar system. Contrib Mineral Petrol 82: 1-12

Parsons I, Brown WL (1984) Feldspars and the thermal history of igneous rocks. In: Brown WL (ed) Feldspars and feldspathoids: D Reidel Publ Co, Dordrecht, 317-371

Parsons I, Brown WL (1988) Sidewall crystallization in the Klokken intrusion: zoned ternary feldspars and coexisting minerals. Contrib Mineral Petrol 98: 431-443

Parsons I, Brown WL (1991) Mechanisms and kinetics of exsolution - structural control of diffusion and phase behavior in alkali feldspars. In: Ganguly J (ed) Diffusion and flow in minerals and fluids: Advances in physical geochemistry, vol 9. Springer-Verlag, Berlin: 306-346

Parsons I, Rex DC, Guise P, Halliday AN (1988) Argon-loss by alkali feldspars. Geochim Cosmochim Acta 52: 1097-1112

Paterson MS, Weiss LE (1961) Symmetry concepts in the structural analysis of deformed rocks. Geol Soc Am Bull 72/ 841-882

Pearce TH (1984) Optical dispersion and zoning in magmatic plagioclase: laser interference observations. Can Mineral 22: 383-390

Pearce TH, Kolisnik AM (1990) Observations of plagioclase zoning using interference imaging. Earth Sci Rev 29: 9-26

Petford N, Bryon D, Atherton MP, Hunter RH (1993) Fractal analysis in granitoid petrology: a means of quantifying irregular grain morphologies. Europ J Mineral 5: 593-598

Price RC, Taylor SR (1980) Petrology and geochemistry of the Banks Peninsula volcanoes, South Island, New Zealand. Contrib Mineral Petrol 72: 1-18

Ramseyer K, Fischer J, Matter A, Eberhardt P, Geiss J (1989) A cathodoluminescence microscope for low intensity luminescence. J Sed Petrol 59: 619-622

Ricci JE (1966) The phase rule and heterogeneous equilibrium. Dover Publ, New York 505 p

Riley GN Jr, Kohlstedt DL (1990) An experimental study of melt migration in an olivine--melt system. In: Ryan MP (ed) Magma transport and storage: Wiley, New York, 77-86

Ringwood AE (1975) Composition and petrology of the Earth's mantle. McGraw-Hill, New York 618 p

Russell JK, Nicholls J (1987) Early crystallization history of alkali olivine basalts, Diamond Craters, Oregon. Geochim Cosmochim Acta 51: 143-154

Schulze DJ (1987) Megacrysts from alkali basalts. In: Nixon PH (ed) Mantle xenoliths: Wiley, Chichester, 433-451

Seki Y, Kennedy GC (1964) The breakdown of potassium feldspar, $KAlSi_3O_8$, at high temperatures and high pressures. Am Mineral 49: 1688-1706

Singer BS, Pearce TH, Kolisnik A, Myers JD (1993) Plagioclase zoning in mid-Pleistocene lavas from the Seguam volcanic center, central Aleutian arc, Alaska. Am Mineral 78: 143-157

Slimming EH (1976) An electron diffraction study of some intermediate plagioclases. Am Mineral 61: 54-59

Smith AL, Carmichael ISE (1968) Quarternary lavas from the Southern Cascades, Western U.S.A. Contrib Mineral Petrol 19: 212-238

Smith JV, Brown WL (1988) Feldspar minerals. vol 1, Springer-Verlag, Heidelberg, 882 p

Smith JV, Stenstrom RD (1965) Electron-excited luminescence as a petrologic tool. J Geol 73: 627-635
Smith KL, McLaren AC, O'Donnell RG (1987) Optical and electron microscope investigation of temperature-dependent microstructures in anorthoclase. Can J Earth Sci 24:528-543
Snow E, Yund RA (1988) Origin of cryptoperthites in the Bishop Tuff and their bearing on its thermal history. J Geophys Research 93:8975-8984
Soldatos K (1965) Über eine monokline kryptoperthitische Natronfeldspat-Modifikation. Z Kristallogr 121: 317-320
Sparks RS, Huppert HE, Koyaguchi T, Hallworth MA (1993) Origin of modal and rhythmic igneous layering by sedimentation in a convecting magma chamber. Nature 361: 246-249
Speer JA, Ribbe PH (1973) The feldspars of the Kiglapait intrusion, Labrador. Am J Sci 273A: 468-478
Stewart DB, Wright TL (1974) Al/Si order and symmetry of natural alkali feldspars, and the relationship of strained cell parameters to bulk composition. Bull Soc franç Minéral Crist 97: 356-377
Stimac JA, Wark DA (1992) Plagioclase mantles on sanidine in silicic lavas, Clear Lake, California: implications for the origin of rapakivi texture. Geol Soc Am Bull 104: 728-744
Stolz, AJ (1986) Mineralogy of the Nandewar volcano, northeastern New South Wales, Australia. Mineral Mag 50: 241-255
Streckeisen A (1976) To each plutonic rock its proper name. Earth Sci Rev 12: 1-33
Tait SR, Huppert HE, Sparks RSJ (1984) The role of compositional convection in the formation of adcumulate rocks. Lithos 17: 139-146
Tiller WA (1977) On the cross-pollenation of crystallization ideas between metallurgy and geology. Phys Chem Minerals 2: 125-151
Toramura A, Fujii N (1986) Connectivity of melt phase in a partly molten peridotite. J Geophys Res 91: 9239-9252
Tsuchiyama A (1985) Dissolution kinetics of plagioclase in the melt of the system diopside--albite--anorthite, and origin of dusty plagioclase in andesites. Contrib Mineral Petrol 89: 1-16
Tsukui M (1985) Temporal variation in chemical composition of phenocrysts and magmatic temperature at Daisen volcano, southwest Japan. J Volcan geotherm Res 26: 317-336
Turbeville BN (1993) Petrology and petrogenesis of the Latera Caldera, Central Italy. J Petrol 34: 77-123
Turcotte DL (1987) Physics of magma segregation processes. In: Mysen BO (ed) Magmatic processes: physicochemical principles: Geochem Soc Spec Publ 1, Penn State Univ, University Park, Penn., USA, 69-74
Tuttle OF, Bowen NL (1958) Origin of granite in the light of experimental studies in the system $NaAlSi_3O_8$--$KAlSi_3O_8$--SiO_2--H_2O. Geol Soc Am Mem 74: 153 p
Upton BGJ (1987) Gabbroic, syenogabbroic and syenitic cumulates of the Tugtutôq Younger Giant Dyke complex, South Greenland. In: Parsons I (ed) Origins of igneous layering: D Reidel Publ Co, Dordrecht, 93-123
Upton BGJ, Emeleus CH (1987) Mid-Proterozoic alkaline magmatism in southern Greenland: the Gardar province. In: Fitton JG, Upton BGJ (eds) Alkaline igneous rocks: Blackwell, Oxford, 449-471
Upton BGJ, Aspen P, Hunter RH (1984) Xenoliths and their implications for the deep geology of the Midland Valley of Scotland and adjacent regions. Trans Roy Soc Edin: Earth Sci 75: 65-70
Upton BGJ, Stephenson D, Martin AR (1985) The Tugtutôq older giant dyke complex: mineralogy and geochemistry of an alkali gabbro--augite syenite--foyaite association in the Gardar Province of South Greenland. Mineram Mag 49: 623-642
Van Kooten GK (1980) Mineralogy, petrology, and geochemistry of an ultrapotassic basaltic suite, central Sierra Nevada, California, USA. J Petrol 21: 651-684
Wager LR, Brown GM (1968) Layered igneous rocks. Oliver and Boyd, Edinburgh, 588 p

Wager LR, Brown GM, Wadsworth WJ (1960) Types of igneous cumulates. J Petrol 1: 73-85
Waldron KA, Parsons I (1992) Feldspar microtextures and the multi-stage thermal history of syenites from the Coldwell Complex, Ontario. Contrib Mineral Petrol 111: 222-234
Waldron K, Parsons I, Brown WL (1993) Solution--redeposition and the orthoclase--microcline transformation: evidence from granulites and relevance to ^{18}O exchange. Mineral Mag (in press)
Waldron K, Lee MR, Parsons I (submitted) The microstructures of perthitic alkali feldspars revealed by hydrofluoric acid etching. Am Mineral
Walker FDL (1990) Ion microprobe study of intragrain micropermeability in alkali feldspars. Contrib Mineral Petrol 106: 124-128
Walker FDL (1991) Micropores in alkali feldspars. Unpub PhD thesis, Univ Edinburgh, 247 p
Wallace P, Carmichael ISE (1989) Minette lavas and associated leucitites from the Western Front of the Mexican Volcanic Belt: petrology, chemistry, and origin. Contrib Mineral Petrol 103: 470-492
Wang Y, Merino E (1993) Oscillatory magma crystallization by feedback between the concentrations of the reactant species and mineral growth rates. J Petrol 34: 369-382
Willaime C, Brown WL (1974) A coherent elastic model for the determination of the orientation of exsolution boundaries: application to the feldspars. Acta Cryst A30: 316-331
Willaime C, Brown WL, Gandais M (1973) An electron-microscopic and X-ray study of complex exsolution textures in a cryptoperthitic alkali feldspar. J Materials Sci 8: 461-466
Worden RH, Rushton JC (1992) Diagenetic K-feldspar textures: a TEM study and model for diagenetic feldspar growth. J Sed Petrol 62: 779-789
Worden RH, Walker FDL, Parsons I, Brown WL (1990) Development of microporosity, diffusion channels and deuteric coarsening in perthitic alkali feldspars. Contrib Mineral Petrol 104: 507-515
Wyers GP, Barton M (1989) Polybaric evolution of calc-alkaline magmas from Nisyros, Southeastern Hellenic Arc, Greece. J Petrol 30: 1-37
Yund RA, Chapple WM (1980) Thermal histories of two lava flows estimated from cryptoperthite lamellar spacings. Am Mineral 65: 438-443
Yund RA, Davidson P (1978) Kinetics of lamellar coarsening in cryptoperthites. Am Mineral 63:470-477

EVOLUTION OF FELDSPARS IN GRANITIC PEGMATITES

P. ČERNÝ
Department of Geological Sciences
University of Manitoba
Winnipeg, MB, R3T 2N2
Canada

ABSTRACT. In most geological categories of granitic pegmatites, the feldspar minerals crystallize under conditions of disequilibrium fractionation facilitated by increasing concentration of some or all of H_2O,F,B,P, and Li. This departure from cotectic-to-eutectic or peritectic termination of ordinary granitic crystallization generates zoning in textures and feldspar assemblages, controlled by melt undercooling and variable rates of nucleation and crystal growth: in general, from granitic borders through graphic K-feldspar + quartz in plagioclase + quartz matrix to blocky K-feldspar in the core margins, associated with cleavelandite and followed by saccharoidal albite. - Trends in bulk composition of the feldspars, as it changes throughout their crystallization sequences can be related, to a degree, to the liquidus-solidus-subsolidus relations in the Ab-An-Or or Ab-Or systems. Minor elements comprise compatible Ba,Sr and Fe^{3+} in early generations and incompatible Li,Rb,Cs,Tl,Pb and NH_4 which fractionate into late primary feldspars, locally to extreme concentrations not encountered outside the granitic pegmatites. Incompatible behaviour is also typical of P,Ga and possibly B, Be. Low-temperature hydrothermal feldspars are very close to end-member compositions, except local occurrences of Rb- or Ba-rich phases. - Perthite exsolution and recrystallization are common in K-feldspar, whereas peristeritic and antiperthitic breakdown of plagioclase and albite seem to be restricted. Perthitization of the disordered monoclinic (K,Na)-feldspar, as originally crystallized from the melt and/or fluid, is accompanied by diverse degrees of tetrahedral ordering of the K-phase, dependent on the specific local conditions in individual pegmatites. In contrast, plagioclase and albite generally achieve very high order, even in low-temperature hydrothermal assemblages which frequently preserve extreme disorder in K-feldspar. - Replacement of K-feldspar by mica + quartz assemblages does not seem to be widespread, and sodic metasomatism is restricted to the margins of otherwise primary albitic units. Corrosion, leaching, regeneration, replacement and late overgrowths are typical of feldspars in miarolitic cavities. Replacement of diverse precursors by microscopic networks of hydrothermal K-, Na- (and Rb- or Ba-enriched) feldspars is relatively widespread.

1. Introduction

Compared to accessory minerals such as garnet, tourmaline, phosphates or oxide minerals of Nb and Ta, the rock-forming feldspars are a Cinderella of pegmatite research. It is perhaps the more intriguing crystal chemistry and petrology of the above minor phases, and the simple means available for industrial evaluation of feldspar quality that conspire against systematic research into pegmatitic feldspars, as well as subsolidus disturbances that render them difficult to interpret in terms of primary magmatic phases. Nevertheless, a smattering of information has accumulated over the last three decades, since the classic optical studies of the early part of this century and after a near-hiatus in activity during the nineteen thirties to sixties. The present attempt at

I. Parson (ed.), Feldspars and Their Reactions, 501–540.

TABLE 1 The four classes of granitic pegmatites.

Class	Family*	Typical Minor Elements	Metamorphic Environment	Relations to Granites
Abyssal	-	U,Th,Zr,Nb,Ti, Y, REE,Mo	(upper amphibolite to) low- to high-P granulite facies; ~4-9 kbar, ~700-800°C	none (segregations of anatectic leucosome
Muscovite	-	Li,Be,Y,REE,Ti, U,Th,Nb>Ta	high-P, Barrovian amphibolite facies (kyanite-sillimanite); ~5-8 kbar, ~650-580°C	none (anatectic bodies) or marginal and exterior
Rare-Element	LCT	Li,Rb,Cs,Be,Ga, Sn,Hf,Nb>Ta,B, P,F	low-P, Abukuma amphibolite facies (andalusite-sillimanite); ~2-4 kbar, ~650-500°C	(interior to marginal to) exterior
	NYF	Y,REE,Ti,U,Th, Zr,Nb>Ta,F	variable	interior to marginal
Miarolitic	NYF	Y,REE,Ti,U,Th, Zr,Nb>Ta,F	shallow to subvolcanic;	interior to marginal

*See Table 3 for explanation;
**Some Russian authors distinguish a rare-element-muscovite class, in all respects intermediate between the muscovite and rare-element classes proper.

generalization of the evolution of feldspars in granitic pegmatites is based not only on published information, but also on a backlog of unpublished data. If this review stimulates new studies of feldspars in pegmatites, which would contribute to understanding the feldspars in general and the process of pegmatite formation in particular, I'll be fully rewarded for writing it.

2. Terms of Reference

"Granitic pegmatite" is a rather general term that encompasses quite a spectrum of geological classes, of paragenetic and geochemical types and subtypes within each class, and of families splitting the classes by geochemical signatures and petrogenetic affiliations. Geological environments separate the classes as shown in Table 1, whereas mineral assemblages and associations of elements characterize the types and subtypes of pegmatites within the rare-element class in Table 2. The split of the rare-element class into petrogenetic families is presented in Table 3. All aspects of this classification are discussed in more detail, and references to representative occurrences are given in Černý (1990, 1991a,b).

Understandably enough, the histories of primary crystallization of feldspars and their subsequent modifications are far from uniform across the above variety of categories. Consequently, *the present review is aimed primarily at the feldspars of the very widespread subaluminous to peraluminous zoned pegmatites of the rare-element class* (generally subsolvus), in which they are most diversified and relatively best known in all respects, particularly so in those of the LCT family (typically enriched in Li,Cs,Ta). Occasional notes also are included on the feldspars of the miarolitic (in part hypersolvus) class of the NYF kindred (typically enriched in Nb,Y,F). In contrast, the abyssal and muscovite classes are not discussed: the former boils down to segregations of anatectic leucosome, the latter probably has a dual origin, and there are no

TABLE 2 Classification of pegmatites of the rare-element class.

Pegmatite type	Pegmatite subtype	Geochemical signature	Typical minerals	
RARE-EARTH	allanite-monazite	(L)REE,U,Th	allanite monazite	(topaz)
	gadolinite	Y,(H)REE,Be, Nb>Ta F(U,Th,Ti,Zr)	gadolinite fergusonite euxenite	(beryl)
BERYL	beryl-columbite	Be,Nb>Ta (±Sn,B)	beryl columbite-tantalite	
	beryl-columbite-phosphate	Be,Nb>Ta,P (Li,F,±Sn,B)	beryl triplite	columbite-tantalite triphylite
COMPLEX	spodumene	Li,Rb,Cs,Be, Ta>Nb (Sn,P,F,±B)	spodumene beryl tantalite	(amblygonite) (lepidolite) (pollucite)
	petalite	Li,Rb,Cs,Be, Ta>Nb (Sn,Ga,P,F,±B)	petalite beryl tantalite	(amblygonite) (lepidolite)
	lepidolite	F,Li,Rb,Cs,Be Ta>Nb (Sn,P,±B)	lepidolite topaz beryl	microlite (pollucite)
	amblygonite	P,F,Li,Rb,Cs Be,Ta>Nb (Sn±B)	amblygonite beryl tantalite	(lepidolite) (pollucite)
ALBITE-SPODUMENE		Li (Sn,Be,Ta,>Nb,±B)	spodumene (cassiterite)	(beryl) (tantalite)
ALBITE		Ta>Nb,Be (Li,±Sn,B)	tantalite beryl	(cassiterite)

meaningful studies of feldspar minerals available for either of them.

As to the general conditions of consolidation, the rare-element granitic pegmatites are treated here as products of magmatic crystallization, transitional into supercritical and ultimately hydrothermal + gaseous regimes with decreasing temperature. However, the pegmatite-forming granitic magmas are volatile-undersaturated through most of their solidification. The most evolved melts can hold spectacular quantities of H_2O in solution: up to 15-20 wt.% H_2O after 75-85% crystallization from an initially H_2O-undersaturated Li,B,P-rich starting composition at 2 kbar (London et al. 1989). The liquidus-solidus span of geochemically primitive magmas may be ~750 - 600°C, but it is extended and downshifted for extremely fractionated magmas to ~650-450°C (London 1986, Chakoumakos and Lumpkin 1990). Intrusion into relatively cool host-rocks and the small size of most pegmatites generally lead to rapid magmatic crystallization (e.g., 1000 years to possibly as short as 200-100 years; London 1986, 1992a; Chakoumakos and Lumpkin 1990). The subsolidus, mainly hydrothermal processes operate in part after thermal equilibration with the host rocks, during regional uplift, and may become open to influx from, or exchange with the wallrock lithologies.

TABLE 3 The three petrogenetic families of rare-element pegmatites.

Family	Pegmatite types	Geochemical signature	Associated granites	Source lithologies
LCT	beryl, complex, albite-spodumene, albite	Li,Rb Cs,Be,Sn,Ga Ta>Nb(B,P,F)	(synorogenic to) late orogenic (to anorogenic); largely heterogeneous; peraluminous;** S,I or mixed S+I types	undepleted upper- to middle-crust supracrustals and basement gneisses
NYF*	rare-earth	Nb>Ta,Ti,Y,Sc, REE,Zr,U,Th,F	(syn-, late, post- to) mainly anorogenic; largely homnogeneous; (peraluminous to) subaluminous to metaluminous (rarely peralkaline); A and (I) types	depleted middle to lower crustal granulites, or un-depleted juvenile granitoids
Mixed	"crossbred"	mixed	(postorogenic to) anorogenic; moderately heterogeneous; subaluminous to slightly peraluminous; mixed geochemical signature	mixed protoliths, or assimilation of supracrustals by NYF granites

* This family also includes pegmatites of the low-pressure miarolitic class (*cf.* Table 1).
**Definitions: Peraluminous, A/CNK>1; subaluminous, A/CNK~1; metaluminous, A/CNK<1 at A/NK>1; subalkaline, A/NK~1; peralkaline, A/NK<1, where A = molecular Al_2O_3, CNK = CaO + Na_2 + K_2O, and NK = Na_2O + K_2O (Černý, 1991c).

Throughout this paper, the term K-feldspar is used for phases of all structural states with Or ≥ Ab ≫ An chemistry, whereas plagioclase is applied to phases characterized by Ab > An (usually < 30) ≫ Or, including the albite end-member. The term albite is used whenever a virtually pure albite phase is the only plagioclase encountered, and in petrographic terminology (e.g., albitic aplite, saccharoidal albite *et sim.*). The non-specific term alkali feldspar is avoided, as it covers two different phases - K-feldspar and albite - commonly associated in numerous pegmatite assemblages.

Nomenclature characterizing the structural state of feldspars follows Smith and Brown (1988; Fig. 1.6, 3.6, chapter 9), as does the terminology characterizing exsolution phenomena in general, and perthitic intergrowths in particular (chapter 19). Triclinicity (obliquity) ▲ is used in the original definition by Goldsmith and Laves (1954): 12.5 [d(131)-d(1$\bar{3}$1)].

The varietal name of cleavelandite is used for curved platy albite extremely flattened parallel to (010), as applied in the literature on pegmatites since 1823 (Fisher 1968). The term adularia is used for low-temperature hydrothermal K-feldspar of typical morphology (cf. Černý and Chapman 1986).

Mineral symbols of Kretz (1983) are used in abbreviated expressions, if available.

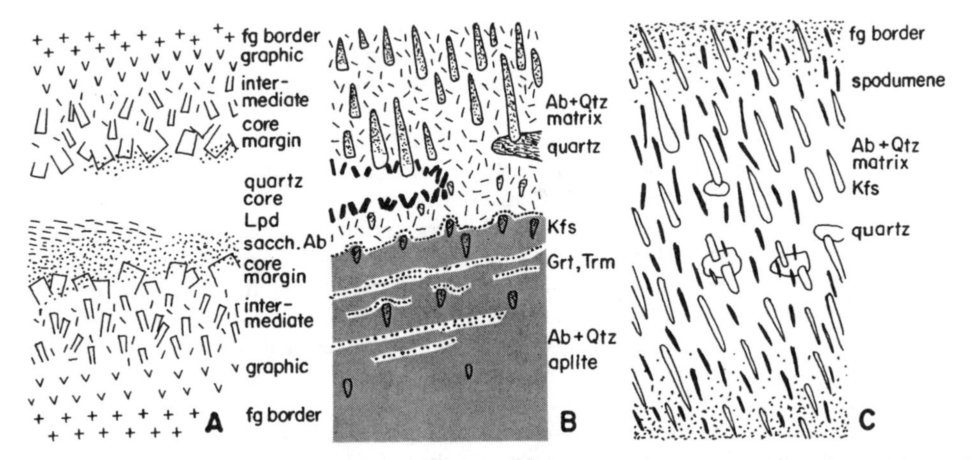

Figure 1. Schematic sections across (A) concentrically zoned and (B) subhorizontal layered complex, and (C) quasi-homogeneous albite-spodumene pegmatites of the rare-element class. Modified from Černý (1991a).

3. Paragenetic and Textural Relationships

3.1. ZONED PEGMATITES

The widespread zoned bodies of rare-element granitic pegmatites of the beryl and complex types (and largely also those of the rare-earth type) display a sequence of units that may have evolved to a different degree in individual dikes (cf. Vlasov 1952, Černý 1982a), and are highly variable among diverse localities even at a very local scale (Cameron et al. 1949, Beus 1960, Norton 1983). However, in concentrically zoned bodies this sequence does follow a fairly uniform basic pattern (Fig. 1A). In its most evolved form, the zonal sequence consists of the following units:

(i) the outermost border zone commonly has a granitic composition, and granitic to aplitic texture; it is usually enriched in plagioclase at the expense of K-feldspar;

(ii) wall zone has similar granitic to Na-rich composition but distinctly different texture; megacrystals of K-feldspar in a coarse-grained matrix of Pl+Qtz(±Trm,Qrt,Bt,Ms) are widespread, and the K-feldspar commonly contains graphically intergrown quartz; graphic Pl+Qtz occasionally substitute for the coarse-grained matrix;

(iii) one or more intermediate zones have variable proportions of Pl and Kfs, but the latter is usually dominant;

(iv) core-margin zone is adjacent to quartz core, with blocky K-feldspar and negligible, if any, plagioclase, but with giant crystals of spodumene, petalite or beryl;

(v) albite-rich units consist of platy cleavelandite or fine-grained saccharoidal albite, with a more or less prominent assemblage of accessory minerals of Nb,Ta,Be,Sn and Zr, among others;

(vi) micaceous units are comprised of medium- to fine-flaked muscovite, lithian muscovite or lepidolite, and usually also are enriched in minerals of rare elements;

(vii) quartz core is usually located at the centre of pegmatite bodies.

From the viewpoint of crystallization of the feldspars, this zonal sequence shows a gradual increase in the proportion of K-feldspar, culminating in the core-margin. The abundance of K-feldspar in the border and wall zone is even more reduced in many highly fractionated pegmatites that also lack the "graphic granite". Albite is the sole feldspar constituting the late cleavelandite or saccharoidal units - an abrupt about-face in the feldspar chemistry.

It must be stressed that subtle to conspicuous deviations from the above sequence are widespread, particularly in the concentricity of the unit distribution. Some of the complex pegmatites exhibit additional units, such as central accumulations of blocky K-feldspar within the quartz "core" (Beus 1960, Solodov 1962, Černý 1982b).

Parallel to the changing proportions of the feldspars, textural relations also evolve in a rather spectacular manner. From the border zone to the core-margin, the K-feldspar shows a remarkable increase in size - up to 6 m in length of individual morphological units. It also displays an inward-oriented growth fabric, with club-shaped crystals or comb-like aggregates, locally with skeletal to dendritic branching of morphological individuals. The switch to crystallization of albite in internal units is marked by dramatic reduction in crystal size - plates of cleavelandite normally <10 cm across and saccharoidal albite in crystals averaging 0.5-2 mm.

The zonal pattern of the internal pegmatite structure, including the typical phase assemblages and textures of consecutive zones, was successfully simulated in laboratory experiments on H_2O-undersaturated bulk compositions. London et al. (1989) used a compositional analog of highly evolved pegmatites rich in Li,Rb,Cs,B,P and F, the Macusani glass. London (1990, 1992) explained the dramatic departure of pegmatite crystallization from cotectic pathways, peritectic reactions or potential final H_2O eutectic or minimum granitic compositions by the influence of the above volatile components and H_2O.

According to London (1990, 1992a), crystallization of pegmatitic textures stems from a disequilibrium growth of feldspars and quartz, facilitated by their supersaturation, as promoted by increasing concentrations of H_2O, Li, B, F and P in the melt. Supersaturation is generated by undercooling, and aided by an increasing degree of disruption of the alkali-alumino-silicate framework of the melt by the above fluxing components. These components also reduce the density of the melt and rate of crystal nucleation (particularly of K-feldspar), increasingly so with their gradual accumulation during progressive crystallization of the melt.

In terms of textural evolution, initial nucleation in the outermost zones is internal (near-homogeneous), and it generates aplitic or granitic textures. However, decreasing rate of nucleation promotes a rapid transition into a heterogeneous, substrate-controlled nucleation style. This leads to the growth of comb-structured, directional fabrics commonly combined with dendritic and graphic features. Inhibited nucleation promotes the growth of giant crystals, also facilitated by the high fluidity of the volatile-rich melts, particularly in the intermediate and central zones.

Components such as B,P and F are not requisite, but they promote zoning in a pegmatite body by expanding the liquidus field for quartz. The activity of albite seems to be more depressed than that of K-feldspar, so that addition of B or F expands the liquidus field of the latter as well (core-margin blocky masses). Consequently, an increase in concentration of H_2O, B,P and F drives the residual melt to a silica-depleted, Na-rich alkaline composition enriched in incompatible elements. It eventually crystallizes as late-stage albitic $\pm$ micaceous units. Consolidation of these units is triggered by stabilization of spodumene, tourmaline or phosphates, which dramatically reduces the solubility of H_2O in the residual melt. Removal of the fluxing components, culminating in exsolution of a supercritical aqueous fluid, increases the solidus temperature and leads to dumping of rapidly nucleating albite and micas in fine-grained aggregates.

Crystallization of graphic intergrowths of feldspar + quartz develops in conjunction with liquidus undercooling and metastable saturation of the pegmatite melt. Local increase in silica activity, in a boundary layer of the melt near the surface of a rapidly growing feldspar, results in periodic crystallization of quartz, in part simultaneous with that of the feldspar. London (1992a)

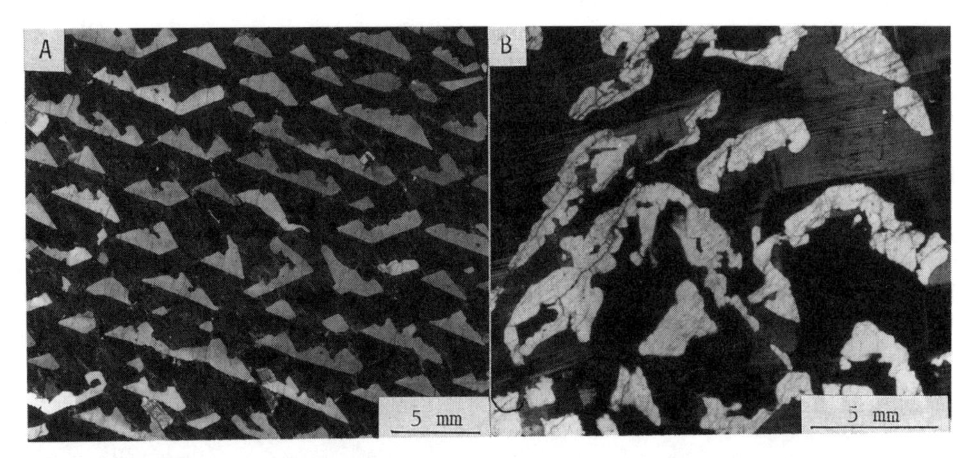

Figure 2. Graphic intergrowth of (A) K-feldspar + quartz from Rožná and (B) albite + quartz from Věžná, both in western Moravia, Czech Republic (optical microphotograph, crossed polars). Note the differences in the shapes and outlines of the quartz ichthyoglypts.

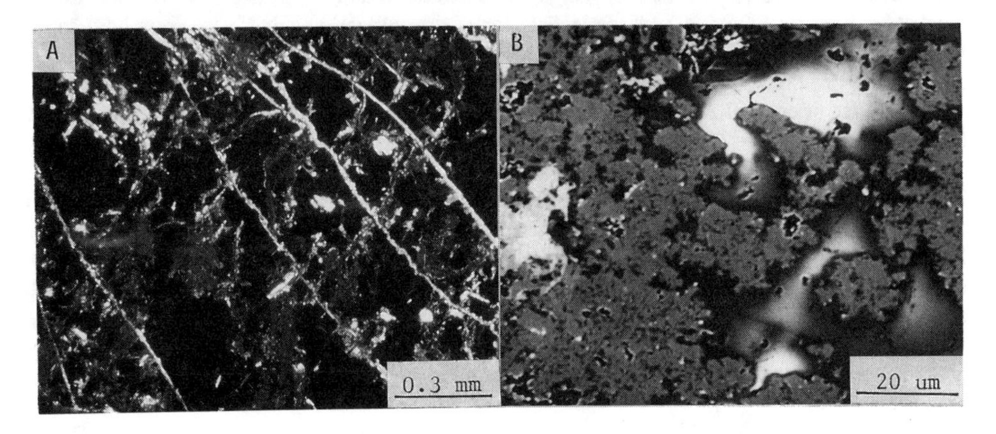

Figure 3. Late K-feldspar (grey) (A) replacing pollucite (black) among micaceous veinlets (white; optical microphotograph, crossed polars, from Teertstra et al. 1993) and (B) replacing analcimized (grey to black) pollucite (white) from Utö, Sweden (BSE image).

questioned the presence of H_2O oversaturation during this process, considered by Fenn (1986) and advocated by Lentz and Fowler (1992). Compositional differences between the K-feldspar- and plagioclase-based graphic intergrowths are pointed out in Černý (1971); textural differences are shown in Figure 2.

To date, London's model of pegmatite consolidation is the most universal, and the only one that explains the course of crystallization of the geochemically and paragenetically most complex bodies. It also accounts for the sequential and textural evolution of the K-feldspar and plagioclase during the main, primary consolidation of pegmatite-generating melts. The model will have to be adjusted, however, to interpret many of the rare-earth pegmatites of the NYF affiliation. They show significant deviations from the structure and rock-forming mineral assemblages of their LCT counterparts, such as plagioclase-rich wall zones followed by intermediate zones and core margins carrying exclusively K-feldspar (e.g., Paul 1984), or giant-size blocky plagioclase coexisting with blocky K-feldspar in the same zones (e.g., some pegmatites in the Grenville Province of Ontario).

The only additional note must be devoted to low-temperature, hydrothermal feldspars locally developed during the subsolidus reactions of primary minerals with residual or host-rock fluids. These "alpine-vein stage" feldspars fill fractures, line cavities and replace other minerals, including primary feldspars (Fig. 3). They are largely microscopic and inconspicuous in the field but locally attain quite a significant role in the overall feldspar (and alkali) budget of the pegmatite bodies. They correspond to Ginsburg's (1960) "second K-stage" of pegmatite evolution, which concludes its overall course.

3.2. OTHER STRUCTURAL TYPES

3.2.1. *Layered Pegmatites* such as those of southern California (Jahns and Tuttle 1963, Foord 1976) pose a problem, together with any other containing extensive aplitic units. London (1990, 1992a) found no satisfactory explanation for their appearance, either in his experiments or in the interpretations of earlier workers, although supersaturation and high rates of nucleation could be involved (London 1992a). A concentric direction of crystallization is indicated from both contacts inwards (Fig. 1B), with aplite consolidating somewhat earlier than the pegmatitic units. However, locally a cyclic, layered pattern is developed (Rockhold et al. 1987, Linnen et al. 1992).

The course of crystallization of layered pegmatites of the lepidolite subtype (e.g., Brown Derby, Colorado; Heinrich 1967) is even more obscure, as are the characteristics of their feldspars.

3.2.2. *Quasi-homogeneous Pegmatites* of the albite-spodumene type are not known in sufficient detail from the viewpoint of their fabric and of patterns of feldspar growth. In some of them, a finer-grained border zone is detectable, and small near-central pods of blocky texture, but most of them lack these features (Fig. 1C). The distribution and orientation of club-shaped megacrystals of K-feldspar locally indicate concentric crystallization, but rod-shaped megacrystals are widespread in many bodies, obscuring growth directions. However, embayments of the fine-grained plagioclase + quartz (± Trm, Grt) matrix into the corroded K-feldspar and spodumene are omnipresent.

3.2.3. *The Albite Type* of rare-element pegmatites is very poorly known, commonly near-homogeneous, and no meaningful data are available on its feldspars.

3.2.4. *Miarolitic Pegmatites* of the NYF-family systems (Fig. 4) are commonly of the low-pressure hypersolvus persuasion, with Na-rich K-feldspar as the only primary feldspar throughout their crystallization. Albitic plagioclase appears only as a subsolidus phase in cavities. Diverse patterns of corrosion, leaching, epitactic overgrowth and regeneration of the vug-lining K-feldspar are very widespread.

4. Compositional Evolution at the Magmatic Stage

The chemical composition of feldspars is rather easy to establish, by a variety of bulk or microbeam methods. However, the coverage of chemistry of feldspars from granitic pegmatites

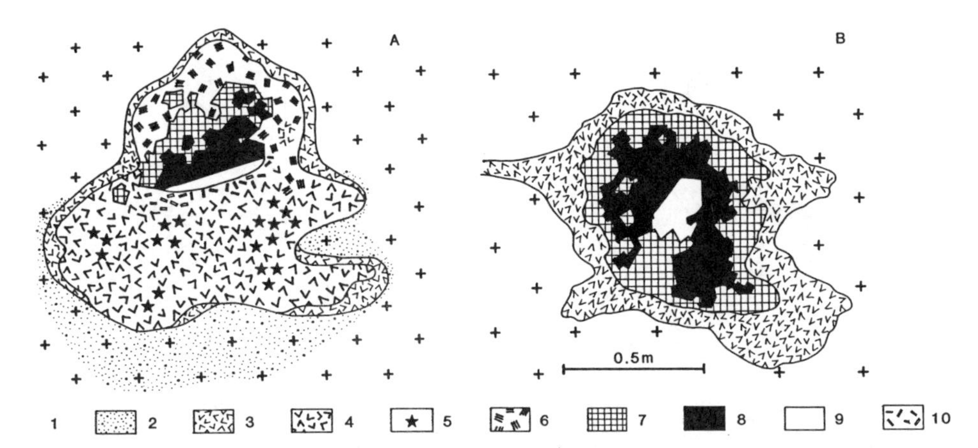

Figure 4. Miarolitic pod-like pegmatites from (A) hypersolvus NYF granites of the Korosten batholith in Ukraine (after Lazarenko et al. 1968) and (B) California (from Jahns 1954). 1 - granite, 2 - altered granite, 3 - medium-grained and 4 - coarse-grained graphic zone, 5 - radial and quasi-graphic unit, 6 - pegmatoid zone, 7 - blocky K-feldspar, 8 - quartz, 9 - crystal-lined vug, 10 - leaching and recrystallization.

is extremely spotty in the literature. The spectrum of trace elements hosted by K-feldspar has long ago attracted attention for geochemical reasons and in exploration, largely at regional scale (e.g., Heier 1962, Correia Neves and Lopes Nunes 1966, Gordiyenko 1971, Makagon and Shmakin 1988). Bulk compositions were studied only exceptionally (e.g., Tan 1966), and attempts at deciphering the overall evolution of feldspar chemistry within individual pegmatites are rare. Bulk composition and trace-element content will be dealt with separately, because of their dramatically different evolution and significance.

4.1. TRENDS IN BULK COMPOSITION

Ideally, three "end-member" cases can be envisaged for crystallization of feldspars in rare-element (and closely related NYF miarolitic) pegmatites:

(i) Relatively primitive, distinctly Ca-bearing granitic pegmatites of the rare-earth and beryl types should crystallize pairs of coexisting K-feldspar and plagioclase along converging paths of the ternary solidus-solvus intersection. The final magmatic two-feldspar assemblage should be controlled by the intersection of the ternary solidus with binary Ab-Or solvus (Fig. 5A), which is to be expected in the presence of quartz at $\geq$ 2.5 kbar (Stewart and Roseboom 1962, James and Hamilton 1969, Černý 1971). The relatively high solidification temperature of the primitive melt would render this intersection relatively narrow, with a rather small gap separating the compositions of the two feldspars.

(ii) Highly fractionated granitic pegmatites enriched in Li,Rb,Cs,B,P and F that solidified from highly hydrous melts are usually extremely poor in Ca, all of which could be strongly partitioned into apatite, other Ca-bearing phosphates and oxide minerals. Also, these pegmatites have much lower solidus than those sub (i). Consequently, their melts should precipitate mainly a pair of Kfs + Ab defined by a broad intersection of the binary Ab-Or solidus and solvus (Fig. 5B), with the feldspars much closer to the end-member compositions than in (i). The first feldspar to crystallize, before the solidus-solvus intersection is reached, would be determined by the relationship of the

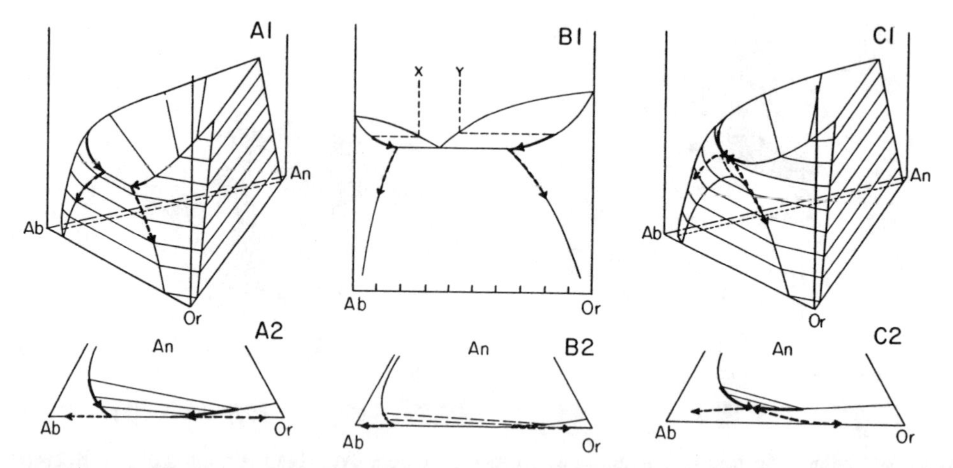

Figure 5. Hypothetical pathways of feldspar crystallization (heavy solid lines) in (A) primitive to moderately fractionated, Ca-bearing, subsolvus pegmatites of the rare-earth or beryl types, (B) highly fractionated Ca-"free" (1) or very Ca-poor (2), subsolvus pegmatites of the complex type, and (C) shallow-seated, hypersolvus NYF pegmatites. Heavy dashed lines schematically indicate the trends of subsolidus re-equilibration of the last-crystallizing magmatic feldspars. In B, X and Y show examples of plagioclase-first and K-feldspar-first examples of crystallization respectively, controlled by the bulk melt composition.

bulk melt composition to the position of the two feldspar loops (Fig. 5B).

(iii) Miarolitic granitic pegmatites of the NYF family, and some rare-earth pegmatites of the same affiliation consolidate at relatively shallow levels. The low-pressure environment compounded by the relatively primitive composition of the melts (and consequent high temperature of solidification) leads to a hypersolvus regime of feldspar crystallization. Theoretically, a brief span of two-feldspar crystallization (with very close compositions) should precede the terminal crystallization of a single feldspar (Fig. 5C). Complications might possibly be inflicted by the staggering array of terminal possibilities in this system (Stewart and Roseboom 1962, Dubrovskyi 1978, Boettcher 1980).

The meagre data available indicate that the three models are not closely followed by actual patterns of feldspar crystallization but, in some cases, rather ignored:

(i) The relatively primitive granitic pegmatites may approach the feldspar trends of Figure 5A, but a true coexistence of Kfs + Pl pairs is rather questionable. Coexistence is well expressed in the early outer zones, but the textural relationships are generally not convincing in the intermediate and central, late-crystallizing units. The K-feldspar follows the path of increasing Ab (Ab_{20} to Ab_{45} at $An_{<2}$), the plagioclase shows decreasing An and decreasing or steady Or (Fig. 6; Černý et al. 1984, Abad Ortega et al. 1993). Relatively high Ab contents that fall within the above range of K-feldspar compositions are reported from other localities in which the sequence of crystallization is not rigorously established (e.g., Correia Neves 1964, Karnin 1980), or is not statistically supported (Shearer and Papike 1985).

(ii) In the highly fractionated pegmatites, coexistence of Kfs and Pl is quite doubtful even in the outer units. Albitic plagioclase usually shows low Ca contents (routinely $An_{<6}$) which decrease from the outer zones inward and suggest crystallization at the very end of fractionation (Fig. 5B2).

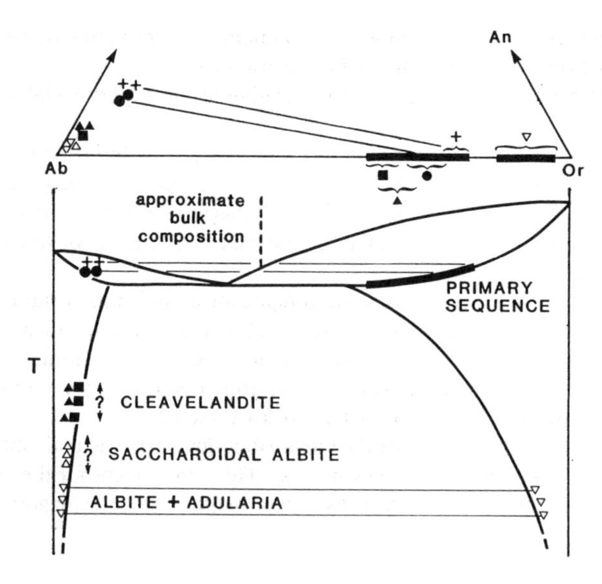

Figure 6. Model of feldspar crystallization in the Věžná I pegmatite, based on the sequence of feldspar assemblages, bulk compositions and limited intersections of the binary solidus and solvus. Border-zone feldspars (+) were contaminated by Ca from the wall-rock serpentinite, shifting the bulk compositions temporarily into the ternary Or-Ab-An system (top triangle). Intermediate-zone (•) and graphic (▲) K-feldspar grade into the core-margin blocky variety (■) with increasing Ab content, the last one associated with albite. Adularia and albite in fissures (▽) conclude the feldspar crystallization at low temperatures. After Černý et al. (1984).

K-feldspar, however, manifests a narrow range of Ab contents *decreasing* from the outer to the inner zones (about Ab_{25} to Ab_{15}; unpublished data of the author). Moreover, Ab contents are commonly reported as low as Ab_5, or even less, and non-perthitic homogeneous microcline was identified as a rock-forming constituent of the central units of some pegmatites (e.g., Norton et al. 1962, 1964).

(iii) The only pegmatites that apparently adhere to the theoretical model are the miarolitic pegmatites of the NYF family, and some of the rare-earth pegmatites of the same affiliation. Initial crystallization of two feldspars transitional into a single Ab-rich K-feldspar is documented at least in terms of phase relationships (Martin 1982, Simmons and Heinrich 1980), but data that would verify the compositional trends are lacking.

It must be stressed again that the primary feldspar crystallization covers rather restricted ranges of bulk composition. Despite of the 150-200°C span from liquidus to solidus of granitic pegmatite consolidation, the bulk composition of the K-feldspar may be virtually constant, as shown by Neiva (1993) on the example of composite aplite-pegmatite dikes. The length of the arrows indicating compositional trends in feldspar evolution in Figure 5 is purely schematic and exaggerated.

All of the above indicates that disequilibrium crystallization advocated by London (1990, 1992a), controlled by the largely unbuffered, increasing abundance of network-modifying components, dominates the precipitation of the feldspars. It is significant in this respect that the only category that seems to adhere to the theoretical expectation - the NYF miarolitic and rare-earth pegmatites - originates from volatile-poor melts largely devoid of Li,B and P, and consequently relatively close to the haplogranitic experimental systems. In contrast, the extremely fractionated LCT pegmatites, which feature the maximum concentrations of network modifiers, show the most extreme deviations from the haplogranitic model.

Additional complications in the natural process relative to the experimental systems should be noted:

(i) Contamination of the pegmatite-forming melts by reaction with host rocks also affects the course of crystallization, as documented by Černý et al. (1984). It may strongly affect the chemistry of plagioclase in the outermost units of otherwise Ca-poor, strongly fractionated pegmatites (e.g., Lenton 1979, unpublished data of the author) and shift coexisting feldspars away from expected compositions (Fig. 6).

(ii) Rubidium attains the level of a substantial component in feldspars from highly fractionated pegmatites: at Red Cross Lake, Manitoba the Rb-f end member reaches 19 mole % in the bulk composition of primary blocky K-feldspar (Černý et al. 1985b). The influence of the rubidian component on the binary and ternary feldspar systems involving K-feldspar has not been experimentally investigated and remains a totally unknown factor.

(iii) Last but not least, subsolidus modifications of K-feldspar are undoubtedly significant, particularly in the highly fractionated LCT pegmatites. This can account for the extremely low Ab content in Kfs of central zones, inexplicable by direct magmatic crystallization (see below sub "Subsolidus Processes").

4.2. MINOR ELEMENTS IN K-FELDSPAR

Three modes of substitution are considered here for K-feldspar: large cations entering the M sites, small cations accepted into the tetrahedral sites, and diverse allocations of different forms of water.

The most prominent minor substituent for K are the compatible Ba and Sr, and the incompatible Li,Rb,Cs,Tl,NH_4 and Pb.

Ba and Sr may reach respectable concentrations in the early generations of K-feldspar of primitive LCT pegmatites (e.g., 13100 and 400 ppm, respectively at Věžná, Černý et al. 1984; 8800 and 500 ppm, in the New York pegmatite, Black Hills; Shmakin 1979). However, their concentrations drop dramatically at the end of fractionation in extremely evolved bodies (42-10 and 130-20 ppm; Shmakin 1979, Shearer and Papike 1985, unpublished data of the author and X.-J. Wang).

Natural K-feldspar attains the maximum concentration of the incompatible alkalis in granitic pegmatites: Li ~ 500 ppm (Černý et al. 1985a), Rb ~ 49000 ppm and Cs 4240 ppm (Černý et al. 1985b), and Tl 650 ppm (Borovik-Romanova and Sosedko 1960). In highly fractionated complex pegmatites, concentrations of Rb routinely attain 1000-2500 ppm, and those of Cs are in the 2000-3000 ppm range (e.g., Černý 1982b, Wang et al. 1986). The substitution is straightforward for Rb,Cs and Tl, but the mechanism of incorporation of small Li is not entirely clarified. Experimental incorporation of Li into sanidine was shown to be explainable by a local lattice-deformation effect (Iiyama and Wolfinger 1976), possibly approaching the collapsed configuration of a Li-exchanged feldspar (Deubener et al. 1991). Otherwise, positional disorder of Li in an oversize cavity must be envisaged. However, the extremely high contents of Li are suspected to be introduced by incipient subsolidus alteration, bound in secondary phases such as micas or clay minerals.

Maximum contents of Pb are shown mainly by amazonite from the NYF pegmatites, ranging up to 10000 ppm (Foord and Martin 1979, who also quote 13500 ppm from the literature). Čech and Mísař (1971) reported green orthoclase from anatectic pegmatites at Broken Hill with 11200 ppm Pb, and Martin (1982) quoted up to 30000 ppm from the same environment. According to

Hofmeister and Rossman (1985), the color of amazonite is generated by electron transitions involving Pb^{3+} or Pb^{+} on irradiation, catalyzed by an associated H_2O molecule. In contrast, Vokhmentsev et al. (1989) proposed five possible mechanisms, the most likely one being a charge-transfer complex Pb^{+}-$(O,OH)^{-}$-Fe^{3+}. However, Petrov et al. (1993) and Hafner (1993) documented $[Pb - Pb]^{3+}$ dimeric centres as the chromophoric agent, an interpretation approached already by Marfunin and Bershov (1970).

The factor responsible for formation of amazonite is the availability of Pb for incorporation into the feldspar structure, which is controlled by the activity of S. Beige to flesh-coloured K-feldspar from amazonite-bearing pegmatites is known to contain dispersed microscopic galena, but amazonite is free of these inclusions (Foord and Martin 1978, Martin 1982). The NYF affiliation of most pegmatite occurrences of amazonite is evidently conditioned by the low concentration of sulphur in this family of granitoid rocks, as opposed to the relatively common occurrences of accessory sulphides in the LCT parageneses. In many cases the primary nature of chromophoric Pb, incorporated during the growth of the green feldspar, is evident (e.g., sector-zoned crystals described by Vokhmentsev et al. 1989). However, Oftedal (1957), Taylor et al. (1960), Zhirov and Stishov (1965) and other authors suggested the possibility of metasomatic introduction of Pb into (or at least its migration within) crystals of K-feldspar. This seems to be supported by streaky patterns of green coloration in some specimens illustrated by Vokhmentsev et al. (1989), and its restriction to exocontacts of late albite veins in other cases.

Ammonia was first discovered in pegmatitic K-feldspar by Honma and Itihara (1981; 0.06 mole % NH^{4+}) and later confirmed in numerous samples from the pegmatite field of southern Black Hills (Solomon and Rossman 1988; up to 0.7 mole % NH_4^{+}). More data are necessary to assess the distribution and quantities of ammonia which represents a petrogenetically significant component of pegmatitic systems.

The most significant substitutents in the tetrahedral sites are Fe^{3+}, Ga, P and, potentially, B and Be. Fe^{3+} behaves in a compatible manner, whereas Ga tends to be, and P distinctly is incompatible. Available data are insufficient to judge the trends of B and Be.

Reliable data on Fe^{3+} contents are not available in any meaningful quantities. Fe^{3+} is known to readily substitute for Al (e.g. Klein et al. 1993), but most of it seems to be present as discrete particles of hematite, introduced by pervasive hydrothermal "hematitization" widespread in some granitoid complexes, or possibly liberated from the feldspar structure (cf. aventurine in "Subsolidus processes" below). In any case, Fe^{3+} is typically enhanced in K-feldspar of the NYF pegmatites, and restricted to only the most primitive members of the LCT family. However, contamination from mafic host-rocks may increase the Fe content of feldspar in outer zones of pegmatite bodies.

The nature of the pegmatites which contain the famous Fe^{3+}-rich orthoclase at Itrongay, Malagasy Republic is uncertain; these bodies possibly represent products of metamorphism in a granulite-facies environment (Martin 1982).

Gallium is omnipresent but highly variable. Considered typical of feldspars from peralkaline systems (Bowden 1964, Gottardi et al. 1972), and introduced by late metasomatic processes (e.g., Severov and Vershkovskaya 1960, Bowden 1964), Ga also was found to be enhanced in peraluminous granite + pegmatite systems but subject to significant regional variations. Černý (1982b) reported 31-116 ppm Ga in K-feldspar from Tanco, but much lower concentrations were found in K-feldspar from the otherwise analogous Bikita (Zimbabwe) pegmatite (unpublished data of the author).

A relative newcomer into the family of tetrahedral substituents is phosphorus, incorporated by the berlinite substitution $AlPSi_{-2}$. Detected first by Corlett and Ribbe (1967), it was found to be widespread in feldspars of peraluminous LCT pegmatites (Pan and Černý 1989, London et al. 1990) and, with a few exceptions, of peraluminous granitic suites in general (London 1992b,

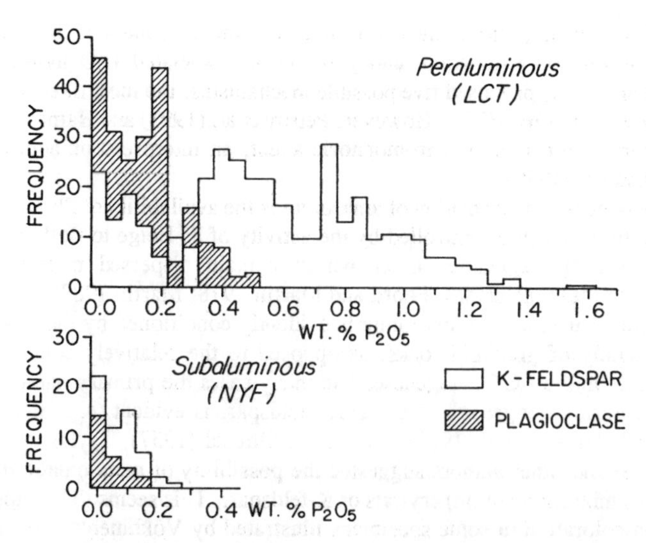

Figure 7. Distribution of phosphorus in K-feldspar and plagioclase of the peraluminous (LCT) and subaluminous (NYF) pegmatites. Data from London et al. (1990), London (1992), Martin et al. (1993), Tomascak and Černý (1993) and unpublished data of the author. Note the preference of P for the K-feldspar structure, and its abundance in the peraluminous environment relative to that of the subaluminous rocks. The columns are superimposed, not stacked.

Tomascak and Černý 1993). Fig. 7 shows a compilation of data on P in K-feldspar of LCT pegmatites (up to 1.62 wt.% P_2O_5) juxtaposed to that of the subaluminous NYF pegmatites, which carry much lower concentrations. The geochemical impact of feldspars as P-carriers is of course enormous, with significant consequences for norm calculations.

High P in feldspars of the LCT pegmatites (and of their parent granites) is generated by anatectic mobilisation of the peraluminous LCT magmas from largely metasedimentary, undepleted protoliths. Early plutonic crystallization of plagioclase decreases the Ca/P ratio, and crystallization of apatite is metastably suppressed in the chemically evolved leucogranitic melts. Excess of Al facilitates the entry of P into the feldspars, also aided by the formation of $AlPO_4$ melt species (London et al. 1990, Černý 1991c). In contrast, the P-poor subaluminous NYF systems consolidate from melts mobilized from depleted granulite-facies protoliths (Collins et al. 1982, Whalen et al. 1987) or I-type meta-igneous lithologies, in part juvenile (e.g., Wilson 1980, Anderson and Morrison 1992). These melts are relatively rich in Ca and F that combine with the low abundances of P to form apatite in early plutonic stages, and their subaluminous (to metaluminous) bulk composition does not provide excess Al required for the berlinite substitution.

In contrast to phosphorus, boron has been so far known in appreciable quantities only in sodic feldspar from peralkaline environments (pegmatitic reedmergnerite of Dusmatov et al. 1968). However, minor contents of B were reported in K-feldspar from peraluminous pegmatites (Shearer and Papike 1986, 10-20 ppm; Foord et al. 1989, up to 290 ppm), and the recent discovery of boromuscovite (Foord et al. 1991), a late-stage hydrothermal mineral, suggests that B-rich feldspars may also be formed in the LCT pegmatites: alkaline conditions are expected to prevail during this stage, promoting tetrahedral coordination of B and its substitution for the deficient Al (Pichavant et al. 1984). Such a regime also is suggested by the recent determinations of B in late but rock-forming muscovite and lepidolite (up to 1.1 and 0.2 wt.% B_2O_3, respectively; Černý et al. 1993).

Data on Be are very limited, and of doubtful value unless collected by modern microbeam methods. In general, Be ranges from 0.X to low XO ppm.

Water is undoubtedly a minor structural constituent of K-feldspar, but no meaningful data are available for feldspar from granitic pegmatites. It can only be assumed that it should contain at least as much H_2O as volcanic sanidine or hydrothermal adularia (hundredths to a few tenths of %; Beran 1986), given the hydrous nature of pegmatite-forming melts and supercritical fluids. The mode of "water" incorporation is diversified: H^+,H_2O and possibly H_3O^+ in the M sites, H^+ as an interstitial cation, and OH^- for briding oxygens; several orientations of H_2O and H_3O^+ were identified (Solomon and Rossman 1979, Hofmeister and Rossman 1985a,b; Beran 1986; Graham and Elphick, this volume).

4.3. MINOR ELEMENTS IN PLAGIOCLASE

Because of the smaller cation in the M site, sodic plagioclase shows only a limited number of substituents. The main one is Sr (up to 600 ppm) and locally also Ba (up to 330 ppm; Černý et al. 1984). The large compatible alkalis Rb,Cs and Tl are present in negligible concentrations (< 10 ppm); higher contents are exceptional and random, strongly suggesting sample contamination by potassic minerals, or association with microfractures rather than M sites. However, the small Li seems to be preferred by albite (up to 67 ppm; Černý et al. 1984, Edgar and Piotrowski 1967); the structure of albite is distinctly closer to that of a Li-exchanged feldspar than to that of K-feldspar (Deubener et al. 1991).

In contrast, substitutions in the tetrahedral sites are as abundant and diversified as in K-feldspar. A few hundred ppm of Fe is fairly common, and Ga is slightly higher than in the associated K-feldspar (e.g. 39-125 ppm in albite of the Tanco pegmatite; Černý 1982b). The potential for massive incorporation of B into late-stage plagioclase seems to be even better than in the case of the K-feldspar; boron shows a slight but distinct preference for the albite structure (3-90 ppm; Černý et al. 1984, Shearer and Papike 1986).

Figure 7 shows that the P content of plagioclase, including albite, is significantly lower than that of K-feldspar; the maximum P_2O_5 recorded to date is 1.06 wt. % (Martin et al. 1993). Experimental results, however, show only a slight (if any) preference of P for the structure of K-feldspar (London et al. 1989). Martin et al. (1993) documented a much more extensive evolution of microporosity in perthitic albite relative to its K-feldspar host in peraluminous granites of the South Mountain batholith, Nova Scotia. They claimed that leaching of P during the evolution of microporosity was responsible for the lower P content of albite. Further research is required to test this mechanism in feldspars of other rock types, including primary plagioclase from granitic pegmatites.

Beryllium tends to be slightly more abundant in albitic plagioclase than in K-feldspar, but it only rarely increases over 100 ppm. Steele et al. (1977) reported ~200 ppm Be in oligoclase from the Quadeville pegmatite, Ontario - a NYF body significantly enriched in Be, Y and REE (Ercit 1993). Beus and Fedorchuk (1955) found increased Be and REE contents in feldspars from Y,REE-enriched pegmatites of Kola Peninsula. These environments could promote the substitution mechanism REE Be $(Ca\ Al)_{-1}$ and REE Be $[(K,Na)Si]_{-1}$, proposed by Beus (1960).

In general, data on plagioclase are rather scarce relative to those available for the K-feldspar, mainly because of the lack of the easy-to-analyze large alkalis in the former that serve as very useful geochemical indicators in the latter. Reviews of geochemistry of granitic pegmatites such as that of Makagon and Shmakin (1988) are based solely on K-feldspar, as far as the feldspar group is concerned. Future studies should rectify this situation.

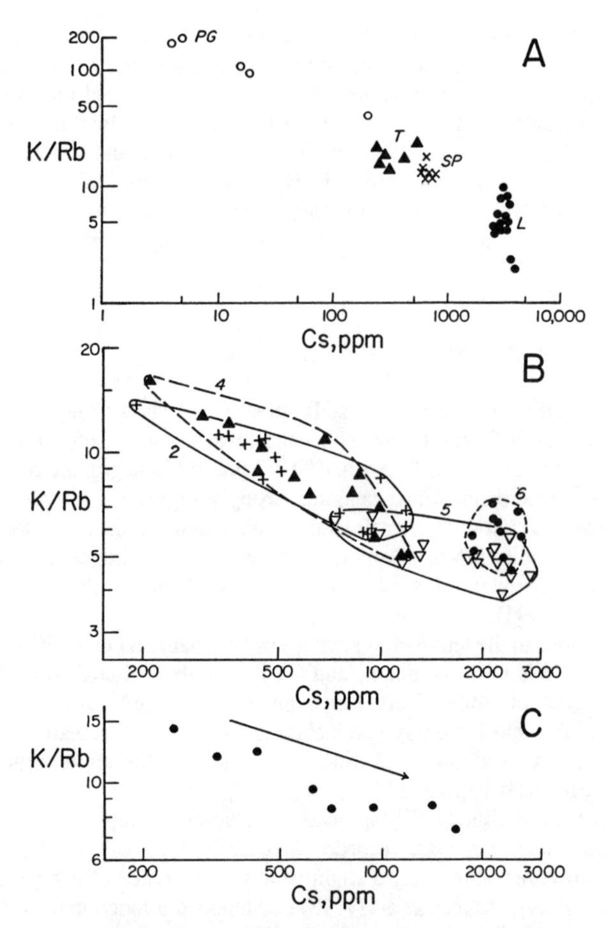

Figure 8. K/Rb vs. Cs in K-feldspar of (A) the Red Cross Lake group (PG - parental pegmatitic leucogranite, T - tourmaline pegmatites, SP - spodumene pegmatites, L - lepidolite-petalite pegmatites; Wang et al. 1986), (B) the Tanco pegmatite, southeastern Manitoba (numbers designate internal zones in the sequence of solidification) and (C) a single blocky crystal, 70 cm long, from the SQ pegmatite at Red Sucker Lake, northeastern Manitoba (arrow indicates the direction of growth; B and C from unpublished data of the author). Note the changes in scale among the three diagrams.

4.4. FRACTIONATION TRENDS OF MINOR ELEMENTS

In contrast to the very limited ranges of bulk compositions that characterize primary feldspar sequences in granitic pegmatites, minor elements show quite spectacular variations in the same feldspar suites. Solidification of granitic pegmatites in general, and within the rare-element class in particular, is the most efficient fractionation process in igneous petrology, as far as minor elements are concerned. Among the rock-forming and accessory minerals contributing to the overall fractionation effect, feldspars play a significant role, and especially the K-feldspar. First

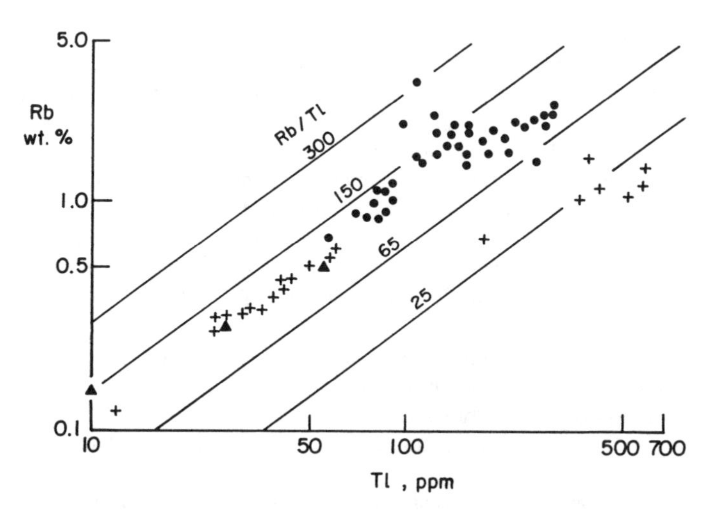

Figure 9. The Rb vs Tl plot of primary K-feldspar from the Kola Peninsula field (+ ; Borovik-Romanova and Sosedko 1960), Mongolian Altai #3 pegmatite (▲ ; Solodov 1962) and the Tanco pegmatite (● ; Černý 1982b). Note the very modest range of fractionation in all three cases. The only trend distinctly crossing the lower limit of fractionation of terrestrial igneous rocks (Rb/Tl ~65) is unique on the worldwide scale.

of all, it is the substituents for K that show "astronomic" ranges in fractionation ratios; substitutions involving the tetrahedral site are rather moderate numerically but also very significant relative to their ranges in plutonic rocks.

Figure 8 demonstrates examples of fractionation ranges at three scales of the pegmatite-generating process, using one of the classic geochemical indicators, K/Rb vs Cs. The Red Cross Lake pegmatite group in northeastern Manitoba consists of parental leucogranites, largely pegmatitic in texture, and a sequence of progressively evolved tourmaline-, spodumene- and lepidolite+petalite-bearing pegmatite dikes (Fig. 8A); the overall ranges of K/Rb and Cs in K-feldspar are 200-1.9 and 4-4300 ppm, respectively (Wang et al. 1986). The Tanco pegmatite in southeastern Manitoba contains primary K-feldspar in four zones (Fig. 8B); its K/Rb ratio varies from 16 in early feldspar down to 4.0 in internal zones, with concomitant increase in Cs from 190 to 2840 ppm (unpublished data of the author). Finally, a longitudinal section along the 70-cm-long growth axis of a single club-shaped crystal of K-feldspar, from the SQ pegmatite at Red Sucker Lake (Chackowsky 1987), shows progressive decrease in K/Rb from 14.5 to 7.3, and increase in Cs from 240 to 1640 ppm (unpublished data of the author; Fig. 8C).

The strong fractionation of Rb from K is closely matched by the behaviour of the K-Cs, Rb-Cs, Ba-K, Sr-K and most remarkably Ba-Rb pairs. In contrast, even the extremely potent fractionation process of crystallization of rare-element pegmatites cannot effectively separate Rb and Tl (Fig. 9). Most of the Rb/Tl ratios of K-feldspar fall within the range typical of terrestrial igneous rocks (de Albuquerque and Shaw 1972, Fung 1978). Reduction of Rb/Tl from ~100 to ~20 is accomplished only across a multitude of pegmatite types, within a pegmatite field which seems to be exceptionally enriched in Tl (Borovik-Romanova and Sosedko 1960). Fractionation trends within individual pegmatites are very shallow (Fig. 9).

The literature teems with data on fractionation in feldspars of a broad variety of pegmatites, the references quoted throughout this section on minor elements being just a representative selection. The fractionation trends show quite a range of scatter (e.g., Černý et al. 1981, 1985a, Černý

1989), and one wonders about the reasons. Besides the possibility of analytical errors, there is a number of objective factors that inevitably contribute to the dispersal of data:

(i) The disequilibrium nature of crystallization of feldspars in granitic pegmatites promotes compositional heterogeneity among crystals within a single zone, at a very short-range scale. Progressive fractionation within a pegmatite body can be meaningfully demonstrated only by analyzing statistically significant numbers of samples per zone (cf. Černý et al. 1984, Fig. 5). If not, interpretations can be seriously distorted as in Kretz (1970).

(ii) Individual crystals are compositionally heterogeneous: concentric zoning (Karnin 1980, Vokhmentsev et al. 1989), sector zoning (Leavens 1972, Long 1978, Vokhmentsev et al. 1989) and micro-patchy distribution of minor elements (London et al. 1990) are well documented in only a few studies, but they must be widespread.

(iii) Long (1978) documented several factors affecting partition coefficients of Rb,Sr, and Ba, such as degree of silica saturation, peraluminous vs. peralkaline chemistry of the melt, growth rate of crystals, concentration of related minor elements, and the Or content of the feldspar. These factors undoubtedly influence the fractionation, as all of them change in the course of pegmatite consolidation.

(iv) Subsolidus reactions of the feldspars with a fluid phase will affect the distribution of elements within individual crystals, disturbing their concentration and, possibly, also their ratios (cf. "Subsolidus Processes").

Although briefly mentioned in some cases, the geochemical differences between the LCT and NYF families and the diverse mechanisms of fractionation are not a direct concern of this review. Further information on these topics is available in the literature (Li, Rb,Cs,Tl,Ba,Sr,Ga - Černý et al. 1985a; P- London et al. 1990, London 1992b; B - Pichavant et al. 1984; Pb - Foord and Martin 1979; LCT vs NYF - Černý 1991a,b,c; 1992).

5. Isotopic Data

Isotopic relationships are commented on here separately from the preceding discussion of magmatic geochemistry and from the following treatment of subsolidus phenomena, as they are influenced by both the primary and deuteric events. Magmatic solidification sets the stage by incorporating all categories of isotopes into the feldspar structures, as they are available in the melt. Since the crystallization, decay of radioactive isotopes generates radiogenic products, and all isotopes are subject to subsolidus reequilibration and migration by diffusion or fluid action. This is happening not only during the postmagmatic thermal equilibration of pegmatite bodies with their host rocks, but most significantly during the subsequent regional uplift and cooling, occasionally disrupted by thermal events of metamorphic or plutonic nature.

In terms of unstable isotopes, feldspar minerals from granitic pegmatites have been utilized in K-Ar, Ar-Ar, Rb-Sr and Pb-Pb studies aimed principally at age dating. However, progress of research revealed numerous problems caused mainly by external disturbances, and the emphasis shifted in many cases from geochronology to other applications of isotope behavior.

The systems involving ^{40}Ar are very easily disturbed, and their current understanding is clouded by "disagreement, controversy, and ambiguity", particularly as far as feldspars are concerned (cf. Foland, this volume). All the problems plaguing Ar in feldspars of plutonic rocks must be inevitably magnified in those of granitic pegmatites, because they are subject to much more intensive post-crystallization disturbances. Thermochronology is complicated by the presence of Ar in fluid inclusions, by dislocations at incoherent perthite boundaries and deuteric microporosity, all of which show protracted histories of evolution (Parsons et al. 1988, Burgess et al. 1992,

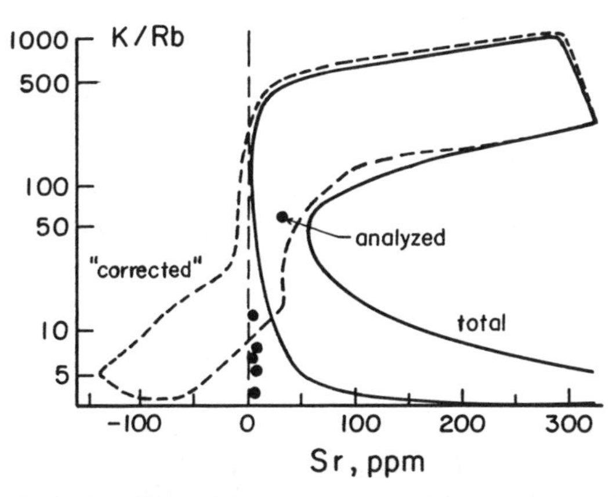

Figure 10. K/Rb vs Sr in blocky K-feldspar of pegmatitic granites and several categories of pegmatites of the Cross Lake field, central Manitoba (after Anderson 1984, Clark and Černý 1987). The C-shaped trend results from plotting total Sr as determined chemically, apparently contradicting the usual trend of Sr depletion. "Correction" of Sr by calculation and subtraction of radiogenic ^{87}Sr for a geologically realistic age leads to negative values of common Sr, because of loss of ^{87}Sr from the K-feldspar. Dots represent common Sr for several samples of isotopically analyzed K-feldspar. The content of common Sr determined by isotopic analysis follows the expected fractionation trend at the time of pegmatite crystallization.

Parsons 1993). Significant precautions must be taken in sample preparation to obtain reliable age spectra and Arrhenius plots (e.g., Harrison et al. 1993).

The Rb-Sr system in feldspars from granitic pegmatites is affected not only by episodic resetting, as commonly encountered in plutonic rocks (e.g., Giletti 1991; this volume), but also by protracted migration of radiogenic ($\pm$ common) Sr which may or may not affect the Rb-Sr systematics in associated minerals (Clark 1982).

In many cases, the decay of ^{87}Rb and mobility of radiogenic ^{87}Sr generate serious disturbances of classic "whole-element" geochemistry. Results of routine analytical determination of Sr in K-feldspar must be treated with caution, particularly in Rb-rich K-feldspar from geologically old pegmatites. Most of the Sr in highly fractionated Archean K-feldspar is radiogenic: for example, a K-feldspar from Cross Lake, Manitoba containing 225.9 ppm total Sr and 28373 ppm total Rb has 216.6 ppm radiogenic ^{87}Sr but only 9.2 ppm common Sr, incorporated during crystallization (Clark and Černý 1987). It is evident that Sr data uncorrected for radiogenic ^{87}Sr must generate fictitious fractionation trends, with initial decrease but a false increase in the most fractionated samples (Fig. 10).

The only valid correction can be achieved via isotopic analysis. Adjustment by calculation of decayed ^{87}Rb and resulting radiogenic ^{87}Sr for a known or assumed age generates negative values of common Sr, -64 ppm in the above example. The reason is the partial loss of radiogenic ^{87}Sr, first recognized in pegmatitic K-feldspar by Brookins et al. (1969) and found to be a widespread feature (e.g., Riley 1970a,b, Penner and Clark 1971, Methot 1973, Baadsgaard and Černý 1993). Complementary to this extensive loss, Rb-poor phases such as albite and phosphate minerals regularly show ^{87}Sr concentrations that are anomalously high, indicative of gain from other minerals. Thus a pegmatite as a whole may represent a closed system, but extensive redistribution takes place among its minerals.

Examination of the Pb-Pb systematics reveals the same relationship. K-feldspar is commonly depleted and plagioclase enriched in radiogenic $^{206,207}Pb$ (e.g., Baadsgaard and Černý 1993) but exceptions do exist: K-feldspar of the Gloserheia pegmatite in southern Norway appears to retain both radiogenic Sr and Pb, whereas plagioclase has Pb undisturbed but Sr excessive, probably gained from micas which lost it (Baadsgaard et al. 1984).

So far, there is no universal explanation available for the above redistribution of radiogenic Sr and Pb in the pegmatitic feldspars. Locally, regional thermal events can be blamed if the losses converge on a geologically realistic age, but the loss-and-gain totals commonly are too high to be interpreted in this fashion, and they locally affect some minerals but not others which should be equally sensitive. Evidently, much is to be done to understand the behavior of radiogenic isotopes in feldspars of pegmatite bodies.

Feldspars also play a substantial role in the study of oxygen isotopes, contributing to the characterization of the source and to the estimates of temperatures of crystallization, collectively with other associated minerals. In the first case, valuable information is usually obtainable. However, subsolidus activity commonly resets the temperature readings distinctly below values petrologically reasonable for primary crystallization. The ubiquitous presence of late fluids, so characteristic of consolidating pegmatite intrusions, is a major factor in this reequilibration, as H^+ is recognized to be the main "water" component promoting oxygen mobility in feldspars and other silicates (cf. Graham and Elphick, this volume). Pervasive flow of regional fluids through solidified pegmatite bodies can even erase the original magmatic signature of $\delta^{18}O$. The reader is referred to Longstaffe (1982), Taylor and Friedrichsen (1983a,b), Walker et al. (1986), and references quoted in these papers, for further information which concerns multiphase assemblages and consequently does not fall within the scope of the present review.

6. Subsolidus Processes

In general, three processes take place after the primary magmatic crystallization of rock-forming feldspars in granitic pegmatites: ordering of their structures, exsolution and recrystallization, and precipitation of low-temperature hydrothermal phases. These phenomena will be dealt with separately for K-feldspar and plagioclase.

6.1. EVOLUTION OF K-FELDSPAR

6.1.1. *Ordering - Structural and Textural Aspects*. Pegmatitic K-feldspar displays a full range of tetrahedral Al-Si order-disorder, generally in an apparently haphazard distribution of structural states (Fig. 11). This indicates locally variable conditions of cooling rate, pressure, deformation, surface area, composition of the feldspar, presence and composition of fluids, which all influence the ordering of the primary, homogeneous and disordered K-feldspar as crystallized from the melt (Martin 1974). In granitic pegmatites, the cooling rate, character of the fluid phase, and local deformation are the most important factors.

In plutonic rocks solidified from relatively H_2O-poor magmas, incipient exsolution and

(Fig. 11 cont.) microcline, in contrast to the late high-sanidine adularia (Černý and Chapman 1984); (F) microperthitic strained orthoclase to vein-perthitic strained intermediate microcline from the Dolní Bory pegmatites, western Moravia, Czech Republic (DB; unpublished data of the author). Numbers in brackets at the locality symbols designate the number of examined samples; the fields of $\underline{b}$ - $\underline{c}$ values circumscribe 2σ error bars of all data.

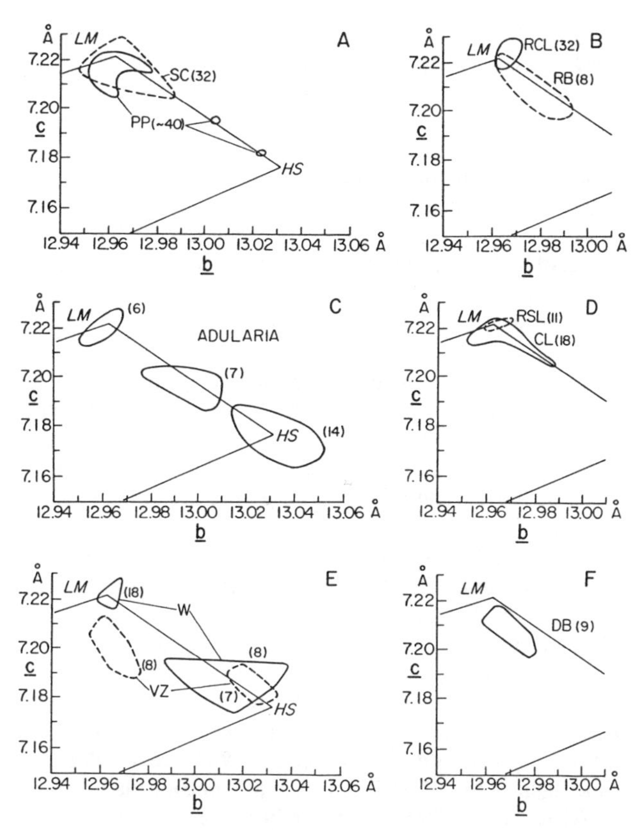

Figure 11. The b - c plot for K-feldspar from several pegmatite populations and/or of specific paragenetic types. (A) Blocky crystals of low microcline with overgrowths of ordered orthoclase or high sanidine from cavities in the miarolitic NYF pegmatites of the Pikes Peak batholith (PP; Martin 1982), and low to intermediate microcline and orthoclase in the walls of pockets in the lepidolite-elbaite pegmatites of southern California (SC; Martin 1982); (B) Blocky K-feldspar from the lepidolite-petalite pegmatites at Red Cross Lake, Manitoba (RCL; Wang et al. 1986, Černý et al. 1985c), with b - c trend extending outside the Na-K-quadrilateral because of high Rb contents; K-feldspar from the border, intermediate and core zones of the Radkovice and Biskupice pegmatites, western Moravia, Czech Republic, grading from strained cryptoperthitic orthoclase through intermediate microcline to maximum low microcline (RB; Černý and Macek 1974); (C) Adularia from a variety of granitic pegmatites, clustering at low microcline, orthoclase (low-sanidine) (with extensive standard errors in b) and at high sanidine (Černý and Chapman 1984); (D) Blocky K-feldspar from the Red Sucker Lake pegmatite field, northeastern Manitoba at the maximum low-microcline corner (RSL; Chackowsky 1987) and blocky K-feldspar from the Cross Lake field, central Manitoba in the same structural state, except a few P-rich samples of orthoclase (CL; Anderson 1984); (E) Adularia and adularia-like K-feldspar from pockets of the NYF pegmatites at Wausau, Wisconsin (W; Černý and Chapman 1984, Martin and Falster 1986) encompassing all structural varieties of K-feldspar; K-feldspar of the border, intermediate and core-margin zones from the Věžná pegmatite, western Moravia, Czech Republic (VZ; Černý et al. 1984) ranges from highly strained cryptoperthitic orthoclase to strained vein-perthitic intermediate

particularly symmetry transformation are relatively slow. Deuteric alteration, which promotes both the symmetry change and coarsening of exsolution textures, may be considerably restricted (cf. Brown and Parsons, this volume). The opposite is true for granitic pegmatites. In virtually all cases, the action of ubiquitous late fluids erases the original structural state and its incipient subsolidus modification. In many instances, the fluids drive the ordering and recrystallization well beyond the stages known from plutonic intrusions.

The original high-sanidine state is preserved only exceptionally. In xenolithic blocks of subvolcanic pegmatites, transported to near-surface levels at Rabb Park, New Mexico, a partially coherent cryptoperthite of bulk composition $Or_{55}Ab_{43}An_2$ is preserved in the high-sanidine structural state (Keefer and Brown 1978, O'Brient 1986). Other occurrences of high sanidine are reported by Martin (1982), but the structural state is secondary in these cases, induced by thermal or shock metamorphism.

Microperthitic, at least partially coherent and considerably strained low sanadine (orthoclase)is found in the Moldanubian pegmatites of the Czech Republic. Best preserved at Věžná (Fig. 11E) and, to a degree, also at Dolní Bory (Fig. 11F) it forms relics transitional into intermediate microcline. In contrast to these localities of the beryl-columbite subtype, the closely related Li-enriched pegmatites at Radkovice and Biskupice have ordering of the K-feldspar extended to near-maximum low microcline (Fig. 11B; Černý and Macek 1974). At all these and other Moldanubian localities, rapid initial cooling after intrusion and different degrees of subsequent action of late (and possibly alkaline) fluids in the central parts of the pegmatites are evidently responsible for the variety and distribution of the structural states.

In contrast, K-feldspar in the Kenoran pegmatites of the Superior and Slave Provinces of the Canadian Shield is almost uniformly close to low microcline. Structural states of K-feldspar of the Red Cross Lake, Red Sucker Lake and Cross Lake pegmatite populations are shown as examples in Figures 11B and D. The reasons for local occurrences of orthoclase, e.g. at Cross Lake (Fig. 11D) and in the Yellowknife pegmatite field (Meintzer 1987) are not clear. The $\underline{b}$ - $\underline{c}$ data of some of the low-microcline samples from Red Cross Lake plot outside the LM corner: this is due to the high content of Rb (Černý et al. 1985c), but at other localities the "excessive" values of $\underline{c}$ may suggest that the standard low microcline is not fully ordered (Martin 1982) or contains significant Fe^{3+}.

Martin (1982) presented data on blocky K-feldspar from the gem-bearing LCT pegmatites (of the lepidolite-to-elbaite subtype) from southern California, and from the miarolitic NYF-affiliated pegmatites of the Pikes Peak batholith. The former contain abundant orthoclase as well as microcline, whereas the latter have dominantly a highly ordered microcline (Fig. 11A); however, a definite sampling bias at both localities favours crystals of K-feldspar from the pocket zones (Martin 1982). In any case, a definite difference is indicated here in terms of cooling rates, extent of fluid action and probably also alkaline fluid chemistry, all of them favouring profound ordering at Pikes Peak.

In some pegmatites, conspicuous changes in colour accompany the progressive stages of Al-Si ordering (and exsolution), notably at the Radkovice and Biskupice localities mentioned above (Černý and Macek 1974). Glassy to white, structurally strained orthoclase of the wall and intermediate zones grades into brownish intermediate microcline, culminating in a core-margin, dark chocolate-brown feldspar tightly clustered at $\blacktriangle = 0.66$. Two consecutive generations of near-maximum low microcline develop at the expense of this 0.66 phase, both with very restricted range of order (Fig. 12A). The chromophoric agents are not known. Some of the late varieties of feldspar show different character of microporosity, which may be responsible for the colour changes but which was so far not systematically investigated in feldspars from granitic pegmatites (cf. Harrison and Parsons, this volume). Three pulses of fluid activity are suggested to generate

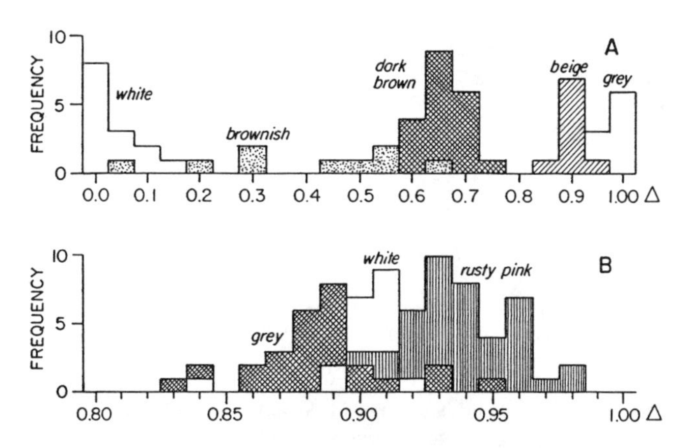

Figure 12. Triclinicity vs colour in hand specimens in (A) K-feldspars from the Radkovice and Biskupice pegmatites, western Moravia, Czech Republic (Černý and Macek 1974), and (B) blocky K-feldspar from zone (5) of the Tanco pegmatite and the corresponding zone of Lower Tanco, Manitoba (Černý and Macek 1972, Ferreira 1984, and unpublished data of the author). Note the transitional nature of the brownish phase but very restricted ranges of all other varieties in (A); the broad and overlapping ranges in (B) result from impure separates, to a degree, but show statistically unambiguous distinction of the three varieties. The columns are superimposed, not stacked.

this spectrum of order levels.

A similar correlation of colour changes with increasing degree of Al-Si order was observed in the Tanco and Lower Tanco pegmatites, Manitoba (Černý and Macek 1972, Ferreira 1984). Giant-size blocks of K-feldspar show a good correlation of triclinicity increasing from grey through white to rusty pink phases within individual blocks, but over a narrow range of Δ values (Fig. 12B). The ranges of Δ overlap among individual colour phases, probably because of unavoidable averaging of the main colour phase (as sampled for X-ray diffraction) with microscopic relics, or incipient evolution, of the other varieties.

Much has been written about the retarding influence of large cations - Rb, Cs, Ba - on the ordering process. The concept was advocated mainly by Gordiyenko and Kamentsev (1967), Afonina and Shmakin (1970), Afonina et al. (1979) and supported by Karnin (1980). However, such a correlation was put in doubt, at least as far as Rb is concerned, by Shmakin (1979), Martin (1982), and Černý et al. (1985b). Figure 13 convincingly demonstrates that intensive action of order-promoting factors can easily overcome any retarding influence of Rb, at least within the range of its contents in primary magmatic K-feldspar.

Microtextural expression of phase transformation, which occurs during the monoclinic-to-triclinic conversion from low sanidine (orthoclase) to microcline, is initially observed only by transmission electron microscopy and related techniques. The development of tweed structure in orthoclase is the first step (e.g., McLaren 1974). This diffuse pattern, which evolves further only with considerable input of energy, coarsens into sharp-bounded tartan microcline, which criss-crosses orthoclase matrix (Eggleton and Buseck 1980, Fitz Gerald and McLaren 1982). On total disappearance of orthoclase, tartan microcline grades into diverse complex textures of M-twinning - a combined albite + pericline twinning, which becomes observable at optical scale as the well-known cross-hatched texture (Fig. 14D). However, the four orientations of lamellae paired by albite and pericline twinning are rarely preserved. In most samples examined to date, pericline

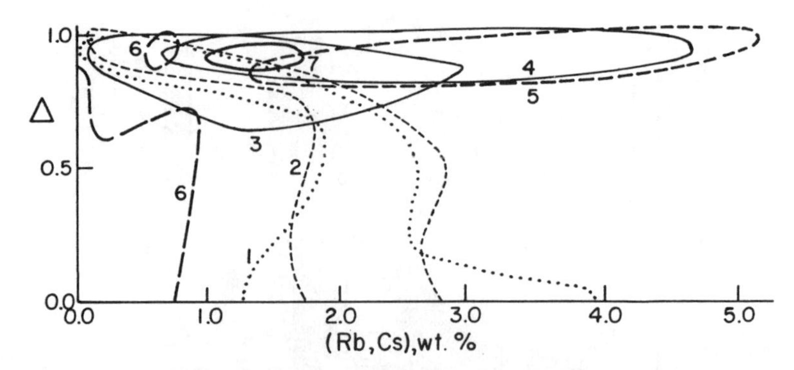

Figure 13. Triclinicity vs the Rb + Cs content of K-feldspar from rare-element pegmatites of 1 - "European part of the USSR", 2 - "one of the Siberian fields", 3 - "another pegmatite field in Siberia" (all from Afonina et al. 1979), 4 - the Tanco and Lower Tanco, southeastern Manitoba (Černý and Macek 1972, Ferreira 1984, unpublished data of the author), 5 - Red Cross Lake, northeastern Manitoba (Černý and Macek 1985c, Wang et al. 1986), 6 - Radkovice, western Moravia, Czech Republic (Černý and Macek 1974, 19 samples), 7 - Buck Claim, Bernic Lake, southeastern Manitoba (Lenton 1979, 12 samples).

lamellae are converted into "secondary", super-fine albite twins. Disregarding the complexity of the ensuing textural patterns, the resulting low-microcline then consists essentially of only two sets of lamellae related to each other by albite law (Fitz Gerald and McLaren 1982, McLaren 1984).

The above textural evolution is commonly observed in K-feldspar of plutonic rocks. What is very rare in this environment, but rather common in granitic pegmatites, is a continuing recrystallization of the cross-hatched (but only albite-twinned) K-feldspar into increasing volumes of single-crystal low microcline. In a process of reduction of surface energy, one set of lamellae grows and coalesces at the expense of the other (Fig. 14E), up to the apparently total obliteration of twinning (Fig. 14F). Volumes of essentially single-crystal low microcline attain up to 15 cm^3 (the grey Radkovice phase of Figure 12; Černý and Macek 1974).

6.1.2 *Perthitic Exsolution and Coarsening.* Incipient strain-controlled cryptoperthite is only exceptionally preserved in pegmatitic K-feldspar, notably in the Rabb Park high sanidine (Keefer and Brown 1978). Relics of cryptoperthite grading into micro- and film perthite are common in the monoclinic K-feldspar of the Moldanubian pegmatites (Fig. 14A,B; Černý and Macek 1974). Structures of these coherent perthites are highly strained, with $a_{obs.}$ - $a_{est.}$ variable from 0.05 to 0.13 Å. Coarsening into vein perthite is widespread at these localities, concomitant with ordering and phase transformation of the potassic component and loss of coherency (Fig. 14B). Strain-free vein perthite with cross-hatched microcline and albite-twinned plagioclase constitutes the most characteristic K-feldspar of granitic pegmatites, commonly visible to the naked eye and occasionally reaching very coarse dimensions (Fig. 14C,D). Association with subordinate film perthite and minor plate perthite is, however, widespread even in the most coarsened intergrowths. Segregation of albite veins into chessboard blebs culminates the process (Fig. 14F).

Exsolution and coarsening of perthitic albite proceed simultaneously with the Al-Si ordering discussed in the preceding section. The two processes are inter-related, to a degree, but exsolution is considerably faster than the rate of ordering. The first involves alkali interdiffusion which is rapid relative to the ordering of Al-Si (cf. Brown and Parsons, this volume). Whereas the temperature of initial exsolution depends on the starting composition of the K-feldspars and the

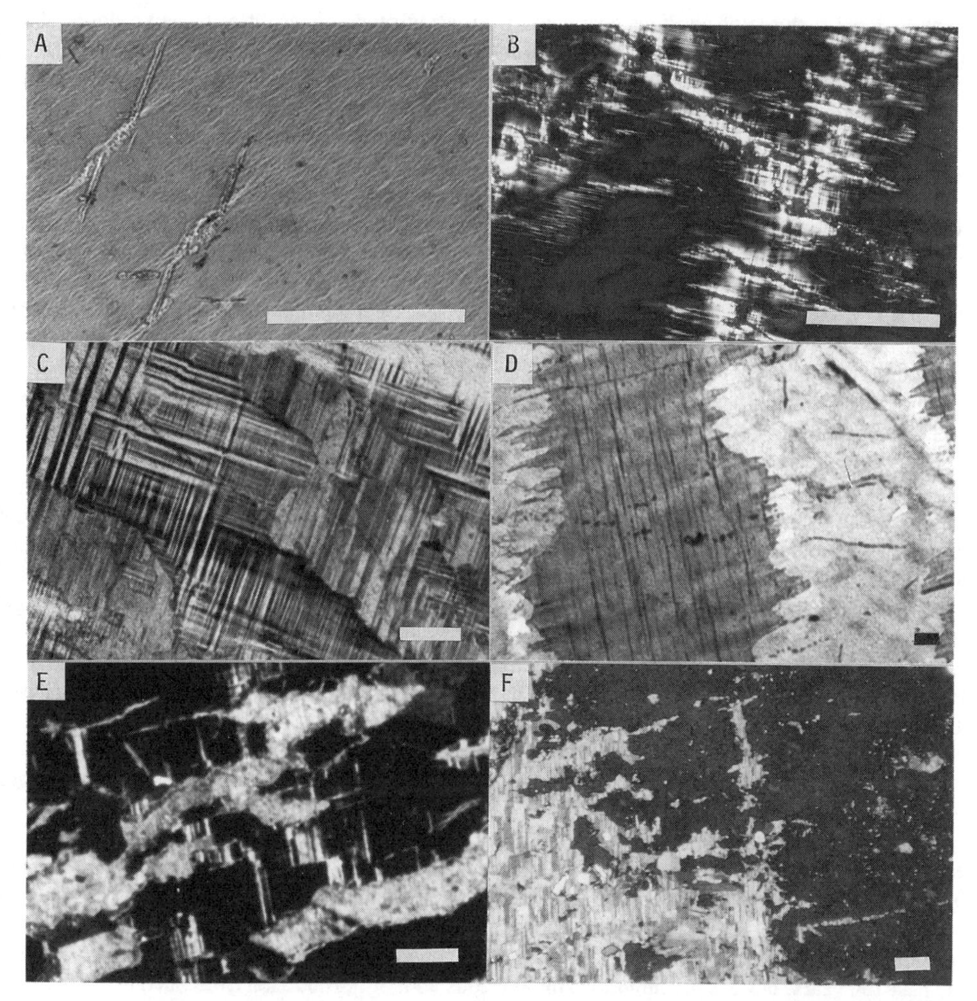

Figure 14. Diversity in pegmatitic perthite: (A) strained orthoclase - microperthite with incipient segregation of albite at deuteric flakes of mica, Věžná I, western Moravia, Czech Republic; (B) transition from cryptoperthite (homogeneous-looking dark patches) into micro- and vein-perthite which develop from micro-fractures, with local film perthite, Biskupice, western Moravia, Czech Republic (from Černý and Macek 1974); (C) well-developed coarse microcline-perthite, Tanco, Manitoba; (D) extremely coarse microcline perthite, with advanced recrystallization of both phases, from Otty Lake near Perth, Ontario; (E) recrystallization of the microcline component into an untwined single-crystal phase, locality unknown; (F) recrystallization of microcline perthite into large volumes of untwinned single-crystal K-phase and patches of chessboard albite, Radkovice, western Moravia, Czech Republic (Černý and Macek 1974). All scale bars are 0.2 mm long.

conditions of its cooling, the process of coarsening can be expected to begin at 450° - 400°C, within the stability field of low microcline which is invariably generated at this stage.

Although the incipient stages of exsolution are rarely encountered undisturbed, the textural and compositional features of perthites correlated from a multitude of localities leave no doubt about a common line of evolution: initial crypto- to micro-scale coherent exsolution in a monoclinic Or-rich matrix, followed by deuteric coarsening, up to virtual elimination of twinning in the K-feldspar and segregation of albite into coarse chessboard patches. The differences among individual localities or regional pegmatite populations stem from the different maximum levels of perthite evolution attained, and from the degree of preservation of the preceding stages.

As pointed out by Parsons (1978), the coherent exsolution does not necessitate intergranular water, but the loss of coherency and coarsening of perthite are strongly dependent on the presence of H_2O. Thus it is not surprising that coarse vein perthite is the most common type of pegmatitic K-feldspar, promoted by the omnipresent residual fluids, and that it initially evolves from its precursor along microfractures, intergranular surfaces and similar channelways (Fig. 14B). According to Graham and Elphick (this volume), it is specifically the high proton activity which facilitates coarsening of perthite and the concomitant ordering of Al and Si.

The coarsening and segregation of perthite components, which proceed much farther than in any other geological environment, are the sources of many petrologic and geochemical frustrations. The redistribution of the potassic and sodic phases may become so extensive and random that the exsolution origin may be legitimately doubted (Roubault 1958). A total removal of the albitic component to the outside of the original homogeneous crystal may be responsible for, or contribute to, the two phenomena mentioned sub "Trends in bulk composition": the presence of non-perthitic K-feldspar in the inner zones of some pegmatites, and the formation of (some of) the late albite, in particular cleavelandite. Last but not least, one wonders whether such a thorough recrystallization preserves the original contents and ratios of minor elements.

Mason (1980, 1982) examined the composition of K-feldspar and plagioclase lamellae in perthite from granitic pegmatites. He found Li,Cs,Ba,Pb,Rb and B partitioned into the potassic phase. This is in general agreement with the behavior of these elements in coexisting K-feldspar and plagioclase, except for Li and B which prefer plagioclase (Smith and Brown 1988). Martin (1982) reported the albitic phase of perthite from several localities as slightly Ca-bearing and largely but incompletely ordered.

Aventurine may be mentioned in conclusion, although it should be placed into a separate category of basically unknown phenomena. Aventurine shows "a play of light and colors" caused by reflection from oriented lamellae in microcline perthite (Andersen 1915). Most of the lamellae consist of iron oxide, probably hematite, but other and so far unidentified species must be present, as some are transparent. Exsolution is just one of several genetic speculations put forward over the past 150 years, besides precipitation simultaneous with or introduction from outside the feldspar (Smith and Brown 1988). Exsolution is an improbable mechanism in this case, because of complex charge-balance requirements by even the simplest model of breakdown of a Fe-feldspar component, and because of the inevitable but unobserved formation of byproducts.

6.2. EVOLUTION OF PLAGIOCLASE

6.2.1. *Ordering.* As in the case of geochemistry of minor elements, our present understanding of the structural state of plagioclase and albite in granitic pegmatites is only sketchy. In pegmatites of those categories which are considered in this review, all plagioclase is very close to being ordered. As shown in a series of experimental studies (e.g. Martin 1969, Mason 1979), ordering in albite (and presumably also in oligoclase exsolved into peristerite) proceeds at a fast rate in an alkaline environment, aided by increased pressure, and there is no analog to the orthoclase "tweed barrier" in the ordering path of albitic plagioclase (Brown and Parsons 1989).

It should be pointed out, however, that the degree of order attained by pegmatitic plagioclase is not necessarily total, as shown by some data in Černý et al. (1984), and as documented for the plagioclase component of some perthite samples by Martin (1982; β^* vs γ^*). Sweeping assumptions of total order in, as well as Ca-free composition of pegmatitic "low albite" are rather misleading, and some petrologically significant information may be revealed by detailed study. However, both perthitic and discrete plagioclase are virtually always much better ordered than the associated K-feldspar (Martin 1982).

6.2.2. *Exsolution.* Three types of heterogeneous assemblages develop in plagioclase at subsolidus conditions: peristerite, antiperthite and aventurine.

Historically, granitic pegmatites yielded the first specimens of peristerite, conspicuous by the mainly bluish-purplish play of colour on the (010) cleavage surfaces, but grading to white, reddish or yellow-green at some localities. Coarse-grained to blocky plagioclase with bulk composition of Ca-poor oligoclase ($\sim An_5$ - $\sim An_{18}$) shows the best optical effects in the NYF-family pegmatites of, e.g., the Grenville Province of Ontario and Quebec, and in the mixed NYF-LCT pegmatites of southern Norway. However, peristerite does not necessarily develop planar lamellae of thickness near the range of light wavelengths. The iridescence is absent in specimens with incipient tweed-like texture, or with lamellae thinner than about 400 Å (e.g., McLaren 1974, Ribbe 1983).

To date, virtually all peristerite has been found to be suboptical in scale, and is revealed only by different single-crystal X-ray diffraction or TEM techniques. Exceptions are extremely rare (Brown 1960). As reviewed by Ribbe (1983) and Smith and Brown (1988), peristeritic intergrowths are well characterized today but their thermal history and conditions of formation are not; experimental and theoretical work presents several lines of contradictory evidence and a variety of models (e.g. Iiyama 1966, Lagache 1978, Carpenter 1981 and this volume, Ribbe 1983).

In contrast, antiperthitic exsolution of K-feldspar from plagioclase tends to be coarser and routinely observable in the light-optical microscope. However, observations of antiperthite in plagioclase from granitic pegmatites are rare, compared to its widespread occurrence in metamorphic rocks. The reason might be the relatively rapid cooling of oligoclase in outer zones of pegmatite bodies, in which this feldspar typically occurs. The cooling regime may be adequate for peristeritic exsolution but insufficient for separation of a potassic phase, which is virtually always very minor in the starting bulk composition (Or_5 or less in most cases, exceptionally up to Or_{15}; Černý 1971, Černý et al. 1984).

Plagioclase-based aventurine is actually much more widespread than its K-feldspar counterpart mentioned sub 6.1.2. The physical aspects, the largely unknown composition of component phases, and the uncertainties about the origin of the aventurine are the same in both cases (e.g. Divljan 1960, Neumann and Christie 1963).

6.3. LOW-TEMPERATURE HYDROTHERMAL PHASES

Besides affecting the structural state and phase composition of the magmatic feldspars, the subsolidus regime also produces late generations of plagioclase and K-feldspar. The plagioclase is very close to the ordered sodium end member (e.g., Černý et al. 1984) and thus rarely examined in any meaningful detail. The following account therefore focusses on the very diversified occurrences and properties of the late K-feldspar, arranged for descriptive purposes by its parent environments.

(i) In the LCT pegmatites, adularia is known from open fissures or leaching cavities, in association with albite, celadonite, fluorite, calcite and other minerals of hydrothermal origin (Černý and Chapman 1984). It forms locally oriented, shingle-like overgrowths on blocky K-feldspar substrate. Chemical composition is very close to that of end-member $KAlSi_3O_8$ but the structural state varies over the full range of order-disorder.

(ii) The same characteristics apply to adularia-type K-feldspar in LCT pegmatites, which replaces pollucite, albite or early generations of primary K-feldspar at many localities (Černý and Chapman 1984). However, Rb-rich K-feldspar grading compositionally to a potassic Rb-feldspar also is found in this association. It forms a two-phase assemblage with the K-feldspar, suggestive of a solvus in the K-Rb feldspar series (Teertstra et al. 1993, Teertstra and Černý 1993). Barium-enriched feldspar was found associated with late products of pollucite alteration, but so far only at a single locality (unpublished data of D.K. Teertstra and the author).

(iii) Early coarse crystals of K-feldspar that line cavities in the LCT miarolitic pegmatites display overgrowths of a late generation of K-feldspar (Martin 1982). The blocky crystals are perthitic low sanidine to microcline but the overgrowths are largely nonperthitic, close to ideal ordered orthoclase $K(Al_{0.5}Si_{0.5})_2Si_2O_8$ (Prince et al. 1973, Horsky and Martin 1977), or less ordered orthoclase (low sanidine).

(iv) Veins and fissure-coatings of late K-feldspar also are known from NYF-family pegmatites. In the Landsverk I pegmatite of southern Norway, brick-red feldspar of this type is a nonperthitic low microcline, poor in Na and minor elements relative to the primary K-feldspar (Taylor et al. 1960). It is found intergrown with saccharoidal albite but also replacing perthitic lamellae of albite, and the primary K-feldspar as well.

(v) A great variety of K-feldspars occurs in the vugs of NYF-affiliated miarolitic cavities. At Wausau, Wisconsin, Černý and Chapman (1984) found the high-temperature variety of adularia, with Zillertal-Fibbia morphology (in monoclinic notation, {110} > {010} > {101} > {001} to {010} ≅ {001} ⋙ {110} > {201} {101}), consisting of cross-hatched non-perthitic low microcline, structurally analogous to the rock-forming blocky microcline-perthite in the substrate. However, Martin and Falster (1986) described a broader spectrum of structural states from this locality: drusy and folded crusts, spherulitic aggregates, euhedral crystals and glazes alike show all degrees of order, from low microcline to high sanidine but rather uniform, near-end-member Or chemistry. The structural variety is equally broad in the vugs of miarolitic pegmatites in the Pikes Peak batholith (Foord and Martin 1979). Crystals of amazonite, themselves well-ordered low microcline, carry more or less selective overgrowths of late albite, ordered orthoclase (low sanidine) and intermediate microcline (Martin 1982). In one case, a nonperthitic untwinned coating of low microcline was identified, in another one an abundant coating of high sanidine covered all cavity minerals (Foord and Martin 1979, Martin 1982).

Despite the above diversity, all occurrences of hydrothermal K-feldspar are controlled by the same array of factors: rate of precipitation, stability of vugs or fracture systems, availability and chemistry of fluid action plus thermal history after crystallization. It is the individual significance and mutual interplay of these factors at specific localities that determine the resulting feldspar.

High sanidine was commonly rapidly nucleated from solution in fine-grained masses by pressure-quenching, and preserved by evacuation of fluids and subsequent sealing of the cavities (Foord and Martin 1979, Černý and Chapman 1984). On the other end of the spectrum of conditions, slow crystallization from alkaline fluids was probably responsible for the untwinned overgrowth of low microcline.

Adularia of intermediate and well ordered structural states is commonly coarse-grained, with Zillertal-Fibbia morphology, turbid, overgrown by other hydrothermal minerals - all of which are the earmarks of relatively high-temperature origin, and protracted history of subsequent interaction with fluids at temperatures that were high enough to advance the ordering. In contrast, the high-

Figure 15. Wavy pattern of layered saccharoidal albite, zone (3) of the Tanco pegmatite, southeastern Manitoba.

sanidine phase is commonly fine-grained, with Felsöbanya-Maderaner morphology (in monoclinic notation, {110} > {10$\bar{1}$} ≫ {001}), waterclear, and free of any minerals of later origin: all of this indicates rapid nucleation at low temperature, a short period of fast growth, and shielding from any order-promoting agents afterwards (Černý and Chapman 1986; cf. Flehmig 1977 for experimental work). This process is so far the only one that produced a natural end-member high sanidine, surpassing in its total Al-Si disorder all synthetic phases as well (Ferguson et al. 1991).

The generally uniform chemistry of the late K-feldspar and associated albite in most of their occurrences, very close to the respective end-member compositions of Or and Ab, corresponds to the low-temperature of crystallization. The anionic chemistry of the parent fluid does not play a significant role in K-feldspar-albite equilibria (e.g., Orville 1963, Fournier 1976, Barton and Frantz 1983). However, the anions do influence the precipitation of one or the other of the two feldspars in processes involving metasomatism (Pichavant 1983), as do variations in pressure and the transition from supercritical fluids to the aqueous fluid plus gas system (Lagache and Weisbrod 1977). The Rb- and Ba-rich feldspars associated with replacement of pollucite evidently require very specific compositonal parameters of their parent fluids, and are currently under study.

7. Replacements Involving Feldspars

Two large-scale phenomena should be mentioned here, both discussed in the literature since the beginning of pegmatite studies: the origin of late albite-rich and micaceous units.

7.1. ALBITE-RICH UNITS

Albite-rich units are a typical late component of moderately to highly fractionated granitic pegmatites, occasionally in more than a single generation. Saccharoidal albite is the main

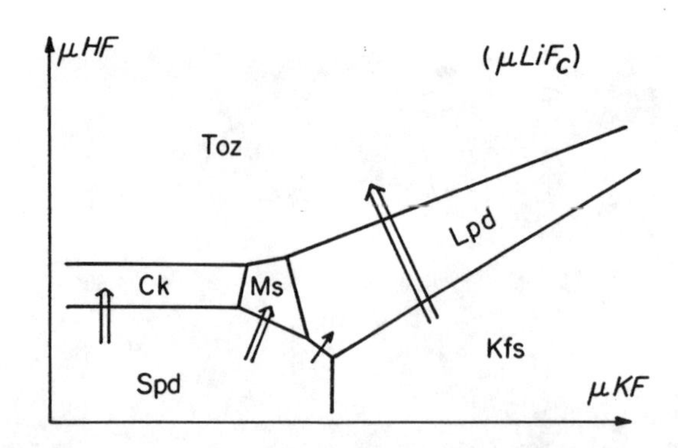

Figure 16. Schematic isobaric-isothermal μHF - μKF phase diagram, at constant μLiF in the system $LiAlO_2$ - SiO_2 - H_2O - HF - LiF - KF, under quartz- and H_2O-saturated conditions within the stability field of spodumene (modified from London 1982); the arrows indicate the directions of mineral reactions commonly observed in complex rare-element pegmatites. CK = cookeite.

constituent, with variable but generally low quantities of quartz, muscovite or lepidolite, commonly in a layered texture marked by fluctuations in the abundance and grain size of individual minerals (Fig. 15).

These units were traditionally considered metasomatic in their entirety, although questions were raised about the provenance of the metasomatizing fluids (from within or outside the pegmatite bodies, with a space problem associated with the former case) and about the fate of components liberated from the replaced assemblages. On the other hand, primary origin could not be justified because of widespread replacement features along contacts with adjacent units, and a bulk composition outside the classic granitic minima.

It is true that the albite-rich units do show extensive reaction with pre-existing mineral assemblages along their margins, especially with K-feldspar and lithium aluminosilicates. Metasomatic replacement of K-feldspar by beryl + albite generated by action of fluorine-bearing fluids was experimentally proven by Beus et al. (1963). However, the bulk volume of the albite-rich units shows primary growth features indicative of crystallization on a solid substrate into fluid-filled space, and slumping or folding structures that also require a fluid or plastic medium (e.g., Norton et al. 1964; Lenton 1979, Černý et al. 1981, Thomas and Spooner 1988, London 1992a).

Fluid-inclusion studies and the experimental work of London (1986, 1987) and London et al. (1989) show that in the presence of Li and B, a $Li_2B_4O_7$-fluxed, sodic, hydrous silicate liquid can evolve during the crystallization of core-margin K-feldspar + quartz. Such a liquid can generate fine-grained albite-rich units upon loss of B and Li, by stabilization of tourmaline and Li-aluminosilicates and consequent rapid separation of aqueous fluid. If this model also works for moderately fractionated pegmatites with low B and Li contents, a universal interpretation would be available, consistent with observations in pegmatites and broader ramifications of the primary vs. metasomatic controversy, mentioned above: late but primary albite units which do inflict replacement "damage" along contacts with pre-existing assemblages.

7.2. MICACEOUS UNITS

The origin of local pods of muscovite + quartz, and of larger-scale lepidolite units, has been so far judged mainly on textural evidence, which apparently favors metasomatism (Norton et al. 1962,

Stewart 1978, Černý 1982b, Thomas and Spooner 1988). However, experimental data indicate that a primary origin is a realistic proposition. In hydrous K,Na-aluminosilicate systems, muscovite ± albite can form instead of K-feldspar at reduced K-salinity, despite the availability of a pre-existing K-feldspar phase (Hemley and Jones 1964, Burt 1981). Consequently, the assemblage of fine-grained muscovite + quartz is no guarantee of pre-existing K-feldspar. Lepidolite ± albite may form instead of K-feldspar because of moderate changes in $\mu LiF/\mu KF$ in an otherwise identical system; metasomatism is possible but not inevitable (Fig. 16; London 1982). Under univariant conditions of assemblage boundaries, K-feldspar + muscovite or K-feldspar + lepidolite may form in equilibrium. Furthermore, lepidolite is found to be stable in silicate melts (Glyuk et al. 1980) as well as under subsolidus conditions (Munoz 1971), similar to muscovite relationships. Consequently, some authors propose primary crystallization from an immiscible liquid (Melentyev and Delitsyn 1969) or from a residual "magma-like phase" (Norton 1983), although they acknowledge the existence of metasomatic lepidolite-bearing assemblages.

The possibility of late but primary micaceous units was confirmed by the experiments of London et al. (1989) which yielded fine-flaked mica aggregates among the latest units of synthetic analogs of zoned pegmatites. However, much is to be done to understand the natural mica-rich units: the bulk compositions, mineral modes, internal heterogeneity and other characteristics of lepidolite units are not even remotely known. It is quite probable that both primary and secondary mica-rich units exist, but so far no simple criteria are available to unambiguously distinguish them.

8. Concluding Remarks

The present overview of the feldspars in granitic pegmatites clearly demonstrates two facts of life: the tremendous diversity of the paragenetic position, of conditions of crystallization , low-temperature modifications, chemistry, structural state and reactions of the feldspars in a broad variety of pegmatite classes, types and sub-types on the one hand, and the extent of our ignorance about many aspects of these feldspars on the other. A series of problems should be addressed in the future:

(i) A thorough and systematic characterization of feldspars from zoned, highly fractionated LCT pegmatites, from the quasi-homogeneous albite-spodumene bodies, and from the NYF-related rare-earth and miarolitic pegmatites is required.

(ii) The extensive migration of radiogenic Sr and Pb needs quantitative analysis and interpretation of its driving force and timing.

(iii) A quantitative approach is also required for the muscovite + quartz and lepidolite units as potential replacement products after feldspars.

(iv) The structural state and chemistry of albitic plagioclase must be documented throughout the crystallization sequences.

(v) Examination of the Rb-feldspar - K-feldspar system is required to complement the ternary Or-Ab-An relationships, which are inadequate for feldspar phases in extremely fractionated pegmatites.

(vi) Examination of factors affecting the crystal-melt and crystal-fluid partition of minor elements is required.

(vii) Last but not least, and difficult as it may be in sampling and interpretations, the bulk composition of perthitic K-feldspar must be analyzed in conjunction with microbeam studies, to facilitate the study of primary bulk compositions at the time of crystallization.

Acknowledgements

I wish to thank many of my colleagues who shared their knowledge of feldspars in general, and pegmatite feldspars in particular, in countless discussions and collaborative work. Special thanks go to R.B. Ferguson, E.E. Foord, M. Lagache, R.F. Martin, the late P.M. Orville, B.J. Paul, H. Pentinghaus, J.V. Smith, and D. K. Teertstra. Extensive support of my research by NSERC of Canada Research and Equipment Grants, DEMR Canada Research Agreements, University of Manitoba and Tantalum Mining Corporation of Canada Ltd. has been much appreciated. A. Lester, P.H. Ribbe and particularly R.F. Martin reviewed the manuscript with meticulous care.

References

Abad-Ortega MDM, Fenoll Hach-Alí P, Martin-Ramos JD, Ortega-Huertas M (1993) The feldspars of the Sierra Albarrana granitic pegmatites, Cordoba, Spain. Canad Mineral 31: 185-202

Afonina GG, Makagon VM, Shmakin BM, Glebov MP, Makrygin AI (1979) Effects of rubidium and cesium on the structural states of potash feldspars from rare-metal pegmatites. Internat Geol Rev 21: 597-604

Afonina GG, Shmakin BM (1970) Inhibition of lattice ordering of potassic feldspar by barium ions. Doklady Acad Sci USSR 195: 133-135

de Albuquerque CAR, Shaw DM (1972) Thallium. In: Handbook of Geochemistry III/3. Springer-Verlag

Anderson AJ (1984) Geochemistry, mineralogy and petrology of the Cross Lake pegmatite field, Manitoba. Unpubl M.Sc. thesis, Univ Manitoba, Winnipeg: 251pp

Anderson JL, Morrison J (1992) The role of anorogenic granites in the Proterozoic crustal development of North America. In: Condie KC (ed) Proterozoic Crustal Evolution. Elsevier, 263-299

Anderson O (1915) On aventurine feldspar. Amer Jour Sci ser 4, 40:351-358

Baadsgaard H, Černý P (1993) Geochronological studies in the Winnipeg River pegmatite populations, southeastern Manitoba. Geol Assoc Canada - Mineral Assoc Canada Ann Meet Edmonton, Prog w Abstr 18: A-5

Baadsgaard H, Chaplin C, Griffin WL (1984) Geochronology of the Gloserheia pegmatite, Froland, southern Norway. Norsk Geol Tidsskr 64: 111-119

Barton MD, Frantz JD (1983) Exchange equilibria of alkali feldspars with fluoride-bearing fluids. Carnegie Inst. Washington Year Book 82: 377-381

Beus AA (1960) Geochemistry of beryllium and the genetic types of beryllium deposits. Acad Sci USSR Moscow, 329 pp (in Russian). Engl. transl. Freeman & Co. 1966, 401pp

Beus AA, Fedorchuk SN (1955) Beryllium clarke in granitic pegmatites. Dokl Acad Sci USSR 104, Earth Sci Sect, 124-129

Beus AA, Sobolev BP, Dikov JP (1963) On the geochemistry of beryllium in high-temperature postmagmatic mineralization events. Geokhimiya 1963: 297-304 (in Russian)

Beran A (1986) A model of water allocation in alkali feldspar, derived from infrared-spectro-scopic observations. Phys Chem Minerals 13:306-310

Boettcher AL (1980) The system albite-orthoclase-water and albite-orthoclase-quartz-water: chemographic phase relationships. Jour Geophys Res 85, B12:6955-6962

Borovik-Romanova TF, Sosedko AF (1960) Relation between thallium and rubidium in minerals from pegmatite veins of the Kola Peninsula from the results of spectrographic analysis. Geochem Int: pp34-42

Bowden P (1964) Gallium in younger granites of northern Nigeria. Geochim Cosmochim Acta 28:1981-1988

Brookins DG, Fairbairn HW, Hurley PM, Pinson WK (1969) A Rb-Sr geochronologic study of the pegmatites of the Middletown area, Connecticut. Contrib Mineral Petrol 22:157-168

Brown WL (1960) X-ray studies in the plagioclases. Part 2. The crystallographic and petrologic significance of peristerite unmixing in the acid plagioclases. Zeit Krist 113:297-344

Brown WL, Parsons I (1989) Alkali feldspars: ordering rates, phase transformations and behaviour diagrams for igneous rocks. Mineral Mag 53:25-42

Burgess R, Kelley SP, Parsons I, Walker FDL, Worden RH (1992) ^{40}Ar-^{39}Ar analysis of perthite microtextures and fluid inclusions in alkali feldspars from the Klokken syenite, South Greenland. Earth Planet Sci Letters 109:147-167

Burt DM (1981) Acidity-salinity diagrams - application to greisen and porphyry deposits. Econ Geol 76:832-843

Cameron EN, Jahns RH, McNair A, Page LR (1949) Internal structure of granitic pegmatites. Econ Geol Monogr 2: 115p

Carpenter MA (1981) A "conditional spinodal" within the peristerite miscibility gap of plagioclase feldspars. Amer Mineral 66:553-560

Čech F, Mísar Z, Povondra P (1971) A green lead-containing orthoclase. Tscherm Min Petr Mitt 15: 213-231

Černý P (1971) Graphic intergrowths of feldspars and quartz in some Czechoslovak pegmatites. Contr Mineral Petrol 30:343-355

Černý P (1982a) Anatomy and classification of granitic pegmatites. In: Cerny P (ed) Granitic Pegmatites in Science and Industry. Mineral Assoc Canada Sh Course Handb 8:1-39

Černý P (1982b) The Tanco pegmatite at Bernic Lake, southeastern Manitoba. In: Cerny P (ed) Granitic Pegmatites in Science and Industry. Mineral Assoc Canada Sh Course Hbk 8:527-543

Černý P (1989) Contrasting geochemistry of two pegmatite fields in Manitoba: products of juvenile Aphebian crust and polycyclic Archean evolution. Prec Res 45:215-234.

Černý P (1990) Distribution, affiliation and derivation of rare-element granitic pegmatites in the Canadian Shield. Geol. Rundschau 79:183-226

Černý P (1991a) Rare-element granitic pegmatites. Part I: Anatomy and internal evolution of pegmatite deposits. Geosci Canada 18:49-67

Černý P (1991b) Rare-element granitic pegmatites, Part II: Regional to global environments and petrogenesis. Geosci Canada 18:68-81

Černý P (1991c) Fertile granites of Precambrian rare-element pegmatite fields: is geochemistry controlled by tectonic setting or source lithologies? Precamb Res 51:429-468

Černý P (1992) Geochemical and petrogenetic features of mineralization in rare-element granitic pegmatites in the light of current research. Appl Geochem 7:393-416

Černý P, Chapman R (1984) Paragenesis, chemistry and structural state of adularia from granitic pegmatites. Bull Minéral 107:369-384

Černý P, Chapman R (1986) Adularia from hydrothermal vein deposits: extremes in structural state. Can Mineral 24:717-728

Černý P, Hawthorne FC (1989) Controls on gallium concentration in rare-element granitic pegmatites. Geol Assoc Canada - Mineral Assoc Canada Ann Meet Montreal, Progr w Abstr

14:A21

Černý P, Macek J (1972) The Tanco pegmatite at Bernic Lake, Manitoba. V. Coloured potassium feldspars. Canad Mineral 11:679-689

Černý P, Macek J (1974) Petrology of potassium feldspars in two lithium-bearing pegmatites. In: Mackenzie WS, Zussman J (eds) The Feldspars. Manchester University Press, pp 615-628

Černý P Smith JV, Mason RA, Delaney JS (1984) Geochemistry and petrology of feldspar crystallization in the Vezná pegmatite, Czechoslovakia. Canad Mineral 22: 631-651

Černý P, Meintzer RE, Anderson AJ (1985a) Extreme fractionation in rare-element granitic pegmatites: selected examples of data and mechanisms. Canad Mineral 3:381-421

Černý P, Pentinghaus H, Macek JJ (1985b) Rubidian microcline from Red Cross Lake, northeastern Manitoba. Bull. Geol. Soc. Finland 57:217-230

Černý P, Stanek J, Novák M, Baadsgaard H, Rieder M, Ottolini L, Chapman R (in press) Geochemical and structural evolution of micas in the Rozná and Dobrá Voda pegmatites, Czech Republic. Mineral Petrol

Chackowsky LE (1987) Mineralogy, geochemistry and petrology of pegmatitic granites and pegmatites at Red Sucker Lake and Gods Lake, northeastern Manitoba. Unpubl M.Sc. thesis, Univ Manitoba, Winnipeg: 157pp

Chakoumakos BC, Lumpkin GR (1990) Pressure-temperature constraints on the crystallization of the Harding pegmatite, Taos County, New Mexico. Canad Mineral 28:287-298

Clark GS (1982) Rubidium-strontium isotope systematics of complex granitic pegmatites. In: Cerný P (ed) Granitic Pegmatites in Science and Industry. Min Assoc. Canada Hbk 8:347-372

Clark GS, Cerný P (1987) Radiogenic ^{87}Sr, its mobility, and the interpretation of Rb-Sr fractionation trends in rare-element granitic pegmatites. Geochim Cosmochim Acta 51:1011-1018

Collins WJ, Beams SD, White AJR, Chappell BW (1982) Nature and origin of A-type granites with particular reference to southeastern Australia. Contrib Mineral Petrol 80:189-200

Corlett M, Ribbe PH (1967) Electron probe microanalysis of minor elements in plagioclase feldspars. Schweiz Mineral Petrogr Mitt 47:317-332

Correia Neves JM, Lopes Nunes JE (1966) Geochemistry of the pegmatite field of Alto Ligonha (Mozambique, Portuguese East Africa). Est Gerais Univ de Mozambique 3, Ser 2, Cienc Geol:1-30

Correia Neves JM (1964) Genese des zonar gebauten Beryll-pegmatits von Venturinha (Viseu, Portusal) in geochemischer Sicht. Beitr Mineral Petrogr 10:357-373

Deubener J, Sternitzke M, Müller G (1991) Feldspars $MAlSi_3O_8$(M=H,Li,Ag) synthesized by low-temperature ion exchange. Amer Mineral 767:1620-1627

Divljan S (1960) The results of field and laboratory studies on aventurine plagioclases from some Norwegian pegmatites. 21st IGC Norden Rept: 94-101

Dubrovskyi MI (1978) The albite-anorthite-orthoclase system. Izvest Akad Nauk USSR 1978, No. 6:20-32

Dusmatov VD, Popova NA, Kabanova LK (1967) The first discovery of reedmergnerite in the USSR. Dokl Akad Nauk Tadzhik. SSR 10:51-53

Edgar AD, Piotrowski JM (1967) $\Delta\theta_{131\text{-}131}$ for albites crystallized in the system $NaAlSi_3O_8$-β-$LiAlSi_2O_6$-H_2O and $NaAlSi_3O_8$-$LiAlSiO_4$-H_2O. Mineral Magazine 36:578-582

Eggleton RA, Buseck PR (1980) The orthoclase-microcline inversion: a high-resolution transmission electron microscope study and strain analysis. Contrib Mineral Petrol 74:123-133

Ercit TS (in press) The geochemistry and crystal chemistry of columbite-group minerals from

granitic pegmatites in the Grenville Province. Canad Mineral
Fenn PM (1979) On the origin of graphic intergrowths. Geol Soc Amer - Mineral Soc Amer Abstr w Progr 11:424
Fenn PM (1986) On the origin of graphic granite. Amer Mineral 71:325-330
Ferguson RB, Ball NA, Cerný P (1991) Structure refinement of an adularian end-member high sanidine from the Buck Claim pegmatite, Bernic Lake, Manitoba. Canad Mineral 29:543-552
Ferreira KJ (1984) The mineralogy and geochemistry of the Lower Tanco pegmatite, Bernic Lake, Manitoba, Canada. M.Sc. thesis, Univ Manitoba, Winnipeg:239ppFisher DJ (1968) Albite, variety cleavelandite, and the signs of its optic directions. Amer Mineral 53:1568-1578
Fitz Gerald JD, McLaren AC (1982) The microstructures of microcline from some granitic rocks and pegmatites. Contrib Mineral Petrol 80:219-229
Flehmig W (1977) The synthesis of feldspars at temperatures between 0°-80°C, their ordering behaviour and twinning. Contrib Mineral Petrol 65:1-9
Foord EE (1976) Mineralogy and petrogenesis of layered pegmatite-aplite dikes in the Mesa Grande district, San Diego Country, California. Unpubl. Ph.D. thesis, Stanford University:326pp
Foord EE, Martin RF (1979) Amazonite from the Pikes Peak batholith. Mineral Record 10:373-384
Foord EE, Martin RF, Fitzpatrick JJ, Taggart JE Jr, Crock JG (1991) Boromuscovite, a new member of the mica group, from the Little Three mine pegmatite, Ramona district, San Diego County, California. Amer Mineral 76:1998-2002
Foord EE, Spaulding LB Jr, Mason RA, Martin RF (1989) Mineralogy and paragenesis of the Little Three pegmatites, Ramona District, San Diego county, California. Mineral Record 20:101-127
Fournier RO (1976) Exchange of Na^+ and K^+ between water vapor and feldspar phases at high temperature and low vapor pressure. Geochim Cosmochim Acta 40:1553-1561
Fung PC (1978) K,Rb and Tl distribution between coexisting natural and synthetic rock-forming minerals. Unpubl. Ph.D. thesis, McMaster University:262p
Ginsburg AI (1960) Specific geochemical features of the pegmatitic process. Int Geol Congress, 21st session Norden, Rep 17:111-121
Giletti BJ (1991) Rb and Sr diffusion in alkali feldspars, with implications for cooling histories of rocks. Geochim Cosmochim Acta 55:1331-1343
Glyuk DS, Trufanova LG, Bazarova SB (1980) Phase relations in the granite-H_2O-LiF system at 1,000 kg/cm^2. Geochem Internat 17:35-48
Gordiyenko VV (1971) Concentration of Li,Rb, and Cs in potash feldspar and muscovite as criteria for assessing the rare metal mineralization in granite pegmatites. Internat Geol Review 13:134-142
Gordiyenko VV (1976) Diagrams for prognostic evaluation of rare-element mineralization in granitic pegmatites utilizing compositional variations in potassium feldspar. Dokl Acad Sci USSR 228:442-444 (in Russian)
Gordiyenko VV, Kamentsev IE (1967) On the nature of the rubidium admixture in potassic feldspar, Geokhimiya 1967, 478-481 (in Russian)
Gottardi G, Burton JD, Culkin F (1972) Gallium. Handb of Geochem, Springer Berlin:Pt 31
Hafner SS (1993) Crystallography of trace elements in feldspars. NATO ASI Feldspars and Their Reactions, Abstracts No. 14
Harrison TM, Lovera OM, Heizler MT (1991) $^{40}Ar/^{39}Ar$ results for alkali feldspars containing

diffusion domains with differing activation energy. Geochim Cosmochim Acta 55: 1435-1448
Heier KS (1962) Trace elements in feldspars - a review. Norsk Geol. Tidsskr 42:415-454
Heinrich EW (1967) Micas of the Brown Derby pegmatites, Gunnison County, Colorado. Amer Mineral 52: 1110-1121
Hemley JJ, Jones WR (1964) Chemical aspects of hydrothermal alteration with emphasis on hydrogen metasomatism. Econ Geol 59:538-569
Hofmeister AM, Rossman GR (1985a) A spectroscopic study of irradiation coloring in amazonite: structurally hydrous, Pb-bearing feldspar. Amer Mineral 70:794-804
Hofmeister AM, Rossman GR (1985b) A model for the irradiative coloration of the smoky feldspar and the inhibiting influence of water. Phys Chem Minerals 12:324-332
Honma H, Itihara Y (1981) Distribution of anmonium in minerals of metamorphic and granitic rocks. Geochim Cosmochim Acta 45:983-988
Horsky SJ, Martin RF (1977) The anomalous ion-exchange behavior of "ordered" orthoclase. Amer Mineral 62:1191-1199
Iiyama JT (1966) Contribution à l'étude des equilibres sub-solidus du système ternaire orthose-albite-anorthite à l'aide des réactions d'échange d'ions Na-K au contact d'une solution hydrothermale. Bull Soc franç Minéral 89:442-454
Iiyama JT, Volfinger M (1976) A model for trace-element distribution in silicate structures. Mineral Mag 40:555-564
Jahns RH (1954) Pegmatites of southern California. In: Geology of Southern California. Calif Div Mines Bull 170 pt 7:37-50
Jahns RH, Tuttle OF (1963) Layered pegmatite-aplite intrusives. - Miner Soc America Sec Paper 1:78-92
James RS, Hamilton DL (1969) Phase relations in the system $NaAlSi_3O_8$-$KAlSi_3O_8$-SiO_2 at 1 kilobar water vapour pressure. Contrib Mineral Petrol 21:111-141
Karnin W-D (1980) Petrographic and geochemical investigations on the Tsaobismund pegmatite dyke, South West Africa/Namibia. Neues Jahrb Mineral, Monatsh:193-205
Keefer KD, Brown GE (1978) Crystal structures and compositions of sanidine and high albite in cryptoperthitic intergrowth. Amer Mineral 63:1264-1273
Klein J, Hafner SS, Pentinghaus H (1993) Crystallography of iron in alkali feldspars. NATO ASI Feldspars and Their Reactions, Abstracts, Poster 6
Kretz R (1970) Variation in the composition of muscovite and albite in a pegmatite dike near Yellowknife. Canad Jour Earth Sci 7:2\1219-1235
Kretz R (1983) Symbols for rock-forming minerals. Amer Mineral 68:277-279
Lagache M (1978) Nouvelles expériences portant sur le système ternaire albite-orthose-anorthite en présence de solutions de chlorures sodipotassiques, dans la domaine de composition des péristerites. Comp Rend Acad Sci Paris 287, Série D:849-952
Lagache M, Weisbrod A (1977) The system: two alkali feldspars-KCl-NaCl-H_2O at moderate to high temperatures and low pressures. Contrib Mineral Petrol 62:77-101
Lazarenko EK, Matkovskyi OI, Pavlishin VI, Sorokin YG (1968) News on the structure and mineralogy of Volynian pegmatites. Dokl Acad Sci USSR 176:171-174 (in Russian)
Leavens PB (1972) Rubidium and cesium in New England pegmatites. IGC Montreal, Section 10, Geochemistry, Abstracts 179
Lenton PG (1979) Mineralogy and petrology of the Buck Claim lithium pegmatite, Bernic Lake, southeastern Manitoba. Unpubl M.S. thesis, Univ. Manitoba:164p
Lentz DR, Fowler AD (1992) A dynamic model for graphic quartz-feldspar intergrowths in granitic

pegmatites in the southwestern Grenville province. Canad Mineral 30:571-585

Linnen RL, Williams-Jones AE, Martin RF (1992) Evidence of magmatic cassiterite mineralization at the Nong Sua aplite-pegmatite complex, Thailand. Canad Mineral 30:739-761

London D (1982) Stability of spodumene in acidic and saline fluorine-rich environments. Carnegie Inst Geophys Lab Ann Rpt 81:331-334

London D (1986) Magmatic-hydrothermal transition in the Tanco rare-element pegmatite: Evidencefrom fluid inclusions and phase-equilibrium experiments. Amer Mineral 71:376-395

London D (1987) Internal differentiation of rare-element pegmatites: effects of boron, phosphorus,and fluorine. Geochim Cosmochim Acta 51:403-420

London D (1990) Internal differentiation of rare-element pegmatites; a synthesis of recent research. In: Stein HF, Hannah JL (eds) Geol Soc America Spec Paper 246:35-50

London D (1992a) Application of experimental petrology to the genesis and crystallization of granitic pegmatites. Canad Mineral 30:499-540

London D (1992b) Phosphorus in S-type magmas: the P_2O_5 content of feldspars from peraluminous granites, pegmatites and rhyolites. Amer Mineral 77:126-145

London D, Černý P, Loomis JL, Pan JC (1990) Phosphorus in alkali feldspars of rare-element granitic pegmatites. Canad Mineral 28:771--786

London D, Morgan GB IV, Hervig RL (1989) Vapor-undersaturated experiments Macusani glass + H_2O at 200 MPa, and the internal differentiation of granitic pegmatites. Contrib Mineral Petrol 102:1-17

Long PE (1978) Experimental determination of partition coefficients for Rb,Sr, and Ba between alkali feldspar and silicate liquid. Geochim Cosmochim Acta 42:833-846

Longstaffe FJ (1982) Stable isotopes in the study of granitic pegmatites and related rocks. In: Černý P (ed) Granitic Pegmatites in Science and Industry. Mineral Assoc. Canada Short Course Hbk 8:373-404

Makagon VM, Shmakin BM (1988) Geochemistry of the main formations of granitic pegmatites. Nauka Novosibirsk:208p

Marfunin AS, Bershov LV (1957) Paramagnetic center(s) in feldspar and their possible crystallochemical and petrological significance. Dok Acad Sci USSR 193, Earth Sci Sect, 129-131

Martin RF (1969) The hydrothermal synthesis of low albite. Contr Mineral Petrol 23:323-339

Martin RF, Falster AU (1986) Proterozoic sanidine in pegmatite, Wausau complex, Wisconsin. Canad Mineral 26:709-716

Martin RF, Kontak DJ, Richard LR (1993) The significance of P enrichment in alkali feldspar, South Mountain batholith, Nova Scotia, Canada. NATO ASI Feldspars and Their Reactions, Abstracts, No. 15

Mason RA (1980) Trace elements in alkali feldspars and element distribution in a perthite by in probe microanalysis. Geol Soc America Ann Meet, Abstr w Progr 12, No. 7:477

Mason RA (1982) Trace element distributions between the perthite phases of alkali feldspars from pegmatites. Mineral Mag 45:101-106

Mason RA (1979) The ordering behaviour of albite in aqueous solutions at 1 kbar. Contrib Mineral Petrol 68:269-273

McLaren AC (1974) Transmission electron microscopy of feldspars. In: Mackenzie WS, Zussman J (eds) The Feldspars. Manchester Univ Press, 378-423

McLaren AC (1984) Transmission electron microscope investigations of the microstructures of microclines. In: Brown WL (ed) Feldspars and Feldspathoids. Reidel Publishing Company

Dordrecht:373-409

Meintzer RE (1987) The mineralogy and geochemistry of the granitoid rocks and related pegmatites of the Yellowknife pegmatite field, Northwest Territories. Unpubl Ph.D. thesis, Univ Manitoba, Winnipeg:708pp

Melentyev GB, Delitsyn LM (1969) Problem of liquation in magma. Dokl Acad Sci USSR, Earth Sci Ser 186:215-217 (transl AGI New York)

Methot RL (1973) Internal geochronologic study of two large granitic pegmatites, Connecticut. Unpubl Ph.D. thesis. Kansas State University:123p

Munoz JL (1971) Hydrothermal stability relations of synthetic lepidolite. Amer Mineral 56:2069-2087

Neiva AMR (submitted) Distribution of trace elements in feldspars of granitic rocks from Alijó-Sanfins, northern Portugal. Mineral Mag

Neumann H, Christie DHJ (1963) Observations on plagioclase aventurines from southern Norway. Norsk Geol Tids 42:389-393

Norton JJ (1983) Sequence of mineral assemblages in differentiated granitic pegmatites. Econ Geol 78:854-874

Norton JJ, Page RL, Brobst DA (1962) Geology of the Hugo pegmatite, Keystone, South Dakota. USGS Prof Paper 297-B:49-128

Norton JJ and others (1964) Geology and mineral deposits of some pegmatites in the southern Black Hills, South Dakota. USGS Prof Paper 297-E:293-341

O'Brient JD (1986) Preservation of primary magmatic features in subvolcanic pegmatites, aplites and granite from Rabb Park, New Mexico. Amer Mineral 71:608-624

Oftedal I (1957) Heating experiments or amazonite. Mineral Mag 31:417-419

Orville PM (1963) Alkali ion exchange between vapor and feldspar phases. Amer Jour Sci 261:201-237

Pan JC, Černý P (1989) Phosphorus in feldspars of rare-element granitic pegmatites. Geol Assoc Canada - Mineral Assoc Canada Ann Meet 1989 Montreal, Progr w Abstr 14:A82

Paul BJ (1984) Mineralogy and geochemistry of the Huron Claim Pegmatite, Southeastern Manitoba. Unpubl M.Sc. thesis, Univ of Manitoba, Winnipeg:108p.

Parsons I (1978) Feldspars and fluids in cooling plutons. Mineral Magazine 42:1-17

Parsons I, Rex DC, Guise P, Halliday AN (1988) Argon-loss by alkali feldspars. Geochim Cosmochim Acta 52:1097-1112

Penner AP, Clark GS (1971) Rb-Sr age determinations from the Bird River area, southeastern Manitoba. In: Turnock AC (ed) Geoscience Studies in Manitoba. Geol Assoc Canada Spec Paper 9:105-109

Petrov I, Mineeva RM, Bershov LV, Agel A (1993) EPR of $[Pb\text{-}Pb]^{3+}$ mixed valence pairs in amazonite-type microcline. Amer Mineral 78:500-510

Pichavant M (1983a) (Na,K) exchange between alkali feldspars and aqueous solutions containing borate and fluoride anions: experimental results at P=1 kilobar. Programme and Abstracts, 3rd NATO A.S.I. on Feldspars, Feldspathoids and Their Parageneses, Rennes:102

Pichavant M (1983b) Melt-fluid interaction deduced from studies of silicate - B_2O_3-H_2O systems at 1 kbar. Bull Minéral 106:201-211

Pichavant M, Schnapper D, Brown WL (1984) Al_B substitution in alkali feldspars: preliminary hydrothermal data in the system $NaAlSi_3O_8$-$NaBSi_3O_8$. Bull Minéral 107:529-537

Prince E, Donnay G, Martin RF (1973) Neutron diffraction refinement of an ordered orthoclase structure. Amer Mineral 58:500-507

Ribbe PH (1983) Exsolution textures in ternary and plagioclase feldspars; interference colors. In: Ribbe PH (ed) Feldspar Mineralogy. Reviews in Mineralogy 2:241-270
Riley GB (1970a) Isotopic discrepancies in zoned pegmatites, Black Hills, South Dakota. Geochim Cosmochim Acta 34:713-725
Riley GH (1970b) Excess Sr^{87} in pegmatitic phosphates. Geochim Cosmochim Acta 34: 727-731
Rockhold JR, Nabelek PI, Glascock MD (1987) Origin of rythmic layering in the Calamity Peak satellite pluton of the Harney Peak granite, South Dakota: the role of boron. Geochim Cosmochim Acta 51:487-496
Roubault M (1958) Observations sur la composition chimique d'un cristal de microcline perthitique. Compt Rend Acad Sci Paris 247:1355-1358
Severov EA, Vershkovskaya OV (1960) The behavior of gallium during the process of albitization of granitoids. Dokl Acad Sci USSR, ser geol 135: 1498-1500 (in Russian)
Shearer CK, Papike JJ (1985) Chemistry of potassium feldspars from three zoned pegmatites, Black Hills, South Dakota: implications concerning pegmatite evolution. Geochim Cosmochim Acta 49:663-673
Shearer CK, Papike JJ (1986) Distribution of boron in the Tip Top pegmatite, Black Hill, South Dakota. Geology 14:119-123
Shmakin BM (1979) Composition and structural state of K-feldspars from some U.S. pegmatites. Amer Mineral 64:49-56
Simmons WB, Heinrich EW (1980) Rare-earth pegmatites of the South Platte district, Colorado. Colo Geol Survey, Resource Ser 11:131pp
Smith JV, Brown WL (1988) Feldspar Minerals I. Crystal Structures, Physical, Chemical and Microtextural Properties (2nd ed.) Springer-Verlag, Berlin
Solodov NA (1962) Distribution of thallium among the minerals of a zoned pegmatite. Geochem Int:738-743
Solomon GC, Rossman GR (1979) The role of water in structural states of K-feldspar as studied by infrared spectroscopy. Geol Soc America Abst w Progr 11:521
Solomon GC, Rossman GR (1988) NH_4^+ in pegmatitic feldspars from the southern Black Hills, South Dakota. Amer Mineral 73:818-821
Steele IM, Hutcheon ID, Smith JV (1980) Ion microprobe analysis of plagioclase fledspar for major, minor and trace elements. Proc. 8th Internat Conf on X-ray Optics and Microanalysis Boston, Pendell Publ. Co., 515-525
Stewart DB (1978) Petrogenesis of lithium-rich pegmatites. Amer Mineral 63:970-980
Stewart DB, Roseboom EH (1962) Lower-temperature terminations of the three-phase region plagioclase - alkali feldspar - liquid. Jour Petrology 3:280-315
Tan L-P (1966) Major pegmatite deposits of New York state. N.Y. State Mus Sci Service Bull 408
Taylor BE, Friedrichsen H (1983a) Light stable isotope systematics of granite pegmatites from North America and Norway. Isotope Geoscience 1:127-167
Taylor BE, Friedrichsen H (1983b) Oxygen and hydrogen isotope disequilibria in the Landsverk I pegmatite, Evje, Southern Norway: evidence for anomalous hydrothermal fluids. Norsk Geol Tidsskr 63:199-209
Taylor SR, Heier KS, Sverdrup TL (1960) Trace element variations in three generations of feldspars from the Landsverk I pegmatite. Norsk Geol Tidsskr 40:133-156.
Teertstra DK, Černý P (1993) Rubidian adularia as an alteration product of pollucite, Morrua Mine, Mozambique. NATO ASI Feldspars and Their Reactions, Edinburgh, Abstracts, Poster 9

Teertstra DK, Lahti SI, Alviola R, Cerný P (1993) Pollucite and its alteration in Finnish pegmatites. Bulletin of the Geological Survey of Finland 368:39pp
Thomas AV, Spooner ETC (1988) Occurrence, petrology and fluid inclusion characteristics of tantalum mineralisation in the Tanco granitic pegmatite, s.e. Manitoba. In: Taylor RP, Strong DF (eds) Granite-Related Mineral Deposits. CIM special Volume 39 :208-222
Tomascak P, Černý P (in press) The geochemistry of phosphorus in alkali feldspars: evidence from granitic pegmatites in the Northwest Territories, Canada. Mineral Petrol
Vlasov KA (1952) Textural-paragenetic classification of granitic pegmatites. Izvest Acad Sci USSR, ser geol 1952:30-44 (in Russian)
Vokhmentsev AY and others (1989) Amazonite. Nedra Moscow: 192pp (in Russian)
Walker RJ, Hanson GN, Papike JJ, O'Neil JR, Laul JC (1986a) Internal evolution of the Tin Mountain pegmatite, Black Hills, South Dakota. Amer Mineral 71:440-459
Wang XJ, Cerný P, Chackowsky LE, Eby R (1986) Evolution of K-feldspar in the Red Cross Lake pegmatitic granite and its pegmatite aureole, northeastern Manitoba, Canada. Internat Mineral Assoc. 14th General Meet Stanford. Abstr w Progr 262
Whalen JB, Currie KL, Chappell BW (1987) A-type granites: geochemical characteristics, discrimination and petrogenesis. Contrib Mineral Petrol 95:407-419
Wilson MR (1980) Granite types in Sweden. Geol För i Stockholm Förh 102:167-176
Zhirov KK, Stishov SM (1965) Geochemistry of amazonitization. Geokhimiya, 32-42

SURFACE CHEMISTRY OF FELDSPARS

JOSEPH V. SMITH
Department of Geophysical Sciences
University of Chicago
5734 South Ellis Avenue
Chicago Illinois 60637
United States of America

ABSTRACT. Feldspars and other silicate minerals must have structural perturbations at the external surface that involve hydroxyl, water and various molecular and ionic complexes. Internal surfaces around voids and at twin boundaries and intergrowths must also have chemical properties that differ from the ideal bulk structure. Quantum chemistry, lattice dynamics, and electrostatics offer theoretical guidance for interpretation of experimental data. Ideally, the chemical type and structural position of each surface species would be measured without any degradation from the analytical probe. Some techniques can be used only in a vacuum, while others cause significant structural and chemical changes from heating, electrical charging, and momentum transfer. Each technique has restrictions on the type of sample, which may require special preparation. The scanning electron microscope/microprobe gives morphological information at sub-micrometer resolution from the low-energy secondary electrons, and chemical information at the micrometer scale from the high-energy scattered electrons and x-rays. Transmission electron microscopy gives structural and some chemical information for thinned specimens down to the nanometer scale. Conventional X-ray photoelectron and Auger electron spectroscopies yield quantitative chemical information from X-rays and photoelectrons, but have limited spatial resolution (horizontal, 0.n μm; vertical, 0.0n μm). Ion bombardment provides micrometer horizontal and subnanometer depth resolution of isotopes from mass spectrometer analysis of secondary ions and neutrals, but quantitative understanding of the destructive sputtering and knock-on processes is incomplete. Rutherford back-scattering and nuclear reaction analysis are useful, especially for light elements. Other techniques, involving infrared absorption, nuclear magnetic resonance and electron spin resonance have specialized applications. Rapidly growing are scanning techniques using a pointed probe which rides over surface species as mechanical, electrical and magnetic forces are measured. New techniques that use a brilliant, tunable and polarized X-ray beam from a synchrotron storage ring at near-surface incidence (absorption spectroscopy, fluorescence analysis, and diffraction) should probe the chemical and physical relations between surface species, with single-layer depth profiling as the aim. Information on particular elements, especially at coherent internal surfaces, is attainable from resonant X-ray scattering. The sparse information on feldspar surfaces is supplemented by observations on zeolites, clays and other oxygen-rich materials. Atomic-scale drawings of hydrated/hydroxylated feldspar surfaces are given to focus discussions of chemical and physical phenomena. Dynamic changes of surface chemistry as a feldspar is subjected to wetting/drying and heating/cooling cycles, to organic acids, and to metal-halide and other complexes are considered in preparation for the plenary lecture on weathering. Surfaces might become more stable as aluminol groups are replaced by silanol, as expected from experience with industrial zeolites. Absorption complexes on feldspar surfaces might have been involved in biological evolution. Feldspars from the Moon, other planets and meteorites should have unusual surfaces in response to high vacuum, intense radiation and particle bombardment. Potential chemical concentrations at twin boundaries and other internal surfaces are

I. Parson (ed.), Feldspars and Their Reactions, 541–593.

discussed. Ideas are given for systematic experimental and theoretical study of surface properties of feldspars, with emphasis on choice of natural and synthetic specimens which optimize the advantages of the techniques, and minimize the weaknesses.

1. Introduction

Surface science is a sturdy infant that is just beginning to toddle. Even the refractory metals, alloys and oxides with simple crystal structures offer severe experimental and theoretical challenges. New techniques show that their surfaces, when cleaned in high vacuum, undergo structural perturbation in response to one-sided electron bonding, and may even adopt a new atomic arrangement. At one extreme, the atoms at a bare surface may retain the connectivity of the bulk structure, and the external layer of atoms merely becomes more tightly bonded to the underlying atoms. At the other extreme, the outer atoms adopt a new structural pattern, as exemplified by the 7 x 7 supercell of silicon metal which is epitaxially related to the internal structure. Controlled adsorption of molecules and ionic complexes induces various types of surface perturbations and reconstructions, some of which are becoming understood qualitatively at the atomic and electronic levels. Quantum-chemical and molecular-dynamical theories for atomic clusters embedded in lattices and surface layers are beginning to yield quantitative predictions to be matched against experimental observations. These observations and theories for simple materials provide some help in thinking about the surface chemistry of feldspars, but must be extrapolated with caution.

Complex aluminosilicate materials like feldspars must display a wide range of surface phenomena of which we have only a few glimpses, mostly from measurements of bulk samples. In addition to the complexity of the atomic bonding, a host of potential molecular and ionic complexes can be expected that are related to fundamental geochemical and geophysical processes. How does a dry feldspar surface differ from a wet one? How do organic compounds and humidified complexes from living organisms adsorb on feldspar surfaces? Do transition elements in the feldspar structure change valence state at the surface in response to a variable environment? What happens to a feldspar surface as it wastes away during geological weathering? How does a feldspar surface change in a soil during the yearly agricultural cycle, and how does it interact with the surface of adjacent clay minerals? Have feldspar surfaces changed over geological time in response to biological evolution and coupled changes in the chemical composition of the atmosphere? What is the chemical and physical interaction across the boundary of a feldspar phenocryst suspended in a magma? Obviously there is an infinity of poorly understood phenomena which pose questions for which there is little current understanding. The best that can be done here is to outline the general features to be expected for feldspar surfaces; to list the current experimental and theoretical techniques, and point out expected developments for the future; and to supplement the sparse observations on feldspar surfaces with whatever is relevant for other materials.

Partly because of commercial applications, much more is known about the surface properties of zeolites and clays than for feldspars. Most of the evidence at the atomic and electronic level has been obtained only in the past few years, and is still subject to debate. The publications are scattered over many journals not read routinely by most feldspar investigators. Hence, this review will gather together relevant experimental data and theoretical interpretations of ion exchange and molecular sorption for these microporous and layered materials.

Current macroscopic measurements of feldspars merely hint at the underlying atomic and electronic nature of the surface species, and there is a wonderful world of *atomic surface geochemistry* waiting for the magic wands (please excuse the humorous hyperbole!) of new experimental and theoretical techniques. Just appearing are new experimental techniques capable of probing individual surface features, including individual atoms and molecules. However it is becoming even more important to consider how much a surface is disturbed by the experimental probe. In general, the finer the desired spatial resolution, and the more sophisticated the desired information, the greater the physical and chemical perturbations. Heating, momentum-transfer and electronic charging must be considered. A pretty picture might be misinterpreted, or an apparent diffusion profile might be an experimental artifact.

A reviewer of the first draft urged me to discuss what is a surface. We could be mathematically precise and draw a strict boundary between those atoms that are topologically connected to the normal bulk structure and those which are not. Thus water molecules would be outside the mathematical surface, and could be specified as part of an absorption layer. But what about the outer layer of framework oxygen atoms which may be turned into hydroxyl by addition of a proton? From the viewpoint of connectivity, the hydroxyl might be considered as part of the normal structure, but from the viewpoint of chemical bonding it is distinct. Consider a crystal of anorthite in contact with an aqueous solution of NaCl: ionic exchange between Ca and Na might occur in the outer layer of anorthite. There is excellent evidence that proton exchange occurs in the outer layers of all feldspars in acidic solutions. Consider the bombardment of a feldspar by atoms and molecules from the atmosphere. Atoms of rare gases would have only trivial interaction with the feldspar surface, but polar molecules such as carbon monoxide might form transient activated complexes. Unstable molecules such as ozone might react with the feldspar surface producing an oxygen molecule and an oxygenated complex. Most feldspar surfaces are not flat, and some are thoroughly honeycombed. Defining a mathematical surface across the edges, ledges, etc. of the irregular surface would be challenging. The import of these remarks is that *a surface is best considered as a gradational zone from an interior which is essentially unaffected by the exterior to an exterior which is essentially unlinked to the interior.* Thus the water molecules of a wet feldspar might range from (i) a first layer strongly hydrogen-bonded to framework oxygens/hydroxyls, via (ii) perhaps a dozen layers with decreasing linkage to inward water molecules, to (iii) the dynamic uncontrolled water molecules distant from the feldspar. Ionic and organic complexes would also show a gradient away from the surface. (Freezing the water would reduce or eliminate the temporal disorder, but not eliminate the spatial disorder.)

Consider further the defects inside feldspars. An array of dislocations might be considered as an internal surface with characteristic chemical properties. Atoms of a trace element might congregate in the dislocation array. A twin boundary is associated with a structural change which might prove favorable to a trace element. Two-feldspar intergrowths have internal surfaces at which chemical changes occur. Because dissolution is faster at defects, internal surfaces become encompassed by external ones. Hence this review will consider both internal and external surfaces.

Classical thermodynamics assumes that each phase is infinite and that the internal energy is proportional to the volume. To cover phases of finite volume, a surface energy between two phases is expressed most simply as proportional to surface area. Thus elementary textbooks consider the nucleation barrier to condensation of a homogeneous liquid droplet from a gas. Heterogeneous nucleation is much more challenging because the interface energy depends on the particular atomic and electronic bonding across each part of the interface, and this is directional. Furthermore, spatial and temporal disorder involves many complications requiring application of statistical thermodynamics.

To conclude, six principles enumerated by Hochella (1990) encapsulate the underlying essence of this review:

"1. *Mineral surface compositions are not representative of the bulk and are laterally heterogeneous.*
2. *Mineral surface microtopography is complex.*
3. *The atomic structure of the top few monolayers of a mineral is not representative of the bulk structure.*
4. *The reactivity of a mineral in aqueous solution depends ultimately on its surface composition, microtomography, and atomic structure, all of which can be considered independent factors.*
5. *The reactivity of a mineral surface will generally increase as its atomic and molecular-scale roughness increases.*
6. *Mineral surfaces are not static; they are dynamic systems sensitive to ambient conditions and local reactions."*

This review lecture must be highly speculative because of the paucity of experimental evidence on atomic, ionic and molecular species at feldspar surfaces. Nevertheless it is upbeat because there are so many techniques now available for study of feldspar surfaces, and there are so many ideas for future experiments. Because weathering of feldspars, reviewed by A. Blum, depends on the adsorption and migration of chemical species at the feldspar surface, comments are given here on the experimental techniques applied to artificially-weathered feldspars. Attention is also given to possible experimental tests of a new idea that natural surfaces opf old feldspars become covered by silanol (Si-OH) groups which are more stable than aluminol (Al-OH) ones.

2. Quantitative Chemical and Physical Theories

2.1. ELECTRONIC THEORIES OF BULK CRYSTAL STRUCTURE

Consider first the regular repetitive atomic packing of feldspar. How well can we predict the atomic positions and electron configurations of silicates from first principles? Reviews by Burdett (1983), Hoffman (1988), Sauer (1989), Lasaga (1992), Bleam (1993) and Catlow et al. (1993) summarize the general principles. Because the fundamental quantum-mechanical equation cannot be solved exactly for many-electron systems, various well-defined approximations using mixing of functions for individual electrons are used in the calculations (Hehre et al. 1986). Appropriate features from these calculations are used to control empirical energy functions for geometrical modeling. Some recent papers for structures similar to, but rather simpler than, the feldspar structure are valuable in pointing the way to future models of greater complexity: quartz and cristobalite (Boisen & Gibbs 1993); high-cristobalite (Liu et al. 1993). Extension to surface properties is exemplified by calculations for atomic relaxations at the (0001) surface of corundum (Manassidis et al. 1993), and physi- and chemisorbtion of CO_2 at surface and step sites of MgO (100) (Pacchioni 1993).

Historically, considerable progress was made by considering the crystal structure as a set of atoms or ions on the vertices of a three-dimensional lattice. Some combination of ionic, covalent and van der Waals bonding was used in the calculation of lattice energy. The simple electrostatic ideas embodied in Pauling's rules were extended to algorithms for fast computers. The simplest "bridge-building" programs for prediction of atomic positions for tetrahedral(T)-oxygen (O) frameworks treat the chemical bonds as stiff rods, and adjust the positions to obtain the least-squares deviation from strict regularity. Target values and statistical weights for T-O distances and T-O-T angles are empirical, or based on *ab initio* quantum-mechanical calculations for atomic clusters. Changing the weights crudely adjusts the ratio of ionic to covalent bonding. More complex computer programs minimize lattice energies by various mathematical procedures which change atomic positions to optimal values. For feldspars, the strong interaction of oxygens with the extra-framework cations must be considered, as demonstrated by the large variations in cell dimensions and atomic positions from the relatively regular Rb- and K-feldspars to the irregular twisted structure of Ca-feldspar (Smith 1974, Smith & Brown 1987: hereafter S & SB). Furthermore, the large changes of cell dimensions and atomic positions for a given feldspar as the temperature and pressure are varied testify to the importance of the thermal and external energies. Megaw's (1973) codification of this "working approach" to the stereogeometry and bonding of feldspars and other complex structures still offers an excellent semiquantitative guide. In the same spirit, Brown (1992) has summarized the chemical and steric constraints in inorganic solids using a combined bond-valence and stereogeometrical approach.

Although the Schrödinger equation cannot be solved exactly for many-electron systems, considerable progress has been made by using assumptions about the hybridization and coupling between the different electronic states of the atoms. An electrically-neutral piece of the crystal structure is carefully selected, and calculations are made using computer programs developed for isolated molecules. For example, a cluster of linked silica tetrahedra can be terminated by hydrogen atoms to maintain charge balance (Gibbs 1982). The bonding geometry for a cluster or a ring (Kramer et al. 1991a) can then be used to derive empirical force fields for algorithms used in computer programs for simulation of crystal structures of framework materials (Kramer et al. 1991b for silicas, aluminophosphates and zeolites).

So far, *ab initio* quantum-mechanical techniques have not been applied in detail to the structural effects of temperature and pressure on the feldspar family of structures. Post & Burnham (1987) made structure energy calculations for albites, and Jackson & Gibbs (1988) predicted the pressure dependence of the feldspar and coesite structures using a modified electron gas model. Purton & Catlow (1990) modeled microcline, albite and anorthite with an ionic Born potential coupled with empirical three-body and polarization terms which represent the covalent bonding. Addition of a short-range modified-electron-gas potential for the K/Na/Ca-O bonding was important for simulation of the O-T-O angles.

Finally, the crystal structure must be considered from the viewpoint of statistical thermodynamics. The vibrational energy is covered by the band theory developed by solid-state physicists. The configurational entropy depends on the order-disorder relations between atoms occupying the same crystallographic site. Engel (1991) observed that interruptions in the T-O-T connectivity of a framework by T-OH are related to the electronic charge distribution between T atoms and extra-framework species. This analysis may provide a guide to thoughts on the interruptions at a surface.

2.2. THERMODYNAMIC CONCEPTS

Parks (1990) reviews the complex thermodynamics of external surfaces. Classically, the surface free energy, g, is expressed as dG/dA, where the Gibbs free energy changes with surface area at constant pressure, temperature and composition. The free energy can be split up into enthalpy and entropy terms. A uniform surface on a uniform bulk phase surface can be described mathematically by a fictional plane. Each crystallographically-distinct face of a crystal must have a different surface free energy. Each terrace, edge and vertex require a local change of surface free energy.

The complex, non-uniform *atomic* nature of a surface requires empirical modification of the classical concept of a surface. The overall surface energy can be broken down into several empirical terms expressing different features of the atomic bonding. Thus the adsorption of an ionic complex on a surface has been represented by the constant-capacitance, double-diffuse-layer and triple-layer models. Sverjensky (1993) decided that they are inadequate for quantitative treatment of ion solvation at a mineral-water interface, and proposed that the Gibbs free energy of sorption be decomposed into three terms specifying Coulombic electrostatic forces, a Born-type solvation, and a specific property intrinsic to each metal cation.

Statistical thermodynamics allows consideration of surface disorder. Until experimental techniques yield estimates of the extent of atomic substitutions in lattice-related atomic sites, speculation on the degree of surface disorder entropy is hardly profitable. Most of the atomic movements at surfaces will be vibrational (cf. fluctuations in zeolite apertures modeled by crystal dynamics: Deem et al. 1992) but topological changes should occur. In general, atomic movements will be greater at higher temperature, and will be smaller for a surface between two dense phases than for a surface in which one phase is mobile (liquid or gas).

Internal surfaces, including twin boundaries, have a surface free energy. Liu et al. (1993) provide an example of a first-principles study of the relaxation of atomic positions at a twin boundary, specifically for high cristobalite. This free energy would require an extra term if the twin plane were decorated by atoms not present in the bulk structure. Modulated structures have an associated energy which can be expressed mathematically in terms of an internal crystal surface across which chemical and physical change back and forth (Price & Ross 1992). An external surface layer, just a few unit cells thick at most, might have a chemical/physical gradient with properties related to those of a modulated structure.

2.3. ACID-BASE BEHAVIOR

The two common definitions of acid-base behavior are:*Brønsted acids and bases* are species which can supply or accept a proton; *Lewis acids and bases* are species which can accept or donate an electron pair. Thus a proton from a surface hydroxyl can take part in a Brønsted acid-base reaction, and a transition-metal with two accessible valence states can be involved in a Lewis-type electron transfer.

The strength of the acid or base depends on the nature of the species associated with the proton or electron pair. Theoreticians are developing methods for estimating the acidity of surface species of zeolites in order to understand catalytic behavior. Some representative papers with implications for feldspars are: (i) Sauer et al. (1989) estimated from ab initio calculations for water-Al-Si-OH clusters that the barrier for a proton jump is ~50 kJ/mol; (ii) Cook et al. (1993) incorporated a Madelung term into the Hamiltonian to allow modeling of the relaxation of atomic positions for substitution of Al for Si in a tetrahedral site; (iii) Kramer & van Santen (1993) & Kramer at al. (1993) used ab initio quantum chemistry to estimate a range of 0.8 eV in proton affinity as the Si/Al ratio changed (3.2).

2.4. DEFECTS

The classical treatment of defects involves omitted-, substituted- and interstitial-species in a mathematically-regular bulk structure. Petrov et al. (1989) found three types of O^{1-} defects in gemmy albite (5.5.6). Point defects may congregate to form line, area and volume defects of various geometries, including spirals and tubes. Feldspars etched in Nature or the laboratory show complex morphological features indicative of non-uniform surfaces with local weak spots (5.2). Hence care is needed in the interpretation of measurements of surface properties by bulk techniques. Particularly indicative of the problems in preparing feldspar surfaces for experimental study are the observations of emission of charged particles during indentation fracture of rocks (Enomoto & Hashimoto 1990) and of alkali atoms from fractured Colorado amazonite (Dickinson et al. 1992).

3 General Ideas on Surfaces

Useful general references are Davis & Hayes (1986), Gonis & Stocks (1992), Hochella (1990), and Stumm (1987, 1992).

3.1. GROWTH MORPHOLOGY.AND SURFACE PROPERTIES

The simplest concept of the external surface is that it is a symmetrical polyhedron with planar faces whose areas are determined by the inverse of the growth rate. The next idea is that the combined surface energy is a minimum. This implies that the surface is terminated at the weakest bonds. Hartman (1979) selected the strongest periodic bond chain (PBC) vectors in the bulk structure, and distinguished between F(lat)-, S(tepped)- and K(inked)-faces, respectively with 2, 1 or 0 PBCs. These simple ideas are generally applicable to many simple inorganic materials, but there are complications reviewed in Hochella (1990).

Some materials have a range of morphological habits which cannot be explained just by simple chemical bonding in the isolated bulk crystal structure. Furthermore, growth of crystals generally proceeds by addition of material at steps and ledges, which may be part of spiral dislocations. Adsorption of ionic or molecular species on particular faces is an obvious possibility for changing the rate of growth. Thus Gratz et al. (1993) have characterized the step dynamics and spiral growth of calcite by atomic force microscopy, and Gratz & Hillner (1993) have observed the change of step shape as phosphonates and phosphates become attached, and poison the growth. Other ideas relevant to feldspars include change of surface energy with the varying bulk composition of a parent magma or hydrothermal fluid, and change of crystal energy in response to order-disorder and symmetry change related to pressure and temperature.

Feldspar shows a spectrum of morphological habits spanning the Carlsbad (prominent {010}, *c*-axis elongation) to the Baveno (*a*-axis elongation) to the adularia (near-rhombohedral {110} & {$\bar{1}$01}) types (S, v.2, p.256; SB, p. 516). In general, the habit appears to depend mainly on the growth temperature, and partly on the chemical environment, but detailed explanations at the atomic level are lacking.

3.2. HYDROXYLS, TERMINAL OXYGEN ATOMS AND WATER MOLECULES

Useful general references are: mineral-water interface geochemistry (Hochella & White 1990); hydrogen bonds in crystalline hydrates (Baur 1965); chemical and steric constraints on hydrogen bonds in inorganic solids (Brown 1992); role of OH and H_2O in oxide and oxysalt minerals (Hawthorne 1992).

By analogy with silica minerals, zeolites and clays, it can be assumed that some of the surface O atoms bonded to only one T atom of a feldspar are actually hydroxyl with the embedded proton pointing outwards. Under hard vacuum, two hydroxyls might condense to release a water molecule, leaving a terminal O atom and an oxygen vacancy. Structural relaxation and a new local topochemistry must ensue. The terminal O atom might reach electrostatic balance by bonding to any available nonframework cation(s), and the distance to the T atom would change. Even under humid conditions, it is possible that some terminal O atoms achieve electrostatic balance without attachment to a proton. A water molecule might be bonded directed to a T atom, especially to Al which has been found to have one or two extra ligands in certain tetrahedral frameworks. Water molecules can also be hydrogen bonded to framework O and OH, and also bonded ionically to extraframework cations.

Even in simple oxide structures, the surface OH and H_2O have complex properties. Variable-temperature ^1H nuclear magnetic resonance spectroscopy (5.5.5) of silica gel dried at 298 & 773K showed reduction of the Si-H_2O resonance as the Si-OH one increased (Kinney et al. 1993). The freezing point of the adsorbed water was ~ 40K lower than for normal water. Fourier-transform infrared spectroscopy (5.5.2) of hydroxylated Cr_2O_3 (Kittaka et al. 1993) showed three types of hydroxyl: two bonded to water, and one not bonded. In hydroxylated ZnO (Kittaka et al. 1992), FTIR revealed detachment of physisorbed H_2O from OH below 233K to form condensed clusters. Even more complexities can be expected for hydroxyl & water at feldspar surfaces. Perhaps the 40K depression of freezing of condensed surface water will also apply to feldspar; however, dissolved cations should also be considered.

Infrared absorption energies show that the Si-OH (*silanol*) group is tightly bonded and stronger than the Al-OH (*aluminol*) group. The computer models for H-O-Si clusters (e. g. Lasaga & Gibbs 1990) give useful guidance towards speculation about the stereogeometry, but the details are complex and must vary from one feldspar to another. The computer modeling should be tested against the atomic positions determined for several H-containing aluminosilicate frameworks, including zeolites (Engel 1991). The clearest experimental data are for ussingite, $Na_2AlSi_3O_8OH$ (Rossi et al. 1974), which has an internal silanol [T(4)-O(8) = 1.626(2) Å, O(8)-H = 0.97(5)] hydrogen-bonded [1.54(5)] to another framework oxygen O(2) with O(2)-H-O(8) = 171(3)°. Protons were not located in bavenite (Cannillo et al. 1966), wenkite (Wenk 1973), roggianite (Galli 1980) and parthéite (Engel & Yvon 1984), but framework hydroxyls are implied by interatomic distances and electrostatic charge calculations. In sarcolite (Guiseppetti et al. 1977) and wadalite (Tsukimura et al. 1993) the one-connected framework O atoms attain electrostatic balance by bonding to nonframework cations.

The crystal structure of hydrogen-feldspar (Paulus & Müller 1988) is particularly relevant to the surface layer of acid-treated feldspar. Eifel sanidine was ion-exchanged with NaCl at 1073 K and H_2SO_4 at 593 K to yield a zoned crystal with composition mainly $(K_6Na_4Ba_1H_{89})AlSi_3O_8$. The very short *a* (8.00 Å) and very long *b* result from strong kinking of the crankshaft chains caused by absence of a shared edge between cation polyhedra. The bond lengths and angles present no evidence for the position of the protons. Several types of framework hydroxyls may be present, perhaps associated with Al atoms in a random pattern.

Assume first that the tetrahedral species in a feldspar have the same pattern at the surface as in the bulk. All surface oxygens of pure ordered anorthite and celsian could, in principle, be bonded to one Al and one Si if the surface boundaries were chosen in a particular way. How would charge balance be maintained? Some combination of hydroxyls and Ca ions could be considered subject to overall electrical neutrality. Under wet conditions these species would be interacting with water molecules. Because of the spectroscopic evidence for fast dynamical changes involving the ions and water molecules in the broad pores of zeolites (nano- to milli-second) it is essentially certain that similar changes are occurring at the surface of wet feldspars

The general aspects of the interaction of water with surfaces are reviewed by Thiel and Madey (1987), but the experimental data refer mainly to simple surfaces of metals and oxides. Water acts as a Lewis base to a surface oxygen to which it transfers a fractional charge. Water molecules near a surface should form hydrogen-bonded clusters unless each molecule is separately bonded to a surface cation. A water molecule may react with a surface oxygen to yield two hydroxyls.

Neutron diffraction studies of zeolites at ~ 15K show that a single water molecule in a small cage has a single center-of-vibration for each proton (e. g. thomsonite: Stahl, Kvick & Smith 1990), whereas water molecules in larger cages have fractional occupancy of multiple centers-of vibration (e. g. laumontite: Artioli, Smith & Kvick 1989). Even near absolute zero, the zero-point energy is sufficient for large vibrations of protons, and jumping from one energy minimum to another can be expected. Hence hydrogen-bonded clusters at the surface of feldspar, especially at elevated temperature, might be ephemeral with only statistical occupancy of low-energy sites at best.

Lasaga's (1992) review of *ab initio* methods relevant to mineral surface reactions emphasizes these points: model clusters must be chosen carefully to obtain a reasonable charge distribution; a *terminal* hydroxyl is more reactive than a *bridging* one (Mortier et al. 1984; Sauer 1987); a bridging hydroxyl is more acidic (Pelmenshchikov et al 1987; Sauer 1987) if linked to Al & Si rather than 2 Si; models show "tilted"geometries for water molecules hydrogen-bonded to surface species, and the bond lengths and angles are perturbed in response mainly to the first three neighbors (Saur 1987); 5-coordinated Si may occur; models are proposed for monovalent cations coupled to framework species and water molecules.

Kramer and van Santen (1993) used both ab initio quantum chemical and classical force field calculations for small tetrahedral rings to show that the proton affinity of an Al-Si-connected bridging hydroxyl is reduced by up to 0.8 eV from a high-silica zeolite to a high-alumina one. The exact values for the zeolites depend on the topological features, and faujasite and ZSM-5 zeolites were taken as illustrative examples. Schöder & Sauer (1993) suggested that in high-silica zeolites the small amount of Al should tend to form small -Al-OH-Si-OH-Al- clusters, rather than being dispersed. Such calculations on zeolites are thought-provoking for feldspars, but caution is needed because of the much more open nature of the former.

Clay mineralogists have developed experimental evidence for a theoretical model in which (a) some water molecules fit into the "holes" where they are strongly bonded to the silicate, (b) other water molecules are "associated" with the surface species, and are only weakly bonded thereto, and (c) yet other water molecules constitute an ill-defined multi-layer with decreasing structural order away from the surface that approximates to the structure of ice or liquid water depending on the temperature (e. g. Davis & Hayes 1986: especially Giese & Costanzo 1986). It is very important that distinction is made between the pure Si-O surface of the *siloxane* layers on the outside of the triple sandwich in mica and on one side of the double sandwich of kaolinite, and the mixed Al,Si-O surfaces that occur on feldspar and other framework materials.

A model for the solubility mechanism of water in albite melt (Sukes & Kubicki 1993) is useful in thinking about hydroxyls at the surface of feldspar phenocrysts growing from magmas.

3.3. INORGANIC IONS AND COMPLEXES.

By analogy with zeolites, clays and oxides, it can be assumed that a feldspar surface is in contact with a wide range of inorganic ions and complexes. A cation will be surrounded by various anions and molecules. An *inner-sphere complex* contains a cation which is directly linked only to anions of the bulk structure (such as K^+ in mica), or to some anions of the bulk structure and some adsorbed species: thus a Ca^{2+} cation might be linked to three surface oxygens of a framework and another three or so water molecules, as is common inside large cavities of zeolites. An *outer-sphere complex* consists of a cation surrounded by anions not belonging to the bulk structure: thus an Mg^{2+} or an Al^{3+} cation might be surrounded by six water molecules or hydroxyls, or a mixture of both, some of which would be weakly attached to framework oxygens or hydroxyls, or both. In general, a cation will be surrounded by 4 to 8 anions and molecules depending on the radius ratio, with many coordinations close to octahedral (e. g. hydrated ions: review by Ohtaki & Radnai 1993). The ionic potential of the cation and the electrical charge distribution across a surface are important determinants of the stereogeometry. Examples are given by Sposito (1984, 1990) and Stumm (1992).

Adsorption of metal ions and complexes on aluminosilicate minerals depends not only on the specific structural nature of the surface, but also on the physicochemical nature of the metal-carrying fluid (Goodman 1986). Some ligands weaken the bonding of metal ions to surface hydroxyls (e. g. of Cu^{2+} on kaolinite), but Cu^{2+} adsorption on allophane is enhanced by phosphate anions, perhaps because of formation of a stable ternary complex. Metal complexes may be polymeric, as in the iron- and aluminum-oxygen-hydrogen clusters found on the surface of clays, or intercalated in the pillared clays. A review of the geochemistry of clay-pore fluid interactions (Manning et al. 1993) discusses many factors relevant to feldspars.

The structural features of interlayer water in Ni- & Na-exchanged vermiculites (Skipper, Soper & McConnell 1991) are relevant to feldspar surfaces. At high humidity, Ni^{2+} lies midway between silicate layers, where it is octahedrally coordinated to six H_2O, each hydrogen--bonded to a silicate-oxygen. In contrast, Na^{1+} appears to be directly bonded to silicate-oxygen, with a complex arrangement of Na & H_2O. At low humidity, the silicate layers approach as water departs, and both protons of each H_2O are bonded to silicate-oxygens of the same layer.

Ion exchange in clay minerals and zeolites is highly selective for one structure type, and for one ion among the various structure types (Maes & Cremers 1986). Interaction with silanol and aluminol groups is important. Based on the hydroxylated models for feldspar, it is likely that ion exchange at the faces depends greatly on the bulk composition and the crystallographic orientation. Because feldspars may be in contact with ions different from those in its bulk structure (e. g. an albite soaked in a K-rich brine), consideration must be given to a dynamic ionic exchange in the outer unit cells.

Aluminum is very important because of the small energy differences between various anionic complexes. Four- and six-coordinated complexes are common, and five-coordination occurs in several crystal structures, and may be important in complexes undergoing a transition.

A wide range of inorganic anions and molecules must be interacting with natural feldspars. All feldspars on the seafloor and in saline lakes are exposed to halogens, some of which will be complexed with metals. Feldspars exposed to air must be bombarded by many types of molecules. Neutral ones will merely bounce off, but those with electrical dipoles (e. g. CO) will form activated complexes with a brief lifetime. Even normal rain carries carbonic acid and nitrogenous species, while acidic rain contaminated by industrial and other emissions from human activities carries a cocktail of chemicals.

Fritz & Mohr (1984) observed plagioclase partly coated with calcite in a spheroidally-weathered boulder. Calcite did not coat a limonite veneer on the plagioclase. Al-rich amorphous material occurred on limonite-free regions. These observations indicate that Fe- and Al-bearing complexes are mutually exclusive on feldspar surfaces, and that CO_2-associated dissolution of plagioclase is blocked by Fe-complexes. Theoretical modeling of these phenomena could take advantage of calculations for physi- & chemisorbed CO_2 on the MgO (100) surface (Pacchioni 1993), and a surface complexation model for carbonate/aqueous solution (Van Capellen at al. 1993).

Absorption of light can drive redox changes (Waite 1986; Hering & Stumm 1990). The literature concentrates on oxide-hydroxides of iron, titanium and manganese. Electronic energy transfer can occur for adsorbed inorganic and organic compounds. Redox changes of Fe and other transition metals may occur directly in the feldspar structure, and also in attached non-feldspar materials such as aventurine inclusions and external metal-rich species such as desert varnish. Junta et al. (1992) found by SEM, AES and XPS that aqueous Mn(II) is precipitated on rough fractures of the {010} cleavage of albite at pH greater than 7.8. The precipitate contained Mn(III), and possibly MnOOH.

3.4. ORGANIC IONS AND COMPLEXES.

Organic cations will interact ionically with a charged aluminosilicate surface, and will compete with inorganic cations. As demonstrated for the internal surfaces of zeolites, bulky organic cations can fit neatly into certain cages (e. g. tetramethylammonium ion in a truncated octahedron of linked tetrahedra - the cage in sodalite). Whether such steric-specific adsorption can occur at the outer surface of a feldspar requires detailed investigation, but might occur at the near-hexagonal rings of tetrahedra expected at the surfaces of homogeneous feldspars, and in the grooves of weathered nano-intergrowths. Co-adsorption of metal ions and organic ligands can yield ternary surface complexes (Schindler 1990).

Under completely dry conditions, saturated organic groups would be attached to an aluminosilicate surface only by the induced dipoles responsible for weak van der Waals forces. Saturated organic compounds are hydrophobic when in contact only with water, but can be adsorbed on an aluminosilicate surface if the hydrogen bonding at the surface is weaker than in water (Curtis et al. 1986). The higher the Si/Al ratio (i. e. the lower the polarity of the surface), the greater is the preference for adsorption of a hydrophobic organic compound, as is evidenced by the silicalite/ZSM-5 series of molecular sieves.

Application of these ideas to the rates of chemical processes in soils (textbook: Sposito 1989) is summarized in Huang and Schnitzer (1986) and Sparks and Suarez (1991). The following organic mixtures and compounds are important in the formation of complexes with soil minerals ((McKeague et al. 1986; Tan 1986):

- *humic acid* - mixture of organic substances precipitated by acid treatment of an alkali extract of soil - has abundant benzenecarboxylic groups; *fulvic acid* - the dissolved residue - has abundant phenols, and is generally rich in COOH and other oxygen-containing functional groups. Both are charged polymers which may bond to metals. *Lignin* is persistent but slowly degrades to humic acid.
- *low-molecular-weight acids and phenols.* - includes formic, acetic, propionic, butyric, oxalic, citric and 2-ketogluconic acids; long-chain fatty acids; phenolic acids including *p*-hydroxybenzoic, vanillic, syringic and ferulic acids.

• *carbohydrates* - including monosaccharides, oligosaccharides and sugars (predominantly glucose).

• *proteins and related compounds* - including amino acids (generally aspartic > glycine > arginine > glutamic > alanine > lysine) which form complexes with Al and other metals.

The relative amounts of these organics vary from one soil type to another, and change greatly during the growing season of plants and trees. Generally, the carbohydrates, proteins and related compounds are rapidly consumed by microorganisms. Lichens produce acids: oxalic and other organic acids form complexes with Al and other metal ions (Welch & Ullman 1993), and calcium oxalate is a common precipitate on mineral surfaces. Silica microspheres indicate dissolution/absorption/excretion by *Acarospora hospitans* (Robert & Berthelin 1986). Bacteria also have specific roles: *Thiobacillus ferrooxidans* occurs on weathered microcline; *Pseudomonas* mobilizes alumina and *Bacillus siliceus* is believed to dissolve silicates. Bacterial colonies are associated with etching of feldspars in a petroleum-contaminated sand/gravel aquifer (Hiebert & Bennett 1992). Fungi produce oxalic and citric acids, whereas bacteria generate acetic and formic acids.

Very little is known about the specific atomic processes involved in attachment of organic compounds to mineral surfaces. Cambier & Sposito (1991) showed that at pH > 7 interlayer Al from synthetic hydroxy-Al montmorillonite formed an external amorphous complex with citric acid.

Organic compounds may polymerize in the presence of metal ions: thus glycine is polymerized on bentonite to pentamers and other oligomers, with Cu^{2+} more effective than Ni^{2+} and Zn^{2+}, and much more so than Na^{+}. Similarly the adsorption of ATP and ADP is higher for Mg- and Zn-montmorillonite than for Na-. In general, the abiotic reactions of hydrolysis, elimination, substitution, redox change and polymerization can be affected by interaction of adsorbed organic species with Lewis and Brønsted sites at a mineral surface (Voudrias & Reinhard 1986). Feldspars and quartz are less effective than other primary minerals in promoting abiotic polymerization of hydroquinone (Wang et al. 1986). From the environmental viewpoint, the adsorption, oxidation and hydrolysis of organic pollutants is particularly germane, but there is no current indication that feldspars would be commercially effective.

3.5. ULTRA-HIGH VACUUM.

Complete removal of all adsorbed species in an ultra-high vacuum may be just a theoretical dream for all terrestrial feldspars. One might imagine pumping off the weakly-adsorbed water molecules, and then any water generated by reaction of two hydroxyls to leave just one framework oxygen. The zeolite literature contains simple drawings in which the aluminosilicate surface adjusts to one-sided bonding: termination by SiO_4 and AlO_3 is one option. None of them are more than conceptual drawings that must be upgraded into 3D to become useful to a crystallographer. It would be valuable to conduct a theoretical exercise in which plausible atomic structures are generated that pay attention to the molecular orbitals and ionic forces. Major topological changes should be considered as well as minor perturbations.

In practice, secondary ion mass spectroscopy and other surface-sensitive techniques demonstrate that terrestrial feldspars contain a cocktail of impurities at the outer surface and internal discontinuities. The SIMS mass spectrum of a sputtered feldspar shows that some water is driven off easily by the primary ion beam, while carbonaceous species are tenacious. Although some water is driven off easily, the persistence of mass peaks related to water indicates that water migrates towards the sputtered surface directly under the the primary ion beam.

Soil from the lunar regolith obtained during the Apollo missions surprised us because it did not clump together like terrestrial powders. Perhaps the surfaces were equilibrated on the Moon in an environment essentially free of hydrogen (though there is unprotected bombardment by the solar wind). Or perhaps the surfaces are coated with vapor deposits ultimately resulting from flash heating at meteorite impacts on the Moon. Study in an ultrahigh vacuum might yield fascinating results.

3.6. HIGH- AND LOW-TEMPERATURE; HIGH PRESSURE.

Most thinking about natural surfaces has been in relation to soils and weathering at the Earth's surface. In contrast, feldspars in volcanic rocks have surfaces that were developing at high temperature, generally in contact with a magma containing reactive species very different from those at the Earth's surface. Hydroxyls, but not free water, can be expected to be present in wet magmas, and possibly carbonate species in carbonatites. Some unusual phenomena may occur at the feldspar surfaces in magma chambers prior to exposure at the Earth's surface to aqueous solutions and atmospheric species. Chapter 17 of SB describes many phenomena pertinent to feldspar crystallization in magmas. Transfer of transition metals from magma to growing feldspar is of particular interest because of the local perturbation of atomic coordination in the feldspar structure, and the competition with coexisting ferromagnesian and oxide minerals.

High pressure should change the surface chemistry of a feldspar. Jackson and Gibbs (1988) modeled the effect of pressure on the bulk structure of silicas with the coesite and feldspar topologies using modified electron gas calculations. A covalent contribution should be added, and consideration given to possible structural changes at the surface. Whatever the details of the resulting models, a key feature will be the enforced reduction of T-O-T angles as the near-rigid tetrahedra swivel to generate a smaller volume. Any water molecules at the surface should be more compressible than the bulk feldspar, and might adopt an ordered or disordered topology related to those in the high-pressure polymorphs of ice. Particularly interesting is the effect of a transient shock wave on a feldspar in contact with another mineral. Absorption of energy will be greater at any surface because of the mismatch of impedance, and localized melting will occur before the bulk feldspar melts.

4. Atomic Packing at Feldspar Surfaces

Lack of time made it impossible to prepare a comprehensive set of 3D graphics on possible atomic packing at feldspar surfaces. The following 2D drawings and concise text merely hint at the wealth of future developments. A detailed paper with 3D graphics of the possible atomic positions at feldspar surfaces is in preparation. The possible positions for OH, H_2O, K, Na and Ca are being considered, as are the distortions of bond angles and distances. Further papers will consider the stereogeometry of inorganic and organic adsorbates.

4.1. HYDROGEN-FREE TERMINATION OF REGULAR STRUCTURE

Cleavage of a pure homogeneous feldspar should instantaneously produce a hydrogen-free surface. Where would the cleavage surface lie in the crystal structure, and what would be the distribution of cations?

Consider mica. Assume that the cleavage surface develops at the outer oxygen atoms of one side of the double-silicate layer. Then, what happens to the simple hexagonal array of K cations in muscovite and phlogopite, and of Ca cations in margarite? For simple electric neutrality, one-half of these cations should go to each of the new surfaces. Each cation has an intrinsic equality of choice, but do the cations flip left or right like a zipper to yield a regular pattern of occupied and vacant sites (still hexagonal, but with double-area cell)? In principle, this could be tested by surface diffraction (5.4). For the simple ionic model, each oxygen adjacent to a cation would receive a double charge contribution, whereas an oxygen next to a vacancy would be underbonded. This electrical misbalance is probably insufficient to drive a topologic change of tetrahedral bonding, and readjustment of bond distances and angles in the tetrahedral groups should occur to balance the local electrical potential. For the quantum-mechanical models, readjustment of the atomic orbitals would ensue, especially for the terminal oxygens and the one-sided K and Ca cations. [Emission of alkali species from fractured minerals and rocks (2.4) may imply that this simple model is inadequate. However, most minerals contain inclusions of volatile-rich materials which would be released upon fracture, and the alkali atoms of "clean" fractures might remain attached to the mineral structure.]

For feldspar, the structural details are more complex than for mica, but similar effects might be expected. Fig. 1-upper is a plan of oxygen (O) atoms that might terminate the (010) face of a feldspar. The parallelogram shows the *ac* cross-section of the unit cell of albite. [Fig. 1-upper is reversed with respect to Fig. 4.2 of SB.] The linked squares show the bifurcated chains of T atoms (hidden by the O atoms) along *c*. Each medium grey circle represents a 1-connected O atom projecting upwards from its T neighbor. Below it, a triangle of three 2-connected O atoms completes the tetrahedral coordination. For clarity, isolated tetrahedra are shown at the left and right with medium and light grey colors. Pairs of T atoms, each attached to a 1-connected O, lie across a diagonal of each 4-ring. Each of the remaining pair of T atoms in the 4-ring lies in a tetrahedron of three 2-connected O (one ring is shown with black and dark grey prototypes) and one 2-connected O projecting downwards (almost hidden). Whatever the chemical type of feldspar, the 1- and 2-connected O atoms define a distorted hexagonal lattice with a bowl-shaped depression inside each 6-ring.

Fig. 2 shows the location of Si and Al in the ordered varieties of anorthite and albite/microcline. [The albite geometry is used in all figures.] In anorthite, the strict alternation of Al (circled star & under the bracelet) and Si (small filled circle & under the large filled circle) generates two types of projecting 1-connected O. The hexagonal pattern of 1-connected O yields two zigzags of Si-O and Al-O. In albite/microcline, the single Al of each 4-ring can be chosen so that all projecting O are bonded to an Si. If the (010) surface is chosen across the mirror plane of the ideal space group C2/m (corresponding to a twin operation), the projecting O are linked alternately to Al and Si.

An atomic-force-microscope (5.2.2) image of the (010) face of Amelia albite in the presence of water (Drake & Hellmann 1991) is shown in Fig. 1-lower, rotated to give a match with the 1-connected O. A more detailed structural drawing of atomic positions is given by Drake & Hellmann, but the essential geometric features are given in Fig. 1. *The good fit between the bumps on the AFM image and the positions of the 1-connected O gives some confidence that the essential topology of the feldspar structure is maintained at the surface*. However, under H-free conditions one must consider the possibility of surface reconstruction.

Fig. 3-lower illustrates the possible oxygen positions on the (001) surface, for which no AFM image has been published. The projecting 1-connected O (medium grey) again yield a distorted hexagonal lattice. Each medium grey O forms a tilted tetrahedron with the three adjacent 2-

connected O (light grey). Two light grey O and one dark grey O plus a 2-connected O (almost hidden) generate a tilted tetrahedron pointing downwards. Fig. 3-upper shows that the 1-connected O form two zigzags of subtypes connected to either Si or Al. Fig. 4 shows the two mirror-related choices for placement of 6 Si and 2 Al in each 8-ring. The occupancy of the T1(O) site by Al enforces a regular alternation of Si and Al around the distorted hexagonal pattern of 1-connected O in Fig. 4-upper. The alternate choice (Fig. 4-lower) has all the 1-connected O linked to Si.

Possible positions for the K, Na and Ca atoms under H-free conditions are under consideration. Please refer to Fig. 4 of SB for the cation positions in the bulk structure. Just as for mica, these cations must be shared between the pair of new surfaces. Hence, from each pair at approx. $z = 0.85$ & 1.15, one would go to each (001) surface. Because one-sided bonding is unfavorable, the cation position which is submerged would remain occupied, and the newly exposed one would become vacant. The removal of the O(A1)-O(A1') shared edge must result in substantial relaxation of atomic positions, and other factors will undoubtedly turn up.

4.2 HYDROGENATED SURFACE

Consider the simplest model in which each 1-connected oxygen is converted into a hydroxyl. Figures 2-4 shows how these hydroxyls might be attached to surface oxygens to yield aluminol (Al-OH) and silanol (Si-OH) groups. To a first approximation, one can ignore the non-framework atoms, and imagine that each proton is approximately 0.1 nm from its framework oxygen pointing approximately perpendicular to the surface. As reviewed by Brown (1992), the proton of a hydrogen bond in silicates is usually about 0.095 nm from one oxygen and 0.18 nm from a second one (see also 3.2). Hence a "bare" hydroxyl projecting normal to the feldspar surface would be unusual. Under ambient conditions, the aluminol and silanol groups would be hydrogen-bonded to water molecules at one side of a film several molecules thick. Presumably the AFM image in Fig. 1-lower results from interaction of the probe with the strongly anchored Al/Si-OH, essentially unaffected by the weak hydrogen-bonding of the water molecules.

This simple model of an ordered surface with projecting hydroxyls is useful for preliminary discussion. Because aluminol is less stable under most conditions than silanol, those feldspar surfaces with the least number of aluminols should be disfavored. All the growth faces of feldspar, and especially the edges between faces, are being modeled to determine the relative potential density of aluminol and silanol. Twin boundaries in $AlSi_3$-feldspars that intersect a surface may enforce a change in the aluminol density.

Figure 1. (next page) Comparison of a structure model of oxygen atoms at (010) face of albite (upper) with atomic-force-microscope image (lower: Drake & Hellmann 1991). The upper diagram shows the outline of the unit cell projected down - *b*, with *a* horizontal. The vertices of each square reveal the positions of T atoms, and the squares are linked into bifurcated chains along *c*. Each surface oxygen atom is shown by a grey-scale coded circle: *medium grey*, connected to one T atom, and pointing upwards; *light grey*, triplet of O which complete the tetrahedron with the medium grey O; *dark grey and black*, triplets of O in tetrahedra with the fourth vertex pointing downwards - note that two out of each triplet could also be coded light grey; *white*, not specified. The lower figure is an AFM image of Amelia albite rotated so that its bumps match the light grey 1-connected O of the upper model. With kind permission of American Mineralogist and Roland Hellmann.

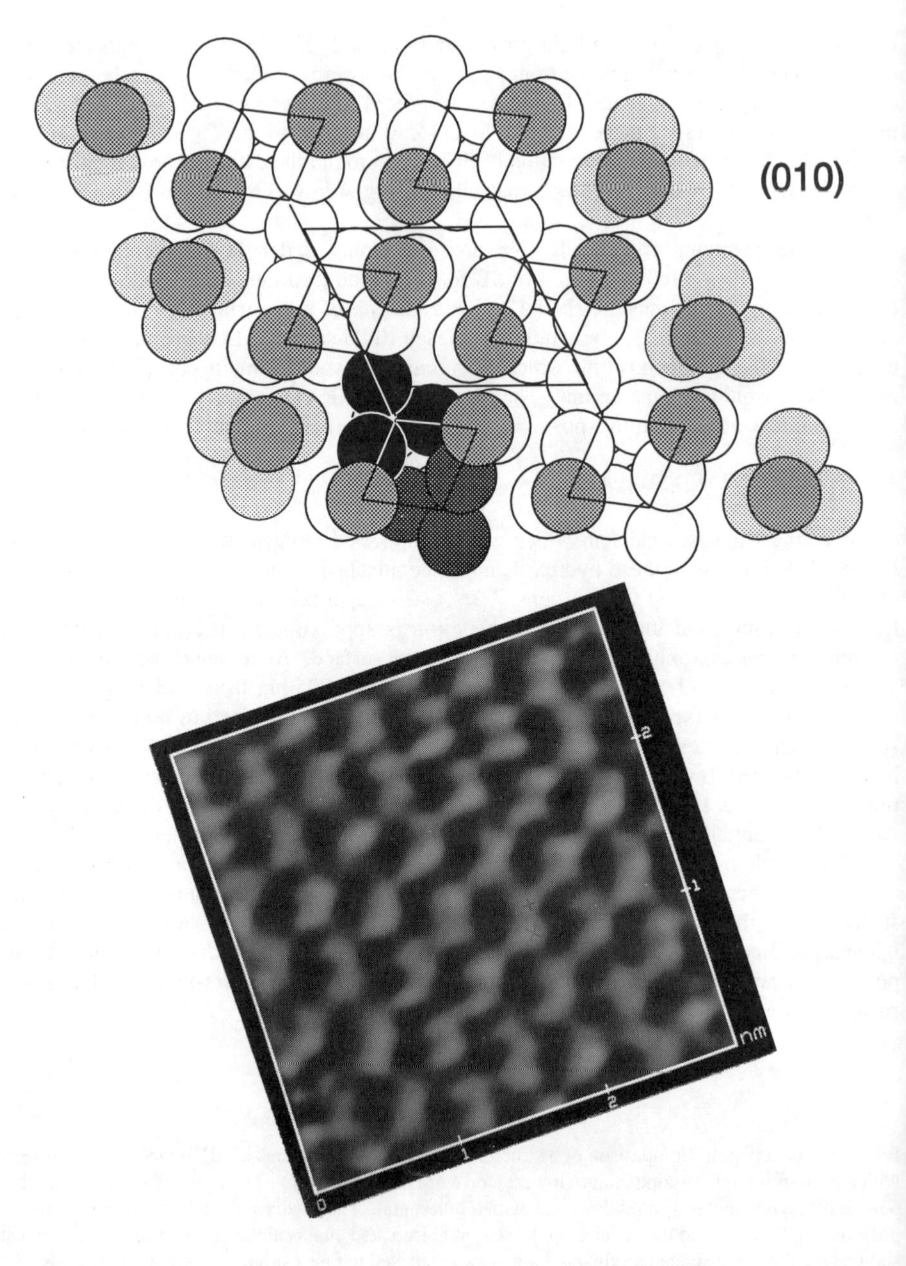

Figure 1. (Caption on previous page)

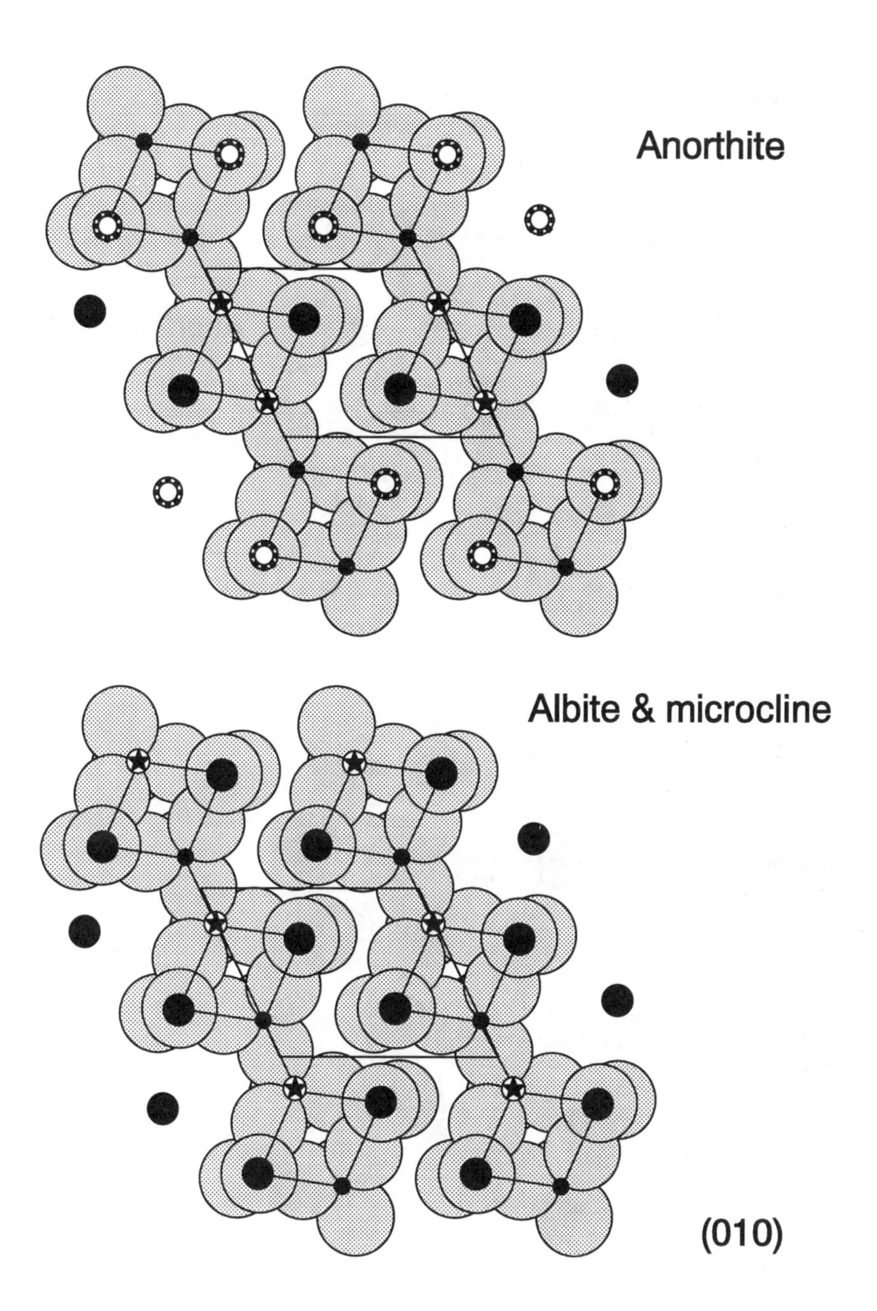

Figure 2. Structure models for the (010) surface of anorthite (upper) and albite/microcline (lower). The O positions match those in Fig. 1-upper. Al & Si in tetrahedra with the fourth O pointing downwards are marked respectively by a circled star and a small filled circle. Al & Si in tetrahedra with an upward-pointing 1-connected O are marked respectively with a bracelet and a large filled circle.

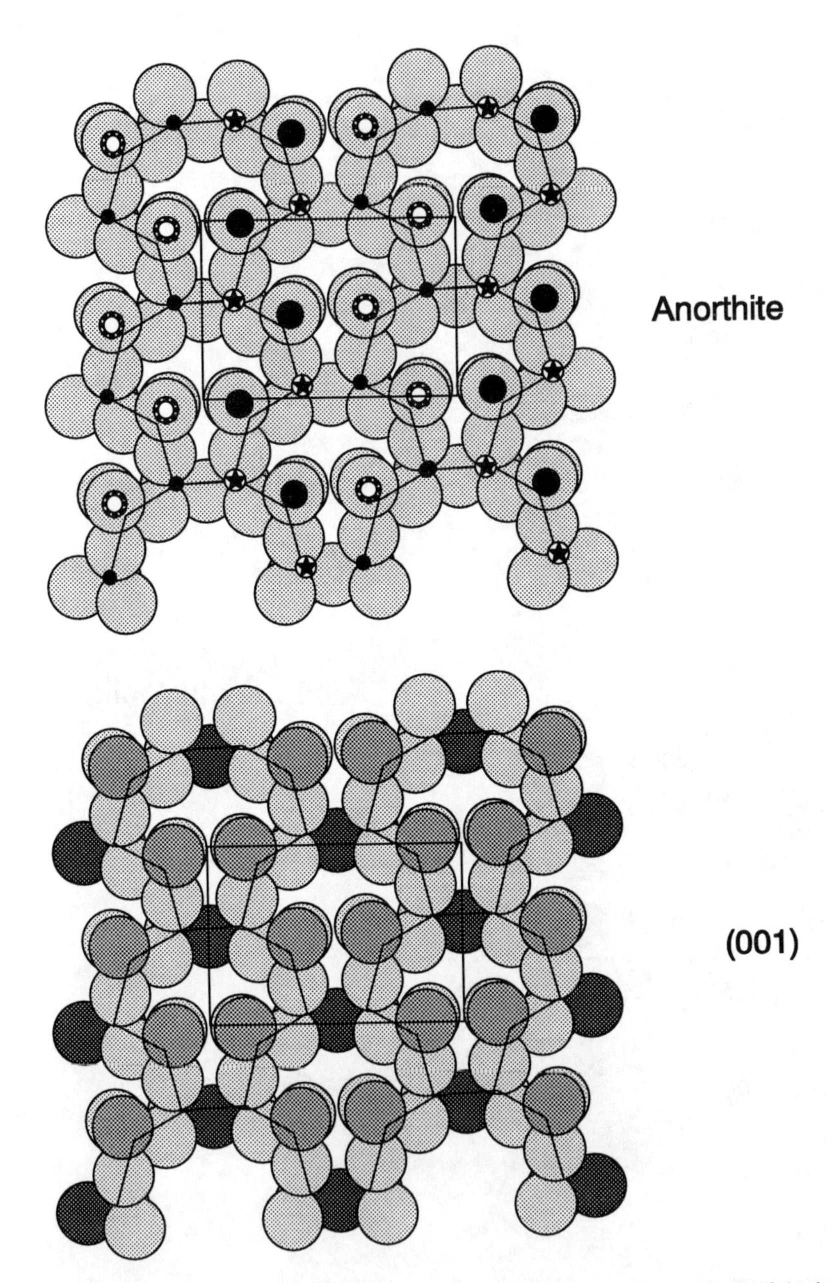

Figure 3. Structure models for the (001) surface of anorthite. Projection down *c* with *b* horizontal. The "mushrooms" show 8-rings of T atoms. Each surface oxygen is shown by a grey-scale coded circle in the lower diagram: *medium grey*, connected to one T atom, and pointing upwards; *light grey*, triplet of O which complete the tetrahedron with the medium grey O; *dark grey*, completes a triplet of O with two light grey O of tetrahedron with fourth O pointing downwards. The upper diagram has the same coding as Fig. 2.

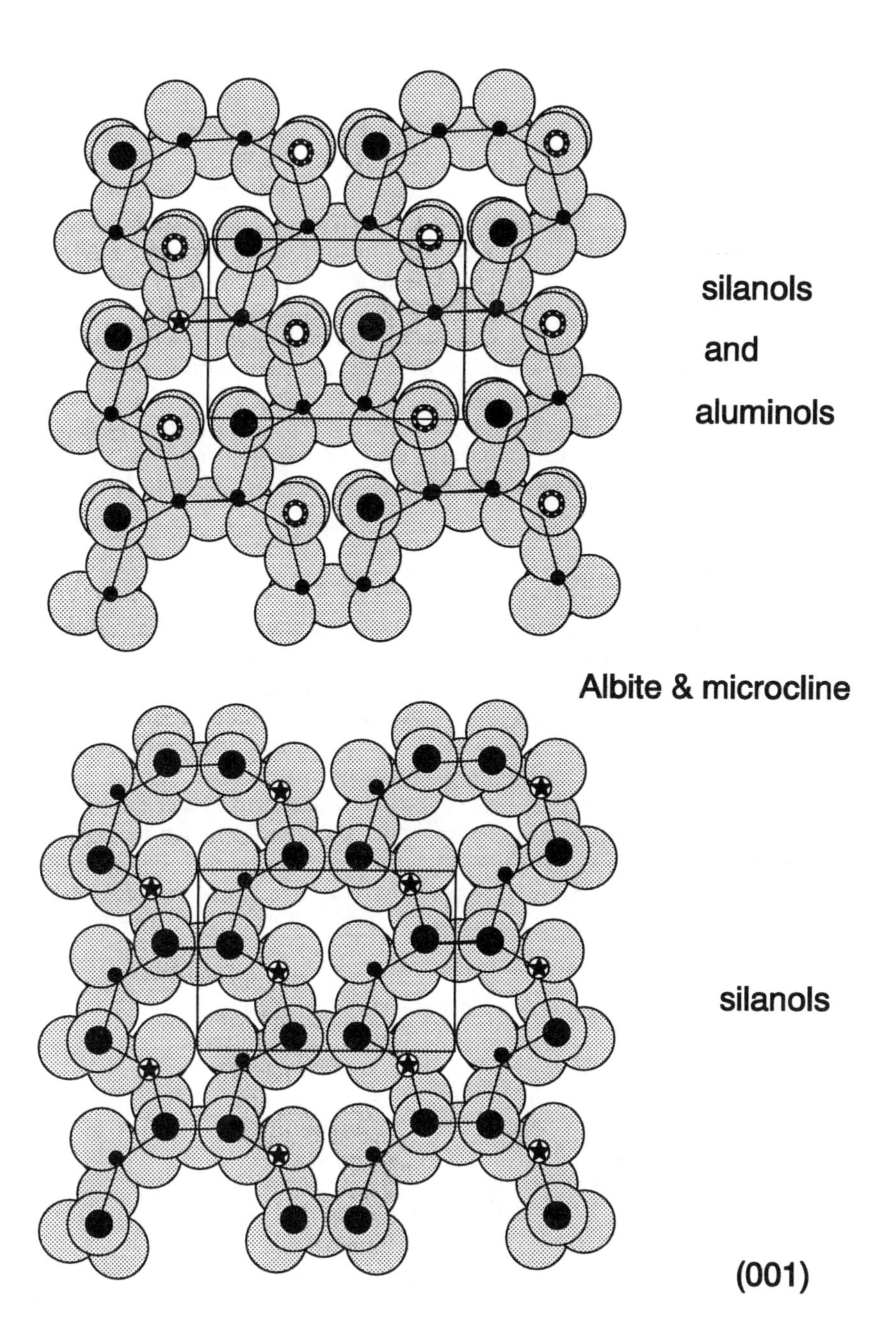

Figure 4. Structure models for the (001) surface of albite/microcline. Two choices for the Al in site T1(0) are depicted, one giving only silanols (lower) and one giving both aluminols and silanols (upper).

4.3 INORGANIC ADSORBATES

The shallow hexagonal dimples on the models in Figs. 1-4 are potential sites for shape-selective adsorption of inorganic complexes. The 1-connected O on (010) are approximately 4 to 5 Å apart, and those on (001) are less regular at 3 to 6 Å. Details must vary greatly from the relaxed Rb- & K-feldspars to the twisted Ca-feldspars (SB, Chapter 4). Simple adsorbates, such as mono-metal ionic complexes are narrower than the dimple diameter. Under strictly H-free conditions, such complexes might nestle into one side of a dimple. Under hydrous conditions, a mono-metal hydrate/hydroxyl/oxide complex might also settle into a dimple, but structure direction might be weak because of the likelihood of a dynamic disorder with the film of water. Poly-metal complexes would be larger, and might have special electronic features which would favor electrostatic interaction with the dangling atoms of the feldspar. Al and various transition metals are likely candidates for such lock-and-key complexes, and indeed there is evidence (3.3) for their presence on natural feldspars. Guidance for modeling the adsorption of inorganic complexes can be obtained from the extensive literature on zeolites, clays and other materials used industrially for catalysis.

4.4 ORGANIC ADSORBATES

What has already been written for inorganic adsorbates is qualitatively pertinent to organic ones. The extensive literature on the shape-selective absorption of organic compounds in zeolites, clays and other microporous materials is particularly relevant. Oxygen atoms of aluminosilicates interact with organic complexes via single or bifurcated hydrogen bonds to unsatisfied functional groups (e. g. organo-ammonium ion), and by weak van der Waals induced interactions with satisfied groups (e. g. methyl). I intend to prepare structural models for some of the organic species listed in 3.4, using the published structural data on microporous materials as a guide. Note that the dimples between the 1-connected O on the feldspar surface are wide enough for stereogeometrical interaction with benzene and derivatives. Further note that almost all structural data for organic species encapsulated in micropores indicate high spatial disorder, and that a similar "sloppiness" is likely at a feldspar surface.

5. Analytical Techniques and Results on Feldspars, Clays, Zeolites, etc.

5.1. INTRODUCTION

Because the surface chemistry is only meaningful in the context of the physical properties, it is necessary to consider a wide gamut of techniques (Woodruff & Delchar 1986; Hawthorne 1988; Brown 1990; Fuoss & Brennan 1990; Hochella 1990; Lagally 1990; Perry, 1990; Skoog 1987; Turner 1990; Walls 1990; Török & Van Grieken 1992; Yates 1992; Howe 1993; McGuire et al. 1993; Roberts 1993). An ideal group of techniques would essentially identify the atoms, ions and molecules in sequence from the outermost adsorbed layer inwards to the bulk feldspar structure, and characterize the chemical bonding between each set of adjacent species. This goal has been achieved for some simple metals and alloys that have been carefully prepared under ultra-clean conditions at very high vacuum. Atoms at the surface have significantly different bonding geometry to those in the interior, and may occur in a different topology (e. g. the 7x7 superstructure of atoms at a clean Si metal surface). Feldspars have not been studied under such extreme controlled conditions. Furthermore, all natural feldspar surfaces are exposed to a wide range of ionic and molecular species which must be involved in complex dynamic arrangements.

To illustrate the problems faced in surface analysis, consider what happens to a feldspar as it is prepared for electron microprobe analysis (Bonsfield 1992). It is sliced with an abrasive-bonded saw, crudely leveled with coarse grit, polished with μm diamond powder, cleaned in an organic solvent, outgassed in a fairly crude vacuum, and coated with a sputtered layer of electrically-conducting carbon. Reflected light from a petrographic microscope commonly reveals traces of a few coarse grooves, various pits of unclear origin, and spots with different reflectivity from mineral inclusions. In the electron microprobe, there is abundant evidence of physical and chemical changes at the region under the electron beam. The impact area can commonly be seen as a white spot under reflected light as a result of some unknown change in the carbon film. In cathodoluminescence, the impacted area can be identified by a change of color indicative of electronic modifications. Finally, as the beam current density is increased, the x-ray yield from the alkali atoms progressively decreases. Some combination of ionization and thermal effects is presumably responsible for the loss of alkali (van der Pluijm et al. 1988). Hence it is absolutely certain that we must consider how the specimen preparation and the analytical probe affect the outcome of the analysis.

In general, low-energy electromagnetic radiation does less damage than electrons and ions (Smith & Walls 1990). However, the latter have special properties that cannot be ignored. Furthermore, we should consider how to choose feldspar specimens that are least damaged by the analytical technique, and still answer important general questions. Most surface-sensitive techniques work best for atomically flat surfaces. Natural surfaces are generally very irregular, at least in part because of dissolution of reactive regions. Even a fresh cleavage is generally uneven and littered with fragments. Furthermore, charged species (probably alkali ions) are emitted during fracturing (Enomoto & Hashimoto 1990). Perhaps we should consider exotic ways of scraping surfaces with a diamond blade, or blasting them with a powder from the same feldspar. A review of the chemical aspects of chemical tribology (Fischer & Mullins 1992) is pertinent. Specimen preparation should be carried out in an environmental chamber. All feldspars exposed to aqueous solutions have textured surfaces, and it is important to determine whether the roughness affects the emitted signal. Thus, emitted low-energy electrons and x-rays are strongly absorbed, and the overall signal at the detector may favor hilltops over valleys. Prolonged low-temperature annealing in high vacuum may be desirable to clean out the junk from cavities. Perhaps lunar feldspar or gemmy sanidine from mantle eclogites (Rossman & Smyth 1990) might yield more uniform surfaces than feldspar from water-rich environments.

Terrestrial feldspars are undoubtedly terminated in part by hydroxyls interacting with water molecules. Ionic species and molecules including CO_2, NO_x and organic acids may well be present in significant amounts. Prior to analysis, it would be prudent to characterize the mobile species that are interacting with the feldspar. For example, a feldspar from a desert may be covered with varnish, and its surface may be quite different under the blast of the midday sun than after an evening shower, especially in a region with urban smog. A feldspar may be covered by lichens or bacteria; it may be in contact with soil enriched in lead and other toxic metals; and so on. Careful sample selection, collection, storage and preparation takes time and money, but will be the key to reliable advances in the chemical analysis of feldspar surfaces.

We can learn a lot from studies that have been made on other hydroxyl-rich surfaces, especially zeolites and clays for which industrial activity has driven a host of technical achievements.Please read Hochella's (1990) review of atomic structure, microtomography, composition, and reactivity of mineral surfaces. I am concentrating only on those features of particular relevance to thoughts on feldspar surface chemistry, and on recent references.

5.2. MICROSCOPY

5.2.1 *General.* In order to understand the chemistry of a feldspar surface it is important to characterize its morphology by microscopy. A flood of new techniques must be considered here.

Classical light optics is limited by the numerical aperture of the objective, and by the strict limits on interferometry. Chapter 17 of SB gives many examples of important textures revealed by light optics, including a picture of growth steps on a hydrothermal synthetic sanidine (p. 517). Some technical advances are occurring, but the spatial resolution will remain limited.

5.2.2. *Atomic force microscopy.* The concept is absurdly simple (Friedbacher et al. 1991; Rugar & Hansma 1990). Run a fine tip across a surface, like a needle over a gramophone/phonograph record, and measure the displacement and force. **AFM** is certainly very useful for imaging specimens at the nanometer level under ambient conditions. The challenge is to go to the atomic level and relate the force profile to chemical identification of individual atoms. Improvements are being sought: reducing the tip diameter from 30 to 10 nm; adjusting the load so that the tip "rolls" over the surface without gouging it or jumping off; varying the relative humidity for runs in air; running in different types of fluid including water, ethanol and ionic solutions. A review of AFM and related techniques based on electronic and other forces (Snyder & White 1992) lists 590 references. Studies of calcite, anhydrite and zeolites are reviewed here to get a "feel" for the achievements of AFM and the challenges facing its practitioners.

Several published images of feldspars and other minerals run in air show terraces at the unit-cell scale, and one image has bumps that match with expected positions of oxygen atoms in the known bulk structure.

Hochella et al. (1990) cleaved Amelia albite in air, and studied the (010) surface by AFM and low-energy electron diffraction (**LEED**). The well-defined spots in the LEED pattern demonstrate that the outer few unit cells have the usual feldspar topology. Still puzzling, however, is the apparent distortion of the lattice (7.22 x 6.51 Å x 60° compared with 7.28 x 6.39Å x 63.5° for the bulk structure). Perhaps the absence of the C-centering and the geometrical deviation might be explained by the the physics of a low-n diffraction event. The AFM images displayed pits and depressions. Mounds ~30 nm high disappeared after immersion in deionized water and drying in air. Atomic resolution images were not obtained on the flat parts.

Drake and Hellmann (1991) obtained images in air for random sites on a single cleavage flake of Amelia albite 3 months apart. The published images were processed with a fast-Fourier transform to reduce the noise while retaining the peaks. Fig. 1-lower shows that the blobs tend to fall near the nodes of "six-sided, en-echelon rings". There was a match with the oxygen atoms of a drawing of a 5 Å thick slice of the bulk structure, and this is demonstrated in a slightly different way in Figs. 1 and 2. One might imagine that the humidified tip interacts weakly with the anchored hydroxyls while several layers of water molecules act as a lubricant. However, this is pure speculation because the blobs have no chemical identification. Hellmann et al. (1992) showed 12 images taken over ten minutes of a freshly cleaved Amelia albite dissolving in aqueous HCl solution of pH = 2. Two fronts coalesced leaving a few islands which subsequently disappeared leaving a step/terrace structure matching the original cleavage surface. Hellmann (pers comm., 9/93) described damage to a feldspar surface from the AFM tip.

To guide future AFM of feldspars, consider the extensive evidence for calcite, whose easy cleavage has attracted AFM practitioners. Calcite has strongly-bonded CO_3 groups ionically bonded to Ca atoms. At first sight, hydroxyls are not needed , and the surface Ca ions can complete their coordination by water. However, the one-sided bonding of the outer CO_3 and Ca species, although neutral overall, is locally unbalanced without some transfer of charge. Hence the chemistry of an ambient surface may be complex.

AFM images of the rhombohedral cleavage of calcite in water show periodic maxima, whose grainy nature was averaged by a Fourier transform algorithm (Rachlin et al. 1992). The lattice repeats match those expected for the bulk structure, but individual atoms were not resolved. Hillner et al. (1992) observed straight growth steps on a perfect piece of cleaved calcite, and dendrites above defective regions. Gratz et al. (1992) concluded that step nucleation is primarily at growth spirals. Gratz & Hillner (1993) concluded from AFM images of calcite growing in a cell from controlled mother liquor that attachment of phosphate and phosphonate species to growth steps inhibited growth, whereas Mg^{2+} had only minor effect, probably because of substitution in Ca sites. Dove & Hochella (1993) observed that phosphate caused irregular shape of surface nuclei, and disrupted steps during spiral growth. These sets of observations on a simple material clearly demonstrate the close linkage between surface chemistry and growth morphology, but the details of atomic bonding are elusive.

Shindo & Nozoye (1992) obtained AFM images in air of the three types of faces of anhydrite immediately after cleaving. High regions match the outer oxygen atoms of the sulfate groups. The ridge pattern on the (100) face was attributed to avoidance of instability from Ca^{2+}-O^{2-} dipoles. Deterioration overnight was attributed to restructuring caused by adsorbed water. Storage in a salt solution preserved the ridges.

AFM images of freshly cleaved muscovite and montmorillonite (Sharp et al. 1993) showed approximations to pseudo-hexagonal arrays of bumps consistent with oxygen rings. They discussed the general theory that AFM depends on the summation of several forces including Coulombic and van der Waals attraction; capillary adhesion from adsorbed water which necks up around the tip; and molecular-orbital-overlap repulsion. The balance of these forces depends on whether the needle is in air or in liquid. Thundat et al. (1993) selected 20-30% relative humidity, a low adhesion force and small tip to obtain a hexagonal pattern with 5.2 Å spacing for freshly cleaved muscovite. Previously, Hartman et al. (1990) had selected 80% relative humidity to obtain similar hexagonal images from montmorillonite and illite. Do these images result from the tip responding to the basal oxygen atoms of the aluminosilicate layer? See also complex results for pillared montmorillonites (Occelli et al. 1993).

Gratz et al. (1991) etched Brazilian quartz in 0.01 molar KOH solution at 421 & 484 K, and observed the ledges of the pits by AFM. The best images were obtained in ethanol. They applied the Burton-Cabrera-Frank model of ledge motion, which should also be tested by AFM of feldspar.

Zeolites are a prime target for AFM because of the extensive technical literature on ion exchange and molecular adsorption. Weisenhorn et al. (1990) presented images of the in-water (010) surface of the Na-rich zeolite clinoptilolite after cleaning in 0.1M NaOH; then with a layer of *tert*-butanol adsorbed from a water-methanol solution; and thirdly with clusters of *tert*-butyl ammonium cations adsorbed from a TBA-chloride solution. Adsorption and desorption were studied directly by AFM. Although the geometry of the AFM image is consistent with the expected lattice repeat of the C-centered unit cell, the in-water image shows considerable deviation from regularity, and each supposed organic species is represented by a single bump. Hence, it is probably prudent to draw a line between a plausible interpretation of the AFM images, and what can be rigorously deduced about the atomic chemistry. MacDougall et al. (1991) imaged twin boundaries at the surface of scolecite, observed terraces and atomic-scale cavities on stilbite (010), and found a rough surface for faujasite which has very large cages and windows. Scandella et al. (1993) display images of the (010) face of the Ca-rich analog of clinoptilolite, heulandite, in 0.1 molar NaOH. A Fourier transform of a grainy image is stated to show a repeating surface unit of 8.4 x 21.0 Å x 115° which is considerably larger than the repeat of 7.4 x 17.9 x 116 for the C-centered unit cell of heulandite in the accepted mineralogical databases. My measurement of the printed image gives dimensions which agree with the mineralogic data; this implies an error involving sin 115° in the original calculation of the lattice repeats. The spotty image shows a tendency to a superstructure in violation of the C lattice repeat, perhaps involving a step of $(a + b)/2$. Has the surface of the Ca-rich heulandite become ion-exchanged to its Na-rich structural relative clinoptilolite?

Currently it appears that AFM study of zeolites is in the typical early stage of exploitation of a new technique in which bright ideas enter print as the complexities are being learnt. Perhaps it will be necessary to choose adsorbed species which become strongly anchored and have identifiable interactions with the probe. Even so, there is no current indication that AFM will yield details of chemical bonding (as distinct from quasi-atomic imaging) at a aluminosilicate surface.

Acid-base properties of surfaces can be characterized by AFM. The Brønsted acid/base behavior of a surface is characterized by the point of zero charge(**PZC**), which is equal to the isoelectric point (**IEP**) for surfaces in contact only with protons, hydroxyls and water. For a pure silica/water surface, two -Si-OH are in reaction equilibrium with $-Si-O^-$ and $-Si-OH_2^+$. Lin et al. (1993) imaged the oxidized surface of a (111) silicon wafer with a 30-nm-radius Si_3N_4 tip in millimolar aqueous KNO_3/KOH solution. The transition from an attractive to a repulsive interaction in a plot of cantilever deflection vs. sample displacement as a function of pH yielded IEP = 6.0 ± 0.4. Application of this technique to feldspar would be complex because of the multi-cation chemistry, but might allow mapping of acid-base properties of chemically modified surfaces.

A comprehensive study under ultra-high vacuum of the (001) surface of magnetite (Tarrach et al. 1993) illustrates complex atomic changes at an oxygen-rich surface. Scanning Tunneling Microscopy (**STM**), which measures the variation of electronic state across a conducting surface, revealed morphological reconstructions at the surface. The crystallographic features were characterized by low-energy electron diffraction, and the chemistry was monitored by analytical electron microscopy, and X-ray and ultraviolet photoelectron spectroscopy. This type of STM study may be possible for metallic and semiconducting oxide inclusions in feldspars.

5.2.3. *Electron microscopy.* This has developed into a sophisticated set of techniques which exploit different aspects of electron scattering.

Scanning Electron Microscopy (**SEM**) utilizes the secondary electrons to image surface tomography with a resolution which can reach a few nanometers for an instrument with a field-ion source (Goldstein et al. 1992). SEM is a routine tool for the study of feldspar surfaces, generally with a fairly coarse resolution of some tens of nanometers. The key observation that natural feldspar surfaces are pitted and even reduced to a "house of cards" is illustrated in the following representative papers: Berner & Holdren (1977, 1979), Al-Shaieb et al. (1980), AlDahan & Morad (1987), Cremeens et al. (1987, 1989, 1992), Blum et al. (1992). Feldspars in soils are coated with clays and other minerals. Bacterial colonies on feldspars were observed by

Hiebert & Bennett (1992). SEM showed that albite hydrolysed at 573K developed a porous rind of boehmite, whose needles were perpendicular to the retreating surface (Hellmann et al. 1989).

That low-energy secondary electrons are much more sensitive to topography than surface chemistry, is nicely illustrated by comparing an SEM image of a manganese precipitate on hematite with Auger electron maps specific to Mn and Fe (Junta et al. 1992). SEM showed that overnight shaking with 2% sodium bicarbonate at pH 9.5 cleans weathered microcline and albite with little apparent damage to the surfaces (Cremeens et al. 1987), but Na-exchange must have occurred at the microcline surface. Ultrasonification and boiling with HCl and H_2O_2 solutions generated etch pits 0.5 µm across.

Transmission Electron Microscopy (**TEM**) and its scanning variant (**STEM**) are particularly valuable for examination of internal surfaces, particularly those between micrometer intergrowths. Submicrometer heterogeneity of an external surface can be imaged by transmission through a very thin slice. Such a specimen can be prepared by thinning a standard thin section with an argon beam in a vacuum chamber. Feldspars degrade easily in an intense electron beam, however, and care is needed in making chemical deductions from the images, and especially from AEM (next section). Techniques for minerals in soils and sediments are described by Smart and Tovey (1982) with many illustrations. Z-contrast STEM offers new opportunity to relate chemical composition to an electron image (Pennycook 1992).

TEM is a challenging technique because images must be collected quickly, and interpretation of contrast involves subtle considerations (McLaren 1991): here are some examples of applications. Fracturing is common in plagioclase from brittle rocks deformed at low temperature. The surfaces are determined by dislocation walls and arrays (Ague 1988), where chemical gradients can be expected. TEM was used to characterize mixed-layer illite-montmorillonite on K-feldspar (Eggleton & Buseck 1980), epitaxial smectite on {001} plagioclase (Page & Wenk 1979), and amorphous Si-rich layers on acid-treated labradorite (Casey et al. 1989b). Partial and complete pseudomorphs of clay minerals after plagioclase and K-feldspar were described by Romero et al. (1992). Twin planes of detrital microcline did not continue into a diagenetic intergrowth (Worden & Rushton 1992); are the dislocations at the interface associated with chemical gradients? Inskeep et al. (1991) showed that Ca-rich lamellae of labradorite were weathered more deeply than Na-rich lamellae producing ridged surfaces. TEM of surface replicas (Fung & Sanipelli 1982) is being replaced by AFM.

TEM of the polytypic synthetic zeolites EMT and FAU (Alfredsson et al. 1993) hints at potential applications to feldspar surfaces. Images of synthetic crystals some 20 nm thick, corresponding to about 10 unit cells, revealed the stacking changes and 12- and 6-ring channels. Surface steps on cubic octahedral faces of FAU are identical from one crystal to another, and computer simulations indicate termination at 4- and 6-rings of an incomplete truncated-octahedral cage of tetrahedra. Unlike cubic FAU, the hexagonal polytype EMT was synthesized with a crown ether as template. Its surface steps are rougher, perhaps because the template causes growth of complete cages. A high-silica product of FAU was obtained by ion-exchange of Na^+ with NH_4^+, heating to drive off ammonia, and acid leaching of released Al. The images showed that the Si-rich product has a regular FAU structure except for a 2 nm amorphous surface layer. Hence, an epitaxial recrystallization occurs during removal of Al, as had been demonstrated by other investigators using other techniques.

The value of Analytical TEM is illustrated by the demonstration that plagioclase and K-feldspars alter to clay minerals without any detectable composition gradient under the feldspar surface (Banfield & Eggleton 1990; Banfield et al. 1991), and that secondary albite grows in parallel orientation in brittle *en echelon* fractures in altered plagioclase (Hirt et al. 1993).

5.2.4. *X-ray Microscopy.* This includes synchrotron soft X-ray microscopy (**SSXM**) and hard X-ray tomographic microscopy (**SHXTM**). Synchrotron microscopy (Howells et al. 1985; Attwood 1992; McNulty et al. 1992; Ade et al. 1992) has not yet been applied to feldspars. The X-ray microscope at the X1 beamline of the National Synchrotron Light Source uses the variation of absorption as a narrow soft X-ray beam is scanned across the sample. The energy range of 2 to 5 nm is particularly suited to the K edges of light elements C to O, but the spatial resolution of >20 nanometer is insufficient for many surface-chemistry problems. Considerable technical advances can be expected with new instruments being designed for third-generation synchrotron sources. Holography and XANES spectroscopy (5.5.4) are important wrinkles on this type of microscopy. Not only can the spatial variation of bulk chemistry be determined, but also the type of chemical bonding. Large metal-organic complexes sorbed on albite might be worth examination.

SHXTM (Kinney & Nichols 1992) in its simplest form uses differential total absorption as the basis for the tomographic inversion. By tuning through an absortion edge, the anomalous absorption can be selected (Flannery et al. 1987). This technique has not been applied to feldspars, but might be useful for studying a surface diffusion profile of a heavy element.

5.2.5. *Angular Distribution Auger Microscopy.* Frank et al. (1990) demonstrated that the angular distribution of Auger electrons emitted from a platinum (111) surface shows "silhouettes" of surface atoms "backlit" by emission from underlying atoms. Monolayers of Ag and I were characterized. The surface must be "clean" and in ultrahigh vacuum, and any application to feldspar would be challenging because of electrical charging (5.5.2).

5.3. CHEMICAL MICROANALYSIS USING A NARROW BEAM

5.3.1 *General.* Some general references are: Potts (1987), Willard et al. (1988), Grasserbauer (1990), Jackson et al. (1991), and Török & Van Grieken (1992).

5.3.2. *Secondary Ion Mass Spectrometry.* Bernius & Morrison (1987), Briggs et al. (1989), Grasserbauer (1990), McGuire et al. (1993), Paul & Vickerman (1990), Sykes (1990), Vickerman et al. (1989), Vickerman (1990) and Benninghoven et al. (1993) reviewed the experimental aspects of dynamic and static secondary ion mass spectrometry (**SIMS**). A primary beam drills a hole into the sample, and the sputtering process produces neutral and charged species (Smith & Walls 1990). In principle, all elements can be mass-analyzed simultaneously by measuring the intensity of each charged species (including ones produced by subsequent ionization of the neutrals), but in practice there are severe limitations. Dynamic-SIMS uses a high-energy intense primary ion beam and secondary ions from a deep crater in order to achieve a bulk analysis. Static-SIMS uses a low-energy weak primary beam which causes emission of polyatomic species as well as monatomic ones from a shallow subatomic crater. These ions contain information pertinent to the surface chemistry and bonding. The primary beam may consists of uncharged atoms. Hence static-SIMS may also be listed under surface sensitive mass spectrometry (**SSIMS**) or fast atom bombardment mass spectrometry (**FABMS**). A broad primary beam is used to analyze a horizontally-uniform sample. Imaging down to the micrometer range is obtained by scanning a focused beam over the area of interest. The greater the horizontal and vertical resolution, the poorer the detection level for the chosen time of the overall analysis.

In reviews of geological applications of dynamic-SIMS, Reed (1989) and Hinton (1990) noted that the sputtering process is complex and not fully understood. Silicates are usually bombarded by negative oxygen ions. The emission of charged ions is inversely related to the work function: thus alkalis yield a much higher ratio of ions to neutrals (several factors of ten) than refractory elements like the noble metals and rare earths. The secondary ion intensity must be calibrated against standards which should be close in composition to the unknown to minimize matrix effects. In addition, the sputtering intensity depends on crystallographic orientation. Even though nonconducting surfaces are generally coated with a thin Au film, which must be sputtered away before analysis can begin, there are problems at the boundaries of minerals with different electrical conductivities. Flooding a sputtered surface with low-energy electrons may help in achieving near-neutrality. Hydrogen-bearing secondary ions are generally observed, even for nominally H-free materials, and transport along the surface and into the crater is the likely cause: a cold finger with large solid angle at the crater may reduce the unwanted signal. Depth profiling is achievable by drilling a hole, and results are improved by rastering the primary beam and eliminating the ions from the ragged edges of the crater by electronic aperturing. The design of the mass spectrometer is important. Combined electric/magnetic focusing is needed to get high mass resolution for separation of adjacent ions, but at the cost of poor transmission. Lower-resolution quadrupole mass spectrometry, which has much higher transmission efficiency, allows higher depth resolution for depth profiling, but only for easily resolved species.

I. M. Steele (pers. comm.) found that the ratio of secondary ions for Si and O did not stabilize on silica-rich surfaces, especially of quartz, until the surface region had been sputtered away. In addition, it was important to make analyses in regions away from grain boundaries, especially those between silicates and dense oxides with different electrical properties. Hence it is legitimate to be sceptical or agnostic about SIMS analyses of the first atomic layers of the outer surface of a feldspar, and at boundaries with other minerals. Great care is needed in SIMS analysis of hydrogen in feldspar because of danger of contamination, even when a cold trap is very close to the specimen. Secondary ions attributable to carbon-bearing species are typically detected during the initial sputtering, and are generally ignored by analysts interested only in the bulk composition.

In spite of legitimate concerns about driving atoms and ions deeper into the sample (knock-on), and about enforced ionic and electronic changes, the dynamic-SIMS analyses of plagioclase and alkali feldspar micro-intergrowths by Miura and Rucklidge (1979) appear sensible in relation to information from diffraction and electron microscopy (SB, Chapter 19). Other dynamic-SIMS studies of feldspars are: diffusion of ^{18}O in dry anorthite (Elphick et al. 1988); reexamination of the role of hydrogen in Al-Si interdiffusion (Graham & Elphick 1990); demonstration that micropores in turbid feldspars are connected in 3D (Walker 1990; Elphick et al. 1991); and depth profiling of altered and weathered plagioclases (Muir et al. 1989, 1990; Nesbitt & Muir 1988; Nesbitt et al. 1991; Shotyk & Nesbitt 1992). Because many labradorites are not chemically homogeneous, care is needed in the interpretation of ion emission during the sputtering process.

Static-SIMS has not been applied to feldspars to the best of my knowledge. It is much less destructive than dynamic-SIMS, and has yielded analyses of propylene on ruthenium, organic lubricants on 'gold', an antibacterial agent on cholesterol, and organic species on fractured alloy-epoxide composite (Paul & Vickerman 1990). Typically a 2 keV Ar^+ beam of very low current density ~ 0.1 $nAcm^{-2}$ generates secondary ions which are analyzed in a quadrupole or time-of-flight mass analyzer. For a simple organic species, the sensitivity can reach 10^{-4} of a monolayer. Careful choice of a feldspar using a carefully selected molecular species might yield useful information. The interpretation must consider the possibility that the organic species undergoes reaction during the bombardment.

Westrich et al. (1989) used dynamic-SIMS to characterize a 100 nm leached layer in labradorite exposed to 0.01 N $HCl-H_2{}^{16}O$ solution followed by 0.01 N $HCl-H_2{}^{18}O$. The heavy oxygen was stable to heating at 393 K and vacuum treatment. Hence exchange at silanol and siloxane was inferred. Would milder isotopic exchange allow characterization of surface silanol and aluminol groups by static-SIMS?

Several technical developments are important for SIMS of surface chemistry. Surface Analysis by Resonant Ionization of Sputtered Atoms (SARISA) uses a tuned laser to selectively ionize one of the neutral species. This results in much higher sensitivity for the chosen element, particularly a refractory one. Thus high sensitivity was obtained for Fe (30 x 10^{-9}) and Ti impurities in a Si wafer (Pellin et al. 1986). A wide-band laser can be used to ionize secondary neutrals in order to obtain better analytical quantitation.

Scanning of the ion beam with a narrow spot can yield a chemical image of the surface. Instruments under development (Bernius & Morrison 1987; Levi-Setti 1990) use an intense Ga^+ ion beam 20-70 nm across, but face the problem of capturing enough secondary ions to achieve sufficient sensitivity for minor and trace elements. Levi-Setti's instrument uses a radiofrequency quadrupole mass filter/transport system which collects 0.2% of the secondary ions. Rapid switching between mass peaks permits matching of chemical images as the crater deepens. The technique has been applied to samples containing micrometer grains of Al-Li alloys, Ag-Br-I emulsions and Y-Ba-Cu-F-O films.

SIMS is unique in its capability to measure isotopic ratios. The vicissitudes of the sputtering process are currently limiting the accuracy of dynamic-SIMS measurement of oxygen isotopic ratios to about 1 in 200, thus making it difficult to search for surface profiles on natural feldspars. Static-SIMS might yield important information on the various types of H- and O-containing species at feldspar surfaces gently exchanged with deuterium and ^{17}O.

5.3.3. *Electron-excited X-ray Emission Analysis.* Electron microprobe analysis (**EMPA**) is widely used in the geosciences, and its strengths and weaknesses are well-known (Goldstein et al. 1992; Reed 1993). Analyses are generally made on a polished thin section some 30 μm thick, at 15 keV. This accelerating voltage is a compromise that yields K spectra of comparable intensity for all elements from Na to Zn in the common rock-formong minerals. Electrons from a focused beam are scattered by electrons bound to atoms in the sample. Much of the kinetic energy of the incoming electrons is dissipated by electrostatic interaction with the bound electrons producing a continuous spectrum of emitted X-rays, generally called the Bremsstrahlung (German: braking radiation). Some of the kinetic energy is transformed into potential energy as a bound inner electron is driven outwards to a higher energy state. The empty quantum state is filled by an electron falling in from one of the outer states according to selection rules and probability functions. The potential energy is used either by a characteristic X-ray photon or an Auger electron which escape from the atom: the latter is favored for low-Z elements. In the energy distribution, the characteristic X-rays occur as sharp spikes sticking out of the continuum. The atomic number (Z) of the atom is revealed by the characteristic energy of the X-ray, and the concentration of the atom type by the intensity. Matrix corrections are needed for the efficiency of generation of the empty quantum level (approximately related to the mean atomic number of the bulk sample), the photoelectric emission efficiency, the absorption of the X-ray before it escapes, and the X-ray fluorescence from X-rays of higher energy which are captured in the sample to generate the same type of empty quantum level. In general, most elements from Na to U can be analyzed within a few percent relative accuracy using simple standards, with a detection level of a few tens to hundreds of parts-per-million limited by the noise from the Bremsstrahlung.

The primary electrons lose energy non-linearly and are scattered irregularly over an onion-shaped volume a few μm across and deep. The characteristic X-rays are generated most effectively when the primary electron has just enough energy to knock out a bound electron. Thus, for the K spectrum, X-rays for a high-Z element will be generated at shallower depth than those for a low-Z element. High-energy X-rays tend to be absorbed less than low-energy ones. Hence even semiquantitative EMP analysis of a sub-micrometer surface layer on a polished thin section is hardly practical. In principle, the acceleration voltage can be reduced to obtain a smaller volume in which X-rays are generated, but at the price of lower sensitivity. Thus for Fe, X-ray photons would be generated only at the surface for electrons accelerated by ~7 keV. For a wide beam a millimeter across, it should be possible to get enough signal for analysis of a layer some nanometers thick, but it would be very difficult to obtain a quantitative analysis. The background scatter makes it difficult to detect elements below the 0.01% level, even for 25 keV incident electrons.

These considerations for a thick sample do not apply to a very thin foil used for TEM. Analytical Electron Microscopy (**AEM**) uses the same principles as the electron microprobe, but the specimen is so thin that very few electrons scatter sideways. Hence the area of analysis can be close to that of the primary electron beam. Because the thickness of the Ar-thinned foil varies from point to point, absolute concentrations are unattainable. Elemental ratios can be obtained to a few percent with careful calibration. AEM is particularly appropriate for internal surfaces and defects, but could give information for an external surface, if the edge of the foil survived Ar-thinning.

It is convenient to list here the technique of Electron Energy Loss Spectroscopy (**EELS**) which can be combined with STEM to yield very high sensitivity approaching the detection of single atoms (Leapman & Newbury 1993). The excitation of the inner electrons by a narrow electron beam is detected directly by measurement of core edges in the energy loss spectrum. The sample must be thinner than ~ 50nm.

5.3.4. *X-ray Fluorescence Analysis*. The high brightness, tunability and strong polarization of X-rays from synchrotron sources(Smith 1993) has rejuvenated the classical bulk technique of X-ray fluorescence (**XRF**) analysis (Jones & Gordon 1989; Rivers et al. 1991; Jones 1992; Sutton et al. 1993). In the classical method, a primary beam of white X-rays from a laboratory source excites the bound electrons in atoms of the sample, and the fluorescent X-rays are analyzed for energy and intensity. The primary X-ray beam is broad (cm) and unpolarized, and the sample is commonly fused and diluted to obtain higher accuracy. Because there is no Bremsstrahlung, XRF can go down to trace levels inaccessible to the electron microprobe. However, Rayleigh and Compton scattering from an unpolarized beam limit the minimum detection level to a few parts per million for many elements. With some difficulty, X-rays from laboratory sources can be focused for small-spot analysis, or polarized for lower detection level, or reflected from surfaces, but synchrotron radiation-induced X-ray emission (**SRIXE**) offers important advantages.

First, the dipole emission of X-rays produced as electrons in the synchrotron storage ring are accelerated radially by a bending magnet is a very narrow jet because of the relativistic transformation from the high-speed electrons to the sample. Hence, focusing with a metallic mirror is very easy. There are enough X-rays in present synchrotrons for analysis of trace elements with a nanometer-diameter beam. Second, the synchrotron X-rays are 99% polarized in the plane of the electron ring, and Compton and Rayleigh scattering can be greatly reduced by collecting the fluorescent X-rays at 90°. This permits a lower detection level for trace elements than for the classical method. Four years ago, analyses of several trace elements in doubly-polished thin sections of feldspar standards with a detection level near the part-per-million level were attained with simple energy-dispersive detection and unfocused radiation limited by crossed slits (Lu et al. 1989). For elements with hard X-rays, the sample can be analyzed in air. For light elements, the sample must be in a vacuum or helium.

Increase of brightness from focusing optics has allowed wavelength dispersive techniques with detection below the part-per-million (ppm) level. A spot diameter of 1 nm yields sufficient signal for current synchrotron sources. New third-generation undulator sources will have enough brilliance for analysis with a 1 μm beam diameter. Replacement of a single bending magnet with a series of magnets of alternate polarity causes the electron beam to undulate down a straight section of the synchrotron ring. For small displacement, the emitted X-ray pulses from each undulation are coherent but out of phase. Interference concentrates the photon energy into narrow harmonics which may be tuned by changing the magnetic field of the undulator. Sweeping through an absorption edge with undulator-generated X-rays should reduce the detection level for many trace elements down to 0.01-0.1 ppm (Smith 1992).

Three-dimensional tomography will allow element analysis on a volume-element as small as a few cubic micrometers. Such tomography, of course, is really a type of 3D chemical microscopy, and will be applicable to internal surfaces and defects. Grazing incidence will permit analysis of the first few layers by **SSXRF** (synchrotron surface X-ray fluorescence) thus greatly extending the capability of laboratory X-ray sources (Klockenhämper et al. 1992). Tuning an undulator harmonic to just greater than the excitation energy for a bound electron should prove useful in "high-lighting" a particular element in a surface layer. In spite of the brilliance, the energy deposition in a sample will be small compared to that from electrons in AEM and protons in PIXE. A detailed review of the science and engineering is given in a proposal to the US National Science Foundation (Rivers 1993).

5.3.5. *Nuclear Analysis.* Several techniques fall here. A nuclear particle such as a proton or a deuteron may undergo a "billiard-ball" collision with a nucleus in which the momentum transfer provides the chemical signature - Rutherford Back Scattering Analysis (**RBA**): Grant (1990). The nuclear particle may knock an electron out of a bound orbit, and the ensuing X-ray photon provides the basis for Particle-Induced X-ray Emission (**PIXE**): Johansson and Campbell (1988). The particle is often a proton (also *P*). The particle may drive an atomic nucleus into an excited state and one or more particles may be emitted to be used for Nuclear Reaction Analysis (**NRA**): emission of a gamma-ray gives **PIGE**.

Fraser(1990), Petit et al. (1990b), Tapper & Malmqvist (1991), McGuire et al. (1993) and Ryan and Griffin (1993) describe the latest applications of the nuclear microprobe to chemically zoned materials. The minimum impact diameter is about 1 μm and nuclear particles are so much heavier than electrons that they essentially maintain a straight path. The beam may penetrate deeper than a typical 30 μm thin section, and the region of maximum conversion to an analyzed photon or secondary particle is generally well below the surface, commonly near 20 μm. Athough there is no current indication that nanometer-scale surface layers will be analyzable, Baumann et al. (1986) were able to distinguish surface and bulk species in zeolites by RBS. Uranyl ions at the surface of Na-zeolites gave a sharp peak in the RBS spectrum. Laursen et al. (1993) used RBS to characterize Cu atoms in an arachidate film on a Si wafer. Heavy ions would similarly be easily detected on a feldspar surface. Dran et al. (1992) used RBS to study the absorption of colloids on silica and muscovite. Light elements can be analyzed by NRA and PIGE. Casey & Bunker(1990) describe the use of Elastic Recoil Detection Analysis for determination of the H profile in leached minerals and glasses. PIXE is good for trace elements because the background is low - the detection level is a few parts-per-million for many elements, and the calibration is accurate if all the protons are stopped in the sample. Two illustrative studies are olivines in a very thin section (7 μm) impacted by a 3 μm proton beam (Bajt et al. 1991) and of brine inclusions in quartz from cassiterite-granite veins (Ryan et al. 1993).

RBS was used by Petit et al. (1987) on ion-implanted albite, amazonite and labradorite, Casey et al. (1988, 1989a,b) on acid- and alkali-treated labradorite, Petit et al. (1990a,b) on hydrated orthoclase, and Cherniak (1992) and Cherniak and Watson (1992) on Pb- and Sr-exchanged alkali feldspars and anorthite. Resonant Nuclear Reaction analysis of H and Na was used by Petit et al. (1989, 1990) to characterize leached orthoclase and glass.

5.4 DIFFRACTION

5.4.1. *General.* The special characteristics of electrons, X-rays and neutrons can be selected to optimize diffraction of surface structure. Neutrons are very useful for study of hydroxyl and water because hydrogen and deuterium have negative and positive coherent scattering lengths respectively (- 3.7 & + 6.2 Fermi), while other elements have useful differences (e. g. Mn - 3.7, Fe + 9.2). Because scattering is so weak, at least a gram of sample is needed. Hence, neutron diffraction is currently confined to internal surfaces in bulk powders of zeolites and clays, and in very rare large single crystals. X-ray diffraction of surfaces is currently limited to simple surfaces (liquid metals, metallic films, etc.) using intense radiation from second-generation synchrotron sources, but the much more intense radiation from the third-generation synchrotrons should open up the study of complex surfaces on minerals, not necessarily in a hard vacuum. Well established is low-energy electron diffraction (**LEED**), which has the restriction that the surface is in a vacuum.

5.4.2. *Electron Diffraction.* Diffraction of low-energy electrons (30-200 eV) is convenient for probing surface geometry (Pendry 1974; Spence & Zuo 1992), but multiple scattering makes it difficult to obtain quantitative diffraction intensities for structure determination. Some progress is being made on structure determination for surfaces of simple metals (Hu & King 1992) and oxides (Tarrach et al. 1993).

Hochella (1990) reproduced LEED patterns for several minerals, including albite(010), without sample degradation. The (010) surface of Amelia albite (Hochella et al. 1990) was cleaned in air and immediately analyzed after pumpdown. The diffractions fitted a 2D cell with 7.22 x 6.57 Å x 60° which indicate substantial distortion from the bulk structure. The symmetry was primitive instead of the face-centering for the bulk structure. LEED patterns result from constructive interference in the outer unit cells, and the diffractions are broad. Hence, the significance of the lower symmetry and apparent geometrical perturbation is unclear. However, consideration should be given to formation of a superstructure, perhaps involving the angular orientation of the terminal T-OH and the position of Na and linked framework species.

Because rough surfaces are unsuitable for LEED, this technique is not promising for study of dissolution of feldspars, or of irregular surfaces on natural specimens. It offers promise for characterization of surface relaxations and reconstructions induced by adsorbed species on good feldspar cleavages.

5.4.3. *X-ray Diffraction.* Surface sensitive X-ray diffraction (Feidenhans'l 1989; Fuoss & Brennan 1990, Coppens 1992; Matsui & Mizuki 1993; Robinson & Tweet 1992; Toney & Wiesler 1993) is being developed as a tool for the study of the outer atomic layers. A brilliant X-ray beam from a synchrotron source (Smith 1992) travels at grazing incidence to a surface so that total reflection occurs. The single-scattering event yields scattering rods from the 2D lattice array of surface species. Background scattering is reduced by placing the detector at 90° to the plane of polarization of the synchrotron X-rays. Surface species of different chemical type to the bulk species can be pinpointed by collecting diffractions above and below an absorption edge. The difference corresponds to the anomalous scattering which can reach about one-quarter of the normal scattering: furthermore, careful tuning around the absorption edge can yield evidence on the electronic state (5.5.4). Jach & Bedzyk (1993) describe how such selective excitation can be coupled with standing waves (Abruna et al. 1988) produced by interference between the transmitted wave and the scattered wave.

All published studies are on simple systems that can be prepared precisely and interpreted with a few structural parameters. For example, the {111} surface of silicon has a 7x7 superstructure, and epitaxial intergrowths show various complex relations. Tweet et al. (1992) observed Ge and Si ordering at the Si/B/Ge_xSi_{1-x} (111) interface by anomalous diffraction. Wang et al. (1991) pinpointed Zn atoms in an arachidate bilayer with Å resolution. Specht & Walker (1993) showed that Cr is trivalent at an interface with alumina. Chiarello et al. (1993) measured the thinning of the water film on a calcite cleavage face as the humidity was reduced.

Berthold & Jagodzinski (1990) used anomalous X-ray scattering from Rb to demonstrate that it is preferentially present in the Albite-twin boundaries of both albite and M-twinned microcline. A single crystal of each was studied with Si-monochromatized radiation from a 12 kW rotating anode generator. The diffraction streak between the pair of 002 diffractions was 20% more intense for 0.79 Å than for 0.82 Å, as expected for Rb, whose K absorption edge is at 0.815 Å. No intensity change was found for the 002 streak of the Pericline-twin, which apparently does not have a concentration of Rb in the twin boundary. No change of intensity of the twin streaks was found for Sr whose edge is near 0.77 Å. This type of experiment would be easier with synchrotron X-rays. It is important to select a diffraction whose intensity is particularly sensitive to the position of the chosen element, in this case 002. Further research is needed to exploit this surface technique which is potentially quite sensitive for checking whether certain trace elements, including Pb, are concentrated at internal boundaries.

5.4.4. *Neutron Diffraction.* The theory of neutron scattering from surfaces was given by Dosch (1992). The transition between "water" and "ice" encapsulated within Vycor glass, Ludox gel and Na-exchanged montmorillonite was studied by small-angle neutron scattering (1-100 nm wavelength) of deuterated samples at 240-280K (Li et al. 1993). A slurry of deuterated feldspar powder in an H-free dispersant might provide useful information on the liquid/solid transition of the hydrous film. Single-crystal neutron diffraction studies of water encapsulated in zeolites are typified by determination of the disordered positions for protons in partially dehydrated laumontite (Artioli et al. 1989). The data are useful for structure modeling of H-containing feldspar, but direct measurement of H in feldspars is currently impractical.

5.5. SPECTROSCOPY AND RESONANCE

5.5.1. *General.* This group of techniques has been applied mainly to bulk samples, but recent developments are permitting study of surfaces. Each technique has its own particular niches, and careful choice of the best technique, rather than the most accessible one, will become the key to future success in characterizing surface chemistry. The collection of review articles in Hawthorne (1988) gives up-to-date details of most of the spectroscopies.

5.5.2. *Infrared, Raman , & Luminescence Spectroscopies.* Nakamoto (1978) provided a general review of infrared and Raman spectra of inorganic and coordination compounds. McMillan and Hofmeister (1988) reviewed applications to mineralogy and geology, and Rossman (1988) covered vibrational spectroscopy of hydrous components. Flanigen (1976) and Ward (1976) summarized the uses of IR and Raman spectroscopy for zeolites, and SB (Chapter 11) assembled applications to feldspars. Luminescence spectroscopy is covered by Waychunas (1988). In IR spectroscopy, infrared radiation is absorbed preferentially at frequency bands corresponding to vibrational transitions. In solids, the quantized vibrational modes are broadened and shifted by complex coupling of near-neighbor interactions, but the individual peaks can be assigned from the frequency to particular modes such as Si-OH stretching. In Raman scattering, a tiny fraction of the infrared radiation is scattered inelastically, and the energy shift of each photon is a measure of vibrational mode with which it transferred energy. Polarization and selection rules are important in teasing out the structural significance of IR and Raman spectra.

Essentially all IR measurements of zeolites have been in transmission with powders mulled with an inert liquid. Generally there are 3 types of absorption bands in the mid-IR for OH stretching: one near 3750 cm^{-1} for hydroxyl projecting from the external surface, one near 3650 for a bridging OH at a large internal surface, and a third near 3550 for an internal surface with strong interaction between OH and an extra-framework cation. Jacobs and Mortier (1982) found a linear relation between the 3650 band and the Sanderson electronegativity of the framework (approximately related to the Si/Al ratio). Fourier-transform IR in the 250-1400 cm^{-1} region is revealing changes induced by protonation of zeolites (Jacobs et al. 1993). The asymmetric and symmetric T-O stretching modes shift to higher wavenumbers, and subtle changes of wavenumber and intensity occur for other modes. The far-IR at 350-10 $cm^{-1}$is now being exploited for information on translatory motions of extra-framework cations, including ammonium and organic species (Ozin et al. 1989). Exchange of hydrogen by deuterium is useful for sorting out IR peaks. Thus Kramer at al. (1993) studied D-H exchange between deuterated methane and protonated zeolites, a trick which could be used for studying the adsorption of an organic cation on a wet feldspar surface.

Still to be exploited fully is IR spectroscopy for characterization of surface species. Two illustrative examples from the materials sciences using conventional techniques are: Fourier-transform transmission IR spectroscopy of water molecules on the surface of a fine powder of Cr_2O_3 (Kittaka et al. 1993), and a combined reflection/absorbance study of a silica-supported Cu catalyst interacting with CO (Xu et al. 1992).

Study of a surface would ideally be done by reflectance from a single crystal using a polarized IR beam. In principle, the hydroxyl and water species at the surface could be characterized in 3D. To obtain enough signal with conventional IR sources (lamps and arcs), about 10 reflections would be needed for a partially hydroxylated silicate surface, which ideally would have strong order. Careful sample preparation would be needed and measurements should be made at low temperature under controlled environment. The Harrick internal-reflection spectroscope (Harrick 1979; Zeltner et al. 1986) with a cylindrical transmitter surrounded by a mulled powder demonstrates the principle of reflectance IR, but a polarized multi-reflectance technique with a TlBr mirror above a crystal face would be more sophisticated.

Such a multi-reflectance procedure with a conventional source may be upstaged by synchrotron IR spectroscopy. Current low-energy synchrotron sources are 100 to 1000 times brighter than a conventional globar source in the far-IR, and the next generation will be even brighter. Maroni et al. (1993) used synchrotron-IR spectroscopy to characterize the molecular absorption of water on a 25 x 6 mm gold plate at 100-140 K. At 0.3 langmuir, the O-H stretching vibration at 3387 cm^{-1} was observed, and other bands demonstrated frustrated rotation. Clusters of $(HOH)_n$ at low coverage transformed into an ordered multilayer with ice-like structure at higher coverage. This level of sensitivity should permit controlled experiments with a feldspar cleavage plate.

Beran (1986) obtained polarized IR spectra of gemmy Eifel sanidine with 0.036 wt. % H_2O. Bands at 3400 and 3050 cm-1 were assigned to asymmetric and symmetric stretching of O-H-O lying in the mirror plane at the M site. A very weak band at 4550 cm-1 was attributed to overtones from (Si,Al)-OH groups. Are these at the outer surface? In contrast to the Eifel sanidine, a sanidine from a mantle eclogite gave no detectable IR signal for structural hydroxyl (Rossman & Smyth 1990).

Petrov, Agel & Hafner (1989) observed six types of O^- defect centers in gemmy Amelia albite: four at oxygens bonded to two Al, and one each involving Na and M^{2+}. The centers were destroyed by heating and regenerated by X-irradiation. Do these observations for the interior of a millimeter sample have any implications for the outer surface?

5.5.3. *Electron- and X-ray-induced Photoelectron & X-ray Spectroscopies.* These techniques are so well known that reliance will be placed here mainly on signalling the review articles (Hochella 1988; Hochella & Brown 1988; Bishop 1990; Christie 1990; Frank et al. 1990; Turner 1990) and relevant papers on zeolites (Barr & Lishka 1986; Barr et al. 1988; Howe 1993) and iron oxides (Harvey & Linton 1981). The importance of reducing surface charging is emphasized by Grünert et al. (1993).

XPS studies of feldspars are discussed extensively by Blum because of their relevance to weathering profiles: Berner & Holdren (1977), altered natural feldspars; Hellmann et al. (1990), leached albite; Hochella et al. (1988a), leached albite; Hochella et al. (1988b), leached labradorite; Inskeep et al. (1991), laboratory weathered labradorite; Junta et al. (1992), Mn oxidation in presence of albite; Muir et al. (1989,1990), altered plagioclase; Petrovic et al. (1975), leached sanidine. Caution in needed in accepting some of the data for rough surfaces because of loss of signal from the valleys.

5.5.4. *X-ray Absorption Spectroscopies.* The availability of brilliant, tunable and polarized X-rays from synchrotron sources (Smith 1992) has revolutionized X-ray absorption spectroscopy (Brown et al. 1988; Koningsberger & Prins 1988; Brown & Parks 1989; Bassett & Brown 1990; Brown 1990;Waychunas & Brown 1990; Ade et al. 1992; Attwood 1992). Using the low-energy side of the formal absorption edge, the electronic state can be determined along with some information about the type of coordination (e. g. tetrahedral vs. octahedral) by X-ray Absorption Near Edge Spectroscopy (**XANES**). From the oscillations at the high-energy side, the radial distribution of neighboring atoms can be determined by Extended X-ray Absorption Fine Structure Spectroscopy (**EXAFS**). Two illustrative papers covering bulk samples are on: Na-Al-Si-O glasses (McKeown et al. 1985a,b); and Fe, Mn, Zn and Ti in staurolite (Henderson et al. 1993). Polarized EXAFS is illustrated by a study of hematite-like clusters in Fe-diaspore (Hazemann et al. 1992). Illustrations of application to surface chemistry are given by EXAFS spectroscopy of a slurry that revealed the stereogeometry of selenium oxyanions on the surface of goethite (Hayes et al. 1987), and of catalyst surfaces in a reactive atmosphere at high pressure (Moggridge et al. 1992).

Surface spectroscopy is expected to become much more sophisticated as even more brilliant synchrotron sources are commissioned. Glancing-angle XAS can probe the surface species, and applications to electrochemistry, metallic overlayers, biological and zeolitic films, etc. are growing rapidly (Smith & Woodruff 1982; Abruna et al. 1988; Sharpe et al. 1990; Stöhr 1992).

XAS of feldspars is just beginning. The valence state of Fe can be estimated from XANES K-edge spectra, and iron in lunar plagioclase is divalent whereas Fe in terrestrial feldspars is a mixture of ferrous and ferric. L-edge spectroscopy (Cressey et al. 1993) gives better energy resolution and somewhat different information than K-edge spectroscopy of Fe and other transition metals. XAS data for Ti in Oregon sunstone have revealed tetrahedral coordination in clear areas, and octahedral coordination in cloudy areas (S. Bajt et al., to be published), and further samples of Ti-bearing feldspar are being studied. All present studies are on bulk samples, but the third-generation synchrotrons will permit surface-XAS of minor and trace elements.

5.5.5. *Ultraviolet Spectroscopy.* See applications to the conduction band of metal particles on zeolite surfaces (Grünert et al. 1993) and the (001) surface of magnetite (Tarrach et al. 1993) for inspiration that might lead to ideas for studying metal particles on the surface of feldspars from reduced meteorites and the Moon. Electrostatic charging of the surface caused by loss of photoelectrons must be compensated by supplying new electrons.

5.5.6. *Nuclear Magnetic Resonance.* High-resolution solid-state NMR (Engelhardt & Michel 1987; Kirkpatrick 1988; Stebbins 1988) is a bulk technique which allows characterization of the bonding environment of atoms with dipole and higher order spin. Howe (1993) describes how FTIR, XPS and NMR are useful for probing catalytic surfaces in 3D. The major advance was the application of sample spinning at the magic angle (cubic 111) which greatly sharpened the spectrum for solid samples. Cross-polarization allows characterization of the spin-coupling of adjacent atoms, and is particularly important for the coupling of a proton to nearby species. New techniques include multi-pulsing, higher magnetic fields and sample rotation inside the magic-angle rotor. Much smaller samples can be used, and surface features should be easier to sort out from internal structural ones. A study of water in silicate glasses by ^{1}H solid-echo and magic-angle-spinning NMR (Eckert et al. 1988) offers useful guidance for thinking about possible study of a wet feldspar slurry. A representative paper on zeolites is characterization of H-bonded complexes and Brønsted sites in H-exchanged ZSM-5 zeolite (White et al. 1992). There is a large expanding literature that can be accessed from the Chemical Abstracts database on Zeolites.

The value of NMR spectroscopy for the study of feldspar surface chemistry is illustrated by Yang & Kirkpatrick (1989). The H-exchanged layer in albite was shown to be <0 3nm thick, and new instruments of higher sensitivity would provide an even stricter limit.

5.5.7. *Electron Paramagnetic Resonance.* EPR is reviewed by Calas (1988). It is generally used for study of low levels of paramagnetic species in the bulk structure of mm-size gemmy single crystals. Thus Petrov and Hafner (1988) studied an Eifel sanidine by single-crystal EPR. In addition to strong resonance lines attributed to Fe^{3+} in T1and T2 sites, a very weak line in the Q band (~35 GHz) at g_{eff} ~6 with **B** parallel to *a* was found. The pseudotrigonal symmetry was consistent with $Fe^{3+}O_3OH$ at T1 and/or T2. Petrov et al. (1989) characterized six defect centers involving O^{1-} in Amelia albite.

Paramagnetic species are also used as surface probes, and wet samples with low concentration of the probe can be studied by electron spin resonance. McBride (1986) described measurements on clay films oriented in the EPR spectrometer. For example, vermiculites bond water and organocations more strongly than smectite. Perhaps very thin cleavage flakes of feldspar could be prepared to mimic the oriented clay films. Cu^{2+} is useful because it can become surrounded by six water molecules in a distorted octahedron, and Cu-containing surface adsorbates on feldspar should yield stereogeometric information of value in economic and environmental sciences. Nitroxide radical cations are useful probes; TEMPAMINE, = 4-amino-2,2,6,6-tetramethyl-piperidine-NO, is an example.

5.5.8. *Ion scattering spectroscopy.* Armour (1990), Rabalais (1990) and McGuire et al. (1993) reviewed ion scattering spectroscopy which is the surface-sensitive analog of Rutherford Back Scattering. Kilovolt ions are scattered from the outer surface, only a few unit cells deep, by a "soft" impact which depends partly on simple momentum transfer and partly on the lattice forces. A pulsed incident beam and a time-of flight energy detector of the secondary ions and atoms provide information on the surface chemistry and the stereogeometry of the chemical bonding. Possible application to feldspar can be assessed from a study of Na- and Ag-exchanged zeolite type Y (Grünert et al. 1993).

5.6 CHEMICAL ANALYSIS USING CLASSICAL METHODS

5.6.1 *Adsorption.* Determinations of adsorption isotherms and of accessible surface area and internal volume are the fundamental bases for characterization of surfaces. General references are Gregg & Singh (1982) and Potts (1987). An example of the importance of determining the accessible area by the BET absorption technique is the demonstration by Anbeek et al. (1992) that naturally weathered surfaces have internal surfaces some hundred times greater than the geometrical envelope of the outer surface.

5.6.2 *Titration.* This classical technique retains its value for characterization of bulk surfaces. Recent applications are Wollast & Chou (1992) on Na^+/H^+ exchange in albite, and Zeltner et al. (1986) on titration calorimetry of surface complexes on goethite.

6. Speculations on Surface Properties of Feldspars in Various Geological Environments and Ideas for Future Research

6.1. ARE SILANOL-RICH SURFACES RESPONSIBLE FOR SLOWER WEATHERING IN NATURE THAN EXPERIMENTAL ALTERATION?

In general, theoretical hydroxylated surfaces of a feldspar would have both silanol and aluminol atoms, except for special choice of surfaces in ordered $AlSi_3$-feldspars (Figs. 2 & 4). Do natural feldspars, however, tend to develop surfaces dominated only by silanols, partly by more rapid erosion of aluminol-rich surfaces, and perhaps partly by slow silicon-aluminum exchange? If so, the evolved surface might become more stable to further weathering by analogy with quartz. Natural weathering rates would then become much less than those in short-term laboratory experiments for which silanol-rich surfaces had not developed: this is consistent with observations (Velbel 1986) - see Blum (next paper). Aluminum can be extracted from zeolites by several processes. Will the same processes work on feldspars studied under controlled conditions? Even more interesting is whether any of these processes can be proven to occur in Nature. There is an extensive literature on the removal of Al from industrial zeolites by a range of processes (Flanigen 1976; McDaniel & Maher 1976), and, by analogy, an auto-catalytic process of replacement of an aluminol by a silanol at a feldspar surface in contact with an Si-bearing phase appears worthy of consideration. Pluth and Smith (1982) showed that a sodium zeolite, Linde Type A, lost Al from the framework upon ion-exchange with divalent cations at room temperature. Exposure to acids and chelating agents also have caused removal of Al from zeolites at low temperature. Remarkable is the epitaxial recrystallization of the surviving framework at elevated temperature, with removal of the defects left after Al migration. Natural feldspars exposed to high temperatures after contact with acidic aqueos fluid, as in sunny deserts after thunderstorms, might be good targets for testing this concept.

Reflectance infrared absorption spectroscopy (6.7) could measure the aluminol/silanol ratio in principle, but might be experimentally difficult because of the roughness of natural feldspar surfaces.

Adsorption on feldspar surfaces of selected organic species, used as gauging molecules for molecular sieves, might provide an evaluation of the ratio of silanol and aluminol species. In general, a high-silica surface is more organophilic than an alumina-rich one, as evidenced by the remarkable selectivity of the synthetic zeolite, silicalite, for benzene over water.

6.2. HOW DO FELDSPARS IN SOILS INTERACT WITH NATURAL ORGANICS AND MICROBES?

A conference proceedings (Huang & Schnitzer 1986) concentrated on clay minerals. Although there are a few references to feldspars, some important analogies can be drawn to guide the selection of organic complexes and microbes for experimental study (3.4). Perhaps feldspar surfaces will be found to have unsuspected specific catalytic interactions with particular organic species. Computer modeling of the potential bonding of organic species in the dimples between 1-connected O at feldspar surfaces should be valuable. One negative prospect is indicated by Wang et al. (1986, p.260) who reported that abiotic polymerization of hydroquinone by rock-forming minerals follows the sequence tephroite (highest) to muscovite-albite-orthoclase-microcline-quartz (approximately equal). Systematic study of feldspars using the techniques developed by the clay mineralogists should reveal important phenomena.

6.3. BIOLOGICAL EVOLUTION: ABSORPTION ON GROOVED SURFACES OF FELDSPARS.

Speculations on the origin of life on Earth have focused on possible catalytic reactions and subsequent replication on clay minerals (Cairns-Smith & Hartman 1986). However, there are no specific structural models at the atomic level. Furthermore, it is not clear how organic species would become associated between the layers of clay minerals, and desorbed after reaction. Adsorption on an external surface may not give a clay mineral any advantage over the many other minerals, including feldspar, with interesting surface geometries. That interesting inorganic-organic interactions did occur in biological evolution is well documented by many research papers in biomineralization. Of particular significance was the experimental demonstration of heterogeneous and epitaxial nucleation of protein crystals on mineral surfaces, including lysozyme on apophyllite (McPherson & Shlichta 1988).

Micro- and nano-intergrowths of feldspars develop grooved surfaces as a result of differential resistance to ionic solutions (S, Chapter 10; Inskeep et al. 1991, Brown and Parsons and frontispiece, this volume). All feldspars containing intimate intergrowths (SB, Chapter 19) have the potential to develop grooved surfaces because of differential weathering, especially between Ca-rich and Na-rich phases. Electron microscopy has revealed essentially a continuum of spacings from the nano- to micrometer scale, and from near-linear boundaries to braided ones. Fascinating chemical gradients should exist across the grooves. Are macro-organic species adsorbed in the grooves? Do catalytic reactions occur, especially during cooling-heating and wetting-drying cycles? Is photochemistry important? Can qualitative arm-waving be replaced by systematic quantitative studies combining surface structural techniques and stereochemical modeling?

6.4. SURFACE PROPERTIES OF FELDSPAR IN HIGH-PRESSURE ROCKS.

What happens to the surface of a feldspar as it is compressed deep in the crust? Boisen & Gibbs (1993) have modeled the structure and compressibility of silica minerals based on the cluster $H_6Si_2O_7$, and their work should be extended to the more difficult problem of the aluminosilicates. Do the hydroxyl and water species at a feldspar surface in a rock bond more closely with atoms of adjacent minerals? Does chemical adjustment occur at the edges of faces where the chemical potential varies rapidly? Is there epitaxi with adjacent non-feldspar minerals? Does coarsening occur at surfaces, or does independent nucleation occur?

6.5. SURFACE PROPERTIES OF EXTRATERRESTRIAL FELDSPARS.

Are lunar feldspars terminated by silanol and aluminol groups? If not, are they generated on Earth by reaction with water? Similarly, what terminates the surfaces of feldspars on the other planets, asteroids, comets and planetary dust? Are ionic and molecular complexes adsorbed on the surface? Do photochemical reactions occur? Are surfaces amorphized by impact of fine particles and ions? Do metal clusters nucleate on anorthite crystals of high-temperature condensates?

6.6. TRANSITION-METAL COMPLEXES ON FELDSPAR SURFACES.

Complexes of first-series transition metals may be adsorbed on feldspar surfaces. Fritz and Moore (1984) found "limonite"-rich and -free areas in microenvironments of a spheroidally-weathered anorthosite boulder. Calcite, presumably formed by interaction with CO_2-charged water, grew only on limonite-free feldspar. Comparison of the surface of the An_{72} plagioclase from the Fe-rich and -poor regions might be instructive. Potential Fe-oxygen clusters on the feldspar surface should be sought using AFM. Xu et al. (1992) provide a guide with their study of silica-supported copper catalysts: note that STM requires a conducting matrix, which Xu et al. obtained by growing a silica film on a molybdenum foil.

Many feldspars from high-temperature rocks contain up to 1 wt. % Fe, and a few go even higher. Annealed feldspars contain a wide range of discrete Fe-rich inclusions (SB, Chapter 20) that reflect the redox conditions. Intermediate states that involve small clusters can be expected. Polarized X-ray absorption spectroscopy using brilliant synchrotron beams offers the opportunity to characterize the electronic state and bonding of the Fe. Particularly interesting would be the clouded feldspars from achondritic meteorites, the Fe,Ni,Co alloys in lunar feldspars, the characteristically-colored feldspars from the various types of granites, and the gemmy feldspars. The technical aspects would be similar to those described by Hazemann et al. (1992) in the study of hematite-like clusters in Fe-diaspore.

6.7. SURFACE INFRARED SPECTROSCOPY OF FELDSPAR.

Section 5.5.2 contains the essential justification for giving this technique the highest priority for technical development in general, and application to feldspar surfaces in particular. Attachment of a dedicated controlled-atmosphere chamber to a beamline of a high-brilliance low-energy synchrotron source is desirable so that the design can be optimized over a period of several years for surface reflectance. The feldspar and other samples should be on a precision stage to permit translation and rotation. The detection system should be optimized. In addition to control of the pressure and temperature of the atmosphere, provision is needed for sputtering. Furthermore, it is desirable to be able to introduce controlled amounts of inorganic and organic species on the surface. The chamber should be detachable to permit transfer to other scientific instruments, including ones for synchrotron surface diffraction/absorption spectroscopy.

6.8. SURFACE X-RAY ABSORPTION SPECTROSCOPY AND DIFFRACTION.

Combined surface X-ray absorption spectroscopy and diffraction using synchrotron radiation is a potentially useful probe for monitoring monolayers of sorbate on a mineral surface. Useful guidance is provided by Moggridge et al. (1992). Controlled adsorption of CO, CO_2 and organic acids on feldspar cleavage faces that have been thoroughly cleaned and equilibrated should yield promising specimens. More exotic would be the study of metal-organic species, including porphyrins. Yet even more challenging would be simultaneous adsorption of two organic species that might react to produce a third: obviously, there is the chance that speculations about the origin of biological evolution at mineral surfaces might be tested. Presence of a first-series transition metal, actually in the feldspar structure below the surface boundary, and a different one in an adsorption complex on the surface, offers further subtle potential. The possibilities are unlimited.

6.9. NEUTRON DIFFRACTION OF HYDROGEN-FELDSPAR.

High-resolution neutron powder diffraction of hydrogen-feldspar should reveal the H positions of the proposed bridging hydroxyls, thus providing a test of theoretical models for stereogeometry and chemical bonding of silanol and aluminol groups. Deuteration would reduce the incoherent background scattering. It would be desirable to increase the H content higher than the 89% achieved by Paulus & Müller (1988) for the center of a single crystal of Eifel sanidine used for single-crystal x-ray study. Because statistical distribution of H at several oxygen sites may be expected for this highly disordered sanidine, H-exchange of highly ordered albite and microcline should be considered. Even for a fully ordered H-feldspar, high spatial disorder can be expected for the H. Hence, data should be collected at very low temperature to minimize the thermal disorder.

6.10. SIMS OF EXCHANGE PROCESSES AT FELDSPAR SURFACES.

SIMS analysis of isotopically-labeled species should allow characterization of exchange processes at a feldspar surface. Cooling of the sample during analysis, and flooding with low-energy electrons, should reduce chemical changes induced by the primary ion beam.

7. Acknowledgements

I am deeply grateful to Joe Pluth for preparing the drawings of surface atoms on feldspar faces; to Ed Nater and Alex Blum for thorough reviews of a seriously incomplete first draft which was cobbled together in odd hours stolen from other activites with higher priority; to Roland

Hellmann for prompt and generous lending of a key AFM image reproduced in Fig. 1; to Ian Parsons for his efficient and friendly direction of the NATO ASI; and to the staff at the University of Chicago libraries who have made it possible to assemble the pertinent literature in so many scientific disciplines. To Mike Hochella, I express sincere thanks for his advice and help, but he should have given this review, and not dropped me into the apple-cart. Finally, I could not have survived the slogging through the surface literature without having had 40 years experience with zeolite science in association with brilliant scientists including Donald Breck, Edith Flanigen and Stephen Wilson. Furthermore, I was forced to bone up on several new types of surface science over the past four years as part of my duties in the Consortium for Advanced Radiation Sources. This organization is managed by the University of Chicago on behalf of large groups of geoscientists, soil mineralogists, environmental scientists, inorganic and physical chemists, materials scientists, biologists and physicists, and is becoming increasingly interdisciplinary. Surface science spans all these disciplines, and gathers strength from synergy among them. Thanks particularly to Stuart Rice, Mark Schlossman, Mark Rivers and Steve Sutton for pertinent scientific assistance with synchrotron matters, and to Nancy Weber for secretarial and managerial support during the pioneering phase of CARS. Finally, I apologize in advance for any errors or infelicities. Because of various unforeseen challenges over the summer, I was unable to ruminate over this review, and it is inevitable that there will be some rough spots and errors. In order to meet Ian Parsons's deadline, it went straight to press without the benefit of detailed review.

8. References

Abruna HD White JH Albarelli MJ Bommarito GM Bedzyk MJ McMillan M (1988) Is there any beam yet? Uses of synchrotron radiation in the in situ study of electrochemical interfaces. J Phys Chem 92: 7045-7052

Addadi L Weiner S (1992) Control and design principles in biological mineralization. Angew Chem Int Ed English 31: 153-160

Ade H Zhang X Cameron S Costello C Kirz J Williams S (1992) Chemical contrast in X-ray microscopy and spatially resolved XANES spectroscopy of organic specimens. Science 258: 972-5

Ague DM (1988) Universal stage measurements and transmission electron microscope observations of fractured plagioclase. J Struct Geol 10: 701-705

AlDahan AA Moran S (1987) A SEM study of dissolution textures of detrital feldspars in Proterozoic sandstones, Sweden. Am J Sci 287: 460-514

Alfredsson V Ohsuna T Terasaki O Bovin J-O (1993) Investigation of the surface structure of the zeolites FAU and EMT by high-resolution transmission electron microscopy. Angew Chem Int Ed Engl 32: 1210-1213

Al-Shaieb Z Hanson RE Donovan RN Shelton JW (1980) Petrology and diagenesis of sandstones in the Post Oak Formation (Permian) southwestern Oklahoma. J Sedim Petr 50: 43-50

Anbeek C (1992) Surface roughness of minerals and implications for dissolution studies. Geochim Cosmochim Acta 56: 1461-1469

Armour WA (1990) Ion scattering spectroscopy. In Walls, 263-298

Artioli G Smith JV Kvick Å (1989) Single-crystal neutron diffraction study of partially dehydrated laumontite at 15 K. Zeolites 9: 377-391

Attwood D (1992) New opportunities at soft-X-ray wavelengths. Physics Today August: 24-31

Bajt S Pernicka E Traxel K (1991) PIXE analyses of olivine grains in Semarkona: microdistribution and correlation of trace elements. Proc Lunar Planet Sci Conf 21: 513-521, Lunar Planet Sci Inst, Houston

Banfield JF Eggleton RA (1990) Analytical transmission electron microscopy studies of plagioclase, muscovite, and K-feldspar weathering. Clays Clay Minerals 38: 77-89

Banfield JF Jones BF Veblen DR (1991) An AEM-TEM study of weathering and diagenesis, Albert Lake, Oregon: I. Weathering reactions in the volcanics. Geochim Cosmochim Acta 55: 2781-2793

Barr TL Chen LM Mohsenian M Lishka MA (1988) XPS valence band study of zeolites and related systems. 1. General chemistry and structure. J Am Chem Soc 110: 7962-7975

Barr TL Lishka MA (1986) ESCA studies of the surface chemistry of zeolites. J Am Chem Soc 108: 3178-3186

Bassett WA Brown GE Jr (1990) Synchrotron radiation in the earth sciences. Ann Rev Earth Planet Sci 18: 387-447

Baumann S Strathman MD Suib SB (1986) The use of Rutherford Backscattering to distinguish surface and bulk species in zeolites. J Chem Soc Chem Comm 308-309

Baur WH (1965) On hydrogen bonds in crystalline hydrates. Acta Cryst 19: 909-916

Benninghoven A Hagenhoff B Niehuis E (1993) Surface MS: probing real-world samples. Anal Chem 65: 630-639A

Beran A (1986) A model of water allocation in alkali feldspar, derived from infrared-spectroscopic investigations. Phys Chem Minerals 13: 306-310

Berner RA (1992) Weathering, plants, and the long-term carbon cycle. Geochim Cosmochim Acta 56 3225-3231

Berner RA Holdren GR Jr (1977) Mechanism of feldspar weathering: Some observational evidence. Geology 5: 369-372

Berner RA Holdren GR Jr (1979) Mechanism of feldspar weathering - II. Observations of feldspars from soils. Geochim Cosmochim Acta 43: 1173-1186

Bernius MT Morrison GH (1987) Mass analyzed secondary ion microscopy. Rev Sci Instr 58: 1789-1804

Berthold T Jagodzinski H (1990) Analysis of the distribution of impurities in crystals by anomalous X-ray scattering. Z Kristallogr 193: 85-100

Bishop HE (1990) Auger electron spectroscopy. In Walls , 87-126

Bleam WF (1993) Atomic theories of phyllosilicates: quantum chemistry, statistical mechanics, electrostatic theory, and crystal chemistry. Reviews in Geophysics 31: 51-73

Blum A White A Hochella M (1992) The surface chemistry and topography of weathered feldspar surfaces. Eos 73 suppl 10/27: 602

Boisen MB Jr Gibbs GV (1993) A modeling of the structure and compressibility of quartz with a molecular potential and its transferability to cristobalite and coesite. Phys Chem Mineral 20: 123-135

Bonsfield B (1992) Surface preparation and microscopy of minerals. John Wiley, Chichester.

Briggs D Brown A Vickerman JC (1989) Handbook of static secondary ion mass spectrometry. Wiley, Chichester

Brimhall GH & 9 others (1992) Deformational mass transport and invasive processes in soil evolution. Science 255: 695-702

Brown GE (1990) Spectroscopic studies of chemisorption reaction mechanisms at oxide-water interfaces. Reviews in Mineralogy 23: 309-363

Brown GE Jr Calas G Waychunas GA Petiau J (1988) X-ray absorption spectroscopy and its applications in mineralogy and geochemistry. Reviews in Mineralogy 18: 431-512

Brown GE Jr Parks GA (1989) Synchrotron-based x-ray absorption studies of cation environments in earth materials. Reviews in Geophysics 27: 519-533

Brown ID (1992) Chemical and steric constraints in inorganic solids. Acta Cryst B48: 553-572

Burdett JK (1983) Theoretical challenges in understanding the structure of solids. J Phys Chem 87: 4368-4375

Cairns-Smith AG Hartman H eds (1986) Clay minerals and the origin of life. Cambridge University Press

Calas G (1988) Electron paramagnetic resonance. In Hawthorne, 513-571

Cambier P Sposito G (1991) Interactions of citric acid and synthetic hydroxy-aluminum montmorillonite. Clays & Clay Minerals 39: 158-166

Cannillo E Coda A Fagnani G (1966) The crystal structure of bavenite. Acta Cryst 20: 301-309

Carozzi AV (1992) Horace-Bénédict de Saussure, a forerunner in 1794-95 of experimental weathering. Clays & Clay Minerals 40: 323-326

Casey WH Bunker B (1990) Leaching of mineral and glass surfaces during dissolution. Reviews in Mineralogy 23: 397-426

Casey WH Westrich HR Arnold GW (1988) Surface chemistry of labradorite feldspar reacted with aqueous solutions at pH = 2, 3, and 12. Geochim Cosmochim Acta 52: 2795-2807

Casey WH Westrich HR Arnold GW Banfield JF (1989a) The surface chemistry of dissolving labradorite feldspar. Geochim Cosmochim Acta 53: 821-832

Casey WH Westrich HR Massis T Banfield JF Arnold GW (1989b) The surface of labradorite feldspar after acid hydrolysis. Chem Geol 78: 205-218

Catlow CRA Gale JD Grimes RW (1990) Recent computational studies in solid state chemistry. J Solid State Chemistry 106: 13-26

Cherniak DJ (1992) Diffusion of Pb in feldspar measured by Rutherford Backscattering Spectroscopy. Eos 73 suppl 10/27: 641

Cherniak DJ Watson EB (1992) A study of strontium diffusion in K-feldspar, Na-K feldspar and anorthite using Rutherford Backscattering Spectroscopy. Earth Planet Sci Lett 113: 411-425

Chiarello RP Wogelius RA Sturchio NC (1993) In-situ synchrotron X-ray reflectivity measurements at the calcite-water interface. Geochim Cosmochim Acta 57: 4103-4110

Christie AB (1990) X-ray photoelectron spectroscopy. In: Walls, 127-168

Cook SJ Chakraborty AK Bell AT Theodorou DN (1993) Structural and electronic features of a Brønsted acid site. J Phys Chem 97: 6679-6685

Coppens P with contributions by Cox D Vlieg E Robinson IK (1992) Synchrotron radiation crystallography. Academic, London

Cremeens DL Darmody RG Jansen IJ (1987) SEM analysis of weathered grains. Pretreatment effects. Geology 15: 401-404

Cremeens DL Darmody RG Norton LD (1992) Etch-pit size and shape distribution on orthoclase and pyriboles in a loess catena. Geochim Cosmochim Acta 56: 3423-3434

Cressey G Henderson CMB van der Laan G (1993) Use of L-edge absorption spectroscopy to characterize multiple valence states of 3d transition metals; a new probe for mineralogical and geochemical research. Phys Cmem Minerals 20 111-119

Curtis GP Reinhard M Roberts PV (1986) Sorption of hydrophobic organic compounds by sediments. In Davis & Hayes 191-216

Davis JA Hayes KF eds (1986) Geochemical processes at mineral surfaces. Am Chem Soc Symp Ser 323
Deem MW Newsam JM Creighton JA (1992) Fluctuations in zeolite aperture dimensions simulated by crystal dynamics. J Am Chem Soc 114: 7198-7207
Dickinson JT Jensen LC Langford SC Rosenberg PE (1992) Fracture-induced emission of alkali atoms from feldspar. Phys Chem Minerals 18: 453-459
Dosch H (1992) Critical phenomena at surfaces and interfaces: evanescent x-ray and neutron scattering. Springer, New York
Dove PM Hochella MF Jr (1993) Calcite precipitation mechanisms and inhibition by orthophosphate: In situ observations by Scanning Force Microscopy. Geochim Cosmochim Acta 57: 705-714
Drake B Hellmann R (1991) Atomic force microscopy imaging of the albite (010) surface. Am Mineral 76: 1773-1776
Dran J-C Moulin V Petit J-C Ramsay JDF Russel P Stefanini A (1992) Mechanisms of colloid sorption on mineral surfaces: a RBS study. Applied Geochemistry Suppl Iss 1: 187-191
Eckert H Yesinowski JP Silver LA Stolper EM (1988) Water in silicate glasses: Quantitation and structural studies by ^1H solid echo and MAS-NMR methods. J Phys Chem 92: 2055-2064
Eggleton RA Buseck PR (1980) High resolution electron microscopy of feldspar weathering. Clays Clay Minerals 28: 173-178
Elphick SC Graham CM Dennis PF (1988) An ion microprobe study of anhydrous oxygen diffusion in anorthite: a comparison with hydrothermal data and some geological implications. Contrib Mineral Petrol 100: 490-495
Elphick SC Graham CM Walker FDL Holness MB (1991) The application of SIMS ion imaging techniques in the experimental study of fluid-mineral interactions. MM 55: 347-356
Engel N (1991) Crystallochemical model and prediction for zeolite-type structures. Acta Cryst B47: 849-858
Engel N Yvon K (1984) The crystal structure of parthéite. Z Kristallogr 169: 165-175
Engelhardt G Michel D (1987) High-resolution solid-state NMR of silicates and zeolites. Wiley, Chichester
Enomoto Y Hashimoto H (1990) Emission of charged particles from indentation fracture of rocks. N 346: 641-643
Feidenhans'l R (1989) Surface structure determination by x-ray diffraction. Surface Structure Reports 10: 105-188
Fischer TE Mullins WM (1992) Chemical aspects of ceramic tribology. J Phys Chem 96: 5690-5701
Flanigen EM (1976) Structural analysis by infrared spectroscopy. Am Chem Soc Monograph 171: 80-117
Frank DG Batina N Golden T Lu F Hubbard AT (1990) Imaging surface atomic structure by means of Auger electrons. Science 247: 182-188
Franke WA (1989) Tracht and habit of synthetic minerals grown under hydrothermal conditions. Eur J Mineral 1: 557-566
Flannery BP Deckman HW Roberge WG D'Amico KL (1987) Three-dimensional X-ray microtomography. Science 237: 1439-1444
Fraser DG (1990) Applications of the high-resolution scanning proton microprobe in the earth sciences: an overview. Chem Geol 83: 27-37
Friedbacher G Hansma PK Ramli E Stucky GD (1991) Imaging powders with the atomic force microscope: from biominerals to commercial materials. Science 253: 1261-1263

Fritz SJ Mohr DW (1984) Chemical alteration in the micro weathering environment within a spheroidally-weathered anorthosite boulder. Geochim Cosmochim Acta 48: 2527-35
Fung PC Sanipelli GG (1982) Surface studies of feldspar dissolution using surface replication combined with electron microscopic and spectroscopic techniques. Geochim Cosmochim Acta 46: 503-512
Fuoss PH Brennan S (1990) Surface sensitive X-ray scattering. Ann Rev Mater Sci 20: 365-390
Galli E (1980) The crystal structure of roggianite, a zeolite-like silicate. Proc 5th Int Conf Zeolites, ed. LVC Rees, 205-213, Heyden, London
Gibbs GV (1982) Molecules as models for bonding in silicates. Am Mineral 67: 421-450
Giese RF Jr Costanzo PM (1986) Behavior of water on the surface of kaolin minerals. In Davis & Hayes, 37-53
Giuseppetti G Mazzi F Tadini C (1987) Further observations on the extra-framework atoms in the crystal structure of sarcolite. Neues Jb Mineral Monat (11): 521-7
Goldstein JI Newbury DE Echlin P Joy DC Romig AD Jr Lyman CE Fiori C Lifshin E (1992) Scanning electron microscopy and X-ray microanalysis. 2nd ed, Plenum, New York
Gonis A Stocks GM eds (1992) Equilibrium structure and properties of surfaces and interfaces. NATO ASI Ser B Physics, Plenum, NY
Goodman BA (1986) Adsorption of metal ions and complexes on aluminosilicate minerals. In Davis & Hayes, 342-361
Graham CM Elphick SC (1990) A re-examination of the role of hydrogen in Al-Si interdiffusion in feldspars. Contrib Mineral Petrol 104: 481-491
Grant WA (1990) Rutherford back-scattering spectrometry. In: Walls, 299-337
Grasserbauer M (1990) Elemental trace analysis of surfaces and interfaces: goals, accomplishments and challenges. Phil Trans R Soc London A 333: 113-132
Gratz AJ Hillner PE (1993) Poisoning of calcite growth viewed in the atomic force microscope (AFM). J Crystal Growth 129: 789-793
Gratz AJ Hillner P Hansma PK (1992) Step dynamics and spiral growth on calcite. Geochim Cosmochim Acta 57: 491-495
Gratz AJ Manne S Hansma PK (1991) Atomic force microscopy of atomic-scale ledges and edge pits formed during dissolution of quartz. Science 251: 1343-1346
Gregg SJ Sing KSW (1992) Adsorption, surface area, and porosity. Academic, NY
Grünert W Schögl R Karge HG (1993) Investigations of zeolites by photoelectron and ion scattering spectroscopy. 1. New applications of surface spectroscopic methods to zeolites by a high-temperature measurement technique. J Phys Chem 97: 8638-8645
Harrick NJ (1976) Internal reflection spectroscopy. Harrick Scientific Corporation, Ossining, New York
Hartman H Sposito G Yang A Manne S Gould SAC Hansma PK (1990) Molecular-scale imaging of clay mineral surfaces with the atomic force microscope. Clays & Clay Minerals 38: 337-342
Hartman P (1979) Zum Verständnis des Mineralwachstums. Fortschr Miner 57: 127-171
Hawthorne FC ed (1988) Spectroscopic methods in mineralogy and geology. Rev Mineralogy 18: 1-698
Hawthorne FC (1992) The role of OH and H_2O in oxide and oxysalt minerals. Z Kristallogr 201: 183-206
Hayes KF Roe AL Brown GE Jr Hodgson KO Leckie JO Parks GA (1987) In situ absorption study of surface complexes: selenium oxyanions on a-FeOOH. Science 238: 783-786

Hazemann JL Manceau A Sainctavit Ph Malgrange C (1992) Structure of the a-$Fe_xAl_{1-x}OOH$ solid solution. I. Evidence by polarized EXAFS for an epitaxial growth of hematite-like clusters in Fe-diaspore. Phys Chem Minerals 19: 25-38

Hehre WJ Radom L Schleyer PVR Pople JA (1986) Ab initio molecular orbital theory. Wiley, New York.

Hellmann R Crerar DA Zhang R (1989) Albite feldspar hydrolysis to 300°C. Solid State Ionics 32/33: 314-329

Hellmann R Drake B Kjoller K (1992) Using atomic force microscopy to study the structure, topography and dissolution of albite surfaces. In: Kharaka YK Maest AS (eds) Water-rock interaction: Balkema, pp 149-151

Hellmann R Eggleston CM Hochella MF Jr Crerar DA (1990) The formation of leached layers on albite surfaces during dissolution under hydrothermal conditions. Geochim Cosmochim Acta 54: 1267-1281

Henderson CMB Charnock JM Smith JV Greaves GN (1993) X-ray absorption of Fe, Mn, Zn, and Ti structural environments in staurolite. Am Mineral 78: 477-485

Hering JG Stumm W (1990) Oxidative and reductive dissolution of minerals. Reviews in Mineralogy 23: 427-465

Hiebert FK Bennett PC (1992) Microbial control of silicate weathering in organic-rich ground water. Science 258: 278-281

Hillner PE Gratz AJ Manne S Hansma PK (1992) Atomic-scale imaging of calcite growth and dissolution in real time. Geology 20: 359-362.

Hinton RW (1990) Ion microprobe trace-element analysis of silicates: measurement of multi-element glasses. Chem Geol 83: 11-25

Hirt WG Wenk H-R Boles JR (1993) Albitization of plagioclase crystals in the Stevens sandstone (Miocene), San Joaquin Basin, California, and the Frio Formation (Oligocene), Gulf Coast, Texas: A TEM/AEM study. Geol Soc Am Bull 105: 708-14

Hochella MF Jr (1988) Auger electron and X-ray photoelectron spectroscopies. Reviews in Mineralogy 18: 573-637

Hochella MF Jr (1990) Atomic structure, microtomography, composition, and reactivity of mineral surfaces. Reviews in Mineralogy 23: 87-132

Hochella MF Jr Brown GE (1988) Aspects of silicate surface and bulk structure analysis using X-ray photoelectron spectroscopy (XPS). Geochim Cosmochim Acta 52: 1641-1648

Hochella MF Jr Eggleston CM Elings VB Thompson MS (1990) Atomic structure and morphology of the albite {010} surface: an atomic-force microscope and electron diffraction study. Am Mineral 75: 723-730

Hochella MF Jr Lindsay JR Mossotti VG Eggleston CM (1988a) Sputter depth profiling in mineral-surface analysis. Am Mineral 73: 1449-1456

Hochella MF Jr Ponader HB Turner AM Harris DW (1988b) The complexity of mineral dissolution as viewed by high resolution scanning Auger microscopy: labradorite under hydrothermal conditions. Geochim Cosmochim Acta 52: 385-394

Hochella MF Jr White AF (1990) Mineral-water interface geochemistry: an overview. Reviews in Mineralogy 23: 1-16

Hoffman R (1988) Solids and surfaces: a chemist's view of bonding. VCH, New York

Howe RF (1993) Surface science in three dimensions: zeolites as model catalysts. Springer Proceedings in Physics 73: 242-245

Howells M Kirz J Sayre D Schmahl G (1985) Soft-x-ray microscopes. Physics Today August: 22-32

Hu P King DA (1992) A direct inversion method for surface structure determination from LEED intensities. Nature 360: 655-8.

Huang PM Schnitzer M eds (1986) Interactions of soil minerals with natural organics and microbes. Soil Science Society of America Spec Publ 17

Inskeep WP Nater EA Bloom PR Vandervoort DS Erich MS (1991) Characterization of laboratory weathered labradorite surfaces using X-ray photoelectron spectroscopy and transmission electron microscopy. Geochim Cosmochim Acta 55: 787-800

Jach T Bedzyk MJ (1993) X-ray standing waves at grazing angles. Acta Crystallogr A49: 346-350

Jackson LL Baedecker PA Fries TL Lamothe PJ (1993) Geological and inorganic materials. Anal Chem 65: 12R-28R

Jackson MD Gibbs GV (1988) A modeling of the coesite and feldspar framework structure types of silica as a function of pressure using modified electron gas methods. J Phys Chem 92: 540-545

Jacobs PA Mortier WJ (1982) An attempt to rationalize stretching frequencies of lattice hydroxyl groups in hydrogen-zeolites. Zeolites 2: 226-230

Jacobs WPJH van Wolput JHMC van Santen RA (1993) Fourier-transform infrared study of the protonation of the zeolitic lattice. Influence of silicon:aluminium ratio and structure. J Chem Soc Faraday Trans 89: 1271-1276

Johansson SAE Campbell JL (1988) PIXE: a novel technique for elemental analysis. John Wiley: Chichester

Jones KW (1992) Synchrotron radiation-induced X-ray emission. In: Van Grieken RE & Markowicz AA (eds) Handbook of X-ray spectrometry: Marcel Dekker, NY, 411-451

Jones KW Gordon BM (1989) Trace element determination with synchrotron-induced X-ray emission. Anal Chem 61: 341-358A

Junta JL Hochella MF Jr Harris DW Edgell M (1992) Manganese oxidation at mineral-water interfaces, a spectroscopic approach. In: Kharaka YK Maest AS (eds) Water-rock interaction: Balkema, pp 163-166

Kinney DR Chuang I-S Maciel GE (1993) Water and the silica surface as studied by variable-temperature high-resolution ^{1}H NMR. J Amer Chem Soc 115 6786-6794

Kinney JH Nichols MC (1992) X-ray tomographic microscopy (XTM) using synchrotron radiation. Ann Rev Mater Sci 22: 121-152

Kirkpatrick RJ (1988) MAS NMR spectroscopy of minerals and glasses. In Hawthorne, 341-403

Kittaka S Sasaki T Fukuhara N (1992) Fourier-transform infrared spectroscopy of H_2O molecules on the ZnO surface. Langmuir 8: 2598-2600

Kittaka S Sasaki T Fukuhara N Kato H (1993) Fourier-transform infrared spectroscopy of H_2O molecules on the Cr_2O_3 surface. Surface Science 282: 255-261

Kittrick JA ed (1986) Soil mineral weathering. Van Nostrand Reinhold, NY

Klockenkämper R Knoth J Prange A Schwenke H (1992) Total-reflection X-ray fluorescence spectroscopy. Anal Chem 64: 1115-1123A

Koningsberger DC Prins R (eds) (1988) X-ray absorption: principles, applications, techniques of EXAFS, SEXAFS, and XANES. John Wiley, New York

Kramer GJ De Man AJM Van Santen RA (1991a) Zeolites versus aluminosilicate clusters: the validity of a local description. J Am Chem Soc 113: 6435-6441

Kramer GJ Farragher NP van Best BWH van Santen RA (1991b) Interatomic force fields for silicas, aluminophosphates, and zeolites: Derivation based on *ab initio* calculations. Phys Rev B 43: 5068-78

Kramer GJ van Santen RA (1993) Theoretical determination of proton affinity differences in zeolites. J Am Chem Soc 115: 2887-2897

Kramer GJ van Santen RA Emeis CA (1993) Understanding the acid behavior of zeolites from theory and experiment. Nature 363: 529-531

Lagally MG ed (1990) Kinetics of ordering and growth at surfaces. Plenum, New York

Lasaga AC (1990) Atomic treatment of mineral-water surface reactions. Reviews in Mineralogy 23: 17-85

Lasaga AC (1992) Ab initio methods in mineral surface reactions. Reviews in Geophysics 30: 269-303

Lasaga AC Gibbs GV (1990) Ab initio quantum-mechanical calculations of surface reactions - a new era? In W Stumm 1987, 259-89

Laursen T Palmer GR Amm DT Johnson D (1993) RBS analysis and Langmuir-Blodgett films. Nuclear Instruments & Methods in Physics Research B 82 125-128

Leapman RD Newbury DE (1993) Trace element analysis at naometer spatial resolution by parallel-detection electron energy loss spectroscopy. Anal Chem 65: 2409-2414

Levi-Setti R (1990) Recent applications of high resolution secondary-ion-mass-spectrometry imaging microanalysis. Vacuum 41: 1598-1600

Li J-C Ross DK Heenan RK (1993) Small-angle neutron-scattering studies of the transition between water and ice in a restricted geometry. Phys Rev B48: 6716-6719

Lin X-Y Creuzet F Arribart H (1993) Atomic force microscopy for local characterization of surface acid-base properties. J Phys Chem 97: 7272-6

Liu F Garofalini SH King-Smith RD Vanderbilt D (1993) First-principles studies on structural properties of b-cristobalite. Phys Rev Lett 70: 2750-2753

Lu F-Q Smith JV Sutton SR Rivers ML Davis AM (1989) Synchrotron X-ray fluorescence analysis of rock-forming minerals 1. Comparison with other techniques 2. White-beam energy-dispersive procedure for feldspars. Chem Geol 75: 123-143

MacDougall JE Cox SD Stucky GD Weisenhorn AL Hansma PK Wise WS (1991) Molecular resolution of zeolite surfaces as imaged by atomic force microscopy. Zeolites 11: 429-433

Maes A Cremers A (1986) Highly selective ion exchange in clay minerals and zeolites. In Davis & Hayes, 254-295

Manassidis I De Vita A Gillan MJ (1993) Structure of the (0001) surface of a-Al_2O_3 from first principles calculations. Surface Science Letters 285: L517-521

Manning DAC Hall PL Hughes CR eds (1993) Geochemistry of clay-pore fluid interactions. Chapman & Hall, London

Martini IP Chesworth W eds (1992) Weathering, soils & paleosols. Elsevier, Amsterdam

Matsui J Mizuki J (1993) Studies of semiconductor interfaces by grazing incidence x-ray diffraction. Ann Rev Mater Sci 23: 295-320

McBride MB (1986) Paramagnetic probes of layer silicate surfaces. In Davis & Hayes, 362-388

McDaniel CV Maher PK (1976) Zeolite stability and ultrastable zeolites. Am Chem Soc Monograph 171: 285-331

McGuire GE & 8 others (1993) Surface characterization. Anal Chem 65: 311-333R

McKeague JA Cheshire MV Andreux F Berthelin J (1986) Organo-mineral complexes in relation to pedogenesis. In Huang & Schnitzer, 549-592

McLaren AC (1991) Transmission electron microscopy of minerals and rocks. Cambridge University Press

McMillan PF Hofmeister AM (1988) Infrared and Raman spectroscopy. Reviews in Mineralogy 18: 99-159

McNulty I Kirz J Jacobsen C Anderson EH Howells MR Kern DP (1992) High-resolution imaging by Fourier transform X-ray holography. Science 256: 1009-1012

McPherson A Shlichta P (1988) Heterogeneous and epitaxial nucleation of protein crystals on mineral surfaces. Science 239: 385-387

Megaw HD (1973) Crystal structures: a working approach. Saunders, Philadelphia

Melendres CA Beden B Bowmaker G Liu C Maroni VA (1993) Synchrotron infrared spectroscopy of H_2O adsorbed on polycrystalline gold. Langmuir 9: 1980-1982

Miura Y Rucklidge JC (1979) Ion microprobe analysis of exsolution lamellae in peristerites and cryptoperthites. Am Mineral 64: 1272-1279

Moggridge GD Rayment T Ormerd RM Morris MA Lambert RM (1992) Spectroscopic observation of a catalyst surface in a reactive atmosphere at high pressure. Nature 358: 658-660

Mortier WJ Sauer J Lercher JA Noller H (1984) Bridging and terminal hydroxyls. A structural and quantum chemical discussion. J Phys Chem 88: 905-912

Muir IJ Bancroft GM Nesbitt HW (1989) Characteristics of altered labradorite surfaces by SIMS and XPS. Geochim Cosmochim Acta 53: 1235-1241

Muir IJ Bancroft GM Shotyk W Nesbitt HW (1990) A SIMS and XPS study of dissolving plagioclase. Geochim Cosmochim Acta 54: 2247-2256

Nakamoto K 1978 Infrared and Raman spectra of inorganic and coordination compounds. 3rd ed. Wiley, NY

Nesbitt HW MacRae ND Shotyk W (1991) Congruent and incongruent dissolution of labradorite in dilute, acidic, salt solutions. J Geol 99: 429-442

Nesbitt HW Muir IJ (1988) SIMS depth profiles of weathered plagioclase and processes affecting dissolved Al and Si in some acidic soil solutions. Nature 334: 336-338

Occelli ML Drake B Gould SAC (1993) Characterization of pillared montmorillonites with the atomic force microscope. J Catal: 142 337-348

Ohtaki H Radnai T (1993) Structure and dynamics of hydrated ions. Chem Rev 93: 1157-1204

Ozin GA Baker MD Godber J Gil CJ (1989) Intrazeolite site-selective far-IR cation probe. J Phys Chem 93: 2899-2908

Pacchioni G (1993) Physisorbed and chemisorbed CO_2 at surface and step sites of the MgO (100) surface. Surface Science 281: 207-219

Page R Wenk HR (1979) Phyllosilicate alteration of plagioclase studied by transmission electron microscopy. Geology 7: 393-397

Parks GA (1990) Surface energy and adsorption at mineral-water interfaces: an introduction. Reviews in Mineralogy 23: 133-175

Paul AJ Vickerman JC (1990) Organics at surfaces, their detection and analysis by static secondary ion mass spectrometry. Phil Trans R Soc London A333: 147-158

Paulus H Müller G (1988) The crystal structure of a hydrogen-feldspar. Neues Jb Mineral Monat 481-490

Pellin MJ Young CE Calaway WF Gruen DM (1986) Trace surface analysis: 30 ppb analysis with removal of less than a monolayer. Fe and Ti impurities in the first atomic layer of Si wafers. Nucl Instr Meth Phys Res B13: 653-657

Pelmenshchikov AG Pavlov VI Zhidomirov GM Beran S (1987) Effects of structural and chemical characteristics of zeolites on the properties of their bridging hydroxyl groups. J Phys Chem 91: 3325-3327

Pendry JB (1974) Low energy electron diffraction. Academic, London

Pennycook SJ (1992) Atomic-scale imaging of materials by Z-contrast scanning transmission electron microscopy. Anal Chem 64: 263A-272A

Perry DL ed. (1990) Instrumental surface analysis of geological materials. VCH Publishers, New York

Petit J-C Della Mea C Dran J-C Magonthier M-C Mando PA Paccagnella A (1990a) Hydrated-layer formation during dissolution of complex silicate glasses and minerals. Geochim Cosmochim Acta 54: 1941-1955

Petit J-C Dran J-C Della Mea G (1990b) Energetic ion beam analysis in the earth sciences. Nature 344: 621-6

Petit J-C Dran J-C Paccagnella A Della Mea G (1989a) Structural dependence of crystalline silicate hydration during aqueous dissolution. Earth Planet Sci Lett 93: 292-298

Petit J-C Dran J-C Schott J Della Mea G (1989b) New evidence on the dissolution mechanism of crystalline silicates by MeV ion beam techniques. Chem Geol 76: 365-369

Petrov I Agel A Hafner SS (1989) Distinct defect centers at oxygen positions in albite. Am Mineral 74: 1130-1141

Petrov I Hafner SS (1988) Location of trace Fe^{3+} ions in sanidine, $KAlSi_3O_8$. Am Mineral 73: 93-104

Petrovic R Berner RA Goldhaber MB (1975) Rate control in dissolution of alkali feldspars. I. Study of residual feldspar grains by X-ray photoelectron spectroscopy. Geochim Cosmochim Acta 40: 537-548

Pluth JJ Smith JV (1982) Crystal structure of dehydrated Sr-exchanged zeolite A. Absence of near-zero-coordinate Sr^{2+}. Presence of Al complex. J Am Chem Soc 104: 6977-6982

Post JE Burnham CW (1987) Structure energy calculations on low and high albite. Am Mineral 72: 507-514

Potts PJ (1987) A handbook of silicate rock analysis. Blackie, Glasgow

Price GD & Ross NL eds (1992) The stability of minerals. Chapman & Hall, London

Purton J Catlow CRA (1990) Computer simulation of feldspar structures. Am Mineral 75: 1268-1273

Rabalais JW (1990) Scattering and recoiling spectrometry: An ion's view of surface structure. Science 250: 521-527

Rachlin AL Henderson GS Goh MC (1992) An atomic force microscope (AFM) study of the calcite cleavage plane: Image averaging in Fourier space. Am Mineral 77: 904-910

Reed SJB (1989) Ion microprobe analysis - a review of geological applications. Mineral Mag 53: 3-24

Reed SJB (1993) Electron microprobe analysis. Second edition. Cambridge University Press

Rivers ML Sutton SR Jones KW (1991) Synchrotron X-ray fluorescence microscopy. Synchrotron Radiation News 4: 23-26

Rivers ML (1993) GeoCARS: A national resource for earth, planetary, soil and environmental science research at the Advanced Photon Source. Proposal submitted to NSF June 14. Copy available from JVS.

Robert M Berthelin J (1986) Role of biological and biochemical factors in soil mineral weathering. In Huang PM & Schnitzer M, 453-495

Roberts KJ (1993) The application of synchrotron X-ray techniques to problems in crystal science and engineering. J Crystal Growth 130: 657-681

Robinson IK Tweet DJ (1992) Surface x-ray diffraction. Rep Progr Phys 55: 599-651

Romero R Robert M Elsass F Garcia C (1992) Evidence by transmission electron microscopy of weathering microsystems in soils developed from crystalline rocks. Clay Minerals 27: 21-33

Rossi G Tazzoli V Ungaretti L (1974) The crystal structure of ussingite. Am Mineral 59: 335-340

Rossman GR (1988) Vibrational spectroscopy of hydrous components. Reviews in Mineralogy 18: 193-206

Rossman GR Smyth JR (1990) Hydroxyl contents of accessory minerals in mantle eclogites and related rocks. Am Mineral 75: 775-780

Rugar D Hansma P (1990) Atomic force microscopy. Physics Today October: 23-30

Ryan CG Griffin WL (1993) The nuclear microprobe as a tool in geology and mineral exploration. Nuclear Instruments & Methods in Physics Research B77: 381-398

Ryan CG Heinrich CA Mernagh TP (1993) PIXE microanalysis of fluid inclusions and its application to study ore metal segregation between brine and vapor. Nucl Instr Methods Phys Res B77: 463-471

Sauer J (1987) Molecular structure of orthosilicic acid, silanol, and $H_3SiOH.AlH_3$ complex: models of surface hydroxyls in silica and zeolites. J Phys Chem 91: 2315-2319

Sauer J (1989) Molecular models in ab initio studies of solids and surfaces: from ionic crystals and semiconductors to catalysts. Chem Rev 89: 199-255

Sauer J Koelmel CM Hill JR Alrichs R (1989) Brønsted sites in zeolitic catalysts. An *ab initio* study of local geometries and of the barrier for proton jumps between neighboring sites. Chem Phys Lett 164: 193-198

Scandella L Kruse N Prins R (1993) Imaging of zeolite surface structures by atomic force microscopy. Surf Sci 281: L331-334

Schindler PW (1990) Co-adsorption of metal ions and organic ligands: formation of ternary surface complexes. Reviews in Mineralogy 23: 282-307

Schröder K-P Sauer J (1993) Preferred stability of Al-O-Si-O-Al linkages in high-silica zeolite catalysts. Theoretical predictions contrary to Dempsey's rule. J Phys Chem 97: 6579-6581

Schwartzman DW Volk T (1989) Biotic enhancement of weathering and the habitability of the Earth. Nature 340: 457-460

Sharp TG Oden PI Buseck PR (1993) Lattice-scale imaging of mica and clay (001) surfaces by atomic force microscopy using net attractive forces. Surface Science Letters 284: L405-410.

Sharpe LR Heineman WR Elder RC (1990) EXAFS spectroelectrochemistry. Chem Rev 90: 705-22.

Shindo H Nozoye H (1992) Structure of cleaved surfaces of anhydrite ($CaSO_4$) studied with atomic force microscopy. J Chem Soc Faraday Trans 88:711-714

Shotyk W Nesbitt HW (1992) Incongruent and congruent dissolution of plagioclase feldspar: effect of feldspar composition and ligand complexation. Geoderma 55: 55-78

Skipper NT Soper AK McConnell JDC (1991) The structure of interlayer water in vermiculite. J Chem Phys 94: 5751-60

Skoog DA (1987) Principles of instrumental analysis. 3rd ed. Saunders College Publishing, Philadelphia

Smart P Tovey NK (1982) Electron microscopy of soils and sediments: techniques: examples. Oxford University Press

Smith JV (1974) Feldspar minerals. 1. Crystal structure and physical properties. 2. Chemical and textural properties. Springer, Berlin

Smith JV (1992) Present status of and future prospects for synchrotron-based microtechniques that utilize X-ray fluorescence, absorption spectroscopy, diffraction and tomography. Inst Phys Conf Ser 130: 605-612

Smith JV Brown WL (1988) Feldspar minerals. 1. Crystal structures, physical, chemical and microtextural properties. Springer, Berlin

Smith R Walls JM (1990) Ion erosion in surface analysis. In Walls, 20-56

Snyder SR White HS (1992) Scanning tunneling microscopy, atomic force microscopy, and related techniques. Anal Chem 64: 116-134R

Sparks DL Suarez DL eds (1991) Rates of soil chemical processes. Soil Sci Amer Spec Publ 27

Specht ED Walker FJ (1993) Oxidation state of a buried interface: near-edge x-ray fine structure of a crystal truncation rod. Phys Rev B 47 13743-13751

Spence JCH Zuo JM (1992) Electron microdiffraction. Plenum, NY

Sposito G (1984) The surface chemistry of soils. Oxford University Press, New York

Sposito G (1989) The chemistry of soils. Oxford University Press, New York

Sposito G (1990) Molecular models of ion adsorption on mineral surfaces. Reviews in Mineralogy 23 261-279

Sposito G Reginato RJ eds (1992) Opportunities in basic soil science research. Soil Science Society of America, Madison WI

Stähl K Kvick Å Smith JV (1990) Thomsonite, a neutron diffraction study at 13 K. Acta Cryst C46: 1370-1373

Stebbins JF (1988) NMR spectroscopy and dynamic processes in mineralogy and geochemistry. In Hawthorne, 405-429

Stöhr J (1992) NEXAFS spectroscopy. Springer

Stumm W (1987) Aquatic surface chemistry. John Wiley, New York

Stumm W (1992) Chemistry of the solid-water interface. Wiley-Interscience, New York

Sukes D Kubicki JD (1993) A model for H_2O solubility mechanisms in albite melts from infrared spectroscopy and molecular orbital calculations. Geochim Cosmochim Acta 57: 1039-1052

Sutton SR Rivers ML Bajt S Jones KW (1993) Synchrotron X-ray fluorescence microprobe analysis with bending magnets and insertion devices. Nucl Instr Methods Phys Res B75: 553-558

Sverjensky DA (1993) Physical surface-complexation models for sorption at the mineral-water interface. Nature 364: 776-780

Sykes DE (1990) Dynamic secondary ion mass spectroscopy. In Walls JM, 216-262

Tan KH (1986) Degradation of soil minerals by organic acids. In Huang PM & Schnitzer M, 1-27

Tapper UAS Malmqvist KG (1991) Analysis, imaging, and modification of microscopic specimens with accelerator beams. Anal Chem 63: 715-725A

Tarrach G Bürgler D Schaub T Wiesendanger R Güntherodt H-J (1993) Atomic surface structure of Fe_3O_4 (001) in different preparation stages studied by scanning tunneling microscopy. Surface Science 285: 1-14

Thiel PA, Madey TE (1987) The interaction of water with solid surfaces: fundamental aspects. Surface Science Reports 7: 211-385

Thundat T Zheng X-Y Chen GY Warmack RJ (1993) Role of relative humidity in atomic force microscopy imaging. Surface Science Letters 294: L939-943

Toney MF Wiesler DG (1993) Instrumental effects on measurements of surface X-ray diffraction rods: Resolution function and active sample area. Acta Cryst A49: 624-642

Török SB Van Grieken RE (1992) X-ray spectrometry. Anal Chem 64: 180-196R

Tsukimura K Kanazawa Y Aoki M Bunno M (1993) Structure of wadalite $Ca_6Al_5Si_2O_{16}Cl_3$. Acta Cryst C49: 205-207

Turner NH (1990) Surface analysis: X-ray photoelectron spectroscopy and Auger electron spectroscopy. Anal Chem 62: 113-125R

Tweet DJ Akimoto K Tatsumi T Hirosawa I Mizuki J Matsui J (1992) Direct observation of Ge and Si ordering at the Si/B/Ge_xSi_{1-x}(111) interface by anomalous X-ray diffraction. Phys Rev Lett 69: 2236-2239

Van Capellen P Charlet L Stumm W Wersin P (1993) A surface complexation model of the carbonate mineral-aqueous solution interface. Geochim Cosmochim Acta 57: 3505-3518

van der Pluijm BA Lee JH Peacor DR (1988) Analytical electron microscopy and the problem of potassium diffusion. Clays Clay Minerals 36: 498-504

Velbel MA (1986) Influence of surface area, surface characteristics, and solution composition on feldspar weathering rates. In: Davis & Hayes, 615-634

Vickerman JC (1990) Static secondary ion mass spectrometry. In Walls, 169-215

Vickerman JC Brown A Reed NM (1989) Secondary ion mass spectrometry - principles and applications. Oxford University Press

Voudrias EA Reinhard M (1986) Abiotic organic reactions at mineral surfaces. In Davis & Hayes, 462-486.

Waite TD (1986) Photoredox chemistry of colloidal metal oxides. In Davis & Hayes, 426-445

Walker FDL (1990) Ion microprobe study of intragrain micropermeability in alkali feldspars. Contrib Mineral Petrol 106: 124-128

Walls JM ed (1990) Methods of surface analysis: techniques and applications. Cambridge Univ Press: Cambridge, paperback

Wang J Bedzyk MJ Penner TL Caffrey M (1991) Structural studies of membranes and surface layers up to 1,000Å thick using X-ray standing waves. Nature 354: 377-380

Wang J Bedzyk MJ Caffrey M (1992) Resonance-enhanced x-rays in thin films: a structure probe for membranes and surface layers. Science 258: 775-778

Wang TSC Huang PM Chou C-H Chen J-H (1986) The role of soil minerals in the abiotic polymerization of phenolic compounds and formation of humic substances. In PM Huang & M Schnitzer, 251-281

Ward JW (1976) Infrared studies of zeolite surfaces and surface reactions. Am Chem Soc Monograph 171: 118-284

Waychunas GA (1988) Luminescence, X-ray emission and new spectroscopies. Reviews in Mineralogy 18: 639-698

Waychunas GA Brown GE Jr (1990) Polarized X-ray absorption spectroscopy of metal ions in minerals. Applications to site geometry and electronic structure determination. Phys Chem Minerals 17: 420-430

Weisenhorn AL Mac Dougall JE Gould SAC Cox SD Wise WS Massie J Maivald P Elings VB Stucky GD Hansma PK (1990) Imaging and manipulating molecules on a zeolite surface with an atomic force microscope. Science 247: 1330-1333

Welch SA Ullman WJ (1993) The effect of organic acids on plagioclase dissolution rates and stoichiometry. Geochim Cosmochim Acta 57: 2725-2736

Wenk H-R (1973) The structure of wenkite. Z Kristallogr 137: 113-126

Westrich HR Casey WH Arnold GW (1989) Oxygen isotope exchange in the leached layer of labradorite feldspar. Geochim Cosmochim Acta 53: 1681-1685

White JL Beck LW Haw JF (1992) Characterization of hydrogen bonding in zeolites by proton solid-state NMR spectroscopy. J Am Chem Soc 114: 6182-6189

Willard HH Merritt LL Jr Dean JA Settle FA Jr (1988) Instrumental methods of analysis. 7th ed. Wadsworth

Wollast R Chou L (1992) Surface reactions during the early stages of weathering of albite. Geochim Cosmochim Acta 56: 3113-21

Woodruff DP Delchar TA (1986) Modern techniques of surface science. Cambridge University Press, New York

Worden RH Rushton JC (1992) Diagenetic K-feldspar textures: A TEM study and model for diagenetic feldspar growth. J Sedim Petrol 62: 779-89

Xu X Vesecky SM Goodman DW (1992) Infrared reflection-absorption spectroscopy and STM studies of model silica-supported copper catalysts. Science 258: 788-790

Yang W-H A Kirkpatrick RJ (1989) Hydrothermal reaction of albite and a sodium aluminosilicate glass: A solid-state NMR study. Geochim Cosmochim Acta 53: 805-819

Yates JT Jr (1992) Surface chemistry. Chem & Eng News March 30: 22-35

Zeltner WA Yost EC Machesky ML Tejedor-Tejedor MI Anderson MA (1986) Characterization of anion binding on goethite using titration calorimetry and cylindrical internal reflection-Fourier transform infrared spectroscopy. Davis & Hayes, 142-161

FELDSPARS IN WEATHERING

ALEX E. BLUM
U.S. Geological Survey
345 Middlefield Rd. MS 420
Menlo Park, CA 94301 USA

ABSTRACT. Feldspar weathering occurs via dissolution of all components into solution, with the subsequent precipitation of secondary minerals from solution, and it is the feldspars dissolution rate which controls the overall rate of feldspar weathering. The rate of feldspar dissolution is controlled by the kinetics of surface reactions at the mineral-water interface, not by mass transfer processes, either in solution or through a protective surface layer. At neutral to basic pH conditions, the entire range of feldspars compositions appears to dissolve nearly stoichiometrically, although a thin Al enriched surface layer (<20Å) may form, and cations, particularly Na^+, may be exchanged with H^+ to depths of several 100 Å. The exchange of cations with H^+ appears to be reversible, with cation occupancy favored in the basic pH region. It is not clear whether the observed Al enrichment on the surface is a consequence of slightly non-stoichiometric dissolution, or readsorption of Al from solution at charged surface sites. At acidic solution pH's, a silica-enriched surface layer 100's to 1000's of Å thick may form. This layer is highly hydrated and disordered, and analogous to an amorphous SiO_2 gel. The silica-enriched surface layer does not provide a diffusional barrier to the transport of Al and cations to solution, and does not appear to effect the destruction rate of the feldspar tetrahedral lattice.

Feldspars of all compositions have an experimental dissolution rate which increases with increasing H^+ activity at pH <6, and increasing OH^- activity above pH 8.5, a pattern typical of many silicate minerals. In the acidic pH region (<pH 6) albite and K-feldspar have nearly identical experimental dissolution rates, with the dissolution rate (R) proportional to $[H^+]^{-0.5}$. This is in contrast to the observation in natural soils that albite weathers much more rapidly than K-feldspar. In the basic pH region, albite dissolution rates have a pH dependence of $R \propto [OH^-]^{-0.3}$, with K-feldspar dissolution rates having a stronger dependence on $[OH^-]$ than albite, and absolute dissolution rates approximately a factor of 10 slower than albite. The plagioclase series has a pH dependence of the dissolution rate of $R \propto [H^+]^{-0.5}$ in the acidic region in the compositional range An_{0-70}. However, at compositions of $An_{>70}$ there is a transition to increasing values in the exponent until $R \propto [H^+]^{-1.0}$ for $An_{>90}$. At a single solution pH, the dissolution rate of the plagioclase series increases gradually with increasing Ca content until ~An_{75}, with a rapid increase in the dissolution rate of plagioclase between the compositions An_{75} and $An_{>90}$.

There is general agreement that feldspar dissolution in the acid region occurs by selective attack on the Al sites in the tetrahedral framework, and the rate limiting process must involve the hydrolysis of the Al-O-Si bond. It is therefore the increasing Al/Si ratio with increasing An content, and not the Na/Ca ratio, which causes the increasing dissolution rate with increasing An content. In the basic pH region, the relative importance of the Al/Si ratio and cation occupancy are more uncertain.

The most comprehensive models for the dissolution of feldspars employ surface speciation models, in which the dissolution rate is proportional to the equilibrium concentration of surface chemical complexes, which are preferential sites for dissolution reactions. Most commonly, the surface species involve the protonation and deprotonation of dangling tetrahedral oxygens to form positively and negatively charged surface sites in the acid and basic regions, respectively. The concentration of charged surface sites is controlled by adsorption isotherms of species in solution. The large effect of solution pH on feldspar dissolution rates is indirect, controlling the equilibrium concentration of surface species through an

I. Parson (ed.), Feldspars and Their Reactions, 595–630.

adsorption process. The surface speciation model has major implications for the effects of temperature and adsorption of impurities on dissolution kinetics, which differ significantly from models based upon homogenous kinetic theory.

Field measurements of plagioclase dissolution rates from watershed mass-balance studies and from changes in the abundance of plagioclase in soils of different ages indicate that feldspar weathering is one to three orders of magnitude slower than predicted from laboratory studies. The leading explanation for this effect is the isolation of a large proportion of the feldspar surface area in isolated micropores. Alternative explanations include; (i) adsorption of inhibitors, of which Al and Fe are the most likely, (ii) the high saturation states of soil solutions, and (iii) experimental artifacts in the dissolution experiments. The reconciling of natural and experimental feldspar dissolution rates is a critical step before feldspar dissolution kinetics can be used reliably in predictive models.

1. Introduction

The weathering of feldspar is a major step in the geological cycling of crustal material, and it has a profound influence on many near surface processes. The overall feldspar weathering process can be simplified as the alteration of feldspar to common clay minerals, such as kaolinite. For example;

$$2NaAlSi_3O_8 + 2H^+ + 9H_2O = Al_2Si_2O_5(OH)_4 + 4H_2SiO_{4(aq)} + 2Na^+ \quad (1)$$

However, it is now widely recognized that this overall reaction for feldspar alteration involves two independent processes: (i) the initial dissolution of feldspar into solution, and (ii) the subsequent precipitation of kaolinite and other secondary phases from solution. These two reactions are linked by the effects of the solution composition on their relative rates, and in a closed system will reach steady-state. However, because the dissolution kinetics of feldspar are very slow, and most near-surface environments are open systems on this time scale, the feldspar dissolution rate should dominate the kinetics of the overall feldspar alteration process. Consequently, the dissolution of feldspar has been the focus of much of the research on mineral alteration and weathering.

Feldspar dissolution kinetics have been studied for many years, but have been the subject of increasing study over the last decade. Some of the motivations for this interest are: (i) understanding the effects of acid precipitation, and the long-term capacity of soils to neutralize anthropogenic acidic rainfall (e.g. Reuss and Johnson, 1986); ii) the importance of silicate weathering in regulating atmospheric CO_2 concentration over geologic time, and consequently in controlling global climatic change (e.g. Walker et al., 1981; Berner, et al. 1983; Lasaga et al., 1985; Volk, 1987); iii) the incorporation of chemical kinetics into reactive transport models, which require rate laws for the dissolution and precipitation of major minerals; and iv) the tremendous advances in surface science and heterogeneous kinetics which have provided both new analytical and theoretical techniques, allowing progress on complex problems of heterogeneous kinetics such as feldspar dissolution.

A major advance in our understanding feldspar dissolution kinetics is recognition of the central importance of the mineral-solution interface. Neither the characteristics of the bulk solution nor the feldspar chemistry and structure control the rate of the dissolution reaction; rather, it is how the solution and mineral surface interact at the mineral/water interface which is the crucial chemical question. In this chapter will (i) summarize some of the experimental data for feldspar dissolution kinetics, (ii) present some of the theoretical approaches used to explain the observed experimental results, (iii) briefly summarize measurements of natural plagioclase weathering rates in soils, and (iv) compare these rates with the experimental results.

2. Experimental Determination of Feldspar Dissolution Rates

2.1. Transport versus Surface Reaction Control

There are three distinct phases observed during experimental feldspar dissolution experiments. There is a rapid exchange of H^+ for cations such as K^+, Na^+ and Ca^{2+}, which occurs within minutes after feldspars are placed in solution, forming a hydrogen feldspar surface layer. This is followed by a period of rapid, sometimes non-stoichiometric dissolution. Finally, the dissolution rate approaches a nearly constant rate after several tens to many hundreds (or even thousands) of hours with a more or less stoichiometric release of elements.

The dissolution of feldspars (and all other materials) must involve two processes: (i) the *chemical reactions* which break bonds, allowing the removal of chemical species from the crystal lattice; and (ii) the *transport* of the reactants and products away from the surface (summaries by Berner, 1978; Lasaga, 1984). Transport control of the dissolution rate may itself occur in two ways: (i) Aqueous diffusion of solutes to and from the surface, and (ii) Solid-state diffusion of species through a protective (or armoring) layer on the surface. The former commonly limits the rate of rapid dissolution reactions, and may be recognized by the influence of stirring the solution on the reaction rate. This is the reason stirring your tea increases the dissolution rate of the sugar. The latter is a common phenomena in the corrosion of metals such as aluminum and iron. Feldspar dissolution rates are so slow that aqueous diffusion does not limit the dissolution rate. This leaves two major classes of theories for the rate control of feldspar dissolution: (i) surface-layer diffusion, and (ii) surface reaction kinetics.

Surface-layer diffusion theory suggests that the dissolution rate is limited by solid-state diffusion through a continuous surface layer with a different structure and composition than the original mineral. In this model, the freshly fractured feldspar surface has no protective layer, and consequently has a higher initial dissolution rate when first exposed to solution. As the dissolution reaction proceeds, a surface layer develops and thickens with time, providing a progressively greater diffusional barrier to both reactants and products of the dissolution reaction. The surface layer may form either by precipitation from solution, or by preferential removal of some components from the feldspar structure. A simple diffusional model predicts that the dissolution rate (R) should decrease as $R \propto t^{½}$, a behavior often referred to as parabolic kinetics. The surface layer must be both tightly adhering and continuous in order to provide an effective diffusion barrier. As the reaction progresses, the surface layer will reach a thickness at which the rate of diffusion of feldspar components though the layer equals the dissolution rate of the layer itself. The overall feldspar dissolution rate will then reach a constant and stoichiometric steady-state value. This kinetic mechanism was proposed in much of the early work on feldspar dissolution (e.g. Wollast, 1967; Helgeson, 1971; Paces, 1973; Busenburg and Clemency, 1976) but is now considered very unlikely, although it cannot be totally discounted.

The surface reaction theories for feldspar dissolution center upon the rates of chemical reactions at the feldspar-solution interface. The important aspect of this approach is that it is the structure and chemistry of the feldspar surface in contact with the solution, and not necessarily the bulk chemistry of the feldspar, which controls the rate of the dissolution reaction. Therefore, surface properties such as the surface physical configuration of the surface, step density and surface roughness, and adsorption reactions effecting surface speciation become critical factors. The surface reaction theories include those based upon transition state theory (TST), surface speciation models, and surface configuration models such as surface nucleation and spiral growth/dissolution (i.e. Burton-Cabrera-Frank (BCF) theory). As we shall see, there is still much disagreement about the details, but most workers now agree it is the chemical and physical characteristics of the feldspar-solution interface which controls the rate of feldspar dissolution.

What we will now do is to look at the nature of the rapid, essentially equilibrium adsorption and exchange reactions on the feldspar surface, and then the period of rapid dissolution, finally finishing with a summary of the steady-state dissolution rates of feldspars and how they vary with solution chemistry and feldspar composition. However, we will first examine the issue of surface area determinations. Because both dissolution rates and surface sorption reactions are expressed per unit surface area, surface area determinations are potentially a major source of error and/or misinterpretation encountered in feldspar surface chemistry and dissolution studies.

2.2. Feldspar Surface Areas

Surface areas (SA) can be estimated geometrically, based on the size and shape of particles. However, this geometrical estimate will not include surface roughness or any internal surface area. Surface areas are most often determined by the BET (Brunauer-Emmett-Teller) method (Brunauer et al., 1938; Gregg and Sing, 1982) which measures the adsorption of a gas on the surface as a function of the partial pressure of the gas. The technique most commonly used adsorbs N_2 or Kr at 77°K, and measures the amount of gas released when the sample is warmed to room temperature. At a low and fixed N_2 or Kr partial pressure, the surface area is proportional to the amount of gas released. Because the diameter of the N_2 and Kr molecules is ~4Å, surface roughness, porosity, and fractures with dimensions >4Å will be included in a BET surface area measurement. The 4Å scale is very similar to the effective size of a water molecule (~3Å), and probably represents a reasonable estimate of the wetted surface area.

Table 1 shows both estimated geometrical surface areas and BET surface areas of freshly ground and washed feldspar particles used in dissolution experiments. The ratio of the BET to geometrical surface areas is called the surface roughness (SR) factor. This follows the surface science nomenclature, although this SR also includes any internal surface area such as etch pits, fractures or any other void and connected microstructures. The SR of freshly ground and unaltered feldspar generally fall in the range of 9±6, and can thus be estimated within a factor of ~3 from the grain size.

Weathered feldspars from soils typically have a SR of several hundreds (Table 1). The interpretation of the large discrepancy between the SA of freshly ground experimental and naturally weathered feldspar grains introduces a large uncertainty in the comparison of natural and experimental feldspar dissolution rates. Surface roughness of weathered feldspars measured by atomic force microscopy (AFM) can explain only a surface roughness of ~2, suggesting that much of the observed increase in surface area results from internal surface area. This agrees with the extent of etching observed by SEM and TEM in weathered feldspars (e.g. Berner and Holdren, 1979; Banfield and Eggleton, 1990). Anbeek (1992a,b) suggested that even "pristine" mineral specimens used in dissolution experiments have extensive internal surface area. This is consistent with the high SR's of the experimental material reported in Table 1. In particular, the Grass Valley anorthite used by Holdren and Speyer (1987) has a surface roughness approaching that of highly weathered feldspar grains. Microporosity has been shown to commonly exist in many unweathered feldspars (Wordon et al.; 1990; Walker, 1990). The nature of microporosity in feldspars and its contribution to both natural weathering and experimental determination of feldspar dissolution kinetics has not been adequately addressed.

2.3. Initial Adsorption/Exchange Reactions

2.3.1. Surface Titrations - Sorption and exchange reactions on mineral surfaces can be determined by surface titration. During surface titrations acid, base and other solutes are added to an aqueous suspension of the mineral, and the composition of the solution is monitored, often with specific-ion electrodes. The amount of a species adsorbed or desorbed from the surface can be determined by the difference between the amount of species added to the system, and the concentration of species in solution (e.g. Stumm and Morgan, 1981; Huang, 1981). Several cautions in the interpretation of surface titrations are warranted. Because we are implying surface chemistry from changes in solution chemistry, we have no direct information on the actual surface species involved in the reactions. We can only guess at the appropriate surface complexes based upon the observed stoichiometry. Furthermore, if the same aqueous species is involved in several different reactions, we may not be able to differentiate between reactions.

Table 1 - Surface areas of feldspar grains

mineral	Grain Size (μm)	Geometric Surface Area (m^2/g)	BET Surface Area (m^2/g)	Surface Roughness (SR)
Microcline[1]	300-600	0.005	0.067	14.0
Microcline[1]	150-300	0.010	0.073	7.6
Microcline[1]	75-150	0.019	0.114	5.9
Microcline[1]	37-75	0.038	0.327	8.5
Microcline[1]	300-600	0.005	0.067	14.0
Microcline[1]	150-300	0.010	0.073	7.6
Microcline[1]	75-150	0.019	0.114	5.9
Microcline[1]	37-75	0.038	0.327	8.5
Microcline[1]	300-600	0.005	0.056	11.7
Microcline[1]	150-300	0.010	0.072	7.5
Microcline[1]	75-150	0.019	0.107	5.6
Microcline[1]	37-75	0.038	0.179	4.7
Alkali feldspar[1]	300-600	0.005	0.039	8.1
Alkali feldspar[1]	150-300	0.010	0.060	6.3
Alkali feldspar[1]	75-150	0.019	0.089	4.6
Alkali feldspar[1]	37-75	0.038	0.277	7.2
Alkali feldspar[1]	300-600	0.005	0.050	10.4
Alkali feldspar[1]	150-300	0.010	0.072	7.5
Alkali feldspar[1]	75-150	0.019	0.108	5.6
Alkali feldspar[1]	37-75	0.038	0.176	4.6
$Ab_{80}K_{20}$[2]	150-300	0.010	0.053	5.5
$Ab_{80}K_{20}$[2]	75-50	0.036	0.108	3.0
Albite[1]	300-600	0.005	0.036	7.5
Albite[1]	150-300	0.010	0.049	5.1
Albite[1]	75-150	0.019	0.066	3.4
Albite[1]	37-75	0.038	0.154	4.0
Albite[3]	104-250	0.012	0.083	7.0
Albite[3]	53-104	0.028	0.106	3.8
Albite[4]	75-125	0.022	0.086	3.9
Albite[4]	75-125	0.022	0.092	4.2
Albite[5]	25-75	0.041	0.405	10.0
Albite[5]	25-75	0.041	0.490	12.1
Albite[6]	100-200	0.014	0.046	3.2
Albite[6]	50-100	0.029	0.075	2.6
Oligoclase[1]	300-600	0.005	0.079	16.5
Oligoclase[1]	150-300	0.010	0.090	9.4
Oligoclase[1]	75-150	0.019	0.128	6.7
Oligoclase[1]	37-75	0.038	0.334	8.7
Oligoclase[5]	25-75	0.041	0.180	4.4
Oligoclase[5]	25-75	0.041	0.645	15.9
Oligoclase[7]	75-150	0.019	0.116	6.1

[1] Holdren and Speyer (1987)
[2] Holdren and Speyer (1985)
[3] Rose (1991)

mineral	Grain Size (μm)	Geometric Surface Area (m^2/g)	BET Surface Area (m^2/g)	Surface Rough. (SR)
Andesine[1]	300-600	0.005	0.119	24.8
Andesine[1]	150-300	0.010	0.134	14.0
Andesine[1]	75-150	0.019	0.210	11.0
Andesine[1]	37-75	0.038	0.386	10.0
Andesine[5]	25-75	0.041	0.405	10.0
Andesine[5]	25-75	0.041	0.350	8.6
Bytownite[1]	300-600	0.005	0.130	27.1
Bytownite[1]	150-300	0.010	0.127	13.3
Bytownite[1]	75-150	0.019	0.214	11.2
Bytownite[1]	37-75	0.038	0.473	12.3
Bytownite[5]	25-75	0.041	0.200	4.9
Bytownite[5]	25-75	0.041	0.270	6.6
Anorthite[1]	300-600	0.005	0.619	129.2
Anorthite[1]	150-300	0.010	0.700	73.0
Anorthite[1]	75-150	0.019	0.757	39.5
Anorthite[1]	37-75	0.038	1.020	26.5
Anorthite[5]	25-75	0.041	0.520	12.8
Anorthite[8]	20-50	0.060	0.500	8.4
Anorthite[8]	50-100	0.029	0.300	10.4
Anorthite[8]	100-250	0.012	0.110	9.2
Average				8.5
Standard Deviation				4.9

Feldspar grains from soils near Merced, Calif.

Mineral	Age (Kyr)	geometric	BET	SR
oligoclase[9]	500-1000	.0014	0.39	271
oligoclase[9]	500-1000	.0014	0.26	180
oligoclase[9]	500-1000	.0014	0.46	320
oligoclase[9]	500-1000	.0014	1.48	1029
Alkali fspr[9]	500-1000	.0014	0.12	83
Alkali fspr[9]	500-1000	.0014	0.26	180
Alkali fspr[9]	500-1000	.0014	0.94	653
Alkali fspr[9]	500-1000	.0014	0.81	563
Alkali fspr[9]	250-500	.0029	0.32	111

[4] Knauss and Wolery (1986)
[5] Casey et al. (1991)
[6] Chou and Wollast (1984)
[7] Mast and Drever (1987)
[8] Amrhein and Suarez (1992)
[9] J. Schulz (*per. comm.*)

Surface titrations also give no information on heterogeneity. They average the behavior of all exposed faces, yet we know that different crystallographic faces may have dramatically different chemical behaviors. For example, the (001) face of quartz dissolves 10^3 times faster than the prismatic faces in HF solution (Ernsberger, 1960), so that a surface titration would be heavily biased toward "unreactive" surface area. We currently have no information on the relative dissolution rates of feldspar surfaces, but the (001) faces of feldspar grains tend to be more heavily corroded. It has also been proposed that a large proportion of dissolution may occur at high energy "reactive" sites such as dislocations and impurities which are only a small fraction of the surface. These features would also not be reflected in surface titrations. Finally, surface titrations are very subject to experimental errors, especially precipitation of secondary phases, errors in electrode measurements, and inaccurate compensation for mineral dissolution and aqueous speciation of solutes.

2.3.2. Surface Complexes on Feldspars - When unaltered feldspar surfaces are first placed in solution, there are three fundamentally different types of rapid surface reactions which may occur. All are thought to reach equilibrium within minutes, and to be governed by classical exchange and partitioning relationships.

1) Exchange and/or sorption reactions involving non-framework cations such as K^+, Na^+, and Ca^{2+}. The proximity of a free surface allows these cations access to the solution to react or exchange. Examples of these reactions include:

$$X-Na \rightleftharpoons X^- + Na^+ \qquad (2)$$

$$X-Na + H^+ \rightleftharpoons X-H + Na^+ \qquad (3)$$

$$X_2-Ca + 2H^+ \rightleftharpoons 2(X-H) + Ca^{2+} \qquad (4)$$

where X represents a cation exchange site, and X^- is an empty exchange site which gives the surface a net negative charge. These exchange reactions result in an immediate decrease in solution $[H^+]$ concentration and release of cations when feldspar are first placed in solution. This exchange reaction was recognized early in the study of feldspar dissolution (Tamm, 1930; Nash and Marshall, 1956a,b; Garrels and Howard, 1959). Table 2 shows the depth of the exchange layer measured on albite, anorthite, and potassium feldspar, calculated from BET surface area

Table 2 - Depth of the Cation Exchange Layer

Reference	Mineral (cation)	sites/nm^2	depth (Å)
Wollast and Chou (1992)[1]	albite (Na)	11.8	21
Blum and Lasaga (1991)[1]	albite (Na)	8.4	15
Amrhein and Suarez (1988)[1]	anorthite (Ca)	14-17	25-30
Schweda (1989)[2]	microcline (K)	21	37
Schweda (1989)[2]	sanidine (K)	48	85

[1] determined by surface titration: total number of exchange sites.
[2] determined from non-stoichiometry of initial dissolution at ~30 min at pH 3. Assumed to represent total exchange.

and assuming a uniform exchange depth. The depth of the exchange layer is a several unit cells in albite, and slightly deeper in anorthite and the K-feldspars. The observed cation exchange depths of several unit cells is greater than can be explained by direct cation exposure at a surface terminating the bulk structure, and suggests higher H^+ and cation mobility near the feldspar surface. Possible explanations include surface relaxation, which may broadly enhance cation mobility in the near surface region, and defects intersecting the surface which may create localized regions of enhanced cation mobility. However, both the cause and distribution of the high cation exchange site densities in feldspars are not well understood.

At pH's <8 and low Na concentrations, albite exchanges essentially all the Na^+ for H^+ in the outermost few unit cells (i.e. eq. 2.3 goes completely to the right). There has been some controversy about the reversibility of the H^+/Na^+ exchange reactions for albite (Chou and Wollast, 1989; Murphy and Helgeson, 1989). Blum and Lasaga (1987, 1991) did not detect any Na^+ uptake during surface titrations of albite at low Na^+ concentrations. However, Wollast and Chou (1992) presented convincing evidence that the exchange of H^+ for Na^+ in albite is reversible, with significant Na^+ occupancy of the exchange sites at pH's >8, and moderate Na^+ concentrations. Thus, in saline environments, the exchange sites on the albite surface may have significant occupancy. This seems consistent with the common observation of secondary albite precipitation in saline diagenetic environments. The exchange of cations on the surfaces of feldspars of other compositions have not been studied in detail, but the common occurrence of diagenetic K-feldspars suggests that K^+ exchange may also be reversible.

2) Adsorption and desorption reactions involving dangling oxygens are created by the termination of the tetrahedral lattice at the surface. These dangling oxygens are rapidly hydrated by atmospheric or aqueous water (Iler, 1979) to form hydroxyl groups. Sorption reactions at these sites result in charged surface species, for which equations and equilibrium expressions can be written. Examples of these reactions include:

$$>S\text{-}O^- + H^+ \rightleftharpoons >S\text{-}OH \qquad K_1 = \frac{a_{[>S\text{-}OH]}}{a_{[>S\text{-}O^-]}\, a_{[H^+]}} \tag{5}$$

$$>S\text{-}OH + H^+ \rightleftharpoons >S\text{-}OH_2^+ \qquad K_2 = \frac{a_{[>S\text{-}OH_2^+]}}{a_{[>S\text{-}OH]}\, a_{[H^+]}} \tag{6}$$

$$>S\text{-}OH + M^{n+} \rightleftharpoons >S\text{-}OM^{(n-1)+} + H^+ \qquad K_M = \frac{a_{[>S\text{-}OM^{(n-1)+}]}}{a_{[>S\text{-}OH]}\, a_{[M^{n+}]}} \tag{7}$$

where >S is a tetrahedral Al or Si on the surface, >S-O is a dangling oxygen on the surface, and M is any other adsorbing ligand. Blum and Lasaga (1991) performed surface titrations of albite and found they could describe eqns. 5 and 6 by a Langmuir adsorption isotherm, with $K = 2.7\times10^4$ and $K_2 = 9.4\times10^4$. In the acid region, they found a site density of 35 μmol/m^2, and in the basic region of 27 μmol/m^2. Blum and Lasaga (1991) interpreted the charged surface species as protonation and deprotonation of dangling Al-OH surface groups to $>Al\text{-}OH_2^+$ and $>Al\text{-}O^-$, respectively. Wollast and Chou (1992) examined only the basic region, but found a negative site density of 21.3 μmol/m^2, in good agreement with Blum and Lasaga (1992). However, Wollast and Chou (1992) attribute the negative charge density to the removal of H^+

from cation exchange sites via eqn. 2. Amrhein and Suarez (1988) performed a surface titration of anorthite and found a total site density of 120 μmol/m^2.

3) Adsorption may also occur at bridging oxygens exposed at the surface. Examples include the adsorption of H^+ or Na^+:

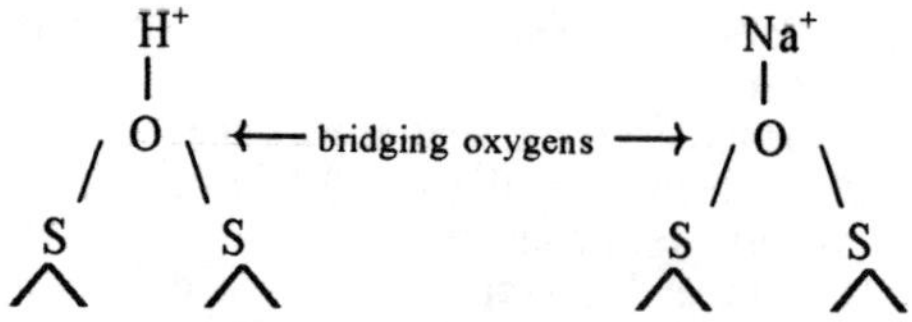

```
       H+                                   Na+
       |                                     |
     / O \  <-- bridging oxygens -->       / O \          (8)
    S     S                               S     S
    ^     ^                               ^     ^
```

Blum and Lasaga (1991) suggested that bridging oxygens on feldspar surfaces tend to be hydrophobic, based primarily on an analogy with quartz and silica surfaces (Iler, 1979). This implies that adsorption of cations at bridging oxygens should be very weak, and only a very small proportion of the sites will be occupied at any specific time. Such low surface coverage would not be detectable by surface titration or any other analytical technique. However, adsorption at bridging oxygens could still have a potentially large effect on the activation energy of breaking the bridging tetrahedral bonds (Lasaga and Gibbs, 1990; Dove and Crerar, 1990), and therefore, could still be important in controlling the kinetics of the overall dissolution rate. In contrast to Blum and Lasaga (1991), Wieland et al. (1988) and Stumm and Wollast (1990) have suggested a dissolution model for feldspars and other oxides, based upon the oxide dissolution kinetics of Furrer and Stumm (1986), which suggests that bridging oxygens behave identically to dangling oxygens during adsorption reactions, and are active adsorption sites.

2.4. Rapid Initial Dissolution Rates

During feldspar dissolution experiments, the rapid surface sorption reactions are followed by a period of rapid dissolution, which decreases with time. Finally, the dissolution rate will approach a nearly constant dissolution rate after several tens to many hundreds (or even thousands) of hours. It is this long term steady-state dissolution rate which is most relevant to most geological systems. The cause of the rapid initial dissolution is generally unknown. Holdren and Berner (1979) suggested two possible explanations, (i) small particles adhering to the surface after sample grinding have a higher surface free energy, and would dissolve more rapidly than large grains, and (ii) disruption of the surface during grinding produces a disordered layer which dissolves more rapidly. Subsequent experiments in which most fine particles were removed prior to dissolution did not completely reduce the high initial rates. Petrovic (1981a,b) supported the idea of a disturbed surface layer produced during grinding, and this is still probably the most widely cited explanation. However, Chou and Wollast (1984, 1985) and many subsequent investigators have also observed a more rapid period of dissolution immediately after changing the solution pH during a continuous experiment. Mast and Drever (1987) also observed a high transient Al release after introducing oxalic acid into oligoclase dissolution experiments. They have suggested that the rapid release of elements may be a consequence of the presence of a thin steady-state surface layer with a non-stoichiometric composition. Changing the solution composition may require establishing a new steady-state surface composition. As the surface reequilibrates, there may be a transient release of those elements which have a lower concentration in the new surface layer. With this interpretation, the rapid dissolution rates observed at the start of the experiments may result from more rapid dissolution before a non-stoichiometric, steady-state, surface composition can evolve.

2.5. STEADY-STATE FELDSPAR DISSOLUTION RATES

2.5.1. Experimental techniques - The experimental techniques used to determine feldspar dissolution rates can be crudely divided into two groups, batch systems and flow through systems. Batch experiments are closed systems, in which the feldspar sample and starting solution are sealed in a closed vessel, usually a polyethylene or teflon bottle. Solutes released or consumed by reaction of the feldspar are allowed to accumulate in solution (free drift), resulting in a changing solution composition with time. Buffers may be used to minimize fluctuations in some solution properties, usually pH and/or ionic strength, and solutions may be changed periodically to prevent the solution from drifting over too large a compositional range. Precipitation of secondary phases can be a serious problem in batch systems, particularly at moderate pH, where the system almost always oversaturates with respect to gibbsite and/or kaolinite, rendering the results a minimum rate only.

Flow-through systems continuously add and remove solution from a well mixed reactor (e.g. Chou and Wollast, 1984). The dissolution rate is calculated from the difference in solute concentrations of the input and output solutions. The continuous changing of solution allows the experiment to be designed so that it is never saturated with respect to secondary phases, or to manipulate the concentration of solutes in a systematic fashion (e.g. Burch et al., 1993). Because the reactor is well mixed, the composition of the output solution is identical to the solution in contact with the mineral. After an initial transitory period, the flow-through system should reach a steady-state at which the output solution reaches a constant composition, allowing the feldspar dissolution rate to be determined under constant chemical conditions.

There are many factors which must be considered in determining and evaluating feldspar dissolution rates. Most emphasis in experimental studies has been on the effects of solution pH, temperature, and feldspar composition on the dissolution rate. Other factors considered include; (i) the catalytic or inhibiting effect of dissolved inorganic and organic solutes (including pH buffers), (ii) the effect of the saturation state of the solution, (iii) the type and concentration of both point and line defects within the feldspar, and (iv) compositional variations within the feldspar crystals. In addition, different studies have treated starting material differently, determined mineral surface areas differently, used mineral names inconsistent with the actual composition, and not always evaluated the possibility of precipitation of secondary phases. Variations in many of these factors between different studies mean that the comparison of results from different studies must be done carefully and critically.

2.5.2. Albite dissolution kinetics - The feldspar with the most intensively studied dissolution kinetics is albite, and its dissolution behavior is typical in most respects of feldspars of all compositions. Solution pH appears to be the most important single solution variable controlling the dissolution rate. This is because both H^+ and OH^- are important participants in feldspar dissolution mechanisms, and because H^+ and OH^- have wider fluctuations in commonly occurring natural concentrations than any other solute. Figure 1 shows a compilation of measured dissolution rates for albite as a function of pH at temperatures near 25°C. The first thing to note in Fig. 1 is that the dissolution rate has a minimum at pH 6 to 8, and increases in both the acid and basic regions. This pattern is common to all feldspars, as well as many other silicates (e.g. Helgeson et al., 1984: Lasaga, 1984; Murphy and Helgeson, 1987), and it strongly suggests at least two different dissolution mechanisms for feldspars; a proton promoted mechanism in the acid region, and a hydroxyl promoted mechanism in the basic region. Many workers (e.g. Murphy and Helgeson, 1987; Chou and Wollast, 1985; Knauss and Wolery, 1986; Mast and Drever, 1987) have suggested that the feldspar dissolution rates are independent of pH in the neutral pH region from ~5 to 8. They have interpreted this as a third dissolution mechanism for feldspars which dominates in the neutral pH region and is pH independent. Other workers (e.g. Blum and Lasaga, 1991; Brady and Walther, 1989) have proposed only two pH dependent mechanisms. However, the determination of the very slow feldspar dissolution rates at near neutral pH's is often problematic. Analytical detection limits of solutes in solution often

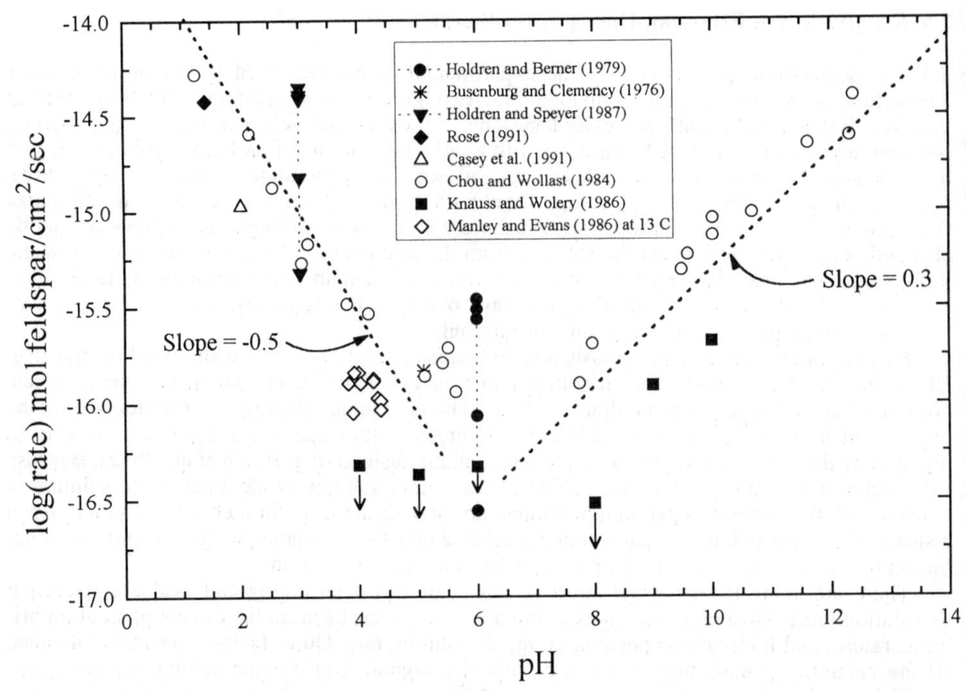

Figure 1 - Compilation of albite dissolution rates as a function of solution pH.

limit the sensitivity of the experiments. Very long reaction times are necessary to dissolve a significant volume of material and, therefore, definitively remove any anomalously high dissolution rate material on the surface. It is still not unambiguously resolved whether the feldspar dissolution kinetics are controlled by only a proton promoted mechanism in the acid region and a hydroxyl promoted mechanism in the basic region, or if a third pH independent mechanism operates in the neutral pH region. The neutral region includes many natural environments including most soils, surface and ground waters, and pH independence in the neutral region would greatly simplify the practical application of feldspar kinetics.

2.5.3. Potassium Feldspar Dissolution Kinetics - Figure 2 compiles K-feldspar dissolution rates as a function of solution pH at ~25°C. Fig. 3 compares the K-feldspar rates reported in fig. 2 with albite rates from fig. 1. At pH <6, the dissolution rates of K-feldspars are indistinguishable from albite, both in their pH dependence ($R \propto [H^+]^{-0.5}$) and in their absolute rate. The major difference between the structures of albite and K-feldspar is the substitution of K^+ for Na^+ and the slight distortion of the tetrahedral lattice which accommodates the cation substitution. In acidic solutions both the albite and K-feldspar surfaces undergo rapid exchange of cations with H^+ to form a hydrogen-feldspar surface layer several unit cells thick. This implies that the structure and composition of the albite and K-feldspar surfaces at the mineral-solution interface are nearly identical, which is consistent with the similar dissolution kinetics of albite and K-feldspar in the acid region. The presence of a hydrogen feldspar surface layer also implies that it is the nature of the tetrahedral framework, and hydrolysis of the bridging oxygen bonds at tetrahedral sites which is controlling the dissolution rate of feldspars under acidic conditions, whereas the nature of the cation is of secondary importance. Calcic-plagioclase also forms a hydrogen-feldspar surface layer in the acidic solutions. By analogy, the differences between alkali feldspar and

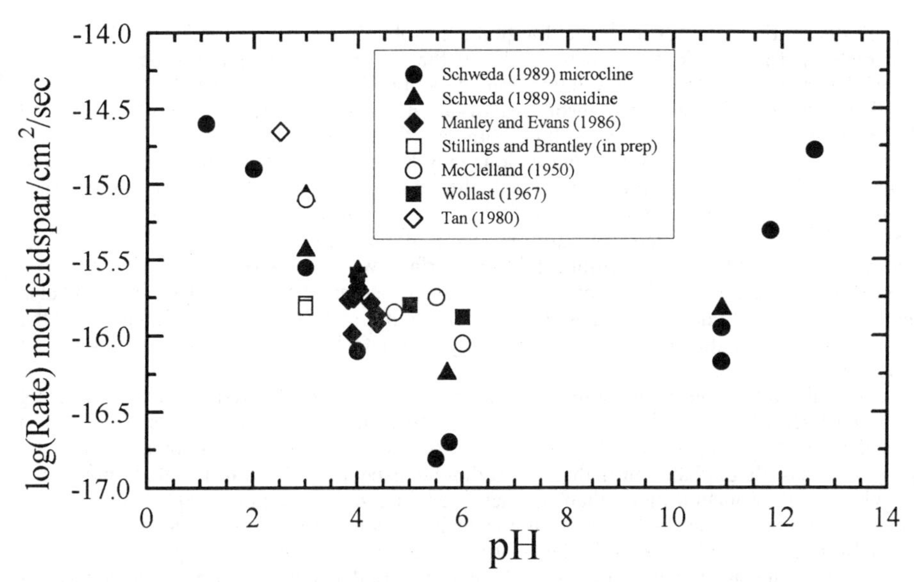

Figure 2 - Compilation of K-feldspar dissolution rates as a function of solution pH.

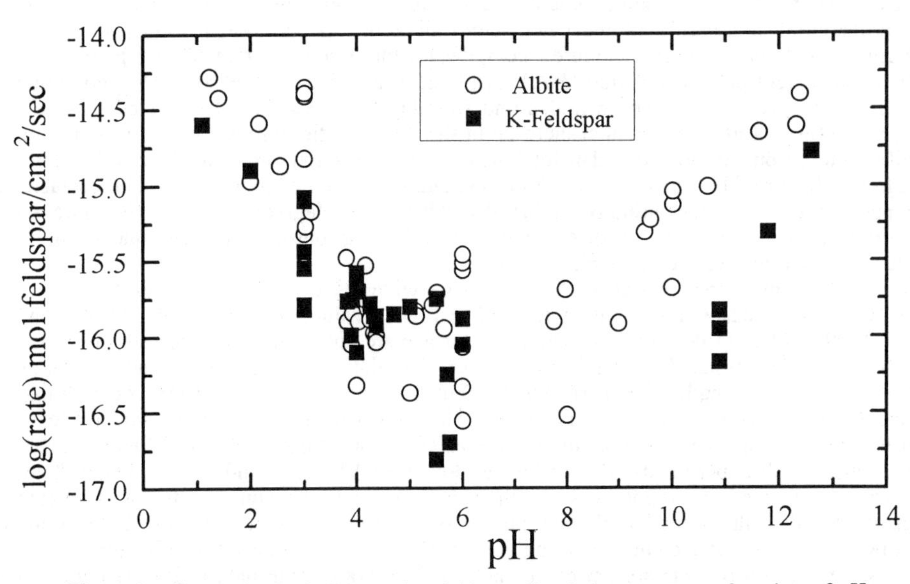

Figure 3 - Comparison of albite and K-feldspar dissolution rates as a function of pH (data from figs. 2 and 3).

plagioclase dissolution kinetics in the acid region are probably a result of the higher Al/Si ratio in the tetrahedral framework, and not the nature of the cation in the original mineral.

In the basic pH region the dissolution rate data for K-feldspar are sparse. However, it appears that K-feldspar has a slower absolute dissolution rate than albite by a factor of ~10, and possibly a greater pH dependence than albite, although the slope is poorly defined. The slower dissolution rate of K-feldspar implies that cations may have an important influence on feldspar dissolution kinetics in basic solutions, presumably due to K^+ readsorption at cation sites stabilizing the surface more strongly than Na^+. This is consistent with the observation of Wollast and Chou (1992) that Na^+ reversibly adsorbs on the albite surface. Since both the exchange constant and the stability of the resulting feldspar surface would be expected to increase with ΔG of adsorption, presumably K^+ exchanges on to the surface more strongly than Na^+. This also implies that dissolution rates of feldspars in the basic region may be strongly influenced by the concentrations of any cations which can effectively compete for cation exchange sites on the feldspar surface.

In most soils, including acidic soils, it is commonly observed that K-feldspars weathering rates are much slower than sodic plagioclase. This is in contrast to the experimental results in fig. 3, which indicates approximately similar experimental dissolution rates for albite and K-feldspar. One possible explanation is that soil solutions, especially those in isolated pores, may reach higher solute concentrations than are achieved in most dissolution experiments. If K-feldspar is more prone to cation occupancy in the exchange sites, high solute concentrations may preferentially retard K-feldspar dissolution relative to albite. However, this is quite speculative, and the cause of the discrepancy between the experimental results and field observations is quite uncertain.

5.2.4. Plagioclase Dissolution Kinetics - It has been recognized for quite some time (e.g. Goldich, 1938) that the feldspar weathering rates in soils vary as calcic plagioclase > sodic plagioclase. There is abundant morphologic evidence for this relationship. A common example is the preferential weathering of the calcic cores of zoned plagioclase crystals before the weathering of the more sodic rims. Figure 4 compiles recent dissolution rate data for plagioclase as a function of feldspar composition (%An) at pH 2, 3, and 5. Casey et al. (1991) compiled similar plots for pH 2 and 3, and Stillings and Brantley (in press) at pH 3. At all three pH values, we observe a near-exponential increase in the feldspar dissolution rate with increasing anorthite content from An_0 to $\sim An_{80}$. The total increase in dissolution rate from albite to anorthite ranges from a factor of $10^{2.5}$ to 10^4, and %An is a strong control on plagioclase dissolution rates. There appears to be a discontinuity in the dependence of plagioclase dissolution rate on composition at $\sim An_{80}$, with anorthite dissolution rates increasing considerably faster than the trend predicted from the less calcic plagioclase data.

The anorthite measurements at pH 5 demonstrate wide disagreements, with differences of almost 10^6 in the measured dissolution rates. The higher values are from Fleer (1982) and Sverdrup (1990). Fleer (1982) used the geometric surface area of a single large crystal, and may have severely underestimated the BET surface area. The data from Busenburg and Clemency (1976) was collected using batch experiments at 1 atm CO_2, and Busenburg (1978) reported that these experiments precipitated secondary phases. The data point is based on the linear release rate of Ca, but still represents only a minimum rate. The data from Amrhein and Suarez (1988, 1992) are also batch experiments which were run for up to 4.5 years, and in which significant secondary precipitates were reported. Thus, these low anorthite dissolution rates are probably the result of either the removal of material by secondary precipitates, or the high solute concentrations allowed to accumulate during the experiment which reduced the reaction rate. In either case, these experiments are not comparable with the other rate data in fig. 4c collected in dilute solutions far from equilibrium, and they have not been included in the indicated trends or in the interpretation. However, these discrepancies of a factor of 10^6 illustrate the difficulties in the comparison and interpretation of feldspar dissolution experiments.

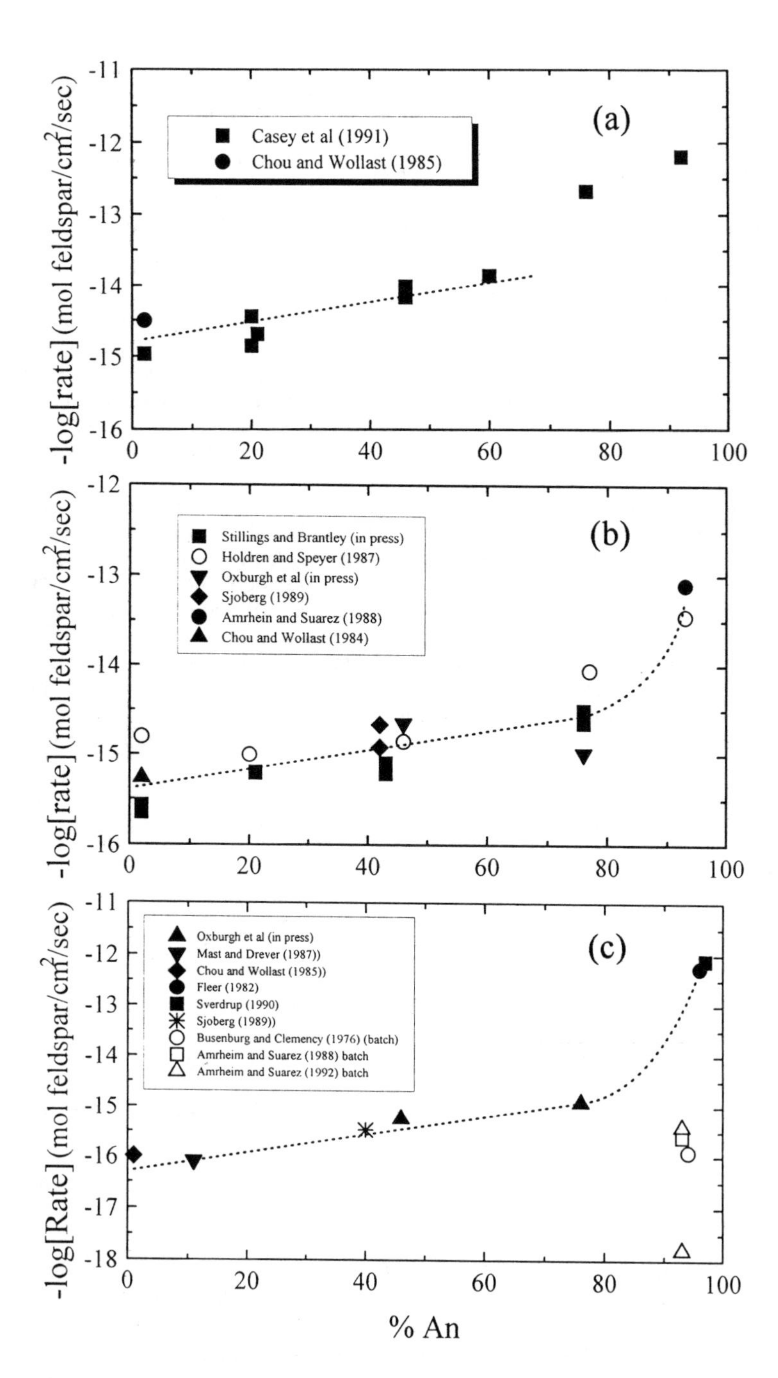

Figure 4 - Plagioclase dissolution rates as a function of anorthite content (%An) at (a) pH 2, (b) pH3, and (c) pH 5.

Fig. 5 shows the dissolution rate for a series of plagioclase compositions as a function of solution pH in the acid region. The plagioclase series from albite to $\sim An_{70}$ (as well as K-feldspar) have a dependence of $R \propto [H+]^{-0.5}$, but the exponent increases to 0.76 for An_{76} (Oxburgh et al., in press), and 1.0 for anorthite (Sverdrup, 1990). This transition appears to occur at $\sim An_{70}$, and corresponds to the observed change in dissolution behavior with %An observed at pH 2, 3 and 5 (fig. 4). There appears to be a critical Al/Si ratio at $An_{70\text{-}80}$ which dramatically effects plagioclase dissolution kinetics. This may be related to a decrease in the average size of linked clusters of Si tetrahedra after preferential attack and removal of tetrahedral Al. In albite with an Al/Si ratio of 3, preferential attack and removal of Al will still leave a fragmented but partially linked framework of tetrahedral Si. Removal of the Si will still require the hydrolysis of Si-O-Si bonds, although the Si centers will probably have several hydroxyl groups. In anorthite (An_{100}) the Al/Si ratio equals 1, and the ordering imposed by the aluminum avoidance principle means that removal of Al will leave completely detached Si tetrahedra, with the result that hydrolysis of Si-O-Si bonds are never required to decompose the anorthite tetrahedral lattice. There may be a critical Al/Si ratio at which the hydrolysis of Si-O-Si is no longer a control on the plagioclase dissolution rate, corresponding to a change in the overall dissolution mechanism.

2.5.5. Reproducibility of feldspar dissolution rates - The reproducibility of feldspar rate determinations by the same laboratory are generally $<\pm 50\%$ and almost always within a factor of 2. For example, Holdren and Speyer (1987) found a reproducibility of individual experiments of $<\pm 15\%$. The agreement between feldspar rate determinations conducted in different laboratories under similar conditions, such as in dilute solutions with no secondary precipitates, is generally within $\sim 10^{0.7}$, maybe a little better at low pH. The differences may reflect subtle differences in composition and microstructure of the mineral specimens, grain size used, solution saturation state, cation concentrations, surface area determinations, and the concentrations of other possible catalysts and inhibitors such as organic buffers and Al^{3+}, respectively.

Holdren and Speyer (1987) measured the dissolution rates of different grain size fractions

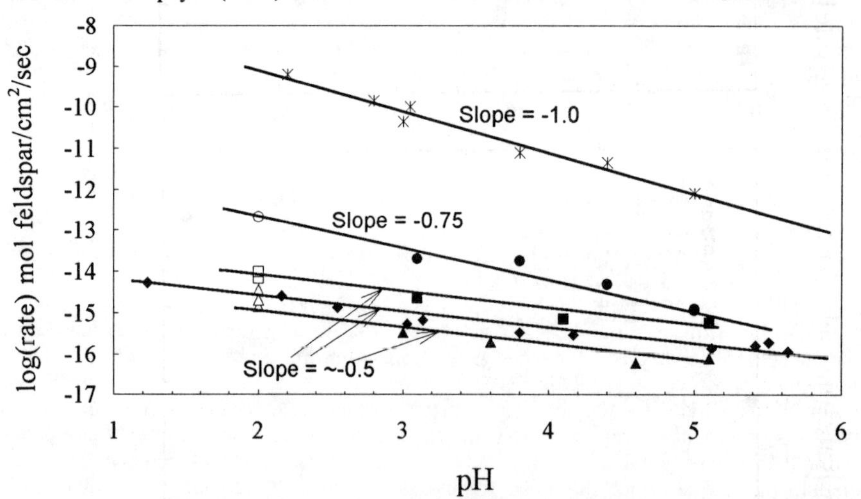

Figure 5 - Dissolution rate as a function of solution pH for plagioclase feldspars of different compositions. * anorthite, Sverdurp (1990); ● bytownite, Oxburgh and et al. (in press); O bytownite, Casey et al. (1991);■ andesine, Oxburgh and et al. (in press); □ andesine, Casey et al. (1991); ▲ oligoclase, Oxburgh and et al. (in press); △ oligoclase, Casey et al. (1991); ◆ albite, Chou and Wollast (1985). (*after* Oxburgh and et al., *in press*)

from nine different feldspar compositions at pH 3, and one sample each at pH 2 and 6. They found a systematic decrease in dissolution rate with decreasing grain size from >600μm to <37μm, with the rates decreasing by a factor from 2.8 to 28.5. Their data for albite is shown on Fig. 1 at pH 3 and 6, and their observed change in rate with grain size is greater than the variability between the rates of most other workers. A similar decrease in dissolution rate with decreasing grain size is reported by Amrhein and Suarez (1992) for anorthite. Holdren and Speyer (1987) proposed that the defect structures at twin boundaries and between exsolution domains are preferential sites for the dissolution of feldspars. Additionally, grinding of feldspar starting material preferentially cleaves the crystals along twin and exsolution boundaries effectively destroying them. Thus, the density of these high energy boundaries exposed on the grain surfaces decreases with decreasing grain size. This continues until the grain size drops below the average spacing of the defect. However, in contrast to the data of Holdren and Speyer (1985, 1987), Mast and Drever (1993) measured the dissolution rates of seven albite fractions from 0.5 μm to 150 μm at pH 4 and found the dissolution rate did *not* vary with grain size. Inskeep et al. (1991) observed differential weathering of the more calcic exsolution lamellae in experimentally dissolved labradorite using TEM, but they did not observe preferential dissolution along the lamellae boundaries.

2.5.6. Evidence for a Leached Layer on the Surface - There is considerable evidence for development of a Si-rich layer on the feldspar surface after dissolution in the acid region (≤3), both from direct measurement by surface analytical techniques (Petrovic et al., 1976; Berner and Holdren, 1979; Casey et al., 1988, 1989; Sjöberg, 1989; Hellmann et al., 1989, 1990; Goossens et al., 1989; Nesbitt and Muir, 1988; Nesbitt, 1990; Muir et al.,1989; Muir and Nesbitt, 1991, 1992) and from non-stoichiometric dissolution (Chou and Wollast, 1984; Holdren and Speyer, 1985; Schweda, 1989; and others). The thickness of the Si-rich layer is typically several thousand angstroms thick at a pH of 1-3, and the layers seem to be thicker and develop at higher pH's the more calcic, and therefore, the more rapid the dissolution rate of the feldspar. When the feldspar is removed from solution, the Si-rich layer has been reported to crack and spall (Casey et al., 1989). The silica-rich layer is highly porous as indicated by TEM imaging and nitrogen porosimetry measurements on labradorite (Casey et al, 1989), and ion beam measurements on albite (Petit et al, 1990), alkali feldspar (Petit et al, 1989), labradorite (Casey et al, 1989) and diopside (Petit et al, 1987; Schott and Petit, 1987). All indications are that the silica-rich layer has a structure similar to that of an amorphous hydrated silica gel, and TEM imaging indicates there is an atomically discrete interface between the silica layer and the feldspar surface (Casey et al, 1989). Diffusion rates though similar porous silica layers on glass and zeolites are very similar to aqueous diffusion rates (Bunker et al, 1984, 1986, 1988), and Westrich et al (1989) demonstrated O^{18} will diffuse into a 1000 Å leached layer on labradorite within hours. Casey et al (1988), Westrich et al (1989), Hellmann et al (1990), Rose (1991) and Armhein and Suarez (1992) all conclude that the dissolution rate of feldspars in acid solutions is not limited by diffusion through the Si-rich leached layer, and formation of the leached layer does not necessarily indicate any change in the rate controlling mechanism of the dissolution reaction which decomposes the feldspar structure.

A conceptual model for the formation of the leached layer is that during the dissolution process Al is preferentially removed from the tetrahedral lattice, leaving small fragments of Si tetrahedra with a few remaining bridging oxygen bonds. If the rate of feldspar dissolution is slower than the rate at which the silica fragments can be dissolved, then no silica layer will form. This is the case above pH ~3 to 5, where feldspar dissolution rates are slow. As feldspar dissolution rates increase below pH 3, the silica fragments start to accumulate near the surface where they can repolymerize to form a continuous, but porous and hydrous silica layer. There is evidence for the repolymerization of silinol groups within the silica layer, based upon hydrogen inventories measured by elastic recoil detection (ERD) by Casey et al, (1988), hydrogen inventories determined by ion beam analysis (Petit and others), and irreversible and rapid incorporation of O^{18} into the silica-rich layer by Westrich et al, (1989).

There is less agreement about the presence and nature of the leached layer above pH 3. Most studies agree that the cation and Al depleted layer reaches thicknesses <20 Å between pH 5 and 8 for all the feldspars. This is comparable to the depth of rapid cation exchange measured by Chou and Wollast (1984) and Blum and Lasaga (1991) for albite and Amrhein and Suarez (1988) for anorthite, and there is a general consensus that no depleted layer forms during feldspar dissolution in the neutral pH region. There is some evidence that the leached layer persists but grows rapidly thinner between pH 3 and 5 (Nesbitt et al, 1991; Hellmann et al 1990; Sjöberg, 1989). However, Inskeep et al (1991) observed with TEM the preferential dissolution of Ca-rich exsolution lamellae in labradorite with a width of 700Å to a depth of 1350Å. They have suggested that much of the evidence for non-stoichiometric dissolution of plagioclase based upon surface analytical techniques and dissolution stoichiometries in the pH 3-5 region can be explained by the preferential dissolution of calcium rich exsolution phases.

There is less information at basic pH. Casey et al. (1988) using Rutherford backscattering (RBS) analysis found no detectable leached layer and very low hydrogen inventories with ERD after labradorite reaction at pH 12, indicating no leached layer. Hellmann et al. (1990) observed a Si enriched layer ~500Å thick on albite at pH 10.8, but at an elevated temperatures of 200°C. Holdren and Speyer (1985) concluded that K-feldspar dissolution remained stoichiometric up to at least pH 9.

Dissolution experiments have a duration of from several hundred to several thousand hours, whereas natural weathering of feldspars may occur over 100,000's of years. With the slow feldspar dissolution rates at neutral pH, there is still some doubt if thick leached layer might form on feldspar surfaces given sufficient time. However, Berner and Holdren (1979) found no evidence of a leached layer >20 Å thick on naturally weathered plagioclase with XPS. Blum et al. (1990) observed no cation depletion of >50 Å for Na and Ca, and >30 Å for K on the (001) faces of naturally weathered alkali and oligoclase feldspars using XPS. There was no indication of a non-stoichiometric Al/Si ratio greater than a few angstroms, with a slight Al enrichment which appears to be on the very near surface. The soils ranged in age from 10K to 800K years old, with soil pH's of 5-7. These results agree with experimental observations in near neutral pH solutions; cations are removed by exchange to depths of several 10's of Å, and silica and aluminum dissolve essentially stoichiometrically. This indicates that the experimental results may reflect the natural weathering processes. However, Nesbitt and Muir (1988) used SIMS to measure surface compositions of 9 oligoclase grains from a 10K year old till in an acidic watershed in Ontario. They observed consistent Al enrichment of the surface to depths of ~500 Å, which contrasts with the dissolution experimental results. The reasons for the disagreement are unclear, but explanations such as the high Al^{3+} and other solute concentrations in acidified soil solutions (Muir and Nesbitt, 1991; Nesbitt et al., 1991), failure to remove secondary precipitates, or errors in the SIMS data interpretation have been suggested.

3. Theoretical Approaches to Surface Reaction Controlled Dissolution Kinetics

There are several different theoretical approaches to dissolution and precipitation kinetics in the fields of chemistry and materials science. These models approach the problem on very different scales and are often appropriate for very different applications. Geochemists are still in the process of selecting the most appropriate conceptual models to describe feldspar dissolution kinetics. There is still some ambiguity in how to relate the different types of approaches, and the description of geological systems may require the development of more comprehensive models. I have somewhat arbitrarily divided the models into three categories, based upon the conceptual scale.

3.1. ATOMISTIC MODELS

In atomic models of dissolution, we are looking at the detailed configuration and energetics of interactions between the mineral surface and the solvent atoms. The most widely applied atomistic approach to kinetics is transition state theory (TST) which has been applied to many studies of mineral dissolution kinetics (e.g. Lasaga, 1981b; Aagaard and Helgeson, 1982, Helgeson et al., 1984; Chou and Wollast, 1985; Hellmann et al., 1990). Geochemists usually write overall reactions, such as eq. 1, which reflect the overall stoichiometry of a chemical process. In contrast, an elementary reaction reflects the exact molecular species actually involved in the reaction at the atomic level, and there may be a large system of elementary reactions involved in a single overall reaction. An elementary reaction involves a collision between reactant molecules (e.g. A+BC), which initially are confined to a well defined potential energy well (fig. 6). As the reactants approach each other they must overcome an energy barrier, which is at a saddle point in the potential energy surface. The molecular configuration in the vicinity of the saddle point is the activated complex (ABC), which will then decompose to either the reaction products (AB+C) or back to the initial reactants. TST calculates the reaction rate by assuming equilibrium between the reactants and the activated complex, and calculating the concentration of the activated complex, and its decomposition rate. Both of these quantities are a function of the energetics of the molecular species, and can be calculated from their partition functions. Lasaga (1981b) gives a simple and concise summary of TST and its application to geochemistry.

Several aspects of transition state theory need to be considered in the application to feldspar dissolution kinetics: (i) The activated complex is a well defined molecular configuration. However, there is no energy barrier for the decomposition of the activated complex along the reaction coordinate, so that the decomposition rate will be on the order of the vibrational

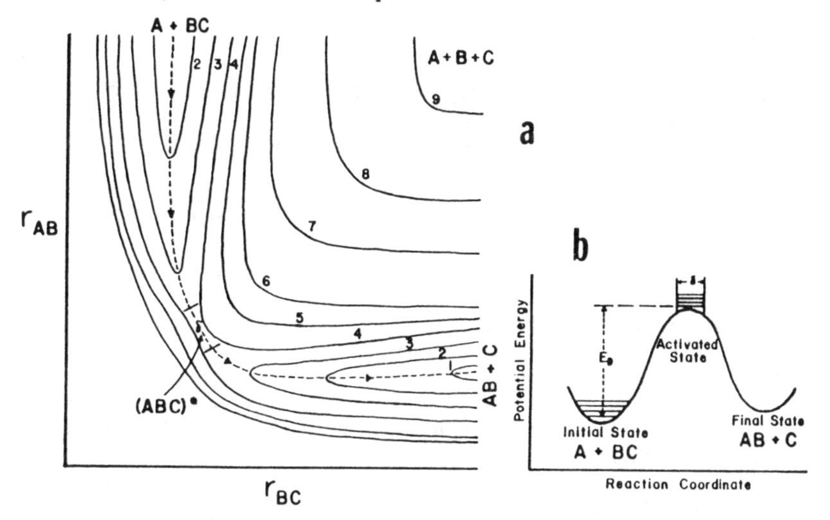

Figure 6 - Transition state theory. a) Schematic diagram of the potential energy as a function of the bond distance for a three atom system. The contours are labeled with arbitrary energy units. The reaction coordinate follows the trough in the energy surface, shown by a dotted line. The transition state is at the saddle point in the trough. This treatment can be extended mathematically into many dimensions for complex systems. b) Typical one dimensional representation of the reaction coordinate, which shows the total system energy as a function of linear distance along the reaction coordinate. *(from Lasaga, 1981b)*

frequency of the molecules (~10^{-14} sec). Thus, the concentration of the activated complex at any specific time is extremely low; (ii) Application of TST to a specific reaction requires, as a minimum, knowledge of the elementary reaction of interest; (iii) Rigorous application of TST requires knowing such quantities as activity coefficient of the activated complex, which can generally be determined only by calculation; and (iv) TST is most successful when applied to homogenous systems with well defined chemistry and structure. It is difficult to apply TST to reactions where larger scale characteristics of the system, such as defect structures, have a large influence on the energetics of the reaction. In these cases, there may be a whole range of subtly different activation complexes at different sites, with a range of geometries and energetics. In this case, the overall reaction kinetics will then be the summation of all these subtly different pathways, and will be very difficult to quantify on an atomic scale. Our understanding of feldspar dissolution kinetics is not yet refined to the point where we can conclusively point to any specific elementary reaction or activated complex in the rate limiting step of the dissolution process. Despite these difficulties, TST is still a powerful conceptual framework which shapes much of our thinking about chemical kinetics.

The rapid advances in ab-initio quantum mechanical calculations hold the prospect of calculating proposed activated complexes, and testing the energetics of the proposed reaction mechanisms. Lasaga and Gibbs (1990a, 1990b) have used ab-initio calculations to investigate the hydrolysis of Si-O-Si bonds. They calculated the reaction trajectory and energetics of the transition state for a proposed mechanism, and applied the results to quartz dissolution kinetics. This computational approach has tremendous potential for testing proposed dissolution mechanisms for feldspar dissolution, and facilitating the quantitative application of TST. However, the application of quantum mechanical modeling to feldspar dissolution still requires making the correct choice of molecules interacting during the rate limiting elementary reaction. This requires insight which can only be obtained from detailed and well designed experiments.

3.2. SURFACE SPECIATION MODELS

Surface speciation models take a large step back in both rigor and scale from TST. Surface speciation models attempt to relate an observed dissolution rate to the surface concentration of specific chemical complexes on the surface. In this model, the dissolution rate is a consequence of a series of elementary reactions.

$$\text{surface + aqueous species} \leftrightarrows \text{surface species} \quad \text{(fast and reversible)}$$

$$\text{surface species + reactant} \rightarrow \text{dissolved species} \quad \text{(slow and irreversible)}$$

One of the early steps in that reaction sequence is a rapid and reversible adsorption reaction (such as described in section 2.3). The configuration of this surface complex then facilitates the next step in the reaction sequence which is a slow and irreversible reaction. In the case of feldspars, the slow step is probably the hydrolysis of bridging tetrahedral oxygen bonds. If the rate limiting reaction occurs with equal probability at each occurrence of a surface complex and only at sites with this specific surface complex, then

$$\text{dissolution rate (R)} = k_{dis}\ [\#\ \text{surface complexes}\ (S_c)] \qquad (9)$$

so that the dissolution rate (R, $mol/m^2/s$) is directly proportional to the number of surface complexes (S_c, mol/m^2), where k_{dis} is the rate constant in s^{-1}. In terms of TST, the presence of the surface complex provides a reaction path with a lower activation energy, and is essentially a stable intermediate in the dissolution reaction mechanism. Several points should be clarified. The actual "activated complex" for the rate limiting step is some unknown configuration in a subsequent slow reaction. Identification of the surface species gives only general insight into the rate limiting step, in that we are one step closer in the reaction mechanism, and *perhaps* the surface species is one of the reactants. However, there may also be additional fast intermediate steps.

Nevertheless, surface complexation models may be extremely powerful. If we can identify the surface complex involved in the dissolution reaction: (i) We can predict dissolution rate using eq. 9 from only a knowledge of the rate constant (k) and the adsorption isotherms for the solutes involved in the surface complex. Determining the adsorption isotherm is in principle a much simpler matter than explicitly measuring the kinetics of slow reactions like feldspar dissolution over a full range of solution compositions; (ii) The activity of both poisons and inhibitors will probably result from competition for the adsorption sites, which are then either blocked or enhanced by the adsorbed species. These effects may also be quantified by surface adsorption measurements, and intelligent guesses as to the effects of solutes may be made even without measurements; and (iii) We are one step closer to the true "activated complex", and may be in a position to propose reasonable molecular clusters for testing using ab-initio quantum mechanical calculations.

3.2.1. Applications of Surface Speciation Models to Feldspar Dissolution Kinetics - The adsorption characteristics of simple oxides have been studied extensively, and surface complexation models have been developed to estimate the protonation reactions of hydroxyls at oxide surfaces, as well as the adsorption of other ligands (see Davis and Kent (1990), Wehrli et al. (1990) and Stumm and Wieland (1990) for summaries).

Furrer and Stumm (1986) and Guy and Schott (1989) found a correlation between the dissolution rate of simple oxides and their surface charge with a reaction order equal to the formal charge of the metal center.

$$\text{Dissolution rate} \propto c^{n} \qquad (10)$$

where c is the surface charge and n is the formal charge of the metal center (i.e. n = 2 for Be, 3 for Al and Fe(III) and 4 for Si). This suggests that dissolution is occurring at metal centers at which n of the oxygens coordinated around that metal center are simultaneously protonated. Wieland et al. (1988) presents a model for this dependence. They have shown statistically that if each metal center has n possible adsorption sites, which are filled in completely random fashion, then the concentration of metal centers with all n sites filled (F) will vary as $F \propto (X_c)^n$ where X_c is the fraction of total number of surface sites which are filled. Blum (unpublished) has numerically verified the validity of this statistical result.

Blum and Lasaga (1988, 1991) measured the change in surface charge of albite as a function of pH by surface titration (fig 7). They did not observe Na^+ uptake during the titrations, and assumed that the protonation and deprotonation of surface hydroxyls (eqns. 5 and 6) were the major reactions controlling the change in surface charge. The inflection point in the surface charge is at pH 6.8, which Parks (1967) estimated to be the zero point of charge for tetrahedral Al. From this, and the depth of the adsorption layer, they concluded that it is protonation of Al sites which is controlling the surface charge of albite as a function of pH. Blum and Lasaga (1991) also observed a strong correlation between the surface charge of albite and the dissolution rate of albite measured by Chou and Wollast (1985) (fig. 8). They conclude that in the acid region (pH <6.8) the albite dissolution rate is directly proportional to the positive surface charge, and in the basic region, to the negative surface charge, with both charged species occurring on Al sites. The rate laws are then:

$$\textit{Dissolution rate} = k_a \, c_{>Al-OH_2^+}^{\sim 1.0} \qquad (pH < 6) \qquad (11)$$

where the reaction constants for the dissolution mechanisms in acid and basic regions are $k_a \approx 10^{-6.5}$ s^{-1} and $k_b \approx 10^{-6.1}$ s^{-1}, respectively. They concluded that the hydrolysis rate of bridging

$$Dissolution\ rate = k_b\, c_{>Al\text{-}O^-}^{-1.0} \qquad (pH > 7.5) \qquad (12)$$

Al-O-Si bonds adjacent to the charged aluminum species is the rate limiting step in the feldspar dissolution process. Wollast and Chou (1992) found a similar linear relation between negative surface charge and the dissolution rate of albite in the basic region ($R \propto c^{-1.0}$). Schott (1990) conducted similar surface titrations of albite and found a good correlation between the dissolution rate and surface charge of albite in the basic region, but with a reaction order of 3.3 ($R \propto c^{-3.3}$). Amrhein and Suarez (1988) found a correlation between the dissolution rate and the surface charge of anorthite with a reaction order of 4 ($R \propto c^{-4.0}$). Reaction orders of 3 to 4 are consistent with the theory of Wieland et al. (1988) (eq. 10) for control of feldspar dissolution by surface species at Al centers. This is in contrast to the reaction order of 1 found by Blum and Lasaga (1991) and Wollast and Chou (1992).

The hydration reaction in which water reacts with the Al-O-Si bridging bonds of the feldspar lattice in the acid region can be schematically represented as:

<u>reactants</u>		<u>transition state</u>		<u>products</u>
≡Si-O		≡Si-O		≡Si-O
\		\		\
≡Si-O-Al-O-Si≡	⇒	≡Si-O-Al-O-Si≡	⇒	≡Si-O-Al-**OH** + **HO-Si≡**
/		/ \		/
OH_2^+ + $\mathbf{H_2O}$		OH_2^+ **O-H**		OH_2^+
		H		

The reaction above has been presented to represent my biases. The Al on the surface is shown with an OH_2^+ surface complex adjacent to a bridging Al-O-Si bond which will be hydrolyzed, as proposed by Blum and Lasaga (1991). The approach trajectory of the water and

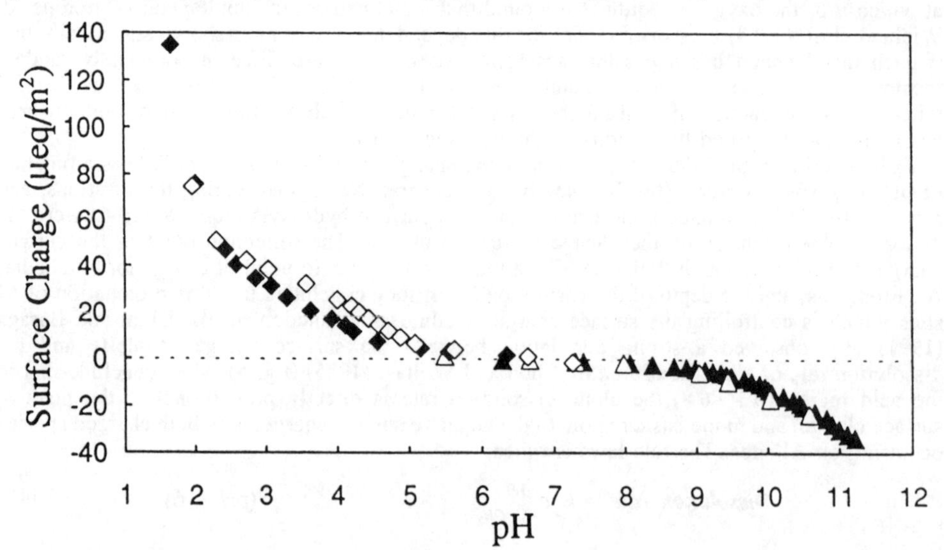

Figure 7 - Change in albite surface charge as a function of solution pH from surface titrations. Open symbols are the initial titrations, and open symbols are the back titration. (*from* Blum and Lasaga, 1991)

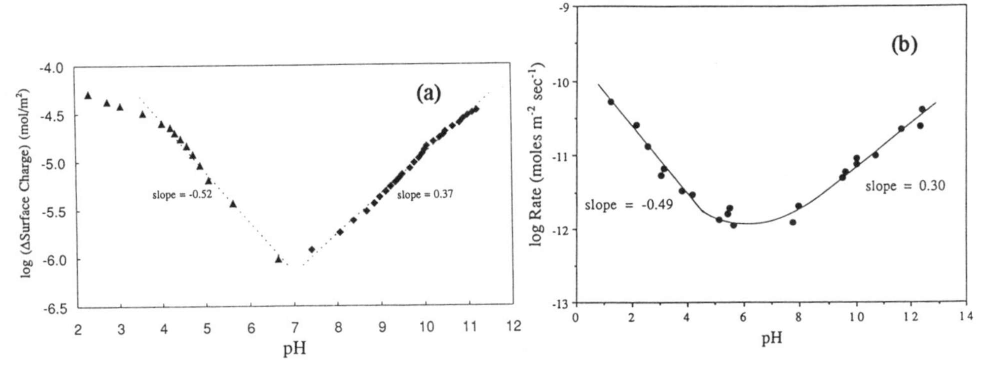

Figure 8 - Correlation between pH dependence of the albite surface charge and dissolution rate. a) Log of the change in positive surface charge in the acid region and of negative surface charge in the basic region as a function of solution pH. b) Data of Chou and Wollast (1985) for albite dissolution rate as a function of pH. (*from* Blum and Lasaga, 1991)

the configuration of the activated complex are analogous to those proposed by Lasaga and Gibbs (1990), based upon quantum mechanical modeling, for the hydrolysis of quartz Si-O-Si bonds. The presence of the positively charged surface species at the Al dangling oxygens may reduce the energy barrier for the approach of the negative oxygen dipole of the water toward the Al (i.e. reducing the activation energy). Wieland et al. (1988) and Schott (1990) suggest that the Al should have three charged groups. This may present a problem. If we envision the protonation of only dangling hydroxyls on the Al, then the mechanism can only apply to Al with only one remaining bridging tetrahedral bond. Those Al with only a single bridging bond should be the easiest to remove from the lattice, and it is unlikely that this is a rate limiting reaction. Alternatively, protonation may occur on bridging oxygens themselves. The model of Wieland et al. (1988) requires equivalent adsorption behavior at all sites in order for the relation $R=kc^n$ to be valid. Si-O-Si bonds have been shown to be highly hydrophobic on quartz and silica, and adsorption site densities on quartz are consistent with protonation of only dangling hydroxyls (Iler, 1979). The behavior of Si-O-Al should be at least crudely similar to Si-O-Si bonds. This does not appear consistent with the requirement of Wieland et al. (1988) that the adsorption characteristics of bridging oxygens and dangling oxygens must be nearly identical.

Brady and Walther (1989) examined the dissolution versus surface charge behavior of a range of oxides and silicates. They concluded that Si hydroxyls become highly deprotonated at basic pH, and it is the detachment of Si, and not Al, which controls the dissolution kinetics of feldspars in the basic region. Obviously, there is still considerable uncertainty concerning the feldspar dissolution mechanism in the basic pH region, with both inconsistent data and different proposed models. Nevertheless, the surface speciation approach holds great promise for a robust model of feldspar dissolution kinetics.

3.2.2. Implications of Surface Speciation on Temperature Effects - If the dissolution rate of feldspars is directly proportional to the surface concentration of surface species, then the traditional application of an Arrhenius plot to determine the activation energy is invalid, even if the reaction mechanism remains unchanged. The simple equation;
where E_a is the apparent activation energy, and A is the pre-exponential frequency factor, assumes that the concentration of the reactants remains approximately constant. Eq. 13 should be valid for surface speciation controlled dissolution so long as the rate is expressed as [(mol reaction) (mol surface complex)$^{-1}$ sec^{-1}]. However, the determination of dissolution rates from

$$rate = A \exp\left(\frac{E_a}{RT}\right) \tag{13}$$

the appearance of solutes in solution typically yields a dissolution rate in [mol cm^{-2} sec^{-1}]. Eq. 13 can be recast as;

$$rate = X_c(T,pH,adsorb)\ A \exp\left(\frac{E_a}{RT}\right) \tag{14}$$

where X_c is the mole fraction of potential sites for surface complex which are actually occupied by the surface complex, and the rate is expressed in [mol cm^{-2} sec^{-1}]. As discussed earlier, the equilibrium concentration of surface complexes may be very sensitive to solution pH and competitive adsorption of other species at potential adsorption sites (*adsorb* in eq. 14). In addition, the equilibrium constant for surface complex formation quite likely temperature dependent, and therefore, X_c is temperature dependent. Consequently, both the linear and exponential terms in eq. 14 have a temperature dependence, and are not independent. Therefore, an Arrhenius plot will not yield a reliable activation energy, even at a constant pH, and even though the elementary reaction mechanism may remain unchanged.

3.3. Macroscopic Models

Macroscopic models are those models based upon the surface configuration of the surface on the molecular scale. These approaches are most useful for dealing with features of the system which are defined only at scales larger than molecular clusters, such as the saturation state of the solution, and the effects of dislocations, impurities and other heterogeneities in the solid.

3.3.1. *Surface nucleation models* - Surface nucleation models were first proposed for growth (Nielsen, 1964; Ohara and Reed, 1973), but apply equally well to dissolution (Blum and Lasaga, 1987). In this discussion we will frame the problem in terms of dissolution kinetics. Figure 9 shows a schematic diagram of a crystal surface. The surface nucleation model takes a basic thermodynamic approach in which there are two competing energy terms to consider; (i) When a unit is removed from the crystal into an undersaturated solution, the energy released is $\Delta\mu = \mu_{liquid} - \mu_{solid}$, where $\Delta\mu$ is the change in chemical potential, which is related to the solution composition by $\Delta\mu = kT \ln(Q/K)$ where k is Boltzmann's constant, Q is the activity product and K is the equilibrium constant; and (ii) Silicate surfaces in aqueous solution have a positive surface free energy (σ), indicating the creation of surface area is thermodynamically unfavorable, and the energy required to increase the surface area by A equals Aσ. For quartz, Parks (1984) suggested a value for σ of 350 ergs/cm^2, and Lasaga and Blum (1986) estimated σ at 500 ergs/cm^2 for feldspars.

The free energy change for the formation of either a cluster on the surface by the precipitation of *n* atoms or a pit formed by the dissolution of *n* atoms is then:

$$\Delta G_n = -n\Delta\mu + A_n\sigma \tag{15}$$

For a cluster or pit with a height of h units, a radius of r units, and an atomic volume υ,

$$n\upsilon = \pi r^2 h \tag{16}$$

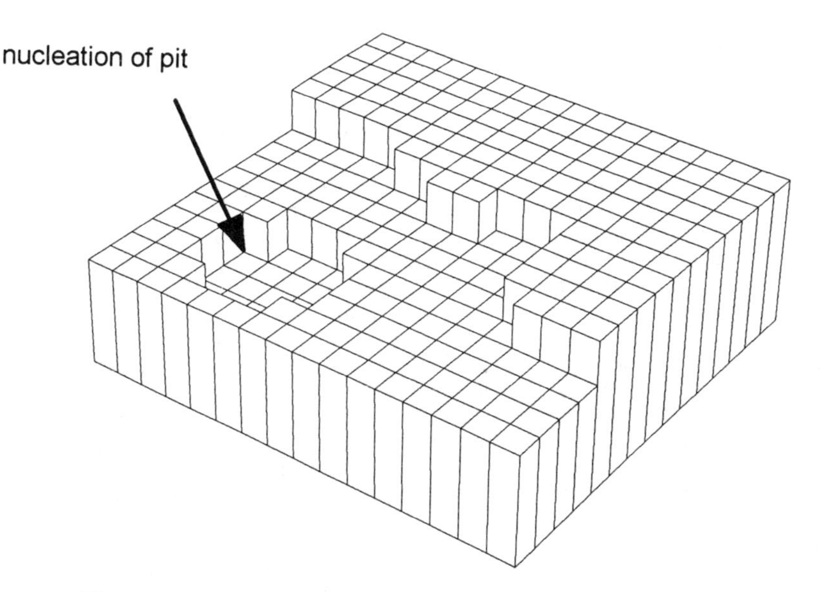

Figure 9 - Schematic diagram of a nucleation pit on the surface.

The *new* surface area created by the formation of the pit (i.e. along the edges) is then:

$$A_n = 2\pi r h \tag{17}$$

Substituting eq. 16 into eq. 17 to solve for *r* and inserting this into eq. 15, we find that ΔG_n initially increases with increasing *n*, and reaches a maximum when

$$n_c = \frac{\pi \nu h}{\Delta\mu^2}\sigma^2 \qquad or \qquad r_c = \frac{\sigma\nu}{\Delta\mu} \tag{18}$$

Thus, there is an energy barrier (ΔG_c) which must be overcome in order to form a dissolution pit of radius r_c. At pit sizes above r_c, continued enlargement of the pit is thermodynamically favorable, and it will continue to expand outward until the entire layer is dissolved. The hardest part in dissolution is getting those first few atoms out of a flat surface. The creation of an incipient dissolution pit creates a lot of new surface area relative to the small amount of material which has been dissolved. At solution compositions common in near-surface environments, r_c for feldspars and other silicates is on the order of several nanometers (Lasaga and Blum, 1986).

The nucleation rate of pits on the surface, *R*, depends exponentially on the energy barrier (ΔG_c) necessary to form a pit with the critical radius:

$$R \propto \exp\left[\frac{-\Delta G_c}{RT}\right] \qquad where \qquad \Delta G_c = \frac{\pi h \nu \sigma^2}{\Delta\mu} \tag{19}$$

The nucleation rate (R) calculated using eq. 19 predicts far too strong a dependence of the dissolution rate on the solution saturation state, and greatly underestimates dissolution and growth rates close to equilibrium. However, the concept of the critical radius is commonly used in dissolution theory.

3.3.2 Dissolution at dislocations - Extensive crystallographically controlled etch pits are a ubiquitous feature in both naturally weathered (e.g. Berner and Holdren, 1979; Velbel, 1986; Banfield and Eggleton, 1990) and experimentally dissolved (e.g. Wilson and McHardy, 1980; Holdren and Speyer, 1985; Knauss and Wolery, 1986) feldspars. The morphology of the etch pits is similar to etch pits shown to form at the intersection of dislocations with the surface. There are several ways in which dislocations may potentially affect feldspar dissolution rates. The intersection of screw dislocations with the surface provides a continuous source of steps on the surface (fig. 10) which circumvents the need to nucleate pits on the surface (eq 19). This is the basis for the Burton-Cabrera-Frank (BCF) theory for spiral dissolution and growth (Burton et al., 1951; see also Ohara and Reed, 1973; Christian, 1975; Bennema, 1984). BCF theory is also completely equivalent for dissolution and growth, with the only difference a change in sign of $\Delta\mu$. In BCF theory, the step produced at a screw dislocation is a preferential site for dissolution and growth. The geometry imposed by the step being fixed at one end causes the propagating step to wind into a large spiral, either as an etch pit during dissolution or a growth hillock during precipitation. The distance between the steps (r_{step}) of the spiral are predicted to be $r_{step} \approx 19r_c$, where r_c is the critical radius given by eq 18. The step velocity is found to be proportional to the saturation state at small values of $\Delta\mu$ and independent of saturation state and at large values of $\Delta\mu$. This yields the relationship that at small values of $\Delta\mu$ (i.e. near equilibrium):

$$R \propto \Delta\mu^2$$

At large values of $\Delta\mu$ (far from equilibrium)

$$R \propto \Delta\mu$$

These relationships are often cited in the literature, and BCF theory has been very successful in describing the behavior of many synthetic systems in the materials literature, although predominately for precipitation. Growth spirals have also been observed in natural kaolinite, beryl, quartz and other minerals.

Dislocations may effect dissolution in a second way. In the vicinity of dislocations, the

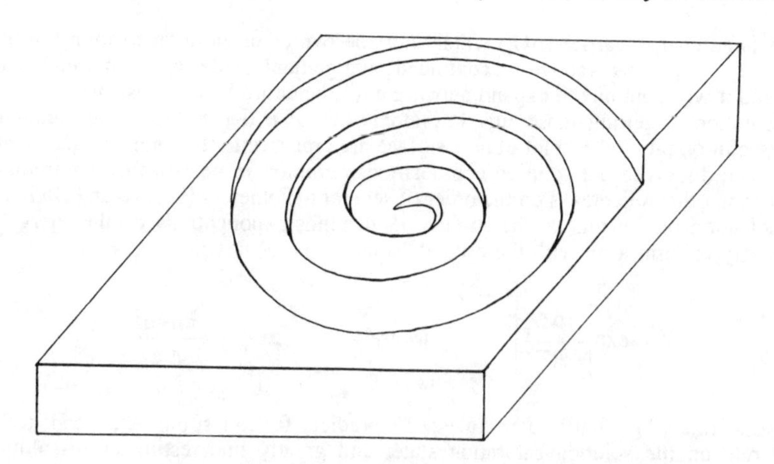

Figure 10 - Schematic diagram of a growth spiral on the surface.

crystal lattice is distorted, introducing strain energy into the lattice. The magnitude of this strain energy is too small to effect the overall thermodynamic properties of the mineral (Helgeson et al., 1984; Wintsch and Dunning, 1985). However, the concentration of strain energy in the vicinity of the dislocation provides a localized area of increased solubility. This provides an additional mechanism for forming etch pits which exceed r_c, and are thermodynamically favored to continue dissolution. Thus, strain energy around dislocations provides an alternative mechanism to surface nucleation or BCF for mineral dissolution (Cabrera and Levine, 1956; Van der Hoek et al., 1982; Lasaga, 1983). Note that strain energy accelerates dissolution, but will marginally inhibit growth. Blum and Lasaga (1987) used Monte Carlo simulations to demonstrate that strain energy is probably the dominant mechanism of etch pit formation at dislocations in quartz and feldspars (fig. 11). They found that far from equilibrium, the strain energy mechanism was ~8 fold faster than the BCF mechanism, with the ratio increasing toward equilibrium. An important prediction of this theory is that there is a critical solution saturation state (ΔG_{bulk}^{crit}) below which macroscopic etch pits will not form and above which the dislocation cores will "open" to form etch pits.

$$\Delta G_{bulk}^{crit} = -\frac{2\pi^2\sigma^2 K_{orient}}{\mu_s b^2}\overline{V} \tag{20}$$

where

$$\Delta G_{bulk} = RT\ln\left[\frac{Q}{K}\right] \tag{21}$$

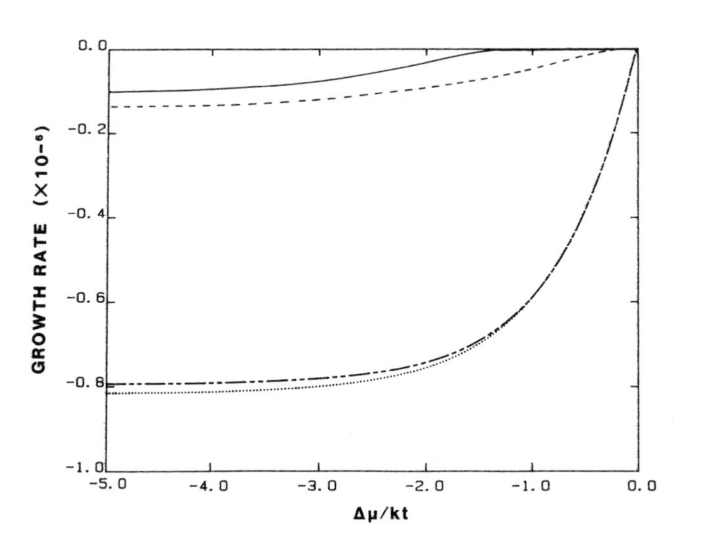

Figure 11 - Dependence of the dissolution rate on the dissolution mechanism as determined by Monte Carlo simulations. The X-axis is the saturation state, with 0 at equilibrium and becoming more undersaturated with more negative values of $\Delta\mu/kT$, and more rapid dissolution rates along the Y axis from 0 to -1. ——— Defect free surface; ------ step at screw dislocation only; -·-·-·-·- strain field around dislocation only; ·········· step and strain field around dislocation. (*from* Blum and Lasaga, 1987)

and μ_s is the shear modulus, b is the magnitude of the dislocation Burger vector, $\bar{V}$ is the molar volume, and K_{orient} is a constant related to the orientation of the dislocation. Lasaga and Blum (1986) suggest that the presence or absence of etch pits may be used as an indicator of solution saturation state in soils and ground water. This effect has been observed experimentally by Burch et al. (1993) for albite and by Brantley et al. (1986) for quartz in soils.

The observation of extreme etch pitting in virtually all feldspars which have undergone extensive dissolution suggests that a large amount of the total dissolution is occurring at etch pits. However, Blum and Lasaga (1987) suggested that dislocation density would have a minimal effect on the overall dissolution rate of quartz and other silicates based upon Monte Carlo model results. Murphy (1989) studied the dissolution rate of sanidine with low and high dislocation densities, and found no difference in the dissolution rates. Similar experimental results were found for quartz (Blum et al., 1990), rutile (Casey et al., 1988) and calcite (Schott et al., 1989). All of these experiments were conducted far from equilibrium. Blum et al. (1990) predicted on theoretical grounds that dislocations should have a larger effect on the overall dissolution rate close to equilibrium.

Burch et al. (1993) conducted albite dissolution experiments at pH 8 and 80°C varying the solution composition over a wide range of saturation states (fig. 12). They observed a dramatic decrease in the dissolution rate of albite as they approach equilibrium, with an abrupt decrease in the dissolution rate at ΔG_r = -6 kcal/mol, which corresponds to a Q/K of $10^{-4.4}$. Thus, the solution saturation state can significantly depress the dissolution kinetics of albite, even in very dilute solutions. They point out that the shape of this curve is inconsistent with the classical TST formulation of the rate law. However, the abrupt transition in dissolution rate occurs approximately at the ΔG_{crit} predicted for the opening of etch pits, and examination of the experimental material with SEM reveals etch pits only in the most undersaturated experiments. A similar dependence of the gibbsite dissolution rate on solution saturation state was observed by Nagy and Lasaga (1992), and also attributed to the opening of dislocations.

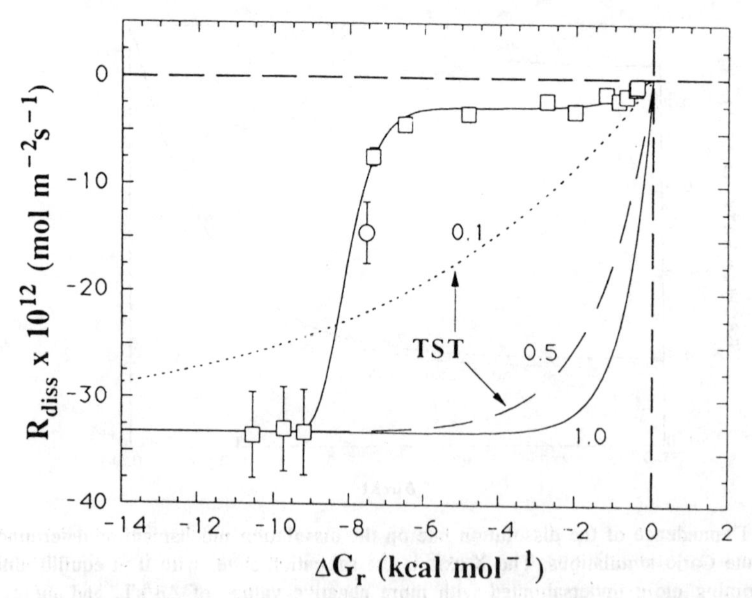

Figure 12 - Experimentally determined dissolution rate of albite as a function of solution saturation state. ΔG_r = 0 at equilibrium and grows more negative at greater undersaturation. More negative values of R_{diss} indicate increasing dissolution rates (*from* Burch et al., 1993).

3.3.3. Numerical Models of Dissolution Kinetics - There is currently a difficulty in relating macroscopic effects, such as dislocations, impurities, and solution saturation state to molecular scale models such as those described by surface speciation and TST models. The most promising approach to including effects at both the molecular and macroscopic scale are numerical simulations. Two approaches have been proposed; Monte Carlo simulations (e.g. Gilmer, 1980; Blum and Lasaga, 1986; Wehrli, 1989) and molecular dynamics simulations (e.g. Lasaga, 1990). The Monte Carlo approach assigns relative probabilities to each molecular reaction which can occur on the surface. The relative reaction probabilities can be estimated from ΔH of reaction (not the activation energy), and can therefore can be easily adjusted for small change in the energetics of different types of sites. The simulation progresses by taking small time steps and allowing reactions to occur randomly at each site on the surface, scaled by the relative probability of each particular type of site reacting. The use of relative probabilities allows the description of a large number of different types of sites on the surface without needing detailed knowledge of the mechanism or activation energy for the elementary reactions. This may include a wide range of subtle differences in the energy, such as in the strain field around a dislocation (Blum and Lasaga, 1986) or small variations in composition. Use of the Monte Carlo method using random probabilities avoids having to solve large systems of simultaneous equations.

Molecular dynamics (MD) simulations model a chemical system by calculating the time averaged motions of atoms to predict the behavior of a macroscopic system. The general MD scheme places several hundred to thousand atoms (limited by computer capacity) with arbitrary initial velocities in a box of finite size but with periodic boundary conditions. All of the interatomic forces between the atoms are calculated, and new velocities and positions are predicted for some small time step in the future. The atoms are then moved to the new positions and the calculation of interatomic forces is repeated. In this manner, the motions of the molecules can be accurately modeled. Reactions occur when the random fluctuations in the localized energy of a molecular cluster exceeds the activation energy of reaction. This often occurs when the translational energy of one cluster is effectively transferred to a vibrational mode of another cluster, providing enough energy to break the bond. In order for MD simulations to accurately reflect these types of atomic interactions, the time step used during the simulations must be less then 1/5 to 1/10 of the vibrational period. For silicates, the time step is typically 10^{-15} sec. Typical simulations utilize several thousand time steps, and therefore simulate a total time of 10^{-11} to 10^{-12} sec. For a dissolution rate of 10^{-15} mol/cm^2/sec (e.g. albite at ~pH 3), the surface retreat rate is ~4×10^{-13} cm/sec. Treating the top 1Å of the bulk crystal as the surface, the average residence time for any individual Al or Si atom on the surface before dissolving into solution is ~10^5 sec. Thus, the odds of observing even one dissolution event during a MD simulation is minuscule (~1 in 10^{15}), and I don't believe MD will be a useful technique for the direct modeling of feldspar dissolution kinetics. However, MD may be extremely useful in quantifying the structure of the hydrated feldspar surface, and perhaps the configuration and kinetics of the rapid exchange and adsorption reactions on the feldspar surface which may proceed the dissolution reaction.

4. Field Measurements of Feldspar Weathering Rates

The motivation for much of the work on feldspar dissolution and weathering rates is to develop the ability to predict feldspar dissolution rates in natural environments, and how variables such as acid rain and climatic change will effect those rates. A critical step in this endeavor is the comparison of field estimates of feldspar weathering rates with laboratory measured dissolution rates. There are two approaches which have been used to quantify natural plagioclase weathering rates in soils. Paces (1973), Velbel (1985,1992), Schnoor (1990) and Swoboda-Colberg and Drever (1993) estimated feldspar weathering rates from mass balance calculations of natural systems. White et al. (1992, in prep) estimated oligoclase weathering rates from the change in abundance of plagioclase in a series of soils of different ages. All the studies agree that natural weathering rates are 10-3000 times slower than predicted from experimentally determined dissolution rates. Reproducibility of experimentally determined rates between studies is generally less than one order of magnitude, suggesting these differences are real.

A major uncertainty in the earlier studies of Paces (1973) and Velbel (1985) was the estimation of feldspar surface areas. White et al. (in prep.) directly determined the BET surface areas of weathered plagioclase. Schnoor (1990) and Swoboda-Colberg and Drever (1993) conducted dissolution measurement on material from the soils, and normalized the results per unit mass, eliminating the uncertainty introduced by surface area determination. In all these cases, significant differences between field and experimental feldspar dissolution rates were observed.

Many explanations have been advanced to explain the discrepancy between field and experimental feldspar dissolution rates. Several of the stronger arguments are:

1) Soil temperatures are typically less than experimental temperatures of 25°C, and lower soil temperatures should depress natural weathering rates. However, the effect of temperature is estimated to be about a factor of 5, and is not sufficient to explain all of the discrepancy natural and field rates.

2) Dissolution experiments are typically run for only a few thousand hours. A small volume of more rapidly solubilized material at high energy sites such as dislocations or impurities may elevate laboratory rates, perhaps for years. In the field, more reactive material has already been removed, usually by thousands of years of weathering, resulting in lower net dissolution rates. In addition, experiments commonly use unusually pure and defect free feldspar samples, which may not accurately represent common feldspars in igneous and metamorphic rocks. However, the data of Schnoor (1990) and Swoboda-Colberg and Drever (1993) in which they used the field material in their dissolution experiments strongly argue against these explanations.

3) The presence of higher solution concentrations of Fe and Al in natural systems may poison the feldspar surface, decreasing field weathering rates. Chou and Wollast (1985) and Amrhein and Suarez (1992) report that increasing Al concentration depresses the dissolution rate of albite and anorthite, respectively. Blum and White (1992, in prep) analyzed weathered feldspar grain surfaces with XPS, and found trace amounts of Fe^{3+} on the surface of all weathered grains as a widely dispersed coating <5Å in thickness. This is consistent with Fe adsorption at charged sites on the surface, which may poison surface sites. There is currently no quantitative evaluation of the effects of Al and Fe adsorption in natural systems on feldspar dissolution rates.

4) Soil solution compositions may be thermodynamically close enough to saturation with respect to feldspars as to reduce the dissolution rate. Velbel (1989) evaluated this effect in terms of a linear dependence between the dissolution rate and $\exp(1-\Delta G_{rxn}/RT)$ (see Nagy et al., 1991 for derivation), and concluded that this effect was negligible. However, Burch et al. (1993) experimentally found a much higher dependence of the albite dissolution rate on saturation state than the thermodynamic effect used by Velbel (1989). Burch et al. (1993) argue that their measured depression of the dissolution rate of albite by a saturation state effect can explain the discrepancy between field and experimental feldspar dissolution rates.

5) Fluid flow in the vadose zone of soils may be extremely heterogeneous (e.g. Glass et al., 1988; Hornberger et al., 1990, 1991; Booltink and Bouma 1991; Gee et al., 1991). The overwhelming proportion of fluid flow in soils may occur along very restricted preferential flow paths (often called macropores). This may leave most of the mineral surface area located in hydrologically isolated micropores, which may not contribute significantly to the mineral dissolution rate. Solutions moving downward through a soil profile and eventually out of the watershed may actually contact only a small proportion of the mineral surface area present in the soil profile, resulting in a much lower weathering flux than predicted based upon the total mineral surface area in the soil profile. This explanation is proposed by Schnoor (1990), Velbel (1993) and Swoboda-Colberg and Drever (1993), and is probably the most widely favored explanation. Velbel (1993) argues that typically 90-99.9% of the mineral surface area is located

in isolated micropores, and are not participating in weathering reactions which contribute solutes to solutions flowing out of the watersheds.

The concept of preferential fluid flow and isolated microporosity in soils may be the dominant process controlling the observed low natural weathering rates, but is not in itself sufficient. Hysteresis in the permeability of unsaturated media may result in transient preferential flow regions (fingering), even in uniform porous media (Hill and Parlange, 1972; Glass et al., 1989a,b,c, 1990). However, this is a thermodynamically unstable situation. In long lived systems such as soils, differential matrix (or capillary) pressure will eventually fill all the micropores, and these micropores will be the last areas to empty of solution during drying of the soil. Micropores, therefore, will almost always remain filled with solutions, although they may be isolated solutions, both hydrologically and chemically. Weathering (i.e. feldspar dissolution and clay precipitation) will progress in these micropores until some mechanism, probably related to either the proton balance or the accumulation of solutes within the micropore, retards feldspar dissolution to a very low rate. The weathering reactions may not proceed again until the solution in that micropore becomes hydrologically connected, and the solution is flushed. However, there must be a chemical mechanism which is retarding the feldspar dissolution rate. The cause of the retardation in feldspar dissolution rate, however, may not be reflected in the chemistry of sampled soil solutions or in the fluxes out of the watershed.

5. Summary

5.1. EXPERIMENTAL DATA

The dissolution of feldspar is controlled by rate of reactions at the mineral-surface interface, not by diffusion through a protective surface layer. A thick Si-rich surface layer does form in strongly acidic solutions, but this layer is too porous and loosely bound to impede the dissolution reaction kinetics. Albite and K-feldspar have similar dissolution kinetics at pH<6, with the dissolution rate $R \propto [H^+]^{-0.5}$. In the basic pH region albite has a pH dependence of $R \propto [OH^-]^{-0.3}$, with K-feldspar dissolution rates having a stronger dependence on [OH-] and absolute rates approximately 10 fold slower than albite. In the acidic region the plagioclase series has a pH dependence of $R \propto [H^+]^{-0.5}$ for An_{0-70}, but there is then a transition of increasing values in the exponent until $R \propto [H^+]^{-1.0}$ for anorthite. This trend is reflected in the dissolution rates of plagioclase of different An content at the same pH. The dissolution rate increases gradually with increasing Ca content until ~An_{75}, with a large increase in the dissolution rate of plagioclase between the compositions An_{75} and $An_{>90}$. The reproducibility of feldspar dissolution rates is generally within $10^{0.7}$ between investigations and within a factor of <2 within the same laboratory.

5.2. FELDSPAR DISSOLUTION MECHANISMS

All the feldspars appear to have two different dissolution rate mechanisms, a proton promoted mechanism in the acid region (pH<5) and a hydroxyl promoted mechanism in the basic region (pH>7.5). It is possible there is a separate pH independent mechanism in the neutral pH region.

5.2.1. Feldspar Dissolution in the Acid Region - There is general agreement that feldspar dissolution in the acid region occurs by selective attack on the Al sites in the tetrahedral framework, and the rate limiting process must involve the hydrolysis of the Al-O-Si bond. There are several observations supporting this conclusion: (i) The presence of a Si-rich layer at low pH indicates the strong preferential removal of Al; (ii) Feldspars of all compositions form cation depleted layers on the surface in acidic solutions, so that the feldspar surface in contact with solution is always hydrogen-feldspar. Albite and K-feldspar have similar Al/Si ratios and have similar dissolution behaviors. The plagioclase dissolution rate increases with increasing anorthite content. The major change in the structure of the plagioclase surface layer in contact with solution with increasing An content is the increasing Al/Si ratio, implying it is Al which is destabilizing the structure; (iii) There is clearly an increase in the positive surface charge of feldspars with increasing $[H^+]$. The zero point of charge for SiO_2 is ~pH 2, implying that very little Si on the feldspar surface should be protonated in the acid region. Al(IV) has a zero point of charge of 6.8, and is the more probable site of adsorption. The change in the surface speciation of Al with pH also points to Al as the critical species in the dissolution mechanism.

5.2.2. Feldspar Dissolution in the Basic Region - Feldspar dissolution in the basic region is less well understood than the acidic region. Blum and Lasaga (1991) proposed a mechanism very similar to the acid region (eq. 12), but with a negatively charged $>Al\text{-}O^-$ surface species rather than a positive charged surface species. However, the difference in dissolution rates between albite and K-feldspar and the evidence of Wollast and Chou (1992) for Na readsorption by albite suggest that cation occupancy in the lattice sites may influence feldspar dissolution kinetics in the basic pH region. Finally, the argument of Brady and Walther (1989) for deprotonation of Si in the basic region controlling the dissolution rate cannot be conclusively refuted. While soils in many of the acid-rain impacted regions of the world happen to be acidic, most surface and ground waters and many soil solutions have a neutral to basic pH, and resolving feldspar dissolution in the neutral and basic pH regions is of considerable practical interest.

5.2.3. Weathering Rates of Feldspars - Field measurements of plagioclase dissolution rates from watershed mass-balance studies and from changes in the abundance of plagioclase in soils of different ages indicate that feldspar weathering is one to three orders of magnitude slower than predicted from laboratory studies. The leading explanation for this effect is the isolation of a large proportion of the feldspar surface area in isolated micropores. Alternative explanations include; (i) adsorption of inhibitors, of which Al and Fe are the most likely, (ii) the high saturation states of soil solutions, and (iii) experimental artifacts in the dissolution experiments, particularly the short duration of the experiments relative to geologic time scales. The reconciling of natural and experimental feldspar dissolution rates is a critical step before feldspar dissolution kinetics can be used reliably in predictive models. Addressing this problem is currently the focus of much of the work in mineral dissolution kinetics, both in field and laboratory studies.

6. References

Aagaard P and Helgeson H C (1982) Thermodynamic and kinetic constrains on reaction rates among minerals and aqueous solution. I. Theoretical consideration. Amer J Sci 282:237-285

Amrhein C and Suarez D L (1988) The use of a surface complexation model to describe the kinetics of ligand-promoted dissolution of anorthite. Geochim Cosmochim Acta 52:2785-2793

Amrhein C and Suarez D L (1992) Some factors affecting the dissolution kinetics of anorthite at 25°C. Geochim Cosmochim Acta 56:1815-1826

Anbeek C (1992a) Surface roughness of minerals and implications for dissolution studies. Geochim Cosmochim Acta 56:1461-1469

Anbeek C (1992b) The dependence of dissolution rates on grain size for some fresh and weathered feldspars. Geochim Cosmochim Acta 56:3957-3970

Banfield J F and Eggleton R A (1990) Analytical transmission electron microscope studies of plagioclase, muscovite, and K-feldspar weathering. Clays and Clay Min 38:77-89

Bennema P (1984) Spiral growth and surface roughening: Developments since Burton, Cabrera and Frank. J Cryst Growth 69: 182-197

Berner, R A (1978) Rate control of mineral dissolution under earth surface conditions. Amer J Sci 5:1252

Berner R A and Holdren G R (1979) Mechanism of feldspar weathering - II. Observations of feldspars from soils. Geochim Cosmochim Acta 43:1173-1186
Berner R A, Lasaga A C and Garrels R M (1983) The carbonate-silicate geochemical cycle and its effect on atmospheric carbon dioxide over the past 100 million years. Am J Sci 283:641-683
Blum A E and Lasaga A C (1987) Monte Carlo simulations of surface reaction rate laws. In Aquatic Surface Chemistry (ed W Stumm) pp 255-292 J Wiley and Sons
Blum A E and Lasaga A C (1988) Role of surface speciation in the low-temperature dissolution of minerals. Nature 4:431-433
Blum A E and Lasaga A C (1990) The effect of dislocation density on the dissolution rate of quartz. Geochim Cosmochim Acta 54:283-297
Blum A E and Lasaga A C (1991) The role of surface speciation in the dissolution of albite. Geochim Cosmochim Acta 55:2193-2201
Blum A E and White A F (1992) The surface chemistry and topography of weathered feldspar surfaces. Trans Amer Geophys Union Abstr 73:602.
Blum A E, Yund R A and A C Lasaga (1990) The effect of dislocation density on the dissolution rate of quartz. Geochim Cosmochim Acta 54:283-297
Booltink H W G and Bouma J (1991) Physical and morphological characterization of bypass flow in a well-structured clay soil. Soil Sci Soc Am J 55:1249-1254
Brady P V and Walther J V (1989) Controls on silicate dissolution in neutral and basic pH solutions at 25°C. Geochim Cosmochim Acta 53:2823-2830
Brantley S L , Crane S R, Crerar D A, Hellman R and Stallard R (1986) Dislocation etch pits in quartz. In: Davis J A and Hayes K F (eds) Geochemical Processes at Mineral Surfaces. Amer Chem Soc pp 639-649
Brunauer S, Emmett P H and Teller E (1938) Adsorption of gases in multimolecular layers. J Amer Chem Soc 60:309-319
Bunker B C, Headley T J and Douglas S C (1984) Gel structures in leached alkali silicate glass. Mater Res Soc Symp 32:226-253
Bunker B C, Arnold G W, Day D E and Bray P J (1986) The effect of molecular structure on borosilicate glass leaching. J Non-Cryst Solids 87:226-253
Bunker B C, Tallant D R, Headley T J, Turner G L and Kirkpatrick R J (1988) The structure of leached sodium borosilicate glass. Phys Chem Glasses 29:106-120
Burch T E, Nagy K L and A C Lasaga (1993) Free energy dependence of albite dissolution kinetics at 80 C and pH 8.8. Chem Geol 105:137-162
Burton W K, Cabrera N and Frank F C (1951) The growth of crystals and the equilibrium structure of their surfaces. Phil Trans Roy Soc London A243:299-368
Busenberg E (1978) The products of the interaction of feldspars with aqueous solutions at 25°C. Geochim Cosmochim Acta 42:1629-1686
Busenberg E and Clemency C V (1976) The dissolution kinetics of feldspars at 25°C and 1 atm CO_2 partial pressure. Geochim Cosmochim Acta 40:41-50
Cabrera N and Levine M M (1956) On the dislocation theory of evaporation of crystals. Phil Mag 1:450-458.
Casey W H, Carr M J and Graham R A (1988) Crystal defects and the dissolution kinetics of rutile. Geochim Cosmochim Acta 52:1545-1556
Casey W H, Westrich H R, Massis T, Banfield, J F and Arnold G W (1989) The surface of labradorite feldspar after acid hydrolysis. Chem Geol 78:205-218
Casey W H, Westrich H R and Holdren G R (1991) Dissolution rates of plagioclase at pH = 2 and 3. Am Min 76:211-217
Chou L and Wollast R (1984) Study of the weathering of albite at room temperature and pressure with a fluidized bed reactor. Geochim Cosmochim Acta 48:2205-2218
Chou L and Wollast R (1985) Steady-state kinetics and dissolution mechanisms of albite. Amer J Sci 285:963-993

Chou L and Wollast R (1989) Is the exchange of alkali feldspars reversible? Geochim Cosmochim Acta 53:557-558
Christian J W (1975) The Theory of Transformation in Metals and Alloys. Part I 2nd Edition, Pergamon Press.
Davis J A and Kent D B (1990) Surface complexation modeling in aqueous geochemistry. Reviews in Mineral 23:177-260
Davis J A and Kent D B (1990) Surface complexation modeling in aqueous geochemistry. Reviews in Mineral 23:177-260
Dove P A and Crerar D A (1990) Kinetics of quartz dissolution in electrolyte solutions using a hydrothermal mixed flow reactor. Geochim Cosmochim Acta 54:955-969.
Ernsberger F M (1960) Structural effects in the chemical reactivity of silica and silicates. J Phys Chem Solids 13:347-351
Fleer V N (1982) The dissolution kinetics of anorthite ($CaAl_2Si_2O_8$) and synthetic strontium feldspar ($SrAl_2Si_2O_8$) in aqueous solutions at temperatures below 100°C: Applications to the geological disposal of radioactive nuclear wastes. PhD thesis, Pa State Univ
Furrer G and Stumm W (1986) The coordination chemistry of weathering. I. Dissolution kinetics of δ-Al_2O_3 and BeO. Geochim Cosmochim Acta 50:1847-1860
Garrels R M and Howard P (1959) Reactions of feldspar and mica with water at low temperature and pressure. Proc Sixth National Conference on Clays and Clay Minerals: pp 68-88
Gee G W, Kincaid C T, Lenhard R J and Simmons C S (1991) Recent studies of flow and transport in the vadose zone. (In: U.S. National Report to International Union of Geodesy and Geophysics 1987-1990) Rev Geophys Suppl 227-239
Gilmer G H (1980) Computer models of crystal growth. Science 208:355-363
Glass R J, Steenhuis T S and Parlange J Y (1988) Wetting front instability as a rapid and far-reaching hydrologic process in the vadose zone. J Contam Hydrol 3:207-226
Glass R J, Parlange J Y and Steenhuis T S (1989a) Wetting front instability; 1.Theoretical discussion and dimensional analysis. Water Resour Res 25:1187-1194
Glass R J, Steenhuis T S and Parlange J Y (1989b) Wetting front instability; 2. Experimental determination of relationships between system parameters and two-dimensional unstable flow field behavior initially dry porous media. Water Resour Res 25:1195-1207
Glass R J, Steenhuis T S and Parlange J Y (1989c) Mechanism for finger persistence inhomogeneous, unsaturated, porous media: Theory and verification. Soil Sci 148:60-70
Glass R J, Cann S, King J, Baily N, Paralange J Y and Steenhuis T S (1990) Wetting front instability in unsaturated, porous media: A three-dimensional study in initially dry sand. Trans Porous Media 5:247-268
Goldich S S (1938) A study in rock weathering. J Geol 46:17-58
Goossens D A, Pijpers A P, Philippareerts J G, Van Tendeloo S, Althaus E and Gijbels R (1989) A SIMS, XPS, SEM, TEM and FTIR study of feldspar surfaces after reacting with acid solutions. In: Proc of the 6th Int Symp on Water-Rock Interaction (ed D L Miles). A A Balkema, Rotterdam 267-270
Gregg, S J and Sing, K S (1982) Adsorption, Surface Area and Porosity. Academic Press, New York, 303 p
Guy C and Schott J (1989) Multisite surface reaction versus transport control during the hydrolysis of a complex oxide. Chem Geol 78:181-204
Helgeson H C (1971) Kinetics of mass transfer among silicates and aqueous solutions. Geochim Cosmochim Acta 35:421-469
Helgeson H C, Murphy W M and Aagaard P (1984) Thermodynamic and kinetic constraints on reaction rates among minerals and aqueous solution. II. Rate constants, effective surface area, and the hydrolysis of feldspar. Geochim Cosmochim Acta 48:2405-2432
Hellmann R, Eggleston C M, Hochella M F and Crerar D A (1989) Altered layers on dissolving albite - 1. Results. In: Proc of the 6th Int Symp on Water-Rock Interaction (ed D L Miles). A A Balkema, Rotterdam 293-269

Hellmann R, Eggleston C M, Hochella M F and Crerar D A (1990) The formation of leached layers on albite surfaces during dissolution under hydrothermal conditions. Geochim Cosmochim Acta 54:1267-1281
Hill D E and Parlange J Y (1972) Wetting front instability in layered soils. Soil Sci Soc Am Proc 36:697-702
Holdren G J and Berner R A (1979) Mechanism of feldspar weathering - I. Experimental studies. Geochim Cosmochim Acta 43:1161-1171
Holdren G R and Speyer P M (1985) pH dependent change in the rates and stoichiometry of dissolution of an alkali feldspar at room temperature. Amer J Sci 285:994-1026
Holdren G R and Speyer P M (1987) Reaction rate-surface area relationships during the early stages of weathering. II. Data on eight additional feldspars. Geochim Cosmochim Acta 51:2311-2318
Hornberger G M, Beven K J and Germann P F (1990) Inferences about solute transport in macroporous forest soils from time series models. Geoderma 46:249-262
Hornberger G M, Germann P F and Beven K J (1991) Throughflow and solute transport in an isolated sloping soil block in a forested catchment. J Hydrol 124:81-99
Huang C P (1981) The surface acidity of hydrous solids. In: Anderson M A and Rubin A J (eds) Adsorption of Inorganics at Solid-Liquid Interfaces. Ann Arbor Sci Pub 183-218
Iler R K (1979) The Chemistry of Silica. John Wiley and Sons, New York
Inskeep W P, Nater E A, Bloom P R, Vandervoort D S and Erich M S (1991) Characterization of laboratory weathered labradorite surfaces using X-ray photoelectron spectroscopy and transmission electron microscopy. Geochim Cosmochim Acta 55:787-800
Knauss K G and Wolery T J (1986) Dependence of albite dissolution kinetics on pH and time at 25°C and 70°C. Geochim Cosmochim Acta 50:2481-2497
Kung K J S (1990) Preferential flow in a sandy vadose zone; 1. Field observations. Geoderma 46:51-58
Lasaga A C (1981a) Rate laws of chemical reactions. in Kinetics of Geochemical Processes (ed. Lasaga A C and Kirkpatrick R J) Reviews Mineralogy 8:1-68
Lasaga A C (1981b) Transition state theory. in Kinetics of Geochemical Processes (ed. Lasaga A C and Kirkpatrick R J) Reviews Mineralogy 8:135-170
Lasaga A C (1983) Kinetics of silicate dissolution. In: Fourth International symposium on Water-Rock Interaction pp269-274 Misasa, Japan
Lasaga A C (1984) Chemical kinetics of water-rock interactions. J Geophy Res 89:4009-4025
Lasaga A C (1990) Atomic treatment of mineral-water surface reactions. in Mineral Water Interface Geochemistry (ed. Hochella M F and White A F) Reviews Mineralogy 23:17-85
Lasaga A C, Berner R A and Garrels R M (1985) An improved geochemical model of atmospheric CO_2 over the past 100 million years. In: Sundquist E T and Broecker W S (eds) The Carbon Cycle and Atmospheric CO_2: Natural Variations Archean to Present, Geophys Monogr Ser AGU Washington DC 32:397-411
Lasaga A C and Blum A E (1986) Surface chemistry, etch pits and mineral-water reactions. Geochim Cosmochim Acta 50:2363-2379
Lasaga A C and Gibbs G V (1990a) Ab initio quantum mechanical calculations of surface reaction - A new era? (In: Aquatic Chemical Kinetics, ed W Stumm),Wiley Interscience 259-289
Lasaga A C and Gibbs G V (1990b) Ab-initio quantum mechanical calculation of water-rock interactions: adsorption and hydrolysis reactions. Am J Sci 290:263-295
Manley E P and Evans L J (1986) Dissolution of feldspars by low-molecular weight aliphatic and aromatic acids. Soil Sci 141:106-112
Mast M A and Drever J I (1987) The effect of oxalate on the dissolution rates of oligoclase and tremolite. Geochim Cosmochim Acta 51:2559-2568
Muir I J, Bancroft G M and Nesbitt H W (1989) Characteristics of altered labradorite surfaces by SIMS and XPS. Geochim Cosmochim Acta 53:1235-1241

Muir I J and Nesbitt H W (1991) Effects of aqueous cations on the dissolution of labradorite feldspar. Geochim Cosmochim Acta 55:3181-3189

Muir I J and Nesbitt H W (1992) Controls on differential leaching of calcium and aluminum from labradorite in dilute electrolyte solutions. Geochim Cosmochim Acta 56:3979-3985

Murphy K and Drever J I (1993) Dissolution of albite as a function of grain size. Trans Am Geophys Union 74:329

Murphy W M (1989) Dislocations and feldspar dissolution. Eur J Mineral 70:163

Murphy W M and Helgeson H C (1989) Surface exchange and the hydrolysis of feldspar. Geochim Cosmochim Acta 53:559

Murphy W M and Helgeson H C (1987) Thermodynamic and kinetic constraints on reaction rates among minerals and aqueous solution. III. Activated complexes and the pH-dependence of rates of feldspar, pyroxene, wollastonite,and olivine hydrolysis. Geochim Cosmochim Acta 51:3137-3153

Nagy K L, Blum A E and Lasaga A C (1991) Dissolution and precipitation kinetics of kaolinite at 80 C and pH 3: The dependence on solution saturation state. Am J Sci 291:649-686

Nagy K L and Lasaga A C (1992) Dissolution and precipitation kinetics of gibbsite at 80°C and pH 3: The dependence on solution saturation state. Geochim Cosmochim Acta 56:3093-3111

Nash V E and Marshall C M (1956a) The surface reactions of silicate minerals, Part 1; Reactions of feldspar surfaces with acid solutions. University of Missouri, College of agriculture, Research Bulletin 613

Nash V E and Marshall C M (1956b) The surface reactions of silicate minerals, Part 2; Reactions of feldspar surfaces with salt solutions. University of Missouri, College of Agriculture, Research Bulletin 614

Nesbitt H W and Muir I J (1988) SIMS depth profiles of weathered plagioclase and processes affecting dissolved Al and Si in some acidic soil solutions. Nature 334:336-338

Nesbitt H W, Macrae N D and Shotyk W (1991) Congruent and incongruent dissolution of labradorite in dilute, acidic, salt solutions. J Geol 99:429-442

Nielsen J W (1964) Kinetics of Precipitation. Macmillan Co

Ohara M and Reed R C (1973) Modeling Crystal Growth Rates from Solution. Prentice-Hall

Oxburgh R, Drever, J I and Sun Y T (in press) Mechanism of plagioclase dissolution in acid solution at 25°C. Geochim Cosmochim Acta

Paces T (1973) Steady-state kinetics and equilibrium between ground water and granitic rock. Geochim Cosmochim Acta 37:2641-2663

Parks G A (1967) Aqueous surface chemistry of oxides and complex oxide minerals. Equilibrium Concepts in Natural Water Systems, ACS Adv Chem Ser 67:121-160

Parks G A (1984) Surface and interfacial free energies of quartz. J Geophys Res 89:3997-4008

Petit J C, Della Mea G, Dran J C, Schott J and Berner R A (1987) Mechanism of diopside dissolution from hydrogen depth profiling. Nature 325:705-707

Petit J C, Dran J C, Paccagenella A and Della Mea G (1989) Structural dependence of crystalline silicate hydration during aqueous dissolution. Earth Planet Sci Let 93:292-298

Petit J C, Dran J C and Della Mea G (1990) Energetic ion beam analysis in the earth sciences. Nature 344:621-626

Petrović R (1981a) Kinetics of dissolution of mechanically comminuted rock-forming oxides and silicates. I. Deformation and dissolution of quartz under laboratory conditions. Geochim Cosmochim Acta 45:1665-1674

Petrović R (1981b) Kinetics of dissolution of mechanically comminuted rock-forming oxides and silicates. II. Deformation and dissolution of oxides and silicates in the laboratory and at the earth's surface. Geochim Cosmochim Acta 45:1675-1686

Petrović R, Berner R A and Goldhaber M B (1976) Rate control in dissolution of alkali feldspars I. Study of residual grains by X-rat photoelectron spectroscopy. Geochim Cosmochim Acta 40:537-548

Reuss J O and D W Johnson (1986) Acid Deposition and the Acidification of Soils and Waters. Springer Verlag, New York 303 p

Rose N M (1991) Dissolution rates of prehnite, epidote, and albite. Geochim Cosmochim Acta 55:3273-3286
Schnoor J L (1987) Kinetics of chemical weathering: A comparison of laboratory and field weathering rates. (In: Aquatic surface chemistry, ed W Stumm) Wiley Interscience 475-504
Schnoor J L (1990) Kinetics of chemical weathering: A comparison of laboratory and field weathering rates. (In: Aquatic Chemical Kinetics, ed W Stumm) Wiley Interscience 475-504
Schott J (1990) Modeling of the dissolution of strained and unstrained multiple oxides: The surface speciation approach. (In: Aquatic Chemical Kinetics, ed W Stumm) Wiley Interscience 337-366
Schott J, Brantley S, Crerar D, Guy C, Borcsik M and Willaime C (1989) Dissolution kinetics of strained calcite. Geochim Cosmochim Acta 53:373-382
Schott J and Petit J C (1987) New evidence for the mechanisms of dissolution of silicate minerals. (In: Aquatic surface chemistry, ed W Stumm) Wiley Interscience 293-315
Schweda P (1989) Kinetics of alkali feldspar dissolution at low temperature. In: Proc of the 6th Int Symp on Water-Rock Interaction (ed D L Miles). A A Balkema, Rotterdam 609-612
Sjöberg L (1989) Kinetics and non-stoichiometry of labradorite dissolution. In: Proc of the 6th Int Symp on Water-Rock Interaction (ed D L Miles). A A Balkema, Rotterdam 639-642
Stillings L L and Brantley S L (in prep) Feldspar dissolution at 25°C and pH 3: The effect of ionic strength and stoichiometry of reaction.
Stumm W and Morgan J J (1981) Aquatic Chemistry. J. Wiley and Sons, New York 780 p
Stumm W and Wieland E (1990) Dissolution of oxide and silicate minerals: Rates depend on surface speciation. (In: Aquatic Chemical Kinetics, ed W Stumm), Wiley Interscience pp. 367-400
Stumm W and Wollast R (1990) Coordination chemistry of weathering: Kinetics of the surface-controlled dissolution of oxide minerals. Reviews of Geophys 28:53-69
Sverdrup H U (1990) The kinetics of base cation release due to chemical weathering. Lund Univ Press, Lund, Sweden
Swoboda-Colberg N G and Drever J I (1993) Mineral dissolution rates in plot-scale field and laboratory experiments. Chem Geol 105:51-69
Tamm O (1930) Experimental Studied Umber die Verwitterung und Tonbildung von Feldspaten. Chem Erde 4:420-430
Van der Hoek B, Van der Eerden J P and Bennema P (1982) Thermodynamical stability conditions for the occurrence of hollow cores caused by stress of line and planar defects. J. Crystal Growth 56:621-632
Velbel M A (1985) Geochemical mass balances and weathering rates in forested watersheds in Southern Blue Ridge. Am J Sci 285:904-930
Velbel M A (1986) Influence of surface area, surface characteristics, and solution composition on feldspar weathering rates. (In: Geochemical Processes at mineral surfaces. J A Davis and K F Hayes eds) Am Chem Soc Symp Ser 323:615-634
Velbel M A (1989) Effect of chemical affinity on feldspar hydrolysis rates in two natural weathering systems. In: Schott J and Lasaga A C (eds) Kinetic Geochemistry. Chem Geol 78:245-253 (special issue)
Velbel M A (1992) Geochemical mass balances and weathering rates in forested watersheds of the Southern Blue Ridge, III Cation budgets in an amphibolite watershed, and weathering rates of plagioclase and hornblende. Am J Sci 292:58-78
Velbel M A (1993) Constancy of silicate-mineral weathering-rate ratios between natural and experimental weathering: implications for hydrologic control of differences in absolute rates. Chem Geol 105:89-99
Volk, T (1987) Feedbacks between weathering and atmospheric CO_2 over the last 100 million years. Am J Sci 287:763-779
Walker F D L (1990) Ion microprobe study of intergrain micropermeability in alkali feldspars. Contrib Mineral Petrol 106:124-128

Walker J C G, Hays P B and Kasting J F (1981) A negative feedback mechanism for the long term stabilization of Earth's surface temperatures. J Geophys Res 86:9776-9782
Wehrli B (1989) Monte Carlo simulations of surface morphologies during mineral dissolution. J Colloid Interface Sci 132:230-242
Wehrli B, Wieland E and Furrer G (1990) Chemical mechanisms in the dissolution kinetics of minerals; the aspect of active sites. Aquatic Sci 52:3-31
Westrich H R, Casey W H and Arnold G W (1989) Oxygen isotope exchange in the leached layer of labradorite feldspar. Geochim Cosmochim Acta 53:1681-1685
Wieland E , Wehrli B and Stumm W (1988) The coordination chemistry of weathering. III potential generalization on dissolution rates of minerals. Geochim Cosmochim Acta 52:1969-1981
White A F, Blum A E, Bullen, T D, Peterson M L, Schulz M S and Harden J W (1992) A three million year weathering record for a soil chronosequence developed in granitic alluvium, Merced, California. In: Kharaka Y K and Maest A S (eds) Proc of the 7th Int Symp on Water-Rock Interaction. A A Balkema, Rotterdam 607-610
White A F and Peterson M L (1990) Role of reactive-surface-area characterization in geochemical kinetic models. Amer Chem Soc Symp Ser 416:461-475
Wilson M J and McHardy W J (1980) Experimental etching of a microcline perthite and implications regarding natural weathering. J Microscopy 120:291-302
Wintsch R P and Dunning J (1985) The effect of dislocation density on the aqueous solubility of quartz and some geologic implications: A theoretical approach J. Geophys Res 90: 3649-3657
Wollast, R and Chou, L (1992) Surface reactions during the early stages of weathering of albite. Geochim Cosmochim Acta 56:3113-3122
Wordon R H, Walker F D L, Parsons I and Brown W L (1990) Development of microporosity, diffusion channels and deuteric coarsening in perthitic alkali feldpars. Contrib Mineral Petrol 104:507-515

SUBJECT INDEX

A

B

C

D

E

F

G

J

K

L

M

N

O

P

Q

R

S

T

U

V

W

X

Z